Lecture Notes in Mechanical Engineering

Lecture Notes in Mechanical Engineering (LNME) publishes the latest developments in Mechanical Engineering—quickly, informally and with high quality. Original research reported in proceedings and post-proceedings represents the core of LNME. Volumes published in LNME embrace all aspects, subfields and new challenges of mechanical engineering. Topics in the series include:

- Engineering Design
- Machinery and Machine Elements
- Mechanical Structures and Stress Analysis
- Automotive Engineering
- Engine Technology
- Aerospace Technology and Astronautics
- Nanotechnology and Microengineering
- Control, Robotics, Mechatronics
- MEMS
- Theoretical and Applied Mechanics
- Dynamical Systems, Control
- Fluid Mechanics
- Engineering Thermodynamics, Heat and Mass Transfer
- Manufacturing
- Precision Engineering, Instrumentation, Measurement
- Materials Engineering
- Tribology and Surface Technology

To submit a proposal or request further information, please contact the Springer Editor of your location:

China: Ms. Ella Zhang at ella.zhang@springer.com
India: Priya Vyas at priya.vyas@springer.com
Rest of Asia, Australia, New Zealand: Swati Meherishi at swati.meherishi@springer.com
All other countries: Dr. Leontina Di Cecco at Leontina.dicecco@springer.com

To submit a proposal for a monograph, please check our Springer Tracts in Mechanical Engineering at https://link.springer.com/bookseries/11693 or contact Leontina.dicecco@springer.com

Indexed by SCOPUS. All books published in the series are submitted for consideration in Web of Science.

Uday S. Dixit · M. Kanthababu · A. Ramesh Babu · S. Udhayakumar
Editors

Advances in Forming, Machining and Automation

Select Proceedings of AIMTDR 2021

Springer

Editors
Uday S. Dixit
Department of Mechanical Engineering
Indian Institute of Technology Guwahati
Guwahati, Assam, India

M. Kanthababu
Department of Manufacturing Engineering
Anna University
Chennai, Tamil Nadu, India

A. Ramesh Babu
Department of Mechanical Engineering
PSG College of Technology
Coimbatore, India

S. Udhayakumar
Department of Mechanical Engineering
PSG College of Technology
Coimbatore, India

ISSN 2195-4356 ISSN 2195-4364 (electronic)
Lecture Notes in Mechanical Engineering
ISBN 978-981-19-3865-8 ISBN 978-981-19-3866-5 (eBook)
https://doi.org/10.1007/978-981-19-3866-5

This Springer imprint is published by the registered company Springer Nature Singapore Pte Ltd.
The registered company address is: 152 Beach Road, #21-01/04 Gateway East, Singapore 189721, Singapore

AIMTDR 2021, Conference Core Organizing Committee

Chief Patron

Shri. L. Gopalakrishnan, Managing Trustee, PSG Institutions

Patrons

Dr. P. Radhakrishnan, Director, PSG Institute of Advanced Studies
Dr. P. V. Mohanram, Secretary, PSG Institute of Technology and Applied Research
Dr. K. Prakasan, Principal, PSG College of Technology
Dr. G. Chandramohan, Principal, PSG Institute of Technology and Applied Research

President (NAC-AIMTDR)

Shri. Vikram Mohan, Managing Director, Pricol Limited, Coimbatore

Vice President (NAC-AIMTDR)

Dr. Uday Shanker Dixit, Professor, IIT Guwahati, India

Chairman

Dr. P. R. Thyla, Professor and Head, Department of Mechanical Engineering, PSG College of Technology

Organizing Secretary

Dr. V. Prabhu Raja, Professor, Department of Mechanical Engineering, PSG College of Technology

Co-organizing Secretary

Dr. N. Saravanakumar, Professor and Head, Department of Mechanical Engineering, PSG Institute of Technology and Applied Research

Joint Organizing Secretaries

Dr. N. Mahendrakumar, Assistant Professor, PSG College of Technology
Dr. C. Shanmugam, Assistant Professor (Sr.Gr), PSG College of Technology
Dr. M. Kalayarasan, Assistant Professor (Sr.Gr), PSG College of Technology
Dr. P. Dhanabal, Assistant Professor (Sr.Gr), PSG College of Technology
Mr. S. Mohanraj, Assistant Professor (Sr.Gr), PSG College of Technology
Dr. V. Rajkumar, Associate Professor, PSG Institute of Technology and Applied Research
Dr. S. Nanthakumar, Assistant Professor, PSG Institute of Technology and Applied Research

Secretaries—Public Relations

Dr. A. Adhiyamaan, Assistant Professor (Sr.Gr), PSG College of Technology
Mr. A. Mohan, Assistant Professor, PSG College of Technology

Treasurer

Mr. D. Martin Suresh Babu, Assistant Professor, PSG College of Technology

Joint Treasurer

Mr. P. Govindaraj, Assistant Professor (Sr.Gr), PSG College of Technology

Members—PSG College of Technology

Dr. K. Ragu, Professor
Dr. A. Ramesh Babu, Professor
Dr. D. Rajenthirakumar, Professor
Dr. R. Jayachitra, Associate Professor
Dr. R. Sridhar, Assistant Professor
Dr. S. Udhayakumar, Assistant Professor
Mr. K. Sadesh, Assistant Professor
Dr. V. Vijayakumar, Assistant Professor (Sr.Gr)
Dr. A. S. Prasanth, Assistant Professor (Sr.Gr)
Mr. D. Shanmuga Sundaram, Assistant Professor
Mr. T. Anantharaj, Assistant Professor
Mr. S. Samsudeensadham, Assistant Professor
Ms. M. Gomathi Prabha, Assistant Professor

Members—PSG Institute of Technology and Applied Research

Dr. R. Ramesh, Professor
Dr. D. Elangovan, Professor
Dr. P. Manoj Kumar, Associate Professor
Dr. G. Rajeshkumar, Associate Professor
Mr. J. Nagarjun, Assistant Professor (Sr.Gr)
Mr. T. Prem Kumar, Assistant Professor (Sr.Gr)
Mr. M. Senthilvel, Assistant Professor
Mr. R. Avinash Kumar, Assistant Professor

National Advisory Committee

Prof. B. B. Ahuja, COEP, Pune
Prof. Amitabha Ghosh, IIEST, Shibur
Dr. R. Balasubramaniam, BARC, Mumbai
Prof. Bijoy Bhattacharyya, Jadavpur University, Kolkata
Prof. Biswanath Doloi, Jadavpur University, Kolkata
Prof. A. K. Chattopadhyay, IIT Kharagpur
Dr. Dasharath Ram Yadav, DRDL, Hyderabad
Prof. V. K. Jain, IIT Kanpur
Prof. P. K. Jain, IIT Roorkee
Prof. Jose Mathew, NIT Calicut
Shri. P. Kaniappan, WABCO, Chennai
Prof. M. Kanthababu, Anna University, Chennai
Prof. N. K. Mehta, IIT Roorkee
Shri. P. J. Mohanram, IMTMA, Bengaluru
Prof. P. V. Mohanram, PSG Institute of Technology and Applied Research, Coimbatore
Dr. Nagahanumaiah, CMTI, Bengaluru
Prof. S. Narayanan, VIT University, Vellore
Shri. Neeraj Sinha, NITI Aayog, New Delhi
Prof. S. S. Pande, IIT Bombay, Mumbai
Dr. D. R. Prasada Raju, MVGR COE, Vizianagaram
Prof. V. Radhakrishnan, IIST, Trivandrum
Prof. P. Radhakrishnan, PSG Institute of Advanced Studies, Coimbatore
Prof. N. Ramesh Babu, IIT Madras, Chennai
Prof. J. Ramkumar, IIT Kanpur
Prof. P. V. Rao, IIT Delhi
Prof. Santosh Kumar, IIT (BHU), Varanasi
Prof. H. S. Shan, IIT Roorkee
Prof. M. S. Shunmugam, IIT Madras, Chennai
Dr. V. Sumantran, Celeris Technologies, Chennai
Dr. V. K. Suri, BARC, Mumbai
Prof. Vinod Yadava, NIT Hamirpur

Foreword

First, I would like to congratulate the editors of five different volumes of proceedings of 8th International and 29th All India Manufacturing Technology, Design and Research Conference (AIMTDR) proceedings being published by Springer. These volumes are very good collection of the research and review papers on the manufacturing processes like Modern Machining processes (Volume-1), Additive Manufacturing and Metal Joining (Volume-2), Simulation, Product Design and Development (Volume-3), Forming, Machining and Automation (Volume-4) and Micro- and Nano-Manufacturing and Surface Engineering (Volume-5). These five volumes are comprehensive collection of the research papers focusing on the most recent research and developments in the area of manufacturing processes.

These subject areas continue to be dominant manufacturing technologies, say, the *technologies of future*, namely 3D printing (additive manufacturing), which generally lacks speed, surface finish and dimensional accuracy. To compensate these weaknesses of 3D printing in the real-life production, I could also see good papers on micro-/nano-manufacturing and nano-finishing. Theoretical analysis, optimization and simulation of manufacturing processes would definitely provide the necessary insights into the physics and mechanisms of these processes, as well as their basic understanding. These five volumes would be invaluable to the researchers working in research laboratories and engineers in industrial organizations working on shop floors for learning, consulting and applying some of the findings deliberated in the conference by the authors of different research papers.

Such conferences encourage the interaction between the research scholars, faculty members and user industries' representatives from different parts of the world. Unfortunately, this could not happen in this hybrid conference to the desired extent due to the pandemic effects across the globe. Apart from these contributed papers, there were many online and offline keynote lectures delivered by the researchers from different countries including India. I am sure that these papers should be of great help to the readers of these proceedings. This proceedings/collection of the papers should be of great help to the academia and industries as well as reference books in different sub-fields of manufacturing processes.

I would like to congratulate the authors for their contributions to all these five volumes of the proceedings and the Editorial Committee Members for their untiring efforts made in bringing out these research papers' collections in five volumes. I will also like to thank the technical committee members in general and ex-vice-president of NAC, Prof. Uday S. Dixit, for inviting me to write this foreword.

Bhopal, India

Dr. V. K. Jain
Professor (Retired), IIT Kanpur

Member, National Advisory
Committee, AIMTDR 2021

Preface

All India Manufacturing Technology, Design and Research (AIMTDR) is a reputed international conference series focused in the domain of manufacturing. The first All India Machine Tool Design and Research conference, also abbreviated as AIMTDR, was organized by Jadavpur University, Kolkata, in the year 1967. Those were the days when the main focus of manufacturing was toward efficient design and utilization of machine tools. PSG College of Technology, Coimbatore, organized the 7th and 15th AIMTDR conferences in 1976 and 1992, respectively. Other institutes who organized this conference before the 15th AIMTDR conference are Central Machine Tool Institute (CMTI), Indian Institute of Technology (IIT) Bombay, IIT Madras, University of Roorkee, IIT Kanpur, Central Mechanical Engineering Research Institute and IIT Delhi, some of them multiple times. After the 15th AIMTDR conference, the necessity of widening the scope of the conference arose. It was decided to encompass entire manufacturing technologies in its fold, rather than just focusing on the technologies related to machine tools. Accordingly, the phrase "machine tool design" was replaced by "manufacturing technology, design." Thus, the conference series was rechristened as All India Manufacturing Technology, Design and Research, without making any alteration to the acronym AIMTDR. Thus, the 16th AIMTDR conference with the new unabridged title was organized at CMTI, Bangalore, in 1994. It is interesting to note that CMTI had renamed itself from Central Machine Tool Institute to Central Manufacturing Technology Institute in 1992, adapting to the current trends in manufacturing.

AIMTDR received its international status in the year 2006, when the first International and 22nd National AIMTDR was organized at IIT Roorkee. Subsequent conferences were held at IIT Madras, Andhra University, Jadavpur University, IIT Guwahati, College of Engineering Pune and Anna University. The recent edition of the conference, viz. the 8th international and 29th national conferences, was jointly organized by the Departments of Mechanical Engineering of PSG College of Technology, Coimbatore, and PSG Institute of Technology and Applied Research, Coimbatore, during December 9–11, 2021, in virtual mode. The theme of AIMTDR 2021 was "*Transformational Changes in Manufacturing*."

After a rigorous review process, about 250 technical papers from academia and industry were accepted for the presentation at the conference. In addition, eight keynote talks on advanced technologies such as the development of high precision machine tools, simulation of material flow, nanostructured surfaces, additively controlled material mechanics, product development platforms, next-generation milling, diamond turning and hybrid machining were delivered by the experts. Further, two case studies from the industry highlighting the innovation practices and challenges in machine tool structure design were presented.

Select papers from the conference are being published by Springer in the series—*Lecture Notes in Mechanical Engineering*, in five volumes—Volume-1: *Advances in Modern Machining Processes*, Volume-2: *Advances in Additive Manufacturing and Metal Joining*, Volume-3: *Advances in Simulation, Product Design and Development*, Volume-4: *Advances in Forming, Machining and Automation* and Volume-5: *Advances in Micro and Nano Manufacturing and Surface Engineering*.

Volume-4 entitled *Advances in Forming, Machining and Automation* covers a wide variety of technical papers related to bulk-forming, incremental forming, laser bending, hydroforming, textured tools, laser-assisted machining, cryogenic grinding, condition monitoring, robotics, intelligent manufacturing, industrial Internet of Things, computer vision, enterprise manufacturing intelligence, etc. These papers clearly indicate the changing trends in the era of Industry 4.0. We hope that researchers, as well as practicing engineers, will find this volume useful.

We sincerely thank the members of the National Advisory Committee of AIMTDR, organizers, reviewers, authors and participants. Special thanks to Springer for publishing the select papers of AIMTDR since AIMTDR 2014. Readers are requested to send us feedback about this volume.

Guwahati, India
Chennai, India
Coimbatore, India
Coimbatore, India

Uday S. Dixit
uday@iitg.ac.in

M. Kanthababu
kb@annauniv.edu

A. Ramesh Babu
arb.mech@psgtech.ac.in

S. Udhayakumar
suk.mech@psgtech.ac.in

Contents

Forming

About the Editors

Dr. Uday S. Dixit is Professor in the Department of Mechanical Engineering, Indian Institute of Technology, Guwahati. He obtained his Bachelor's degree in Mechanical Engineering from the Indian Institute of Technology, Roorkee in 1987. He obtained his M.Tech. and Ph.D. in Mechanical Engineering from the Indian Institute of Technology, Kanpur in 1993 and 1998, respectively. He has more than two decades' experience in carrying out research in the area of metal forming and machining. Apart from FEM, he uses fuzzy set theory and neural networks in his research. Before taking up a research career, he worked as a machine tool designer in HMT Ltd. Pinjore, India. He has published 139 peer reviewed international journal papers and published 17 books. e-mail: uday@iitg.ac.in

Dr. M. Kanthababu is Professor in the Department of Manufacturing Engineering in Anna University, Chennai, India and Director of the Centre for Intellectual Property Right and Trade Mark, Anna University. He obtained his B.E. from Bharathiyar University, M.S. and Ph.D. from the Indian Institute of Technology, Madras. Prof. Kanthababu's research interests include manufacturing technology, composite materials and machining, and automation in manufacturing. He has published 32 peer reviewed international journal papers and two books, and holds one patent. e-mail: kb@annauniv.edu

Dr. A. Ramesh Babu is Professor in the Department of Mechanical Engineering, PSG College of Technology, Coimbatore. He obtained his B.E. from Osmania University, M.Tech. from University of Mysore and Ph.D. from the Indian Institute of Technology, Madras. He pursued his Post-Doctoral Research at Lehrstuhl für Qualitätsmanagement und Fertigungsmesstechnik, Universität Erlangen-Nürnberg, Germany. He worked for Amada Co. Ltd., Japan (at Amada Soft India), world's leading machine tool manufacturers specilized in sheet metal cutting and bending machinery. His research interests include CAD/CAM, Artificial Intelligence in Manufacturing,

Robotics, Augmented Reality, Image processing, Industrial IoT and Product development. He has published 20 technical papers in various reputed international journals and international conferences and contributed for book chapters. e-mail: arb.mech@psgtech.ac.in

Dr. S. Udhayakumar is Assistant Professor in the Department of Mechanical Engineering, PSG College of Technology, Coimbatore. He obtained his B.E. from Bharathiar University, M.E. and Ph.D. from Anna University, Chennai. His major areas of research include manufacturing automation, mechatronics and energy harvesting. He has published more than 60 papers in reputed International and National Journals/Conferences. He has received the 'Institution of Engineers—Young Engineer' award, 'PSG Teacher of the Year' award and 'Institution of Engineers—Young Achiever' award. He has been involved in two technology transfers and has filed four patents. e-mail: suk.mech@psgtech.ac.in

Forming

A Novel Method for Identification of Mechanical Properties During Impact Forming of SS 304L Sheet

S. K. Barik, R. Ganesh Narayanan, and N. Sahoo

Abstract The present investigation focuses to utilize a shock tube facility to perform dynamic forming of a 1-mm-thick SS 304L sheet at an intermediate strain rate. The strain rate evolution is acquired experimentally by mounting a strain rosette on the specimen. During this analysis, the rate-dependent flow stress–strain properties are evaluated at the measured strain rate by the tensile test of the deformed sheet after the shock tube-based experiment. The rate-dependent material properties are identified by testing the deformed tensile sample in the universal testing machine at a crosshead speed of 1 mm/min. The tensile test data of the deformed sheet illustrates that the yield stress and the ultimate tensile stress of the material increase as compared to the results obtained from the base sheet. Moreover, the rate-dependent stress–strain data is validated with the flow stress curve obtained from the Cowper-Symonds flow stress model. Both the flow stress curves demonstrate a good correlation with a slight over prediction. The new method for determining the rate-dependent mechanical properties is reliable and can be implemented in other high strain rate forming processes.

Keywords Shock tube · Tensile test · Strain rate · Impact loading · The flow stress model

1 Introduction

Stainless steel is commonly used in automobile industries because of its good mechanical properties, higher formability and increased resistance to corrosion [1]. Several research activities confirm that during transportation application, the material is exposed to different strain rates in the range of 10^2–10^3 [2]. Further, the mechanical properties of the stainless steel are strongly dependent on the loading rates [3]. Thus, it is important to characterize the deformation behaviour of the stainless steel sheets at different strain rates.

S. K. Barik (✉) · R. G. Narayanan · N. Sahoo
Department of Mechanical Engineering, Indian Institute of Technology Guwahati, Assam 781039, India
e-mail: saibal@iitg.ac.in

U. S. Dixit et al. (eds.), *Advances in Forming, Machining and Automation*, Lecture Notes in Mechanical Engineering, https://doi.org/10.1007/978-981-19-3866-5_1

Generally, the quasi-static tensile test and the split Hopkinson pressure bar (SHPB) test have been used in most of the investigations to describe the material's mechanical properties at low and high strain rates, respectively. It is difficult to attain intermediate strain rates with the normal test set-up. In the SHPB test set-up, the lowest strain rate of 600 s^{-1} is reported [4]. Recently, Grolleau et al. [5] modified the SHPB into a dynamic bulge testing device and performed a dynamic material test at intermediate strain rates. Further, Ramezani and Ripin [6] conclude that this approach is limited due to strain inhomogeneity because of the increased complexity in the setup.

In the last two decades, high energy rate forming devices have been widely used to obtain the bi-axial forming behaviour of the materials over a wide range of strain rates [7–9]. The shock tube facility has recently been used on a lab scale to study the dynamic response of thin sheets [10–13]. It is generally used to establish an impulsive loading environment for a small duration. The uniformity in loading and ease of handling make the shock tube facility advantageous to utilize as a dynamic testing device. Different levels of shock loading provide different loading conditions that help to study the dynamic mechanical behaviour of the material. Stoffel [10] used a shock tube during the investigation of the dynamic behaviour of the sheet metals. Justusson et al. [11] utilized the shock tube during the bi-axial forming of the aluminium sheets and validated it with the FE model results. Barik et al. [12] investigated the forming behaviour of aluminium alloy sheets using a shock tube and validated the forming outputs to the results obtained from FE simulation. They obtained the rate-dependent material parameters by the tensile test of the sheet deformed using the shock tube and incorporated during FE simulation. The predicted results matched quite well with the experimental outputs.

The tensile test of the deformed sheet is new and reliable to determine the rate-dependent mechanical properties. There has never been an attempt to use the shock tube to acquire the mechanical properties of the SS 304L sheet in the intermediate strain rate range. As a result, the shock tube is used in this study to perform impact loading on SS 304L sheet. During the experiment, a strain rosette is fixed on the sheet to capture the strain evolution during the deformation. The stress–strain data under that strain rate is obtained by the tensile test of the sample cut from the sheet deformed using a shock tube. Further, the tensile test data is validated with the flow stress curve obtained from the Cowper-Symonds model.

2 Experimental Methodology

2.1 Mechanical Properties of the Material

In the current investigation, SS 304L sheet of 1 mm thickness having chemical composition Cr %: 18.20, Ni %: 8.01, Mn %: 1.00, Si %: 0.36, P %: 0.031, C %: 0.056, S %: 0.002 is considered for the analysis. The tensile properties of the base sheet are characterized along 0°, 45° and 90° to the rolling direction in a universal testing

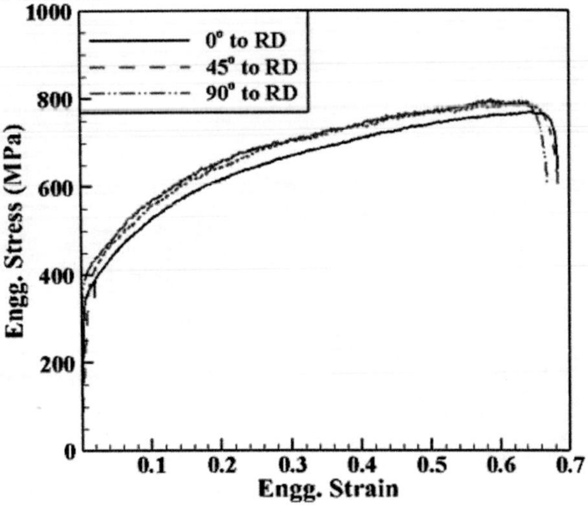

Fig. 1 Engineering stress and strain curves for SS 304L base sheet

Table 1 Mechanical properties of the base sheet

Material	RD	σ_{ys} (MPa)	σ_u (MPa)	K (MPa)	n	e_u (%)	e_t (%)	r
SS 304L	0°	341 ± 2	746 ± 3	1484 ± 3	0.39	64.8 ± 3.4	68.5 ± 3.2	1.02
	45°	384 ± 3	790 ± 3	1562 ± 3	0.38	61.7 ± 4.8	68.5 ± 4.8	0.99
	90°	378 ± 3	798 ± 4	1498 ± 4	0.36	58.7 ± 3.2	65.4 ± 4.1	0.98

Gauge length: 25 mm; σ_{ys}: Yield stress; σ_u: Ultimate tensile stress; K: Strength coefficient; n: Strain hardening coefficient; r: Plastic strain ratio

machine (UTM) at 1 mm/min crosshead speed as per ASTM-E8 and represented in Fig. 1. ASTM-E517 is used to identify the plastic strain ratios (r) of the sheets. Table 1 depicts the mechanical properties of the base sheet.

2.2 *Shock Tube Experiment*

In the present investigation, a shock tube facility as illustrated in Fig. 2 has been utilized to perform dynamic loading on sheets. The details about the experimental facility can be found elsewhere [12, 13]. The pressure difference created between the driver and the driven section of the shock tube helps during the sudden rupture of the diaphragm. It generates a shock wave, which propagates at a high velocity towards the end of the shock tube. After imparting the end of the shock tube, the shock wave reflects and generates higher pressure than the incident shock wave. The high-pressure field zone is created inside the shock tube for a short period. This impulsive environment can be used to deform a sheet in bi-axial mode at different strain rates by placing it at the end of the shock tube.

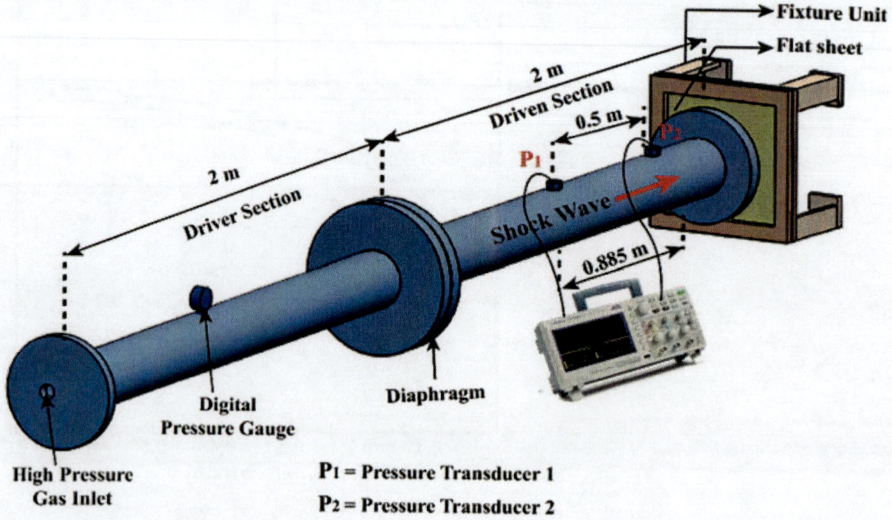

Fig. 2 Illustration of the shock tube experimental facility

During the shock tube experiment, 31.02 ± 0.68 bar of bursting pressure is attained. The detailed instrumentation in the shock tube is mentioned in the previous work of Barik et al. [12]. The pressure–time signals acquired from the pressure transducers positioned in the driven part of the shock tube are illustrated in Fig. 3. The experimentally obtained Mach number (M_s), incident pressure and reflected pressure is 1.94 ± 0.02, 4.32 ± 0.82 bar and 19.02 ± 0.55 bar, respectively.

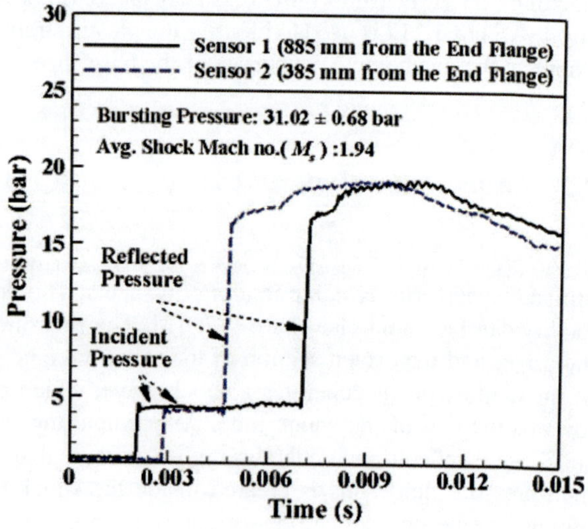

Fig. 3 Pressure–time history obtained from the experiment

It is observed that SS 304L becomes rate-dependent when the rate of loading reaches a threshold limit [3]. SHPB has been used in many studies to characterize the mechanical properties of SS 304L sheets at high strain rates [14]. During the high strain rate forming, Cowper-Symonds (CS) flow stress model is also used, which takes into account the rate-dependent effect during material forming. The Cowper-Symonds (CS) flow stress model is given by

$$\sigma = \sigma_{qs}\left(1 + \left(\frac{\dot{\overline{\varepsilon}}}{C_m}\right)^{\frac{1}{p}}\right) \tag{1}$$

where σ_{qs} is the quasi-static constitutive behaviour of the sheet, σ is the dynamic flow stress, $\dot{\overline{\varepsilon}}$ is the effective strain rate, C_m and P are the strain rate factors that scale the material's initial quasi-static stress to represent the dynamic mechanical behaviour. In this study, the dynamic mechanical properties of the sheet are identified by the tensile test of the sheet deformed using a shock tube. The tensile sample is cut from the deformed sheet's mid location along 0° to the RD (Fig. 4), and then it is tested in UTM at 1 mm/min crosshead speed.

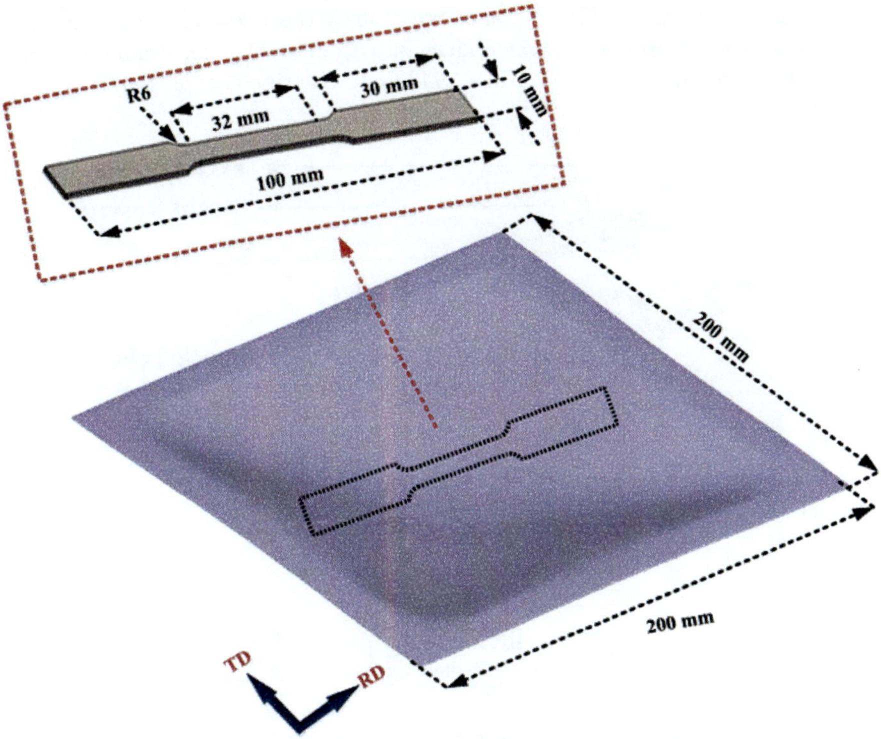

Fig. 4 Tensile test sample obtained from the sheet deformed using the shock tube

2.3 Strain Rate Evolution Measurement

During this study, the strain rosette is used to measure the strain rate evolution during the shock wave-based deformation analysis. Generally, in many high strain rate forming experiments, a high-speed 3D digital image correlation (DIC) system has been utilized to determine the transient variation of the forming parameters [11]. However, it is witnessed that the strain gauge can also be used to measure the strain rate and the results have a good agreement with the results obtained from DIC [15]. Thus, a strain rosette is attached at the centre of the sheet (Fig. 5) to quantify the in-plane strain as well as strain rate along 0°, 45° and 90° to the RD. Details about the strain rosette can be found elsewhere [12]. The strain rosette is connected by the Wheatstone quarter bridge circuit (Fig. 5). When the pressure developed by the shock wave generates impact loading on the sheet, the strain gauge's resistance changes and it causes instability in the Wheatstone bridge.

The output voltage from the Wheatstone bridge circuit is obtained in millivolts. It is difficult to predict the strain outputs from the measured signal. Thus, the voltage outputs are amplified in INA 128, DC voltage amplifier and then captured in an oscilloscope. The voltage outputs can be converted into strain rate signals. The strain rate signals obtained from the strain rosette along 0°, 45° and 90° to the rolling direction are represented in Fig. 6. The effective strain rate ($\dot{\bar{\varepsilon}}$) can be obtained using Hill's 1948 yield criterion [16], which takes into account the peak strain rate along 0° and 90° during the calculation. Hill's 1948 yield criterion is expressed as

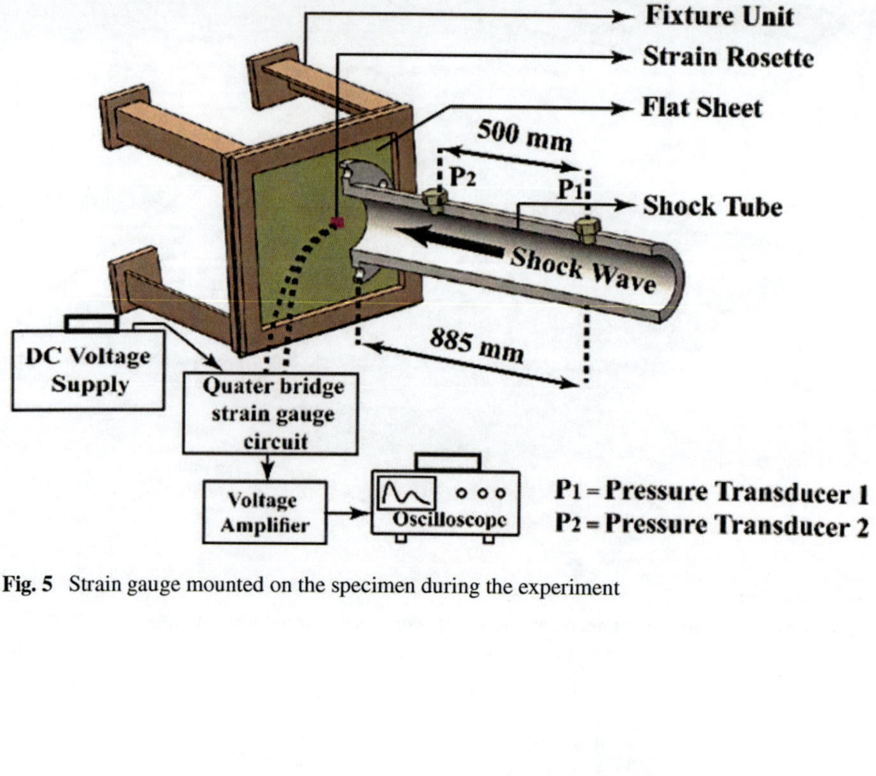

Fig. 5 Strain gauge mounted on the specimen during the experiment

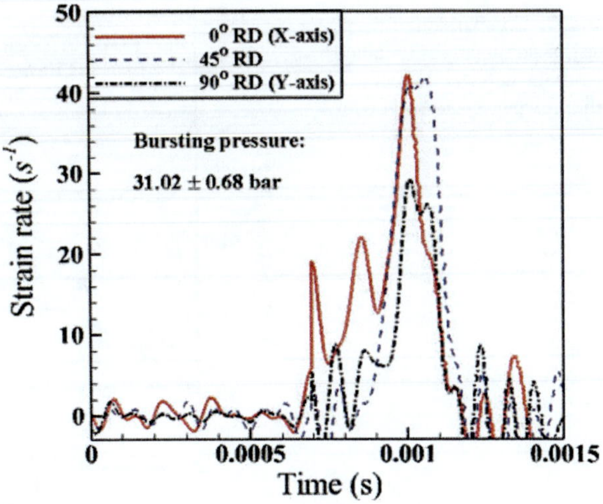

Fig. 6 Strain rate developed during the shock tube-based forming of flat sheet

$$\dot{\bar{\varepsilon}}^2 = \frac{(G+H)}{(FG+FH+GH)^2}\left[F^2(G+H)\dot{\varepsilon}_1^2 + G^2(F+H)\dot{\varepsilon}_2^2 + H^2(F+G)\dot{\varepsilon}_3^2\right] \quad (2)$$

where $\dot{\bar{\varepsilon}}$ is the effective strain rate, $\dot{\varepsilon}_1$ is the strain rate along 0° to RD, $\dot{\varepsilon}_2$ is the strain rate along 90,° and $\dot{\varepsilon}_3$ is the strain rate along the thickness direction. Identification of $\dot{\varepsilon}_3$ is difficult, and it is neglected during $\dot{\bar{\varepsilon}}$ calculation. Hill's constant (*F*, *G* and *H*) can be correlated to the plastic anisotropy parameters (r_0 and r_{90}) by the relations $F = r_0$; $G = r_{90}$; $H = r_0 r_{90}$ [16]. The values of r_0 and r_{90} are identified experimentally as mentioned in Table 1. The effective strain rate ($\dot{\bar{\varepsilon}}$) calculated experimentally is 34.26 s^{-1}.

3 Results and Discussion

3.1 Determination of Rate-Dependent Mechanical Properties

The tensile test results obtained from the deformed sheet are compared to the base sheet (Fig. 7). Table 2 depicts the mechanical properties of the deformed sheet. It is observed that both the yield stress and the ultimate tensile stress of the material rise as compared to the base sheet. The material's strength coefficient (*K*) increases significantly as a result of the action of strain hardening. However, after the deformation, the strain hardening coefficient (*n*) does not vary as much.

Furthermore, the identified rate-dependent tensile properties are validated with the dynamic stress–strain curve obtained from the CS model. During the calculation, the quasi-static flow stress data (σ_{qs}) obtained from the base sheet along 0° to RD

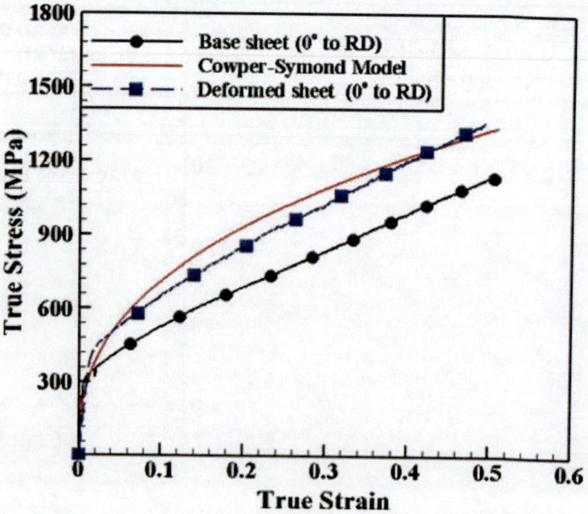

Fig. 7 Comparison of the stress–strain curves obtained from the deformed sheet to the Cowper-Symonds model

Table 2 Comparison of tensile properties of the deformed sheet with the base sheet

Tensile properties (0° to RD)	Base sheet	Deformed sheet
σ_{ys}(MPa)	341 ± 2	386 ± 3
σ_u(MPa)	746 ± 3	828 ± 3
n	0.39	0.39
K	1484 ± 3	1730 ± 5

is considered. The values of $C_m = 17{,}772\ \text{s}^{-1}$ and $P = 3.16$ for SS 304L sheets are obtained from literature [17]. The effective strain rate ($\dot{\bar{\varepsilon}}$) of deformation is calculated experimentally as 34.26 s^{-1} (reported in Sect. 2.3). The results illustrate that rate-dependent flow stress data obtained from the CS model slightly deviated from the data obtained from the tensile test of the deformed sheet. The correlation coefficient (R) between both the flow stress curves is obtained as 0.963, which is an acceptable limit. It confirms that the approach of obtaining the dynamic mechanical properties by the tensile test of the deformed sheets is reliable and can be used to predict the dynamic forming behaviour of the sheet. A similar approach can be implemented in another rate-dependent forming processes to identify the rate-dependent mechanical properties.

4 Conclusions

The present work aims to identify the rate-dependent mechanical properties of a 1-mm-thick SS 304L sheet. For this analysis, a tensile sample is cut from the mid location of the sheet deformed using the shock tube and tested in UTM. The rate-dependent flow stress–strain data is also validated with the Cowper-Symonds flow stress model. The following conclusions are drawn from the results.

i. The tensile test data obtained from the deformed sheet has a reasonable correlation with the Cowper-Symonds model. Thus, the same procedure can be followed during the identification of mechanical properties in other high strain rate forming processes.

ii. The yield strength and ultimate tensile strength obtained from the deformed sheet are higher than the base sheet. The strength coefficient (K) of the material raises significantly after deformation because of the strain hardening phenomenon. However, there is less variation in the strain hardening coefficient.

iii. During the shock tube-based experiment, the strain rosette can able to acquire the strain rate of deformation. The peak strain rate confirms the rate of loading during the experiment.

References

1. Campos HB, Butuc MC, Gracio JJ, Rocha JE, Duarte JM (2006) Theatrical and experimental determination of the forming limit diagram for the AISI 304 stainless steel. J Mater Process Technol 179:56–60. https://doi.org/10.1016/j.jmatprotec.2006.03.065
2. Vaz-Romero A, Rodríguez-Martínez JA, Arias A (2015) The deterministic nature of the fracture location in the dynamic tensile testing of steel sheets. Int J Impact Eng 86:318–335. https://doi.org/10.1016/j.ijimpeng.2015.08.005
3. Rodríguez-Martínez JA, Pesci R, Rusinek A (2011) Experimental study on the martensitic transformation in AISI 304 steel sheets subjected to tension under wide ranges of strain rate at room temperature. Mater Sci Eng A 528(18):5974–5982. https://doi.org/10.1016/j.msea.2011.04.030
4. Smerd R, Winkler S, Salisbury C, Worswick M, Lloyd D, Finn M (2006) High strain rate tensile testing of automotive aluminium alloy sheet. Int J Impact Eng 32(1–4):541–560. https://doi.org/10.1016/j.ijimpeng.2005.04.013
5. Grolleau V, Gary G, Mohr D (2008) Biaxial testing of sheet materials at high strain rates using viscoelastic bars. Exp Mech 48(3):293–306. https://doi.org/10.1007/s11340-007-9073-5
6. Ramezani M, Ripin ZM (2010) Combined experimental and numerical analysis of bulge test at high strain rates using split Hopkinson pressure bar apparatus. J Mater Process Technol 210(8):1061–1069. https://doi.org/10.1016/j.jmatprotec.2010.02.016
7. Oliveira DA, Worswick MJ, Finn M, Newman D (2005) Electromagnetic forming of aluminium alloy sheet: Free-form and cavity fill experiments and model. J Mater Process Technol 170(1–2):350–362. https://doi.org/10.1016/j.jmatprotec.2005.04.118
8. Dariani BM, Liaghat GH, Gerdooei M (2009) Experimental investigation of sheet metal formability under various strain rates. Proc Inst Mech Eng Part B J Eng Manuf 223(6):703–712. https://doi.org/10.1243/09544054JEM1430
9. Balanethiram, V. S., Daehn, G. S. (1992). Enhanced formability of interstitial free iron at high strain rates. Scripta Metallurgica et Materialia 27(12):1783–1788. https://doi.org/10.1016/0956-716X (92)90019-B
10. Stoffel M (2004) Evolution of plastic zones in dynamically loaded plates using different elastic-viscoplastic laws. Int J Solids Struct 41(24–25):6813–6830. https://doi.org/10.1016/j.ijsolstr.2004.05.060

11. Justusson B, Pankow M, Heinrich C, Rudolph M, Waas AM (2013) Use of a shock tube to determine the bi-axial yield of an aluminium alloy under high rates. Int J Impact Eng 58:55–65. https://doi.org/10.1016/j.ijimpeng.2013.01.012
12. Barik SK, Narayanan RG, Sahoo N (2020) Prediction of Forming of AA 5052–H32 Sheets under Impact Loading and Experimental Validation. J Mater Eng Perform 29:3941–3960. https://doi.org/10.1007/s11665-020-04884-w
13. Barik SK, Narayanan RG, Sahoo N (2020) Forming response of AA5052–H32 sheet deformed using a shock tube. Trans Nonferrous Met Soc China 30:603–618. https://doi.org/10.1016/S1003-6326(20)65239-6
14. Jin H, Sanborn B, Lu WY, Song B (2021) Mechanical characterization of 304L-VAR stainless steel in tension with a full coverage of low, intermediate, and high strain rates. Mech Mater 152:103654. https://doi.org/10.1016/j.mechmat.2020.103654
15. Louar MA, Belkassem B, Ousji H, Spranghers K, Kakogiannis D, Pyl L, Vantomme J (2015) Explosive driven shock tube loading of aluminium plates: experimental study. Int J Impact Eng 86:111–123. https://doi.org/10.1016/j.ijimpeng.2015.07.013
16. Wagoner RH, Chenot JL (1996) Fundamentals of metal forming. John Wiley & Sons Inc., New York
17. Mancini E, Campana F, Sasso M, Newaz G (2012) Effects of cold rolling process variables on final surface quality of stainless steel thin strip. Int J Adv Manuf Technol 61:63–72. https://doi.org/10.1007/s00170-011-3698-3

An Experimental Study of Forced Cooling in Single-Scan Laser Bending

Ramsingh Yadav, Dhruva Kumar Goyal, and Ravi Kant

Abstract Laser bending offers a controlled deformation of sheets and generates a small and precise bend angle. In this paper, experimental study of laser bending of duplex stainless steel at forced and natural cooling conditions has been carried out. The forced cooling is applied at the opposite surface to the laser beam irradiation throughout the width simultaneously. Bend angle is analyzed by varying laser power, scanning speed, and line energy during both cooling conditions. An increase in laser power results in the increase of bend angle in the beginning and then remains approximately constant, whereas it continuously increases with line energy and decreases with scanning speed. At most of the parametric conditions, it is observed that the bend angle decreases with the application of forced cooling.

Keywords Laser bending · Forced cooling · Line energy

1 Introduction

Laser bending is a recently developed forming process to get bending with high precision and accuracy. It is a thermo-mechanical process, where thermal stresses are generated by laser heating for the plastic deformation of the worksheet. There is no external mechanical force and hard tool required for this process, leading to negligible spring back effect in laser bending. A laser beam is used as a heat source for this process, irradiating along a predefined path. The worksheet gets heated along the irradiated path, and thermal expansion takes place. The surrounding cooler region restricts the thermal expansion and leads to the generation of thermal stresses. These thermal stresses are compressive in the heated region. At the same time, due to high temperature, the flow strength of the heated region is reduced drastically. As soon as the compressive thermal stresses exceed the flow strength of the heated region, plastic deformation will take place in that particular region. The compressive plastic

R. Yadav · D. K. Goyal · R. Kant (✉)
Department of Mechanical Engineering, Indian Institute of Technology Ropar, Rupnagar, Punjab 140001, India
e-mail: ravi.kant@iitrpr.ac.in

U. S. Dixit et al. (eds.), *Advances in Forming, Machining and Automation*, Lecture Notes in Mechanical Engineering, https://doi.org/10.1007/978-981-19-3866-5_2

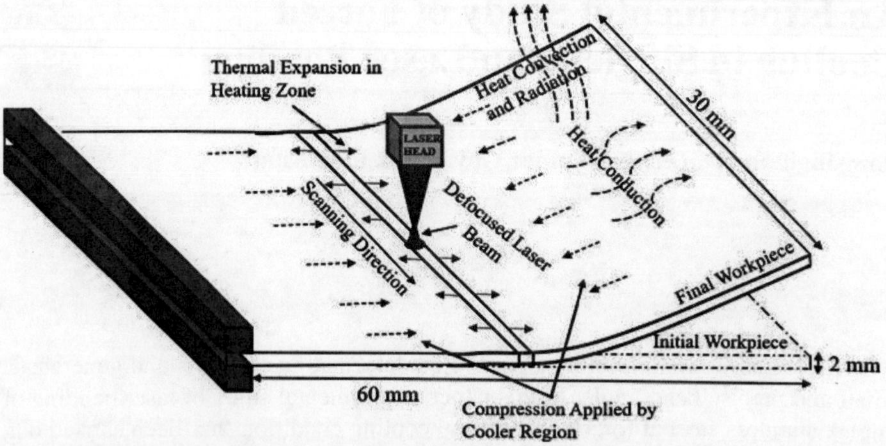

Fig. 1 Dimensions and schematic of the laser bending of duplex stainless steel

deformation in the heating region is responsible for the bending of the worksheet. Figure 1 depicts the schematic representation of the laser bending process.

Various studies have been performed over the globe to get a better understanding of the mechanisms, effect of controlling factors, and improvement of the process efficiency. Geiger and Vollertsen introduced the three mechanisms of laser forming, i.e., temperature gradient, buckling, and upsetting [1]. Some numerical studies have been conducted for a better insight into the mechanisms [2, 3]. Laser power, no. of scans, scanning speed, workpiece geometry, beam diameter, material properties, and cooling conditions are the most prominent factors which affect the process significantly [4]. Some groups have studied that the bend angle increases with laser power continuously and decreases with the scanning speed [5–8], whereas Kant and Joshi investigated the bending of magnesium alloy and reported that bend angle first increases with both laser power and scanning speed, then attains a peak and starts decreasing [9, 10]. Many researchers reported that the bend angle is continuously increasing with no. of scans, although the rate of increase in bend angle per scan is decreasing [11, 12]. Some studies report that the bend angle continuously decreases with an increase in beam diameter [7, 9]. On the contrary, Maji et al. found that there should be some optimum value of beam diameter to get higher bend angles [5]. Maji et al. also explored the effect of worksheet geometry on the bend angle and reported that bend angle has a direct relation with the width of the workpiece and inverse relation with sheet thickness, whereas there is the almost negligible effect of the length of the worksheet on bend angle [5]. Yanjin et al. analyzed the effect of various material properties on bend angle and reported that the bend angle decreases with an increase in specific heat, density, thermal conductivity, yield strength, and Young's modulus, whereas it increases with thermal expansion coefficient[13].

Laser bending has been recognized as a suitable method for bending hard-to-form materials. This process can achieve a small and precise bend angle with a minimal heat-affected zone because of its good controllability. This process is most convenient for the job and batch-type production systems. The major limitation of the process is the poor efficiency in mass production and generating large bend angels. Multi-scanning is used to get a larger bend angle, but a long cooling time is required between the consecutive scans [14]. Chang and Yao investigated the effect of forced cooling between consecutive scans during multiscan laser bending and concluded that forced cooling significantly reduces production time [15]. They also found that forced cooling at the bottom surface is beneficial for increasing the bend angle, whereas simultaneous forced cooling at both top and bottom surfaces, and forced cooling only at top surface are more favorable. Shen et al. performed a numerical study to see the effect of forced cooling and reported that bend angle produced in a single scan could be increased with forced cooling [16]. Lambiase et al. reported that the productivity could be increased with passive water cooling by decreasing the waiting time between two successive scans, but it is not affecting the bend angle [17]. Paramsivan et al. reported that bend angle could be increased by applying forced cooling at the bottom of the worksheet [18].

All the previous studies focused on forced cooling in multiscan and tried to reduce the cooling time between the consecutive scans. Most of the studies are based on numerical simulations. This paper presents an experimental study of forced cooling in single-scan laser forming. The study is focused to analyze the change in bend angle by the forced cooling at the bottom surface. The parametric study of the process in both forced and natural cooling conditions is conducted.

2 Experimental Details

In the present study, experiments have been performed using a fiber laser of power rating 1 kW. The experimental setup for natural and forced cooling conditions is shown in Figs. 2 and 3, respectively. First, the experiments are performed for the natural cooling condition, and bend angle is measured using a laser displacement sensor. Then, for experiments of forced cooling conditions, water is applied by multi-jets opposite to the scan line simultaneously throughout the width of the worksheet.

The experiments are conducted for a range of 250–750 W laser power and 15–25 J/mm line energy with a 4 mm beam diameter. As line energy is the ratio of laser power and scanning speed, the scanning speed is taken accordingly. The material used for the experiments is "Duplex stainless steel" having the dimension of 60 mm × 30 mm × 2 mm.

3 Results and Discussion

Figure 4 shows the variation in bend angle with laser power at different line energies for both natural as well as forced cooling conditions. It is observed that the bend

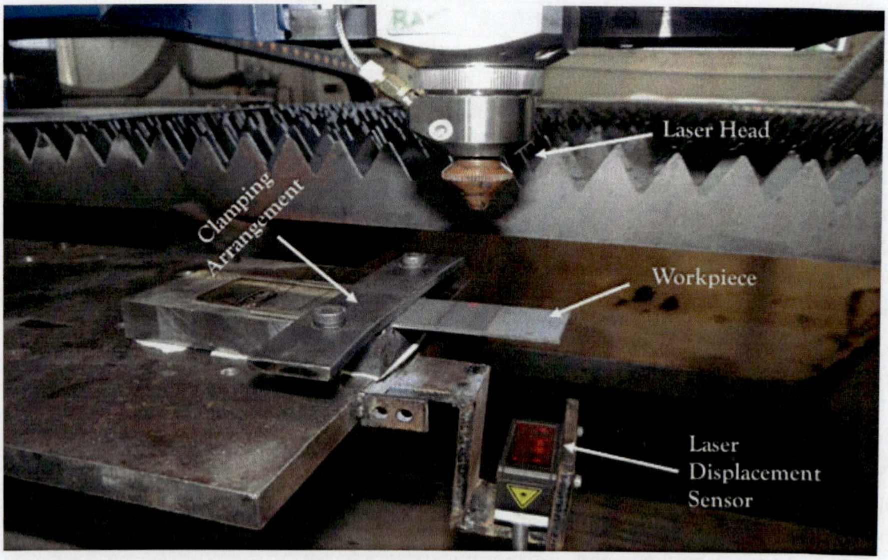

Fig. 2 Experimental setup for natural cooling conditions

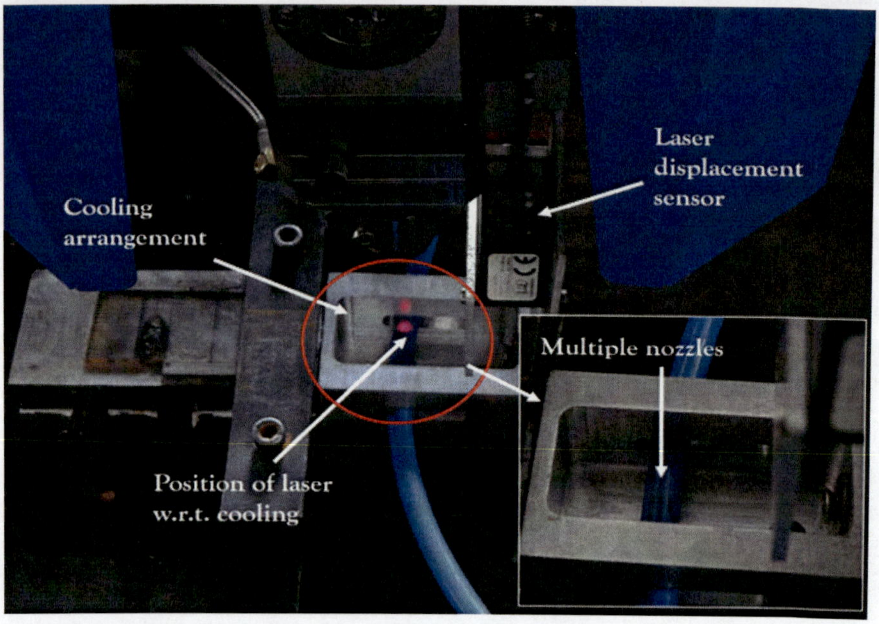

Fig. 3 Experimental setup for forced cooling condition

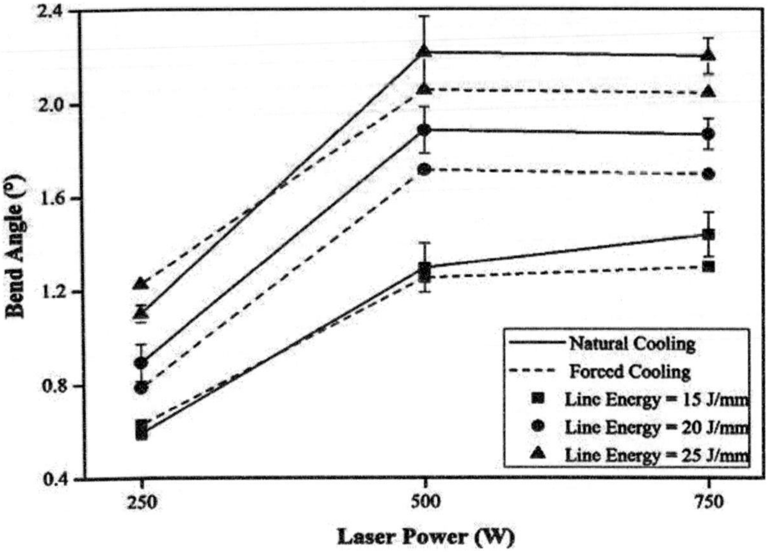

Fig. 4 Variation of bend angle with laser power at different line energies for forced and natural cooling conditions

angle first increases with laser power, then it becomes almost constant with laser power at higher line energy for natural cooling conditions. The rate of increment in bend angle is lower at higher laser powers for given line energy than at low laser powers. A similar trend is observed for forced cooling conditions also. Although at most parametric conditions, a smaller bend angle is observed with forced cooling than natural cooling. The decrement in bend angle due to forced cooling is more dominating at higher laser powers. It may be because, at the constant line energy, the scanning speed is also increasing with laser power, which reduces the laser interaction time with the worksheet. The forced cooling further reduces the energy available for the deformation of the worksheet.

Figure 5 shows the variation in bend angle with line energy at different laser powers for both forced and natural cooling conditions. The results show that the bend angle is continuously increasing with line energy for both the cooling conditions. The possible reason may be continuous increment in the energy available for the deformation of the worksheet. Due to forced cooling, this energy is reduced, which leads to a lower bend angle. At 25 J/mm line energy and 250 W laser power, the bend angle is increased with forced cooling. At low laser power and high line energy, the scanning speed is lower, which leads to a lower temperature gradient, and forced cooling increases the temperature gradient.

The variation of bend angle with scanning speed at different laser power for forced and natural cooling conditions is shown in Fig. 6. It is found that the bend angle continuously decreases with scanning speed in both cooling conditions. It may be due to the continuous decrease in the sheet-laser interaction time with an increase

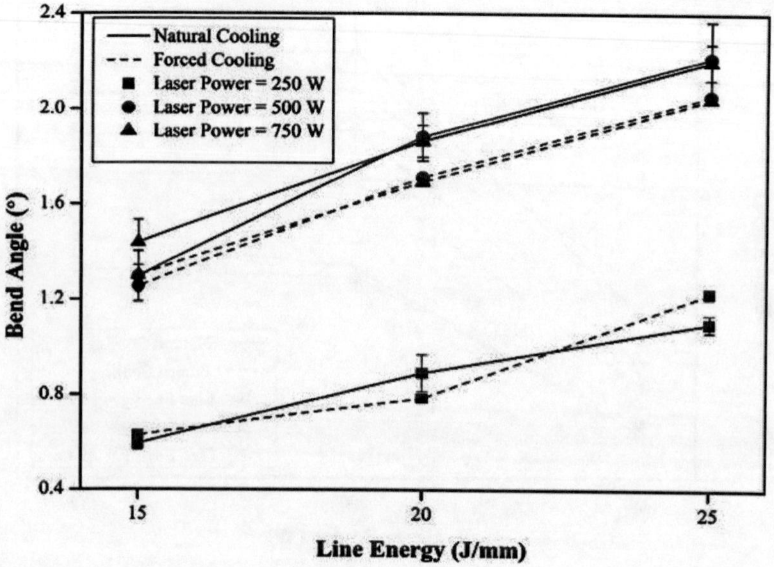

Fig. 5 Variation in bend angle with line energy at different laser powers for forced and natural cooling conditions

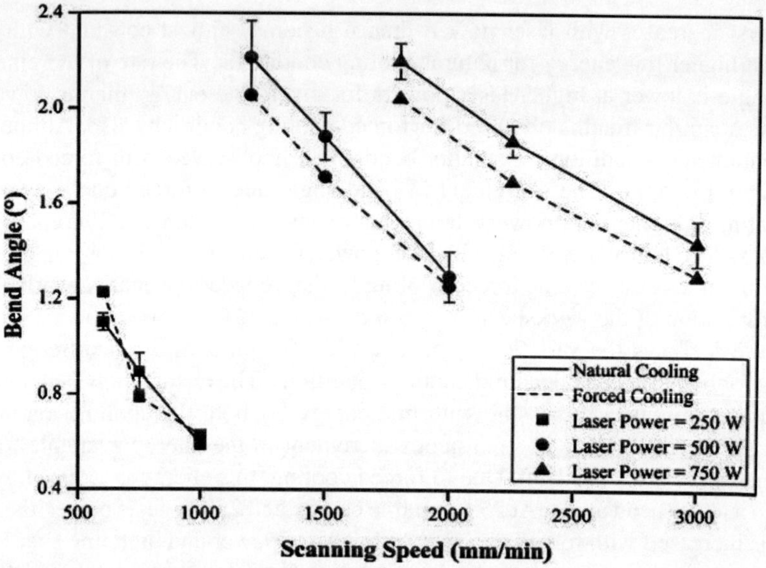

Fig. 6 Variation of bend angle with scanning speed at different laser power for forced and natural cooling condition

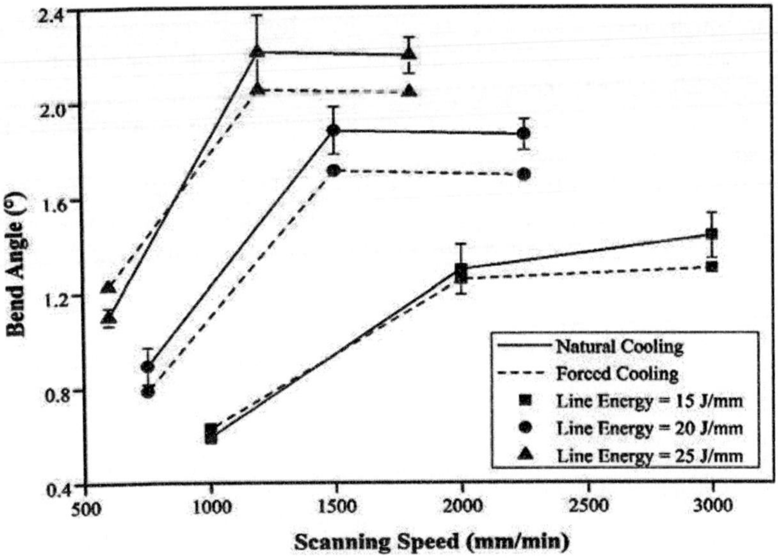

Fig. 7 Variation of bend angle with scanning speed at different line energies for forced and natural cooling conditions

in scanning speed, so the energy available for the bending is also reduced. In two different cooling conditions, the bend angle obtained is lower in forced cooling than natural cooling. It may be because the temperature required for plastic deformation is high, but it reduces due to forced cooling and decreases the bend angle.

In Fig. 7, the bend angle is plotted with scanning velocity at different line energies for forced and natural cooling conditions. It is observed that initially, the bend angle increases with scanning speed for all the line energies. The rate of change in bend angle increases with line energy. But on further increase in scanning speed, the change in bend angle at higher line energies is negligible, whereas, at lower line energy, a considerable increment in bend angle is observed. In forced cooling conditions, the bend angle shows an almost similar trend of bend angle with scanning speed, but the magnitude is decreased for most of the parametric conditions. It is because forced cooling leads to more heat losses.

4 Conclusion

An experimental study has been conducted to analyze the effect of forced cooling for single-scan laser bending. The effect of laser power, scanning speed, and line energy on bend angle for forced and natural cooling conditions is also discussed. Initially, the bend angle increases up to certain laser power, and then, it remains nearly constant. Bend angle continuously decreases with scanning speed and increases with line energy, although the variation of bend angle with various process parameters follows a similar trend in both the cooling conditions. The forced cooling negatively impacts the single-scan laser bending process as the bend angle achieved is less than that in natural cooling conditions. At low line energy, the reduction in bend angle at high laser power is more than that at lower laser power. It is also observed that at high line energies and low laser power, forced cooling is effective in getting a higher bend angle. It can be concluded that forced cooling should be preferred at high line energy with low laser power, as the trend shows that bend angle will increase for high line energy.

References

1. Geiger M, Vollertsen F (1993) The mechanisms of laser forming. CIRP Ann Manuf Technol 42:301–304. https://doi.org/10.1016/S0007-8506(07)62448-2
2. Shi Y, Yao Z, Shen H, Hu J (2006) Research on the mechanisms of laser forming for the metal plate. Int J Mach Tools Manuf 46:1689–1697. https://doi.org/10.1016/j.ijmachtools.2005.09.016
3. Paramasivan K, Das S, Marimuthu S, Misra D (2019) Numerical study of the mechanism of laser forming process. Int J Laser Sci 1:347–382
4. Dixit US, Joshi SN, Kant R (2015) Laser forming systems: a review. Int J Mechatron Manuf Syst 8:160–205
5. Maji K, Pratihar DK, Nath AK (2016) Experimental investigations, modeling, and optimization of multi-scan laser forming of AISI 304 stainless steel sheet. Int J Adv Manuf Technol 83:1441–1455. https://doi.org/10.1007/s00170-015-7675-0
6. Bellisario D, Quadrini F (2012) Laser bending of 5005 aluminum alloy sheets. Int J Manuf Mater Mech Eng 2:1–15. https://doi.org/10.4018/ijmmme.2012070101
7. Fetene BN, Kumar V, Dixit US, Echempati R (2018) Numerical and experimental study on multi-pass laser bending of AH36 steel strips. Opt Laser Technol 99:291–300. https://doi.org/10.1016/j.optlastec.2017.09.014
8. Lambiase F, Ilio A, Ilio AD (2013) A closed-form solution for thermal and deformation fields in laser bending process of different materials. Int J Adv Manuf Technol 69:849–861. https://doi.org/10.1007/s00170-013-5084-9
9. Kant R, Joshi SN (2016) Thermo-mechanical studies on bending mechanism, bend angle and edge effect during multi-scan laser bending of magnesium M1A alloy sheets. J Manuf Process 23:135–148. https://doi.org/10.1016/j.jmapro.2016.05.017
10. Kant R, Joshi SN (2016) Numerical and experimental studies on the laser bending of magnesium M1A alloy. Lasers Eng 35:39–62
11. Siqueira RHM, Carvalho SM, Kam IKL, Riva R, Lima MSF (2016) Non-contact sheet forming using lasers appiled to a high strength aluminum alloy. J Mater Res Technol 8:55
12. Edwardson SP, Abed E, Bartkowiak K, Dearden G, Watkins KG (2006) Geometrical influences on multi-pass laser forming. J Phys D Appl Phys 39:382–389. https://doi.org/10.1088/0022-3727/39/2/021
13. Guan Y, Sun S, Zhao G, Luan Y (2005) Influence of material properties on the laser-forming process of sheet metals. J Mater Process Technol 167:124–131. https://doi.org/10.1016/j.jmatprotec.2004.10.003
14. Lambiase F, Ilio AD, Paoletti A (2016) Productivity in multi-pass laser forming of thin AISI 304 stainless steel sheets. Int J Adv Manuf Technol 86:259–268. https://doi.org/10.1007/s00170-015-8150-7
15. Cheng J, Yao YL (2001) Cooling effects in multiscan laser forming. J Manuf Process 3:60–72. https://doi.org/10.1016/S1526-6125(01)70034-5

16. Shen H, Hu J, Yao Z (2011) Cooling effects in laser forming. Mater Sci Forum 663–665:58–63. https://doi.org/10.4028/www.scientific.net/MSF.663-665.58
17. Lambiase F, Ilio AD, Paoletti A (2013) An experimental investigation on passive water cooling in laser forming process. Int J Adv Manuf Technol 64:829–840. https://doi.org/10.1007/s00170-012-4072-9
18. Paramasivan K, Das S, Marimuthu S, Misra D (2018) Increment in laser bending angle by forced bottom cooling. Int J Adv Manuf Technol 94:2137–2147. https://doi.org/10.1007/s00170-017-1035-1

Determination of the Flow Stress of Material Based on a Friction-Independent Test on a Simple Geometry

G. Safiur Rahiman, N. Muthu, Uday S. Dixit, and Pavel A. Petrov

Abstract This work uses a simple dumbbell-shaped specimen for the compression test to reduce interfacial friction and determine the flow stress as a function of strain. The well-known Hollomon's relation was used to model the strain hardening behavior of the material. The strength coefficient (*K)* and the strain hardening exponent (*n*) were determined inversely using the finite element method by minimizing the error between the experimental and simulated load–displacement curves. The procedure was validated with the experiments based on a test conducted on the standard cylindrical specimens of Al6061-T6 alloy. A good agreement between estimated flow stress and experimental flow stress (instantaneous value of stress required for continuing the plastic deformation of material) was observed. The maximum error was 5% at low strains (<0.2); at higher strains, the error was negligible. This methodology will be useful for estimating the flow stress of small-sized specimens (having a diameter less than 10 mm).

Keywords Friction · Flow stress · Compression test · Finite element method · Aluminum alloy

G. Safiur Rahiman (✉) · N. Muthu · U. S. Dixit
Department of Mechanical Engineering, Indian Institute of Technology Guwahati, Guwahati, India
e-mail: s.g@iitg.ac.in

N. Muthu
e-mail: nelsonm@iitg.ac.in

U. S. Dixit
e-mail: uday@iitg.ac.in

P. A. Petrov
Department of Material Forming and Additive Technologies, Moscow Polytechnic University, Moscow, Russia
e-mail: p.a.petrov@mospolytech.ru

U. S. Dixit et al. (eds.), *Advances in Forming, Machining and Automation*, Lecture Notes in Mechanical Engineering, https://doi.org/10.1007/978-981-19-3866-5_3

1 Introduction

The uniaxial compression loading is one of the practical methods to find the flow stress of metallic materials. The flow stress–strain curve can be easily determined from the load versus displacement curve if the specimen deforms homogeneously, i.e., the cross-section does not vary along with axial (loading) direction. Most of the specimens used in the uniaxial compression test are cylindrical. The friction at the specimen-tool interface results in a non-homogeneous plastic deformation, which increases the load, thereby increasing the apparent flow stress. Further, the interfacial friction causes the barreling effect, and residual stresses are generated upon unloading [1]. Even the most effective lubricants cannot eliminate the interfacial friction but can only reduce it. Therefore, it becomes important to consider the effect of interfacial friction in a uniaxial compression test [2].

With researchers paying increased attention to micro-manufacturing, there is a need to determine mechanical properties accurately for proper modeling purposes. Mechanical properties at micro-scale can be quite different from those at macro-scale due to the so-called "size effect" [3]. Here, the size effect refers to the deviation of output parameters of the process from proportionality during the scaling of geometrical dimensions [4]. A general understanding is that interfacial friction force is several times greater in metal forming of small-sized components than in conventional ones. Vollertsen [5] classified the size effect into three categories—density, shape, and microstructure size effect. The density size effect (valid in approximately 1–10 mm size range) leads to increased strength due to a reduction in the material defect.

On the contrary, the shape size effect (valid in approximately 0.1–1 mm size range) reduces the material flow strength due to the increased surface-to-volume ratio. Engel and Eckstein [6] explained this behavior with a surface layer model. Chen and Tsai [7] conducted experiments on two cylindrical specimens with dimensions—(i) 5 mm diameter, 7.5 mm height, and (ii) 2 mm diameter and 3 mm height, and observed that the flow stress was lesser for the smaller specimen. The microstructure size effect (valid in the 10–100 μ m size range) increases material strength due to the influence of the spatial gradient of strain.

Lubricants are used to reduce interfacial friction. However, during the compression test of small-sized components, wet friction increases tremendously. This is due to the increased share of open pockets on the specimen's surface [8, 9]. Hence, there is a need to develop a compression test procedure that is friction-proof. Geometries such as hyperbolic-shaped specimens were proposed for the uniaxial compression test to alleviate the effect of friction and obtain the flow data at the higher strains [10–12]. The hyperbolic-shaped specimens pose difficulty during fabrication in laboratory settings. In the present work, a dumbbell-shaped cylindrical specimen that can be easily fabricated has been proposed. Due to the unique design of the specimen, friction plays an insignificant role during the compression process. Finally, the flow stress versus strain is obtained using an inverse analysis with a finite element model (FEM) as a direct model.

2 Proposed Method for Determining the Flow Stress

A dumbbell-shaped specimen of Al6061-T6 metal, as shown in Fig. 1, was used for carrying out the uniaxial compression test. Its middle section was of small diameter, and the other two sections were of larger diameter. The specimen was compressed between the die in a universal testing machine (the maximum reduction in the height of the specimen was 50%). Due to the unique shape of the specimen, only the middle section undergoes significant plastic deformation. Therefore, the energy dissipation due to friction can be neglected. When the load is applied, as the middle section undergoes non-uniform deformation, the state of stress is triaxial. The load–displacement curve for the compression test was obtained. Assume that the flow stress follows Hollomon's relation given by [13],

$$\sigma = K\varepsilon^n, \tag{1}$$

where σ is the effective flow stress, ε is the effective plastic strain, K is the strength coefficient, and n is the strain hardening exponent. The parameters K and n can be determined by an inverse model of the compression test from the FEM.

The axisymmetric modeling of the compression test was performed in the commercial FEM package—ABAQUS version 6.14-1. Due to the symmetry of the mid-horizontal plane, only half the specimen was modeled. Here, the discrete rigid surface represents the die. The density, Young's modulus, and the Poisson's ratio were taken as 2.7×10^3 kg/m^3, 69 GPa, and 0.33 [14]. The automatic increment time was chosen for the dynamic explicit analysis in the step module. A penalty-based friction formulation was used in the interaction module. A general contact algorithm available in ABAQUS was used.

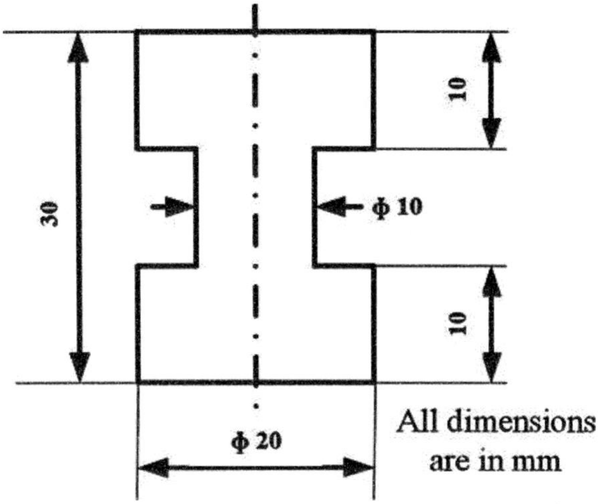

Fig. 1 Geometry of the dumbbell-shaped specimen

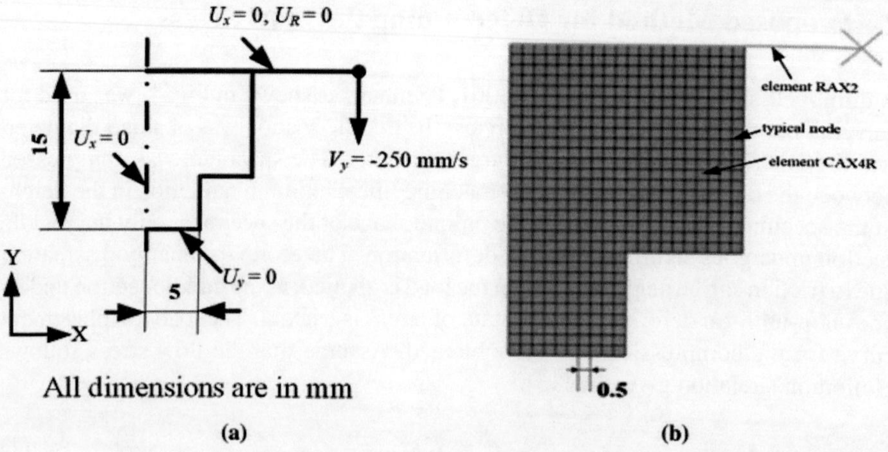

Fig. 2 Schematic of compression model **a** loading and boundary conditions and **b** meshing of the generating plane

The specimen was discretized with a 4-node bilinear axisymmetric quadrilateral, reduced integration, hourglass control (CAX4R) element. A quad element shape with a medial axis algorithm was used to mesh the specimen with an element size of 0.5 mm × 0.5 mm, as shown in Fig. 2b. This mesh was chosen based on the mesh sensitivity analysis, as shown in Fig. 2b. The die was discretized with 2-node linear axisymmetric rigid link (RAX2) elements. Figure 2a shows the Dirichlet boundary conditions where U_x and U_y refer to the displacement along *X*- and *Y*-axes, respectively, and U_R is the rotational displacement about the *Z*-axis.

2.1 Optimization Procedure

The first task in any optimization procedure is to choose a reasonable guess. For an initial guess value of *K* and *n*, the load–displacement curve was obtained through the FEM simulation. The values of *K* and *n* were adjusted to minimize the error between the experimental and simulated curves. The overall procedure is outlined in the form of a flowchart in Fig. 3.

For aluminum alloys, the *n* value lies between 0 and 0.3 [15]. The procedure to estimate a suitable *n* value is as follows. Initially, a guess value of *K* is taken. The optimization is carried out to find out a suitable value of *n*. For this purpose, a parameter *R* is defined as

$$R = \frac{F_{\text{exp}}}{F_{\text{sim}}}, \tag{2}$$

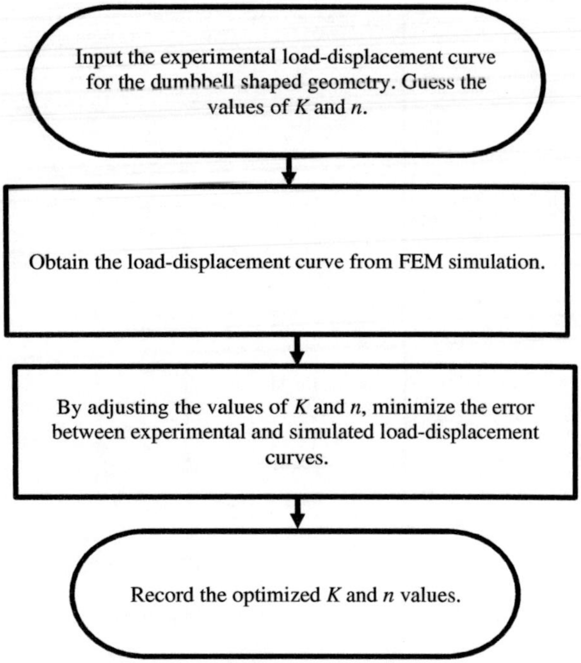

Fig. 3 Flow chart of the proposed method

where $F_{\exp}$ and F_{sim} are the experimental and simulated compression load. This parameter is calculated for nine compression displacements, and its standard deviation is calculated. The optimized value of n corresponds to the minimum standard deviation. When this happens, the experimental and simulated load–displacement curves will be almost parallel, as shown in Fig. 4. It will be easier to reduce the deviation between them by adjusting the K. The optimization was done using the interval halving method [16]. The interval was bracketed using an exhaustive search method.

For the optimized n value, at the various K values, load–displacement curves can be generated. The root mean square (E_{rms}) is calculated for a K value. The optimized K value is obtained using the interval halving method [16]. The E_{rms} between the experimental and simulated loads is defined as

$$E_{\text{rms}} = \sqrt{\frac{\sum_{i=1}^{N} \left(F_{i\exp} - F_{i\text{sim}}\right)^2}{N}}, \tag{3}$$

where N is the number of equally spaced discrete values in the load–displacement curve, $F_{i\exp}$ is the ith discrete load, and $F_{i\text{sim}}$ is the corresponding simulated load.

The interval halving method requires an interval in which the local E_{rms} minimum value lies. This interval was obtained using an exhaustive search method by taking uniformly spaced discrete values of K in a wide interval. The interval halving method is implemented in the following manner:

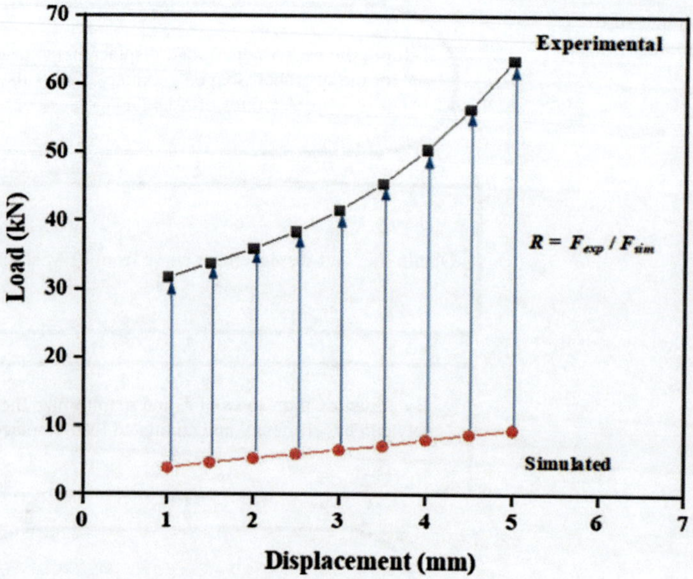

Fig. 4 Experimental and simulated load versus displacement curves

- **Step 1**: Choose a lower and an upper bound of K values, viz., K_a and K_b. The mean value is $K_m = 0.5(K_a + K_b)$. The interval length is $L = K_b - K_a$. Run the FEM simulation for optimized n value and K_m.
  Let the root mean square error of the K_m be $(E_{\text{rms}})_m$.
- **Step 2**: Set $K_1 = K_a + L/4$ and $K_2 = K_b - L/4$. Carry out simulations at these two values and obtain root mean square error as $(E_{\text{rms}})_1$ and $(E_{\text{rms}})_2$.
- **Step 3**: If $(E_{\text{rms}})_1 < (E_{\text{rms}})_m$, then $K_b = K_m$ and $K_m = K_1$; go to Step 6.
- **Step 4**: If $(E_{\text{rms}})_2 < (E_{\text{rms}})_m$, then $K_a = K_m$ and $K_m = K_2$; go to Step 6.
- **Step 5**: Set $K_a = K_1$ and $K_b = K_2$.
- **Step 6**: $L = K_b - K_a$. If the magnitude of L is smaller than 1 MPa, convergence is obtained.

A similar procedure was used to find out the optimum value of n before finding out the optimum value of K. However, the function to be minimized was the standard deviation of R based on data for nine displacements.

2.2 Details of Experiment

The Al6061-T6 specimens were subjected to the uniaxial compression test in the SHIMADZU-Static UTM of 100 kN capacity. The strain rate was of the order of 10^{-3} s^{-1}. The load–displacement curve was automatically obtained from the machine with the help of a data acquisition system. A cylindrical specimen was also

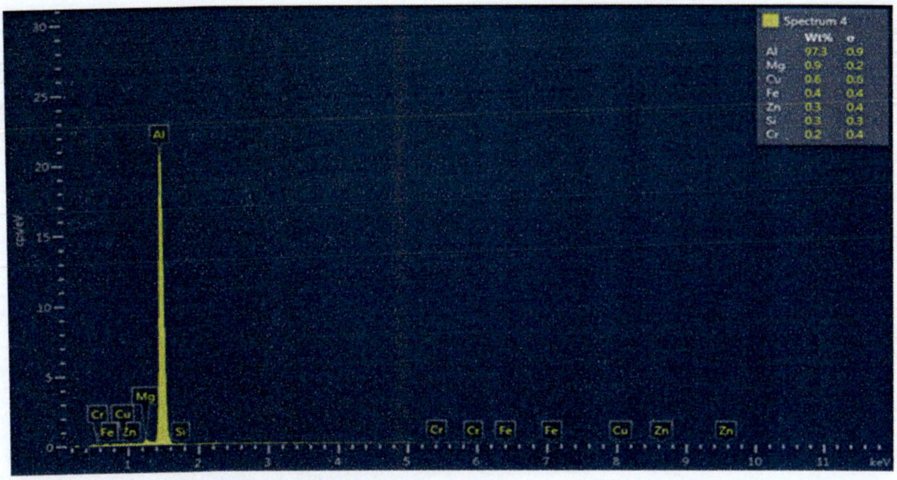

Fig. 5 EDX spectrum chart of the Al6061 specimen

Table 1 Chemical composition of Al6061 alloy

	Si%	Fe%	Cu%	Mg%	Cr%	Zn%	Al%
Composition	0.3	0.4	0.6	0.9	0.2	0.3	97.3

compressed to validate the procedure where the lubricating oil was used to reduce interfacial friction during the testing. The cylindrical specimen was of 10 mm diameter and 10 mm height. The main objective was to demonstrate that the proposed procedure provides almost the same flow stress as that obtained by compression of a cylindrical specimen. The material used in the compression test was Al6061-T6 grade, and its composition was obtained from energy dispersive X-ray (EDX) analysis (Fig. 5) conducted in the Zeiss Sigma FESEM. Composition data obtained is shown in Table 1. The actual sample geometry before deformation and after the deformation are shown in Fig. 6.

3 Results and Discussions

Although the cylindrical specimen was adequately lubricated, it was impossible to make Coulomb's coefficient μ as zero. This was apparent from the small amount of bulging noted. Hence, it was necessary to correct the flow stress values μ estimated from the bulge measurement using the FEM simulation.

Firstly, the engineering stress versus strain data obtained from the compression test of the cylindrical sample was converted to true stress versus strain data and plotted as shown in Fig. 8. This data was used in the subsequent FEM simulation of

Fig. 6 Specimen geometry **a** before deformation and **b** after deformation

the cylindrical specimen subjected to compression loading. The computed bulge B depends significantly on the coefficient of friction during the simulation of simple compression loading of the cylindrical specimen. FEM simulations were performed for two different coefficients of friction values, as shown in Fig. 7, to confirm the effect of μ on B, which is given by

$$B = R_{\max} - R_{\min} \tag{4}$$

Where $R_{\max}$ is the maximum radius at the center and $R_{\min}$ is the minimum radius at the top of the cylindrical specimen.

The experimental bulge was recorded, and the μ value was estimated through different values of μ supplied in the FEM simulation. From Table 2, it is observed

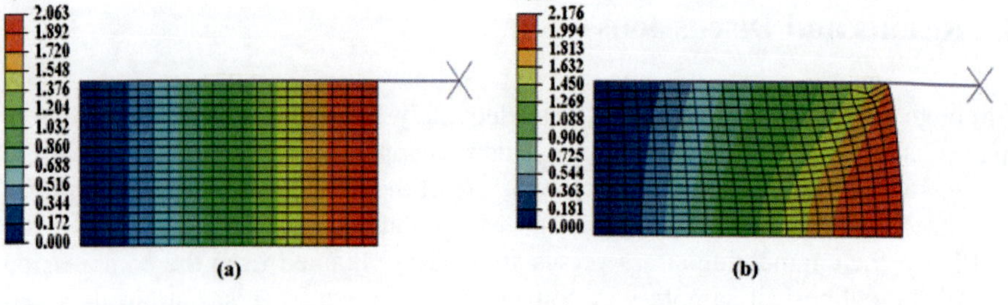

Fig. 7 Effect of friction on bulging for 50% height reduction (where U represents displacement in mm along x-direction) at **a** without friction ($\mu = 0$) and **b** $\mu = 0.05$

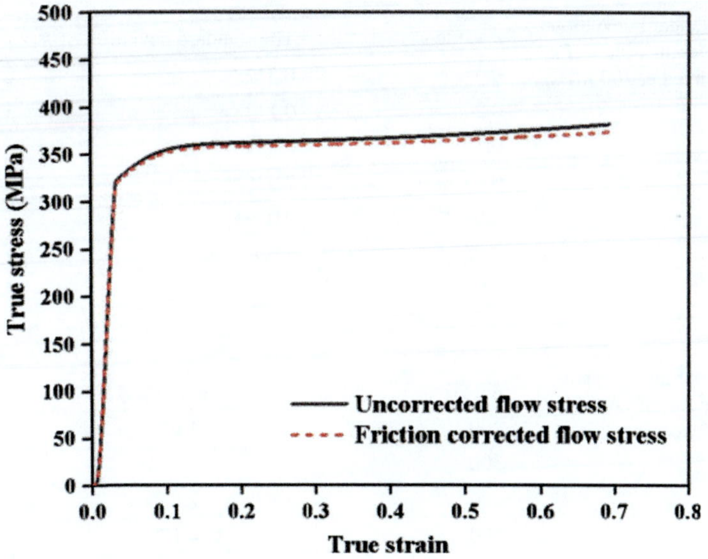

Fig. 8 Corrected and uncorrected flow stress of the cylindrical specimen

Table 2 Simulated bulge values at different friction coefficients

Input friction coefficient (μ)	B_{sim} (mm)	% Error $100\left(\frac{B_{\text{sim}}-B_{\text{exp}}}{B_{\text{exp}}}\right)$
0.02	0.15	−25
0.03	0.23	15

that as the sign of error changes between $\mu = 0.02$ and $\mu = 0.03$, the μ value for the experimental bulge corresponding to $B = 0.2$ mm, can be calculated by the secant method. Accordingly, the final value of μ was found to be 0.026.

The frictional correction on the obtained experimental flow stress was done by the following equation [17]:

$$\sigma = \frac{\sigma_\mu}{1 + \frac{2\mu r}{3h}} \tag{5}$$

where σ is friction corrected stress and σ_μ is uncorrected flow stress, r is the instantaneous radius, and h is the instantaneous height during the compression. The uncorrected and friction corrected flow stress are shown in Fig. 8.

The same value of $\mu = 0.026$ was used in the FEM simulations of the compression of the dumbbell-shaped specimens, although any other value of μ also produces the same results. Following the optimization procedure in Sect. 2.1, for a fixed K value of 200 MPa, R was estimated at different n values. Some typical values obtained during exhaustive search and interval halving method are tabulated in Table 3.

Table 3 Standard deviation of R for different values of n, keeping K fixed at 200 MPa

n value	The standard deviation of R
0.3	0.3128
0.2	0.1133
0.1	0.0912
0.05	0.0361
0.04	0.0293
0.03	**0.0242**
0.02	0.0251

Table 4 Root mean square error values for different values of K, keeping n fixed at 0.03

K (MPa)	E_{rms} (kN)
200	21.07
300	9.138
400	2.755
500	14.170

The minimum value of standard deviation was recorded at $n = 0.03$. For this optimized value of $n = 0.03$, at different K values, the E_{rms} of the load was calculated as shown in Table 4. It is observed that the E_{rms} decreases gradually and then increases. By interval halving method, the optimized K value at minimum E_{rms} can be found in the interval between $K = 300$ and $K = 500$ MPa, respectively. The final optimized value calculated by the interval halving method is $K = 380$ MPa with $E_{rms} = 0.604$ kN.

The load versus displacement curve for the dumbbell-shaped specimen at the optimized values of K and n value was plotted and contrasted with the experimental load versus displacement curve as shown in Fig. 9. There is a close agreement between ABAQUS simulated and experimental load versus displacement data. A FEM simulation was done corresponding to the optimized values of K and n to demonstrate no effect of interfacial friction during the simple compression of the dumbbell-shaped specimen. Figure 10 compares the deformed geometry of the dumbbell-shaped specimen under two different conditions. It was observed that both deformed geometries are identical, proving that friction does not play a role in this testing procedure. Figure 11 shows the stress–strain curve in a uniaxial compression test by taking the inversely obtained values of parameters K and n. This stress–strain curve was compared with the experimentally obtained stress–strain curves. It is observed that except at low strain, there is a very close matching between the two curves. Even at the low strain, the deviation between the simulated and experimental stresses is less than 5%.

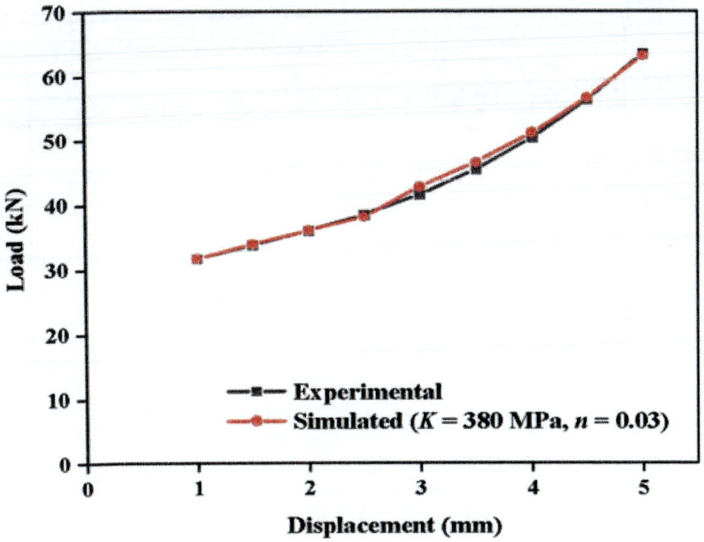

Fig. 9 Experimental and simulated (in ABAQUS) load versus displacement curve of dumbbell-shaped specimen

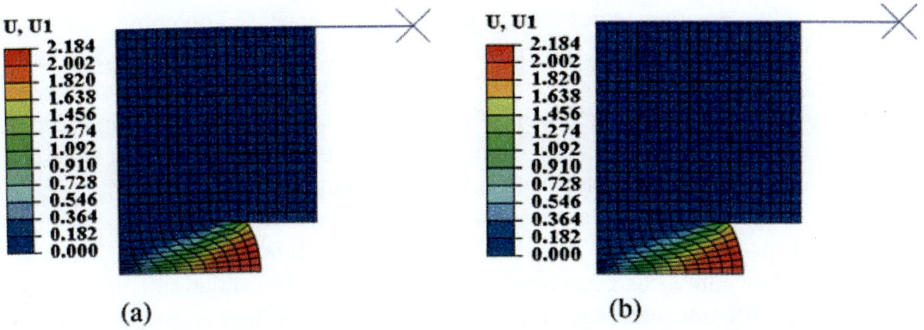

Fig. 10 Simulated deformed geometry of the dumbbell-shaped specimen at 50% height reduction **a** without friction ($\mu = 0$) and **b** $\mu = 0.05$

4 Conclusions

In this work, a new design of the dumbbell-shaped specimen was proposed to overcome interfacial friction during the compression test. This design is simple as compared to the designs used by previous researchers. It was confirmed from the FEM simulation that the bulge was absent in the region above the throat, and the deformation was concentrated in the throat region. A procedure to estimate the hardening parameters, *K* and *n*, by inverse analysis using FEM simulation was outlined. The proposed method is very simple but requires further validation for different materials

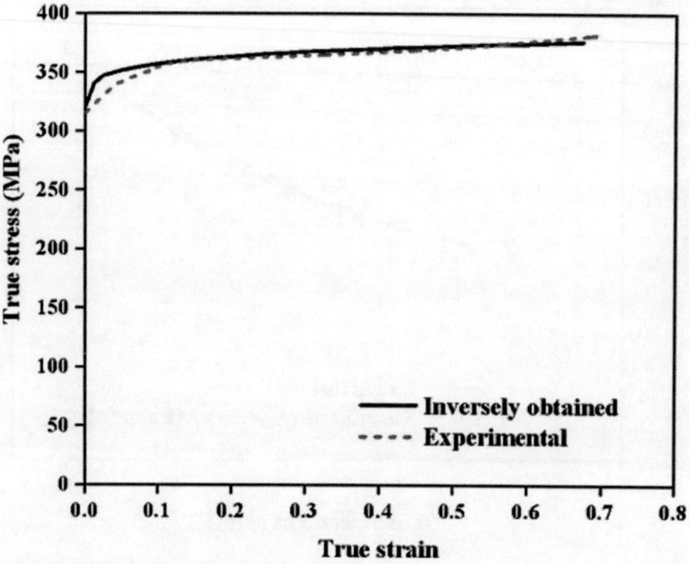

Fig. 11 Flow stress of both cylindrical and dumbbell-shaped specimens

and specimen dimensions, which will be carried out later. Based on compression of the dumbbell-shaped specimen, the following parameters are obtained: $K = 380$ MPa and $n = 0.03$. With these values, the stress–strain curve was obtained for uniaxial compression. The curve matched well with the experimentally obtained flow curve based on a uniaxial compression test on a cylindrical specimen; the significant error was observed only at the low strains (<0.2) and the maximum error was 5%.

Although some earlier researchers have proposed a similar test procedure, the present work proposes a test with simpler geometry. Moreover, a novel heuristic optimization procedure has been used. In the future, more validation tests will be conducted on different materials and specimen geometries. It is envisaged that the method will be useful for compression tests of the small-sized specimen (having a diameter less than 10 mm). The test can be useful in the proper estimation of flow behavior in micro-forming.

Acknowledgements This work is a part of a joint Indo-Russian Project entitled "Experimental and numerical research on contact friction in the process of plastic deformation using compression with torsion." It is funded by the Russian Foundation for Basic Research (RFBR) and the Department of Science and Technology (DST) according to the Research Project №. 19-58-45020 and Grant INT/RUS/RFBR/388, respectively. The authors gratefully acknowledge the support of the funding agencies.

References

1. Mungi MP et al (2003) Residual stresses in cold axisymmetric forging. J Mater Process Technol 142(1):256–266
2. Puchi-Cabrera ES et al (2020) Friction correction of austenite flow stress curves determined under axisymmetric compression conditions. Exp Mech 60(4):445–458
3. Peng L et al (2010) Friction behavior modeling and analysis in micro/meso scale metal forming process. Mater Des 31(4):1953–1961
4. Vollertsen F et al (2009) Size effects in manufacturing of metallic components. CIRP Ann 58(2):566–587
5. Vollertsen F (2008) Categories of size effects. Prod Eng Res Dev 2:377–383
6. Engel U, Eckstein R (2002) Microforming—from basic research to its realization. J Mater Process Technol 125–126:35–44
7. Chen FK, Tsai JW (2006) A study of size effect in micro-forming with micro-hardness tests. J Mater Process Technol 177:146–149
8. Vollertsen F et al (2006) State of the art in micro forming. Int J Mach Tools Manuf 46(11):1172–1179
9. Chan WL et al (2011) Study of size effect in micro-extrusion process of pure copper. Mater Des 32(7):3772–3782
10. Usami M, Oya T (2014) Estimation of work-hardening curve for large strain using friction-free compression test. Proc Eng 81:371–376
11. Chen J et al (2021) Methods for measuring friction-independent flow stress curve to large strains using hyperbolic shaped compression specimen. J Strain Anal Eng Des
12. Chen J et al (2020) A novel method to determine full-range hardening curve for metal bar using hyperbolic shaped compression specimen. J Market Res 9(3):2704–2716
13. Dixit PM, Dixit US (2014) Plasticity: fundamentals and applications. CRC Press, Boca Raton
14. ASM Aerospace Specifications Metals Inc. http://asm.matweb.com/search/SpecificMaterial.asp?bassnum=MA6061T6. Retrieved on 24 August 2021
15. Hosford WF (2010) Mechanical behavior of materials, 2nd edn. Cambridge University Press, New York
16. Deb K (2012) Optimization for engineering design, 2nd edn. PHI Learning, New Delhi
17. Torrente G (2018) Numerical and experimental studies of compression-tested copper-proposal for a new friction correction. Mater Res 21

Development of Flexible Mechanism for Two-Point Incremental Forming

S. Pratheesh Kumar, S. Elangovan, R. Mohanraj, R. Naveen Anthuvan, and V. Nithin

Abstract In incremental sheet forming (ISF), sheet metal is formed practically without the use of dies, which makes it ideal for Industry 4.0. ISF minimizes setup and unit costs but is only suitable for batch manufacturing. Despite recent improvements in ISF such as hybrid ISF, industries will not show interest to this method because of its inflexibility in forming complex shapes. Two-point incremental forming (TPIF) has better dimensional control and can make more complex shapes than single point incremental forming (SPIF) process, but lacks flexibility in deep intricate designs due to sheet failure. An aerospace component created using the developed TPIF fixture is evaluated for dimensional accuracy, surface quality, and profile correctness.

Keywords Fixture design · Two-point incremental forming · Finite element analysis

1 Introduction

Incremental sheet forming is an exciting technology that has been mostly overlooked by researchers over the last few decades. Typically, sheet metal production sectors employ a variety of forming processes that are heavily reliant on the usage of dies and punches. These techniques are feasible for large-scale manufacturing because the high costs associated with die and punch fabrication are distributed across numerous components [1]. As a result, standard dies and punches-based technologies are not economically viable in small-scale and batch production. To address this, incremental

S. Pratheesh Kumar (✉) · S. Elangovan · R. Mohanraj · R. Naveen Anthuvan · V. Nithin
Department of Production Engineering, PSG College of Technology, Coimbatore 641 004, India
e-mail: spratheeshkumarth@gmail.com

S. Elangovan
e-mail: sel.prod@psgtech.ac.in

R. Mohanraj
e-mail: rmr.prod@psgtech.ac.in

R. Naveen Anthuvan
e-mail: rna.prod@psgtech.ac.in

U. S. Dixit et al. (eds.), *Advances in Forming, Machining and Automation*, Lecture Notes in Mechanical Engineering, https://doi.org/10.1007/978-981-19-3866-5_4

sheet forming may be a feasible alternative to cost-effective sheet metal forming in batch production sectors such as prototype, automotive, aerospace, and biomedical, among others [2, 3]. The incremental forming process begins with the development of a tool path based on the component shape generated. Numerical control (NC) code is generated based on the tool path movement and is used to control the form tool movement in a computer numerical control (CNC) machine. The form tool advances across the clamped sheet along the predetermined tool path, forming the appropriate shape [4]. Without the use of a specific die, sheet metal is locally deformed to obtain the desired shape due to frictional contact and down feed movement of the forming tool [5]. This improves the process's suitability for cost-effective part manufacture. On the other hand, the incremental forming process is intrinsically slower than its conventional equivalent, making it suitable for smaller batch sizes of items. Additionally, this method has drawbacks in terms of surface finish and geometric accuracy [6]. Due to these constraints and the unique characteristics of this technology, overcoming its limitations will need a sustained scientific effort [7].

Significant effort has been made by researchers to optimize the incremental sheet forming process. The following sections cover some of the most significant achievements in fixture design by researchers from around the world. The design of a fixture for two-point incremental sheet metal forming was devised and validated [8]. They generated five concepts and chose scissors and the three-pillar concept based on Pugh's approach for a qualitative rating of design matrices. The results indicated that the scissors design lacked sufficient lateral rigidity, the three-pillar concept with nylon bushes was chosen as the final design. Throughout validation, the three-pillar concept demonstrated strain in other sections of the sheet, indicating that the fixture had moved unevenly during the sheet's formation.

Min et al. [9] examined incremental sheet formation using metallic foam as flexible die support. The results indicate that double point incremental forming (DPIF) outperforms single point incremental in terms of geometric precision. In the lowest part of the L-specimen, a significant pillow effect was noticed. The DC04 sheet was compressed locally between the forming tool and the foam, and the compressive force rose proportionately to the foam's compression amount, resulting in severe scratches and early sheet breakage during forming. Paniti et al. [10] created a C-frame fixture for TPIF sheet formation. The blank holder is secured to the base plate via the support rod. The upper forming tool is moved in the X, Y, and Z directions by the tool path defined in the software. The bottom forming tool is copying the higher forming tool's X and Y movements. This is accomplished by the use of a mechanical movement copying mechanism. The C-frame is stabilized in two ways: rationally and translationally. It has a limited working area and prevents sheet flow into the deformation zone. Zhang et al. [11] developed a hybrid approach that incorporates multi-point forming (MPF) and incremental sheet metal forming. MPF is used to create a preform that permits materials in the blanking zone to flow into the deformation region. (a) Avoidance of excessive sheet thinning in MPF. (b) Increased process adaptability. (c) Increased formability. (d) Efficiencies are increased. Araghi et al. [12] proposed a novel hybrid technique combining AISF and stretch forming. This procedure results in a more uniform thickness distribution with less maximum thinning. A significant reduction in manufacturing time is realized when compared to asymmetric ISF. Process planning with finite element analysis (FEA) appears to be more feasible than asymmetric ISF. The primary takeaways from the numerous literature cited are as follows: No material flow is conceivable with the fixtures now available, necessitating the development of a novel design that allows for flow. All known hybrid approaches improve the process's dimensional accuracy and surface finish. While electrically aided, heat assisted, multipoint forming, and stretch forming incremental forming techniques add some flexibility, they all lack material flow.

2 Challenges in ISF

Despite its superior dimensional control and capacity to construct complex designs over single point incremental sheet metal forming, the two-point incremental sheet metal forming (TPIF) technology lacks flexibility when creating complex shapes [13]. Dimensional precision is required for industrial acceptance [14]. Currently, industries prefer single point incremental sheet metal (SPIF) forming over two-point incremental sheet metal (TPIF) forming for job order production and prototyping. There is a need for a process that satisfies industry-specific dimensional standards while minimizing setup and unit costs and increasing flexibility; however, in the TPIF technique, the inability of material to form complex shapes results in premature tearing of the sheet before forming the complete shape as that of the die [15, 16]. Thus, the purpose of this project is to design, build, and validate a novel two-point incremental sheet metal forming fixture that enables material movement into the deformation zone during forming.

3 Fixture Design

A. Requirements of the fixture

The fixture must allow material to flow into the deformation zone when forming. The fixture must travel downward to make contact with the die and to maximize material use. The fixture must be modular in design and maximize the usage of available space. The fixture must be flexible enough to generate complicated shapes while maintaining a low setup and unit cost.

B. Concept generation

The specifications for the fixture have been defined in light of the issues with the existing fixture. Numerous concepts have been produced in response to these parameters, and the most suitable concept is chosen using Pugh's chart. The chosen concept is transformed into the final design following the specifications.

With a pin in the center, the scissor design lowers the sheet holder as shown in Fig. 1. Four rollers travel in a straight line and allow the sheet holder to adjust vertically while maintaining a constant horizontal position. An examination of the idea prototype revealed that it lacked lateral rigidity.

Although various revisions were made to the initial design, these modifications reduced, rather than eliminated, the fixture's lateral movement, or play. These trials established that idea 4 possessed an intrinsic decreased lateral stiffness.

Steel and wooden dies of various shapes were created to verify the prototype's kinematic and mechanical behavior. The dies depicted in the illustration are modular and may be simply replaced to create symmetrical and asymmetric shapes. Additionally, the fixture might be used with 3D-printed polymer dies as shown in Fig. 2. To allow for unfettered movement of the fixture on the three columns, nylon bushes were employed rather than linear rolling-element bearings. Strain in different sections of the sheet would indicate that the fixture moved unevenly during the sheet's formation.

Fig. 1 Model of scissors type

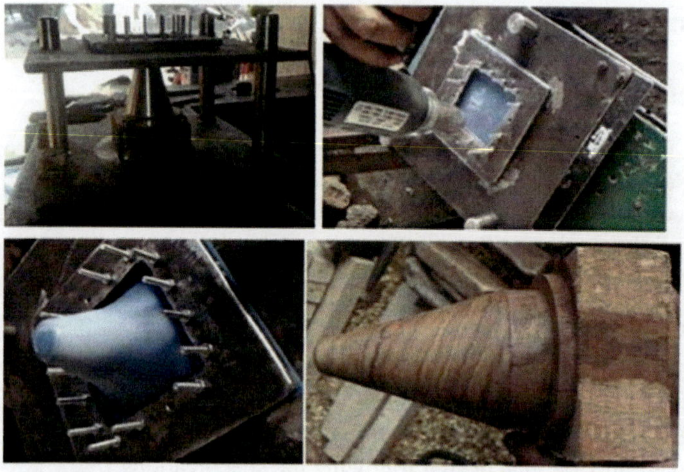

Fig. 2 Model of pillar type

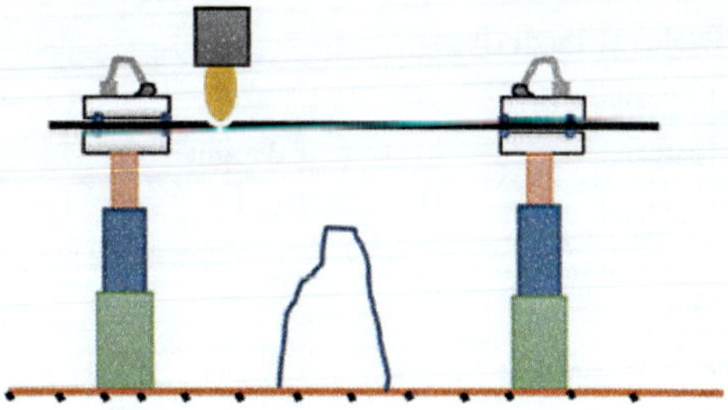

Fig. 3 Expanding cylinder concept

This design incorporates rollers between the blank holders to maintain point contact with the sheet during clamping as shown in Fig. 3. When the material is required, the clamping mechanism is loosened and the material flows in. While forming, the setup must move down to contact to die efficiently, which is why expanding cylinders are utilized.

This configuration allows for both material flow and downward movement, but it lacks effective utilization of the machine's working space and is more expensive than the hydraulic configuration. This concept is deemed unacceptable for two primary reasons.

This design is identical to the previous one, but instead of expanding cylinders, a lead screw configuration is utilized as shown in Fig. 4, which reduces the cost of the fixture and allows for easy control of the fixture's download movement using a gear unit, unlike the previous design. This system corrects the prior design's flaws.

C. Selection of concept

The generated concepts are chosen using Pugh's weighted technique. The design's primary goal is to ease material flow into the deformation zone, and the design must be modular. As a result, these two factors are weighted more heavily in the design as given in Table 1. Additionally, other characteristics like stability, cost inclusion, ease of production, and convenience of use are ranked and weighted.

The lead screw model is chosen for quantitative analysis as a result of its qualitative ranking in the design matrix. For this concept, three-dimensional computer-aided design (CAD) models were built.

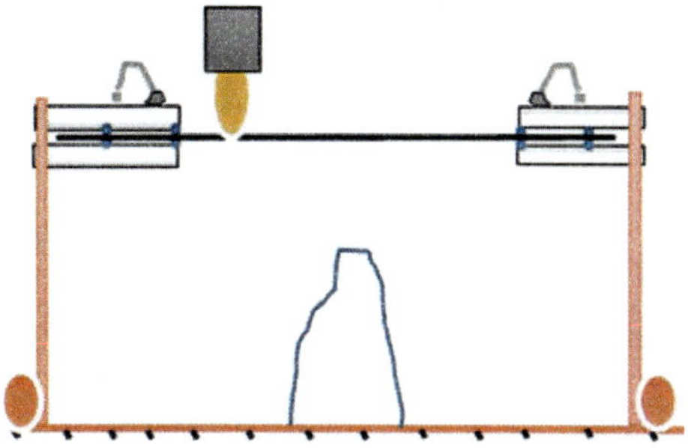

Fig. 4 Lead screw model

Table. 1 Pugh chart

Description		Concepts				
Criteria	Weight	1	2	3	4	5
Material in-flow	5	−	−	−	0	+
Modular design	5	−	−	−	0	+
Stability	4	−	+	+	−	+
Cost inculcated	3	+	0	0	−	−
Ease of manufacturing	3	+	+	+	+	0
Ease of usage	2	+	+	+	−	0
+		8	9	9	3	14
0		0	3	3	10	6
−		14	10	10	7	3
Net score		−6	2	2	−4	11

4 Development of Design

The fixture is meant to accomplish the purpose of eradicating the issues associated with existing fixtures. The created fixture is described in Figs. 5 and 6. The fixture's components are roughly categorized into two types.

1. Top frame assembly
2. Bottom frame assembly.

A. **Top frame assembly**

The top frame assembly establishes the path for material to enter the deformation zone. As the upper frame continues to build, it slides downward. The supporting frame that is attached to the lead screw is capable of moving sideways in response to the rotation of the lead screw.

Fig. 5 Top frame assembly

Fig. 6 Bottom frame assembly

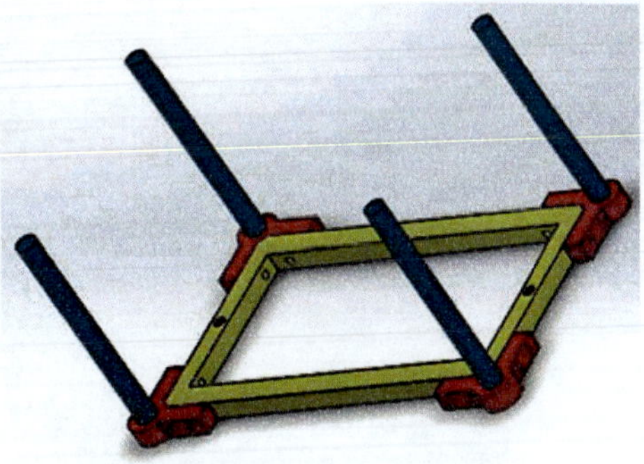

The sheet is secured to the supporting frame in one direction and a clamp rod in the other two. When forming begins, the supporting frame moves sideways in response to the movement of the lead screw, and material flows into the deformation zone. The rotation of the lead screw is proportionate to the step depth of the tool being formed.

B. **Bottom frame assembly**

The bottom frame assembly is made up of L-joint bottom, shafts, and bottom frame. Four L-joints are secured to the bottom frame's four corners. The shaft is inserted into the L-internal joint's provision.

The design comprises two supporting frames guided by lead screws that are attached to the outer frame and used to clamp the sheet metal as shown in Fig. 7.

C. **General assembly**

The new design will allow material to migrate into the zone of deformation while producing complicated shapes. The lead screw is attached to the stepper motor, which provides rotational control. While forming, the angle of rotation is proportional to the step depth. The supporting frame will gravitate toward one another, allowing the material to flow in. The sheet's other two sides are held together with a rod.

Fig.7 Developed design

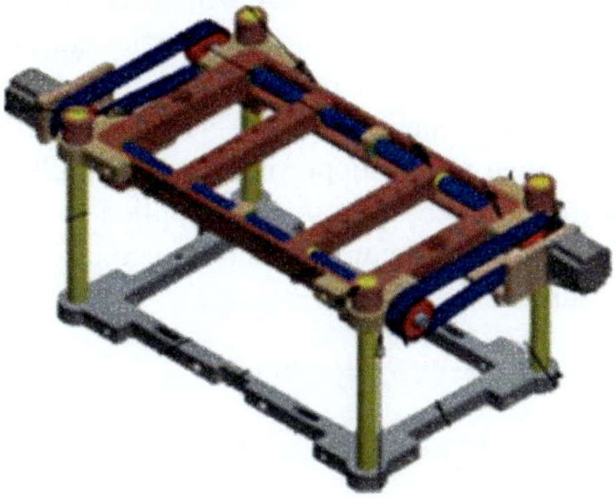

Fig. 8 Fabricated fixture

As a result, as the material is created, it moves in all directions. The outer frame is attached to the shaft on all four sides by L-joints. While forming, the outer frame will slide down the shaft until the die is completely formed. The design is modular in nature, with the outer frame being interchangeable with another.

All components are made of mild steel. The fixture will be constructed to the specified tolerances. Figure 8 shows the fabricated fixture. Experiments will be conducted to fabricate an aircraft component utilizing the fixture designed.

5 Results and Discussion

A. FEM analysis of fabricated fixture

The finite element analysis (FEA) approach is a computer-aided technique for predicting how a product will react to forces, vibration, heat, and other physical influences encountered in the actual condition. It indicates whether a product will break, wear out, or perform as intended. Ansys workbench 15.0 is used to analyze fixture deformation. A 2000 N load is imparted to the supporting frame. 2000 N force is applied to the supporting frame, which bears the majority of the load during the forming process. Pressure is applied to the frame, and the equivalent deformation of the other structures is determined, as well as the equivalent von-Mises stress and strain.

B. Fabricated prototype

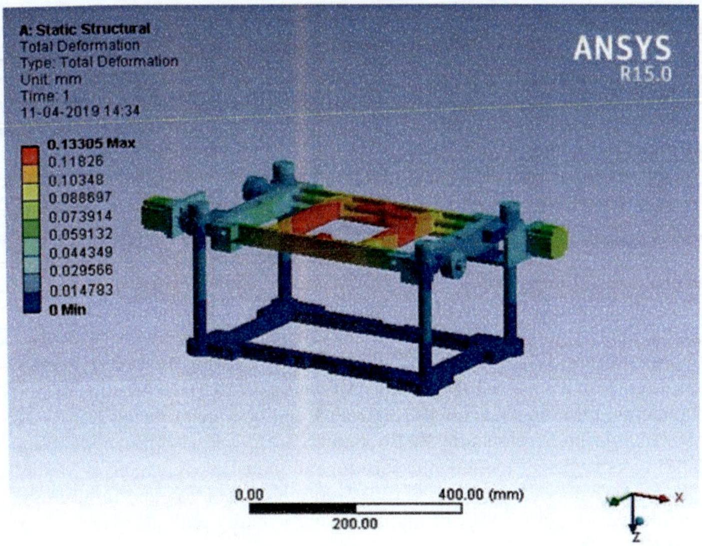

Fig. 9 Load analysis of fabricated fixture

The intended fixture is manufactured to the necessary dimensions and tolerances. Since this is a prototype model, mild steel was chosen as the basic material. Generally, semi-automatic vertical milling machines, turning machines, and jig boring machines are employed. For some components, a lathe was also used. As a prototype model, the primary focus was on the fixture's functional elements.

Due to cost and other considerations, only semi-automatic and traditional machinery were employed. The fabrication work took around three weeks to complete. The 3D model of the fabricated fixture is shown in Fig. 9.

6 Conclusion

The fixture was designed and analyzed. The dimension and material of the fixture are used as the starting point for the FEM analysis. According to the analysis results, the stress is 44 MPa, which is less than mild steel's yield strength (250 MPa). As a result, the design is safe and the fixture allows the material to flow into the deformation zone while generating the complicated shape that is the primary goal of the operation. The fixture is used to create a component for the aerospace industry.

References

1. Allwood JM, Braun D, Music O (2010) The effect of partially cut-out blanks on geometric accuracy in incremental sheet forming. J Mater Process Technol 210(11):1501–1510
2. Lu H, Kearney M, Wang C, Liu S, Meehan PA (2017) Part accuracy improvement in two point incremental forming with a partial die using a model predictive control algorithm. Precis Eng 49:179–188
3. Ai S, Lu B, Chen J, Long H, Ou H (2017) Evaluation of deformation stability and fracture mechanism in incremental sheet forming. Int J Mech Sci 124–125:174–184
4. Pratheesh Kumar S, Elangovan S (2020) Optimization in single point incremental forming of Inconel 718 through response surface methodology. Trans Can Soc Mech Eng 44(1):148–160
5. Lu B, Chen J, Ou H, Cao J (2013) Feature-based tool path generation approach for incremental sheet forming process. J Mater Process Technol 213(7):1221–1233
6. Kumar SP, Pratheesh Kumar S, Elangovan S, Mohanraj R (2020) Experimental study on single point incremental forming of Inconel 718. Trans Can Soc Mech Eng 44(2):179–188
7. Kumar SP, Elangovan S, Mohanraj R, Boopathi S (2021) Real-time applications and novel manufacturing strategies of incremental forming: an industrial perspective. Mater Today
8. Raju C, Haloi N, Sathiya Narayanan C (2017) Strain distribution and failure mode in single point incremental forming (SPIF) of multiple commercially pure aluminum sheets. J Manuf Process 30:328–335
9. Min J, Kuhlenkötter B, Shu C, Störkle D, Thyssen L (2018) Experimental and numerical investigation on incremental sheet forming with flexible die-support from metallic foam. J Manuf Process 31:605–612
10. Paniti I, Viharos ZJ, Harangozó D, Najm SM (2020) Experimental and numerical investigation of single point incremental forming of aluminium alloy foils. ACTA IMEKO 9(1):25
11. Zhang H, Lu B, Chen J, Feng S, Li Z, Long H (2017) Thickness control in a new flexible hybrid incremental sheet forming process. Proc Inst Mech Eng Pt B J Eng Manuf 231(5):779–791
12. Araghi BT, Manco GL, Bambach M, Hirt G (2009) Investigation into a new hybrid forming process: incremental sheet forming combined with stretch forming. CIRP Ann Manuf Technol 58(1):225–228
13. Reddy NV, Lingam R, Cao J (2015) Incremental metal forming processes in manufacturing. In: Handbook of manufacturing engineering and technology. Springer, London, pp 411–452
14. Mohanraj R, Elangovan S (2020) Thermal modeling and experimental investigation on the influences of the process parameters on warm incremental sheet metal forming of titanium grade 2 using electric heating technique. Int J Adv Manuf Technol 110:255–274
15. Pratheesh Kumar S, Elangovan S, Mohanraj R, Boopathi S (2021) A comprehensive review in incremental forming on approaches of deformation analysis and surface morphologies. Mater Today 43:3129–3139
16. Peng W, Ou H, Becker A (2019) Double-sided incremental forming: a review. J Manuf Sci Eng 141(5):050802

Effects of Multi-axis Forging on Mechanical and Microstructural Properties of AA6061-T6 Aluminum Alloy

Bappa Das, Uday S. Dixit, and Biranchi Narayan Panda

Abstract Severe plastic deformation (SPD) is one of the popular techniques to enhance mechanical properties in most metals through grain refinement. In this study, the mechanical behavior and grain refinement of aluminum AA6061-T6 through multi-axis forging (MAF) have been investigated at room temperature. Microstructural investigation through an optical microscope shows a reduced average grain size of 23 μm after 3 cycles of MAF in comparison with the initial size of 53 μm. The increased hardness and compressive strength of 131 HV and 424 MPa, respectively, are observed due to the combined effects of strain hardening and grain boundary strengthening. Minimum wear depth and mean Coulomb's coefficient of friction of 67 μm and 0.32, respectively, are observed after 3 cycles of MAF due to strengthening effects.

Keywords Multi-axis forging · Grain refinement · Severe plastic deformation · Hardness · Wear

1 Introduction

Aluminum (Al) and its alloys are extensively used in aerospace, railway, automobiles, and defense sections due to their lightweight, enhanced ductility, and non-corrosive nature. However, they show a very limited wear resistance and strength after age hardening [1–5]. These problems can be solved by reducing the grain size to an ultra-fine state, which in turn enhances the performance of the materials. In this regard, it has been extensively proven that severe plastic deformation (SPD) is the most efficient

B. Das (✉) · U. S. Dixit · B. N. Panda
Department of Mechanical Engineering, Indian Institute of Technology Guwahati, Assam 781039, India
e-mail: bappadas@iitg.ac.in

U. S. Dixit
e-mail: uday@iitg.ac.in

B. N. Panda
e-mail: pandabiranchi@iitg.ac.in

U. S. Dixit et al. (eds.), *Advances in Forming, Machining and Automation*, Lecture Notes in Mechanical Engineering, https://doi.org/10.1007/978-981-19-3866-5_5

method for the reduction of coarse grain to an ultra-fine state without changing the chemical composition of the parent material; this causes grain boundary (GB) strengthening and strain hardening [6, 7]. One prominent method of accomplishing SPD is multi-axis forging (MAF), which has been carried out on aluminum alloy [8], titanium alloy [9], and ferroalloy.

Several widely used SPD techniques have been developed for strengthening metals and their alloys, and a few among them are high-pressure torsion [10], equal channel angular pressing [11], and constrained groove pressing [12]. The processes using all SPD technology mainly involve the application of shear stress to the workpiece to increase the dislocation density and form dislocation walls, which results in the development of low-angle grain boundary [13].

A variety of metals and alloys have been successfully manufactured through different SPD methods. The existing literature has proved that the SPD process can form ultra-fine grains with sub-grains. However, with the use of different SPD technology and parameters such as strain and temperature, the final shapes and sizes of the grains formed will be different. Sitdikov et al. [14] reported the reduction of average grain size from 20 to 9 μm in 7475 Al alloy manufactured through MAF. Domizio et al. [15] also reported a reduction in average grain size from 40 to 1.3 μm using MAF in Ni-20Cr alloy. Share et al. [16] used the equal channel angular pressing (ECAP) SPD method in Al-7075 at various temperatures. The average grain refinement from 60 to 1 μm was observed in the room temperature processed through ECAP. Joshi et al. [17] carried out MAF of Al-2014 alloy at cryogenic temperature with fine grain refinement from 250 to 100–400 nm and subsequent increase of hardness from 103 to 171 HV. Cryo-forging promotes the development of ultra-fine grains through suppression of dislocation cross slips by introducing high dislocation density compared to room temperature. In the present work, the AA6061-T6 aluminum alloy samples were repetitively forged in a universal testing machine (UTM) machine at room temperature. The microstructural refinement and enhanced mechanical properties due to strengthening mechanisms are discussed.

2 Experimental Procedure

Al samples for this work were obtained with dimensions of 12 mm × 12 mm × 18 mm. The compression test was conducted using SHIMADZU-Static UTM (1–100 kN) machine at the strain rate of 0.001 s^{-1} at room temperature. In each pass, the samples were forged to 10% of the height in all directions. The MAF schematic technique and experimental setup are shown in Fig. 1. After completion of each cycle, the samples were polished in METCO BAINPOL grinding/polishing machine using silicon carbide (SiC) emery paper followed by diamond polish maintaining 160 revolutions per minute (rpm) to maintain a smooth surface for optical microscopy. The polished samples before optical microscopy were etched in standard Keller's reagent for 10 s. The microstructure of the forged samples was evaluated using an upright optical microscope (ZEISS). The average grain size was calculated using the

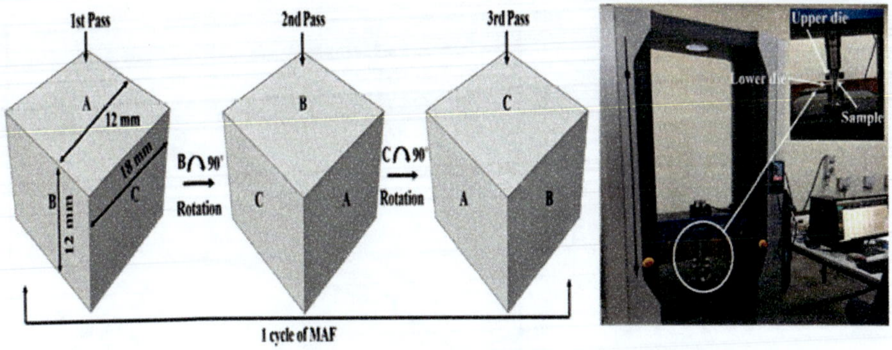

Fig. 1 Schematic of MAF technique and experimental setup

mean line intercept method with the help of Image J software. Vickers microhardness test was evaluated using a microhardness tester (Omnitech, Make: India) with a 200 gf load and 20 s dwell time.

A cylindrical shape of 7 mm diameter was taken out using a wire cut EDM machine, and each forged sample was maintained to get an aspect ratio of 1.5. The compression test was carried out using a UTM at a strain rate of 0.001 s^{-1} for the estimation of compressive strength. The wear test was conducted for different forged samples on a pin on a disc machine (Model: TR-201, Make: Ducom, India) at 20 N load and 365 rpm for 30 min. The surface morphology after the wear test was observed using a field emission scanning electron microscope (FESEM) (Model: GEMINI, Make: ZEISS).

3 Results and Discussion

In this section, some results are discussed. First, the compression load-displacement diagrams are described. After that, microstructure and microhardness are discussed. The wear test was also carried out and results are discussed.

3.1 Compression Test

The load versus displacement plots for different forged samples are shown in Fig. 2. The results show that for a given displacement, the load after 1 cycle increased significantly and gradually increases in the subsequent cycles. This can be partly due to higher resistance to crack propagation by SPD and a lower rate of work hardening [18]. Deformation increases strength. The reduction in work hardening due to SPD is attributed to dislocation entanglement, which is the cause of accelerated dislocation

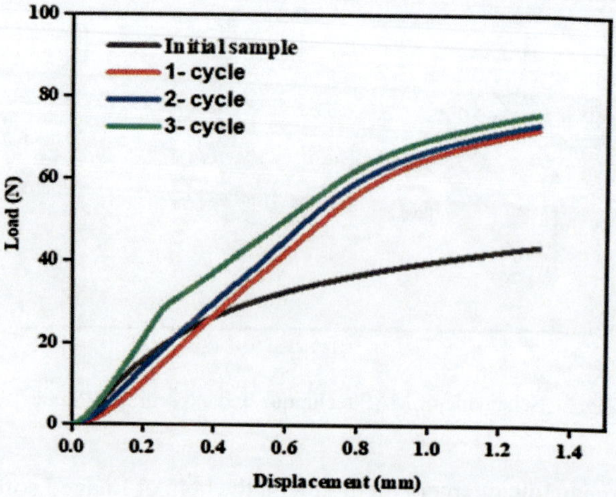

Fig. 2 Compression load versus displacement plot of Al samples under different MAF cycles

propagation kinetics [19]. The difference in strength results from grain boundary and dislocation strengthening. The grain boundary strengthening is the primary strengthening mechanism. The refined grains and grain boundaries can prevent dislocation movement. The enhanced strength is the result of grain refinement structure and is governed by the Hall–Petch relation given by [20]

$$\sigma_{GB} = \sigma_0 + kd^{-0.5}, \tag{1}$$

where σ_0 and k are material constants and d is the average gain diameter.

The load to deformation increased greatly after the 1 cycle of MAF, then relatively at the small rate in further cycles as evident from Fig. 2. The values at lower displacement are not reliable due to low data and the accuracy of direct measurement of the strain. Generally, in SPD by MAF, after a certain number of forging cycles, dynamic recovery and annihilation will dominate, which will reduce the rate of load increase. When a series of dislocations dominate the structure more, the strength will keep enhancing [21]. However, when dynamic recovery dominates, the strength will begin to decline. A similar trend was also reported by researchers during SPD of other metals [22, 23].

3.2 *Microstructural Study*

Figure 3 shows the Al microstructure treated with MAF for up to 3 cycles. The microstructure shows the coexistence of fine and elongated grain structures. Due to grain refinement, the MAF process forms elongated grains along the axis particular to the forging direction. The average grain size of the initial samples was 53 μm, and it was subjected to a maximum of 3 cycles of MAF. After 3 cycles of MAF, the grain

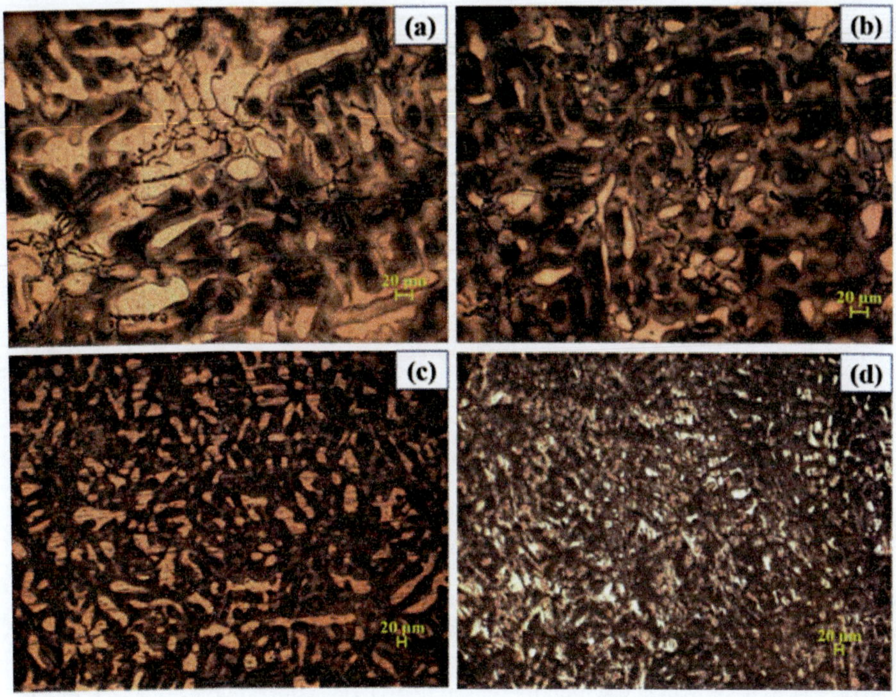

Fig. 3 Optical micrograph (X20) of aluminum samples (**a**, **b**, **c**, and **d**) as-received, 1st, 2nd and 3rd cycles, respectively

boundaries are hardly distinguishable and a large number of sub-grains formed as a result of higher dynamic recrystallization evident in the refinement of grains shown in Fig. 3d. For 1 cycle of MAF, an extended shear deformation zone is observed and as the MAF cycle increases to 3 cycles, dislocation accumulation occurs, and it is very tough to distinguish individual grains as shown in Fig. 3c, d [24]. Increasing the MAF cycles will narrow the distance between bands due to SPD. After 3 cycles of MAF, the average grain size of 23 μm was observed in the microstructure.

Figure 4 shows the dependence of the MAF cycles on the average grain size calculated over the entire area. It is evident that the average grain size gradually decreases as the number of MAF cycles increases. The degree of homogeneity was observed to increase with increasing MAF cycles (results omitted for brevity). Chen et al. [25] also reported similar results for MAF processed as-cast AZ61 magnesium alloy.

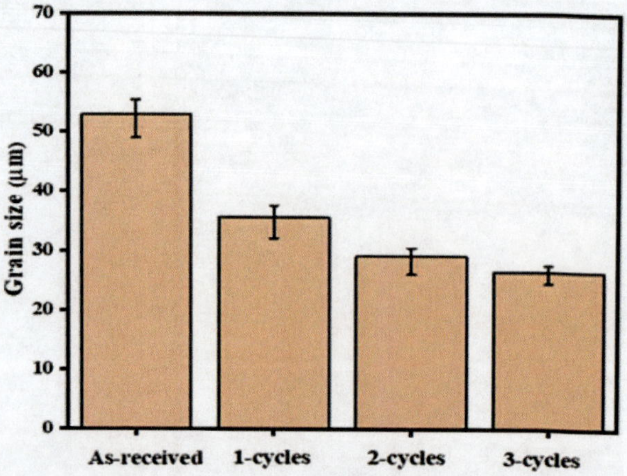

Fig. 4 Grain size of aluminum samples in different cycles

3.3 *Hardness Study*

Variation of Vickers microhardness of Al for different MAF cycles is shown in Fig. 5. To maintain accuracy, the microhardness was obtained at the surface of different cross sections, and an average was taken. The measured microhardness of the initial sample was observed to be 102 HV which increased to 131 HV after 3 cycles of MAF due to strain hardening effects. The sudden increase in hardness after 1 cycle is attributed to working hardening mechanisms thereafter leading to higher dislocation density. The combined effects of elevated dislocation density and grain refinement enhanced the hardness at higher passes. The increase in microhardness of the initial cycle is owing to the strengthening of grain boundary and dislocation, but its effect is less in the subsequent cycle [26].

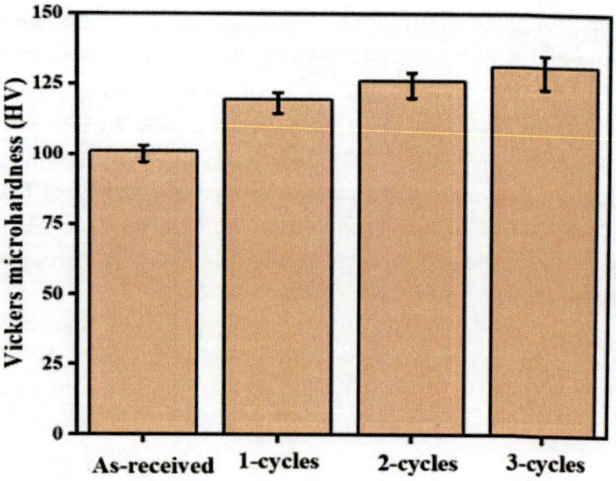

Fig. 5 Microhardness after different MAF cycles

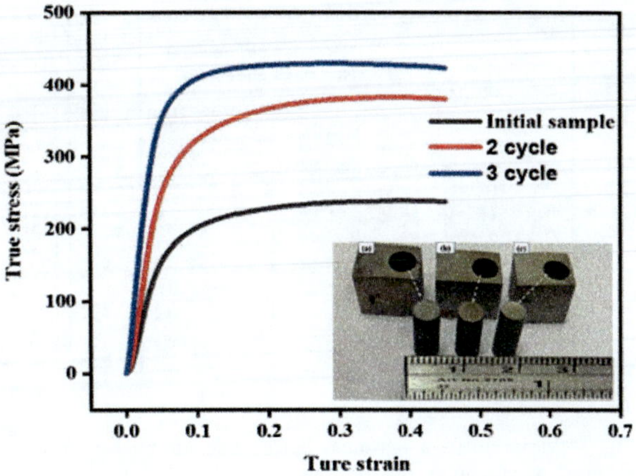

Fig. 6 Flow stress curves for different cycles of MAF (only values in the plastic range are reliable)

3.4 *Approximate Flow Curves*

The compression tests were performed on the Al sample for different MAF cycles at the strain rate of 0.001 s^{-1} shown in Fig. 6. Here, cycle 1 results are not studied due to low significance. The compression strength of the initial samples was 240 MPa. It increased to 383 MPa and 424 MPa after 2 and 3 cycles of MAF, respectively. The sudden increases in the compressive strength are attributed to dislocation pile up in the grain resulting in new sub-grains formation. As the MAF cycles continue, there is a short time heating in the samples and static recrystallization, which accelerates the dislocation pileup even at room temperature. Sharath et al. [27] also observed similarly in results during the MAF of Cu alloy.

3.5 *Wear Study*

The mechanical properties of metals are intensively influenced by the microstructure. The additional dislocation was generated as the MAF cycle continued. Results of the first cycle are not depicted due to low impact. As the dislocation density increases, the dislocation movement becomes more difficult. Hindering the movement of dislocation gives the materials elevated strength. Therefore, the increase in dislocation density and refinement of crystal grains leads to an increase in strength. The material properties such as strength, hardness, and strain hardening solely influence the wear behavior. Increasing the hardness can reduce plastic deformation and wear. In general, the hardness is inversely proportional to the wear volume in adhesive and abrasive wear mode is evident from Fig. 7a, b. The traditional Archard equation of wear also reflects the same [28].

Figure 8 shows wear depth as a function of sliding time for the different MAF cycles. High hardness is usually associated with better wear resistance as is evident

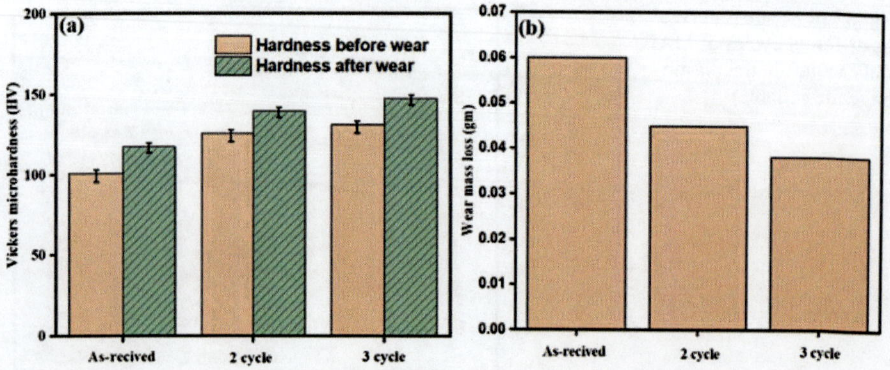

Fig. 7 Variation of **a** hardness before and after wear on the surface and **b** wear mass loss for different MAF cycles, respectively

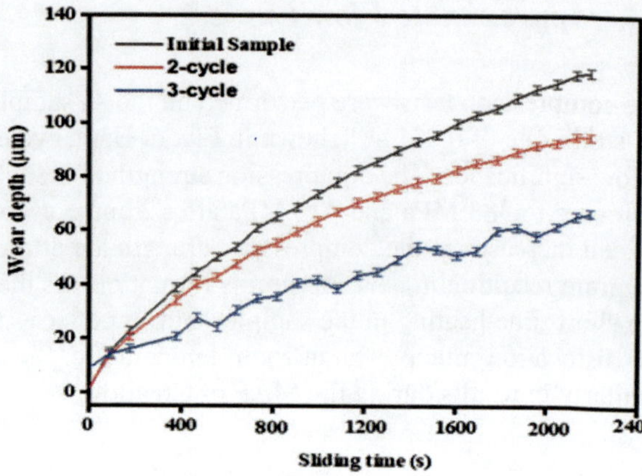

Fig. 8 Variation of wear depth with sliding time after different MAF cycles

from Figs. 5, 7, and 8. The surface hardness of wear samples is more as compared to the hardness before wear test due to work hardening during wear as evident from Fig. 7a. Liu et al. [29] also observed a similar phenomenon while MAF of aluminum bronze alloy.

Due to the SPD with MAF, there is a reduction in wear. Due to increased hardness, the coefficient of friction is low for forged samples. Similar observations were obtained by other researchers [30, 31]. MAF strengthening increases hardness. The coefficient of friction will reduce with the reduction of molecular force and real contact between surfaces as evident from Fig. 9 and Table 1. The wear surface morphology and wear debris of the different forged samples are shown in Figs. 10 and 11, respectively. The scratches are deep and wider, also the average debris size is larger in as-received samples compared to 3 cycle forged samples due to variation in hardness and grain size refinement.

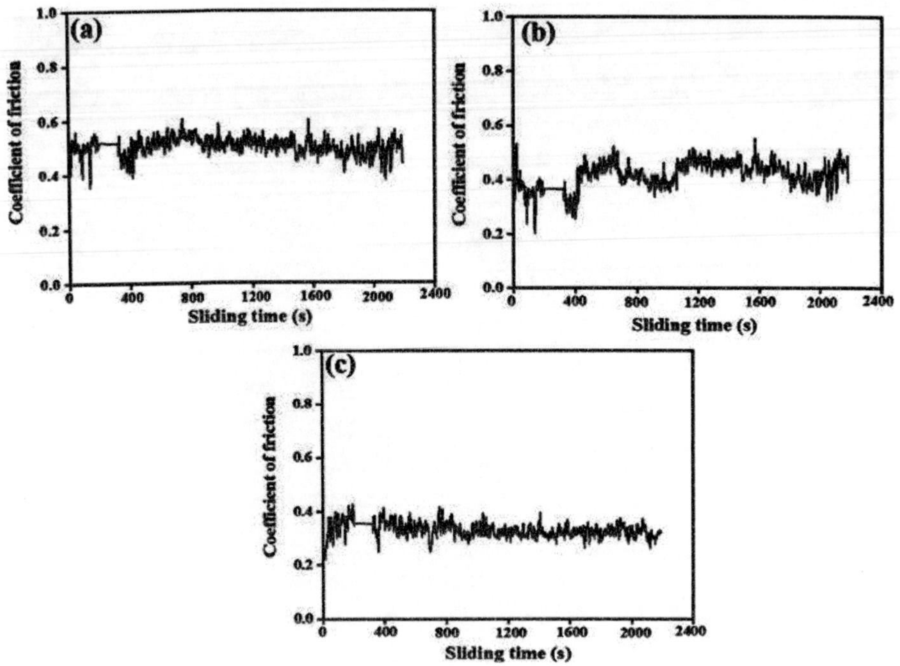

Fig. 9 Variation of friction coefficient with sliding time for different MAF cycles: **a** initial, **b** 2 cycle, and **c** 3 cycle, respectively

Table 1 Representative wear analysis of the Al sample at different MAF cycles

S. No	MAF cycle	Wear depth (μm)	Mean friction coefficient
1	As-received	119	0.50
2	2	95	0.41
3	3	67	0.32

The properties of the Al alloys have been summarized in Table 2 and these properties are compared with the existing literature [32–36]. Table 2 shows that the MAF in this study significantly improved the mechanical properties.

4 Conclusion

AA6061-T6 alloy has been produced through MAF. The average fine grain of 23 μm was observed after 3 cycle MAF. The hardness increased to 131 HV from 101 HV due to grain refinement. The maximum compressive strength of 424 MPa was attained after 3 cycles of MAF due to dislocation pile-up and grain boundary strengthening.

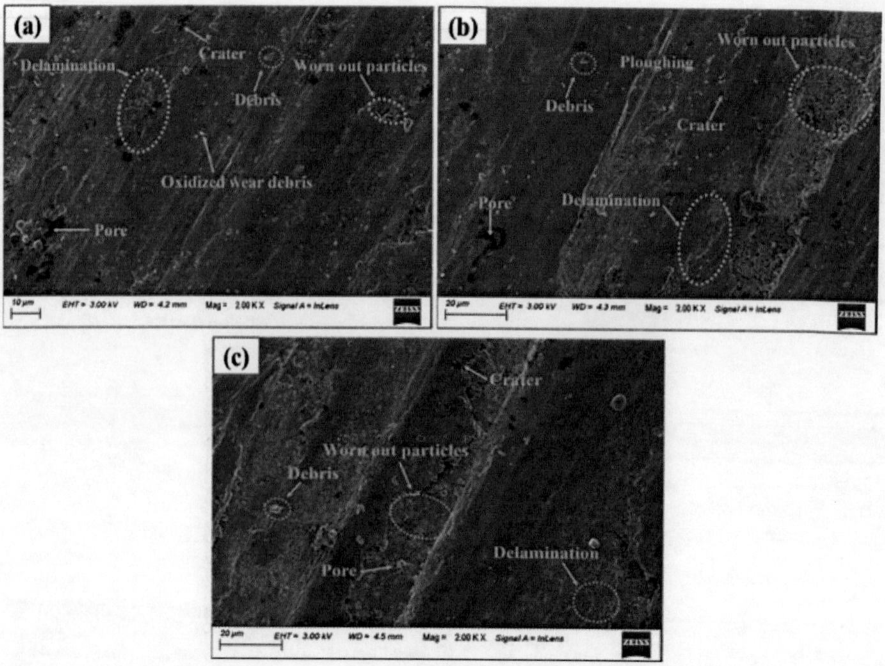

Fig. 10 Wear FESEM micrograph (X2000) of sliding surface **a** initial sample, **b** 2 cycle and **c** 3 cycle, respectively

Due to increased hardness and grain refinement, the wear depth and friction coefficient of 67 μm and 0.32, respectively, were observed after MAF. Further work will be directed toward obtaining accurate flow curves after MAF. Also, an attempt will be made to develop a similar process for bigger samples, so that the process can be used in industry for making bigger size components. The severe plastic deformation process produces mechanical properties, which far exceed the properties attainable by alloying. The present-day limitation is that the process can be employed for smaller size components only. Also, further research is needed to study the machinability after severe plastic deformation.

Fig. 11 Wear FESEM micrograph (X200) of wear debris **a** initial sample, **b** 2 cycle and **c** 3 cycle, respectively

Table 2 Summary of investigated alloys and comparison with present data

Alloy	Consolidation method and parameters	Grain size	Yield strength (MPa)	Compression strength (MPa)	Hardness (HV)	Wear depth (μm)	Ref.
Al–Mg–Mn–Sc–Zr	MAF, room temperature, 3 cycles	25 μm	697 (tensile test)	–	–	–	[32]
2024 aluminum	MAF, room temperature, 1 cycle	–	–	454	148	–	[33]
Al 6082	MAF, room temperature, 3 cycles	500 nm	334 (tensile test)	–	118	–	[34]
Al–Zn–Mg–Cu	MAF, cryo-temperature, 3 cycles	-	612 (tensile test)	–	168	-	[35]
Al–Mg–Mn	MAF, 200–500 °C, 4 cycles	5 μm	200 (tensile test)	–	–	–	[36]
AA 6061-T6	MAF, 3 cycles, room temperature	23 μm	–	424 (compression test)	131	67	Present study

References

1. Duan Y et al (2017) Excellent high strain rate superplasticity of Al-Mg-Sc-Zr alloy sheet produced by an improved asymmetrical rolling process. J Alloy Compd 715:311–321
2. Pereira P et al (2017) Influence of grain size on the flow properties of an Al-Mg-Sc alloy over seven orders of magnitude of strain rate. Mater Sci Eng, A 685:367–376
3. Buranova Y et al (2017) Al3(Sc, Zr)-based precipitates in Al-Mg alloy: Effect of severe deformation. Acta Mater 124:210–224
4. Jiang J et al (2019) Effects of annealing under fixed temperature and cyclic temperature on strength and microstructure of Al-MgMn-Sc-Zr alloy. Mater Sci Eng A 764:138–275
5. Kumar A et al (2017) Effect of Process parameters on microstructural evolution, mechanical properties and corrosion behavior of friction stir processed Al 7075 alloy. J Mater Eng Perform 26:1122–1134
6. Kumar N, Mishra R (2013) Additivity of strengthening mechanisms in ultrafine-grained Al-Mg-Sc alloy. Mater Sci Eng A 580:175–183
7. Zhemchuzhnikova D, Kaibyshev R (2015) Mechanical behavior of an Al-Mg-Mn-Sc alloy with an ultrafine grain structure at cryogenic temperatures. Adv Eng Mater 17:1804–1711
8. Cherukuri B et al (2005) A comparison of the properties of SPD processed AA-6061 by equal channel angular pressing (ECAP), multi-axial compressions/forgings (MAC/F) and accumulative roll bonding (ARB). Mater Sci Eng A 410:394–397
9. Salishchev G et al (1993) Formation of submicrocrystalline structure in the titanium alloy VT8 and its influence on mechanical properties. J Mater Sci 28:28–29
10. Castro M et al (2019) A magnesium-aluminium composite produced by high-pressure torsion. J Alloy Compd 804:421–426
11. Gopi K et al (2016) Investigation of microstructure and mechanical properties of ECAP-processed am series magnesium alloy. J Mater Eng Perform 25:3737–3745
12. Yadav P et al (2016) microstructural inhomogeneity in constrained groove pressed Cu-Zn alloy sheet. J Mater Eng Perform 25:2604–2614
13. Xue Q et al (2007) Mechanisms for initial grain refinement in OFHC copper during equal channel angular pressing. Acta Mater 55:655–668
14. Sitdikov O et al (2005) Grain refinement in coarse-grained 7475 Al alloy during severe hot forging. Phil Mag 85:1159–1175
15. Domizio R et al (2006) Fine-grained structure formation in Al-Cr based alloys during severe plastic deformation. Mater Sci Forum 503:793–798
16. Shaeri M et al (2016) Effect of ECAP temperature on microstructure and mechanical properties of Al–Zn–Mg–Cu alloy. Prog Nat Sci 26:182–191
17. Joshi A et al (2016) Mechanical properties and microstructural evolution in Al 2014 alloy processed through multidirectional cryoforging. J Mater Eng Perform 25:3031–3045
18. Meyers M et al (2006) Mechanical properties of nanocrystalline materials. Prog Nat Sci 51:427–556
19. Kumar S, Raghu T (2013) Mechanical behaviour and microstructural evolution of constrained groove pressed nickel sheets. J Mater Process Technol 213:214–220
20. Cordero Z et al (2016) Six decades of the Hall-Petch effect—a survey of grain-size strengthening studies on pure metals. Int Mater Rev 61:495–512
21. Montazeri-Pour M et al (2015) Microstructural and mechanical properties of AA1100 aluminum processed by multi-axial incremental forging and shearing. Mater Sci Eng A 639:705–716
22. Shin D et al (2002) Constrained groove pressing and its application to grain refinement of aluminum. Mater Sci Eng A 328:98–103
23. Krishnaiah A et al (2005) Applicability of the groove pressing technique for grain refinement in commercial purity copper. Mater Sci Eng A 410:337–340
24. Sharath P et al (2017) Effect of multi-directional forging on the microstructure and mechanical properties of Zn-24 wt% Al-2 wt% Cu Alloy. Trans Indian Inst Met 70:89–96

25. Chena Q et al (2012) Grain refinement in an as-cast AZ61 magnesium alloy processed by multi-axial forging under the multitemperature processing procedure. Mater Sci Eng A 541:98–104
26. Miura H et al (2012) Microstructure and mechanical properties of multi-directionally forged Mg–Al–Zn alloy. Scripta Mater 66:49–51
27. Sharath P et al (2017) Effect of multi axial forging on the microstructure and mechanical properties of Zn-24wt% Al-2wt% Cu alloy. Trans Indian Inst Met 70:89–186
28. Archard J (1953) Contact and rubbing of flat surfaces. J Appl Phys 24:981–988
29. Liu X et al (2020) Multi-directional forging and aging treatment effects on friction and wear characterization of aluminium-bronze alloy. Mater Charact 167:110511
30. Zhang D et al (2003) Research on the fatigue and fracture behavior due to the fretting wear of steel wire in hoisting rope. Wear 255:7–12
31. Zhang Y et al (2006) Friction and wear behaviors of the nanocrystalline surface layer of pure copper. Wear 260:942–948
32. Huang H et al (2021) Microstructural and mechanical behavior of Al-Mg-Mn-Sc-Zr alloy subjected to multi-axial forging at room temperature. Mater Charact 171:110734
33. Nouri S (2020) Flow stress of 2024 aluminum alloy during multi-directional forging process and natural aging after plastic deformation. Mater Chem Phys 254:123446
34. Kumar N et al (2019) Al 6082 alloy strengthening through low strain multi-axial forging. Mater Charact 115:109761
35. Ramesh S et al (2021) Microstructural and mechanical characterization of Al-Zn-Mg-Cu alloy processed by multi-directional cryo-forging. Mater Today Proc 26:5752–5756
36. Kishchik M et al (2018) Effect of multidirectional forging on the grain structure and mechanical properties of the Al–Mg–Mn alloy. Materials 11:2166

Experimental Investigation of Multi-stage Robot-Assisted Single Point Incremental Sheet Forming of Al 6061 Sheet

Ravi Prakash Singh, Santosh Kumar, Pankaj Kumar Singh, and Md. Meraz

Abstract Incremental sheet forming is a die-less forming method to deform the sheet in a stepwise manner to get the final product. In the current work, the Al 6061 sheet has been deformed using single point incremental sheet forming methodology on 6-axis industrial robot to produce a single wall angle and double wall angle conical shape. The wall angle of 68.2° has been obtained using a multi-stage forming strategy. Two types of tools, namely roller ball tool and fixed ball tool have been deployed to accomplish the process. The deformed sheet has been investigated for sheet thinning and surface finish. The quality of the product in terms of forming depth, greater wall angle, and surface finish is better for multi-stage incremental sheet forming.

Keywords Sheet metal forming · Incremental sheet forming · Al6061 · Sheet metal thinning · Industrial robot

1 Introduction

Incremental sheet forming (ISF) is the method of forming sheets into a programmed shape without using any dedicated dies [1, 2]. The patent for this process was granted to Laszek under the patent titled 'Apparatus and process for incremental die-less forming' [3]. Since then, lots of work has been done in ISF due to different advantages of the process, making it a potential candidate for the next generation forming process. One of the most important advantages of such a process is to achieve a high forming angle which is a direct measure of high sheet formability. The high formability in

R. P. Singh (✉) · S. Kumar · P. K. Singh · Md. Meraz
Department of Mechanical Engineering, IIT (BHU), Varanasi 221005, India
e-mail: ravips.rs.mec16@itbhu.ac.in

S. Kumar
e-mail: santosh.kumar.mec@iitbhu.ac.in

P. K. Singh
e-mail: pankaj.krsingh.rs.mec18@itbhu.ac.in

Md. Meraz
e-mail: mdm.rs.mec16@itbhu.ac.in

U. S. Dixit et al. (eds.), *Advances in Forming, Machining and Automation*, Lecture Notes in Mechanical Engineering, https://doi.org/10.1007/978-981-19-3866-5_6

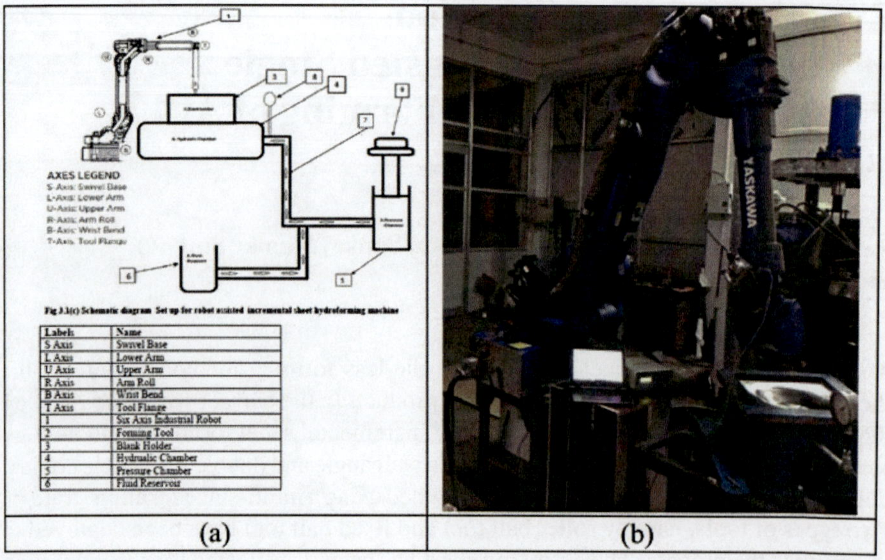

Fig. 1 **a** Labeled schematic diagram of the robotic ISF and **b** the setup developed at IIT (BHU) [1]

the sheet is attributed to the presence of highly localized deformation because of the presence of hydrostatic pressure in a confined deformation zone. Necking in the sheet is either suppressed [4, 5]. Some other benefits of ISF are as follows: (a) as the process does not require dedicated dies, the setup cost can be largely reduced [6]; (b) the degree of flexibility of the process is quite high as different shapes can be achieved by changing the tool path [7–9]. Merely changing the programming for tool paths can lead to the production of new shapes; (c) the process produces a very strong candidature for small batch production in auto-motive, aerospace, and biomedical sectors; (d) the process of ISF forms the basis for rapid prototyping [10–12] (Fig. 1).

In the early advancements of the process, the CNC machine was used for the programming tool path [13]. But in the recent time, robotic industrial arms are being used for the process [14, 15]. In 2003, Timo Tuominen patented the process of ISF using robotic arms under the application titled 'Method and apparatus for forming three-dimensional shapes in a sheet metal' [16]. Deployment of robots can increase the process capabilities in terms of increased speed, feed rate, more work volume, etc.

Forming tool can be mounted on the driving machine and can be programmed to achieve a certain shape. Different types of farming tools can be used depending upon the process requirement. The profile of the tool could be circular, parabolic, angled, etc., and can be made of stainless steel (SS), cubic boron nitride (CBN), etc. [17]. Once the required tool is mounted, the tool path is programmed depending upon the requirement. This is the most crucial thing in the whole ISF process. The accuracy of the formed components depends on the acquired tool path and the strategy. The tool path generation can be done in the following ways: (i) pocketing with constant step depth and (ii) pocketing with constant 'scallop height' [18]. The sheet can be deformed in one stage using a series of forming passes; however, deformation of sheets to attain a high forming angle can be achieved by using multi-stage ISF [19, 20].

2 Experimental Setup

In the current work, Al6061 sheet of initial thickness 1.1 mm was deformed using 6-axis industrial robot. The mechanical properties of the material are enlisted in Table 1.

The initial work dimension of the blank was 20 × 20 mm. Two types of tools are used as shown in Fig. 2a, (i) rollerball tool of tool diameter 15 mm and (ii) fixed ball tool having tool diameter 15 mm. Two types of programming strategies were followed: single-stage forming with constant depth increments and multi-stage forming with constant depth increments in individual stages. In single-stage forming, an increment of 0.5 mm in the negative radial direction responsible for reducing the size of the circle was provided; simultaneously, increment of 0.5 mm in the negative Z-direction was given, pushing the tool to go down. Hence, the cumulative wall angle will be 45°.

In multi-stage forming, the sheet was deformed using more than one stage. In this work, an increment of 0.5 mm in radial and downward directions was chosen for a theoretical wall angle of 45°. Then, to achieve a steeper wall angle, an increment of 0.4 mm in the radial direction and 0.5 mm in the downward direction was provided, thus making the wall angle go on to 51° and so on. In the final step, the increment

Table 1 Mechanical properties for material used for experimentation

Specification of material used for experimentation (Al6061 *T*6) (tensile test)					
Dimension	Proof load	Proof stress	Ultimate load	UTS	Elongation
50 × 50 cm ×0.85 mm	25.82 kN	265.4 MPa	39.240 kN	315.2 MPa	12.6%

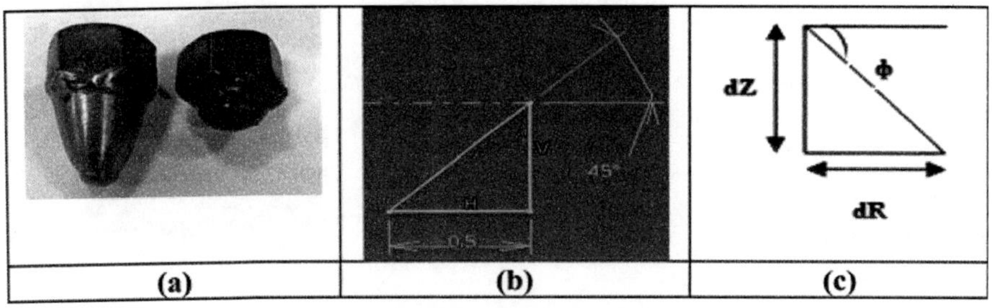

Fig. 2 **a** Different tools used **b** and **c** geometry of the increment

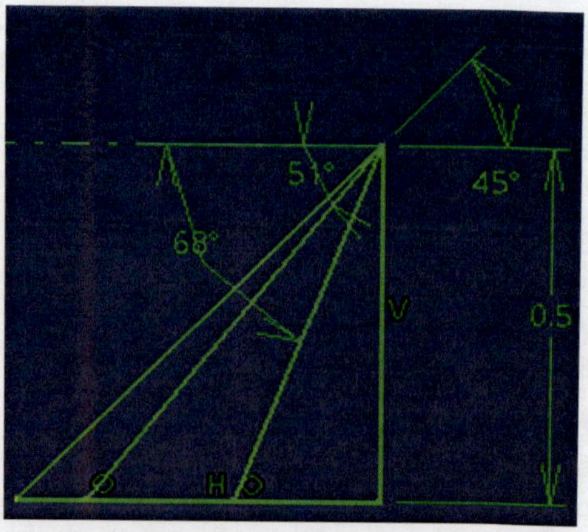

Fig. 3 Increment geometry for multi-stage ISF

of dr = 0.20 mm and dZ = 0.5 mm in the downward direction was given to achieve the theoretical wall angle of 68.1° (Fig. 3).

3 Experimental Parameters

Once the programming for the conical path was accomplished on 6-axis industrial robot, the robotic arm was allowed to follow the programmed path for the hollow cone and deform the Al6061 sheet of initial sheet thickness 1.1 mm in the desired path. However, the actual depth achieved was less than the theoretical depth accounting for the spring back occurred. Here are the details of the experimental parameters for the set of experiments done as tabulated in Tables 2, 3, 4, and 5 for workpieces 1, 2, 3, and 4, respectively. The tool speed was kept at 200 mm/sec for every experiment.

Table 2 Experimental parameters for workpiece 1 and the final shape

Workpiece 1		
Tool used	Rollerball tool	
Tool diameter	15 mm	
Constant step size	$\Delta r = 0.5$ mm, $\Delta z = 0.5$ mm	
Theoretical wall angle	45°	
No. of the cycles run	54	
Anticipated depth	54 * 0.5 = 27 mm	
Actual depth reached	24 mm	
Spring back occurred	11%	
No. of the stages performed	1	
Final status	Product failed after 54 cycles	

Table 3 Experimental parameters for workpiece 2 and the final shape

Workpiece 2	
Tool used	Fixed ball tool
Tool diameter	15 mm
Constant step size	$\Delta r = 0.5$ mm, $\Delta z = 0.5$ mm
Theoretical wall angle	45°
No. of the cycles run	134
Anticipated depth	134 * 0.5 = 67 mm
Actual depth reached	60 mm
Spring back occurred	10.44%
No. of the stages performed	1
Final status	Product successfully formed

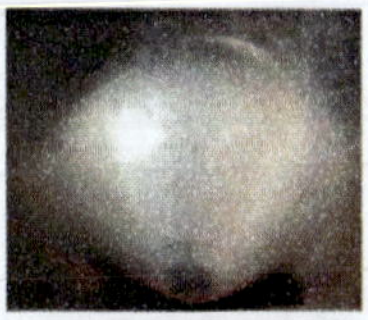

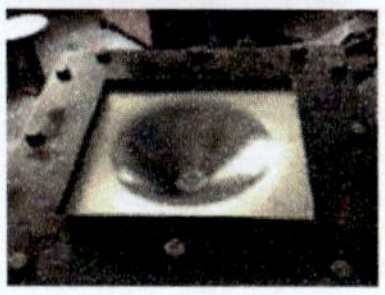

Table 4 Experimental parameters for workpiece 3 and the final shape

Workpiece 3	
Tool used	Fixed ball tool
Tool diameter	15 mm
No. of stages	2
Stage 1	
Constant step size	$\Delta r = 0.5$ mm, $\Delta z = 0.5$ mm
Theoretical wall angle	45°
No. of the cycles run	120
Anticipated depth	120 * 0.5 = 60 mm
Actual depth reached	53 mm
Spring back occurred	11.67%
Final status	Product successfully formed
Stage 2	
Constant step size	$\Delta r = 0.4$ mm, $\Delta z = 0.5$ mm
Theoretical wall angle	51°
No. of the cycles run	60
Anticipated depth	60 * 0.5 = 30 mm
Actual depth reached	26 mm
Spring back occurred	13.33%
Final status	Product successfully formed

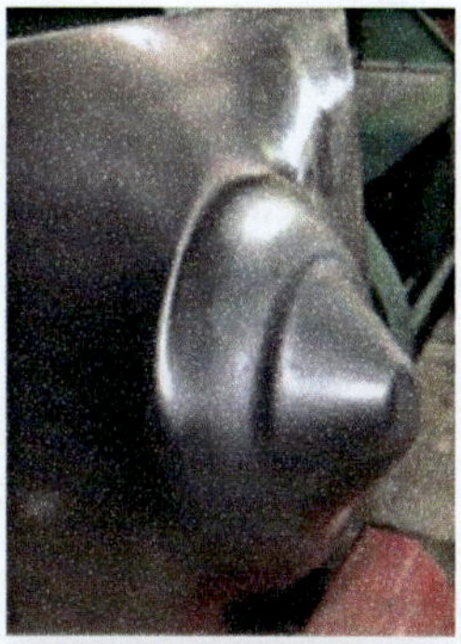

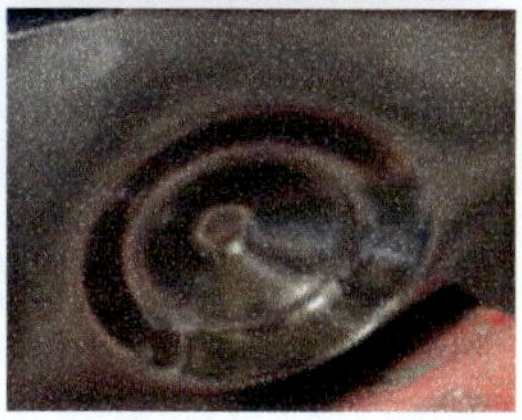

Table 5 Experimental parameters for workpiece 4 and the final shape

Workpiece 4		
Tool used	Fixed ball tool	
Tool diameter	15 mm	
No. of stages	3	
Stage 1		
Constant step size	$\Delta r = 0.5$ mm, $\Delta z = 0.5$ mm	
Theoretical wall angle	45°	
No. of the cycles run	90	
Anticipated depth	120 * 0.5 = 45 mm	
Actual depth reached	41 mm	
Spring back occurred	8.9%	
Final status	Product successfully formed	
Stage 2		
Constant step size	$\Delta r = 0.3$ mm, $\Delta z = 0.5$ mm	
Theoretical wall angle	59.26°	
No. of the cycles run	60	
Anticipated depth	60 * 0.5 = 30 mm	
Actual depth reached	27 mm	
Spring back occurred	10%	
Final status	Product successfully formed	
Stage 3		
Constant step size	$\Delta r = 0.2$ mm, $\Delta z = 0.5$ mm	
Theoretical wall angle	45°	
No. of the cycles run	60	
Anticipated depth	60 * 0.5 = 30 mm	
Actual depth reached	28 mm	
Spring back occurred	6.67%	
Final status	Product successfully formed	

4 Results and Discussion

While performing the experiments with a roller ball tool and the same with a fixed ball tool, more deformation depth was achieved in the case of fixed ball tool, which can be because of more friction in the case of fixed ball tool, which leads to localized heating of the sheet at tool sheet interface and ultimately can be the reason behind increased sheet formability.

Once the sheet was deformed, they were checked for sheet thickness in the region of deformation. The graphical representation of thickness versus distance along the deformation length has been shown in Fig. 4.

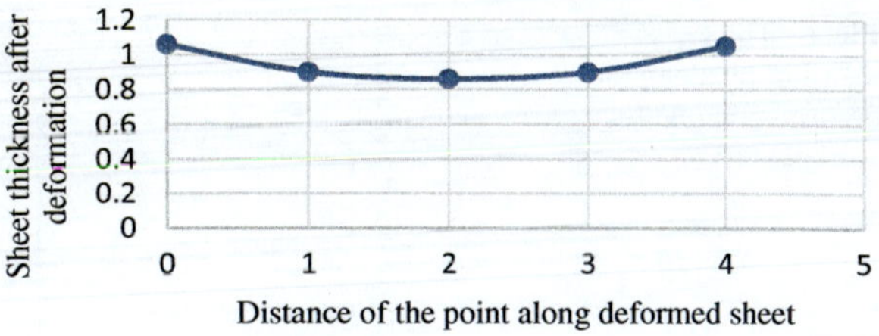

Fig. 4 Thickness distribution along the deformation line for workpiece 1

As shown in Fig. 4, in the case of deformation using a rollerball tool, the workpiece failed after 54 steps, and the product was not being able to be formed completely. The figure of the failed product is shown in Table 2. The deformed portion of the sheet was divided into three categories. Region-1 was near the undeformed portion of the sheet, region-2 being in the middle of the sheet, and region-3 at the bottom of it. Region-2 was of minimum thickness, and the sheet was found to be thin in the middle region. The sheet can be hydroformed to overcome this problem [1].

As shown in Fig. 5, the experiment was carried out using a fixed ball tool. After 134 runs, the product was successfully built; hence, the sheet deformability is better in the case of the fixed ball tool. This can be attributed to increased friction, as in the case of fixed ball tools owing to kinetic friction at the interface. Again the middle region was found to be thinnest.

Once the tool was fixed as a fixed ball tool, a steeper wall angle was achieved. To accomplish the same, a multi-stage ISF was performed, and a stepped cone was successfully generated, the wall angle in the upper region being 51°. The thickness in either of the region was the same, concluding that different regions do not undergo abruptly thin as can be seen in Fig. 6.

Again, for the three-stage ISF, a wall angle of 68.2° has been achieved. The region of steeper wall angle had a decreasing trend of sheet thickness, and in this region, the sheet thinning (Fig. 7) was slightly more; however, silver lining is that the region of steeper wall angle which underwent more number of forming passes had better surface finish.

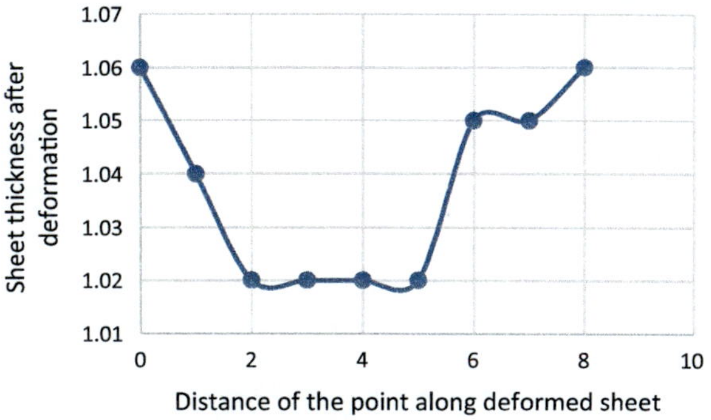

Fig. 5 Thickness distribution along the deformation line for workpiece 2

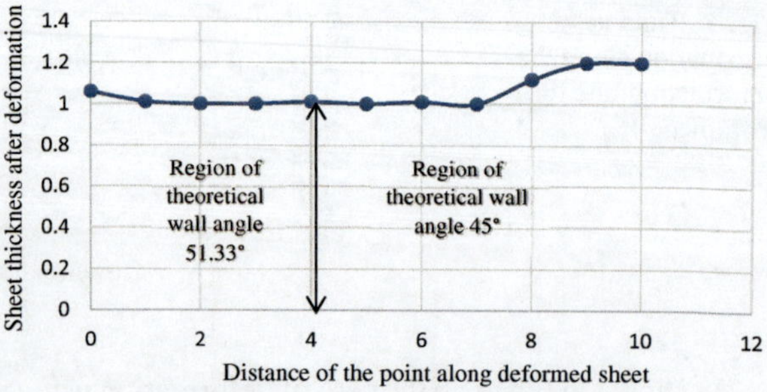

Fig. 6 Thickness distribution along the deformation line for workpiece 3

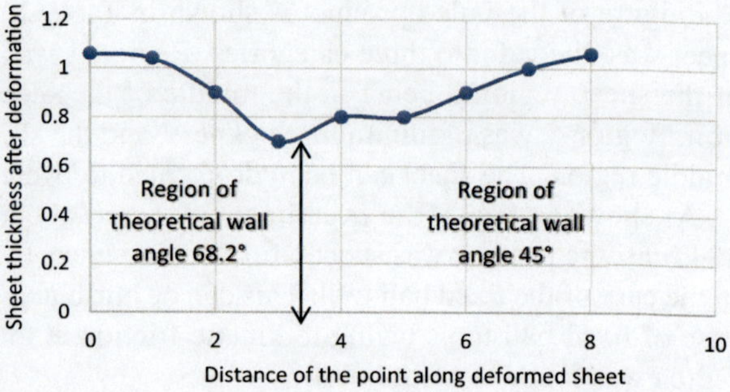

Fig. 7 Thickness distribution along the deformation line for workpiece 4

5 Conclusion

The present work aimed to have an experimental study of comparing single-stage and multi-stage incremental sheet forming under different tool arrangements on the Al6061 sheet. The fixed ball tool was found to be a better candidate for ISF so far better formability is concerned. Furthermore, steeper wall angles can be obtained using multi-stage ISF. Surface quality was better in the region with more forming passes.

Acknowledgements The authors express their sincere thanks to the *Department of science and technology, SERB, for sponsoring the work under Grant no. GP/LT/ME/14-15/08.*

References

1. Singh RP, Gupta SK, Singh PK, Kumar S (2021) Robot-assisted incremental sheet forming of Al6061 under static pressure: preliminary study of thickness distribution within the deformation region. Mater Today Proc 47:2737–2741
2. https://www.routledge.com/Incremental-Sheet-Forming-Technologies-Principles-Merits-Limitations/Ajay-Mittal/p/book/9780367276744. Last accessed 2021/11/15
3. https://patents.google.com/patent/US3342051A/en. Last accessed 2021/11/15

4. Hirt G, Junk S, Bambach M, Chouvalova I (2003) Process limits and material behaviour in incremental sheet forming with cnc-tools. In: THERMEC 2003, international conference on processing & manufacturing of advanced materials - processing, fabrication, properties, applications
5. Pandivelan C, Jeevanantham A (2015) Formability evaluation of AA 6061 alloy sheets on single point incremental forming using CNC vertical milling machine. J Mater Environ Sci. 6:1343–1353
6. Singh Y, Kumar S (2015) Incremental sheet forming (ISF). Adv Mater Form Join, 29–46
7. Meier H, Buff B, Laurischkat R, Smukala V (2009) Increasing the part accuracy in dieless robot-based incremental sheet metal forming. CIRP Ann 58(1):233–238
8. Störkle D, Altmann P, Möllensiep D, Thyssen L, Kuhlenkötter B (2019) Automated parameterization of local support at every toolpath point in robot-based incremental sheet forming. Proc Manuf 29:67–73
9. Wen T, Chen X, Zheng J, Qing J, Tang Z (2017) Multi-directional incremental sheet forming—a novel methodology for flexibly producing thin-walled parts. Int J Adv Manuf Technol 91(5):1909–1919
10. Ambrogio G, Filice L, Gagliardi F (2012) Formability of lightweight alloys by hot incremental sheet forming. Mater Des 34:501–508
11. Kumar A, Kumar D, Kumar P, Dhawan V (2020) Optimization of incremental sheet forming process using artificial intelligence-based techniques. Nat-Inspired Optim Adv Manuf Process Syst, pp 113–130
12. Peter I, Fracchia E, Canale I, Maiorano R (2019) Incremental sheet forming for prototyping automotive modules. Proc Manuf 32:50–58
13. Hirt G, Ames J, Bambach M, Kopp R (2004) Forming strategies and process modelling for CNC incremental sheet forming. CIRP Ann 53(1):203–206
14. Emmens WC, Sebastiani G, van den Boogaard AH (2010) The technology of incremental sheet forming—a brief review of the history. J Mater Process Technol 210(8):981–997
15. Wang D, Li M, Cai Z (2014) Research on forming precision of flexible rolling method for three-dimensional surface parts through simulation. Int J Adv Manuf Technol 71(9–12):1717–1727
16. https://data.epo.org/gpi/EP1560668A1-Method-and-Apparatus-for-forming.three-Dimensional-Shapes-in-a-Sheet-Metal. Last accessed 2021/11/15
17. Adams DW (2014) Improvements on single point incremental forming through electrically assisted forming, contact area prediction and tool development. ProQuest, UMI Dissertations Publishing
18. Attanasio A, Ceretti E, Giardini C (2006) Optimization of tool path in two points incremental forming. J Mater Process Technol 177(1–3):409–412
19. Singh Y, Kumar S (2019) Experimental and analytical evaluation of incremental sheet hydro-forming strategies to produce high forming angle sheets. Heliyon 5
20. Swagatika M, Regalla SP, Rao YVD (2015) multi-stage and robot-assisted incremental sheet metal forming: a review of the state of art and comparison of available technologies. AIR '15. In: Proceedings of the 2015 conference on advances in robotics

Finite Element Prediction of Deep Drawability of Tailor-Welded Blanks Using Non-associated Flow Rule and Anisotropy of Sheet Metal

Shamik Basak, Kaushik Bandyopadhyay, and Sushanta Kumar Panda

Abstract Improved formability prediction is recommended during finite element (FE) simulation of deep drawing tests and for accurate design and manufacturing of lightweight sheet metal components without necking and fracture failures. In the present study, FE models were developed for laboratory-scale deep drawing experiments of tailor-welded blanks (TWBs) made of two different dual-phase sheets of steel, namely DP600 and DP980. In FE models, associated flow rule (AFR) and non-associated flow rule (NAFR) approaches were coupled with the anisotropic Hill48 quadratic yield equation. The deep drawing experiments of the TWBs were conducted successfully, and the experimental data such as thinning development, cup edge profile, and weld line movement were compared to validate the FE predicted results. It was found that FE simulation implementing the NAFR approach predicted deep drawing behavior more accurately compared to that of the AFR approach. Moreover, the non-uniform material flow of the deep-drawn TWB cup was predicted the first time implementing the NAFR-based FE simulation, and the results of the present work demonstrated the improved capability in the prediction of anisotropy of sheet metal. Therefore, it is worthy to implement the NAFR-based approach in FE simulation as far as the accuracy of the predicted results is concerned during forming of anisotropic sheet metal.

Keywords Finite element modeling · Non-associative flow rule · Deep drawing · Tailor-welded blank · Dual-phase steels · Weld line movement

---

S. Basak (✉)
Department of Mechanical and Industrial Engineering, Indian Institute of Technology Roorkee, Roorkee 247667, India
e-mail: shamik.basak@me.iitr.ac.in

K. Bandyopadhyay
Department of Mechanical Engineering, Indian Institute of Technology Bhilai, Raipur 492015, India

S. K. Panda
Department of Mechanical Engineering, Indian Institute of Technology Kharagpur, Kharagpur 721302, India
e-mail: sushanta.panda@mech.iitkgp.ac.in

U. S. Dixit et al. (eds.), *Advances in Forming, Machining and Automation*, Lecture Notes in Mechanical Engineering, https://doi.org/10.1007/978-981-19-3866-5_7

1 Introduction

Accuracy in finite element (FE) modeling is a primary concern to sheet metal researchers as it can help to save energy and resources during component manufacturing by minimizing the failure in the production line. To perform a FE simulation, different material parameters, including failure models and constitutive models, are essential. Modeling of plastic deformation should be precise for a reliable FE simulation. Therefore, appropriate calibration of the hardening laws, flow rules, and yield models is important for characterizing the plastic deformation of anisotropic sheet materials. The first-ever proposed yield criterion for anisotropic materials was the Hill48 model [1]. Afterward, there were many yield models, including Hosford 1979 [2], Barlat Yld89 [3], etc., proposed considering the anisotropy of the sheet materials. However, the main drawback of these yield functions is that the variation in plastic strain anisotropy (r-value) and yield strength anisotropy (σ-value) cannot be predicted simultaneously. Accordingly, several advanced yield models were proposed to consider both the plastic strain ratio and strength anisotropy simultaneously. However, the mathematical complexity increases significantly with these advanced yield models. Therefore, industry researchers are sometimes reluctant to use these advanced models to avoid the increase in the computational cost during the FE simulation. With the advancement in the yield functions, the definition of flow rules was also improvised. Subsequently, a non-associated flow rule (NAFR) was proposed over the classically associated flow rule (AFR) for better accuracy in the FE simulation results. In the case of NAFR, the plastic potential function and the yield function are not assumed to be identical as in the case of AFR. Also, the inclusion of two different functions increases the scope of incorporating more material parameters [4]. One of the pioneering researches on NAFR was published by Stoughton [5]. It was based on the Hill48 quadratic yield function, and the proposed approach was able to predict the variations in r-values and yield strengths simultaneously. Recently, a NAFR model based on Hill48 and Yld2000-2d yield functions was developed to successfully estimate the numbers of ears in the case of cylindrical deep-drawn cups of AA2090-T3 and AA5042 [6]. Furthermore, various other researches were reported to extend the application of NAFR for modeling of the anisotropic hardening [7] and prediction of thinning during the hole expansion test [8].

The tailor-welded blanks (TWBs) are one of the important developments for automotive industries in achieving fuel efficiency by reducing car body weight. TWBs are composed of sheet metals of different grades, thicknesses, and surface conditions by welding to optimize material usage. The application of TWBs enables weight reduction, cost-effectiveness, and better passenger safety [9]. Over the last two decades, researchers have addressed the challenges for the simulation of TWB forming using various anisotropic yield models. There is a tendency to apply the classical and advanced anisotropic yield functions for the simulation of the forming processes of TWBs. However, the number of mechanical tests is increasing as in TWB, more than one material is involved.

Deep drawing is an important metal forming process to fabricate cup-shaped thin-walled sheet components. The formability index is measured using a limiting drawing ratio (LDR) in the case of the deep drawing test. LDR is mathematically quantified as the ratio of two diameters, *D*/*d*. Here, *D* is the maximum circular blank diameter that can be safely drawn into the die cavity, and *d* is the punch diameter. Many researchers used FE tools as a virtual experiment to estimate the LDR and deep drawability of the TWBs. However, the incorporation of the NAFR-based approach coupled with the anisotropic yield function in the FE simulation for the estimation of the forming behavior of TWBs was not addressed to date in the available literature. Consequently, in the present study, FE simulations were conducted incorporating both the AFR and NAFR-based approaches coupling with the Hill48 anisotropic material model for predicting the deep drawing behavior of TWB. Experiments were conducted to validate the FE models of TWB consisting of two different dual-phase (DP) steels, viz. DP600 and DP980. Experimentally obtained thinning development, waviness in the cup edge profile, and weld line movement of the LDR blank were analyzed thoroughly. It was found that FE simulation results with the NAFR approach can predict the non-uniform material flow precisely compared to that of the AFR-based approach.

2 Experiments

2.1 *Material Selection*

Two different grades of DP steels of different strength, DP600 and DP980 of 1.2 mm nominal thickness, were selected. Mechanical properties of DP steel sheets were determined using uniaxial tensile testing experiments along three orientations, including rolling direction (RD), diagonal direction (DD or 45° to RD), and transverse direction (TD or 90° to RD) as per ASTM E8M standard. Also, the anisotropy of the sheet materials in terms of Lankford anisotropy coefficients (*r*-values) was calculated using the ASTM E517 standard. Mechanical properties, including ultimate tensile strength (UTS) and Lankford anisotropy coefficients (*r*-values) along with various directions, are shown in Table 1 for both the parent materials.

Table 1 Ultimate tensile strengths and *r*-values of the parent materials

Materials	σ_0 (MPa)	σ_{45} (MPa)	σ_{90} (MPa)	r_0	r_{45}	r_{90}
DP600	544.1	550.7	561.9	0.962	0.926	1.054
DP980	985.8	1002.8	986.7	0.853	0.699	0.723

2.2 Fabrication of Tailor-Welded Blanks

The laser butt welding process was used for the fabrication of the TWBs. The sheets with dimensions 200 × 100 mm was cut from the cold-rolled sheets by the WEDM process. For ensuring better alignments between the two cut pieces, the cut edges were polished with emery papers. The welding was carried out with a laser power source of 3 kW and a scan speed of 3000 mm/min. Shielding gas (Argon) was used to avoid atmospheric contamination. Also, during welding, the weld line is always kept perpendicular to the RD of the parent materials.

2.3 Deep Drawing

Cylindrical deep drawing tests were performed to evaluate the formability of the fabricated TWBs. Experimental deep drawing tooling arrangements with the hydraulic press setup are shown in Fig. 1a. A closer look at the setup is depicted in Fig. 1b, which comprises 60 mm diameter circular punch with upper and lower dies to hold the TWBs. The punch was attached with the main ram of the hydraulic press, whereas the upper die was connected with the auxiliary ram. The blank holding pressure (BHF) was given through the upper die, as shown in the figure. The lubrication was added on the surface of the blanks at the contact region of both the dies. Also, no lubrication was applied at the punch surface, and this strategy was done to enhance the formability during deep drawing. The load–displacement curves were recorded and also closely navigated. The test was terminated, while a sharp fall in the load curve was observed. This point is an indication of the failure in terms of necking and/or fracture during experiments.

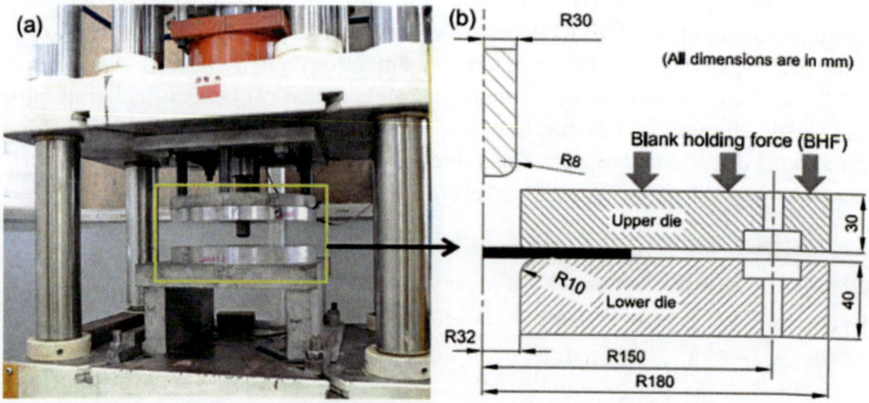

Fig. 1 **a** Experimental arrangement for the deep drawing test with **b** schematic diagram of the setup

3 Material Modeling Using Non-associative Flow Rule

An efficient FE model requires proper identification of the material models. The material model generally consists of three important attributes: yield models, flow rules, and damage/fracture models. In this work, the Hill48 anisotropic yield function ($\sigma(\boldsymbol{\sigma})$) was used during FE simulation. The effective stress calculated with Hill48 yield model for plane stress condition can be represented as

$$\sigma(\boldsymbol{\sigma}) = \left[(G + H)\sigma_{xx}^2 + (F + H)\sigma_{yy}^2 - 2H\sigma_{xx}\sigma_{yy} + 2N\sigma_{xy}^2\right]^{0.5} = \bar{\sigma} \quad (1)$$

Here, σ_{ij} denotes the different stress components of the Cauchy stress tensor. The material constants (F, G, H, N) can be evaluated from both the r-value-based approach and σ-value-based approach as

$$\frac{H}{G} = r_0, \quad \frac{H}{F} = r_{90}, \quad \frac{N}{(F+G)} - 0.5 = r_{45} \quad \text{and} \quad G + H = 1 \quad (2)$$

$$F + H = \left(\frac{\sigma_0}{\sigma_{90}}\right)^2, \quad F + G + 2N = \left(\frac{2\sigma_0}{\sigma_{45}}\right)^2, \quad F + G = \left(\frac{\sigma_0}{\sigma_{bb}}\right)^2 \quad \text{and} \quad G + H = 1 \quad (3)$$

These evaluated constants can be represented with Eqs. (2) and (3) for both approaches, respectively.

The concept of the flow rules is established on yield as well as plastic potential surfaces. The yield surface controls the magnitude of various stress states experienced by the same plastic work. However, the potential surface controls the direction of the increment of the plastic strain. In AFR, the yield function and the plastic potential function are both considered to be the same. In this work, r-value-based Hill48 (Eq. (2)) was used for both the plastic potential function and the yield function for the AFR model. Hence, only a few material parameters (r-values only) can be incorporated into the AFR model. To avoid this drawback, the NAFR model is introduced in the present work. In NAFR, the r-value-based Hill48 was used as yield function ($\sigma_y(\boldsymbol{\sigma})$), and σ-value-based Hill48 was used as the plastic potential function ($\sigma_p(\boldsymbol{\sigma})$) to evaluate the direction of the plastic strain rate. Thus, more material anisotropy information can be incorporated while performing the FE simulations.

For both the AFR and NAFR, the yield function can be expressed as

$$\Phi = \sigma_y(\boldsymbol{\sigma}) - \sigma_{iso}\left(\bar{\varepsilon}_p\right) \quad (4)$$

where Φ is the yield criterion, $\sigma_y(\boldsymbol{\sigma})$ is the yield function of the stress tensor ($\boldsymbol{\sigma}$), and $\sigma_{iso}\left(\bar{\varepsilon}_p\right)$ is the hardening rule connecting the effective stress and strain. Further, the plastic strain increment ($d\varepsilon^P$) is calculated as

$$d\varepsilon^{P} = d\lambda \frac{\partial \sigma_{p}(\sigma)}{\partial \sigma} \tag{5}$$

Here, $\sigma_p(\boldsymbol{\sigma})$ is the plastic potential function and $d\lambda$ is the plastic multiplier which is equivalent to the effective plastic strain increment for the AFR approach. However, for the NAFR approach, the plastic multiplier can be derived as

$$\sigma_{y}(\sigma) d\bar{\varepsilon}^{P} = \sigma : \ d\varepsilon^{P} = \sigma : \ d\lambda \frac{\partial \sigma_{p}(\sigma)}{\partial \sigma} = d\lambda \sigma_{p}(\sigma) \tag{6}$$

Hence, for the NAFR approach, the plastic multiplier can be written as

$$d\lambda = \frac{\sigma_{y}(\boldsymbol{\sigma})}{\sigma_{p}(\boldsymbol{\sigma})} d\bar{\varepsilon}^{P} \tag{7}$$

4 Finite Element Modeling

FE simulations of the deep drawing process of TWBs were conducted to predict the drawability and thinning development during the deformation process. All the forming tools, including upper die, lower die, and punch, were assumed as rigid bodies. The surface-to-surface contact algorithm was assumed at the contact regions between the tools and blank. A predetermined velocity profile was imposed on the punch during deformation. In the preprocessor, the blank was meshed using a mesh size of 1.2 mm. However, in the soft zone and hard zone, the mesh size was approximately 0.5 mm × 1.2 mm due to the smaller width of the regions. The mesh size of the blanks and the details of the different zones in TWB can be found in a previous study by the author [10]. Furthermore, a FORTRAN code was developed to incorporate Hill48 in the form of NAFR along with the hardening model. The desired material parameters were incorporated in LS-DYNA through the MAT50 material card. The incorporation of UMAT in LS-DYNA is explained in detail in DYNA help manual appendix [11]. Also, considering the half-symmetric model, the BHF specified in the simulation was half of the BHF actually applied during the experiment. The numerical value of the coefficient of friction between the blank and the tools was reasonably presumed by iteratively comparing simulated load–displacement curves with experimentally obtained results. It was found that for using the friction coefficients between blank-die, blank-punch, and blank-binder as 0.15, 0.2, and 0.22, respectively, and the FE predicted load progression curve matched the experimental data with sufficient accuracy, as shown in Fig. 2.

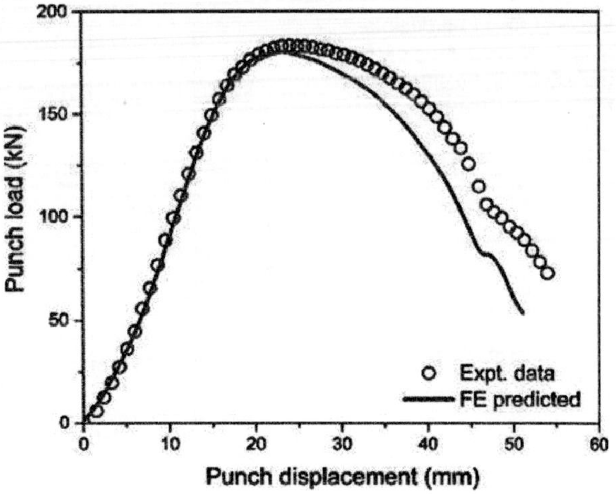

Fig. 2 Comparison of experimental and predicted load progression curves

5 Results and Discussion

5.1 *Parameter Identification for the Combined Hardening Model*

Along with the yield model and flow rules, it is equally important for modeling the work hardening performance of sheet materials. The usual method is to use some well-known hardening models such as the Swift model and Voce model to estimate the strain hardening behavior of sheet material. However, in this work, a combination of the Swift and Voce model, termed as Swift–Voce (SV) model, was used to demonstrate the capability to predict the work hardening behavior of the base materials and TWB. The mathematical expression of the SV hardening model can be expressed as

$$\sigma_{\text{iso}}\left(\bar{\varepsilon}_{\text{p}}\right) = \bar{\sigma} = R \times \left\{K\left(\varepsilon_0 + \bar{\varepsilon}_{\text{p}}\right)^n\right\} + (1 - R) \times \left\{A + B\left(1 - \exp\left(-C \times \bar{\varepsilon}_{\text{p}}\right)\right)\right\} \tag{8}$$

The first part of Eq. (8) is the Swift hardening part, and the second part is the Voce hardening part. Also, R indicates a linear factor of the Swift and Voce models, and the numerical value of R lies between 0 and 1. The fitted curves are shown in Fig. 3 only for the TWB miniature specimen. It can be observed that the Swift curve over-predicts the experimental curve, whereas the Voce law under-predicts the same. The linear combination of these two laws, the SV hardening law, can predict the hardening behavior with 99.6% accuracy. All the hardening parameters are shown for the parent materials and TWB in Table 2.

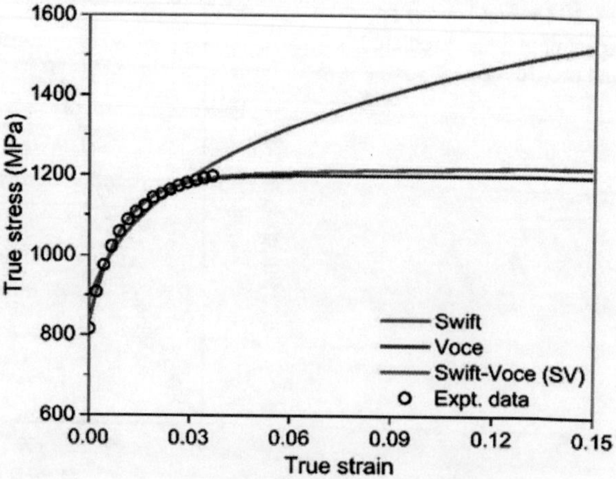

Fig. 3 Effect of hardening models on prediction of the true stress–strain curve of the DP600-DP980 miniature tensile specimen

Table 2 Evaluated hardening parameters for the sheet materials

Materials	*K* (MPa)	ε_0	*n*	*A* (MPa)	*B* (MPa)	*C*	*R*
DP600	1020	0.004	0.20	354.3	277.4	28.19	0.310
DP980	1772	0.005	0.17	679.8	430.9	59.66	0.150
TWB	2086	0.0066	0.169	833.9	365.2	101.1	0.069

5.2 *Prediction of Thinning Profile*

During LDR evaluation, the selection of the proper blank dimension is critical to avoid the wastage of the material. Hence, some initial trials of deep drawing experiments were performed in FE simulation to fix the initial blank diameter. During the experiment, it was observed that 125 mm was the highest circular welded blank (TWB) diameter that could be effectively deep drawn into a deformed cup. Accordingly, the LDR of the DP600-DP980 TWB was fixed as 2.08. Further, the thinning profile or the thickness distribution of the LDR cup was estimated using both the AFR and NAFR approaches in the FE simulation. The middle point of the cylindrical cup bottom area was considered to be the base point during the measurement of the thinning pattern. The experimental data points with the FE predicted data are depicted in Fig. 4 for comparison purposes. It can be found that the thickness distribution is not uniform with respect to the center of the cup. This is because of the two different material properties involved. DP600 material is weaker than DP980 material. Moreover, the weld line shifted toward the stronger material side (DP980) during deep drawing. Due to the excess pulling, the localized thinning (19% of the sheet thickness) on the weaker material (DP600) occurs very near the weld line, as can be observed in Fig. 4.

Interestingly, another secondary thinning is found in the cup corner area on both materials. This is due to the excessive stretching of the material by the cylindrical

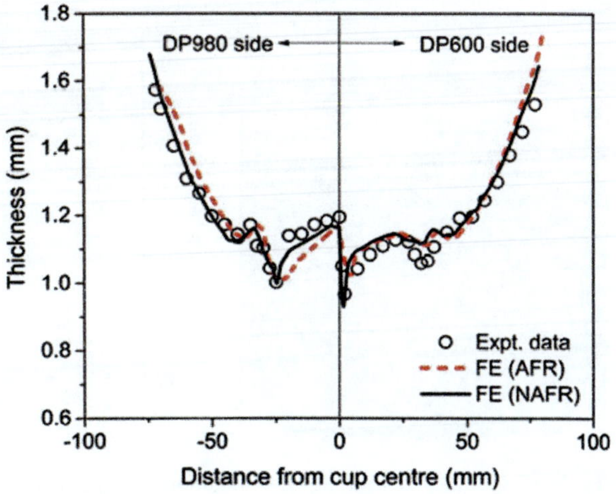

Fig. 4 Prediction of the experimental thinning profile with AFR and NAFR approach

punch corner during the deep drawing experiments. However, the amount of thinning is more in the case of DP980 material (16% of the sheet thickness) compared to that of DP600 (12% of the sheet thickness) side. This is because of the difference in r-value of the parent materials. As it is known that the lower r-value promotes thinning, and DP980 material has a significantly lower r-value compared to DP600 material, as can be observed from Table 1. The thickening of the parent materials is found at the flange area of the TWB drawn cup. The prediction of the thickness profile is relatively accurate with the NAFR approach as it takes account of all the anisotropic material properties, including the r-value and σ-value variations which help the simulation to capture the exact thinning pattern during deformation.

5.3 *Effect of Non-uniform Material Flow on the Cup Edge Profile*

It is found that a strong non-uniform material flow results in an asymmetric thinning pattern for the TWB drawn cup. The cup edge profile is further investigated to understand the material flow behavior during deep drawing. The undulation of the cup edge profile is known as earing during the deep drawing experiments. Plastic anisotropy in the metal results in directional flow of the materials during deep drawing, and it causes the earing in the full drawn cups. The TWB LDR cup with the earing formation is shown in Fig. 5a. The earing heights can be correlated to the respective planer anisotropy (Δr) of the metals. The cup edge profile of the TWB was measured with a coordinate measuring machine (CMM). The cup edge profile is plotted in Fig. 5b for an angular span of 180 degree as the TWB is the half-symmetric cup. It is observed that the weaker material always experiences a sudden increase in the cup height than the DP980 side near the weld junction. This is commonly known as the edge effect

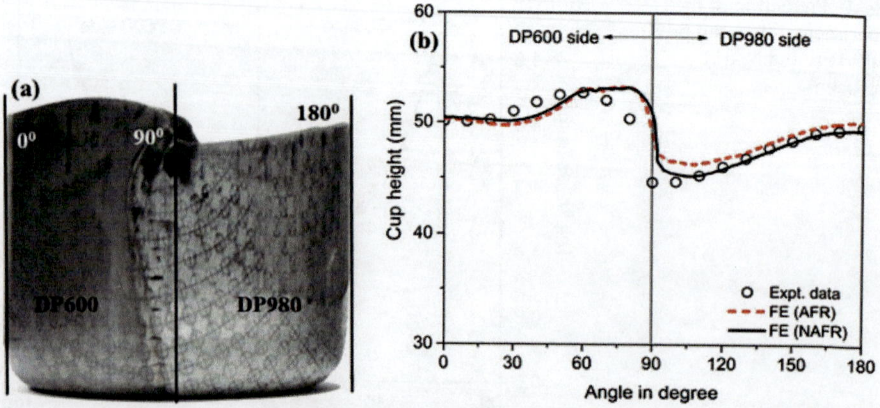

Fig. 5 **a** Experimental TWB at LDR condition with the **b** prediction of the non-uniform material flow with AFR and NAFR approach

during the deep drawing of TWBs. The yield strength of the DP980 is higher than that of DP600. Therefore, DP980 always resists deformation more than its weaker DP600 part. While, during deep drawing, the stress state in the flange region is tensile toward the radial direction and compressive along the circumferential direction. DP980 can withstand more compressive stress, and hence, more stress is induced to the weaker material. This excessive compressive stress causes the weaker material to stretch radially outwards near the weld line as the thickening of the material is resisted by blank holding force. Therefore, the material flow in the outwards direction causes the sudden increase in cup height in the weaker material side near the weld zone, as evident in Fig. 5. Away from the weld zone, tensile force is more predominant, which draws weaker material into the die cavity, causing lower height. On the DP980 side, cup height slightly increases in the region away from the weld due to its resistance to draw into the die cavity. Therefore, severe undulation at the cup edge is observed in the case of TWBs. The cup edge profile is predicted with both the AFR and NAFR approaches. It is found from Fig. 5b that the numerical value of the maximum cup height and the location of the maximum cup height is predicted well with the AFR approach. However, the AFR approach cannot predict the variation in the cup height as accurately as NAFR. It under-predicts the cup height at the DP600 side and over-predicts at the DP980 side.

On the contrary, the NAFR approach reasonably predicts the maximum cup height and the variation of cup heights on both sides. This is mainly because the AFR approach only takes account of the r-value variation of the material, whereas the NAFR approach includes both the r-value and σ-value variations as the yield and potential function, respectively. Hence, as the NAFR approach contains more material properties, this method can estimate the non-uniform material flow during the deep drawing of the TWB. Therefore, the accuracy of the FE simulation through the NAFR approach is comparatively better than that of the AFR approach.

5.4 Weld Line Movement

The strength difference in both the materials results in a non-uniform material flow during deep drawing of the TWBs. Therefore, a strong shift in the weld line is also observed in the deep-drawn cup. The weld line movement was measured using a CMM at a fixed interval of 5 mm and plotted in Fig. 6. As DP980 is the stronger part, the weld line is being pulled toward the stronger side, which results in excess stretching of the DP600 part near the weld line at the bottom of the cup. The weld line shifts toward the weaker material at the cup wall as DP980, being more resistive to the compressive stress, pushes the weld line toward the DP600 side. Also, the weld line takes a turn toward the DP980 side at the cup edge. Material of D600 side expands toward outside at the cup edge near the weld as explained in the previous section. This excess material pushes back the weld toward the DP980 as no material is present at the opposite side to hold out against. So, a curvilinear weld movement is observed in the case of deep-drawn cups from TWBs. It was also observed from Fig. 6 that the weld displacement is more at the cup center and near the top edge of the cup. It was already discussed in the previous section that the non-uniform material behavior is predicted accurately with the NAFR approach. Hence, the weld line movement for the TWB is predicted with FE simulation with the NAFR approach, and it is observed that numerically estimated data corroborated well with the experimental weld line shift data.

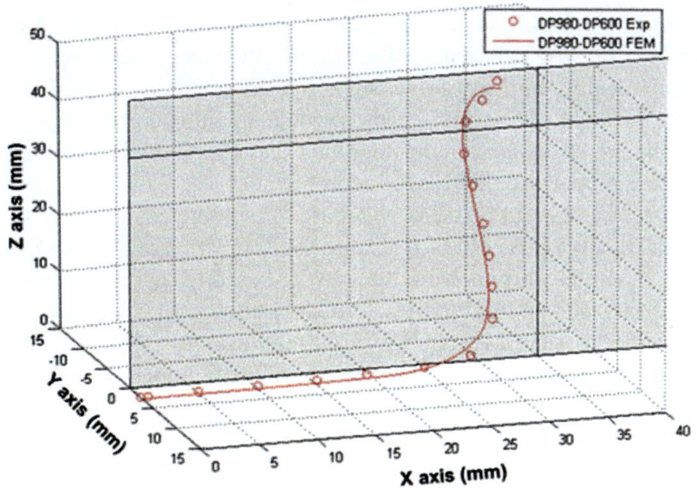

Fig. 6 Prediction of the weld line movement for deep-drawn TWB cup

6 Conclusions

The deep drawing behavior of DP600-DP980 TWB was predicted by FE simulation implementing both the AFR and NAFR approaches in this study. In the AFR approach, the yield and potential functions were both the same and calibrated with the Hill48 r-based anisotropic function. Conversely, in the NAFR approach, the yield function was calibrated with the Hill48 r-based anisotropic function, and the potential function was calibrated with the Hill48 σ-based anisotropic function. Following are the key conclusions obtained from the study.

- Two different thinning locations were observed during the deep drawing of the TWB. The primary thinning location is on the weaker DP600 side near the weld line due to excessive pulling of the weld toward the strong material side. The secondary thinning location is at the cup corner location of both the material side. However, the amount of second thinning is more in the case of DP980 material because of the lesser r-value of the material compared to DP600 material.
- The non-uniform material flow was observed due to the strength difference of the parent materials. Due to the non-uniform material flow, the waviness in the cup edge profile and the weld line movement was observed. FE simulation with the AFR model can predict the maximum cup height reasonably, but it fails to estimate the undulation of the cup height very accurately. However, the FE simulation with the NAFR method is able to predict both the maximum cup height and the variation of the cup height of the TWB deep-drawn cup.
- More number of anisotropic material parameters can be used in the FE simulation with the help of NAFR based on the Hill48 anisotropic yield function. Thus, the mathematical complexity for using so-called advanced yield functions can be avoided successfully. Moreover, the predicted results corroborated well with the experimental observation indicating the excellent prediction capability of the developed FE model with the NAFR approach.
- It has been demonstrated that the NAFR-based FE predicted data showed a prominent improvement over AFR-based results in deep drawing of TWBs. Hence, using the same mathematical framework, the NAFR-based FE approach can be implemented for precise forming and fracture assessments during other sheet metal forming processes such as stretch forming, multistage stamping, bending, and punching/trimming so as to fabricate very complex lightweight components.

References

1. Hill R (1948) A Theory of the yielding and plastic flow of anisotropic metals. Proc R Soc A Math Phys Eng Sci 193:281–297. https://doi.org/10.1098/rspa.1948.0045
2. Hosford WF (1979) On yield loci of anisotropic cubic metals. Proc Seventh North Am Met Work Conf SME 191–197

3. Barlat F, Lian K (1989) Plastic behavior and stretchability of sheet metals. Part I: A yield function for orthotropic sheets under plane stress conditions. Int J Plast 5:51–66
4. Safaei M, Lee M, Zang S, De WW (2014) An evolutionary anisotropic model for sheet metals based on non-associated flow rule approach. Comput Mater Sci 81:15–29
5. Stoughton TB (2002) A non-associated flow rule for sheet metal forming. Int J Plast 18:687–714
6. Park T, Chung K (2012) International journal of solids and structures non-associated flow rule with symmetric stiffness modulus for isotropic-kinematic hardening and its application for earing in circular cup drawing. Int J Solids Struct 49:3582–3593
7. Lee EH, Stoughton TB, Yoon JW (2017) A yield criterion through coupling of quadratic and non-quadratic functions for anisotropic hardening with non-associated flow rule. Int J Plast 99:120–143
8. Kim JJ, Pham QT, Kim YS (2021) Thinning prediction of hole-expansion test for DP980 sheet based on a non-associated flow rule. Int J Mech Sci 191:106067
9. Basak S, Katiyar BS, Orozco-Gonzalez P, Baltazar-Hernandez VH, Arora KS, Panda SK (2019) Microstructure, forming limit diagram, and strain distribution of pre-strained DP-IF steel tailor–welded blank for auto body application. Int J Adv Manuf Technol 104
10. Bandyopadhyay K, Lee M-G, Panda SK, Saha P, Lee J (2017) Formability assessment and failure prediction of laser welded dual phase steel blanks using anisotropic plastic properties. Int J Mech Sci 126
11. Hallquist JO, others (2007) LS-DYNA keyword user's manual. Livermore Softw Technol Corp 970

Investigation on Single-Point Incremental Forming Process of SS316 Sheets at Elevated Temperatures

R. Sridhar, D. Shanmugasundaram, and D. Rajenthirakumar

Abstract This paper aims to investigate the formability of difficult to form stainless steel material by the application of heat in the single-point incremental forming [SPIF] process. The forming tool is made up of EN36 case hardening steel and it is coated with the magnetron sputtering method. A modular fixture was developed with a heating setup, and in this work, stainless steel SS316 grade 1 mm thick sheet was used for forming. Suitable work geometry was selected and the tool path was generated by using MATLAB software. The experiments are carried out with suitable process parameters in LV45 LMW make vertical machining centre (VMC). The output parameters such as forming depth, thickness distribution, surface roughness, and forming limit diagram (FLD) are compared between room temperature forming and with the hot forming condition. The results show that hot forming in the SPIF process will make a constructive influence on the results this could infer from the forming limit diagram (FLD). FLD shows that heating the sheet has made it bear the more longitudinal strain, which allows for smooth forming of the sheets along with postponing the failure to an extended depth, in comparison with that of the room temperature formed sheets. It is found that heating of SS sheet while forming itself gives better results. The coating of the tool does not make any difference in these experiments. The microstructural study has also proved that hot-formed stainless steel parts retained the austenite structure, whereas the room temperature formed parts converted to tempered martensite.

Keywords Incremental forming · Forming limit diagram · Hot forming · Microstructural study

1 Introduction

SPIF is a vital area for researchers in the area of metal forming over the past few years. With the development of new technologies and methods, researchers are aiming to

R. Sridhar (✉) · D. Shanmugasundaram · D. Rajenthirakumar
Department of Mechanical Engineering, PSG College of Technology, Coimbatore 641004, India
e-mail: sri.mech@psgtech.ac.in

U. S. Dixit et al. (eds.), *Advances in Forming, Machining and Automation*, Lecture Notes in Mechanical Engineering, https://doi.org/10.1007/978-981-19-3866-5_8

use this technology for forming difficult to form metals such as nickel and chromium alloys by using this process. Research is aiming to clear understanding of forming mechanism and improve its process capacities. Single-point incremental forming is a sheet metal forming technique with great potential for use in prototyping and custom manufacture [1]. A schematic sketch of (SPIF) has been shown in Fig. 1.

In the incremental sheet forming process, the sheet is formed through step-by-step plastic local deformation to form the final geometry. The movement of the tool is controlled by the cutter location coordinates so that the proper trajectory of the required shape is achieved [2]. It is necessary to develop a forming apparatus to apply the incremental forming technique to steels or thick plates [3]. The thickness reduction of the cup was greater when drawn at higher temperatures but without fracture [4]. The stress analysis proves that the fracture occurs initially on the outer side of the wall at the transition area [5]. The friction between the tool/sheet interface helps to improve the formability but it decreases the surface quality [6]. In stainless steel, it was found that the formability of the sheet is increased with increasing the spindle speed, decreasing the vertical step down and feed rate [7]. SPIF process was feasible enough to produce pyramids in micro-scale with higher forming limits than conventional forming in stainless steel foils [8]. The formability of SS304 steel is initially the shear and compressive stresses were dominating [9]. More starching of the sheet can be possible since the minimum thinning is achieved in both FEM simulation and experiment results [10]. This process inferred with the advantages such as no die is required as in a conventional forming technique and it reduces the cost per piece of the parts made. It will reduce the lead time for low production runs as the need for manufacturing the die is eliminated. Several researchers have found that the formability of metals is better in incremental forming than that of conventional deep drawing. In incremental forming, the geometric deviation can be controlled between the range 1–2 mm, still additional methods are required to make it perfect, and optimization of the tool path, additional support, and hot incremental

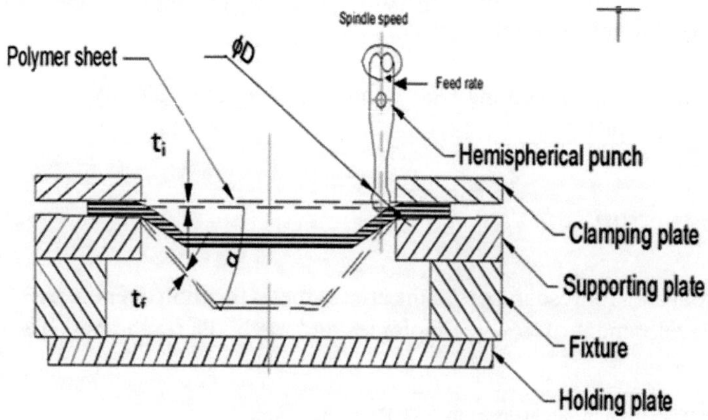

Fig. 1 A schematic sketch of single-point incremental forming (SPIF)

forming is the most suitable and efficient methods for improving the accuracy of parts [11]. The main objective of this paper is to assess and compare the effect of heating in SPIF and the consequent results on the formed part to assess and compare the effect of tool coating.

2 Geometry Selection and Feasibility Analysis

The geometry selected was a truncated hyperbola, and dimensional values of the geometry are top diameter 80 mm, bottom diameter 30 mm, and height 50 mm. The tool path for this geometry was generated by MATLAB software and the generated geometry was shown in Fig. 2.

This geometry was selected to check the thickness distribution and maximum depth formation during forming of increased wall angle conditions. To check the feasibility of incremental forming of difficult to form stainless steel 316-grade material, a specific methodology was employed to simulate the process in Abaqus FEA software.

The properties of the material SS316 were given as input to the software as material properties. Stainless steel 316 (SS316) is an austenitic chromium–nickel stainless steel containing excess molybdenum which aids in increasing the corrosion resistance and especially improves its pitting resistance to chloride ion solutions, and it has wide applications in cooking utensils and automobile interiors.

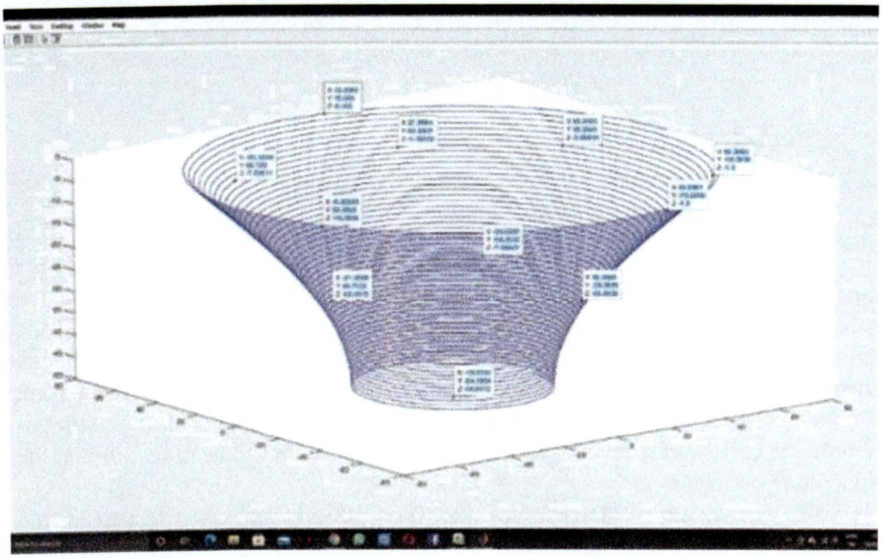

Fig. 2 Output geometry

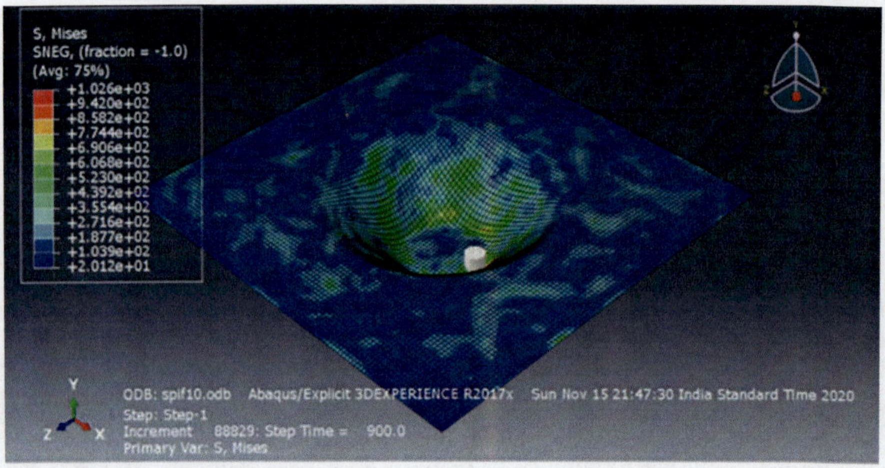

Fig. 3 Stress plot of the SS316 part

The finite element analysis result shows that the desired part with a depth of 60 mm came out to be a failure in Abaqus FEA software. The failure is attributed to the higher value of stress, at a depth greater than 50 mm but in the selected geometry, the depth of the part is to be 50 mm. The stress plot has been obtained for the values amplitude values which are fed. The results show that the SPIF of the desired part geometry for SS316 is feasible. The failure of the sheets is attributed to the excessive stress in the lower region of the part as seen in Fig. 3. It is inferred that the failure of the sheets occurred in the region where stress is higher (yellow region).

3 Experimental Setup

3.1 Experimental Strategy

After the successful feasibility study of this process, a well-formulated experimental procedure was followed. A dedicated experimental setup was developed and the experiments were carried out. Shi et al. found that local heating in incremental forming by the application of electric hot incremental forming in low carbon steel DC04 has enhanced the dimensional accuracy of the formed part [12]. Based on this method, the SPIF experimental setup was developed with the heating coil and the temperature controller.

For this experiment SS316 sheets of size 300 × 300 × 1 mm was selected, and the backside of the sheet was marked with a circular grid of diameter 10 mm, and space between the circles was 10 mm by using the laser method. This was employed to determine the formability of the sheet during the forming process. The hemispherical

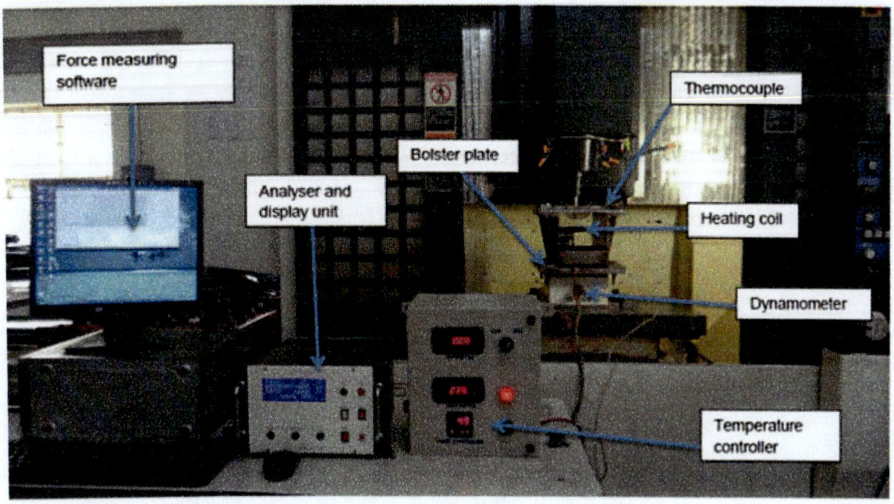

Fig. 4 Hot forming setup

tool of diameter 12 mm was made up of EN36 material used and it was coated with titanium nitride to improve the wear resistance and oxide formation. This process was carried out by using all the combinations involved with elevated temperature and coated and uncoated tools. All four sheets have been measured for various parameters like elongation, sheet thickness, and depth of failure to conclude the effects of heat setup and coating of the tool.

The experimentation has been carried out in a vertical machining centre (VMC) LV45. The codes generated in mastercam have been fed into the machine, and the optimum arrangements are made to experiment. An experimental setup for SPIF with the local heating arrangement is shown in Fig. 4. The SS316 sheet was clamped rigidly to the fixture, and the experimentations were carried out at the forming parameters of speed 1200 rpm, feed 500 mm/min, step size 0.05 mm, and plunge feed 50 mm/min. By using the same parameters, the following four combinations of experiments are carried out. By using an uncoated tool and coated tool at room temperature and elevated temperatures of 750°, the formed parts measured for thickness variations, depth of failure, and forming limit diagram (FLD) are drawn and compared.

3.2 *Forming of Parts*

Based on the designed CAD model, punch motion data (PMD) was controlled by the CNC machine. A square SS316 sheet blank was clamped and formed with the spherical head tool. Once the crack was noted at the bottom of the part, the machine tool Z-axis was stopped manually. The depth at which the part was cracked was noted from the machine tool itself, considering the sheet blank top surface as a reference. It was observed that the part could not reach the final designed depth (50 mm), before it reaches the maximum depth it was cracked. The formed parts and their measured depth are shown in Table 1.

To determine the thickness variations, the formed sheets are cut into half using an EDM machine, from this measured the thickness of parts at different depths and maximum depth formed, and measured the major strain and minor strain in the grid circles at the back of the sheet. The successfully formed shape and a half removed part are shown in Fig. 5.

Table 1 Formed parts and depth measured

	Uncoated tool	Coated tool
Room temperature forming	 [Maximum Depth: 33.572mm]	 [Maximum Depth :35.246mm]
Hot forming at elevated temperature [750 °C]	 [Maximum Depth:37.617mm]	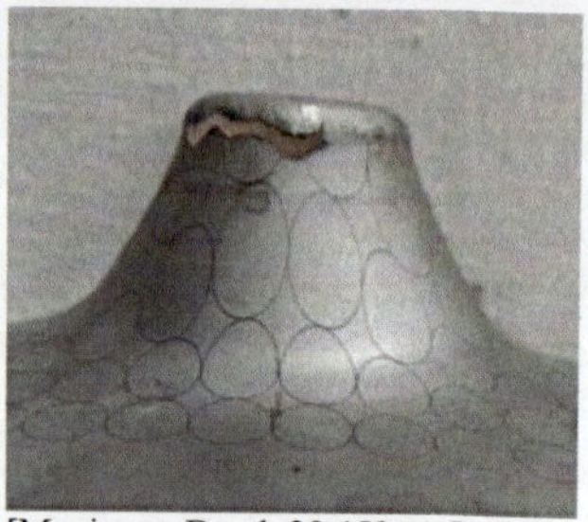 [Maximum Depth:38.459mm]

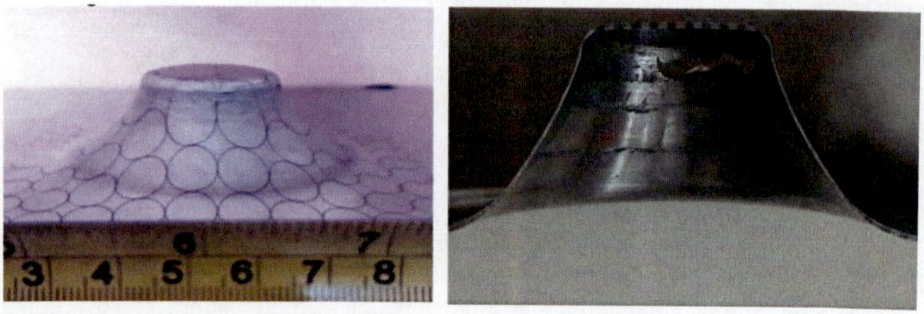

Fig. 5 Formed part and a half removed part

4 Results and Discussion

4.1 Forming Limit Diagram

FLD are most often used in the sheet metal industry to measure and evaluate major strain and minor strain in the surface of the sheet on the onset of necking failure. Major strain is defined in the direction of higher deformation. Minor strain is determined in the direction of lesser deformation.

The FLD has been created in the MINITAB software using regression with groups. The elongations of the bubbles of all the plates were measured in both x and y directions, and the values are fed into the MINITAB software to obtain the FLD of the plates. The forming limit of ISF is much higher compared to other conventional forming processes due to the localization of the plastic deformation, it is shown in Fig. 6 [13].

The formulas which are used are, longitudinal strain (major strain)—y axis $= \ln(d_1/d_0)$ where, $d_1 = y$ axis diameter of the elongated bubble, $d_0 =$ initial diameter of the bubble and lateral strain (minor strain)—x axis $= \ln(d_2/d_0)$ where $d_2 = x$ axis diameter of the elongated bubble, $d_0 =$ initial diameter of the bubble.

As the formability of the SS316 sheet in incremental forming is mostly concerned in the longitudinal direction, minor strain (strain in the x-direction) is negligible when compared to the major strain. The obtained FLD at different conditions and their maximum and minimum values of longitudinal strain are given in Table 2. It was similar to the forming limit diagram obtained by McAnulty et al. for the incremental forming process [13].

From the FLD, it was inferred that only non-coated tools and non-heated setup fall in the failure region, which led to the early failure of the sheet so that lesser depth of the part was obtained. From the values of strain, it is evident that heating the sheet will accept more strain than that of an unheated sheet. As the sheet could bear more strain, the formability of the sheet is increased when the sheet is heated.

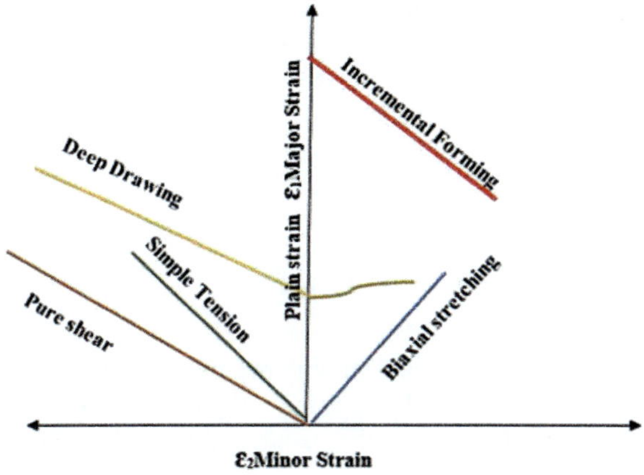

Fig. 6 Forming limiting diagram (FLD) comparison of conventional and incremental forming method

Table 2 Forming limit diagram at different conditions

	Uncoated tool	Coated tool
Room temperature forming	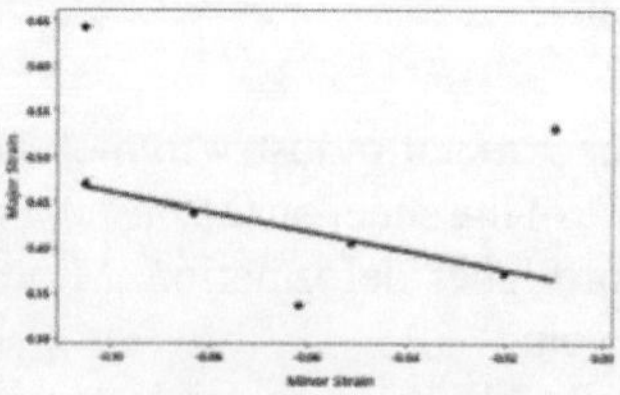 longitudinal strain [Max.0.47, Min 0.36]	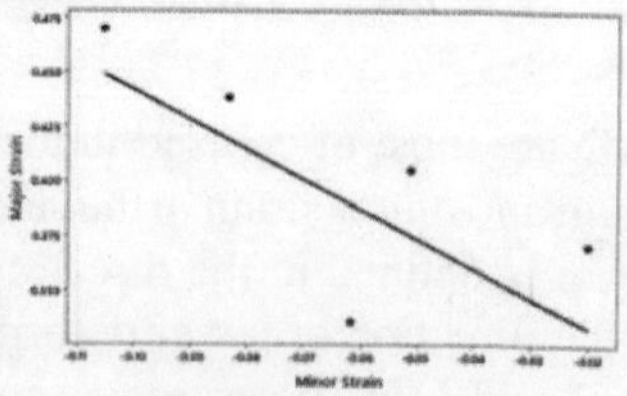 longitudinal strain [Max.0.4, Min 0.33]
Hot forming at elevated temperature [750 °C]	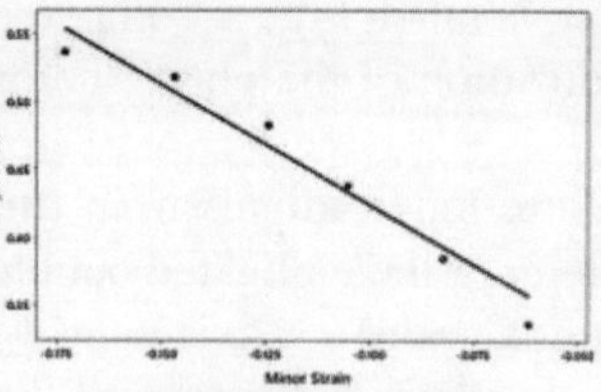 longitudinal strain [Max.0.56, Min 0.38]	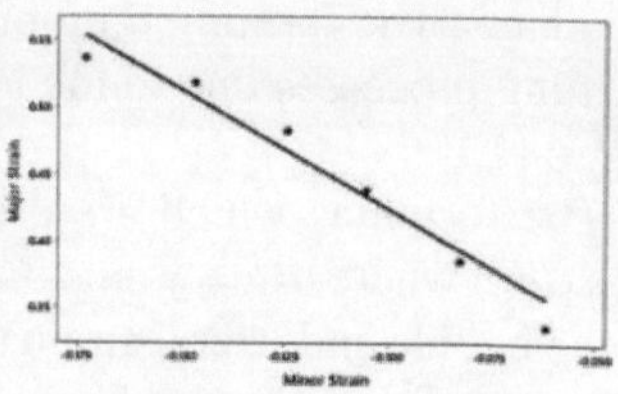 longitudinal strain [Max.0.56, Min 0.35]

These results are obvious, as the depth of forming of heated sheets is more during the experiment, heating the sheet provides for smooth forming of the sheets.

4.2 Depth of Failure

The failure depths at various conditions are compared and inferred that the coated tool pre-formed better than that of the uncoated tool and it is shown in Fig. 7. The forming depths of the coated tool are greater than the uncoated tool in both heated and room temperature setups. Heating the sheets has implied a significant effect on the formability of the sheets as the forming depth has increased. Thus, both coated tool and heated sheet have a constructive effect on the formability of the sheet and can be implied in any SPIF to have better forming depths and to avoid earlier failures.

4.3 Thickness Distribution

The thickness distribution has been measured along with all the sheets at certain depths of the formed part using a digital Vernier calliper. The thickness distribution graphs at different conditions are compared it is shown in Fig. 8. Though there are

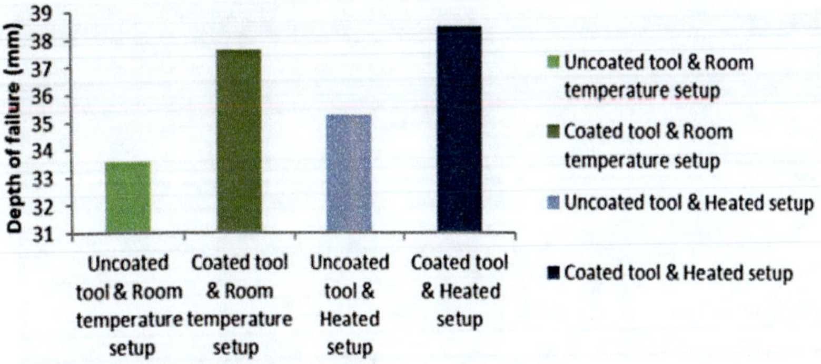

Fig. 7 Depth of failure

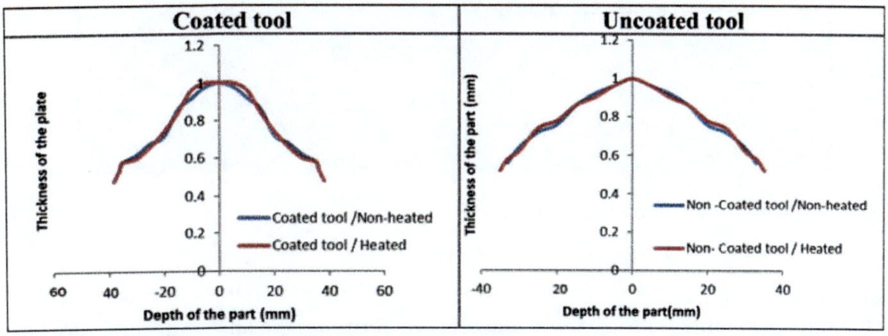

Fig. 8 Thickness distribution in formed parts

very minor differences in the thickness along with the depth of the part, at certain depths, there could be noticeable differences in all the workpieces. The thickness of the sheet does not vary much due to the coating of the tool. However, the heat setup played a major role in altering the thickness of the sheet along the course of the tool path.

4.4 *Microstructural Study*

The microstructure of SS316 sheet at different conditions such as raw material, room temperature formed, and hot-formed is studied by using an optical microscope at 400× magnification, and it was shown in Fig. 9.

From this, it is inferred that the microstructure of the unformed sheet sample, which is at an austenitic state, has been retained as austenite during heat forming in SPIF. Although the austenitic phase is retained during heat forming, elongation of grain size could be seen in heat-formed parts. From this, we could infer that heat

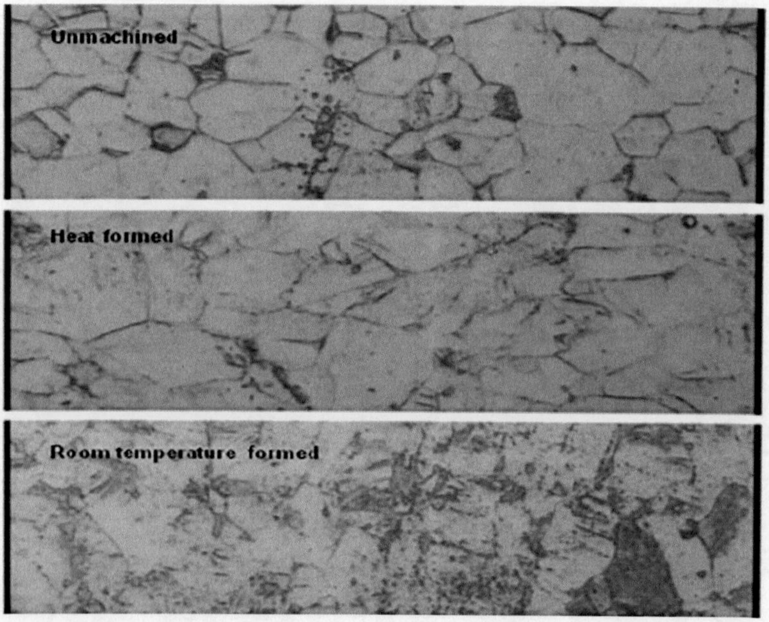

Fig. 9 Microstructure at a magnification of 400×

forming allows for more elongation of the sheets without undergoing more changes in the microstructure. During room temperature forming, the austenitic structure has transformed into a tempered martensitic structure, which has occurred due to rapid cooling of the part than that of the heated forming. This contributed to a major change in the microstructure of SS316.

Martensitic stainless steels are more brittle than that austenitic stainless steel because of the carbon content. Since the part formed during heat forming is of austenitic structure, it could be concluded that heat forming has a constructive effect on (SPIF) as the hardness of the austenitic structure is higher than that of the martensitic structure.

5 Conclusion

This paper aimed at comparing the effect of a coated tool along with the effect of heating on sheets in SPIF. Various parameters like the thickness of the sheet, depth of the formed part, the strain induced in the sheet, and the microstructure are compared. It was concluded that coating the tool with titanium nitride has not proved any significant results in the output, except in the case of depth of the formed part, where the depth of the formed part made using the coated tool is slightly better than that of the part made using the non-coated tool.

The following are the specific conclusions from the investigations:

1. The FLD shows that heating the sheet has made it bear more longitudinal strain, which allows for smooth forming of the sheets along with postponing the failure to an extended depth.
2. The thickness of the sheets during heating is also in cohesion with the findings from the forming limit diagram, as the failure of the heated sheets occurred at a lower thickness with an incredibly larger margin between the room temperature formed sheets. This shows that forming occurs smoother in the heated sheets and heated sheets allow for more depth as well.
3. A finding also shows that heated sheets failed at a greater depth.
4. The microstructural study has also proved to be significant in the positive influence of heating as heated sheets retained the austenite structure, whereas the room temperature formed sheets converted to tempered martensite. It was found that heating the sheets resulted in a part that is much harder than that of normal forming. It could be concluded that heating the sheets is beneficial to SPIF of the SS316 sheets.

References

1. Adams D, Jeswiet J (2014) Design rules and applications of single-point incremental forming. Proc Inst Mech Eng Part B J Eng Manuf 229:754–760
2. Kurra S, Rahman NH, Regalla SP, Gupta AK (2015) Modeling and optimization of surface roughness in single point incremental forming process. J Mater Res Technol 4(3):304–13
3. Park J-J, Kim Y-H (2003) Fundamental studies on the incremental sheet metal forming technique. J Mater Process Technol 140:447–453
4. Hussaini SM, Singh SK, Gupta AK (2014) Experimental and numerical investigation of formability for austenitic stainless steel 316 at elevated temperatures. J Mater Res Technol 3:17–24
5. Fanga Y, Lua B, Chena J, Xua DK, Oub H (2014) Analytical and experimental investigations on deformation mechanism and fracture behavior in single point incremental forming. J Mater Process Technol 214:1503–1515
6. Kim YH, Park JJ (2002) Effect of process parameters on formability in incremental forming of sheet metal. J Mater Process Technol 130–131:42–46
7. Chennakesava Reddy A (2017) Experimental and numerical studies on formability of stain-less steel 304 in incremental sheet metal forming of elliptical cups. Int J Sci Eng Res 8(1)
8. Vignesh G et al. (2012) Forming, fracture and corrosion behaviour of stainless steel 202 sheet formed by single point incremental forming process. Mater Res Express 6(12)
9. Muthusamy T et al. (2018) Formability and force analysis of steel foils in single point micro-incremental forming (SPMIF). Mater Today: Proc 5(5):12772–12781
10. Zeradam Y, Krishnaiah A (2020) Numerical simulation and experimental validation of thickness distribution in single point incremental forming for drawing quality steel. Int J Appl Eng Res 15(1):101–107

11. Cheng Z, Li Y, Xu C, Liu Y, Ghafoor S, Li F (2020) Incremental sheet forming towards biomedical implants: a review. J Mater Res Technol 9(4):7225–7251
12. Shi X, Gao L, Khalatbari H, Xu Y, Wang H, Jin L (2013) Electric hot incremental forming of low carbon steel sheet: accuracy improvement. Int J Adv Manuf Technol 68(1):241–247
13. McAnulty T, Jeswiet J, Doolan M (2017) Formability in single point incremental forming: a com-parative analysis of the state of the art. CIRP J Manuf Sci Technol 16:43–54

Numerical and Experimental Study on Hydroforming of Thin Metallic Sheets

C. Pradeep Raja, T. Ramesh, P. Paavai, and M. Amal Jerald Joseph

Abstract Hydroforming uses controlled fluid pressure to form the sheet metal to the shape of the die. With reduced friction, precise tolerance and single operation allowing a high drawing ratio make it much more preferable than other forming processes. It is also much suited for small irregular contoured parts to be manufactured for prototype and experimentation purposes. In this work, a miniaturized experimental set-up of the hydroforming process has been fabricated by designing, fabrication of a die, and the experimental set-up. Hydroforming was performed on Al6063-*T*6 sample sheets of 95 × 95 mm dimension and 0.5 mm thickness. The pressure was applied using a commercial-grade hand pump, in increments of 10 bars until failure occurred. The obtained sample profiles were analysed using a Zeiss Contura *X*2 CMM (coordinate measure machine) and profile projector. Additionally, a finite element (FE) model of the hydroforming set-up has been developed using 2D simulation in COMSOL Multiphysics. The 2D model has been preferred due to its greatly reduced computational expense compared to a full 3D model. The FE model has been used to obtain deformation across the sheet, depth of compression, and percentage formability for the various pressure of the fluid, until failure. Measurements from the numerical simulation and experiments were correlated, and it was found that the samples were only able to withstand a pressure of 170 bar and beyond that ductile–brittle mixed type, fracture occurred. A good range of formability was observed around 120 bar pressure.

Keywords Hydroforming · Aluminium · Fluid pressure · Die design · Finite element method

1 Introduction

The origin of the hydroforming process started with the interest of the automotive and aircraft industry in hydroforming process 100 years back. The main reason for

C. Pradeep Raja · T. Ramesh (✉) · P. Paavai · M. Amal Jerald Joseph
Department of Mechanical Engineering, National Institute of Technology, Tiruchirappalli, Tamil Nadu 620015, India
e-mail: tramesh@nitt.edu

U. S. Dixit et al. (eds.), *Advances in Forming, Machining and Automation*, Lecture Notes in Mechanical Engineering, https://doi.org/10.1007/978-981-19-3866-5_9

their interest in this process is because they were able to accomplish manufacturing of parts cost effectively as they identified this process as a new important tool to produce parts with high strength to ratio, better rigidity, higher reliability, easy assembly, and fewer post-process operations with new opportunities of flexibility in design and tooling when compared with other similar manufacturing processes [1]. The effectiveness of this process lies in limiting the need for post-process such as redrawing, trimming, and welding to finalize the finished part, and in this process, the blank is drawn in a single cycle to produce the required part. Generally, in hydroforming, a female die is replaced by the pressurized liquid cavity and rubber diaphragm, hence, it is less expensive in the tool set-up than conventional drawing/stamping operations [2]. Furthermore, the hydroforming process has several advantages over other manufacturing processes such as it is overall a less costly process and can save lots of time in producing parts. The overall costs include the development cost, operation cost, and finishing cost. Recently, one of the most important necessary steps in manufacturing is prototyping which can be easily carried out using the hydroforming process. Changes in material, material geometry (thickness), and parameters can also be easily accommodated through the hydroforming process without the need for new tooling. Therefore, development costs can be reduced significantly. Reduced operation/press time due to the single hydroforming cycle we can achieve a reduction of around 60–70% of costs compared to other manufacturing processes. Parts produced from the hydroforming process are of good quality and surface finish and eliminate surface blemishes and polishing. Imperfections such as draw marks, wrinkling, and tearing are mostly eliminated, and this process serves as an ideal manufacturing process for producing parts of irregular contours and complex shapes [3].

Hydroforming can be divided into three groups, tube hydroforming, sheet hydroforming, and shell hydroforming. All three types of hydroforming use fluid pressure to produce high-quality parts with distinct and complex shapes. In tube type, the tube is filled with the liquid and pressurized by two axial rams on either side until the tube is expanded into a required geometry/shape. Sheet type of hydroforming can be divided into two subcategories in which one uses a punch resembling the deep drawing process (hydro-mechanical deep drawing) and the other which uses only the fluid pressure to draw the metal blank into the cavity. In the case of shell type, the metal blanks are welded or joined, enclosed to form a hollow shell and fluid pressure is applied to it. Expansion of the blanks takes place in free space and no die is used in this type of hydroforming [4]. A large number of parametric studies have been made on hydroforming mainly focusing on fluid pressure, lubrication conditions, metal blank thickness, and the use of composite blanks. Early studies made on the effect of hydrostatic pressure on sheet metal by McClintock [5], Tracey and Rice [6] showed that by increasing hydrostatic pressure there was a drastic decrease in fracture ductility. Also, Lang et al. [7] demonstrated that due to the lubrication effect of the fluid, there is a reduced frictional force acting on the blank and the blank can be drawn to a greater depth and drawing ratio. Abedrabbo et al. [8] studied the effect of process parameters such as the pressure of the liquid used and blank holding force (BHF) to determine their role on tearing and wrinkling of 6111-$T4$ aluminium alloys in sheet hydroforming process and concluded with the possibility of achieving hemispherical cups with minimal wrinkling in the flange and die corner area before tearing occurs, and the reason for wrinkling was due to the error in pressure monitor or controller and loss of pressure. Groche et al. [9] combined hydroforming and warm deep drawing to predict the formability of Al–Mg–Si alloy, which are not good at cold forming operations. They applied counter pressure and provided flange heating to form the blanks. It was observed that when formed at elevated temperature, there was a drastic increase in the drawing ratio. Hence, we can conclude that counter fluid pressure plays a very important role in determining the formability of the metal sheets during the hydroforming process.

Computer simulation using the finite element method of hydroforming process helps to predict the possibility of wrinkling and rupture. This also eliminates the number of trial and error experiments of the physical process and helps to save time and cost. Hence, most of the hydroforming processes are simulated using the finite element method and results are predicted to avoid waste of time and material. With the advancement of technologies, the hydroforming process is also emerging by generating lots of advantages when compared to another manufacturing process since it has a lot of potential applications that are being implemented [10, 11]. A lot of research work has been carried out using hydroforming and numerical simulation of the process which has been correlated but a miniaturized prototype experimental set-up to study the effects of process parameters before actual manufacturing and production has been seldom carried out. Since the hydroforming process is flexible and gives lots of opportunities for new design and optimization opportunities, we can now design the process much earlier than the actual manufacturing process to predict the possible outcomes and effects of process parameters that can be optimized for best results. Process parameters were optimized to achieve better results even in the case of fibre metal laminates (FMLs) [12], double-layer sheets [13], tube hydroforming [14], and loading paths [15]. A similar type of miniaturized hydroforming process was carried out in this research work to study the influence of process parameters and their importance.

2 Tooling and Experiment

In this present work, a miniaturized hydroforming set-up was developed by designing and fabricating a die as shown in Fig. 1, and performed hydroforming experiments on Al6063 *T*6 thin sheets. The geometry of the thin specimen experimented on was $95 \times 95 \times 0.5$ mm ($L \times B \times T$). A commercial-grade hand pump was used to apply pressure in increments of 10 bar until failure of the specimen occurred. A numerical model of the hydroforming process using finite element method (FEM) simulation in COMSOL Multiphysics 2D simulation was done. Results from the FE model were to obtain deformation across the sheet, depth of compression, and elongated length for the various pressure of the fluid, until failure. After the failure of the specimen, they were analysed using Zeiss Contura *X*2 CMM (coordinate measure machine)

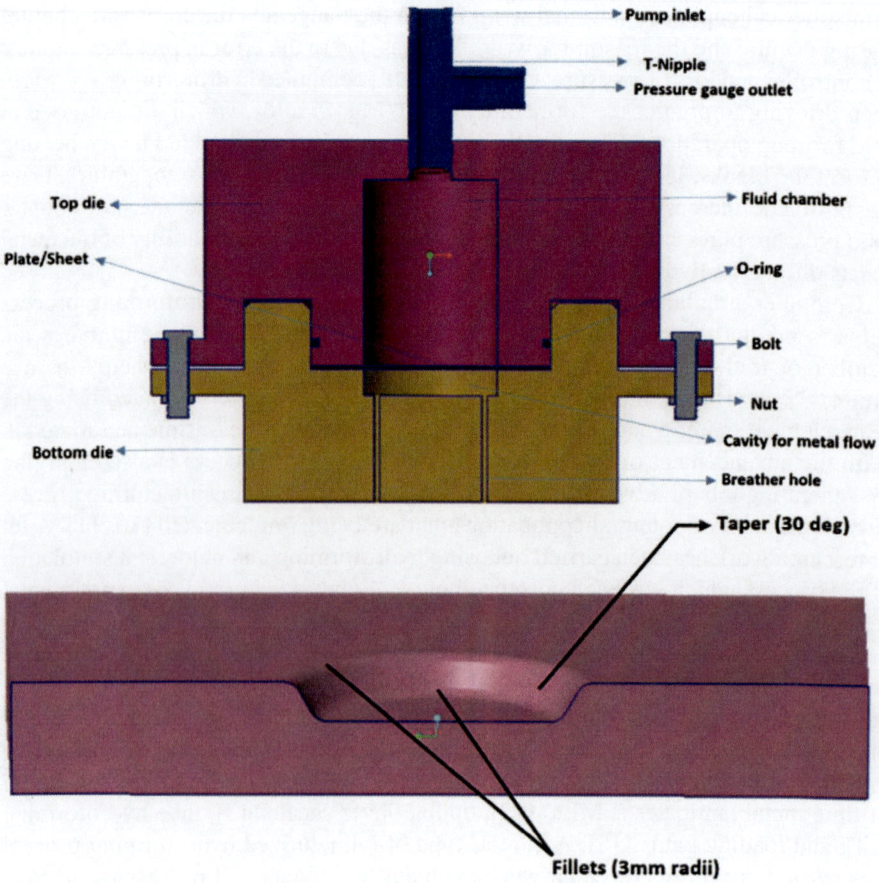

Fig. 1 Design of hydroforming die setup and cut section of the die cavity

and profile projector. Correlations between FEM simulations and experimentally obtained results were done.

2.1 Design of the Die

The dimensions for the experimental set-up were chosen to miniaturize the hydroforming process. This makes the experiment more feasible and suitable for research applications. Also, it reduces the load and pressure requirements and makes the measurement of formed dimensions easier. Commercially available sheet metal Al 6063-*T*6 of thickness 0.5 mm was chosen for the application as thin samples tend to form better. To avoid tearing due to thinning, the die depth was limited to 7 mm. More the taper angle in a contour more the formability. Hence, the die cavity was given a taper angle of 30° to enable better flowing of the material during the forming process as shown in Fig. 1. Also, right angles are impossible to achieve in a forming process. Another advantage of including the taper angle was that it aids easier ejection of the sheet metal from the die cavity after the forming process is complete. As shown in Fig. 1, the die was given a fillet radius of 3 mm at the upper edge and the landing edge to relieve stress concentration, improve manufacturability, aid material flow, and avoid tearing at corners. Also, filleted corners allow continuity in elements thereby promoting better correlation between experimental and numerical results. Allen bolts were chosen for the application as they are generally made of high-grade carbon and stainless steel with increased robustness. Also, Allen bolts are more serviceable.

Hence, two $M12$ Allen bolts were mounted diagonally for the application (Fig. 2). Further, the weight of the die also acts against the upward thrust and holds the sheet in place. Dowel pins were added at the remaining diagonal corners (Fig. 2). The inclusion of dowel pins was to ensure proper positioning of the die and the holder and to prevent any misalignment during the forming process. The O-ring was included

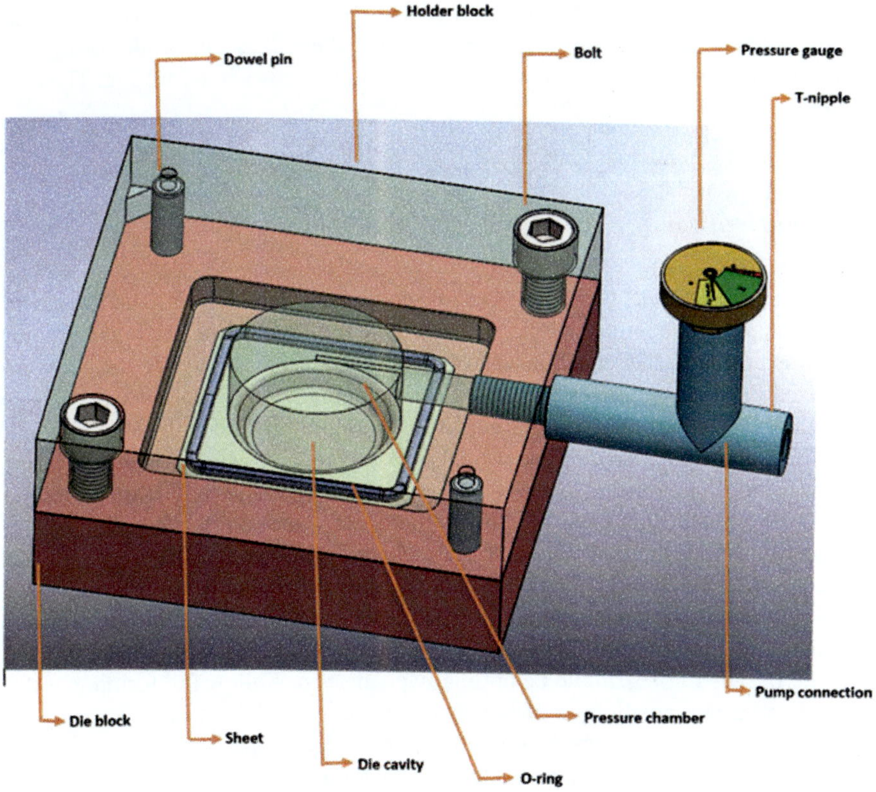

Fig. 2 3D diagram of die setup for hydroforming operation

Table 1 Load calculations used during the design of die for hydroforming experiments

S. No	Load calculations	Calculated values
1	Pressure chamber diameter	53.88 mm
2	Maximum pressure applied	300 bar
3	Area of pressure acting	0.002 m^2
4	Force acting on the holder	68.4 kN
5	Number of bolts used	2
6	Factor of safety	1.5
7	Clamping force rating of bolt	61.3 kN
8	The clamping force required per bolt	34.2 kN

to act as a seal, to prevent the fluid from escaping the fluid chamber. Also, the *O*-ring acts as the point of contact between the sheet and the holder through which the clamping force is applied (Fig. 2). The *O*-ring selected was 3 mm in diameter (wire diameter or WD). The ground width of the *O*-ring = WD + 0.1(for clearance), i.e. 3.1 mm depth of the *O*-ring groove = WD – 0.25(compression allowance) + 0.1(for clearance), i.e. 2.85 mm. Breather holes of diameter 0.5 mm were provided at the bottom of the die cavity to allow the air below the sheet to escape while forming as shown in Fig. 1. It also prevents the formation of air pockets while forming. The load calculations are shown in Table 1.

2.2 *Fabrication of Die*

Billets of material EN18 were procured and faced to required sizes. The die cavity and fluid chamber are initially turned using a lathe. Following this, the outer contour is machined using a CNC machine. The CAD file is given as input into the software EDGECAM which optimizes the path for machining. After this, the path data is fed to the CNC controller, and the machine is programmed. The finished die and holder from the CNC machine are drilled. Two diagonal holes are drilled to insert the dowel pins. The other two diagonal holes are drilled and tapped to fasten the bolts. The hole is drilled from the holder's surface to the fluid chamber to form a passage connecting the pump and the fluid cavity as shown in Figs. 2 and 3. The finished part of the die, i.e. the holder and the die block is shown in Fig. 3. Three breather holes are made on the holder block for better formability and to stop further deformation of the Al 6063-*T*6 samples once the required pressure of the fluid is supplied. Breather holes should be optimally positioned to allow full forming of the metal. Appropriate positioning of breather holes may lead to the formation of air pockets. Once air pockets are formed, the external hydraulic pressure is balanced by the internal air pockets. At this stage, the material will not undergo any further deformation and will not fail as it takes no stress at this stage.

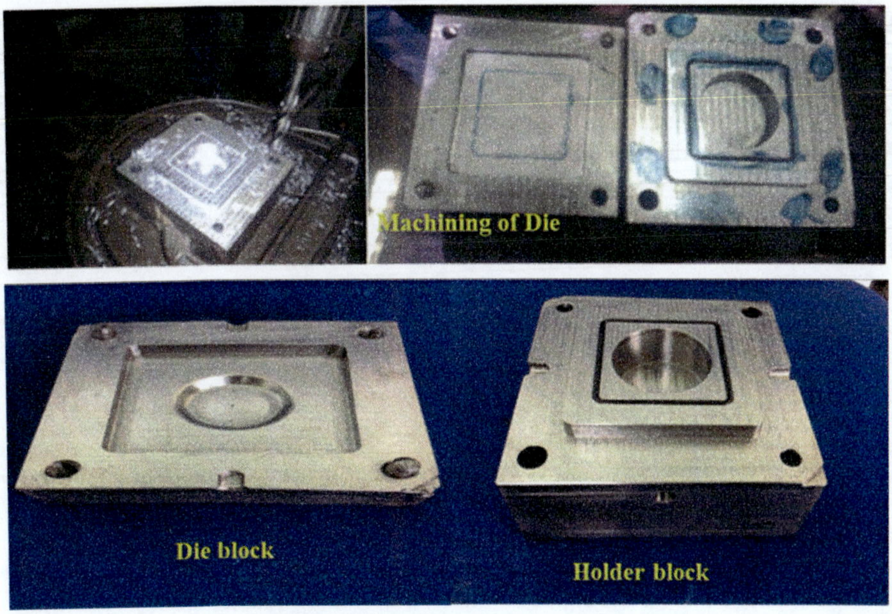

Fig. 3 Machining of die with holder and die block

3 Hydroforming Experimental Activity

Hydroforming experiments were conducted by placing a sample blank in the die set-up and application of hydraulic pressure utilizing a commercial-grade hand pump. The connecting joints die fasteners and hydraulic seals are carefully checked to avoid any leakage of fluid. The experiments were conducted from 0 bar pressure until failure of the sample, in increments of 10 bar. One end of the hosepipe and the pressure gauge is connected to the pump and the other end of the hosepipe is connected to the holder and sufficient fluid is filled into the pump. To start the experimental procedure, the fluid chamber is filled with fluid. The die surface is greased and the sample is placed. The die is inverted over the holder and both the blocks are bolted tightly together, at this stage, the O-ring compresses well and acts as a seal to prevent leakage. The needle valve is closed and the fluid is pumped to the required pressure. The pump is held at that pressure for a minute to let the sample form, the needle valve is removed and pressure is released. Then, the bolts are removed and the formed sample is taken and cleaned with kerosene to remove the grease. And this procedure is repeated for all samples until failure at a particular pressure.

After each cycle of the experiment, the profile of the deformed sample was measured using a Zeiss Contura *X*2 coordinate measuring machine. It has a sensitive probe that traces the profile of the sample and yields output in the form of points. The points are plotted in CAD software to obtain the profile curve. The CMM was initially taught to measure the readings and later programmed enabling it to repeat the process for multiple samples. The output data from CMM was obtained in the form of coordinates. These values were plotted in CAD software to obtain the profile. The process was repeated for all the samples. The curves were then analysed and measured to obtain the upper and lower fillet radius, depth, total drawn length, and the taper angle of the hydroformed sample. Similarly, the line graphs showing the deformation of the sheets were used to plot the profile of the samples under different load conditions. These curves represent the profiles obtained through numerical simulation. Again, these curves were measured for the die dimensions to correlate with the experimental data.

4 Numerical Simulation

The finite element method (FEM) was used to perform the numerical simulation of the hydroforming process in COMSOL Multiphysics. A two-dimensional half model was used to perform the simulation to reduce the computational expense. Appropriate symmetry conditions were applied at the die and sheet boundaries along the central line. To allow sliding of the sheet, frictional contact boundary conditions were assigned between the bottom of the sheet and top of the die then another between the top of the sheet and the bottom of the blank holder. Additionally, contact pairs were defined between the bottom of the sheet and the upper surface of the die to prevent penetration of the sheet through the die as deformation occurs. The hydraulic fluid pressure was given as a boundary load on the top of the sheet. The pressure was applied to the partitioned region. The pressure was defined as a linear function that varied as a function of a solution parameter. To simplify the effects of time-varying equations, a stationary solver was used to compute the FEM model. An auxiliary sweep was incorporated that performed iterations from various values of pressures satisfying the function. In this simulation, a solution parameter-para was varied in increments of 1 MPa (10 bars) starting from 0 until the required maximum pressure. To minimize the variables, the physics-controlled mesh was used with the default (normal) mesh size in COMSOL. To achieve mesh convergence, meshes varying from 'coarser' to 'finest' were carried out and it was found that there was not much variation in the result when the mesh size was increased from 'normal' to 'fine' meshes. Hence, 'normal' was taken to be the optimum mesh for the analysis carried out.

5 Result and Discussion

To validate the numerical simulation, it is necessary to show a correlation between numerical and experimental results. The variation in parameters such as taper angle, depth, upper fillet radius, lower fillet radius, and elongated length with varying pressure was used to correlate numerical simulation with experimental observation.

Experiments were conducted at various pressure values ranging from 10 to 180 bar in intervals of 10 bar and the deformation was measured in each case. The deformed shape at each pressure value as obtained from experimentation and simulation was compared to validate the simulation.

5.1 *Physical Observations at Different Hydroforming Pressures*

Experimental activity on the die started with 10 bar pressure and the result was convincing since the aluminium sheet had undergone deformation and formed according to the dimension/cavity of the die. The hydroforming process was simulated using COMSOL software and obtained result shows the line graph of displacement curve and the deformed profile of the sample was traced using a CMM probe. Later, when pressure was increased to 30 bar, the sheet had undergone deformation and a significant change that can be noted was that the sheet was touching the die base and additionally wrinkles start to appear on the edge of the sheet. COMSOL simulation at 30 bar pressure with line graph for load-displacement curve and CMM profile is shown in Fig. 4. At 160 bar, a small tear mark was visible (Fig. 9) and the aluminium sheet is seated at the base of the die and the hydraulic pressure is balanced by the pressure in the air pockets and further deformation is halted. Then, at 180 bar pressure, the sheet fails by tearing, and the results from COMSOL software and CMM probe are shown in Fig. 5. Hence, the sheet can only handle 170 bar pressure and after which it fails considering the above dimension and condition of the experiment. Further experiments were conducted to study the effect of strain rate, etching (laser and chemical), and also a change of material to verify the working of the build experimental setup on an alternate material with the different mechanical properties.

5.2 *Elongated Length Versus Pressure*

During the hydroforming process, the metal blanks are subjected to biaxial stresses. The nature of stress varies depending on the area of the blank. The part of the blank which forms the wall of the cup experiences radial and circumferential stress. Radial stress along the wall of the cup aids in stretching the blank material around the forming die cavity and circumferential stress aids in pulling the material into the die cavity. Elongation of Al6063-*T*6 sample in the experiment and during numerical simulation is shown in Fig. 6. In this experiment, elongation is a direct measure of deformation until failure and increase in length. We can observe from Fig. 6 that elongation of the samples in both cases of experiment and simulation directly resembles a typical stress–strain curve. Plastic deformation starts around 50 bar pressure in the

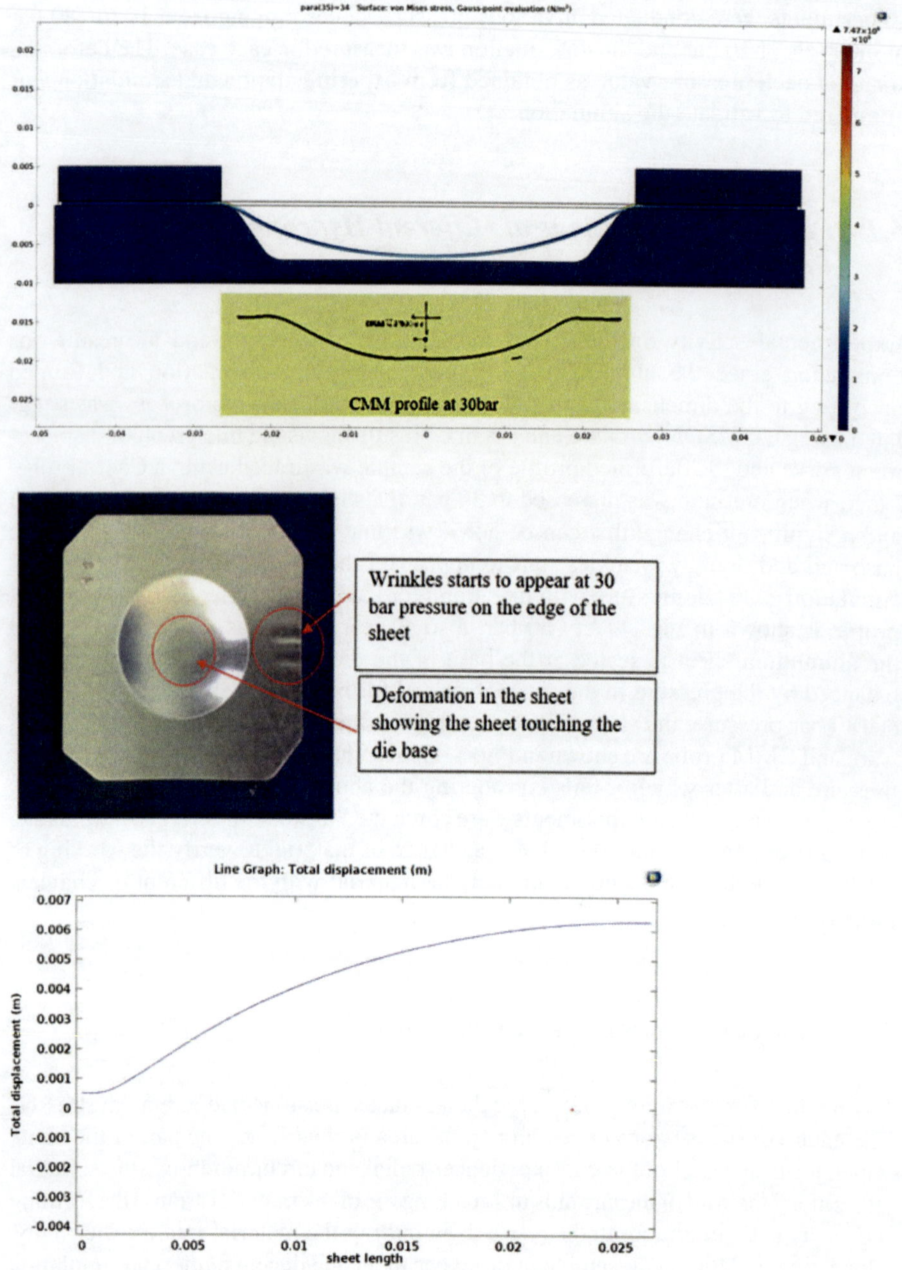

Fig. 4 Deformation result (contour) in COMSOL, CMM profile, line graph, and deformed Al 6063-*T*6 sample at 30 bar pressure

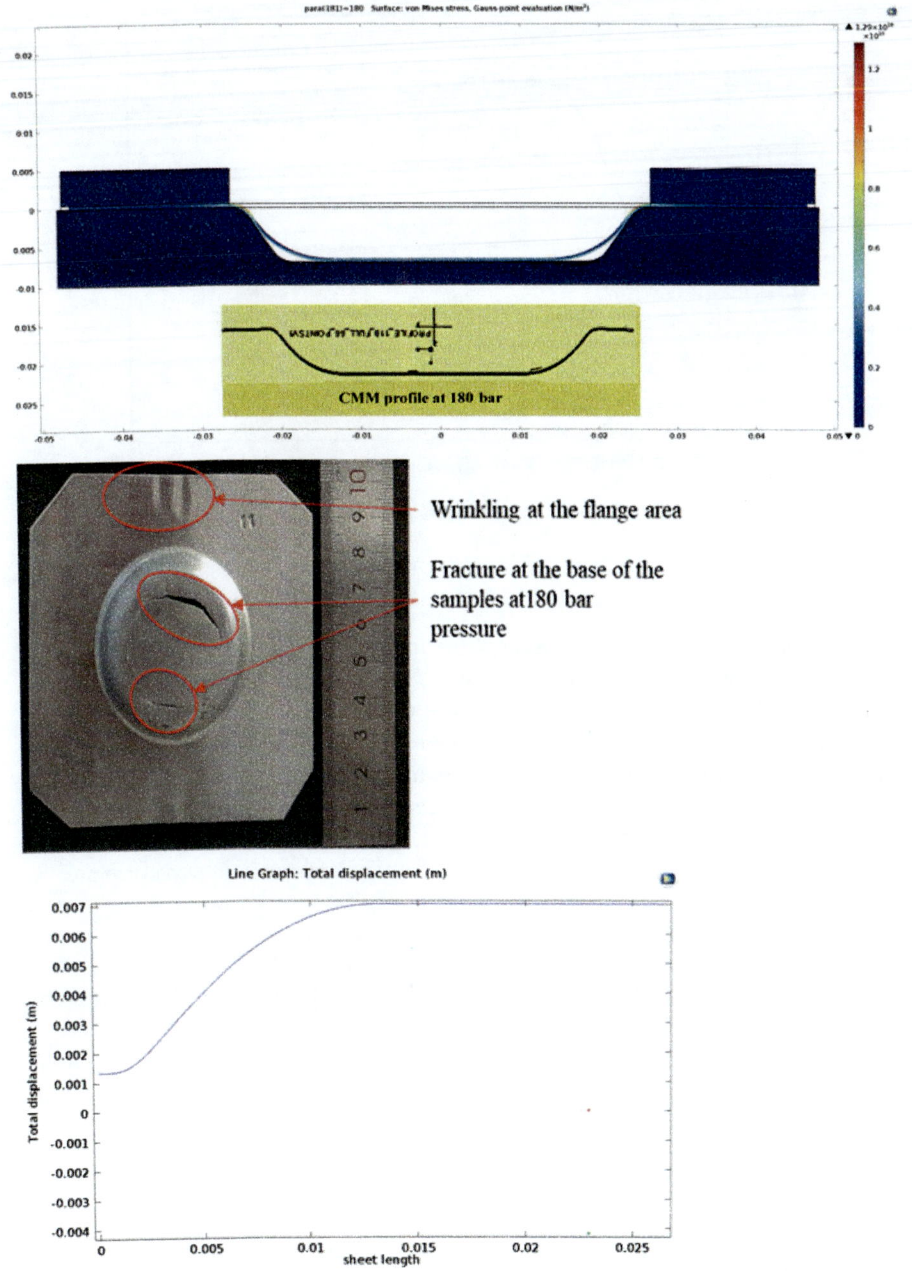

Fig. 5 Deformation result (contour) in COMSOL, CMM profile, line graph, and deformed Al 6063-*T*6 sample at 180 bar pressure

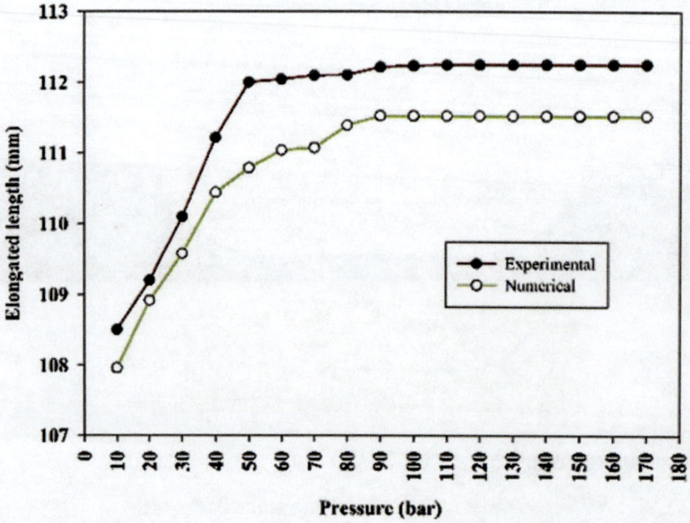

Fig. 6 Comparison of elongated length of Al 6063-*T*6 samples obtained from hydroforming

experimental case and around 60 bar pressure in the case of simulation. The Al6063-*T*6 sample is elastic to the plastic state from 10 to 50 bar pressure and shows definite permanent deformation after 50 bar pressure (yield strength). The samples are in a uniform elongation zone from 50 to 170 bar due to absorption of energy during plastic deformation and then fracture occurs after total elongation at 170 bar. The maximum elongated length of the cups achieved by hydroforming experiment and simulation was 112.252 mm and 111.526 mm, respectively. Here, the experimental and numerical simulation results were almost similar and were able to correlate.

5.3 *Depth Versus Pressure*

The diameter of the pressure chamber is 53.88 mm and the depth of the die cavity is 7 mm. Therefore, a maximum of 7 mm depth can be achieved during the experiment. Figure 7 shows that the samples started to deform from 10 bar pressure itself but were in between the elastic limit till 40–50 bar pressure and thereafter at 80 bar were able to achieve the maximum depth of 7 mm. From the numerical simulation, we can observe that the Al6063-*T*6 sample started deforming from the beginning of the experiment itself and the tip of the sample started touching the base of the cavity around 30 bar pressure. But a flat surface or a proper deformed cup shape was observed only after 70–80 bar of pressure because of the elastic nature and spring back effect of metallic blank once the counter pressure was removed. After 80 bar pressure, a flat stage occurs and the flow stress starts to increase, and a proper cup is formed due to the flow of material into the die cavity. Thickness distribution along the

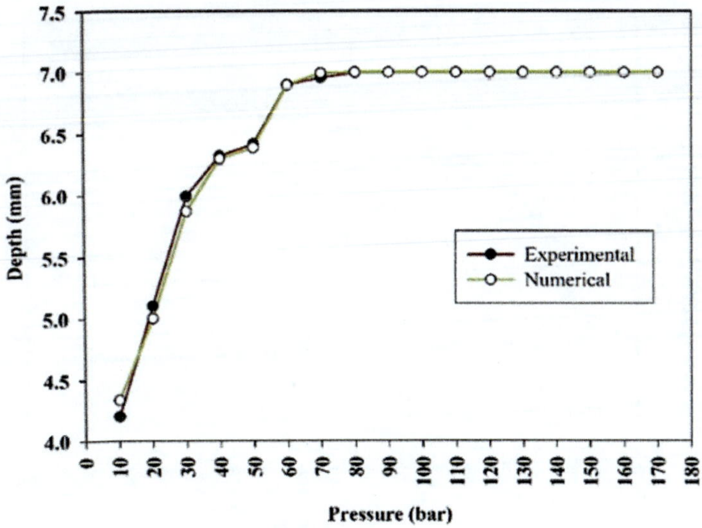

Fig. 7 Comparison of depth of Al 6063-*T*6 samples obtained from hydroforming

wall and base of the cup is uniform in nature because of the uniform deformation of the Al6063-*T*6 samples through fluid pressure. With a consistent observation, we can note that here plastic deformation happened in a very uniform manner throughout the length of the sample and then finally failure (tearing/fracture) due to extreme tensile stress at high pressure beyond 170 bar.

5.4 Taper Angle Versus Pressure

The die was designed with a taper angle of 30° to provide better flow to materials during the forming process and more the taper angle in a contour more the formability can be achieved. Also, taper angle aids in an easier ejection of metals once the forming process is completed. The part of the cup which lies in the taper angle is the wall of the cup which is subjected to continuous circumferential stress. One can observe a clear stretch of the blank and flow of material due to flow stress at this region. Initially, during the lower levels of pressure from 10 to 50 bar, there was only a bulging effect on the samples due to fluid pressure acting on them and it was not much of a wall-like structure around the cup, and the taper angle of the deformed samples were around 4°–20° in both experiment and numerical simulation as shown in Fig. 8. After 50 bar pressure, the flow of material was smooth and the Al 6063-*T*6 metallic samples started gaining strength due to strain hardening and due to even distribution of strains in this stage of fluid pressure, we were able to delay strain localization and premature failure/fracture. Deformed samples achieved a maximum taper angle of 26.04° and 26.23° from experiment and numerical simulation. Starting from 10 bar

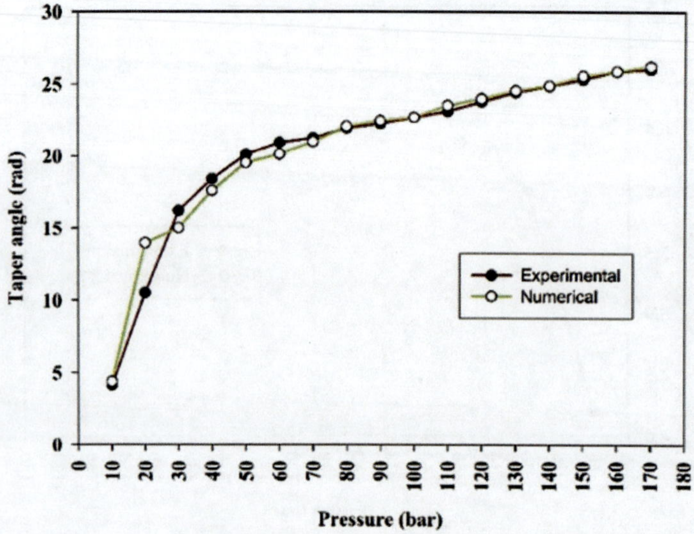

Fig. 8 Comparison of the taper angle of Al 6063-*T*6 samples obtained from hydroforming

pressure till 170 bar there was a uniform distribution of deformation along the wall, and the thickness of Al 6063-*T*6 cups were uniform from one end to the other end until fracture.

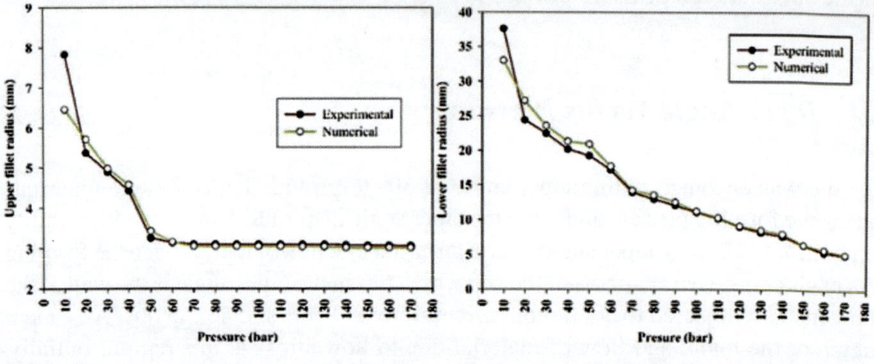

Fig. 9 Comparison of the upper fillet radius and lower fillet radius of Al 6063-*T*6 samples obtained from hydroforming

5.5 *Upper Fillet Radius and Lower Fillet Radius Versus Pressure*

The purpose of designing a fillet in the die was to relieve stress concentration and facilitate a good flow of material and avoid tearing of material especially at corners during hydroforming. Also, filleted corners allow continuity in elements thereby promoting better correlation of experimental and numerical results. A significant factor that can affect the flow of material during hydroforming is the fillets on the upper and lower side of the die cavity. Hydroforming is well known for its sensitivity to the curvature radius even during hydro-mechanical deep drawing which uses punch and fluid pressure together to deform the metallic samples. Even a small strain was found to be an initiator of fracture on the samples. During the hydroforming process, the flange area is under radial tensile stress and tangential compressive stress. And the uniform uninterrupted flow of material along the wall is monitored by the upper fillet radius which also prevents early fracture around the radius area since it is an area at maximum risk. The area above the upper fillet radius is the flange area of the sample which is held between the die block and the holder block. Usually, this is the area that is provided with optimum blank holding force to prevent wrinkling and buckling during the hydroforming process. If an optimum blank holding force is provided, then a uniform wall of cup starts to form starting from the upper fillet radius till lower fillet radius. A steady increase in pressure will lead to the formation of even thickness around the wall and fillet region of the cup and prevent too much accumulation of material around this region. Fluctuations in applied pressure may lead to large scatter in the flow of material and uneven thickness of the samples.

From Fig. 9, we can observe that in the initial stage of the experiment, during the bulging phase, the upper fillet radius was maximum around 7.82 mm and 6.45 mm and kept reducing once we proceeded with higher pressure levels and settled around 3 mm in the end till fracture in both the experiment and numerical simulation case, respectively. And in the case of lower fillet radius, maximum value was around 37.65 mm and 33.03 mm during the bulging phase and settled around 5 mm towards the end in the experiment and numerical simulation, respectively. During the bulging stage in the experiment and numerical simulation, the samples were smoothly deforming under a steady and uninterrupted flow of materials due to the effects of various tensile stress conditions. This stress condition determines the forming operation of the aluminium samples. And once the pressure value crosses the 50 bar pressure, the samples start to touch the base and materials start to fill the base portion uniformly and beyond this stage due to excessive radial stress and compressive stress, there is wrinkling around the flange region and necking starts at the base leading to fracture. And this is also the reason upper fillet and lower fillet radius are large during the initial phase and get reduced and settle to a uniform value starting from the uniform elongation zone (50–170 bar). Reasonable fluid pressure can help in balancing the shape accuracy and drawability in this experiment.

5.6 *Effect of Strain Rate*

To check the effect of strain rate acting on the Al 6063-*T*6 samples, the pressure was applied quicker, and hence the samples deformed faster and failed at 120 bar itself (Fig. 10). At high strain rates, the Al 6063-*T*6 sample fails at a pressure lesser than the maximum it can handle (170 bar). When higher strain rates were applied, elongation was reduced by about 29% and samples were not able to withstand pressure beyond 120 bar and fracture occurred near the corner of the cup. Hence, in our experimental case, we can arrive at a strong conclusion that deformation in the Al 6063-*T*6 samples depends on the strain rates and uniform deformation takes place under steady increments of pressure and also, we can conclude that a higher strain rate reduces formability.

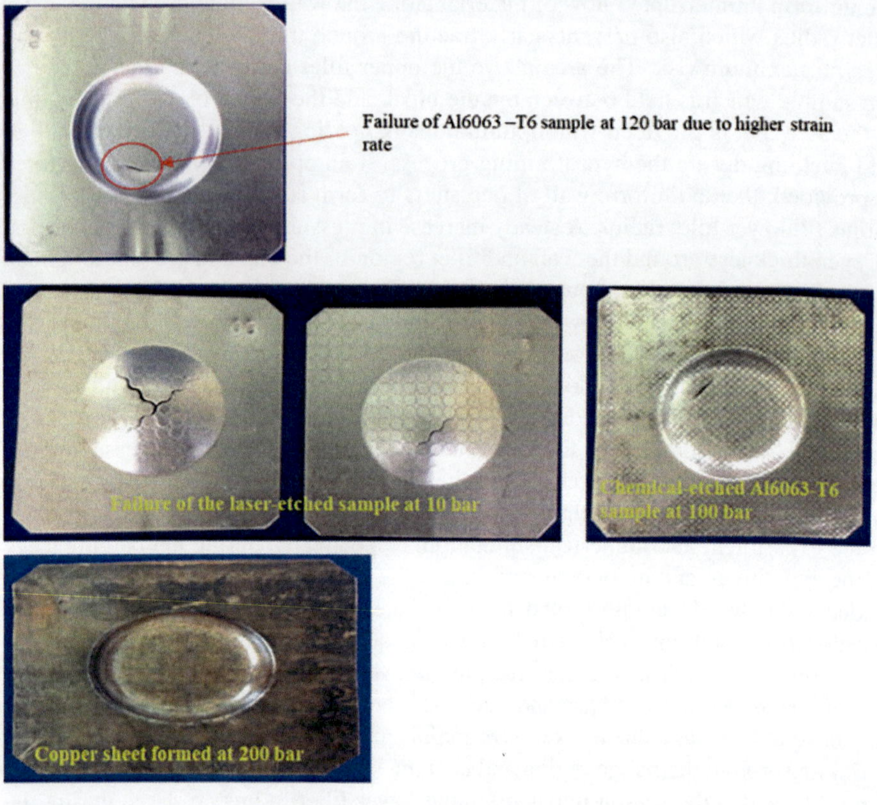

Fig. 10 Effect of strain rate, etching on Al6063-*T*6 samples, and formability of copper sample in the hydroforming die

5.7 Effect of Etching

To study the effect of etching, the samples were both laser and chemically etched. Laser etching on the surface of the sample hardened, where the pattern is printed and increased its strength and reduced its ductility. Hence, the sample failed at a very low pressure of 10 bar Fig. 10. Although chemical etching does not harden the material, it corrodes away a small layer at the locations of the pattern. This leads to geometric discontinuity at the grain levels and aids crack propagation. A chemically etched sample fails at 100 bar when the maximum handling pressure for the sample is 170 bar Fig. 10.

5.8 Experiment with an Alternate Material Sample

To study the variation and working nature of the designed die for the hydroforming experiment, a copper sheet was used with the same dimension of Al6063-$T6$. Copper being more malleable than aluminium was capable of handling a pressure of 200 bar and better traced the contour of the die cavity as shown in Fig. 10. Therefore, the designed die was capable of experimenting with various other metals and ideal for the creation of small prototypes with unique and special shapes with precise tolerances for various manufacturing industries such as defence, health care, and aerospace.

6 Conclusion

The hydroforming process shows a lot of potential in the field of technology as well as industry. The resultant geometrical freedoms for components are enormous, and hence, the process provides far reachable extensions. A miniaturized experimental set-up was made to perform hydroforming on the 0.5 mm thick, 95 $\times$ 95 mm, Al6063-$T6$ samples until failure. The following are the observations that were made from the experiment.

1. A crack was found near the cup corner at 180 bar of pressure and the nature of the fracture was a ductile–brittle mixed fracture.
2. The range of formability was low during the initial stages of applied fluid pressure till 50 bar and wrinkling was also observed during this stage. A medium and good range of formability were observed from 60 to 160 bar pressure (working zone) with stable and uniform deformation and thickness of the metallic blanks.
3. In this experiment, an optimum level of pressure can be suggested as 120 bar pressure, where a good range of formability was achieved considering the process parameters and effects of pressure on depth, elongation, taper angle, upper and lower fillet radius.
4. Effect of strain rates proved that deformation was in dependence of strain rate.

5. Numerical simulation of the whole experiment was carried out on Comsol multiphysics, and the results of the FEM simulation and the experiments have been compared and a very close correlation has been observed, showing the success of the numerical simulation.

References

1. Zhang SH, Danckert J (1998) Development of hydro-mechanical deep drawing. J Mater Process Technol 83(1–3):14–25. https://doi.org/10.1016/S0924-0136(98)00039-9
2. Koç M, Cora ON (2008) Introduction and state of the art of hydroforming. Hydroforming Adv Manuf 1–29. https://doi.org/10.1533/9781845694418.1
3. Singh H (2003) Fundamentals of hydroforming 219
4. Bell C, Corney J, Zuelli N, Savings D (2019) A state of the art review of hydroforming technology. Int J Mater Form 13(5):789–828. https://doi.org/10.1007/S12289-019-01507-1
5. McClintock FA (1968) A Criterion for ductile fracture by the growth of holes. J Appl Mech 35(2):363–371. https://doi.org/10.1115/1.3601204
6. Rice JR, Tracey DM (1969) On the ductile enlargement of voids in triaxial stress fields. J Mech Phys Solids 17(3):201–217. https://doi.org/10.1016/0022-5096(69)90033-7
7. Lang L, Danckert J, Nielsen KB (2004) Investigation into the effect of pre-bulging during hydromechanical deep drawing with uniform pressure onto the blank. Int J Mach Tools Manuf 44(6):649–657. https://doi.org/10.1016/J.IJMACHTOOLS.2003.11.004
8. Abedrabbo N, Zampaloni MA, Pourboghrat F (2005) Wrinkling control in aluminum sheet hydroforming. Int J Mech Sci 47(3):333–358. https://doi.org/10.1016/J.IJMECSCI.2005.02.003
9. Groche P, Huber R, Dörr J, Schmoeckel D (2002) Hydromechanical deep-drawing of aluminium-alloys at elevated temperatures. CIRP Ann 51(1):215–218. https://doi.org/10.1016/S0007-8506(07)61502-9
10. Bell C, Corney J, Zuelli N, Savings D (2020) A state of the art review of hydroforming technology. Int J Mater Form 13(5):789–828. https://doi.org/10.1007/s12289-019-01507-1/Published
11. Hwang Y-M, Manabe K-I (2021) Latest hydroforming technology of metallic tubes and sheets. Materials 11(9):1360. https://doi.org/10.3390/MET11091360
12. Blala H, Lang L, Khan S, Li L, Alexandrovs (2021) An Analysis of process parameters in the hydroforming of a hemispherical dome made of fiber metal laminate. Appl Compos Mater 28(3):685–704. https://doi.org/10.1007/S10443-021-09884-0
13. Zhou B-J, Xu Y-C, Zhang Z-C (2020) Research on the selection principle of upper sheet in double-layer sheets hydroforming. Int J Adv Manuf Technol 109(5):1663–1669. https://doi.org/10.1007/S00170-020-05615-0
14. Feng Y, Liu Z, Luo Z, Wu Q (2021) Application of RSM in optimization of bi-layered *X*-type tube hydroforming. Int J Adv Manuf Technol 1–19. https://doi.org/10.1007/S00170-021-08140-W
15. Zhang C et al. (2021) Process analysis of biconvex tube hydroforming based on loading path optimization by response surface method. Int J Adv Manuf Technol 112(9) 2609–2622. https://doi.org/10.1007/S00170-020-06411-6

Porosity of Al–Cu–Ni Alloy with Addition of FeNb Through Sand and Stir Casting Routes

Kumara Swamy Pulisheru, Anil Kumar Birru, and Uday Shanker Dixit

Abstract This work investigates the porosity of aluminum metal matrix composites (AMMCs) prepared through sand and stir casting methods with varied content of reinforcing ferroniobium (FeNb) particles in Al–5Cu–2Ni alloy. Theoretical and experimental densities were determined by rule of mixture and Archimedes' principle, respectively. Microstructure analysis of the composites was performed with an optical microscope; it revealed the presence of porosity at grain boundaries and uniform distribution of reinforcement particles in composites. It was observed that the porosity of the composite increased with increase in weight fraction of FeNb in both casting methods. A porosity value of 3.3% was observed in stir casting and 13% in sand casting. Thus, the limited experiments clearly pointed out the superiority of stir casting to sand casting as far as control of porosity is concerned.

Keywords Stir casting · Sand casting · Microstructure · Porosity · AMMCs

1 Introduction

Aluminum alloys have a high strength-to-weight ratio, superior mechanical properties and high thermal conductivity. They find applications in the field of aerospace, automobile and structural industries [1]. Recently, aluminum-based composites are also finding applications [2]. The addition of hard ceramic particles into aluminum alloy enhances the mechanical properties, especially wear and corrosion resistance [3–5]. Widely used casting techniques for fabrication of aluminum-based composites are stir casting, squeeze casting, compo-casting and ultrasound-assisted casting [6–9]. Among these, the stir casting is economical and easy to adapt [10]. It is the

K. S. Pulisheru · A. K. Birru
Department of Mechanical Engineering, National Institute of Technology Manipur, Imphal 795 004, India
e-mail: anilbirru@gmail.com

U. S. Dixit (✉)
Department of Mechanical Engineering, Indian Institute of Technology Guwahati, Assam 781 039, India
e-mail: uday@iitg.ac.in

U. S. Dixit et al. (eds.), *Advances in Forming, Machining and Automation*, Lecture Notes in Mechanical Engineering, https://doi.org/10.1007/978-981-19-3866-5_10

most eminent process for industrial production because of uniform distribution of reinforcement particles, fine microstructure due to fast cooling of liquid metal, less porosity, and good bonding between matrix alloy and reinforcement particles [11].

Porosity is a prominent casting defect in aluminum-based composites. Many researchers studied the effect of porosity on mechanical properties, microstructure, corrosion behavior and wear resistance of aluminum-based composites with reinforcement of Al_2O_3, ZrO_2, Ti, SiC and TiC in various casting processes [12–17]. Ahmad et al. [18] studied the incorporation of SiC into A356 alloy by stir casting. It was concluded that the stirring speed had a major influence on porosity formation due to air bubble entrapping due to vortex formation. Bahlooli et al. [19] pointed out that higher solidification time reduced the gas porosity formation during the solidification process. Amirkhanlou and Niroumand [20] highlighted a number of parameters causing porosity in casting, e.g., gas entrapment during stirring, air bubbles present in the slurry, water vapor present on the surface of the reinforcing particles and solidification shrinkage. Kok [21] studied the effect of Al_2O_3 particle reinforcement in AA 2024; it was observed that increasing the weight fraction of Al_2O_3 particle reinforcement increased the porosity of AA 2024 composite. Mishra et al. [22] studied the effect of fly ash and garnet reinforced in Al 7075 composite; it was observed that density and porosity increased with increasing weight fraction of the reinforcement.

Sahu and Sahu [23, 24] explained the importance of the matrix and reinforcement in the fabrication of aluminum metal matrix composites (AMMCs) as well as hybrid metal matrix composites (HAMMCs) to obtain desired mechanical properties such as wear resistance, creep resistance, fatigue strength and high specific stiffness for advanced applications. The study was grouped into different categories based on the synthetic ceramics (B_4C, SiC, Al_2O_3, SiO_2, etc.) and industrial waste (fly ash, red mud, etc.), in which, the main aim was to reduce the cost, enhance the effective utilization of industrial waste and enhance the properties of the material.

James et al. [25] investigated the effect of cooling rate, hydrogen content and grain refiner on formation of micro porosity in Al A356 alloys. Savas and Kayikci [26] investigated the factors affecting the formation of micro porosity in the casting process. The dissolved hydrogen leads to formation of micro porosity. The mold filling and solidification time may also be the reason for the formation of porosity in the casting process. The mostly used post-processing techniques are heat treatment, forming, rolling, laser surface melting and friction stir processing. For example, Sahu et al. [27] carried out upsetting of metal matrix composite to improve the properties.

The main aim of this study is to compare the effects of two casting methods from the viewpoint of porosity in the manufacture of aluminum-based composite. The casting methods chosen were sand casting and stir casting. The material was composite Al–5Cu–2Ni with the reinforcement of FeNb in different weight fractions.

2 Experimental Procedure

The experimental procedure includes manufacturing of composite by two casting routes. After that, porosity is estimated based on density measurement. Micrographs are also studied. Overall methodology is illustrated through Fig. 1. For manufacture of aluminum metal matrix composites (AMMCs), aluminum (Al) with the purity of 99.97%, copper (Cu) with the purity of 99% and nickel (Ni) were taken as matrix material.

The ferroniobium (FeNb) is taken as a reinforcement element in powder form with a mesh size of 250 μm with varying weight fractions as depicted in Table 1.

2.1 *Manufacturing of AMMCs*

In the sand casting process, aluminum ingot with 99.97% purity was melted up to 800 °C; 5 wt% copper (Cu) and 2 wt% nickel (Ni) were added. The temperature was raised to 1000 °C, and molten metal was allowed to reach a semi-solid state at 650 °C ± 5 °C. The magnesium in ingot form approximately 0.1 wt% was incorporated for wettability of matrix alloy [6, 10, 28, 29]. The reinforcement particles of ferroniobium (FeNb) were wrapped in the Al foil as per standard industrial practice. The wrapped

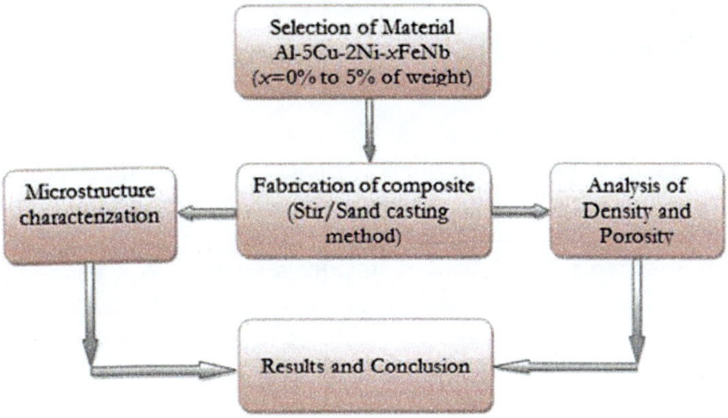

Fig. 1 Schematic view of the experimental study

Table 1 Weight percentage of FeNb in materials

Material	Wt% of FeNb
Al–5Cu–2Ni	0
Al–5Cu–2Ni–1FeNb	1
Al–5Cu–2Ni–3FeNb	3
Al–5Cu–2Ni–5FeNb	5

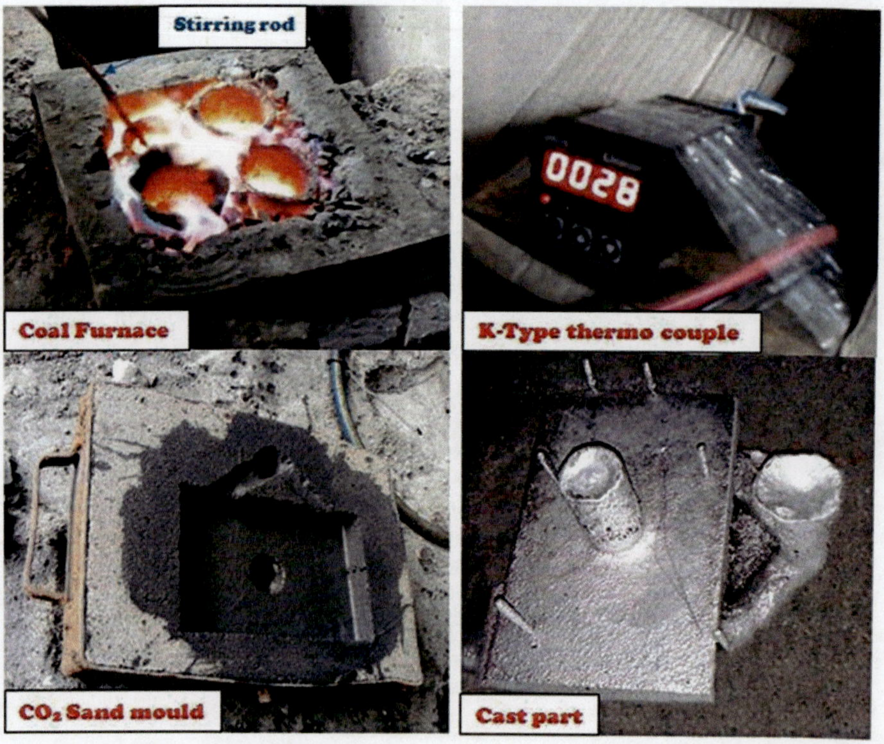

Fig. 2 Sand casting photographs

Al foil was inserted in the melt pool and stirred. The conventional stirring was done for uniform mixing of the reinforcement particles. The temperature was measured by a K-type thermocouple with an accuracy of ±5 °C in the range of 600–1200 °C. Figure 2 shows the photos of sand casting.

The mold was prepared with 95.20% of silica sand and 4.76% of sodium silicate (Na_2SiO_3); for bonding the silica sand, CO_2 gas was injected. The prepared molten metal was placed into the mold cavity and allowed to solidify to get designed casting part as shown in Fig. 2. Sand casting was carried out at Hyderabad Nonferrous Foundries, Balanagar, Hydearabad, in Telangana.

Stir casting method was also used for fabrication of Al–5Cu–2Ni/FeNb composite with varying weight fraction of the ferroniobium (FeNb) reinforcement element with the mesh size of 250 μm. The bottom pouring stir casting setup [10] was used as shown in Fig. 3. Stir casting method offers a better matrix particle bonding due to automated mechanical stirring, which was used for uniform mixing of particles in the composite. Proper selection of process is very crucial [11, 28]. In this work, the parameters of the stir casting were as follows: single impeller stirrer with 30° blade angle, size of the impeller 0.25 *D* (where *D* is the crucible size) and stirring speed 700 revolution per minute rpm.

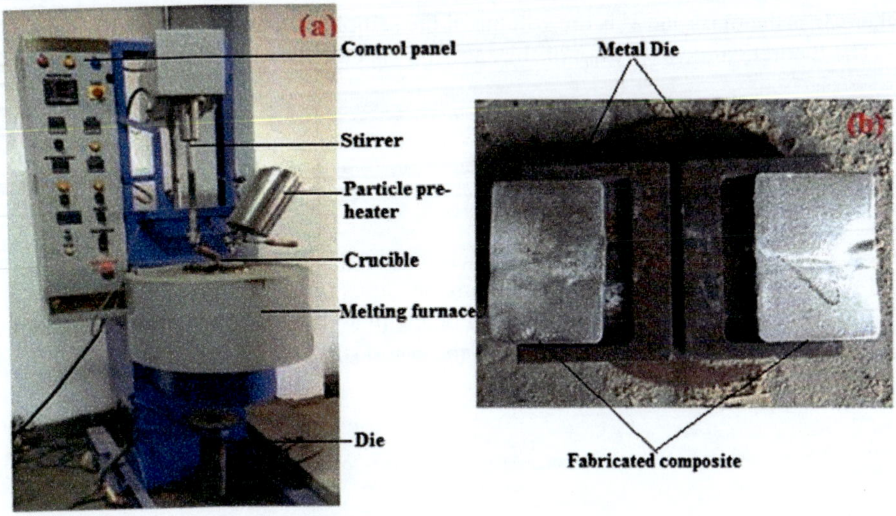

Fig. 3 Stir casting: **a** bottom pouring type experimental setup, **b** composite with metal die

The magnesium in ingot form, in approximately 0.1 wt%, was incorporated for wettability of matrix alloy [6, 10, 28, 29], and the degassing tablets were used for releasing of impurity gases from the liquid metal. The reinforcement particles were pre-heated at a certain temperature before mixing in the crucible; it released all the moisture and trapped air between the particles. The Al–5Cu–2Ni matrix alloy was cut in the form of ingot and placed in crucible. The temperature was raised up to 800 °C for melting of the matrix alloy. The molten metal was allowed to cool in the furnace to a semi-solid state at a temperature of 620–650 °C. The reinforcement particles were added in the semi-solid liquid and stirred for uniform distribution of the reinforcement particles. The molten metal was allowed to solidify after putting in the reinforcement. Stir casting was carried out at Vision Castings in Hyderabad, Telangana.

2.2 *Density and Porosity Measurement of AMMCs*

The experimental density (d_{ex}) of the composite can be calculated by water displacement technique (Archimedes' principle) by weighing the test specimen with a high precision electronic weighing balance having 0.1 mg of accuracy. The density was calculated by dividing the mass with the volume of water displaced [15], i.e.,

$$d_{ex} = \frac{M}{V}, \quad (1)$$

where M is the mass and V is the volume of the sample (amount of water displaced).

The theoretical density (d_{th}) of the specimen is calculated by the rule of mixture as

$$d_{th} = \frac{1}{\frac{w_1}{\rho_1} + \frac{w_2}{\rho_2} + \frac{w_3}{\rho_3} + \frac{w_4}{\rho_4}}, \quad (2)$$

where w_1, w_2, w_3 and w_4 are the weight fractions of Al, Cu, Ni and FeNb, respectively, and ρ_1, ρ_2, ρ_3 and ρ_4 are the corresponding densities of the composite.

The porosity (p) of the composite can be estimated by the ratio of difference of theoretical and experimental densities of the composite to theoretical density of the composite from the following equation:

$$p = \frac{d_{th} - d_{ex}}{d_{th}} \times 100\%. \quad (3)$$

2.3 Microstructure of AMMCs

Material morphology of the base alloy and composite was characterized by optical microscope. The samples were sectioned as per specifications of test procedure in the form of cube of size 10 mm × 10 mm × 10 mm. Samples were polished with abrasive papers of different grit sizes, from rough to fine grade (up to 1500 grit size), to achieve fine surface. Finally, velvet cloth polishing was done with 3 μm diamond paste. The samples were etched with Keller's reagent which is a widely used etchant for aluminum alloys [6, 10]. The Olympus optical microscope was used to study the surface morphology of the specimens.

3 Results and Discussion

In this section, results are discussed. Section 3.1 presents microstructures. Section 3.2 presents quantitative result of porosity.

3.1 Microstructure of AMMCs

The microstructure of the composite with varying reinforcement of FeNb content in Al–5Cu–2Ni has been investigated. Uniform dispersion of reinforcement particles along the grain boundary with some particle clustering and agglomeration is seen

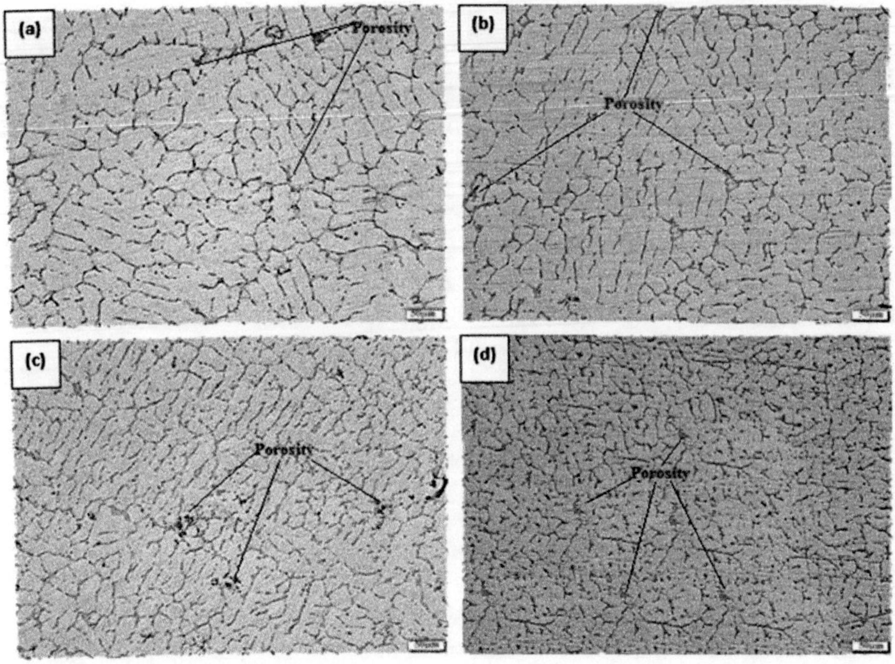

Fig. 4 Sand casting microstructure of Al–5Cu–2Ni with **a** 0% of FeNb; **b** 1% of FeNb; **c** 3% of FeNb; **d** 5% of FeNb

in Fig. 4. The dendrite structure is observed in micrographs of Figs. 4 and 5; it may due to super cooling leading to α–Al dendrites [6]. There are two types of phases at grains, viz. a dark phase having script type morphology and a white phase Al_2Cu along the grain boundary [3, 6]. The dark block region indicates the FeNb particles. The particle segregation may be due to the Al dendrites solidifying first during the solidification of composite [5]. The reinforcement of FeNb content causes grain refinement [6]. Stir casting microstructure shows improved uniform distribution of reinforcement particles due to the stirring mechanism. Micro porosities are observed in sand casting micrographs as shown in Fig. 4. Stir casting provides negligible amount of porosities (Fig. 5).

3.2 *Density and Porosity Analysis*

The theoretical and experimental densities of the composites were determined by rule of mixture (ROM) and water displacement technique (Archimedes' principle) [6, 7, 10]. From the obtained results, the density variations are presented in Fig. 6.

The theoretical density of the composite was greater than the experimental density in both the cases due to presence of porosity in the composites. Theoretical and

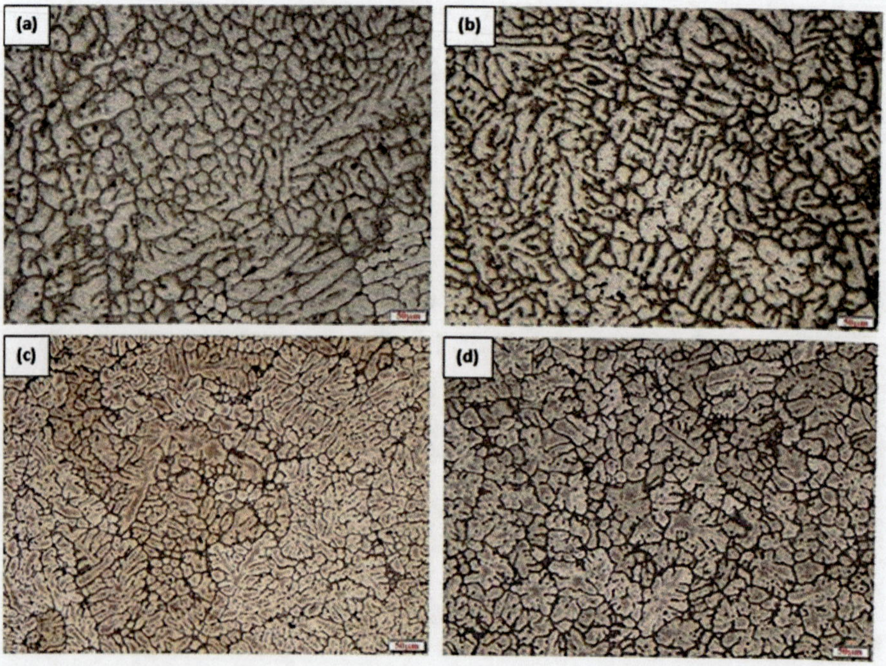

Fig. 5 Stir casting microstructure of Al–5Cu–2Ni with **a** 0% of FeNb; **b** 1% of FeNb; **c** 3% of FeNb; **d** 5% of FeNb

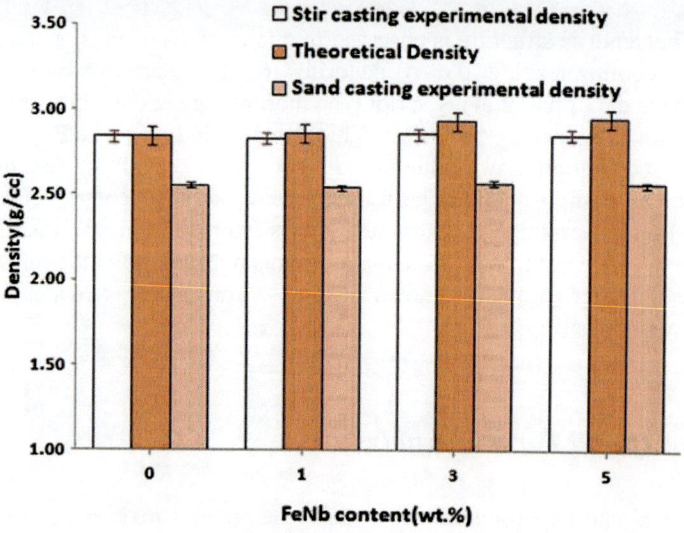

Fig. 6 Density of Al–5Cu–2Ni with FeNb content

experimental densities of the stir casting composite did not have much variation, indicating less porosity. The experimental density of sand casting composite showed significant variation with theoretical density, indicating significant porosity.

The porosity values of the sand casting and stir casting composites are listed in Table 2. The porosity of the composite increased with increased weight fraction of reinforcement of FeNb as shown in Fig. 7. The porosity of the composites was observed due to the entrapped gasses from the environment during the solidification process [6, 10]. Valibeygloo et al. [12] reported that porosity formation is due to environment gasses entrapped into the molten metal during the solidification process and clustering of reinforcement particles. The porosity leads to micro cracks and failure of the lattice structure in the composite [13].

The highest porosity of 3.3% was observed in stir casting and 13% in sand casting. It shows that stir casting obtained lower porosity when compared with the sand casting method. In both the cases, the porosity of the composite was significantly larger than that of the base metal because of the moisture content present in reinforced particles [10]. Among the two casting processes, stir casting is the preferable method for manufacture of composite.

Table 2 Porosity variation of sand casting and stir casting

Wt% of FeNb	Material	Sand casting porosity (%)	Stir casting porosity (%)
0	Al–5Cu–2Ni	10.2	0.1
1	Al–5Cu–2Ni–1FeNb	11.1	1.1
3	Al–5Cu–3FeNb	12.4	2.7
5	Al–5Cu–2Ni–5FeNb	13	3.3

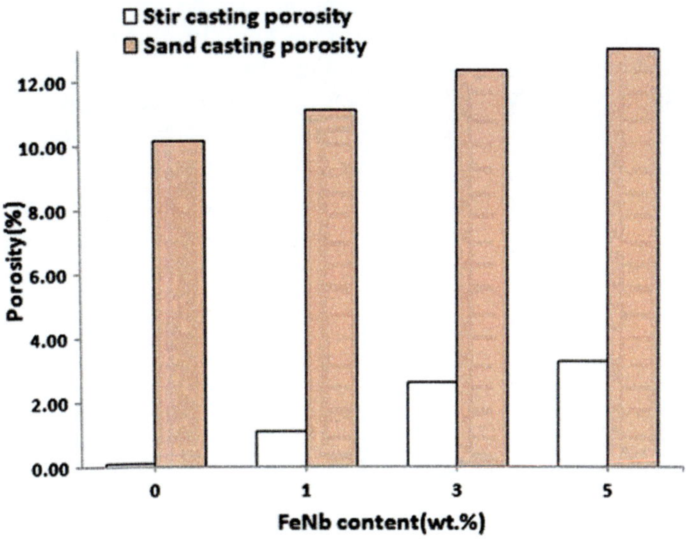

Fig. 7 Porosity of Al–5Cu–2Ni with FeNb content

There are some other factors causing the porosity. The porosity of the composite increased with increasing weight fraction of FeNb because of pore nucleation at the FeNb particle surfaces resulting in the reduction of liquid metal flow. Sajjadi et al. [8] studied A356 aluminum alloy/Al_2O_3 composites fabricated by stir and compo-casting processes. Increasing Al_2O_3 led to increase the porosity of the composite because of the pore nucleation at the Al_2O_3 particle surfaces causing a reduction in liquid metal flow. Porosity depends on the casting route. Stir casting produces less porosity as compared to sand casting because variation in the gravity pressure in metal flow leads to variation in the solidification process.

The gravity pressure is more on metal die compared with sand mold resulting in the faster metal in stir casting compared with sand casting. Xiong et al. [14] studied A356-10%SiC particle composite castings at different solidification pressures; it was observed that the squeeze casting has lower porosity compared with the gravity casting because of pressure variations in the casting process.

4 Conclusions

The following conclusions are drawn from density, porosity and microstructures analysis of the composite with reinforcement of FeNb content:

- The manufacture of Al–5Cu–2Ni/FeNb is possible with sand and stir casting methods.
- Optical micrographs showed the uniform distribution of the particles with presence of porosities.
- The porosity formation is due to entrapped gasses from the environment and moisture content present in reinforcement particles.
- The lower porosity of stir and sand castings are 0.1% and 10.2%, respectively. The highest porosity of stir and sand castings are 3.3% and 13%, respectively.
- Stir casting is the preferable method for fabrication of aluminum metal matrix composite.

Acknowledgements Authors would like to express their gratitude to the NPIU-New Delhi, TEQIP-III and Indian Institute of Technology, Guwahati, for financial support and facilities for carrying out the experimental work.

References

1. Heinz A et al (2000) Recent development in aluminium alloys for aerospace applications. Mater Sci Eng A 280:102–107
2. Miller WS et al (2000) Recent development in aluminium alloys for automotive industry. Mater Sci Eng A 280:37–49

3. Mandal PK, Robi PS (2018) Influence of micro-alloying with silver on microstructure and mechanical properties of Al–Cu alloy. Mater Sci Eng A 722:99–111
4. Panagopoulos CN et al (2009) Corrosion and wear of 6082 aluminum alloy. Tribol Int 42:886–889
5. El-Aziz KA et al (2015) Wear and corrosion behavior of Al–Si matrix composite reinforced with alumina. J Bio Tribo Corros 5:1–10
6. Bannaravuri PK, Birru AK (2018) Strengthening of mechanical and tribological properties of Al–4.5%Cu matrix alloy of with the addition bamboo leaf ash. Results Phys 10:360–373
7. Onat A et al (2007) Production and characterization of silicon carbide particulate reinforced aluminium–copper alloy matrix composites by direct squeeze casting method. J Alloy Compd 436:375–382
8. Sajjadi SA et al (2012) Comparison of microstructure and mechanical properties of A356 aluminium alloy/Al_2O_3 composites fabricated by stir and compo-casting processes. Mater Des 34:106–111
9. Wang K et al (2019) Effects of $TiC_{0.5}N_{0.5}$ nanoparticles on the microstructure, mechanical and thermal properties of $TiC_{0.5}N_{0.5}$/Al–Cu nanocomposites. J Mater Res Technol 1197:1–10
10. Bannaravuri PK, Birru AK (2017) Microstructure and mechanical properties of aluminum metal matrix composites with addition of bamboo leaf ash by stir casting method. Trans Nonferrous Met Soc China 27:2555–2572
11. Sahu MK, Sahu RK (2017) Optimization of stirring parameters using CFD simulations for HAMCs synthesis by stir casting process. Trans Indian Inst Met 10:2563–2570
12. Valibeygloo N et al (2013) Microstructural and mechanical properties of Al–4.5wt% Cu reinforced with alumina nanoparticles by stir casting method. Int J Miner Metall Mater 20:978–985
13. Ravi Kumar K et al (2018) Mechanical properties and characterization of zirconium oxide (ZrO_2) and coconut shell ash (CSA) reinforced aluminium (Al 6082) matrix hybrid composite. J Alloy Compd 1–22
14. Xiong Y et al (2021) Effects of porosity gradient pattern on mechanical performance of additive manufactured Ti6–Al–4V functionally graded porous structure. Mater Des 1–45
15. Dong P-y et al (2013) Microstructures and properties of A356–10%SiC particle composite castings at different solidification pressures. Trans Nonferrous Met Soc China 23:2222–2228
16. Huabing Y et al (2018) Microstructure and mechanical properties at both room and high temperature of in-situ TiC reinforced Al–4.5Cu matrix nanocomposite. J Alloy Compd 767:601–606
17. Jing S et al (2020) Effect of Cu element on morphology of TiB_2 particles in TiB_2/Al–Cu composites. Trans Nonferrous Met Soc China 30:1148–1156
18. Ahmad SN et al (2005) The effects of porosity on mechanical properties of cast discontinuous reinforced metal matrix composite. J Compos Mater 39:451–466
19. Bohlooli V et al (2013) Effect of ablation casting on microstructure and casting properties of A356 aluminium casting alloy. Acta Metall Sin (Engl Lett) 26:85–91
20. Amirkhanlou S, Niroumand B (2010) Synthesis and characterization of 356–SiC_p composites by stir casting and compocasting methods. Trans Nonferrous Met Soc China 20:788–793
21. Kok M (2005) Production and mechanical properties of Al_2O_3 particle–reinforced 2024 aluminium alloy composites. J Mater Process Technol 161:381–387
22. Mishra S et al (2019) Physico–mechanical characterization of garnet and fly ash reinforced Al7075 hybrid composite. Materialwiss Werkstofftech 50:731–741
23. Sahu MK, Sahu RK (2020) experimental investigation, modeling, and optimization of wear parameters of B_4C and fly-ash reinforced aluminum hybrid composite. Front Phys 8:1–14
24. Sahu MK, Sahu RK (2020) Aluminum based hybrid metal matrix composites: a review of selection philosophy and mechanical properties for advanced applications. Int J Mech Prod Eng Res Dev 10:8–28
25. James G et al (2000) Modeling the effects of cooling rate, hydrogen content, grain refiner and modifier on micro porosity formation in Al A356 alloys. Mater Sci Eng A 285:49–55

26. Savas O, Ramazan K (2007) Application of Taguchi's methods to investigate some factors affecting micro porosity formation in A360 aluminium alloy casting. Mater Des 28:2224–2228
27. Sahu MK et al (2014) Multi-objective optimization of upsetting parameters of Al–TiC metal matrix composites: Agrey Taguchi approach. J Eng Manuf 228(11):1501–1507
28. Sahu MK, Sahu RK (2018) Fabrication of aluminum matrix composites by stir casting technique and stirring process parameters optimization. Adv Cast Technol 111–116
29. Sahu MK, Sahu RK (2020) Synthesis, microstructure and hardness of Al7075/B4C/fly-ash composite using stir casting method. Mater Today Proc 27:2401–2406

State-of-the-Art Manufacturing of Metal Foams and Processing—A Review

A. Changdar and S. S. Chakraborty

Abstract Metal foam recently has grabbed a significant amount of interest due to its promising prospects in fulfilling the crisis led by steel and its counterparts. The recent studies indicated commercialization of metal foam which is disrupted mainly because of high cost and lack of design knowledge. Industries had expressed the need of more detailed studies, in order to the amicability of the material to machining, joining and other manufacturing processes. Thus, this article, in brief, clubs all the major processing routes of metal foams along with their limitations. It also discusses how secondary processing methods can contribute in tailoring metal foam for different applications. The article shall help future researchers to bridge the existing gaps which restricts metal foam utility in concerned industries.

Keywords Metal foam · Lightweight structure · Shaping metal foam · Manufacturing metal foam · Processing metal foam

1 Introduction

Metal foams are cellular materials, wherein voids are introduced in a solid base metal, to acquire desired properties. With time, the material emerged as a potential solution for various applications due to its low density coupled with extraordinary mechanical, structural and physical properties [1]. They are porous cellular structure containing air voids within the solid base metal. The material displays high strength-to-weight ratio and potential to absorb shock and noise. The structural reaction of these materials remains same under compressive loads, because the material exhibits plateau stress under such loads. Its potential of acoustic absorption makes it ideal for soundproof construction. The material is also ideal for heat exchangers owing to its substantial surface area and higher cell wall conduction. To have such a unique combination of features within a single conventional material seems non-viable [2, 3]. With time, the material has emerged as a potential solution for various structural and

A. Changdar (✉) · S. S. Chakraborty
Materials Processing and Microsystems Laboratory, CSIR—Central Mechanical Engineering Research Institute, Durgapur 713209, India
e-mail: anirbancmeri04@gmail.com

U. S. Dixit et al. (eds.), *Advances in Forming, Machining and Automation*, Lecture Notes in Mechanical Engineering, https://doi.org/10.1007/978-981-19-3866-5_11

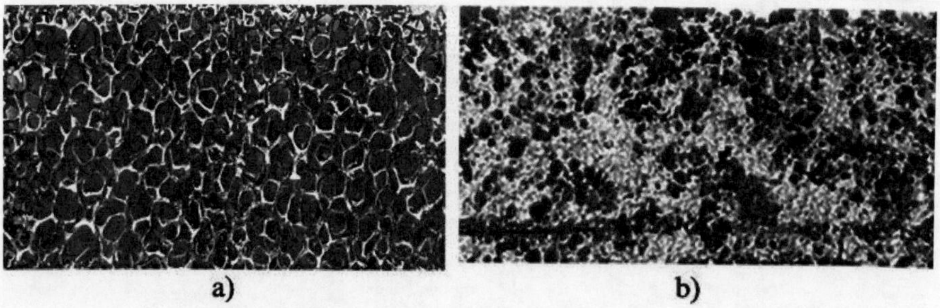

Fig. 1 **a** Open cell metal foam, **b** closed cell metal foam

functional applications. Aluminium-based metal foams out of the rest were exploited the most, and its utilization is expected to rise in near future [4, 5].

The material is classified based on its cellular topology as shown in Fig. 1, and most of its significant properties depend on the same. Foams in which cells are interconnected are known as open cell, whereas when the cells are separated by distinct cell wall, it is known as closed cell [3]. The understanding of the classification is significant as it determines the kind of application, a metal foam is suited for. ASTM standard D2662 quantifies the porosity percentage of closed cell and open cell foams. Closed cell metal foams owing to its lightweight and high stiffness are ideal for structural applications. However, open cell foams due its reticulated structure and larger surface area are mostly applicable in functional areas, like filters and heat exchangers.

Metallic foams exhibit a certain number of unique properties, difficult to obtain using other materials. The structural properties of these materials are of great significance, and its uniqueness lies in the stress–strain response mainly under compression. These materials do not display catastrophic failure, when under deformation. Instead, the material exhibits plateau stress under compressive deformation, because of cell collapse and densification. The reason behind such deformation behaviour is owing to the fact that elastic modulus due to cell collapse obtained during initial loading is less than the actual elastic modulus (E). Apart from this, the deformation characteristics of these materials also depend on the relative density and the ductility of the base metal.

Instead of such wide bandwidth of prospective applications, several factors exist, which can challenge the usability and market growth of aluminium metal foams and its sandwich panels (Banhart et al. 2017). A recent survey suggested, the two major reason which restricts the widespread application of these materials are high cost and a lack of design knowledge. The survey also suggested for increase in literature investigating the amicability of the material to machining and joining process [6, 7]. Growth of aviation and automotive industries has been very prominent due to rapid urbanization and globalization of late. These industries mostly rely on metals and alloys for manufacturing components. The recent tendency of these industries to reduce material weight from their value chains has led to the advent of many

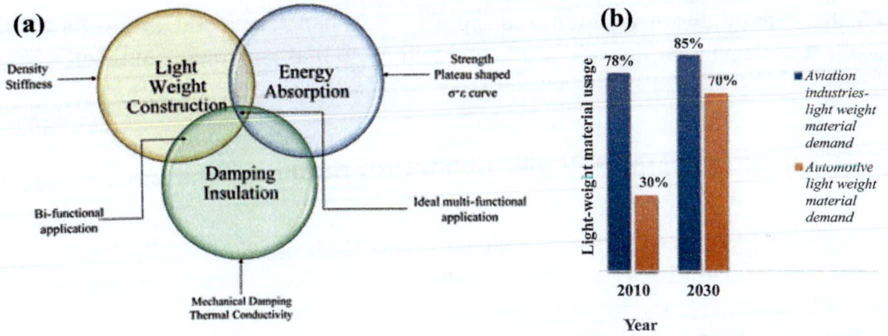

Fig. 2 **a** Multiple applications of metallic foam, **b** survey result on lightweight material usage of aviation and automotive industries

lightweight materials. The current design intent of most industries is biased towards the use of lightweight materials with a wide span of properties. Metal foam, thus stands out to be an ideal material of choice for such industries due its multifunctional capabilities as shown Fig. 2a. Market reports suggest usage of lightweight material share for aviation industries may increase by 10%. However, in automotive industries: a rise up to 70% in 2030 is expected as shown in Fig. 2b. Thus, it is very important to reduce the existing knowledge gap between these materials and industries, in order to scale up its industrial adoption and usability.

Currently, existing processing methods lack the potential to generate parts tailored to industrial requirement. Thus, a more economical way is to process metal foam in bulk in generic shapes like slabs and plates, using available processing methods. Subsequently, the material in their generic form is shaped as desired by introducing a secondary manufacturing process route like forming, welding or cutting. But undergoing conventional manufacturing processes stand out to be challenging for these materials due to their fragility under tensile stresses, and several alternative secondary manufacturing processes like laser or thermal forming of these materials are gaining popularity. Therefore, it can be well-anticipated that a proper process route definition in order to generate practical parts and components out of these materials is of great significance. Currently, several literatures exist enveloping process mechanisms and applications of the material. But a single article confronting all the process routes along with its limitations and describing the possible secondary manufacturing process routes hardly exists.

Henceforth, this article clubs the major processing routes of metal foams along with their limitations. As the properties of metal foam mostly depends on its structure, a reliable fabrication process for continuous production is of prior importance. Unavailability of such a process is the major drawback of the material and thus imposes limitations to its full-fledged industrial adoption [1, 3]. The article also discusses the further processing methods of metal foams, which includes forming, welding and cutting. These processes are quite significant when it comes to tailoring of foam/foam sandwich as per desired application. When a foam core is covered with the sheet of the similar material at top and bottom, it makes a foam sandwich.

2 Processing Methodologies of Metal Foam

Though metal foam's widespread application was realized very recently, the history of metal foam dates back to twentieth century. Since then, the methodologies and materials used for processing of foam underwent many adaptations, as proclaimed in the existing literature. With advancement of technologies and to obtain tailored to application properties, fabrication of metal foams has evolved to next level. Irrespective of ongoing advances, extensive adoption of metal foam is impeded due to inadequacy of production methodologies to process variety of materials in highly effective and economical method [8–10]. The current advancements in processing approaches are quite convincing and certainly show a promise to conquer the existing hindrances in the way of widespread acceptance of metal foam. An exhaustive representation of processing methodologies on porous material is shown in Fig. 3.

2.1 *Liquid-State Foaming Methods*

In liquid-state processing, porosity is introduced in a liquid or semi-solid metal matrix. Currently, this method is widely used for commercial production owing to its higher foaming rate and capabilities to produce greater volume of porous material.

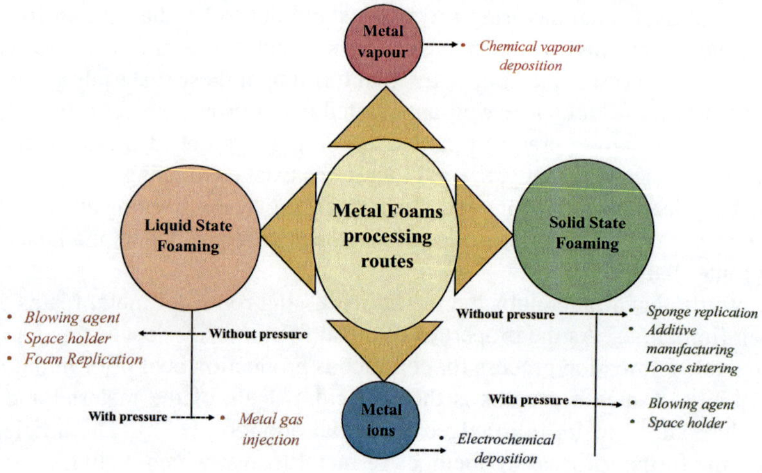

Fig. 3 Processing routes of metal foams

Liquid-state foaming being a traditional method has been studied a lot and detailed understanding of the process can be found in adequate literature. The liquid-state foaming is usually classified as, direct and indirect foaming [10–12] (Tianjian 2002).

Direct foaming methods

The method is achieved by direct introduction of either, a gas (air, nitrogen or argon) or any secondary substance (TiH_2, ZrH_2, MgH_2, $CaCO_3$, etc.) in to the melt pool. Introducing of secondary substance in molten metal matrix is also known as foaming by phase decomposition. The process is achieved by heating the secondary substance in the melt above its decomposition temperature as to evolve gas. In both cases, gas bubbles are formed on the surface of the melt, which are captured and cooled to create voids. CYMAT, a Canada-based company, and HYDRO ALUMINIUM in Norway use direct gas injection method; whereas, Shinko-wire, a Japan-based company, uses phase decomposition method for manufacturing of closed cell aluminium foam. Aluminium foams of Shinko-wire are popularly known as ALPORAS. Both methods are capable of processing close cell aluminium foam with a relative density ranging from 0.04 to 0.4 and cell size varying between 0.5 and 15 mm. The process can be advantageous, as it is capable of producing large volumes of foam along with reduced densities consistently. Foams resulting from these methods are thus probably less expensive compared to other porous materials. The ultimate necessity for cutting off the foam is a disadvantage of this sort of processing technique as it leads to opening of the cells. Additionally, due to the addition of silicon carbide, aluminium oxide or magnesium oxide particles in the melt for enhancing the viscosity lead to the brittleness of the gas injected foam. Reinforcement of ceramic particles within cell wall is one of the complications which is not desired. In order to avoid the effects of additives into melts, foaming of pure metallic melts with gases was suggested. The foaming process is usually established at temperatures near to melting point, as to maintain a uniform viscosity. This is sometimes carried out by bubbling gas through a melt that can be chilled down at a continuous casting process. Further, a foam like structure is formed, by capturing the bubbles in the solidifying molten metal liquid. To circumvent the disadvantages mentioned, foam processing went biased towards indirect foaming in which molten metal is not directly foamed rather they are processed by investment casting [11, 13–15].

Indirect foaming methods

In case of indirect foaming methods, molten metal is not directly foamed. One of the known processing is investment casting using a polymer foam. Polymer foam like polyurethane is employed initially. Polymer foam needs to be converted into an open cell type with reticulation therapy in case it is closed cell type primarily. Further, the polymer foam is filled with adequate substances like easy plaster, mullite or phenolic resin in the form of slurry. The plastic foam is later removed by thermal treatment once the plastic foam has been properly cured and is cast into the consequent open voids to replicate cellular arrangement of the primary polymer foam. After eliminating the mould material, metallic foam is obtained that is a precise replica of the initial polymer foam. ERG company as already described previously manufactured metallic foams using this method known as DUOCEL. Advantages of this process include fabrication of complex shaped foam structures by preforming the polymer foam. The sense of controlled manufacturing is predominant in this method as the pore morphology and density of the metallic foam is determined by polymer foam precursor. Other than aluminium, this process is also capable of producing foams of copper and magnesium which certainly make this process more reliable compared to direct foaming process. Indirect foaming can also be established, by casting molten metal around hollow spheres of low density, and the same can also be done using organic and inorganic granules. The vice versa of the aforementioned method is also possible, wherein the granules are either washed away using proper treatment after casting or are allowed to remain within the cast as to form syntactic foam. This process is also capable of processing diversified materials like aluminium, magnesium, zinc, lead and tin. Mould as desired can be designed as to obtain functional shaped metal foam and porosity up to 98%. Osprey process is another method of liquid-state processing also known as spray forming. This process allows processing of different distinct metals and alloys. Metallic melt is atomized, and a spray of fast flying little metal droplets is made. The resulting droplets are collected to create a dense deposit on a substrate. The process results in formation of dense deposit in generic shapes like billet and sheet. The materials usually exhibit low oxide content with better grain size [11, 13]. Some of the examples of metal foams produced by various liquid-state processes are illustrated in Fig. 4.

Fraunhofer processing method developed at Fraunhofer Institute, Germany, follows powder metallurgical route for development of metal foam. The processing technique involves heating of foaming agent and metal powder mixture at a temperature exceeding melting temperature of the base metal, hence, it can be categorized under liquid-state foaming. The manufacturing process starts with the mixing of metal dyes: basic metallic powder or metallic powder mixes using a blowing agent, and the mixture is compacted to produce a compact, semi-finished item. In principle, the compaction could be carried out by any method that makes sure that the blowing agent is inserted into the metallic matrix with no previously occurring porosities. Selection of the compaction technique (hot uniaxial or isostatic compression, pole extrusion or powder rolling) is dependent upon the necessary form of the precursor substance. The production of this precursor needs to be performed very carefully since any remaining porosity or other flaws will cause bad results in additional processing. The precursor is further heated to temperatures comparable to melting point of the base metal as a result blowing agent start decomposing. The released gas as a result starts expanding the precursor material which consequently results in formation of metallic foam. The foaming agent depends on the sintering and melting temperature of matrix material. For example, TiH_2 and ZrH_2 are used to foam Al and Zn, respectively, $MgCO_3$ and $SrCO_3$ can be used for foaming steels, $BaCO_3$ for copper, and $PbCO_3 \cdot Pb(OH)_2$ and MgH_2 for Pb. The decomposition speed, heating rate and also the firmness of cell wall construction ascertain the final density of the foam generated. This method occasionally also includes heating of the compacted metal powder mixture with additives as to activate exothermic reaction to melt the mix. Overall, the process is very flexible and versatile and can foam a wide range of

Fig. 4 Examples of metallic foams processed by liquid-state processes

materials in near net shape. Proper selection of blowing agent and process parameters allows foaming several metals like aluminium, aluminium base alloys, tin, zinc and brass. On more judicious selection of parameters materials like, lead and gold can also be foamed. Aluminium foams processed using the method are termed as "Foaminal" and "Alulight". More detailed understanding of liquid-state processing is described in several literatures with its current applications and state-of-the-art innovations [8, 9] (Banhart 2013).

Liquid-state forming with progression of time has evolved a lot and even gained capabilities of processing near net shape products. Aluminium foams are the most commercialized product of these processes. However, processing foams of other materials using these processes hardly crossed the prototype stage. Even processing aluminium foams with virtue of these processes faces many challenges, and aluminium melt being less viscous needs to be stabilized using various additives as to ensure porosity. Additives like BaO_3, SiO_2, SiC, Ca and Al_2O_3 along with some other materials are added in the melt to manipulate the surface tension and decrease the viscosity to obtain desired morphology of the foam. Though powder metallurgical replication process by Fraunhofer exhibits a prominent promise but high production cost and limited commercialization has certainly been the barrier to its extensive growth. Owing to fact that metal foam still lacks widespread commercialization, development of more reliable processing techniques is still a research issue of many organizations. Certainly, just the development of processes at laboratory scale is not going to solve the issue. Certain surveys clearly illustrate conquering the psychological constraints will also play a key role in extensive acceptance of these material [14, 15].

2.2 *Solid-State Foaming Processes*

Potential crisis of liquid-state foaming has led to the inception of solid-state foaming techniques. Solid-state processing involves introducing porosity into a solid metal using advanced techniques like additive manufacturing and vapour deposition method. Though solid-state processing is new but it is quite flexible when it comes to foaming of wide range of metals which apparently is challenging using liquid-state processing. With solid-state processing, control over manipulating pore size and relative density of metal foam is easier to achieve, assuring better quality of products in terms of functionalities. As mentioned by Atwater et al. [13], solid-state foaming can be defined as those processes which create metal foams in such a way that the processing temperature remains within the melting temperature of the base metal. The processing of metal foam using this process is done either by introducing a blowing agent directly into the blended solid metal or by integrating metal powder around a pre-processed temporary template having voids. Metal powder is either compacted, sintered or electro-deposited on the templates. The porosity developed as a result of first method is known as intrinsic porosity, whereas the latter is known as extrinsic porosity [13] (Banhart and Seelinger 2006).

The extrinsic porosity methods are quite similar to the liquid-state foaming techniques but the operating temperatures are below the melting point of the matrix metal. Loose powder sintering is a kind of solid-state processing and is done by using a sacrificial template mostly with few exceptions. Most commonly used sacrificial template method includes scaffold technique and space holder technique. Both of these processes can again be subdivided depending on the processing conditions. In scaffold technique, metal powder is introduced in porous polymer foam by adopting methods like coating, deposition or by mixing the metal with an expandable precursor also known as in situ foaming (Banhart and Seelinger 2006).

Grumman et al. [16] in their work have exhibited how metal coating can be used for creating foam out of Ni–Ti shape memory alloy. Metal powder mixed with suitable binders was used for coating a polymer foam with pre-defined pore sizes followed by sintering. Once the required strength is achieved, binders and the polymer foam were removed thermally. Yang et al. [17] also adopted the same method for processing foam out of Tantalum (Ta) particles by coating polyurethane foam with Ta powder and PVA binder mix. Prior to sintering Ta at 1950 °C, the polymer and binders were eliminated at a temperature of about 300 °C. The process is very adaptable in terms of materials but decreasing pore size can be a challenge as smaller pores restricts flow of metal in and out of the sacrificial pattern. Deposition of metal powders on the sacrificial foam is usually done using two routes, one of which is chemical vapour deposition (CVD) and is currently used commercially. International Nickel Company (Inco) currently a subsidiary of Brazilian mining company, Vale, developed Ni foam INCOFOAM using CVD at large scale. The process utilizes Ni carbonyl ($Ni(CO)_4$), which is heated under vacuum to modest temperature, where it decomposes and provides Ni vapour to coat the substrate. The porosity can be controlled with a single template geometry by varying the deposition amount, ranging from 74 to 98% and being comprised of pores from 450 to 3200 micron [13]. The other method of deposition is electroplating but sacrificial foam being non-conductive electro-less plating is used for deposition. Electro-less plating has been successfully utilized for Ni material and few literatures report Cu too. There is always a chance of oxidation while removing the precursor material using deposition method. Hence, the process of removal should be done under vacuum or else a suitable heating range should be adopted as to avoid formation of gaseous bye products. Excessive gas products can result fracturing of metallic shells. In case of in situ foaming, blending of polymer and metallic phase is done before foaming. This process is established by mixing polyurethane precursors with metal powders and letting the mixture to react to form bubbles or additional heat is provided to instigate the reaction [13]. Gauthier et al. [18] and Xie and Evans [19] used this process to foam Ni and Cu, respectively. This process has also been used for processing Ti and Fe foams with nearly 90% porosity. Again, this process too faces the challenges in terms scaffold removal, and many studies have elaborated the proper hierarchy to be followed as to remove the precursor material properly. The intricacies involved make the process very challenging for mass production or batch production. To maintain a suitable production rate, multiple parts should be produced at the same time using this process [13].

In space holder method, metal powder is blended with space holding materials like hollow polymer spheres, ceramic particles, grains of polymer, metals or salts which is removed later to create pores. The physical properties of the space holding material like shape and size along with its volume fraction in the mixture determine the properties of the resultant foam. The mixture is compacted in room temperature or at higher temperature if the material is heat resistant, for better compaction and to initialize the sintering process. The space holder material is either kept as a secondary phase in the final structure to create Syntactic foam, or else it is removed using vaporization or dissolution methods. Dissolution methods are common when the used space holding material is capable of dissolving in an appropriate solvent. NaCl is one such material which can be dissolved and removed. The method is convenient with open cell structure as it is easier for solvent material to flow through it. Vaporization method follows removal of space holder material thermally. The added advantage of the process is removal of the template material and porous metal sintering is combined in one step [13]. Freeze casting method is another template-based foaming method recently applied to metal foam processing. The method involves formation of a slurry of fine nanoscale particles of selected metal using a binder and dispersant in water or other liquid that is appropriate, for example camphene. The mix goes through freeze casted and then dried for a certain time period (24–48 h), and the binder is burned out subsequently. The particles are either sintered or reduced under hydrogen after removal of binder. The process is in its early stages has been applied on diverse range of materials like tungsten, copper, nickel, titanium and iron. Porosity up to 75% can be achieved and foam morphologies can be controlled by selection of proper process parameters. Another method is to sinter hollow metal spheres together to form porous structures. The metallic spheres are either obtained by electro-deposition or chemical deposition of an individual metal onto polymer spheres that are eliminated in a next step, or from coating polymer spheres (e.g., of polystyrene) using a binder/metal powder and then sintering the alloy to acquire a dense metallic shell while the polystyrene is eliminated. There are many other processes to obtain the hollow metallic spheres and can be found in the existing literature. The obtained hollow spheres are subsequently joined together by sintering into desired open cell or closed cell configuration. Titanium, nickel and Inconel foams have been processed using this method. More recent template-based methods include deposition of metal films around bubbles such as dynamic hydrogen bubble template (DHBT). The process is gaining increased attention for processing of battery electrodes and catalysis owing to higher surface area of processed foams. The field is very new and a few available literature [13, 15] should be referred for in-depth understanding of the process. Figure 5 illustrates few of the foams developed by solid-state foaming methods.

Intrinsic porosity methods include gas entrapment technique, wherein, adequate metal powders are compressed in a dense precursor material while a gas is entrapped within it. The precursor material is then heated which leads to expansion of the metal due to the internal pressure created by the entrapped gas which leads to formation of pores. The process is popularly known as Kearns process after the name of Michael Kearns who was the inventor and was first to apply the same at Boeing the US-based aerospace company. Boeing implemented it by filling titanium powder into a

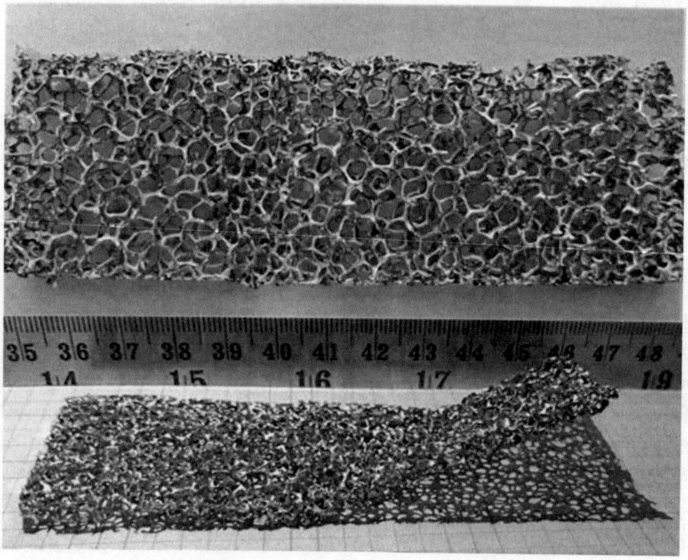

Fig. 5 **a** Examples of metallic foams processed by solid-state processes

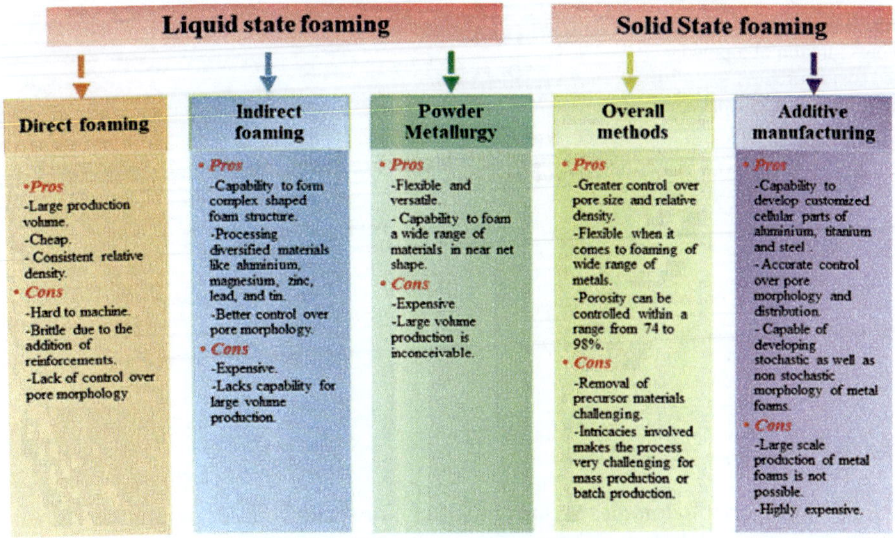

Fig. 6 Diagrammatic representation of pros and cons of various metal foam processing methods

metal can that was outgassed after evacuation. The can was again introduced with pressurized inert gas (Argon) at about 3–5 ATM and then sealed. The structure was densified by applying hot isostatic pressing at a temperature of 950 °C and 1000 ATM for 4 h. The argon gas was entrapped during the compaction and forms uniformly dispersed pores when the compacted material is finally annealed at a temperature of 1240 °C for 65 h. The process is even capable of processing near net shape products and has been widely used for processing commercially pure titanium and its alloys. The other method of intrinsic porosity formation is very new and still have a long way to go. In this process, pores are created in the powder materials unlike other methods, where pores are created in between the powder particles. It is achieved by mixing of metal powders thoroughly with oxides and then subsequent reduction of those oxides leads to formation of steam which starts expanding. The process is very new but many processing routes are being suggested for commercialization of this process and can be found in dedicated literature [12, 13, 15]. Figure 6 schematically illustrates the pros and cons of various foaming methods in processing metal foams.

2.3 Additive Manufacturing

Additive manufacturing technique offers a unique podium for generating, near net shape porous structures, for biomedical and aerospace applications. According to the ASTM standard F2792, it is defined as "the process of joining materials to make objects from 3D model data, usually layer upon layer, as opposed to subtractive

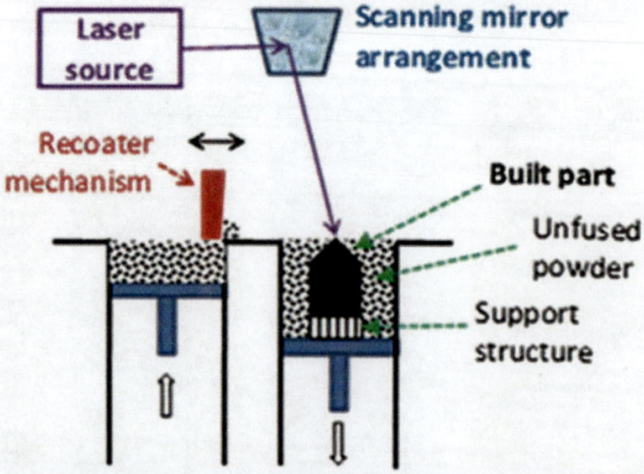

Fig. 7 Schematic representation of a powder bed fusion process. Reprinted with permission from Changdar and Chakraborty [1]

manufacturing methodologies". Currently, SLM or selective laser melting as shown in Fig. 7 is used mostly for manufacturing of porous part. The method is standardized as ISO/ASTM 52,900 and falls under powder bed fusion process [20].

Geometrically defined lattice structure porosity (GDLSP) and Geometrically undefined porosity (GUP) were the two approaches adopted for generating porous parts out of a SLM process. The part fabricated from GDLSP was entitled as non-stochastic cellular materials, whereas the parts out of GUP approach were termed as stochastic porous structure. Stochastic foams were mostly applied as filters and other functional applications, whereas the non-stochastic parts were applied biomedical utilities and heat exchangers [1, 21]. The process was quite efficient in developing customized cellular parts of aluminium, titanium and steel but was not efficient enough for processing alloys of the same. The process is still amateur, and its application for large-scale production of metal foams seems to be a distant dream [21].

3 Further Processing of Metal Foams for More Practical Manufacturability

Powder metallurgical route and additive manufacturing as discussed above are the two processes capable enough, to manufacture metal foam components in near net shape, while both the process offers flexibility in terms of foam structure and choice of the base metal, but are limited to small parts and tiny production volumes. Moreover, both the process involves high production cost, and are mostly limited to laboratory trials till date. A more practical approach of manufacturing metal foam is further processing of metal foams produced initially in the form of sheets and slabs. The foams can be further processed into the desired aesthetics either by cutting, joining or welding, as to tailor it in accordance to application. The approach apart from being more cost-effective, also guarantees better cell distributions and consistent densities.

3.1 Forming of Porous Metals

Capability of a material to form into desired shapes is very crucial to establish its firm foothold in industries. It has been reported that metal foams can resist fracture at severe deformation, which indicates their capabilities of getting formed into desired shape. Several studies successfully attempted forming of wide range of metal foams which include, aluminium foam (Alporas), porous nickel and lotus-type porous copper, plates. Cold extrusion was also applied to process porous metal by applying solid phase bonding of dissimilar metal wires. Though the possibility of mechanical forming exists but its yet not versatile. Studies revealed that corrugation of cell walls during mechanical forming leads to the formation of curvature at extrados. The curvature instigates development of very high tensile stresses, which eventually exceeds the tensile strength of metal foam [22]. Inability of metal foams to withstand such tensile stresses leads to its failure [1, 23, 24]. Limitations of mechanical forming of metal foam instigated utilization of laser forming, for bending metal foams. The process, utilizes non-uniform heating by laser to form a material by inducing non uniform strain within it. It eliminates the threat of mechanical stresses due to hard tooling instead relies on thermal stresses to produce the requisite deformation [1]. The recent studies are also affirmative and suggest that the process is capable of developing large bending deformations in the foam, without any detrimental effect on its structural robustness [3].

3.2 Joining Methodologies of Metal Foams

Several processes have been studied, in order to discover an optimal methodology for joining metal foam to another foam or a solid counterpart. Arc welding posed severe challenges when it comes to welding of metal foams. In order to successfully weld metal foam, heat must be confined to the thin struts and membranes of foam material, which is nearly impossible to achieve using arc welding [24]. As compared to arc welding, diffusion bonding served better results but its extreme processing time limits its usability. Diffusion bonding though delivered efficient joining [25]. Soldering was also attempted in several studies, but considering the porous structure of foam, wetting of soldering faces with molten solder was hard to achieve. As, a result the process failed to provide desired weld strength for any practical applications [24, 26]. Though use of adhesives were successful in joining metal foams with solid sheets, but the major drawback is it restricts utility of such part at higher operating temperature. Mechanical joints were also found not to be compatible. Attempt to rivet and mechanically fasten metal foams led to deformation and pre mature failure of the same [26]. Advanced joining methods like friction stir welding (FSW) and laser welding were found to produce better results in joining metal foams as compared to conventional methodologies. FSW was found to join metal sheets and foam at a single pass if the thickness of the foam was maintained within desired limits. On the other hand, laser welding due to its capabilities of confining heat input within a narrow region, stand out to be the most versatile process for welding metal foams. Laser welding process was found to offer the feasible solution to welding of metal foams on cautiously selecting the process parameters [1].

3.3 Cutting of Metal Foams

The application of metal foam can be enhanced if an appropriate method of machining it can be recognized. Several investigations were reported, wherein all the common methods were exploited. The processes include utilization of circular saw, band saw and disc grinding. But all of the processes tend to create localized plastic deformation and damaging of the surface. Wire electrical discharge machining (WEDM) and laser cutting were found to be more efficient process for machining metal foams. EDM was also found to exhibit higher rate of cutting metal foams, without any adverse effect on the structure of the metal foam. Whereas, laser cutting enabled obtaining cut surfaces which were burr free and with very narrow kerf [1, 25, 27].

4 Conclusions

The article enlisted the currently existing processing routes available for metal foams along with their prospects and limitations. The major insights exhibited by the article are as follows:

- Processes like additive manufacturing and powder metallurgical methods are capable of producing near net shape parts of metal foam, but their large-scale production prospects are limited due to low volume productivity and high cost. Products out of these processes are ideal for biomedical applications, dental implants or other sophisticated applications requiring less volume of part supply and high quality and precision.
- Solid-state processing of metal foam is gaining popularity as they are capable of developing controlled morphology metal foam and at standard volumes. Chemical vapour deposition (CVD) has already been applied for generating nickel foams at large volumes and are being applied by aerospace and automobile industries.
- Liquid-state processing is the most commercialized process, with least control over the morphology of the pores. The process is capable of producing metal foams at large volumes in generic shapes very cheaply. Currently, this processing method is the only sustainable method for large volume production. In order make the

process more versatile, a secondary manufacturing route must be added following the processing of metal foam to shape it as desired in an application. Processes like forming, cutting and welding are vastly studied for their compatibility with metal foam, without disturbing the structure of the material.

- The fragile structure of the material offers several limitations to the currently existing secondary processing methodologies. As this material is limited to deformation under compression, application of mechanical loads in any form to shape it tends to its failure. Similarly, welding, riveting or any other known joining processes turns out to be a challenge for these materials due their lesser contact surface. Laser-based processing methods turned out to be a success for this material and are currently being explored for complete process control.

Acknowledgements Financial support from the Science and Engineering Research Board (SERB) under project grant (no. ECR/2017/002505) is gratefully acknowledged.

References

1. Changdar A, Chakraborty SS (2021) Laser processing of metal foam—a review. J Manuf Process 61:208–225
2. Srivastava VC, Sahoo KL (2006) Metallic foams: current status and future prospects. IIM Metal News 9(4):9–13
3. Bucher T, Yao YL (2018) Advances in laser forming of metal foam: mechanism, prediction and comparison. Int J Mechatron Manuf Syst 11(2–3):250–273
4. Dehnavi A, Ebrahimi GR, Golestanipour M (2020) Effect of SiC particles on hot deformation behavior of closed-cell Al/SiC_p composite foams. J Braz Soc Mech Sci Eng 42(11):1–8
5. Mansuri H, Murugesan G (2018) Experimental study of composite foam sandwich structures for aerospace applications
6. Florek R, Simančík F, Harnúšková J, Orovčík Ľ, Dvorák T, Nosko M, Tekel T (2014) Injection molded plastics with aluminum foam core. Proc Mater Sci 4:323–327
7. Hommel P, Roth D, Binz H (2020) Deficits in the application of aluminum foam sandwich: an industrial perspective. In: Proceedings of the design society: DESIGN conference, vol 1. Cambridge University Press, pp 927–936
8. Körner C, Singer RF (2000) Processing of metal foams—challenges and opportunities. Adv Eng Mater 2(4):159–165
9. Lefebvre LP, Banhart J, Dunand DC (2008) Porous metals and metallic foams: current status and recent developments. Adv Eng Mater 10(9):775–787
10. Alhusseny ANM, Nasser A, Al-zurfi NM (2018) High-porosity metal foams: potentials, applications, and formulations. In: Porosity: process, technologies and applications, p 181
11. Monno M, Negri D, Mussi V, Aghaei P, Groppi G, Tronconi E, Strano M (2018) Cost-efficient aluminum open-cell foams: manufacture, characterization, and heat transfer measurements. Adv Eng Mater 20(8):1701032
12. Banhart J (2001) Manufacture, characterisation and application of cellular metals and metal foams. Prog Mater Sci 46(6):559–632
13. Atwater MA, Guevara LN, Darling KA, Tschopp MA (2018) Solid state porous metal production: a review of the capabilities, characteristics, and challenges. Adv Eng Mater 20(7):1700766

14. Zhao B, Gain AK, Ding W, Zhang L, Li X, Fu Y (2018) A review on metallic porous materials: pore formation, mechanical properties, and their applications. Int J Adv Manuf Technol 95(5–8):2641–2659
15. Jung A, Diebels S (2019) Micromechanical characterization of metal foams. Adv Eng Mater 1900237
16. Grummon DS, Shaw JA, Gremillet A (2003) Low-density open-cell foams in the NiTi system. Appl Phys Lett 82(16):2727–2729
17. Yang H, Li J, Zhou Z, Ruan J (2013) Structural preparation and biocompatibility evaluation of highly porous Tantalum scaffolds. Mater Lett 100:152–155
18. Gauthier M, Lefebvre LP, Thomas Y, Bureau MN (2004) Production of metallic foams having open porosity using a powder metallurgy approach. Mater Manuf Process 19(5):793–811
19. Xie S, Evans JRG (2004) High porosity copper foam. J Mater Sci 39(18):5877–5880
20. Calignano F, Cattano G, Iuliano L, Manfredi D (2017) Controlled porosity structures in aluminum and titanium alloys by selective laser melting. In: International conference on additive manufacturing in products and applications. Springer, Cham, pp 181–190
21. Nazir A, Abate KM, Kumar A, Jeng JY (2019) A state-of-the-art review on types, design, optimization, and additive manufacturing of cellular structures. Int J Adv Manuf Technol 1–22
22. Andrews E, Sanders W, Gibson LJ (1999) Compressive and tensile behaviour of aluminum foams. Mater Sci Eng A 270(2):113–124
23. Ersoy E, Özçatalbaş Y (2015) Deformation of metallic foams with closed cell at high temperatures. Int J Mater Metall Eng 9(7):789–792
24. Nowacki J, Sajek A (2019) Trends of joining composite AlSi-SiC foams. Adv Mater Sci 19(1):70–82
25. Matz AM, Kammerer D, Jost N, Oßwald K (2016) Machining of metal foams with varying mesostructure using wire EDM. Proc CIRP 42:263–267
26. Bernard T, Bergmann HW, Haberling C, Haldenwanger HG (2002) Joining technologies for Al-foam—Al-sheet compound structures. Adv Eng Mater 4(10):798–802
27. Yilbas BS, Akhtar SS, Keles O (2015) Laser cutting of small diameter hole in aluminum foam. Int J Adv Manuf Technol 79(1–4):101–111

VIKOR Method-Based Selection of the Most Suited Bioplastic for the Injection Moulding Process Amongst Potential Alternatives

PL. Ramkumar, Nikita Gupta, Akshay Kumar, and Aman Shukla

Abstract Injection moulding is a widely used process for the manufacturing of plastic products. Numerous plastics can be used in this process. Most of these plastics that are being used for this process are non-biodegradable, which further leads to an increased amount of plastic build-up on earth. Bioplastics are susceptible to be the solution to this worldwide problem. Currently, there are copious bioplastics with diverse properties that are available to us. And from these potential alternatives of bioplastics, selecting the most suited bioplastic for the injection moulding process is undoubtedly tough. In the present research, an attempt has been made to select the most suited bioplastic for the injection moulding process using a decision-making method, VlseKriterijumska Optimizacija I Kompromisno Resenje (VIKOR) method. The outcome of the VIKOR method deduces that polyglycolic acid (PGA) is the most suited bioplastic for the injection moulding process.

Keywords Injection moulding · Bioplastic · MCDM method · VIKOR

1 Introduction

Usage of bioplastics in the manufacturing sector has immensely increased due to the increased climate awareness and the need to reduce non-biodegradable plastic waste. And consequently, research for the use of bioplastics in the manufacturing sector has an emerging requirement [1–3]. The introduction of sustainable engineering has also fuelled this need for research for the use of bioplastics in the manufacturing

PL. Ramkumar · N. Gupta · A. Kumar (✉) · A. Shukla
Institute of Infrastructure Technology Research and Management, Ahmedabad, Gujarat 380008, India
e-mail: akshay.kumar.17m@iitram.ac.in

PL. Ramkumar
e-mail: pl.ramkumar@iitram.ac.in

A. Shukla
e-mail: aman.shukla.17m@iitram.ac.in

U. S. Dixit et al. (eds.), *Advances in Forming, Machining and Automation*, Lecture Notes in Mechanical Engineering, https://doi.org/10.1007/978-981-19-3866-5_12

of various plastic products. Bioplastics have shown great results in their biodegradability and environment friendliness. The use of bioplastics as raw material instead of non-biodegradable alternatives in manufacturing processes has significantly reduced the amount of greenhouse gas emissions. Therefore, determining suitable bioplastics for various manufacturing processes is necessary [4, 5]. At present, numerous researches are going on the properties of bioplastics, which will eventually help in the understanding of the behaviour of the bioplastic materials in different ambient conditions [6, 7].

Currently, for the manufacturing of plastic products, there are plentiful processes present for the manufacturers. And injection moulding process is one of the extensively used manufacturing processes. It is a process to manufacture plastic-moulded products by injecting molten base plastic material into a mould and then solidifying it by cooling [8–12]. This process involves six major steps as listed below [13–15]:

(1) Clamping the mould
(2) Injection of the molten base plastic material into the mould
(3) Dwelling
(4) Cooling the mould
(5) Opening the mould
(6) Removing the final product from the mould.

The physical and mechanical properties of the obtained final plastic product are largely dependent on the selected base material for the injection moulding, so an appropriate selection of the base material is quite necessary before the start of the process [16–18]. At present, linear low-density polyethylene (LLDPE) and high-density polyethylene (HDPE) are vastly used polymers for the manufacturing of plastic products but are extremely resistant to biodegradation [19]. So, worthwhile research is required for the selection of a suitable base material that can be incorporated in an injection moulding process, have similar properties as LLDPE and HDPE, and is also biodegradable. Bioplastics can be the solution to this problem. Bioplastics can be obtained naturally or can be made synthetically. Each bioplastic shows unique properties and chemical constitution. Every property or attribute of the bioplastic should be analysed before incorporating that bioplastic in the injection moulding process. As a result, increasing the number of bioplastics (alternatives) and properties (attributes) has made the selection process even more complex and time-consuming. Currently, there is a number of methods that are used by researchers for this type of selection problem, amongst all of them multi-criteria decision-making (MCDM) emerged as the most adept and decisive method. VIKOR method, which is VlseKriterijumska Optimizacija I Kompromisno Resenje method, is one of the MCDM methods that are easy to comprehend and incorporate in complex systems like this amongst other MCDM methods [20, 21].

In the current study, selection of the most suited bioplastic for the injection moulding process from a list of bioplastics, which are obtained from the valued past research, that comprises PVA, PLLA, PHA, PHB, PGA, PCL, PLA, PBS, and TPS, is carried out (Table 1).

Table 1 Abbreviation of the bioplastics

Bioplastic	Abbreviation
Polyvinyl alcohol	PVA
Poly L-lactic acid	PLLA
Polyhydroxyalkanoates	PHA
Polyhydroxybutyrate	PHB
Polyglycolic acid	PGA
Poly ϵ-caprolactone	PCL
Polylactic acid	PLA
Poly-butylene succinate	PBS
Thermoplastic starch	TPS

The properties of these bioplastics are also obtained from past research, and further, the data are analysed using the VIKOR MCDM method that provides a robust note on the bioplastic that can be used as a base material for the injection moulding process.

2 Properties of Bioplastics (Attributes) Examined for the Selection Process

Every bioplastic has its own unique mechanical and physical properties. For the selection of the most suited bioplastic, it is vital to consider every possible individual property (attribute) of each bioplastic (bioplastic). Impact strength, degradation temperature, cost, and tensile strength are a few of the important properties that are needed to be considered for bioplastic.

Table 2 provides information regarding various properties of the bioplastics (attributes) that are considered in this study [22–25].

3 Values of the Properties (Attributes) of Bioplastics (Alternatives)

For the examination process, various bioplastics and the values of their various properties are taken from the valued research done by various researchers. VIKOR method is used to select the most suited and most capable to process bioplastic for incorporating in the injection moulding process (Table 3).

Manufacturers according to the requirement of their manufactured final product can include or discard the properties of the bioplastics.

Table 2 Properties of bioplastics

Property	Explanation
Tensile strength	To consider the load-carrying capacity of the final product, tensile strength that is the maximum load a material can sustain without breaking when pulled, is considered
Density	To consider the weight of the final product, density is required to be considered
Melting point	To consider the temperature resistance of the final product, the melting point is considered
Degradation temperature	To consider the maximum temperature to which the final product can work before degradation, degradation temperature is considered
Flexure strength	To consider flexibility, that is the ability of the material to bend without being broken, flexure strength is considered
Impact strength	To consider the suddenly applied force to the final product, the impact strength is considered
Vicat softening point	To consider the heat-related properties and patterns, Vicat softening point, that is the temperature at which a flat needle of cross-sectional area 1 mm^2 goes into the material to a depth of 1 mm, is considered
Shore D hardness	To consider the stiffness or strength of the material, shored hardness is considered
Cost	To consider the total expense in manufacturing the final product, the cost is considered

4 Bioplastic Selection Using MCDM Method

4.1 Criteria Weighting

In the present study, entropy method, because of its greater reliability in information gauging and precision, is implemented to determine the weight of the attribute [26]. A problem matrix in multi-criteria decision-making (MCDM) method can be depicted in below equation.

$$F = \begin{matrix} F_1 \\ F_2 \\ \vdots \\ F_j \end{matrix} \begin{pmatrix} f_{11} & f_{12} & \vdots & f_{p1} \\ f_{21} & f_{21} & \cdots & f_{p1} \\ \vdots & \vdots & \ddots & \vdots \\ f_{1j} & f_{2j} & \cdots & f_{ij} \end{pmatrix} \tag{1}$$

where $F_1, F_2, F_3, \ldots F_j$ is the 'j' number of alternatives that are needed to be ranked according to their 'i' number of attributes. W_i is the weight of the attribute F_i.

The values of the weight for our problem matrix are shown in Table 4, calculated using the entropy method.

Table 3 Bioplastics (alternatives) and value of their properties (alternative)

Bioplastics	Tensile strength (MPa)	Density (g/cm^3)	Melting point (°C)	Degradation temperature (°C)	Flexural strength (MPa)	Impact strength (J/m)	Vicat softening point (°C)	Shore D hardness	Cost (Rs/kg)
PLA	61	1.24	171	328.6	111	36	71	77	81
TPS	21	1.47	161	322	44	34	71	63	222
PCL	16.2	1.15	61	374.48	24	81	45.3	52	261
PBA	25.5	1.25	116	394	41	8	102	61.1	286
PGA	114	1.4	226	261	181	201	122	44	451
PLLA	71	1.24	186	241	118	67	166	91	257
PHB	29.5	0.97	173	221	17	88	97	68	121
PVA	79	1.17	201	351	51	161	61	41	201
PHA	25	1.2	163	178	32	25	123	34	101

Table 4 Weight of the attributes

Attributes	Tensile strength	Density	Melting point	Degradation temperature	Flexure strength	Impact strength	Vicat softening point	Shore D hardness	Cost
W_a	0.1852	0.0068	0.0435	0.0278	0.2411	0.2776	0.0655	0.0392	0.1125

4.2 The VIKOR Method

For the optimization of complex systems that has multiple criteria, the VIKOR method was developed. It was developed by Serafim Opricovic for the decision problems that are conflicting criteria or had criteria that have different units. In this method, compromisation is done assuming that it does not affect the resolution much. The solution to the problem is regarded as a compromise. This method finds such a compromise that is closest to the ideal solution. Steps involved in the VIKOR method to solve a decision problem are as follows:

Step 1. Calculation of normalised decision matrix from equation below (R_{ij})

$$R_{ij} = \frac{f_{ij}^* - f_{ij}}{f_{ij}^* - f_{ij}^-} \tag{2}$$

f_{ij}^* and f_{ij}^- are the best and worst values of each alternative that has its criteria.

Step 2. Calculation of weighted normalise decision matrix from equation below (W_{ij})

$$W_{ij} = w_j \frac{f_{ij}^* - f_{ij}}{f_{ij}^* - f_{ij}^-} \tag{3}$$

Step 3. Calculation of utility measure from equation below (S_i)

$$S_i = \sum_{j=1}^{n} w_j \left[w_j \frac{f_{ij}^* - f_{ij}}{f_{ij}^* - f_{ij}^-} \right] \tag{4}$$

Step 4. Calculation of regret measure from equation below (R_i)

$$R_i = \text{Max}\left[w_j \frac{f_{ij}^* - f_{ij}}{f_{ij}^* - f_{ij}^-} \right] \tag{5}$$

Step 5. Calculation of priorities from equation below (Q_i)

$$Q_i = \frac{1}{2}\left[\frac{S_i - S^-}{S^+ - S^-} \right] + \frac{1}{2}\left[\frac{R_i - R^-}{R^+ - R^-} \right] \tag{6}$$

S^+ and S^- are the best and worst values of utility measure; R^+ and R^- are the best and worst values of regret measure.

Step 6. Calculation of rank (R).

Arrange the values of Q_i in ascending order and give their ranks

5 Results and Discussion

Using the steps of the VIKOR method, we obtained the normalised decision matrix, weighted normalised decision matrix, utility measure, regret measure, priorities value, and their ranks which are depicted in Tables 5, 6, 7, 8, and 9.

From Table 10, we can conclude that polyglycolic acid (PGA) can be a potential bioplastic that can be incorporated in the injection moulding process. These are the initial results based on the specific application of the final product with specific applications. Values of weights changes as values of the properties of our desired final product change. Manufacturers according to their needs can change the values of weights of the attributes and get the desired ranking for their decision problem. Manufacturers can also use other multi-criteria decision-making (MCDM) methods like TOPSIS and PROMETHEE to conclude the solution more robustly.

6 Conclusion

Due to the increased usage of non-biodegradable plastic products, there has been an increase in climatic concerns. As this plastic is non-biodegradable, they pollute the environment in a non-avoidable manner, by increasing the greenhouse gas emission and increasing the toxicity of the soil and water reservoirs that eventually impacted the lives of humans and as well as humans. Using bioplastic materials instead of non-biodegradable materials like LLDPE and HDPE turns out to be one of the solutions to this worldwide problem. In this paper, an attempt has been made to select the most suited bioplastic that can be incorporated as a base material for the injection moulding process. VIKOR one of the MCDM methods is used to determine the most suited bioplastic to be as base material. Performing VIKOR method on this problem, following things were concluded:

- Polyglycolic acid (PGA) got the least priority value of approximately 0
- Poly L-lactic acid (PLLA) got the second least priority value of approximately 0.459
- Thus, PGA got ranked 1 and concluded to be most suited bioplastic for rotational moulding process followed by PLLA with rank 2.

Table 5 Normalised decision matrix

Base resin	Tensile strength (MPa)	Density (g/cm^3)	Melting point (°C)	Degradation temperature (°C)	Flexural strength (MPa)	Impact strength (J/m)	Vicat softening point (°C)	Shore D hardness	Cost (Rs/kg)
PLA	0.570	0.519	0.333	0.300	0.452	0.845	0.771	0.218	0.000
TPS	1.000	0.981	0.394	0.339	0.871	0.855	0.787	0.509	0.381
PCL	1.052	0.308	1.000	0.090	1.000	0.611	1.000	0.709	0.486
PBA	0.952	0.538	0.667	0.000	0.903	1.000	0.530	0.544	0.559
PGA	0.000	1.000	0.000	0.619	0.000	0.000	0.373	0.818	1.000
PLLA	0.484	0.519	0.242	0.711	0.394	0.694	0.000	0.000	0.481
PHB	0.930	0.000	0.321	0.803	1.045	0.575	0.572	0.382	0.108
PVA	0.398	0.404	0.152	0.206	0.839	0.207	0.870	0.909	0.324
PHA	0.978	0.615	0.382	1.000	0.948	0.902	0.340	1.000	0.054

Table 6 Weighted normalised decision matrix

Base resin	Tensile strength (MPa)	Density (g/cm^3)	Melting point (°C)	Degradation temperature (°C)	Flexural strength (MPa)	Impact strength (J/m)	Vicat softening point (°C)	Shore D hardness	Cost (Rs/kg)
PLA	0.106	0.004	0.015	0.008	0.109	0.235	0.051	0.009	0.000
TPS	0.185	0.007	0.017	0.009	0.210	0.237	0.052	0.020	0.043
PCL	0.195	0.002	0.044	0.002	0.241	0.170	0.066	0.028	0.055
PBA	0.176	0.004	0.029	0.000	0.218	0.278	0.035	0.021	0.063
PGA	0.000	0.007	0.000	0.017	0.000	0.000	0.024	0.032	0.113
PLLA	0.090	0.004	0.011	0.020	0.095	0.193	0.000	0.000	0.054
PHB	0.172	0.000	0.014	0.022	0.252	0.160	0.038	0.015	0.012
PVA	0.074	0.003	0.007	0.006	0.202	0.058	0.057	0.036	0.037
PHA	0.181	0.004	0.017	0.028	0.229	0.250	0.022	0.040	0.006

Table 7 Utility measure and regret measure

Base resin	S_i	R_i
PLA	0.535	0.235
TPS	0.781	0.237
PCL	0.802	0.241
PBA	0.824	0.278
PGA	0.194	0.113
PLLA	0.465	0.193
PHB	0.685	0.252
PVA	0.478	0.202
PHA	0.777	0.250

Table 8 Best and worst values of utility measure and regret measure

	S_i	R_i
Worst	0.193553	0.1126
Best	0.823746	0.2777

Table 9 Priorities value

Base resin	Q_i
PLA	0.640
TPS	0.844
PCL	0.872
PBA	1.000
PGA	0.000
PLLA	0.459
PHB	0.812
PVA	0.497
PHA	0.880

Table 10 Ranking

Base resin	Rank
PLA	4
TPS	6
PCL	7
PBA	9
PGA	1
PLLA	2
PHB	5
PVA	3
PHA	8

References

1. Ramkumar PL, Kulkarni DM, Chaudhari VV (2014) Parametric and mechanical characterization of linear low-density polyethene (LLDPE) using rotational moulding technology. Sadhana 39(3):625–635
2. Gupta N, Ramkumar PL, Sangani V (2020) An approach toward augmenting materials, additives, processability and parameterization in rotational moulding: a review. Mater Manuf Process 35(14):1539–1556
3. Chen W-Y, Suzuki T, Lackner M (eds) (2017) Handbook of climate change mitigation and adaptation. Springer International Publishing
4. Greco A, Ferrari F, Maffezzoli A (2019) Processing of super-tough plasticized PLA by rotational moulding. Adv Polymer Technol 2019
5. Vroman I, Tighzert L (2019) Biodegradable polymers. Materials 2(2):307–344
6. Hareesh A, Muruli MS, Ramesha A, Ranganath N, Panchakshari HV (2018) 7:89–96
7. Teymoorzadeh H, Rodrigue D (2015) Biocomposites of wood flour and polylactic acid: processing and properties. J Biobased Mater Bioenergy 9(2):252–257
8. Ramkumar PL, Kulkarni DM, Chaudhari VV (2019) Fracture toughness of LLDPE parts using rotational moulding technology. Int J Mater Prod Technol 58(4):305–322
9. Gupta N, Ramkumar P (2020) Polym Polym Compos 096739112095324
10. King FL, Kumar AAJ (2018)
11. Rodrigue D, Robledo-ort JR (2016) Polym Compos 37:915–924
12. Yussuf AA, Massoumi I, Hassan A (2010) J Polym Environ 18:422–429
13. Moghaddamzadeh S, Rodrigue D (2018) Prog Rubber Plast Recycle Technol 34:200–220
14. Sheshmani S (2013) Carbohydr Polym 94:416–419
15. Belhassen R, Boufi S, Vilaseca F, López JP, Méndez JA, Franco E, Pèlach MA, Mutjé P (2009) Polym Adv Technol 20:1068–1075
16. Ramkumar P, Panchal Y, Panchal D, Gupta N (2020) Mater Today Proc
17. Gupta N, Ramkumar PL (2020) 599–606
18. Ramkumar PL, Kulkarni DM (2016) Int J Mater Eng Innov 7:159
19. Sun Z et al (2017) Mechanical and thermal properties of PLA biocomposites reinforced by coir fibers. Int J Polym Sci 2017
20. Brauers WKM, Zavadskas EK, Peldschus F, Turskis Z (2008) Transport 23:183–193
21. Raju SS, Murali GB, Patnaik PK (2020) J Reinf Plast Compos 39:721–732
22. Turco A et al (2015) A magnetic and highly reusable macroporous superhydrophobic/superoleophilic PDMS/MWNT nanocomposite for oil sorption from water. J Mater Chem A 3(34):17685–17696
23. González-López ME, Pérez-Fonseca AA, Cisneros-López EO, Manríquez-González R, RamírezArreola DE, Rodrigue D, Robledo-Ortíz JR (2019) J Polym Environ 27:61–73
24. Quiles-Carrillo L, Montanes N, Pineiro F, Jorda-Vilaplana A, Torres-Giner S (2018) Materials (Basel) 11:2138
25. Abdullah AHD, Putri OD, Fikriyyah AK, Nissa RC, Intadiana S (2020) Polym Technol Mater 5:1250–1258
26. Çalışkan H, Kurşuncu B, Kurbanoğlu C, Güven SY (2013) Mater Des 45:473–479

Machining

3D Finite Element Simulation of Orthogonal Cutting of Ti–6Al–4V Using Textured Tool

Tejanshu Sekhar Sahu, Allan George, Basil Kuriachen, Jose Mathew, P. B. Dhanish, and K. Vipindas

Abstract Ti–6Al–4V alloy is widely used in various industries due to its significant properties. However, due to its poor machinability characteristics, Ti–6Al–4V is classified as a difficult to cut material. The incorporation of surface texturing on the tool faces improves the tribological properties which in turn enhances the overall machinability of the alloy. In this paper, 3D finite element simulation of the orthogonal cutting of Ti–6Al–4V workpiece performed using innovative texture patterns to improve the machinability has been adopted. The stresses induced at the work surface while machining with various textured tools as well as conventional tools have been investigated and compared with the ANSYS WORKBENCH software. The results showed a significant reduction in stress values induced at the work surface while machining with tools having combined textures of both dimples and linear groove patterns.

Keywords Machinability · Orthogonal cutting · Texturing · Von Mises stress

T. S. Sahu · A. George · B. Kuriachen (✉) · J. Mathew · P. B. Dhanish
National Institute of Technology Calicut, Calicut, Kerala, India
e-mail: bk@nitc.ac.in

T. S. Sahu
e-mail: tejanshu2908@gmail.com

A. George
e-mail: allan_p190117me@nitc.ac.in

J. Mathew
e-mail: josemat@nitc.ac.in

P. B. Dhanish
e-mail: dhanish@nitc.ac.in

K. Vipindas
Design and Manufacturing, Indian Institute of Information Technology, Kurnool, Andhra Pradesh, India
e-mail: vipindas.k@iiitk.ac.in

U. S. Dixit et al. (eds.), *Advances in Forming, Machining and Automation*, Lecture Notes in Mechanical Engineering, https://doi.org/10.1007/978-981-19-3866-5_13

1 Introduction

Ti–6Al–4V has found numerous applications in various fields of science owing to its outstanding inherent properties. Titanium alloys exhibit exceptional resistance to corrosion, high-strength capacity at lower weights, and superior strength at high temperatures [1]. Due to its excellent properties, Ti–6Al–4V finds various applications in aviation, biomedical sciences, and automobile sector [2]. However, low thermal conductivity of the titanium alloy combined with its high chemical affinity and work hardening characteristics results in the alloy getting classified as difficult to cut material [3]. Moreover, machining of titanium alloys with a conventional tool with plain rake surface escalates friction at the interface of tool and workpiece which generates higher heat and produces enormous stress at the tool rake surface.

Tool surface modification by surface texturing may be viewed as an effective method to improve the machinability characteristics of the alloy. Sughiara and Enomoto used laser surface texture (femto second laser) to develop patterns on the rake face of WC–Co tool of dimple shape of diameter ranging from 30 μm to 70 μm with a depth of 5 μm [4]. They performed a face milling operation on medium carbon steel under dry conditions and noted that the dimple pattern having a diameter 70 μm performed superior to the other tools in reducing the average crater wear depth. Darshan et al. fabricated dimple patterns of diameter 50 μm on the rake surface of uncoated carbide tool using multi-diode pump fiber laser [5]. They performed dry machining on Inconel 718 and analyzed tool wear, cutting force, and surface finish and observed that dimple textured tool showed best results in minimizing the tool wear, cutting forces, and surface roughness of the specimens. They added that the stabilized builtup edge formation and lesser contact area were the reason for the better performance in case of tools with texture patterns. Rajbongshi et al. used micro-EDM to develop linear grooves of 200 μm length and 50 μm spacing on the flank face of the coated carbide tool [6]. They performed dry machining of AISI D2 steel and studied its effect on cutting temperature and tool wear. They noticed that tools with textures on flank faces performed well and were superior as compared to conventional tool. Manikandan et al. used the wire EDM to develop linear groove patterns of width 100 μm of different orientations such as parallel to cutting edge, perpendicular to cutting edge, and crossed on the rake face of tungsten carbide tool [7]. They conducted orthogonal cutting of Ti–6Al–4V with molybdenum disulfide as the lubricant and analyzed tool wear, cutting temperature, machining forces, and chip morphology characteristics. They noted that cross-patterned tool performed better than all other tools in improving machinability characteristics of the alloy.

The above literature survey showed that numerous researches have been performed to study the different cutting performances such as cutting temperature, tool wear, machining forces, and chip morphology. Studies on the finite element investigation of stress generation in orthogonal cutting with textured tools are few and limited. In this paper, 3D finite element investigations of Mises stress generated on the work surface have been conducted with various tool texture patterns.

2 3D Finite Element Analysis

To investigate the stress generation at the workpiece surface, ANSYS Workbench (2020 R1 version) has been used. The Lagrangian formulation has been adopted for a formulation of finite elements, as it is preferable in the unconstrained flow of materials and provides faster calculation. The stresses generated at the work surface while machining with various tool textures patterns have been discussed.

2.1 *Material Model*

Johnson–Cook model is one of the widely used robust thermo-mechanical models, which relates plastic, viscous, and thermal characteristics observed while performing orthogonal machining operations. The flow stress of this model is represented by the following Eq. (1):

$$\sigma = \left[A + B\varepsilon^n\right]\left[1 + C\ln\frac{\dot{\varepsilon}}{\dot{\varepsilon}_0}\right]\left[1 - \left(\frac{T - T_{\text{room}}}{T_{\text{melt}} - T_{\text{room}}}\right)^m\right] \tag{1}$$

where the 'σ' represents equivalent flow stress, 'ε^n' represents equivalent plastic strain, '$\dot{\varepsilon}$' represents equivalent plastic strain rate, '$\dot{\varepsilon}_0$' initial plastic strain rate, 'T_{room}' represents the workpiece room temperature, and 'T_{melt}' melting point temperature of workpiece. The terms *A*, *B*, *n*, *C*, and *m* are material constants and represent yield strength, hardening modulus, strain hardening exponent, strain rate sensitivity, and thermal softening coefficient, respectively. The first, second, and third parenthesis are elastic–plastic, viscosity, and thermal softening terms, respectively. The J–C parameters used for Ti–6Al–4V in the simulation have been given in Table 1.

The separation of chip or plastic deformation of workpiece and damage model is essential to define a numerical model for orthogonal cutting operation. To initiate the damage, Johnson–Cook law is considered and given by the following Eq. (2):

$$\varepsilon_f = \left[d_1 + d_2 \exp d_3\left(\frac{-p}{\overline{\sigma}}\right)\right]\left[1 + d_4\ln\left(\frac{\dot{\varepsilon}}{\dot{\varepsilon}_0}\right)\left[1 + d_5\left(\frac{T - T_{\text{room}}}{T_{\text{melt}} - T_{\text{room}}}\right)\right]\right] \tag{2}$$

The Johnson–Cook damage criteria parameters for the Ti–6Al–4V are reported in the Table 2.

Table 1 J–C parameters considered for study [8]

Yield stress in MPa (*A*)	Hardening modulus in MPa (*B*)	Strain rate sensitivity (*C*)	Hardening coefficient (*n*)	Thermal softening coefficient (*m*)	Reference strain rate in $\sec^{-1}(\dot{\varepsilon}_0)$
997.9	653.1	0.0198	0.45	0.7	1

Table 2 Johnson–Cook parameters for Ti–6Al–4V for damage criteria [8]

Parameters	d_1	d_2	d_3	d_4	d_5
Values	−0.09	0.27	0.48	0.14	3.87

2.2 *Workpiece/Tool Model*

The workpiece and tool adopted for the simulation are Ti–6Al–4V and tungsten carbide (WC), respectively, whose user-defined thermo-physical and mechanical properties are reported in Tables 3 and 4. The workpiece is fixed with dimensions of 7.5 × 1.5 × 3 mm^3. The geometry of the tungsten carbide tool is CNMA 120,408. Default 4 nodes tetrahedral element and 8 nodes hexahedral element have been adopted for meshing of tool and workpiece, respectively. The inherent low thermal conductivity of Ti–6Al–4V can significantly affect the thermal softening, which affects the induced stress at the workpiece. Hence, the temperature-dependent properties (Table 4) are given as input for better results.

The hybrid-patterned tool consisting of linear grooves and dimples has been made on the rake surface of the tool insert with SOLIDWORKS 2020 as shown in Fig. 1. The diameter of the dimple pattern is considered as 200 μm, while the length and width of the linear groove are 800 μm and 200 μm, respectively. The depth of each pattern is considered as 10 μm.

Table 3 Thermal and mechanical properties of workpiece and tool materials [9]

Properties	Workpiece	Tool
Density in kg/m^3	4430	14,900
Coefficient of thermal expansion in $°C^{-1}$	9.1×10^{-6}	4.8×10^{-6}
Young's modulus in Pa	1.13×10^{11}	6.45×10^{11}
Thermal conductivity in $W\,m^{-1}°C^{-1}$	$K(T)$	91
Poisson ratio	0.34	0.24
Specific heat in $J\,kg°C^{-1}$	$C_p(T)$	206

Table 4 Temperature-dependent properties of Ti–6Al–4V [9]

Properties/temperature (°C)	0	250	500	750	1000	1250	1500
$K(T)\ W\,m^{-1}°C^{-1}$	6.651	6.996	8.591	11.43	15.03	19.10	22.28
$C_p(T)\ J\,kg°C^{-1}$	48.3	535.8	588.3	640.8	600.1	645.2	690.2

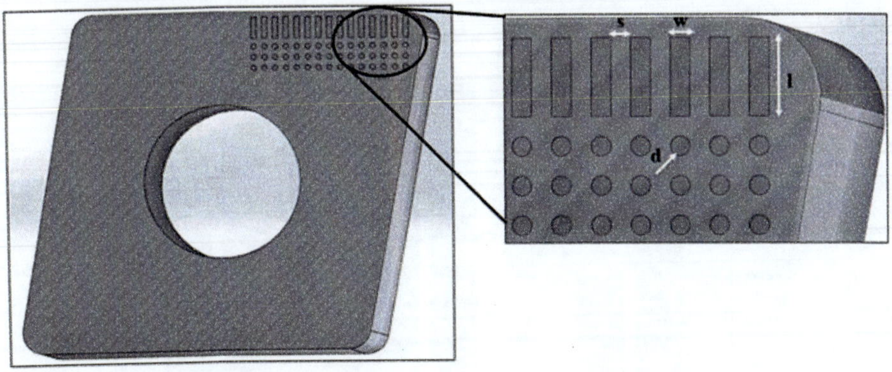

Fig. 1 CAD model of the textured tool

3 Results and Discussion

Micro-slots or patterns are fabricated on the faces of the tool to reduce the contact area at the tool-chip interfaces. Moreover, textures also facilitate effective heat dissipation by reducing the area of contact and help in better entrapment of the wear debris. As the contact area and contact pressure acting along the interface reduces, the workpiece specimen will be subjected to reduce von Mises stress which is an indication of better machinability characteristics. According to the von Mises yield criterion, if the von Mises stress of the material is equal to greater than the yield limit of the material, then the material will undergo yielding. The yielding occurs as a result of both mechanical deformation and also due to thermal loads within the material that accelerate the yielding phenomenon. However, extreme yielding is not preferable as it can result in poor surface integrity and poor machinability characteristics. As may be observed from Figs. 2 and 3, von Mises stress was effectively reduced in the textured tool at both cutting speeds at all conditions of depths of cut, which is an indication of better machinability features. It also may be observed that as the depths of cut increased the von Mises criterion also increases considerably in both the tools. At 0.3 mm depth of cut, the load acting on both the tools was considerably lower. However, due to effective reduction in contact region at the secondary shear zone, the von Mises stress reduced better (1300.3 MPa) in the case of the textured tool as compared to the conventional tool (1315.6 MPa). However, as the depth of cut increases, the load on the tool also increases which resulted in an increase in machining temperature as well as the von Mises stress in both cases.

The von Mises stress observed at a depth of cut of 0.55 mm was observed to be 1372.8 MPa in the specimen machined by a tool with textures, while the specimen machined by conventional tool recorded a stress of 1388.2 MPa. The stress increased further due to even higher contact pressure generated at 0.8 mm and was found to reach 1382.6 MPa in the specimen machined by tool with textures, while it was 1410.4 MPa in the case of conventional tool. Figure 3 shows the stress generation on the workpiece at 100 m/min in both the tool types. Since cutting speed is the

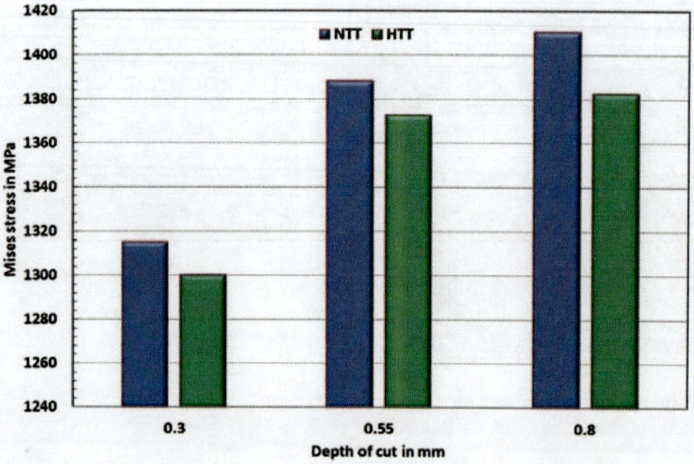

Fig. 2 Comparison of stress generation on workpiece due to different textured tools at 50 m/min

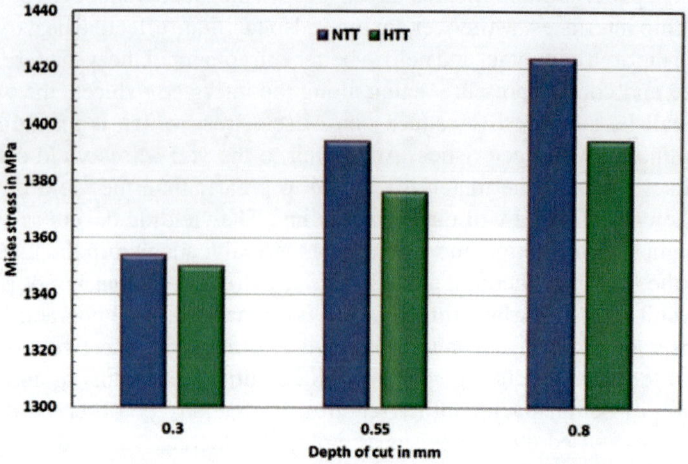

Fig. 3 Comparison of stress generation on workpiece due to different textured tools at 100 m/min

major process parameter that contributes to extreme tool wear due to extremely high temperatures generated at the cutting interface, von Mises stresses will be significantly higher in both the tools. However, due to the presence of micro-patterns that acts as fins for effective heat transfer, the stresses developed in the specimen machined by the tool with texture patterns were comparatively lower as compared to the case of conventional tool. The von Mises stress was observed to be 1350 MPa in the workpiece machined by textured tool, while the stress value was 1354.6 MPa in the case of conventional tool at 0.3 mm. As discussed in previous cases, the von Mises stress was observed to be 1376.3 MPa in the case of textured tool, while the stress was found to be 1394.3 MPa in the case of the conventional tool at 0.55 mm as the contact pressure increased with an increase in depth of cut. Similarly, an even higher von Mises stress value of 1394.4 MPa was recorded in textured tool, while the highest stress value of 1423.4 MPa was observed in the case of the conventional tool.

4 Conclusion

In this paper, the 3D finite simulation of orthogonal cutting of Ti–6Al–4V is discussed at various cutting conditions. The rake face of the tool was modeled by incorporating hybrid texture patterns, which consist of arrays of both dimple and linear groove patterns. The induced stresses at the work surface have been compared and analyzed for tools with hybrid patterns and conventional tools at a combination of two speeds, i.e., 50 m/min and 100 m/min and three depths of cuts, i.e., 0.3 mm, 0.55 mm, and 0.8 mm. The results showed that the stress induced at the workpiece surface while machining with tools with texture patterns is less than that of conventional tools. This can be attributed to the tool-chip contact area, which is reduced due to the presence of textures. The reduced contact area reduces the friction and thereby reduces heat generation. Subsequently, thermal load and contact pressure get diminished. This shows that the texturing on the tool faces effectively reduces the stress generated on the work surface and improves the performance of cutting.

References

1. Ji Z, Yang H, Li H (2015) Predicting the effects of microstructural features on strain localization of a two-phase titanium alloy. Mater Des 87:171–180
2. Khanna N, Davim JP (2015) Design-of-experiments application in machining titanium alloys for aerospace structural components. Meas J Int Meas Confed 61:280–290
3. Li C et al (2020) Electrical discharge-assisted milling for machining titanium alloy. J Mater Process Technol 285:116785
4. Sugihara T, Enomoto T (2017) Performance of cutting tools with dimple textured surfaces: a comparative study of different texture patterns. Precis Eng 49:52–60
5. Darshan C, Jain S, Dogra M, Gupta MK, Mia M (2019) Machinability improvement in Inconel-718 by enhanced tribological and thermal environment using textured tool. J Therm Anal Calorim 138:273–285
6. Rajbongshi SK, Annebushan Singh M, Kumar Sarma D (2018) A comparative study in machining of AISI D2 steel using textured and non-textured coated carbide tool at the flank face. J Manuf Process 36:360–372
7. Manikandan N, Arulkirubakaran D, Palanisamy D, Raju R (2019) Influence of wire-EDM textured conventional tungsten carbide inserts in machining of aerospace materials (Ti–6Al–4V alloy). Mater Manuf Process 34:103–111

8. Kugalur-Palanisamy N, Rivière-Lorphèvre E, Ducobu F, Arrazola PJ (2020) Influence of the choice of the parameters on constitutive models and their effects on the results of Ti6Al4V orthogonal cutting simulation. Proc Manuf 47:458–465
9. Mishra SK, Ghosh S, Aravindan S (2018) 3D finite element investigations on textured tools with different geometrical shapes for dry machining of titanium alloys. Int J Mech Sci 141:424–449

A Study on Ultrasonic Vibration and Laser-Assisted Turning of Aluminum Alloy

N. Deswal and R. Kant

Abstract Ultrasonic vibration-assisted turning (UVAT) and laser-assisted turning (LAT) are widely used advanced machining processes. In this study, a new machining method of ultrasonic vibration and laser-assisted turning (UVLAT) is proposed by simultaneous interaction of UVAT and LAT processes. Experimental analysis is carried out to explore the machinability of the aluminum alloy during conventional turning (CT), UVAT, LAT, and UVLAT. The effect of laser power on machining performance is analyzed for UVLAT and compared with CT, UVAT, and LAT in terms of machining forces and surface roughness. Lower cutting, radial, and feed forces are observed for UVLAT compared to CT, UVAT, and LAT. However, higher surface roughness is obtained for UVAT and UVLAT compared with CT and LAT. Results showed that the UVLAT could be an excellent process to enhance the machinability of aluminum alloys if the vibration is applied in the appropriate direction and will be better than CT, UVAT, and LAT.

Keywords Hybrid machining · Vibration-assisted turning · Laser-assisted turning · Laser and vibration-assisted turning · Machining forces · Surface roughness · Aluminum alloy

1 Introduction

UVAT is a non-conventional machining process in which the cutting tool is provided vibrations through an ultrasonic unit. The ultrasonic unit consists of a generator, transducer, and booster. The cutting tool is separated periodically in the cutting region during each vibration cycle [1]. Compared to CT, some advantages of UVAT

N. Deswal · R. Kant (✉)
Department of Mechanical Engineering, Indian Institute of Technology Ropar, Rupnagar, Punjab 140001, India
e-mail: ravi.kant@iitrpr.ac.in

N. Deswal
e-mail: 2017mez0022@iitrpr.ac.in

U. S. Dixit et al. (eds.), *Advances in Forming, Machining and Automation*, Lecture Notes in Mechanical Engineering, https://doi.org/10.1007/978-981-19-3866-5_14

are lower machining forces, higher tool life, higher surface finish, lower built-up edge, minimal burr formation, etc. [2].

Kim and Lee [3] observed lower cutting forces and lower surface roughness in UVAT than CT during the machining of carbon fiber reinforced plastics. Nath et al. [4] worked on low alloy steel and observed lower cutting forces and surface roughness in UVAT compared to CT. Patil et al. [5] studied Ti6Al4V and reported lower cutting forces and lower surface roughness in UVAT compared with CT. Khajehzadeh et al. [6] used AISI 4140 steel as the workpiece material and observed lower cutting forces in UVAT than that of CT. Khanna et al. [7] investigated Inconel 718 and reported lower surface roughness for UVAT than CT.

To improve the turning operation, some researchers have utilized both vibration and heat energies simultaneously and it is termed as hot ultrasonic-assisted turning (HUAT). Muhammad et al. [8] used a band-resistance heater on Ti-15333 along with vibrations to the cutting tool. They reported lower cutting forces and lower surface roughness in HUAT than CT and hot conventional turning (HCT). Sofuoglu et al. [9] utilized a tunnel furnace on Hastelloy-X and vibrations on the cutting tool. They observed lower surface roughness in HUAT than CT and UVAT.

Numerous heating assisting methods have been applied with turning operation to enhance the machinability such as plasma-assisted, gas torch-assisted, furnace-assisted, band-resistance heating-assisted, induction-assisted, electric-current-assisted, and laser-assisted [10]. Among the heating-assisted methods, LAT is most commonly used due to its localized, rapid, and controlled heating at the workpiece surface [11]. In LAT, a high-intensity laser beam is used to heat the workpiece, usually above the recrystallization temperature to ease material removal through a conventional cutting tool. Due to heating of the workpiece, thermal softening occurs, improving the machinability of the materials and accounts for lower cutting forces, higher tool life, higher surface finish, etc. [12, 13].

Wang et al. [14] investigated on Al_2O_3 particle reinforced aluminum matrix composite and reported lower surface roughness and lower cutting forces in LAT compared with CT. Skvarenina and Shin [15] worked on compacted graphite iron and showed no significant improvement in cutting forces and surface finish during LAT than CT. Venkatesan [16] analyzed Inconel 718 and observed lower cutting forces and lower surface roughness in LAT compared to CT. Song et al. [17] studied fused silica and observed lower cutting forces and lower surface roughness in LAT than that of CT. Zhai et al. [18] used SiCp/2024Al and reported lower surface roughness in LAT than CT.

From the literature survey, it can be concluded that the simultaneous interaction of ultrasonic vibration and laser heat energies is yet to be explored. Therefore, a new machining technique of ultrasonic vibration laser-assisted turning (UVLAT) has been proposed by integrating UVAT and LAT. Therefore, the present work aims to study the machinability of the aluminum alloy in terms of machining forces and surface roughness during UVLAT and compare it with CT, UVAT, and LAT.

2 Experimental Procedure

2.1 Experimental Setup of UVLAT Process

The experimental setup of the integrated ultrasonic vibration and laser-assisted turning process is shown in Fig. 1. Figure 1a shows the laser head fixture to hold the laser head, transducer fixture to fix the transducer, and dynamometer to measure the machining forces. In Fig. 1b, a zoomed view of the cutting area shows fiber laser spot, aluminum alloy material, and carbide tool. The experiments are executed on a high-speed lathe machine with a maximum of 2040 rpm. The lathe machine is modified for mounting the transducer and laser head fixture, and fixtures are so synchronized that it provides efficient movement in the feed direction during the machining operation.

The frequency generator generates the electric signal, and these signals are changed into ultrasonic vibration by the transducer. A booster enhances the vibration amplitude and sends these vibrations on the tool. The cutting tool is fixed at the bottom of the booster by a screw. The vibratory unit is mounted in the radial direction to the workpiece axis. The vibratory unit can be mounted in the feed and tangential direction. A frequency of 20 kHz and an amplitude of 20 μm are provided to the cutting tool. The maximum vibration speed (V_{max}) of the cutting tool can be calculated as $V_{max} = 2\pi af = 150.8$ m/min [19], where a and f represent the vibration amplitude and frequency, respectively.

A continuous-wave (CW) fiber laser having a maximum power of 1 kW and wavelength of 1080 $\pm$ 5 nm is used to preheat the workpiece material. The laser head is situated so that the laser beam is always applied on the workpiece material but not on the cutting tool. Circumferentially, the laser beam is 45° before the cutting tool, i.e., laser approach angle. The average laser beam diameter is measured as 1 mm. The distance between the workpiece surface and the laser nozzle is 18 mm. The axial distance between the center of the laser spot and the cutting tool is taken as 4 mm. It remains constant during the machining operation to preheat the workpiece

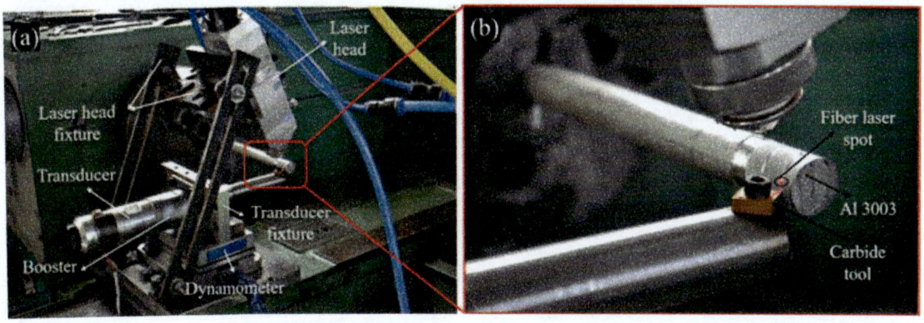

Fig. 1 Experimental setup of UVLAT process

surface. Compressed air is used to protect the focusing lens from extreme heat and debris that originated during the machining operation.

2.2 *Material and Cutting Tool*

The experiments have been performed on the aluminum 3003 alloy cylindrical bar of a diameter of 20 mm and length of 150 mm. The tungsten carbide inserts CNMG 120,408 have been used as the cutting tools during all the experiments.

2.3 *Experimental Methodology*

Before each experiment, an initial cut has been performed to check the roundness of the workpiece material during the machining operation. The experiments have been conducted under similar conditions. The cutting has been performed in the feed direction. The cutting length has been kept as 50 mm for each experiment. Machining parameters, vibration parameters, and laser parameters have been considered as process parameters. Laser power has been varied during the machining operation. Further, all other parameters remain unchanged during the machining operation. The CT and UVAT operations have been performed at room temperature. However, during LAT and UVLAT operations, the power (*P*) has been changed. The process parameters during the machining operation are shown in Table 1. The machining forces during the machining operation have been measured by a Kistler 9257 B dynamometer. The machining force data is analyzed using Dynoware software. The surface roughness has been measured by a Zeiss Handysurf surface roughness tester.

Table 1 Ultrasonic vibration laser-assisted turning process parameters

Machining parameter		Vibration parameter		Laser parameter	
Cutting speed	15 m/min	Frequency	20 kHz	Laser power	100, 150, 200 W
				Distance between laser spot and tool edge	4 mm
Feed rate	0.1 mm/rev	Amplitude	20 μm	Laser approach angle	45°
				Average laser spot diameter	1 mm
Depth of cut	0.1 mm	Direction	Radial	Stand-off distance between laser head and workpiece	18 mm

3 Results and Discussion

3.1 Machining Forces

Machining forces are one of the important factors to know the performance of any machining operation. Cutting, radial, and feed forces are measured during UVLAT and compared with CT, UVAT, and LAT. The average cutting, radial, and feed forces obtained during the experimental study of CT, UVAT, LAT, and UVLAT process are shown in Figs. 2, 3, and 4, respectively.

It can be seen that the average machining forces obtained during the UVLAT operation are found to be lower than CT, UVAT, and LAT. Further, with the increase in laser power the machining forces are found to be reduced for both LAT and UVLAT. The possible reason for lower machining forces in UVLAT is due to the combined effect of ultrasonic vibration and laser heating. Due to the ultrasonic vibration, the cutting tool is periodically separated, and due to the separation, the aerodynamic lubrication comes into effect which reduces friction between the cutting tool and workpiece. Moreover, laser preheats the workpiece surface, thermal softening of the workpiece material occurs, and material loses its yield strength. Therefore, the simultaneous application of the ultrasonic vibration and laser during the turning operation results in lesser machining forces during UVLAT than CT, UVAT, and LAT.

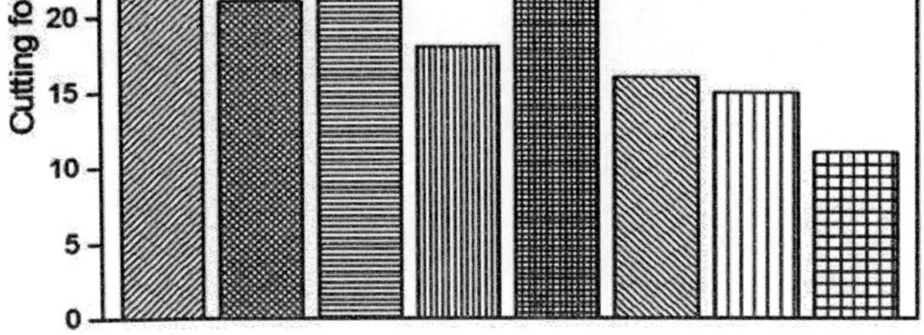

Fig. 2 Cutting force during CT, UVAT, LAT, and UVLAT

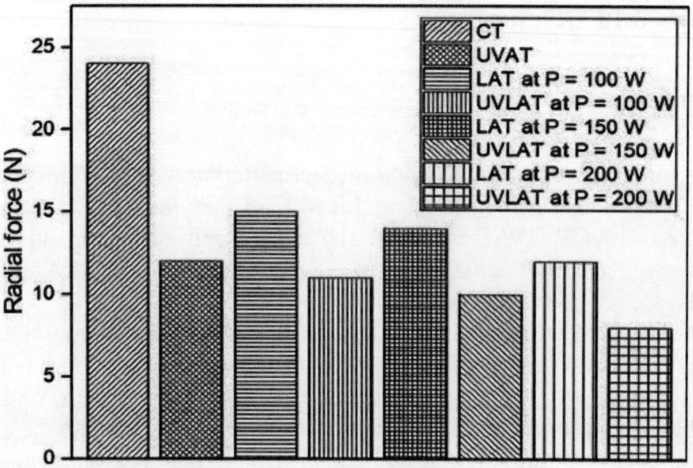

Fig. 3 Radial force during CT, UVAT, LAT, and UVLAT

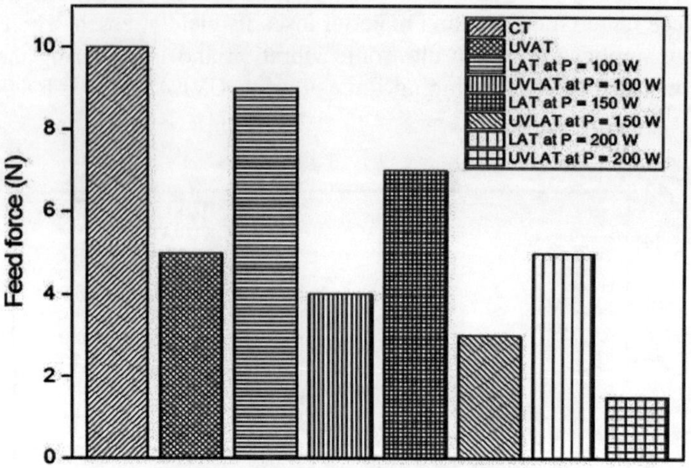

Fig. 4 Feed force during CT, UVAT, LAT, and UVLAT

3.2 Surface Roughness

Surface roughness is another significant factor in determining the quality of the machined surface component. The surface roughness measured during the CT, UVAT, LAT, and UVLAT process is shown in Fig. 5.

It is observed that the higher surface roughness is obtained in UVAT and UVLAT compared to CT and LAT. Besides, surface roughness is found to be decreased with the increase in laser power for both LAT and UVLAT. The possible reason for higher surface roughness in UVAT and UVLAT may be due to the vibration direction. In the

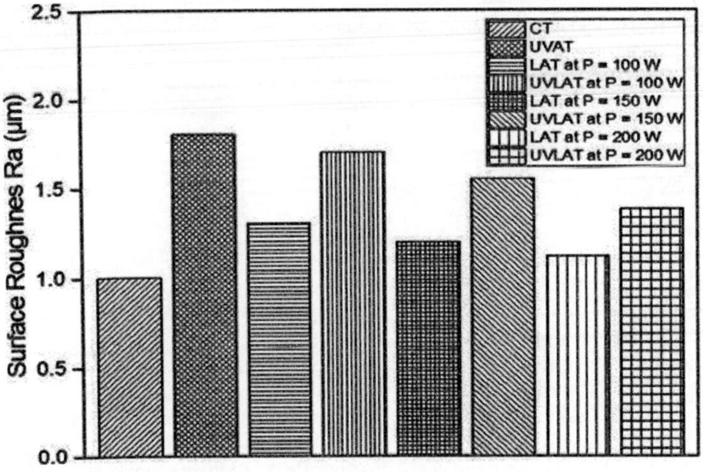

Fig. 5 Surface roughness during CT, UVAT, LAT, and UVLAT

radial direction, the cutting tool may have suffered higher damage as it repeatedly hits on the workpiece surface which leads to higher surface roughness.

4 Conclusions

In the current study, a new machining method of ultrasonic vibration laser-assisted turning (UVLAT) was proposed by the simultaneous interaction of UVAT and LAT. The proposed method was compared with CT, UVAT, and LAT by carrying out experiments on aluminum alloy. Machining forces and surface roughness were investigated during CT, UVAT, LAT, and UVLAT process by varying laser power.

Cutting, radial, and feed forces were found to be reduced in UVLAT compared to CT, UVAT, and LAT. The cutting, radial, and feed forces were found to be reduced with an increase in laser power for both LAT and UVLAT. However, higher surface roughness was obtained in UVAT and UVLAT when compared with CT and LAT. Apart from that, surface roughness was found to be reduced with an increase in laser power during LAT and UVLAT.

UVLAT can be a better machining process compared to CT, UVAT, and LAT if the direction of vibration is appropriately applied to the cutting tool. Additionally, the laser heating method is better compared to other heating methods due to its localized, rapid, and controlled heating at the workpiece surface, less heat-affected zone, noise-free operation, low power requirement, and easy installation. Therefore, it can be concluded that UVLAT is an effective method for enhancing the machinability of metals and alloys.

Acknowledgements The laser system used for this study was established through the Department of Science and Technology, India under project number DST/TDT/AMT/2017/026.

References

1. Sharma V, Pandey PM (2016) Recent advances in ultrasonic assisted turning: a step towards sustainability. Cogent Eng 3:1222776
2. Dixit US et al (2019) Ultrasonic-assisted machining processes: a review. Int J Mechatron Manuf Syst 12:227–254
3. Kim J, Lee E (1996) A study of ultrasonic vibration cutting of carbon fibre reinforced plastics. Int J Adv Manuf Technol 12:78–86
4. Nath C et al (2007) A study on ultrasonic vibration cutting of low alloy steel. J Mater Process Technol 192–193:159–165
5. Patil S et al (2014) Modelling and simulation of effect of ultrasonic vibrations on machining of Ti6Al4V. Ultrasonics 54:694–705
6. Khajehzadeh M et al (2020) Finite element simulation and experimental investigation of residual stresses in ultrasonic assisted turning. Ultrasonics 108:106208
7. Khanna N et al (2020) Inconel 718 machining performance evaluation using indigenously developed hybrid machining facilities: experimental investigation and sustainability assessment. Int J Adv Manuf Technol 106:4987–4999
8. Muhammad R et al (2018) Hybrid machining process: experimental and numerical analysis of hot ultrasonically assisted turning. Int J Adv Manuf Technol 97:2173–2192
9. Sofuoğlu MA et al (2018) Experimental investigation of machining characteristics and chatter stability for Hastelloy-X with ultrasonic and hot turning. Int J Adv Manuf Technol 95:83–97
10. Kalantari O et al (2021) Surface integrity in laser-assisted machining of Ti6Al4V. Proc Inst Mech Eng C J Mech Eng Sci 235:5009–5016
11. Sun S et al (2010) Thermally enhanced machining of hard-to-machine materials—a review. Int J Mach Tools Manuf 50:663–680
12. Tadavani AS et al (2017) Pulsed laser-assisted machining of Inconel 718 superalloy. Opt Laser Technol 87:72–78
13. Nadim N et al (2019) Preheating and thermal behaviour of a rotating cylindrical workpiece in laser-assisted machining. Proc Inst Mech Eng Part B J Eng Manuf 234:559–570
14. Wang Y et al (2002) An investigation of laser-assisted machining of Al2O3 particle reinforced aluminum matrix composite. J Mater Process Technol 129:268–272
15. Skvarenina S, Shin YC (2006) Laser-assisted machining of compacted graphite iron. Int J Mach Tools Manuf 46:7–17
16. Venkatesan K (2017) The study on the force, surface integrity, tool life, and chip on laser-assisted machining of Inconel 718 using Nd: YAG laser source. J Adv Res 8:407–423
17. Song H et al (2019) Experimental analysis and evaluation of the cutting performance of tools in laser-assisted machining of fused silica. Precis Eng 56:191–202
18. Zhai C et al (2020) The study on surface integrity on the laser-assisted turning of SiCp/2024Al. Int J Optomechatronics 14:29–43
19. Nath C, Rahman M (2008) Effect of machining parameters in ultrasonic vibration cutting. Int J Mach Tools Manuf 48:965–974

Advancement of Electrochemical Discharge Machining Process—A Review

S. Santra, B. R. Sarkar, B. Doloi, and B. Bhattacharyya

Abstract Electrochemical discharge machining (ECDM) process, which is referred to as electrochemical spark machining, was known to the world due to the invention by Kurafuji and Suda in 1968. Research papers of the last few years have been reviewed to comprehend the process mechanism and its development up to the milling operation on electrically non-conductive materials. The basic mechanism and development of the ECDM process in terms of gas film stabilization, enhancement of MRR, and improvement of surface quality and dimensional accuracy have been discussed in this paper. The application of electrochemical discharge for milling operation and its development has also been incorporated in this paper. Further, the effects of the use of various electrolytes, tool electrode materials, tool electrode shape and application of vibration, magnetic field as well as the addition of various extra features to the basic process and the future scope of research have been discussed.

Keywords Electrochemical discharge machining · MRR · Microchannel · Micro-milling

1 Introduction

Machining is an essential manufacturing process toward the development of technology and its applications in various fields. Conventional machining ruled the field of manufacturing since stone from stone weapons to the development of CNC and artificial intelligence. Due to the presence of new and advanced materials, the development of machining technologies is also improving. In the age of non-traditional machining, there are a lot of processes, which have been used for machining hard, brittle, and difficult to machine materials like glass, ceramics, metal matrix, etc. Further, the development of a micro-electro-mechanical system (MEMS) pushed forward the micro-machining as well as non-traditional machining processes. Among

S. Santra (✉) · B. R. Sarkar · B. Doloi · B. Bhattacharyya
Production Engineering Department, Jadavpur University, Kolkata 700032, India
e-mail: sudip16santra@gmail.com

U. S. Dixit et al. (eds.), *Advances in Forming, Machining and Automation*, Lecture Notes in Mechanical Engineering, https://doi.org/10.1007/978-981-19-3866-5_15

various kinds of non-traditional machining, there are some thermal, chemical, electrochemical, and mechanical machining processes depending on their material removal mechanisms. Every non-traditional machining process consists of advantages and limitations. The electrical discharge machining process can machine electrically conducting materials with a less hard tool than a workpiece. This process can be used to machine hard materials through the formation of recast layer, HAZ, and other thermal effects on the machined surface is the major drawbacks of this process. The electrically non-conducting materials cannot be machined by the EDM process [1]. Electrochemical machining processes are effective during machining of electrically conductive materials and surface finishing, deburring as well as micropattern can be developed through this process but only in electrically conductive materials [2]. The hybrid machining processes were introduced to overcome the limitations of the separate process. The addition of ultrasonic vibration, powder additives in the EDM process improved the surface finish and material removal rate in the EDM process [3]. Electrochemical discharge machining overcame the limitations present in ECM and EDM separately and was able to machine electrically non-conductive materials. ECDM process proved its effectiveness for machining alumina, borosilicate glass, etc. [4]. The process of electrochemical discharge was first discovered in Japan by Kurafuji and Suda though they did not use the term ECDM rather they proposed it as electrical discharge drilling [5]. After the 1990s lot of works were carried out to understand the process and its usability in the machining of electrically non-conductive materials like glass, ceramics, quartz, etc. [6]. Jain et al. used the term electrochemical spark machining (ECSM) in 1990 and developed a second-order model of responses using the CADEAG-1 software package [7]. In 1996, Singh et al. used cathode wire to machine partially conductive materials and found that material removal rate increased with the increase of supply voltage and electrolyte concentration within the range of machine conditions [8]. Research on the mechanism of spark generation was first carried out by Basak and Ghosh in 1996 [9, 10]. They developed a theoretical model and experimented to verify the model. Jain et al. in 1999 prepared a model for material removal rate by finite element modeling using 3D unsteady state heat conduction problem and observed that the computed MRR was higher than the experimental results [11].

2 Fundamentals of ECDM

Since the first invention of electrochemical discharge machining, researchers around the world have tried to understand the fundamental concept or behavior of the process. Till now it has not been fully understood phenomenon as the removal of materials, gas film formation, or like such things are used to happen during machining which is uncontrollable. It has been observed from the development of ECDM that material removal took place due to the combined effects of electrochemical reaction and electrical spark discharge [12]. During the investigation on non-conducting ceramics materials, Bhattacharyya et al. analyzed the electrochemical reaction that occurred when voltage was applied between the two electrodes immersed into the electrolyte [13]. In ECDM, two electrodes are used which are immersed into the electrolyte and among them, one electrode is a cathode generally used as a tool and another electrode is an anode used as an auxiliary electrode. The size of the auxiliary electrode is much bigger than the tool electrode. The electrochemical reaction that happens in the system can be divided into two parts like reaction on cathode where the evolution of hydrogen gas occurs due to evaporation of electrolytes and reaction on anode where the evolution of oxygen gas occurs.

Reaction at cathode

$$2H^{+} + 2e^{-} \rightarrow H_2 \uparrow \text{ (in acidic electrolytic solution)} \quad (1)$$

$$2H_2O + 2e^{-} \rightarrow 2(OH)^{-} + H_2 \uparrow \text{ (in alkaline solution)} \quad (2)$$

Reaction at anode

$$2H_2 \rightarrow O \rightarrow O_2 \uparrow +4H^{+} + 4e^{-} \text{ (in acidic solution)} \quad (3)$$

$$4(OH)^{-} \rightarrow 2H_2O + O_2 \uparrow +4e^{-} \text{ (in alkaline solution)} \quad (4)$$

Figure 1 shows the schematic diagram of the experimental setup of the electrochemical discharge machining process.

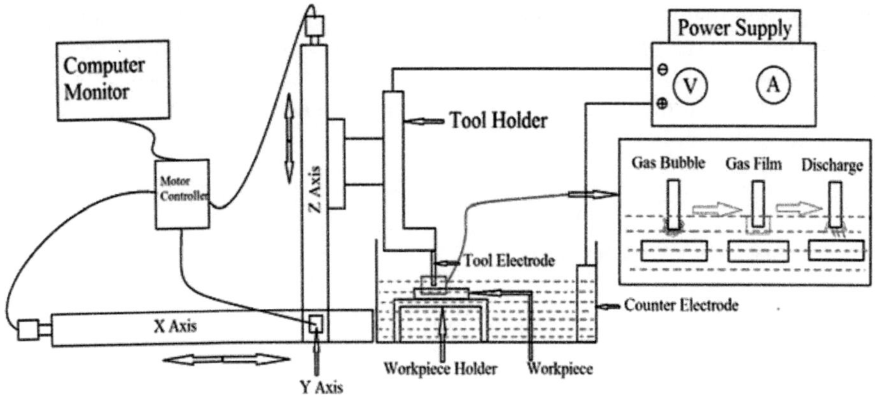

Fig. 1 Schematic diagram of the experimental setup of ECDM process

2.1 Fishbone Diagram

The size of the auxiliary electrode has to be around 100 times larger than the tool electrode. During the application of the voltage between two electrodes, the hydrogen gas bubbles generated by the evaporation of electrolytes accumulate around the tool electrode or cathode. These hydrogen gas bubbles are used to form gas film to cover the open area to the electrode with the increase of voltage. At a certain critical voltage, this gas film collapsed and discharge can be seen in the form of light emission. This discharge did not occur between the two electrodes rather it occurred between cathode and electrolyte [14]. Any electrically non-conductive materials placed in that discharge area are exposed to high thermal energy which melts and vaporized the materials from the workpiece. During the drilling of glass, Wüthrich et al. have increased material removal rate by twice than without vibration [15]. Some researchers have claimed that many phenomena can be monitored like local electrolyte temperature, presence, and absence of gas film by analyzing the current signal during the process [16]. Gas film thickness can be changed by improving the wettability of the tool in the electrolyte [17]. In direct polarity, material removal takes place by melting and vaporization, whereas in reverse polarity chemical reactions also happen with melting and vaporization. Although in reverse polarity overcut, tool wear and surface roughness are higher than direct polarity [18]. The thermal conductivity of the tool can also affect material removal. Mousa et al. [19] found that a tool with higher thermal conductivity can increase the material removal process due to its ability to vaporize more electrolytes in the discharge regime whether hydrodynamic regime material removal decreases due to higher viscosity of molten workpiece. Besides those, many research works have been done and till now development of this process is going on to understand the actual phenomenon that happens during electrochemical discharge and to stabilize it for commercial applications. Figure 2 shows the different effective factors which can influence the ECDM process, and Fig. 3 shows the application of ECDM in different fields.

3 Advancement in ECDM

3.1 Control of Gas Bubbles and Gas Film

Surface active agent or surfactant has the property to reduce the surface tension of the system. Surfactant SDS and CTAB with 25 wt% NaOH and 25 wt% KOH solution can decrease surface tension up to Critical Micelle Concentration (CMC). Due to decreased value of surface tension, the thin gas film produced as a large bubble is unable to be stuck with the tool electrode [20]. Lin et al. have added ethanol to the electrolyte which helped to reduce the angle between electrodes and as bubble, and magnetic field organized the charged particle due to Lorentz force and stabilized the generated flow around the electrode. Besides that minimum level of electrolyte,

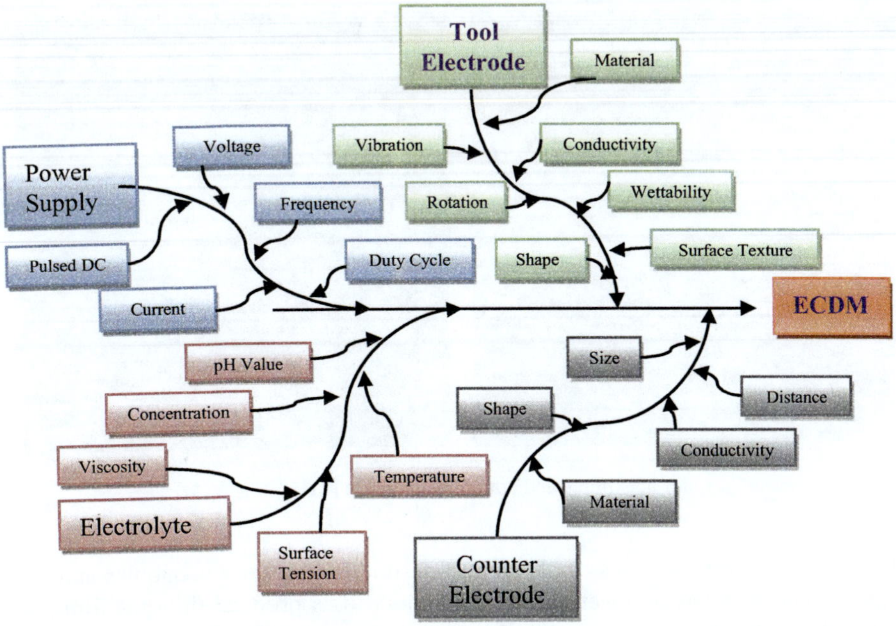

Fig. 2 Cause and effect of different factors in ECDM

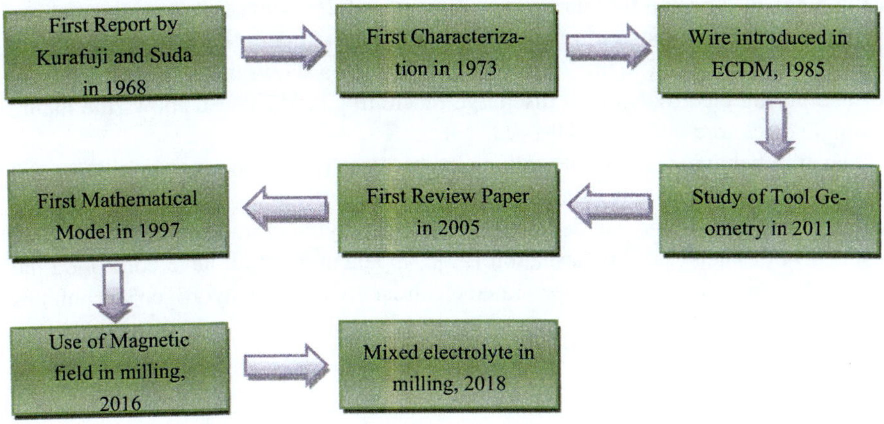

Fig. 3 Development of ECDM

the minimum concentration of electrolyte and maximum possible distance between electrodes is suitable for stable gas film formation. Figure 4 shows the generated gas film under extreme and optimum conditions [21].

As the buoyancy force is a problem in a high level of electrolyte and more concentration of electrolyte accelerates the electrochemical process which enhanced the instability of gas film due to increased buoyant force. Arab et al. have documented

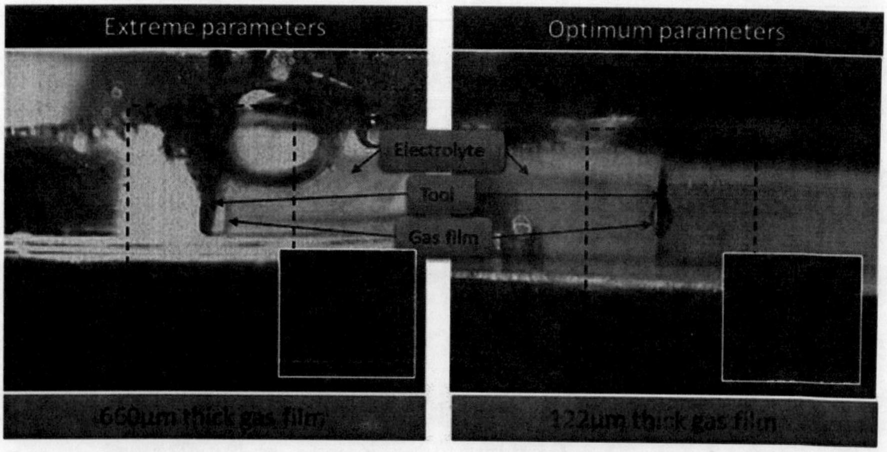

Fig. 4 Generated gas film in extreme and optimum condition [21]

that higher surface roughness tools are responsible for thicker gas bubbles and gas film thickness, whereas lower surface roughness tools produce thin gas film and lower HAZ [22]. Higher surface roughness tool can capture bigger gas bubbles due to the presence of dipper valley which produced the thicker gas film. Behroozfar et al. have reported that in 30 wt% NaOH solution with 35 V applied voltage and using 500 μm tungsten tool an average diameter of 260 μm of plasma channel can be obtained [12]. Assisted ultrasonic vibration of 10 μm amplitude to the tool electrode has reduced gas film thickness by 65%. Figure 5a shows the craters which were formed during electrochemical discharge machining, and Fig. 5b shows the nature of single discharge during ECDM [23, 24].

Gas film behavior is not only influenced by electrolyte, tool, or vibration in tool electrodes but also closely related to the gas bubble evaluation. The balance force acting on the gas bubble can be manipulated by applying extra Lorentz force which resulted in thinner gas film formation [25]. Appalanaidu et al. have controlled the gas film shape with the help of generated electrostatic force on hydrogen gas bubbles [26].

3.2 *Enhancement of MRR*

The application of ultrasonic vibration of 10 μm amplitude has resulted in 5.48 μg of material removal which is greater than without vibration. It has also been observed that a large amount of amplitude destroyed the effect of vibration [24]. Due to the reduction of gas film thickness through ultrasonic vibration, the discharge activities affect more than without vibration which leads to improvement of MRR [23]. Square

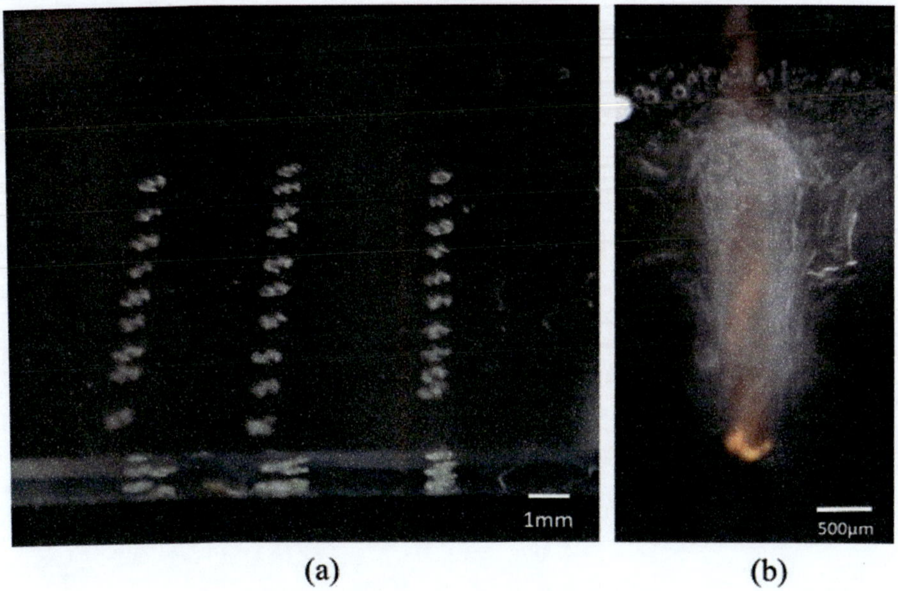

Fig. 5 Craters on a workpiece (**a**) and single discharge (**b**) [23]

waveform voltage to the actuator has increased MRR by 40% in the case of a cylindrical rod and 20% increase of MRR found when sinusoidal waveform applied [27]. Bellubbi et al. have observed that higher electrolyte concentration and machining time can increase MRR [28]. The electrolyte is a very important factor of ECDM because its properties like conductivity, viscosity, density, and temperature influence the gas film making process which leads to proper machining. A mixer of NaOH and KOH has provided more electrical conductivity than separate NaOH and KOH. The depth of the microchannel increased by 19% in the case of 25 wt% mixed electrolyte (NaOH + KOH) than 25 wt% KOH and increased by 31% than 25 wt% NaOH. Figure 6 shows the comparison of current signal among KOH, NaOH, and mixed electrolyte and visible that in case of mixed electrolyte of NaOH, KOH provides more current than other two [29].

By dispersing nanoparticles, Elhami et al. have improved hole depth improved by 21.1% for Al_2O_3 and 18.7% for Cu nanoparticles [30]. Application of electrolyte injection provides sufficient electrolyte even at a deeper hole to maintain the material removal mechanism. It also provides drag forces to detach smaller diameter bubbles from the cathode surface. It has been observed that the use of electrolyte injection increases the hole depth by 70% with a 1.7 mm diameter tool electrode [31]. Researchers also reported that continuous replenishment of vaporized electrolytes at the machining zone stabilized the discharging process. With this stable process, a noticeable amount of increase in hole depth was found [32]. The spherical head tool electrode made less contact with the workpiece and allows the electrolyte to flow in the machining zone which produce stable gas film and lead to an

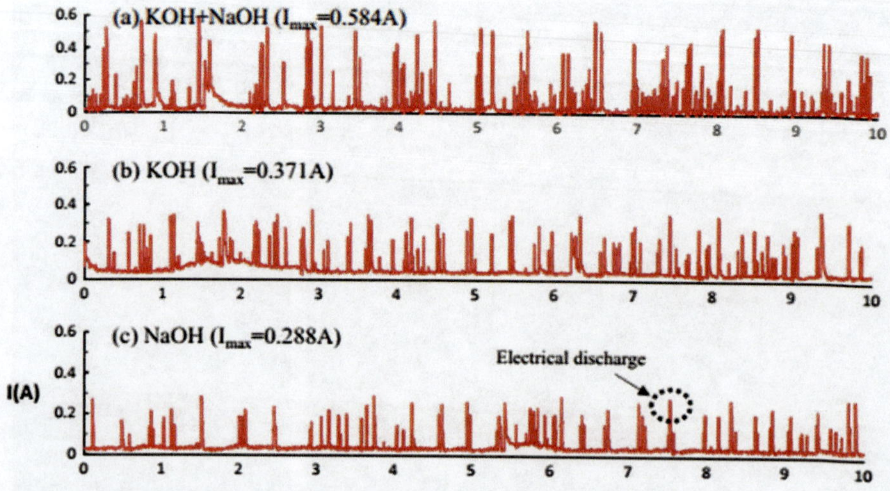

Fig. 6 Comparison of current signal between individual and mixed electrolytes [29]

increase in the machining depth by increasing MRR [33]. Ratan et al. have examined and observed that the presence of a magnetic field in the ECDM process generates Lorentz force onto the ion and helps them to move which creates magnetohydrodynamic convection. Due to MHD, convection electrolytes are used to flow through the small distance between electrodes and workpiece and more frequent sparking occurs which increases the MRR [34].

3.3 Improvement of Surface Quality and Dimensional Accuracy

The optimum tool-workpiece gap has a great impact on the surface quality and dimensional accuracy [35]. Experiments have shown that lower viscosity and higher ion mobility are the two major factors for a better quality of through-hole. In the case of KOH, the quality of the through-hole improved a lot [35]. Chenxiang et al. have found more tapering effect and bigger size hole diameter at higher conductivity solution of $NaNO_3$ [36]. To get more energy from electrochemical discharge, the concentration of electrolyte needs to be increased but at higher concentrations, the machined surface is damaged due to higher thermal conductivity. The mobility and thermal conductivities of potassium ions are greater than sodium ions, but the mixture of KOH and NaOH at 15 and 25% give higher thermal conductivity than separate KOH and NaOH which provide higher depth and better-machined quality [29]. A greater number of $(OH)^-$ ion in the electrolyte have reduced the surface tension and produced the thinner gas film. The presence of $(OH)^-$ ion also enhanced the chemical etching process and improves the quality of machined surfaces [37]. The mobility and

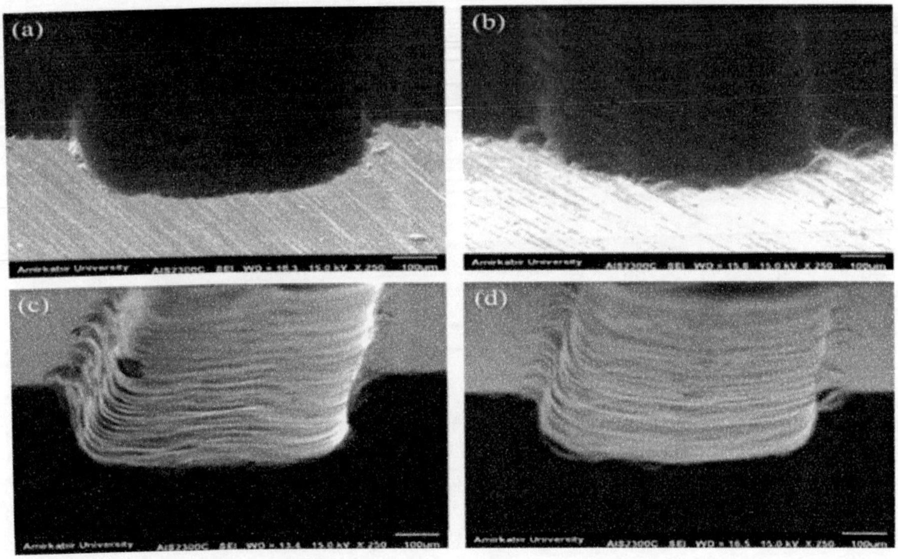

Fig. 7 Different machined surfaces at different electrolyte concentrations of 10–30% NaOH [38]

electrical conductivities of potassium ions are larger than sodium ions which helps the potassium ion to generate more power than sodium ion and consequently heat-affected zone is higher in KOH solution. Figure 7 shows different surface qualities at different electrolyte concentrations [38].

Lower tool electrode feed rate provides more interaction with the workpiece and time for the flow of electrolyte at the machining zone. It has been found by researchers that tool feed rates higher than 5 μm/s produced nonuniform electrochemical discharge. If the tool feed rate is lower than 5 μm/s, then uniform electrochemical discharge can be obtained which resulted in a smaller size of top hole and bottom hole diameter which was 300 μm and 70 μm, respectively at 3–4 μm/s feed rate [39].

3.4 MicroChannel and Milling with Electrochemical Discharge

The research work in the field of channeling and milling is very less compared to the drilling process. Some researchers experimented to find out the capability of ECDM in channeling and milling. Using surfactant in electrolytes, the surface quality of microchannels was improved due to the increase in the viscosity of electrolyte and decrease in the thermal conductivity [20]. Sabahi et al. have increased the depth of microchannel in case of a mixed electrolyte by 19% than KOH and 31% than NaOH solution, and the overcut values for NaOH, KOH, and mixed electrolyte

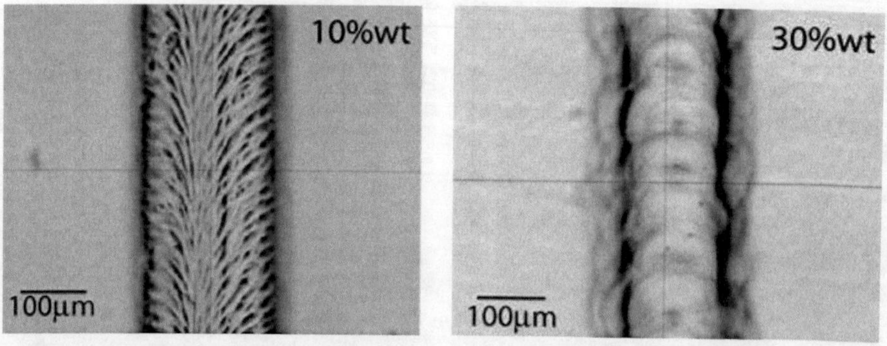

(a) Featherly like texture at 10wt% NaOH (b) Spongy like texture at 30% NaOH

Fig. 8 Texture of machined surfaces at different electrolyte concentrations. **a** Featherly like texture at 10 wt% NaOH and **b** spongy-like texture at 30% NaOH [40]

are 655 μm, 638 μm, and 541 μm, respectively [29]. Electrolyte concentration has a prominent effect on the channel and milling process done by electrochemical discharge machining. Jana et al. observed that at 10wt% NaOH the surface of the machining zone became featherly like texture while by increasing the electrolyte concentration from 10 to 20 to 30 wt%; the featherly like texture became spongy-like texture. Figure 8a shows featherly like texture at 10 wt% NaOH, and Fig. 8b shows the spongy-like texture at 30 wt% NaOH solution [40].

Electrolyte temperature increases the energy of ions at the electrolyte surface which leads to more powerful discharge and a faster channeling process [41]. Higher electrolyte concentration increases the MRR by decreasing inter-electrode resistance but at higher electrolyte concentration, the sludge gets accumulated and side sparking increases [42]. The textured tool has resulted in an increase in MRR and depth of microchannel by 19.27% and 64.81%, respectively. In the case of microchannel, if the tool feed rate increased, then the channel depth decreased such as at lower feed rate channel depth was obtained 23 μm which was 11 μm in case of higher speed [40]. Low feed rate leads to longer interaction time between tool and workpiece which opens up the opportunity for chemical action and increased MRR [42, 43]. The large distance between tool and workpiece increased the kinetic energy as it travels more distance to collide with one another which deteriorates surface quality and reduces the area of groove [44, 45]. The use of the magnetic field in the micro-milling of glass by electrochemical discharge machining was very helpful as stated by Hajian et al. [46]. They published that at lower electrolyte concentration, the viscosity was lower which helped to remove the etched glass from the machining zone. The Lorentz force generated by the magnet controls the motion of gas bubbles at the inter-electrolyte area which improved the MRR. It has been observed that with the increase in applied voltage and electrolyte concentration, the tool feed rate was greatly increased [46, 47]. Another study has shown that by increasing electrolyte concentration, the channel depth can be increased but the surface quality will deteriorate in case of higher electrolyte concentration [47].

4 Significance

Machining a material with the help of non-conventional machining processes like electric discharge machining (EDM), electrochemical machining (ECM), abrasive water jet machining (AWJM), laser beam machining (LBM), etc., can be done but with certain limitations. Electrically non-conducting materials cannot be machined with ECM and EDM. Miniature product is difficult to fabricate with WAJM as overcut is a major problem here and skillful operators are required to operate the machine with the available setup. Transparent materials cannot be easily machined with the help of LBM. These disadvantages of the machining process enhance the demand for micro-electrochemical discharge machining process since conductive materials as well as electrically non-conductive, which are very difficult to machine can be easily machined by ECDM. No isolated chamber is required for the setup. No mask is needed and just as in traditional machining, the desired structure can be machined directly. Electrochemical discharge machining (ECDM) can machine materials irrespective of the electrical and mechanical properties with small thermal effect, i.e., formation of heat-affected zone of workpiece materials, and the area of machining zone can be varied from several millimeters to microns without large force as found in all common conventional machining processes.

5 Summary

After reviewing the research papers on the ECDM process, it has been observed that the process is very complex and uncertain. Lots of phenomena happen during the machining process, and most of the researchers agreed that electrochemical and electrical discharge machining process both occurs in the ECDM process. In this machining process, if a high-temperature etching process happens during machining, then the quality of machining or the machined surface became smooth surface. During the milling tool-workpiece distance and tool-workpiece interaction, time plays a vital role in the smooth milling surface. The addition of surfactant to the electrolyte reduces the surface tension which provides thinner gas film besides that adding vibration to the tool electrode also provides thinner gas film. Different types of tool electrodes, electrolytes, and the addition of magnetic field, vibration may improve the total machining efficiency. From the above discussions, the following conclusions can be made.

- The use of surfactant reduces the surface tension of the medium and the addition of vibration to the tool electrode leads to thinner gas film, and thinner gas film can produce a smooth surface. Using surfactant in electrolyte, the surface quality of microchannels was improved due to the increase in the viscosity of electrolyte and decrease in the thermal conductivity.

- The evaluation of gas bubbles is also an important factor that can be triggered by using a magnetic field to the system where the Lorentz force helps to organize the generated gas bubbles.
- Higher surface roughness tools produce bigger bubbles due to the presence of peak and valley, and the minimum electrolyte concentration and maximum inter-electrode distance may produce the suitable gas film.
- A mixed alkaline solution can produce a better microchannel in glass materials.
- Lower tool electrode feeding provides more time for tool-workpiece interaction which leads to high-temperature etching and smooth machined surface.

6 Future Scope

The complex phenomenon of ECDM is not still well understood. Drilling of non-conductive materials like glass, ceramics, quartz, etc., has been carried out by applying different techniques. Still lot of research is required to make the process commercialized. Very few research papers are available regarding channeling or milling. The process is highly dependent on various factors like workpiece materials, tool electrodes, power supply, electrolytes properties, tool feed rate, etc. So, many factors need to optimize properly to get smooth and accurate machining.

Control of gas film:

- Analysis of electrolyte properties like concentration, viscosity, surface tension by addition of abrasives, mixed electrolyte, surfactant, etc.
- Application of vibration and use of a lower surface textured tool to control the size of the gas bubble.
- Application of magnetic field to organize the gas bubble.

Enhancement of MRR:

- The continuous flow of electrolytes or application of proper flushing techniques.
- Use of appropriate tool electrodes so that availability of electrolyte can make at the machining zone.
- Analysis of voltage application needed at the different conditions to increase MRR without compromising the machining quality.

Improvement of machined quality:

- Generation of the lower gas film reduced the HAZ and tapering effect and the gas film can be reduced by different combinations of effective parameters.

Continuous maintaining of machining gap and tool rotation can also provide a smooth machined surface.

References

1. Ho KH, Newman ST (2003) State of the art electrical discharge machining (EDM). Int J Mach Tools Manuf 43(13):1287–1300
2. Bhattacharyya B, Munda J, Malapati M (2004) Advancement in electrochemical micro-machining. Int J Mach Tools Manuf 44(15):1577–1589
3. Abbas NM, Solomon DG, Bahari MF (2007) A review on current research trends in electrical discharge machining (EDM). Int J Mach Tools Manuf 47(7–8):1214–1228
4. Jain VK, Choudhury SK, Ramesh KM (2002) On the machining of alumina and glass. Int J Mach Tools Manuf 42(11):1269–1276
5. Kurafuji H, Suda K (1968) Electrical discharge drilling of glass. Ann CIRP 16:415–419
6. Gouda M, Sharmaa AK, Jawalkarb C (2016) A review on material removal mechanism in electrochemical discharge machining (ECDM) and possibilities to enhance the material removal rate. Precis Eng 45:1–17
7. Jain VK, Tandon S, Kumar P (1990) Experimental investigations into electrochemical spark machining of composites. J Eng Ind 112(2):194–197
8. Singh YP, Jain VK, Kumar P, Agrawal DC (1996) Machining piezoelectric (PZT) ceramics using an electrochemical spark machining (ECSM) process. J Mater Process Technol 58(1):24–31
9. Basak I, Ghosh A (1996) Mechanism of spark generation during electrochemical discharge machining a theoretical model and experimental verification. J Mater Process Technol 62(1–3):46–53
10. Basak I, Ghosh A (1997) Mechanism of material removal in electrochemical discharge machining a theoretical model and experimental verification. J Mater Process Technol 71(3):350–359
11. Jain VK, Dixit PM, Pandey PM (1999) On the analysis of the electrochemical spark machining process. Int J Mach Tools Manuf 39(1):165–186
12. Behroozfar A, Razfar MR (2016) Experimental and numerical study of material removal in electrochemical discharge machining (ECDM). Mater Manuf Process 31(4):495–503
13. Bhattacharyya B, Doloi BN, Sorkhel SK (1999) Experimental investigations into electrochemical discharge machining (ECDM) of non-conductive ceramic materials. J Mater Process Technol 95:145–154
14. Wüthrich R, Fascio V (2005) Machining of non-conducting materials using electrochemical discharge phenomenon—an overview. Int J Mach Tools Manuf 45(9):1095–1108
15. Wüthrich R, Despont B, Maillard P, Bleuler H (2006) Improving the material removal rate in spark-assisted chemical engraving (SACE) gravity-feed micro-hole drilling by tool vibration. J Micromech Micro Eng 16(11):N28–N31
16. Wüthrich R, Spaelter U, Bleuler H (2006) The current signal in spark-assisted chemical engraving (SACE) what does it tell us. J Micromech Micro Eng 16(4):779–785
17. Wüthrich R, Hof LA (2006) The gas film in spark assisted chemical engraving (SACE)—A key element for micro-machining applications. Int J Mach Tools Manuf 46(7–8):828–835
18. Jain VK, Adhikary S (2008) On the mechanism of material removal in electrochemical spark machining of quartz under different polarity conditions. J Mater Process Technol 200(1–3):460–470
19. Mousa M, Allagui A, Ng HD, Wüthrich R (2009) The effect of thermal conductivity of the tool electrode in spark-assisted chemical engraving gravity-feed micro-drilling. J Micromech Micro Eng 19(1):1–7
20. Sabahi N, Razfar MR, Hajian M (2017) Experimental investigation of surfactant-mixed electrolyte into electrochemical discharge machining (ECDM) process. J Mater Process Technol 250:190–202
21. Kolhekar KR, Sundaram M (2018) Study of gas film characterization and its effect in electrochemical discharge machining. Precis Eng 53:203–211
22. Arab J, Kannojia HK, Dixit P (2019) Effect of tool electrode roughness on the geometric characteristics of through-holes formed by ECDM. Precis Eng 60:437–447

23. Elhami S, Razfar MR (2017) Analytical and experimental study on the integration of ultrasonically vibrated tool into the micro electro-chemical discharge drilling. Precis Eng 47:424–433
24. Elhami S, Razfar MR (2018) Effect of ultrasonic vibration on the single discharge of electrochemical discharge machining. Mater Manuf Process 33(4):444–451
25. Xu Y, Chen J, Jiang B, Liu Y, Ni J (2018) Experimental investigation of magnetohydrodynamic effect in electrochemical discharge machining. Int J Mech Sci 142–143:86–96
26. Appalanaidu B, Dvivedi A (2020) On controlling of gas film shape in electrochemical discharge machining process for fabrication of elliptical holes. Mater Manuf Processes 36(5):558–571
27. Razfar MR, Behroozfar A, Ni J (2014) Study of the effects of tool longitudinal oscillation on the machining speed of electrochemical discharge drilling of glass. Precis Eng 38:885–892
28. Bellubbi S, Naik R, Sathisha N (2021) An experimental study of process parameters on material removal rate in ECDM process. Mater Today: Proc 35(3):298–302
29. Sabahi N, Razfar MR (2018) Investigating the effect of mixed alkaline electrolyte (NaOH + KOH) on the improvement of machining efficiency in 2D electrochemical discharge machining (ECDM). Int J Adv Manuf Technol 95:643–657
30. Elhami S, Razfar MR (2020) Application of nano electrolyte in the electrochemical discharge machining process. Precis Eng 64:34–44
31. Mehrabi F, Farahnakian M, Elhami S, Razfar MR (2018) Application of electrolyte injection to the electro-chemical discharge machining (ECDM) on the optical glass. J Mater Process Tech 255:665–672
32. Arya RK, Dvivedi A (2019) Investigations on quantification and replenishment of vaporized electrolyte during deep micro-holes drilling using pressurized flow-ECDM process. J Mater Process Tech 266:217–229
33. Yang CK, Wu KL, Hung JC, Lee SM, Lin JC, Yan BH (2011) Enhancement of ECDM efficiency and accuracy by spherical tool electrode. Int J Mach Tools Manuf 51(6):528–535
34. Ratan N, Mulik RS (2017) Improvement in material removal rate (MRR) using magnetic field in TW-ECSM process. Mater Manuf Process 32(1):101–107
35. Arab J, Mishra DK, Dixit P (2021) Measurement and analysis of the geometric characteristics of microholes and tool wear for varying tool-workpiece gaps in electrochemical discharge drilling. Mesurement 168:108463
36. Arab J, Mishra DK, Kannojia HK, Adhale P, Dixit P (2019) Fabrication of multiple through-holes in non-conductive materials by electrochemical discharge machining for RF MEMS packaging. J Mater Process Tech 271:542–553
37. Zhang C, Xu Z, Hang Y, Xing J (2019) Effect of solution conductivity on tool electrode wear in electrochemical discharge drilling of nickel-based alloy. Int J Adv Manuf Technol 103(1–4):743–756
38. Yang CP, Wu KL, Mai CC, Wang AC, Yang CK, Hsu YS, Yan BH (2010) Study of gas film quality in electrochemical discharge machining. Int J Mach Tools Manuf 50(8):689–697
39. Sabahi N, Hajian M, Razfar MR (2018) Experimental study on the heat-affected zone of glass substrate machined by electrochemical discharge machining (ECDM) process. Int J Adv Manuf Technol 97:1557–1564
40. Arab J, Dixit P (2020) Influence of tool electrode feed rate in the electrochemical discharge drilling of a glass substrate. Mater Manuf Process 35(15):1749–1760
41. AbouZiki JD, Didar TF, Wuthrich R (2012) Micro-texturing channel surfaces on glass with spark assisted chemical engraving. Int J Mach Tools Manuf 57:66–72
42. Torabi A, Razfar MR (2021) The capability of ECDM in creating effective microchannel on the PDMS. Precis Eng 68:10–19
43. Yang CK, Cheng CP, Mai CC, Wang AC, Hung JC, Yan BH (2010) Effect of surface roughness of tool electrode materials in ECDM performance. Int J Mach Tools Manuf 50(12):1088–1096
44. Singh T, Dvivedi A (2018) On performance evaluation of textured tools during micro-channeling with ECDM. J Manuf Process 32:699–713
45. Changjian L, An G, Meng L, shengyi Y (2012) The micro-milling machining of pyrex glass using the electrochemical discharge machining process. Adv Mater Res 403–408:738–742

46. Hajian M, Razfar MR, Movahed S (2016) An experimental study on the effect of magnetic field orientations and electrolyte concentrations on ECDM milling performance of glass. Precis Eng 45:322–331
47. Lin MY, Tsai TH, Hourng LW, Wang WK (2019) The effects of magnetic field and ethanol addition on the electrochemical discharge machining. Int J Adv Manuf Technol 105(5–6):2461–2467
48. Hajian M, Razfar MR, Movahed S, Etefagh AH (2018) Experimental and numerical investigations of machining depth for glass material in electrochemical discharge milling. Precis Eng 51:521–528

An Experimental Comparison Study of Dry and Air-Cooled Turning of AISI 1040 Steel

D. J. Hiran Gabriel, B. Vaishnavi, and M. Parthiban

Abstract The presented work is a comparative study between dry turning and air-cooled turning process of machining AISI 1040 steel. Parameters like feed, cutting speed, and depth of cut were taken in different combinations for defining each experimental run. For the comparison between dry turning and air-cooled turning, parameters such as cutting force, cutting temperature, and surface roughness of the components are considered. The experiments were conducted and results were obtained. From the results, it is evident that air-cooled turning produces machined components with a better surface finish than components obtained during dry turning, especially during higher cutting speeds. Also, air-cooled turning exhibits reduced cutting temperature and cutting force than dry turning for all experimental runs. Thus, air-cooled turning will be a promising and environmentally friendly option.

Keywords AISI 1040 steel · Dry and air-cooled turning · Green machining

1 Introduction

Metal machining is a controlled material removal technique that cuts metal to a specified final form and size. The machining industry has been subjected to various and uninterrupted modifications in technology [1]. Many industries such as automotive, aerospace depend on the metal cutting process for manufacturing of required components. To accomplish a machining process, relative motion between the tool and the work is necessary. Heat builds up in the tool work contact region, causing the temperature to rise to the point where tool damage occurs. Cutting fluids or metalworking

D. J. Hiran Gabriel (✉) · B. Vaishnavi · M. Parthiban
Department of Mechanical Engineering, PSG College of Technology, Coimbatore, Tamil Nadu 641004, India
e-mail: victorrajan1996@gmail.com

B. Vaishnavi
e-mail: bvi.mech@psgtech.ac.in

M. Parthiban
e-mail: mpn.mech@psgtech.ac.in

U. S. Dixit et al. (eds.), *Advances in Forming, Machining and Automation*, Lecture Notes in Mechanical Engineering, https://doi.org/10.1007/978-981-19-3866-5_16

fluids act as carriers to remove the heat produced at the interface of chip tool during machining. Cutting fluids aid in the evacuation of heat from the machining zone. Around 17% of the total production cost is expended on storage, usage, and disposal of these cutting fluids in the manufacturing industry [2, 3]. Metal cutting fluids were formerly made from crude oil or petrochemical extracts. Toxic chemical reagents were added in coolants to attain favorable properties such as lubrication, chemical stability, and corrosion-resistant properties. These hazardous materials lead to health problems while machining and disposal of coolants. Mineral oil-based cutting fluids are prohibited by federal regulations since they contaminate land, air, surface, and groundwater [2–4]. While operating metalworking machines, operators were immediately exposed to the cutting fluid. Many studies have indicated that direct exposure and long-term interaction with these metalworking fluids and their by-products cause skin disorders, respiratory ailments, and even cancer. Thus, manufacturing industries aim to eliminate the usage of cutting fluid and improve machining characteristics of dry machining [5–7].

In contrast to other technologies such as the MQL, wet cooling, and cryogenic cooling dry machining does not need the use of cutting fluid or its delivery system. Dry machining cannot be accomplished simply by turning off the coolant system; it requires a thorough examination of the tool, material, and machining process. Advanced cutting tools, coated inserts, current machining techniques, better tool design, and modern upgraded machine tools might indeed aid to achieve dry machining. Dry machining has several environmental and economic advantages, including shorter cycle times, lower machine tool and production costs, cleaner chips, and almost no risks to the environment or the operator's health [8–10].

The action of the chips rubbing over the top rake face has been proven in several studies to prevent conventional liquid coolant from reaching the tool interface. Thus, there is a need for a sustainable form of coolant that can penetrate the cutting zone. Gases are substances that can penetrate any region other than solids and liquids due to their freely packed molecular structure. Because of the smaller molecular structure of gas, cold air may have a better penetration at the tool contact. Air cooling is a method of heat dissipation that works based on surface area expansion or increase of airflow over the object to be cooled. In all these cases, air that is cooler than the object is needed, from where the heat is to be removed. This concept is basically from the second law of thermodynamics which states that heat will move continuously only from a hot body to a cold body [8, 11, 12].

In this study, the machining performance comparison of air cooling and dry turning on AISI 1040 was carried out. The experimental comparison includes response parameters such as surface roughness, cutting force, and cutting temperature. The provision of air cooling showed satisfactory performance in that range. The investigation proved the effectiveness of air as a cutting fluid. The usage of chilled air in machining industries as a cutting fluid would lead to a sustainable dry machining process.

2 Experimental Design

The experimental run for the turning experiment was designed considering feed (mm/rev), speed (m/min), and depth of cut (mm). The experiments were designed such that the effect of each parameter on machining can be distinguished. Table 1 shows the series of experimental runs designed for comparison.

2.1 Experimental Setup

A cylindrical rod of diameter 50 mm made up of AISI 1040 steel is used as a workpiece for the turning process. The parameters such as cutting force, surface roughness, and temperature were measured and considered for the comparison of conventional dry turning and compressed air-cooled turning processes. The cutting force during the turning process was measured using a Kistler 9257B dynamometer. To measure the surface roughness of the component after machining, Mahr TR100 surface roughness tester of type Mahr Surf GD 25 was used. Amid the many parameters that are measured in the surface roughness tester, roughness average (Ra) is considered for the study. It is the arithmetic mean of surface heights estimated over a surface. Temperature measurement during the turning process was made using a K-type thermocouple.

The complete experimental setup is shown in Fig. 1. For the experiment, a nozzle of 3 mm diameter shown in Fig. 2 is used. The air supply is given from a compressor at a pressure of 5 bar. A stand was designed to hold the nozzle such that the nozzle points toward the cutting zone. The nozzle was maintained at 15 cm from the cutting zone so that the pressurized air diverges into the cutting zone and cools it. Since the compressor cannot maintain the pressure for a longer period, the metal cutting duration is set for 2 min.

Metal cutting experiments were conducted on a Kirloskar Turn master-35 lathe. A cutting tool consisting of a tool holder PSBNR2525M12 by TaeguTec was used

Table 1 Parameters used in the investigation for each run of the experiment

Run	Speed (m/min)	Feed (mm/rev)	Depth of cut (mm)
S1	80	0.10	0.25
S2	80	0.15	0.50
S3	80	0.20	1.00
S4	160	0.10	0.25
S5	160	0.15	0.50
S6	160	0.20	1.00
S7	240	0.10	0.25
S8	240	0.15	0.50
S9	240	0.20	1.00

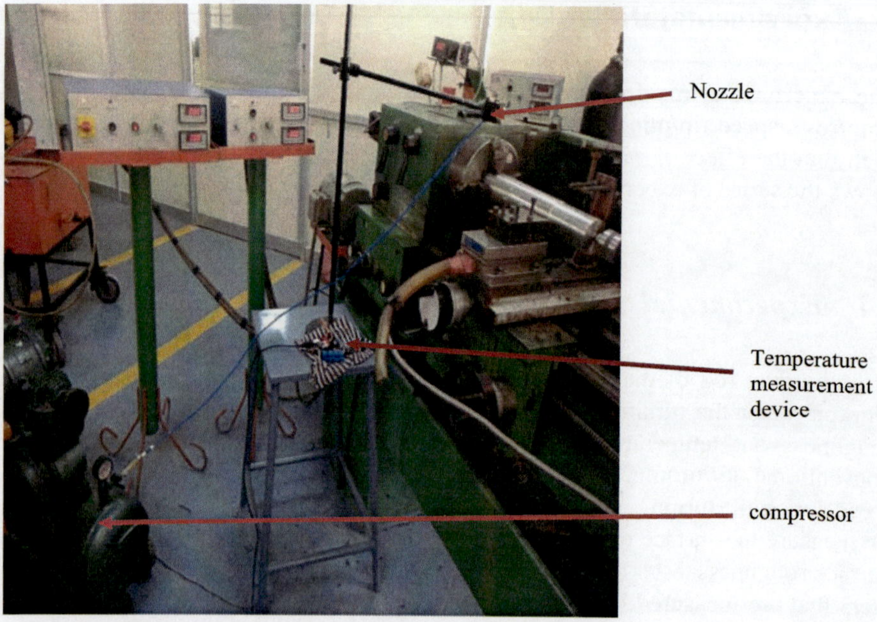

Fig. 1 Experimental setup

Fig. 2 Nozzle setup

in the experiment. Tungsten carbide inserts with a specification SNMG 120408 was mounted on the holder for the experimental purpose. The dynamometer was mounted on the tool post with the tool as shown. A K-type thermocouple was placed in the insert as shown in Fig. 3 such that the temperature of the cutting zone can be measured.

Fig. 3 Thermocouple setup

3 Results and Discussion

The results obtained are tabulated and compared. The comparison includes parameters such as surface roughness, cutting force, and cutting temperature.

From Table 2, it is evident that air-cooled turning is superior to dry turning in terms of all the above-mentioned parameters.

The obtained cutting forces for each experimental run are plotted in the graph as shown in Fig. 4. Reduced cutting forces during the air-cooled turning process show that the performance of cutting has increased due to the application of air in the cutting zone. From Fig. 5, the introduction of air as a coolant in the primary cutting

Table 2 Observation during the experiments

Run	Cutting temperature (°C)		Cutting force (N)		Surface roughness (mm)	
	Dry turning	Air cooling	Dry turning	Air cooling	Dry turning	Air cooling
S1	88	54.00	145.82	140.22	1.724	1.632
S2	96	61.25	131.66	117.42	1.925	1.321
S3	98	76.00	158.29	146.14	3.457	2.984
S4	72	59.75	150.67	140.12	2.765	1.695
S5	85.5	69.25	149.02	145.61	2.556	2.138
S6	97.2	79.75	184.74	160.56	3.145	2.457
S7	83.2	67.25	159.81	152.21	1.734	1.221
S8	89.2	73.50	148.69	140.34	2.747	1.127
S9	96.3	75.25	163.53	150.21	3.117	1.471

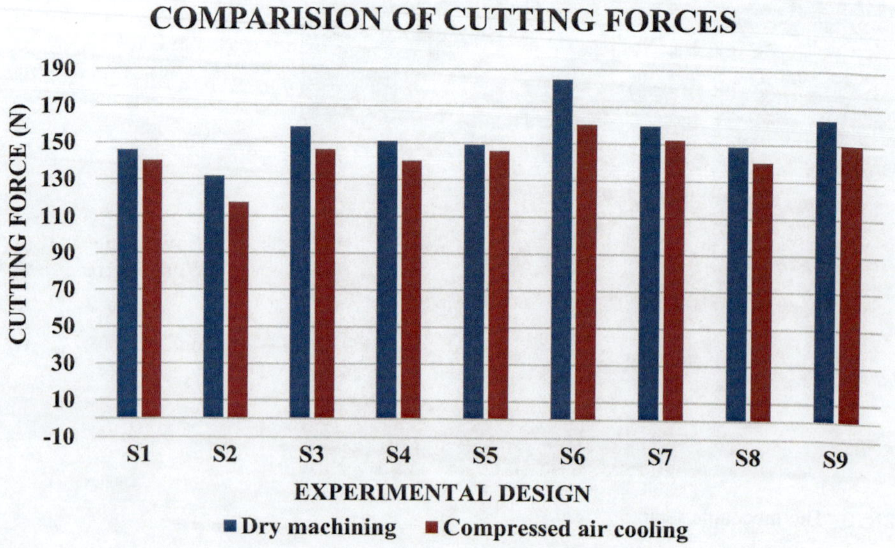

Fig. 4 Comparison of cutting forces

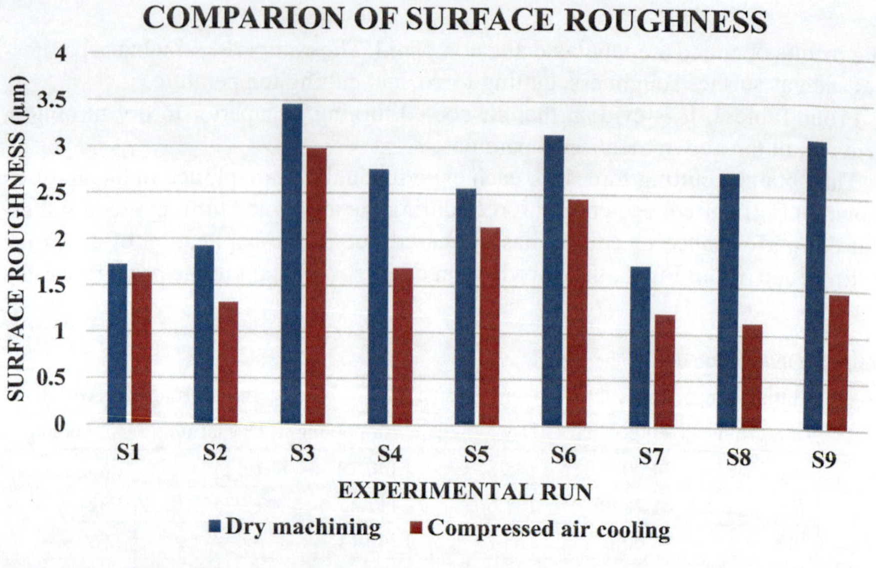

Fig. 5 Comparison of surface roughness

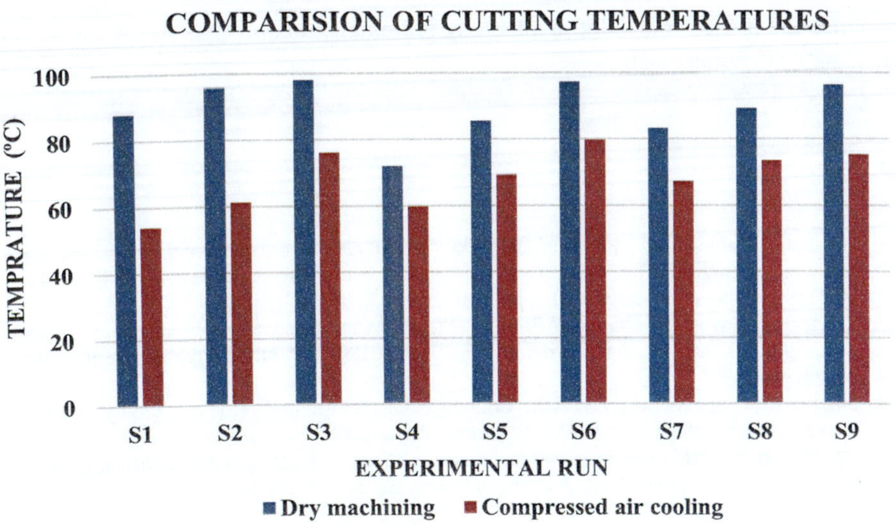

Fig. 6 Comparison of cutting temperatures

zone also improved the finish of the component especially during cutting speeds of 240 m/mi. The results were in accordance with the findings of Tasdelen et al. [12]. It is also observed that during higher speeds of machining, air effectively removes heat from the work material. The surface roughness is improved because of a severe drop in temperature at the metal cutting zone. This can be attributed to well reduced temperature at the cutting zone may lead to slightly affecting the regular shearing process Thus, during higher speeds of turning, air cooling shows increased efficiency in comparison with lower speeds. The temperature of the cutting zone is reduced with the application of air which can be noticed in the comparison shown in Fig. 6. This reduction of temperature aids in the reduction of tool wear, thus enhancing the tool life.

4 Conclusions

For the turning process understudy, a comparative study of parameters such as cutting force, temperature, and surface roughness was made. The comparative study proved that air-cooled turning is superior to dry turning especially at higher speeds of machining. Improved surface finish is obtained during the air-cooled turning. There is a considerable amount of reduction in cutting force and temperatures during air cooling in comparison with the conventional dry turning process. Since air is used as the cooling medium, there is no pollutant involved. Also, air is available freely in the atmosphere and there is no requirement for safekeeping and maintenance of it. Since the process does not involve any pollutants, this process paves a way for sustainable manufacturing.

References

1. Hosseini Tazehkandi A, Pilehvarian F, Davoodi B (2014) Experimental investigation on removing cutting fluid from turning of Inconel 725 with coated carbide tools. J Clean Prod 80:271–281. https://doi.org/10.1016/j.jclepro.2014.05.098
2. Skerlos SJ, Hayes KF, Clarens AF, Zhao F (2008) Current advances in sustainable metal-working fluids research. Int J Sustain Manuf 1(1–2):180–202. https://doi.org/10.1504/IJSM.2008.019233
3. Sharma VS, Dogra M, Suri NM (2009) Cooling techniques for improved productivity in turning. Int J Mach Tools Manuf 49(6):435–453. https://doi.org/10.1016/j.ijmachtools.2008.12.010
4. Hannu T et al (2013) Occupational respiratory and skin diseases among finnish machinists: findings of a large clinical study. Int Arch Occup Environ Health 86(2):189–197. https://doi.org/10.1007/s00420-012-0754-8
5. Koh D, Leow YH, Goh CL (2001) Occupational allergic contact dermatitis in Singapore. Sci Total Environ 270(1–3):97–101. https://doi.org/10.1016/S0048-9697(00)00787-7
6. Marksberry PW (2007) Micro-flood (MF) technology for sustainable manufacturing operations that are coolant less and occupationally friendly. J Clean Prod 15(10):958–971. https://doi.org/10.1016/j.jclepro.2006.01.006
7. Goindi GS, Sarkar P (2017) Dry machining: a step towards sustainable machining–challenges and future directions. J Clean Prod 165:1557–1571. https://doi.org/10.1016/j.jclepro.2017.07.235
8. Carou D, Rubio EM, Lauro CH, Davim JP (2014) Experimental investigation on finish intermittent turning of UNS M11917 magnesium alloy under dry machining. Int J Adv Manuf Technol 75(9–12):1417–1429. https://doi.org/10.1007/s00170-014-6215-7
9. Dasch JM et al (2009) The effect of free-machining elements on dry machining of B319 aluminum alloy. J Mater Process Technol 209(10):4638–4644. https://doi.org/10.1016/j.jmatprotec.2008.11.041
10. Weinert K, Inasaki I, Sutherland JW, Wakabayashi T (2004) Dry machining and minimum quantity lubrication. CIRP Ann Manuf Technol 53(2):511–537. https://doi.org/10.1016/S0007-8506(07)60027-4
11. Sreejith PS, Ngoi BKA (2000) Dry machining: machining of the future. J Mater Process Technol 101(1):287–291. https://doi.org/10.1016/S0924-0136(00)00445-3
12. Hiran Gabriel DJ (2020) Experimental investigation on heat pipe-assisted cooling during milling process of AISI 1040. In: Trends in manufacturing and engineering management. Springer, Singapore

Combined Compromise Solution for Machining Performance Optimization of Modified Polymer Composite

Kuldeep Kumar and Rajesh Kumar Verma

Abstract This paper presents the machining efficiency evaluation of the glass fiber reinforced polymer (GFRP) composites modified by multiwall carbon nanotube (MWCNT). The milling experimentations were performed according to Taguchi-based experimental design. The three distinct machining constraints, such as spindle speed (*S*), feed rate (*F*), depth of cut (*D*), and wt% of MWCNT, are controlled to achieve the desired machining performances. A comparatively new multi-criteria decision-making technique (MCDM) is used by the combined compromise solution (CoCoSo) module for optimal parametric conditions. Analysis of variance (ANOVA) technique on CoCoSo assessment score is applied to determine the variance and adequacy level of variables used during machining of the nanocomposite. The experimental result shows that the depth of cut (*D*) and MWCNT wt% are found to be the most significant input parametric constraints, while spindle speed (*S*) and feed (*F*) obtained to be insignificant. The experimentation also revealed that the proposed CoCoSo optimization approach is a very effective tool for multi-objective optimization of milling parameters.

Keywords Milling · Surface roughness · Cutting force · CoCoSo

1 Introduction

Fiber reinforced polymer (FRP) composites are widely used in various engineering applications, such as aerospace and automotive sports. It provides special qualities of fiber, such as high corrosive resistance and impact properties compared to traditional composites [1, 2]. Glass fibers reinforced polymer (GFRP) composite products as specialized manufacturing materials have risen in automobile, aerospace, building, and sporting goods due to their significant robustness and lightweight application [3, 4], while glass fiber reinforced polymer (GFRP) has served many properties. Still, to some degree of usability, these properties do not reach a sufficient level of

K. Kumar · R. K. Verma (✉)
Materials and Morphology Laboratory, Department of Mechanical Engineering, Madan Mohan Malaviya University of Technology, Gorakhpur 273010, India
e-mail: rkvme@mmmut.ac.in

U. S. Dixit et al. (eds.), *Advances in Forming, Machining and Automation*, Lecture Notes in Mechanical Engineering, https://doi.org/10.1007/978-981-19-3866-5_17

required characteristics, particularly the fragile nature of GFRP. To strengthen the strengthening of GFRP-based composites, some other secondary reinforcement is used to overcome the shortcomings of properties of GFRP to a better extent. The nanofiller-infused matrix provides improved thermal, mechanical, and other properties [5, 6]. Particles produced with nano-level dimensional accuracy from allotropes of carbon can be employed as nanofillers. Multiwall carbon nanotube (MWCNT) has been used in the recent years due to its outstanding thermal and mechanical properties [7, 8]. Several researchers observed that the composite features enhance when considerable MWCNT dispersion in the matrix solution is maintained. The machining operations of fiber composites are one of the most challenging operations. It necessitates real-time applications that require a compact near-net structure with high-dimensional accuracy. Milling is a basic machining operation widely utilized in numerous sectors, most likely for assembly part development and complex structural design-based applications [9, 10]. Milling is primarily an inevitable process, and it generally evades unwanted tolerances during the fabrication of a section part of the assembly design. FRP milling process is distinct from metal matrix composites (MMC) due to a difference in both materials' fabrication methodology and morphological characteristics [11, 12]. During the milling operation, flaws such as debonding, fiber pull-out, fiber breaking, microcracking, stress concentration, and interlaminar crack propagation can occur [13]. By choosing suitable materials and their compositions, the above defects can be reduced. Among those defects, during machining, surface quality mostly affects the shape of the milled surface. Therefore, it is not enough to select the appropriate material composition to obtain a proper milling operation. While machining the FRP-based composites, it is important to identify the suitable milling parameters, such as spindle speed, feed rate, depth of cut, drill diameter, and tool design. Some authors used different techniques to analyze process parameters and optimize them. The mathematical model is employed to determine the impact of process parameters on the response. The milling performances obtained during the machinability study of FRP materials include cutting speed, depth of cut, delamination, and drill geometry [14]. Nowadays, multi-objective optimization techniques, i.e., technique for order of preference by similarity to ideal solution (TOPSIS), gray relational analysis (GRA), CoCoSo, desirability, etc., play a significant role in the manufacturing sector due to the requirement of optimization in quality control during production phase [15–17]. In series, Thakur et al. [18] studied and revealed that the weight percentage of nanofiller (graphene) highly effective parameter for delamination and roughness of the surface. According to the authors, a low feed rate and high spindle speed with a high graphene volume fraction within the parameter value resulted in reduced surface roughness and delamination damage. Kumar et al. [19] had fabricated the nanocomposite materials used to hand lay method with reinforcement by incorporating graphene with 1, 1.51, 2, 2.5, and 3%. The researcher stated that increasing graphene content improved properties, including hardness, bending resistance, and senility. The composite material with a 2.5% graphene fraction of weight was found to have superior properties over other denominators of weight. Davin et al. [20] reported that milling is the most useful machining procedure for reducing excess material for producing a high quality and well sheet. The feed rate was the most significant factor for both materials, with the larger delamination factor (Fd) increasing as the cutting speed and feed rate were increased. Mathivanan et al. [21] carried out milling operations with specific feed speeds, velocity, and cutting speed. Compared to the machining capabilities of CFRP laminates, the machining capacity of GFRP laminates was found to be quite tiny at the same feed rate during the experimental examinations. Parida et al. [22] explored a multi-response optimization approach to optimize the machining parameters in turning GFRP composites. Taguchi L_9 orthogonal array (OA) was used for machining the workpiece. The author also employed a method to analyze multiple performance characteristics. They applied this approach and revealed that it increases machining performance characteristics, including MRR and surface roughness.

In this article, a multi-response optimization of milling processes of MWCNT incorporated GFRP/epoxy nanocomposite is proposed. The surface roughness (SR) and cutting force (Fc) are optimized through a combined compromise solution (CoCoSo) coupled with the Taguchi approach by considering the varying factors, such as spindle speed (*S*), feed rate (*F*), depth of cut (*D*), and reinforcement filler content (MWCNT wt.%). This research aims to estimate the optimum level of process parameters and reduce the effect of surface roughness (SR) and cutting force (Fc) on composites produced. ANOVA is used to examine the most significant parameters which affect in milling of nano-polymer composite. Therefore, the CoCoSo is used to specify appropriate process parameters to obtain a quality machined portion of the MWCNT incorporated GFRP/epoxy nanocomposite milling operation.

2 Materials and Experimentation

To fabricate MWCNT incorporated GFRP/epoxy nanocomposite, the selected glass fiber reinforced polymer (GFRP) has been taken from CF composites Delhi Pvt. Ltd. India. The MWCNT was used as a filler material to improve the interfacial adhesive properties of advanced composites. The properties of MWCNT and characteristics (SEM-XRD) are shown in Table 1 and Fig. 1. During manufacturing, the filler content (wt% MWCNT) is mixed with Lapox resin L-12, and hardener K-6 is poured in a 10:1 ratio. The filler contents materials are obtained from AD Nanotechnology Pvt Ltd, Karnataka. The samples are fabricated by hand layup method with varying and different wt% of MWCNT (1, 2, and 3%) with a size of $100 \times 100 \times 5$ mm.

Table 1 Properties of MWCNT

S. No.	Factors	Feature
1	Form	"Multiwall carbon nanotube"
2	Purity	> 99%
3	Average diameter	10–15 nm
4	Average length	5 μm

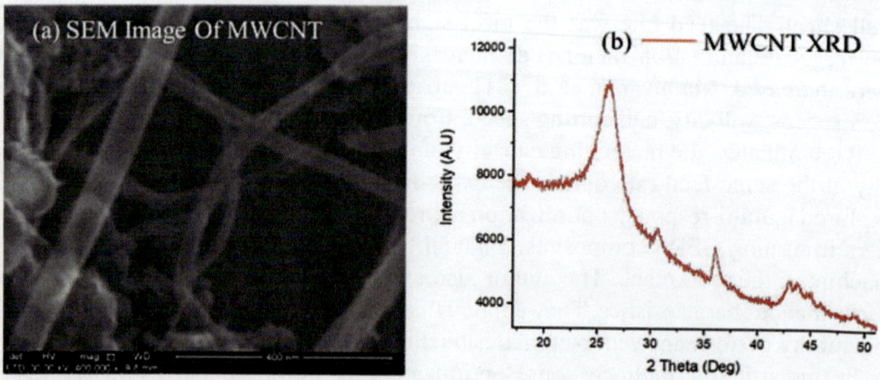

Fig. 1 MWCNT characteristics. **a** SEM image of MWCNT, **b** MWCNT XRD

Table 2 Process parameters

Parameters	Nomenclature	Levels			Unit
		L-1	L-2	L-3	
Spindle speed	*S*	710	1400	2000	Rpm
Feed rate	*F*	56	160	315	mm/min
Depth of cut	*D*	0.8	1.6	2.4	Mm
MWCNT (wt.%)	MWCNT	1	2	4	–

The machining was performed on an automated vertical milling machine setup (Model no BMV35 TC20) with attached electric dynamometers (Kistler) to measure the thrust force and torque. According to Table 2, the milling has been performed using an L_9 orthogonal array with four factors and three different levels [23].

The milling operation of the nanocomposite was carried out with an 8 mm solid carbide end mill tool. The input parameter and corresponding response value of surface roughness (SR) and thrust force (Th) are described in Table 3.

3 Methodology

The main objective of this experimentation is to determine the best optimal solution during milling of MWCNT/GFRP nanocomposite. The process of maximizing the performance of a process or product with two or more outputs, with or without contradicting outputs occurring simultaneously, is referred to as the multi optimization approach. The current study minimizes the milling process's surface roughness (SR) and cutting force (Fc).

Table 3 Experimental array and corresponding response

Exp. No.	Input parameters				Output response	
	S (rpm)	F (mm/min)	D (mm)	MWCNT (wt.%)	SR (μ)	Fc (N)
1	710	56	0.8	1	2.831	45.766
2	710	160	1.6	2	3.175	38.440
3	710	315	2.4	3	3.790	23.772
4	1400	56	1.6	3	3.727	44.276
5	1400	160	2.4	1	2.250	37.500
6	1400	315	0.8	2	3.498	27.246
7	2000	56	2.4	2	2.927	43.477
8	2000	160	0.8	3	3.953	35.733
9	2000	315	1.6	1	2.609	25.817

3.1 Combined Compromise Solution (CoCoSo)

The CoCoSo method is an optimization technique developed by integrating the two most popular multi-criteria decision modeling (MCDM), i.e., "simple additive weighting" (SAW) and "exponentially weighted product" (EWP). This method consists of three mathematical phases and is implemented after accumulating the criteria weights. It uses two approaches to create a comparability sequence (usual multiplication rule and weighted power of the distance). This method necessitates the use of aggregated SAW and EWP methodologies. To solve the CoCoSo decision problem, the following steps are used.

Step-1 First step is determining the initial decision-making matrix, which is represented as:

$$X_{ij} = \begin{bmatrix} x_{11} & x_{12} & \cdot & x_{1n} \\ x_{21} & x_{22} & \cdot & x_{2n} \\ \cdot & \cdot & \cdot & \cdot \\ x_{m1} & x_{m2} & \cdot & x_{mn} \end{bmatrix} \quad (1)$$

Step-2 Normalized values of variables are calculated using the below formula:

for benefit criterion;

$$r_{ij} = \frac{x_{ij} - \min x_{ij}}{\max x_{ij} - \min x_{ij}} \quad (2)$$

for cost criterion;

$$r_{ij} = \frac{\max x_{ij} - x_{ij}}{\max x_{ij} - \min x_{ij}} \quad (3)$$

Step-3 Determining weighted sequence and power of weight of comparability sequences S_i and P_i,

$$S_i = \sum_{j=1}^{n} (w_j r_{ij}) \quad (4)$$

$$P_i = \sum_{j=1}^{n} (r_{ij}) \wedge w_j \quad (5)$$

Step-4 Three different appraisal scores for a weighted sequence are derived using Eqs. (3), (4) and (5):

$$\mathrm{K1} = \frac{P_i + S_i}{\sum_{i=1}^{m} (P_i + S_i)} \quad (6)$$

$$\mathrm{K2} = \frac{S_i}{\min S_i} + \frac{P_i}{\min P_i} \quad (7)$$

$$\mathrm{K3} = \frac{\lambda(S_i) + (1 - \lambda)(P_i)}{\lambda \max S_i + (1 - \lambda) \max P_i} \quad (8)$$

Step-5 According to Eq. 9, the ranking of alternatives is established from higher to lower:

$$k_i = (K1 \times K2 \times K2)^{\frac{1}{3}} + \left(\frac{1}{3}\right)(K1 + K2 + K3) \quad (9)$$

For this study, the values of weights considered are equal weightage for each response as 0.3, and the value of lambda is generally taken as 0.5.

4 Result and Discussion

The considerable results obtained by the CoCoSo methodology are demonstrated in this section with maintaining equal response weight distribution. The output response, viz. SR and Fc is evaluated step by step using the CoCoSo equations from (1–9) for obtaining the assessment score. Table 4 displays the entire range of findings from the CoCoSo. The CoCoSo for all of the experimental trials is provided, along with complete data on normalized values, K_i, and ranking order.

Table 4 Final aggregation (K_i) and ranking of obtained assessment values

Exp. No.	Normalized value		*S*	*P*	K1	K2	K3	K_i	Rank
	SR (μ)	Fc (N)							
1	0.659	0.000	0.554	0.704	0.083	8.639	0.456	3.467	5
2	0.457	0.333	0.437	1.357	0.118	7.955	0.651	3.479	4
3	0.096	1.000	0.239	1.139	0.090	4.920	0.500	2.290	7
4	0.133	0.068	0.122	0.835	0.063	2.873	0.347	1.420	8
5	1.000	0.376	0.901	1.856	0.181	15.055	1.000	6.221	1
6	0.267	0.842	0.359	1.303	0.109	6.794	0.603	3.037	6
7	0.602	0.104	0.523	1.351	0.123	9.133	0.680	3.903	3
8	0.000	0.456	0.073	0.883	0.063	2.254	0.347	1.203	9
9	0.789	0.907	0.808	1.804	0.171	13.702	0.948	5.717	2

Table 4 revealed that experiment number 5 scored the highest value of K_i (6221), which could be considered the optimum conditions for the CoCoSo module, obtained as 1400 spindle speed, 160 feed rate, 2.4 depth of cut, and 1 wt% of MWCNT. The analysis of variance (ANOVA) has been performed, and evaluations are illustrated in Table 5, which demonstrated that depth of cut and wt% of MWCNT are the most significant parameters and the rest are insignificant.

Based on this statistical analysis results, the output *R*-square value = 97.97% and *R*-square (adj) = 95.93% indicates the model's accuracy has good compatibility with the experimental data. The normal probity plot for K_i is shown in Fig. 2, demonstrating that residuals are often straight lines, indicating that errors are normally distributed.

Table 5 ANOVA for K_i

Source	DF	Seq SS	Contribution (%)	Adj SS	Adj MS	*F*-Value	*P*-Value
Regression	4	23.2161	97.97	23.2161	5.8040	48.20	0.001
S	1	0.4350	1.84	0.4350	0.4350	3.61	0.130
F	1	0.7422	3.13	0.7422	0.7422	6.16	0.068
D	1	3.6920	15.58	3.6920	3.6920	30.66	0.005
MWCNT (wt.%)	1	18.3469	77.42	18.3469	18.3469	152.36	0.000
Error	4	0.4817	2.03	0.4817	0.1204		
Total	8	23.6977	100.00				

R-sq = 97.97%, *R*-sq (adj) = 95.93%

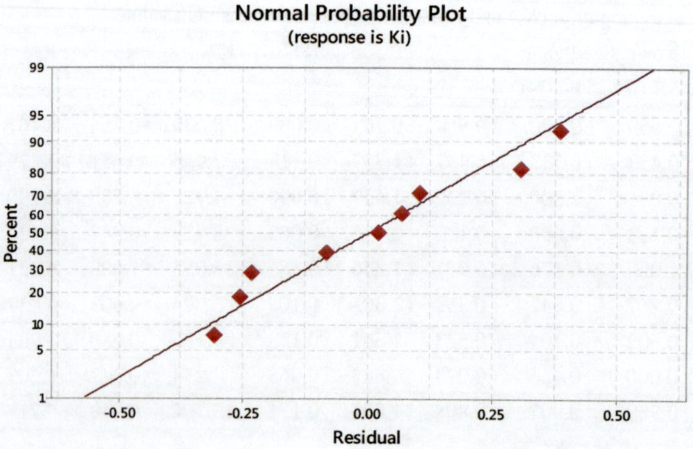

Fig. 2 Normal probity plot for K_i

5 Confirmatory Test

The confirmation studies are conducted (Table 6) to verify the optimum set obtained using the CoCoSo method is within an adequate range concerning obtained experimental set and responses. From Table 4, it can be observed that experiment no. 5 has the highest K_i value (6.22), indicating that the optimum setting for milling operation to be performed on the composite is S2-F2-D3 MWCNT wt%1. Taking the initial experimental one value, the optimal condition for milling came out to be S1-F1-D1-MWCNT wt%1. The CoCoSo experiments show that the smoothness of the milled surface is improved by 20.49%, and the value of Fc declines up to a significant amount (17.99%).

Hence, it can be drawn that CoCoSo optimization has significantly improved the working condition. Compared to the optimal and confirmatory test values, SR improved from 2.831 to 2.25, and Fc significantly decreased from 45.766 to 17.99, showing the effectiveness of the proposed CoCoSo approach in the machining environment.

Table 6 Confirmatory test result

Optimal setting	Taguchi orthogonal array design S1-F1-D1-MWCNT wt.%1	CoCoSo design S2-F2-D3 MWCNT wt.%1	% Improvement
SR	2.831	2.25	20.49
Fc	45.766	37.500	17.99

6 Conclusion

The optimization of the milling performance behavior of MWCNT/GFRP nanocomposites is estimated by using the CoCoSo approach. The experimental results can be concluded as follows:

a. CoCoSo approach improves the estimations of surface roughness up to (20.49%) and 17.99% of cutting force than expected average results. The investigated report depicts the higher application compatibility of the CoCoSo approach in the optimization of milling processes.
b. Based on the rank obtained from the CoCoSo coefficient values, the 5th experimental run gives the highest assessment value ($K_i = 6.22$) is obtained for optimal milling setting for SR and Fc.
c. Proper process parameters control is highly required through the CoCoSo method to improve milling quality on the manufactured composites.
d. The input factor effect analysis K_i for each factor demonstrates that depth of cut and wt% of MWCNT are the most significant input parameter in milling of MWCNT incorporated GFRP/epoxy nanocomposite.

Acknowledgements The authors would also like to thank o/O DC (Handicrafts), Ministry of Textiles, Government of India, New Delhi, INDIA, for their kind assistance to perform this study.

Funding This research work is financially supported by o/O DC (Handicrafts), Ministry of Textiles, Govt. of INDIA, Under Project ID: K-12012/4/19/2020–21/R and D/ST.

References

1. Panchagnula KK, Kuppan P (2019) Improvement in the mechanical properties of neat GFRPs with multi-walled CNTs. In: J Mater Res Technol. 366–376. Brazilian Metallurgical, Materials and Mining Association
2. Navagally RR (2017) Composite materials-history, types, fabrication techniques, advantages, and applications. Int J Mech Prod Eng 82–87
3. Pallapothu H, Kumar A, Laxminarayana P (2020) Micro drilling of glass fibre reinforced polymer composites. Mater Today Proc 1–5. https://doi.org/10.1016/j.matpr.2020.01.545
4. Anand G, Alagumurthi N, Elansezhian R, Palanikumar K, Venkateshwaran N (2018) Investigation of drilling parameters on hybrid polymer composites using grey relational analysis, regression, fuzzy logic, and ANN models. J Brazilian Soc Mech Sci Eng 40:1–20. https://doi.org/10.1007/s40430-018-1137-1
5. Kumar DS, Shukla MJ, Mahato KK, Rathore DK, RKP, BCR (2015) Preparation and mechanical properties of graphene/carbon fiber-reinforced hierarchical polymer composites. In: Effect of post-curing on thermal and mechanical behavior of GFRP composites. p 75
6. Garcia EJ, Wardle BL, Hart AJ, Yamamoto N (2008) Fabrication and multifunctional properties of a hybrid laminate with aligned carbon nanotubes grown in situ. 68:2034–2041. https://doi.org/10.1016/j.compscitech.2008.02.028

7. Kharwar PK, Verma RK, Mandal NK, Mondal AK (2020) Swarm intelligence integrated approach for experimental investigation in milling of multiwall carbon nanotube/polymer nanocomposites. Arch Mech Eng 67. https://doi.org/10.24425/ame.2020.131698
8. Yu Q, Alvarez NT, Miller P, Malik R, Haase MR, Schulz M, Shanov V, Zhu X (2016) Mechanical strength improvements of carbon nanotube threads through epoxy cross-linking. Materials (Basel) 9:1–12. https://doi.org/10.3390/ma9020068
9. Meenakshi R, Suresh P (2020) WEDM of Cu/WC/SiC composites: development and machining parameters using artificial immune system. J Exp Nanosci 15:12–25. https://doi.org/10.1080/17458080.2019.1708331
10. Ramulu M, Kramlich J (2004) Machining of fiber reinforced composites: review of environmental and health effects. Int J Environ Conscious Des Manuf 11:1–19
11. Raj SSR, Dhas JER, Jesuthanam CP (2020) Challenges on machining characteristics of natural fiber-reinforced composites–a review. J Reinf Plast Compos. https://doi.org/10.1177/0731684420940773
12. Kumar J, Verma RK (2020) Experimental investigation for machinability aspects of graphene oxide/carbon fiber reinforced polymer nanocomposites and predictive modeling using hybrid approach. Def Technol. https://doi.org/10.1016/j.dt.2020.09.009
13. Shahabaz SM, Shetty N, Shetty SD, Sharma SS (2020) Surface roughness analysis in the drilling of carbon fiber/epoxy composite laminates using hybrid Taguchi-Response experimental design. Mater Res Express 7. https://doi.org/10.1088/2053-1591/ab6198
14. Zhang X, Yu T, Zhao J (2020) An analytical approach on stochastic model for cutting force prediction in milling ceramic matrix composites. Int J Mech Sci 168:105314. https://doi.org/10.1016/j.ijmecsci.2019.105314
15. Mondal N, Mandal S, Mandal MC (2020) FPA based optimization of drilling burr using regression analysis and ANN model. Meas J Int Meas Confed 152:107327. https://doi.org/10.1016/j.measurement.2019.107327
16. Gokulkumar S, Thyla PR, ArunRamnath R, Karthi N (2020) Acoustical analysis and drilling process optimization of camellia sinensis/ananas comosus/GFRP/epoxy composites by TOPSIS for indoor applications. J Nat Fibers 1–18. https://doi.org/10.1080/15440478.2020.1726240
17. Barua A, Jeet S, Bagal DK, Satapathy P, Agrawal, PK (2019) Evaluation of mechanical behavior of hybrid natural fiber reinforced nano sic particles composite using hybrid Taguchi-Cocoso method. Int J Innov Technol Explor Eng 8:3341–3345. https://doi.org/10.35940/ijitee.J1232.0881019
18. Thakur RK, Singh KK, Sharma D (2019) Modeling and optimization of surface roughness in end milling of graphene/epoxy nanocomposite. Mater Today Proc 19:302–306. https://doi.org/10.1016/j.matpr.2019.07.213
19. Arun GK, Sreenivas N, Reddy KB, Krishna Reddy KS, Shashi Kumar ME, Pramod R (2018) Investigation on mechanical properties of graphene oxide reinforced GFRP. IOP Conf Ser Mater Sci Eng 310. https://doi.org/10.1088/1757-899X/310/1/012158
20. Davim JP, Reis P, António CC (2004) A study on milling of glass fiber reinforced plastics manufactured by hand-lay up using statistical analysis (ANOVA). Compos Struct 64:493–500. https://doi.org/10.1016/j.compstruct.2003.09.054
21. Rajesh Mathivanan N, Mahesh BS, Anup Shetty H (2016) An experimental investigation on the process parameters influencing machining forces during milling of carbon and glass fiber laminates. Meas J Int Meas Confed 91:39–45. https://doi.org/10.1016/j.measurement.2016.04.077
22. Parida AK, Bhuyan RK, Routara BC (2014) Multiple characteristics optimization in machining of GFRP composites using grey relational analysis. Int J Ind Eng Comput 5:511–520. https://doi.org/10.5267/j.ijiec.2014.8.001
23. Asghar A, Raman AAA, Daud WMAW (2014) A comparison of central composite design and Taguchi method for optimizing fenton process. Sci World J 2014. https://doi.org/10.1155/2014/869120

Comparative Study on the Analytical Models for the Workpiece Temperature Distribution During Laser-Assisted Machining

U. Muruga Prabu, Pramod Kuntikana, and Afzaal Ahmed

Abstract Laser-assisted machining (LAM) has been proved to be a suitable process for improving the machinability of difficult-to-cut materials like Inconel 718, Ti–6Al–4 V, etc. Conventional machining processes of such materials led to higher cutting forces, poor surface finish and lesser tool life leading to poor machinability. Laser heating before machining reduces the tensile strength of the material which makes the cutting process easier leading to enhanced machinability. There are various analytical models available in the literature to predict the temperature distribution of the workpiece during laser heating. In this study, the available analytical models are compared with experimental data. The analytical models were implemented with Python code by giving material properties and laser parameters as inputs. The analytical model developed by Shang et al. [1] predicted the peak temperature (with a deviation within 0.15%) more accurately. The same model was used to obtain critical insights such as the effect of the laser feed speed on the laser heating process and power requirement of different materials such as Inconel 728, Ti–6Al–4 V, AISI 1045 and P550. Out of the four materials, P550 was found to have the highest power requirement.

Keywords Laser-assisted machining · Analytical modelling · Thermal modelling

U. M. Prabu (✉) · P. Kuntikana · A. Ahmed
Department of Mechanical Engineering, Indian Institute of Technology Palakkad, Palakkad, Kerala 678557, India
e-mail: 132114005@smail.iitpkd.ac.in

P. Kuntikana
e-mail: pramodkuntikana@iitpkd.ac.in

A. Ahmed
e-mail: afzaal@iitpkd.ac.in

U. S. Dixit et al. (eds.), *Advances in Forming, Machining and Automation*, Lecture Notes in Mechanical Engineering, https://doi.org/10.1007/978-981-19-3866-5_18

1 Introduction

The proliferating usage of superalloys likc Inconel 718 and Ti–6Al–4 V in the oil and gas, aerospace, automotive and biomedical industries brings with it an intrinsic challenge in machining these superalloys with conventional techniques due to their high hardness and strength. Using conventional machining processes on these materials has led to low productivity because of the rapid tool wear. High cutting forces and inferior surface finish are other disadvantages associated with conventional machining of superalloys. In order to overcome these disadvantages, researchers have used heat assistance to ease the machining process by lowering the strength and hardness of the workpiece before its machining. Heat sources like gas flame, induction coil, plasma and lasers are used in assisting turning and milling of such difficult-to-cut superalloys. Amongst these heat sources, lasers have been the most preferred because of their high energy density and ease of control. In course of time, heat-assisted machining (HAM) has evolved as a more industry-friendly process with the optimization of process parameters.

The first major research in laser-assisted machining (LAM) was conducted in the 1980s. Rajagopal et al. [2] worked on a DARPA-sponsored research project and established the benefits of LAM on aerospace alloys. The research on LAM did not interest many in those times because of the high cost and space occupied by the laser equipment. In the 2000s, researchers slowly started to turn towards LAM because of the evolution of portable laser systems. Anderson et al. [3] did an economic analysis of the LAM process on Inconel 718 and reported a 66% decrease in cost compared to conventional carbide machining. Shi et al. [4] numerically modelled the LAM process using DEFORM 3D simulation package and predicted the cutting force and maximum chip thickness in LAM with good accuracy.

In the 2010s, optimized input parameters for obtaining the best process performances were established. The impact of the input parameters on the process performance was studied [5, 6]. One of the hybrid LAM processes was also developed during this phase by combining LAM with cryogenic cooling of the tool [7]. Navas et al. [8] made a significant breakthrough by discovering the elimination of work hardening by laser heating as one of the mechanisms that enhance the machinability during LAM. LAM studies on AISI 1045 steel [9, 10] started due to its increasing industrial applications. AISI 1045 is known for its high wear resistance and strength. Ayed et al. [11] found that the distance between the laser beam axis and the tool rake face is a major factor affecting cutting forces and showed a cutting force reduction of more than 50%. They established the model considering the following deformation behaviours, viz. work hardening and dynamic recovery behaviour and dynamic softening behaviour. Parida and Maity [12] considered tool nose radius as an influencing parameter and found that cutting force increased with an increase in nose radius. The bending of the workpiece as a consequence of LAM was reported for the first time [13]. Parida [14] conducted a study on the chip geometry in hot machining using SEM analysis. They observed an inverse proportionality relationship between heat temperature and equivalent chip thickness; the degree of segmentation; and serration frequency. Oh et al. [15] developed and analysed a hybrid LAM + trochoidal milling process and showed a maximum of 41% decrease in specific cutting energy. Shang et al. [1] made significant headway in LAM by developing a spatial and a temporal (S and T)-controlled LAM process for cutting workpieces with higher widths of cut. In this study, we will be introducing the various analytical models available in the literature for predicting the temperature distribution of the workpiece during laser heating. Further, critical insights will be derived considering the effect of the laser feed speed and material properties.

2 Analytical Models to Predict the Workpiece Temperature Distribution

2.1 *Model Proposed by Woo and Cho [16]*

Woo and Cho [16] developed an analytical model considering a rectangular uniformly distributed heat source and convective top and bottom surfaces. The governing equation of the problem is given by Eq. (1).

$$\frac{1}{\alpha}\frac{\partial \theta}{\partial t} = \frac{\partial^2 \theta}{\partial x^2} + \frac{\partial^2 \theta}{\partial y^2} + \frac{\partial^2 \theta}{\partial z^2} \tag{1}$$

The boundary conditions are given as follows:

$$-k\frac{\partial \theta(X, Y, 0, t)}{\partial Z} + h_1(\theta(X, Y, 0, t)) = 0 \tag{2}$$

$$-k\frac{\partial \theta(X, Y, d, t)}{\partial Z} + h_2(\theta(X, Y, d, t)) = 0 \tag{3}$$

$$\lim_{r \to \infty} T(X, Y, Z, t) = T_0 \tag{4}$$

Here, $r = \sqrt{(X - X_1)^2 + (Y - Y_1)^2 + Z^2}$, h_1 = forced convection coefficient at upper surface, h_2 = forced convection coefficient at lower surface. From Fig. 1, X, Y, Z denotes the coordinate system fixed at O, a point on the surface of the workpiece, while x, y, z denotes the moving coordinate system with its origin fixed at the laser beam centre $(X_c, Y_c, 0)$.

The initial condition is

$$T(X, Y, Z, 0) = T_0 \tag{5}$$

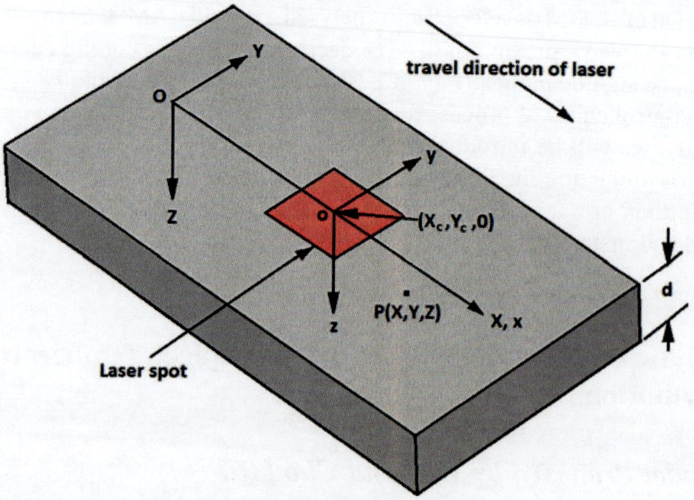

Fig. 1 Coordinates of the workpiece for the thermal analysis

The solution using Green's function method is

$$\theta(x, y, z, t) = \int_0^t \frac{\alpha q(\tau)}{8k\mathrm{d}RS} \sum_{n=0}^{\infty} A_n \exp\left[-\mu_n^2(t-)\right] \times \left[\cos\left(\frac{\mu n Z}{\sqrt{\alpha}}\right) + \frac{\beta_1 \alpha}{\mu n} \sin\left(\frac{\mu n Z}{\sqrt{\alpha}}\right)\right]$$
$$\times \left\{\mathrm{errf}\left[\frac{R + X_C(\tau) - x - X_C(t)}{\sqrt{4\alpha(t-\tau)}}\right] + \mathrm{errf}\left[\frac{R - X_C(\tau) + x + X_C(t)}{\sqrt{4\alpha(t-\tau)}}\right]\right\}$$
$$\times \left\{\mathrm{errf}\left[\frac{S + Y_C(\tau) - y - Y_C(t)}{\sqrt{4\alpha(t-\tau)}}\right] + \mathrm{errf}\left[\frac{S - Y_C(\tau) + y + Y_C(t)}{\sqrt{4\alpha(t-\tau)}}\right]\right\} \mathrm{d}\tau \quad (6)$$

2.2 *Model Proposed by Boo and Cho [17]*

Boo and Cho developed a model that assumes a Gaussian heat source distribution and convective boundary condition at the top and bottom surfaces. The governing equation is given by Eq. (1). The boundary conditions are given by Eqs. (2), (3) and (4). The coordinate system of the model is given in Fig. 2. The Gaussian heat source distribution is shown in Fig. 3.

The equation for the Gaussian heat source distribution is given by

$$Q(x, y, t) = \frac{q(t)}{2\pi\sigma^2} \exp\left\{-\frac{(x^2 + y^2)}{2\sigma^2}\right\} \quad (7)$$

where σ is the distribution parameter. The solution of temperature distribution from this model is given by

$$\theta(x, y, z, t) = \int_0^t \frac{q(\tau)}{\pi\rho c} \sum_{n=0}^{\infty} A_n \exp\left[-\mu_n^2(t-\tau)\right] \times \left[\cos\left(\frac{\mu_n Z}{\sqrt{\alpha}}\right) + \frac{\beta_1\sqrt{\alpha}}{\mu_n}\sin\left(\frac{\mu_n Z}{\sqrt{\alpha}}\right)\right] \times \exp\left[-\frac{\{x + X_C(t) - X_C(\tau)\}^2 + \{y + Y_C(t) - Y_C(\tau)\}^2}{2\sigma^2 + 4\alpha(t-\tau)}\right]$$

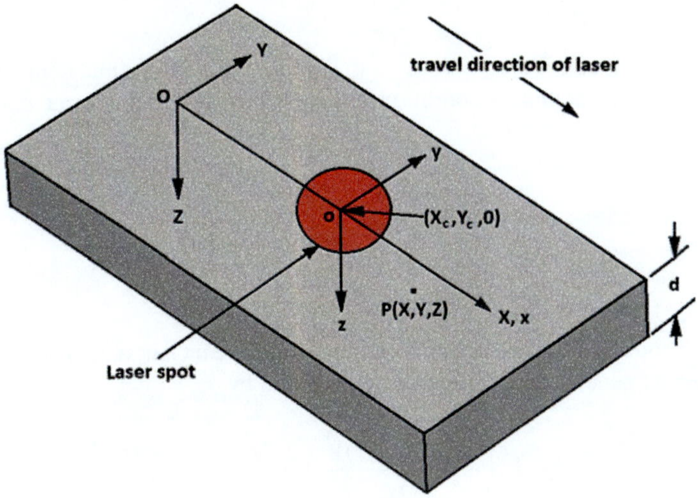

Fig. 2 Coordinates of the workpiece for the thermal analysis

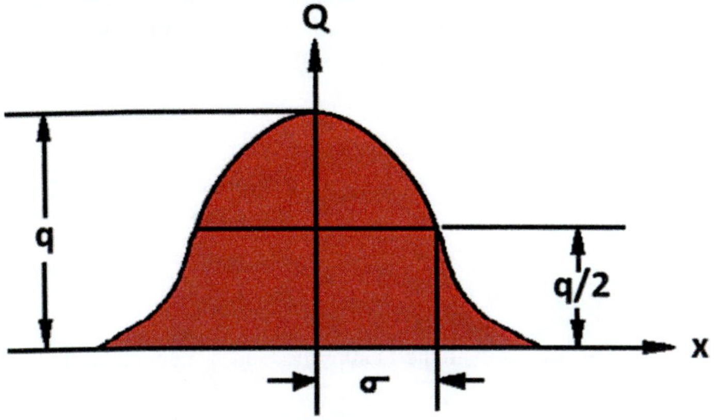

Fig. 3 Schematic of the Gaussian heat source distribution

$$\times \frac{1}{\sigma^2 + 2\alpha(t-\tau)} \mathrm{d}\tau \tag{8}$$

2.3 Model Proposed by Shang et al. [1]

Shang et al. [1] developed an analytical model considering a 3D transient conduction problem. The governing equation is

$$\frac{1}{\alpha}\frac{\partial \theta}{\partial t} = \frac{\partial^2 \theta}{\partial x^2} + \frac{\partial^2 \theta}{\partial y^2} + \frac{\partial^2 \theta}{\partial z^2} + \frac{g(x, y, z, t)}{k} \tag{9}$$

The medium is considered infinite in the x- and y-directions and semi-infinite in the *z*-direction. The boundary condition at the top surface and the initial condition is given by

$$\frac{\partial \theta(x, y, 0, t)}{\partial z} = 0 \tag{10}$$

$$\theta(x, y, z, 0) = 0 \tag{11}$$

The heat source of the laser beam is assumed to have a Gaussian distribution, and hence, the heat generation term *g(x, y, z, t)* is given by

$$g(x, y, z, t) = g_0 \exp\left(-\frac{(x - x_b(t))^2}{r_{\mathrm{bx}}^2}\right) \exp\left(-\frac{(y - y_b(t))^2}{r_{\mathrm{by}}^2}\right) \delta(z) \tag{12}$$

where $g_0 = \eta P / \left(\pi r_{\mathrm{bx}} r_{\mathrm{by}}\right)$, $(x_b(t), y_b(t))$ is the transient position of the laser spot, r_{bx} and r_{by} are the semi-length of the major axis and the minor axis of the elliptical laser spot, respectively, η is the laser heat absorption ratio of the workpiece material and P is the laser power. The temperature distribution obtained by Green's function method is given by

$$\theta(x, y, z, t) = \frac{2\alpha}{k} g_0 \int_{\tau=0}^{t} \frac{1}{(4\pi\alpha(t-\tau))^{\frac{3}{2}}} \frac{4\pi}{\sqrt{\frac{4}{r_{\mathrm{bx}}^2} + \frac{1}{\alpha(t-\tau)}}\sqrt{\frac{4}{r_{\mathrm{by}}^2} + \frac{1}{\alpha(t-\tau)}}} \times \exp\left(-\frac{(x - x_b(t))^2}{r_{\mathrm{bx}}^2 + 4\alpha(t-\tau)}\right)$$

$$\times \exp\left(-\frac{(y - y_b(t))^2}{r_{\mathrm{by}}^2 + 4\alpha(t-\tau)}\right) \times \exp\left(-\frac{z^2}{4\alpha(t-\tau)}\right) \mathrm{d}\tau \tag{13}$$

The integral solution 13 is evaluated using the Gauss–Legendre quadrature.

Table 1 Properties of Inconel 718 [3]

Sl. No.	Property	Value
1	Density	8193–8220 kg/m^3
2	Melting range	1260–1336 °C
3	Specific heat capacity	$10^3(0.0002 \times T + 0.4217)$ J/(kg K)
4	Thermal conductivity	$0.015 \times T + 11.002$ W/(m K)

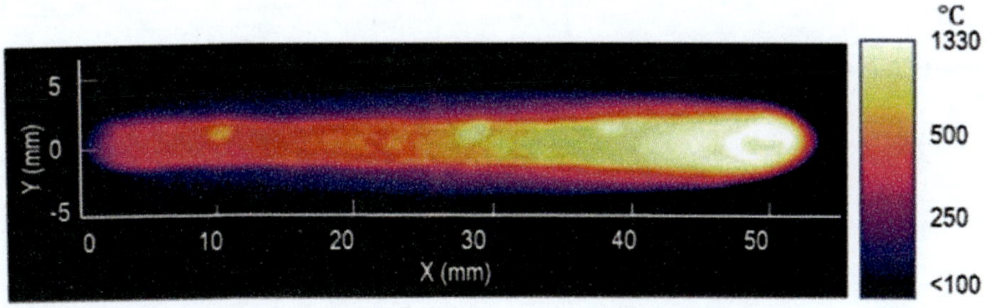

Fig. 4 Thermal image of the laser linear scan captured by the IR camera [1]

3 Experimental Procedure

The experimental details were taken from Shang et al. [1] to validate the analytical models. The linear laser scanning experiments were performed by the authors on square bar samples of precipitation-hardened Inconel 718 with properties shown in Table 1. A laser power of 2000 W and a laser feed speed of 900 mm/min, respectively, were selected. A 10 kW diode laser system with a laser spot radius of 1.5 mm was used as the laser heat source, and a 2-axis laser beam deflection unit was attached to the spindle to control the laser location. An infrared (IR) camera was used to measure the surface temperature. The thermal image of the linear laser scan captured by the IR camera is shown in Fig. 4.

4 Results and Discussion

The models discussed above were compared with the experimental data. The value of the laser heat absorption ratio of Inconel 718 used in the models was 0.345. From Fig. 5, the deviation of peak temperature of the models from the experimental peak temperature can be seen. The percentage deviation is shown in Table 2. The model proposed by Woo and Cho [16] has a deviation of 54.85% when the convection effect was considered. The same model showed a slightly less deviation of 52.28% when convection effects were ignored. The model proposed by Boo and Cho [17] showed a deviation of 45.55% when convection effects were considered and slightly

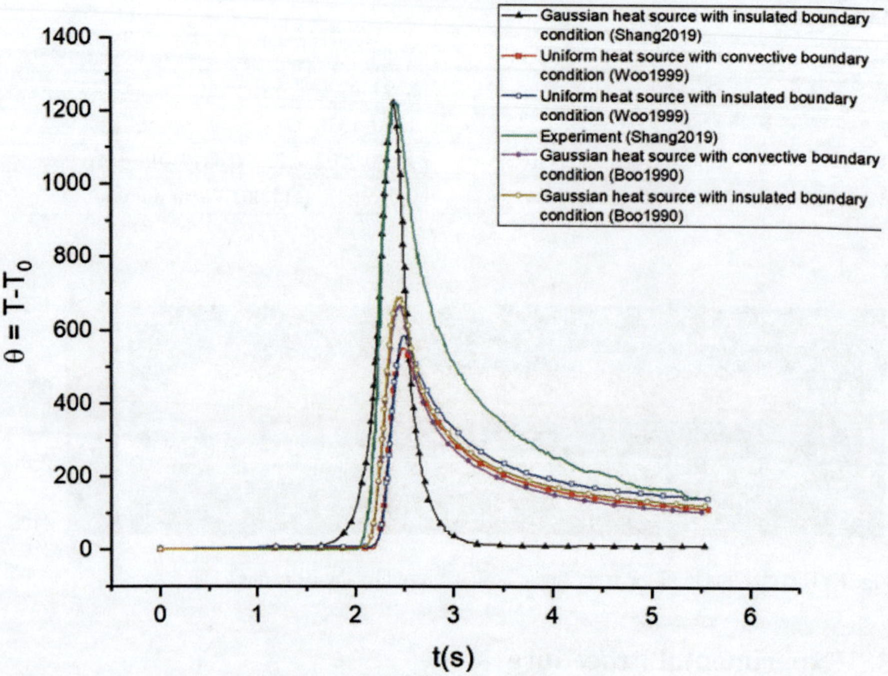

Fig. 5 Comparison of experimental results and the analytical models

Table 2 Analytical models and their percentage deviation from the experimental plot

Sl. No.	Model type	Authors	Analytical peak temperature difference (°C)	Experimental peak temperature difference (°C)	Percentage deviation from experimental peak temperature difference
1	Uniform heat source with convective boundary condition	Woo and Cho [16]	551	1221 [1]	54.85
2	Uniform heat source with insulated boundary condition	Woo and Cho [16]	583	1221 [1]	52.28
3	Gaussian heat source with convective boundary condition	Boo and Cho[17]	665	1221 [1]	45.55
4	Gaussian heat source with insulated boundary condition	Boo and Cho [17]	690	1221 [1]	43.52
5	Gaussian heat source with insulated boundary condition	Shang et al. [1]	1220	1221 [1]	0.15

less deviation of 43.52% when convection effects were neglected. Boo and Cho [17] model showed lesser deviation compared to Woo and Cho [16] model because the former considered a Gaussian distribution of the laser heat source while the latter considered a uniform distribution of the laser heat source. Laser beams act closer to a Gaussian beam because both show a constant divergence behaviour at large distances along the direction of propagation. Shang et al. [1] proposed a better model with a Gaussian heat source distribution and insulated boundary condition. This model showed the least deviation from the experimental peak temperature (0.15% deviation).

Considering its high accuracy, the Shang model was applied to three other materials, namely Ti–6Al–4 V, AISI 1045, P550, whose properties are given in Table 3. The power required for each material to obtain their respective temperature distribution at laser feed speed = 900 mm/min with and without graphite coating (with laser absorption ratio = 0.8) is shown in Fig. 6. As the temperature differences are

Table 3 Properties of the materials

Sl. No.	Material	Melting point (K)	Density (kg/m^3)	Specific heat (J/(kg K))	Conductivity (W/(m K))	Laser heat absorption ratio
1	AISI 1045	1796	7817	658	31	0.4
2	Ti–6Al–4 V	1933	4428	759	28	0.3
3	P550	1783	7800	762	60	0.4

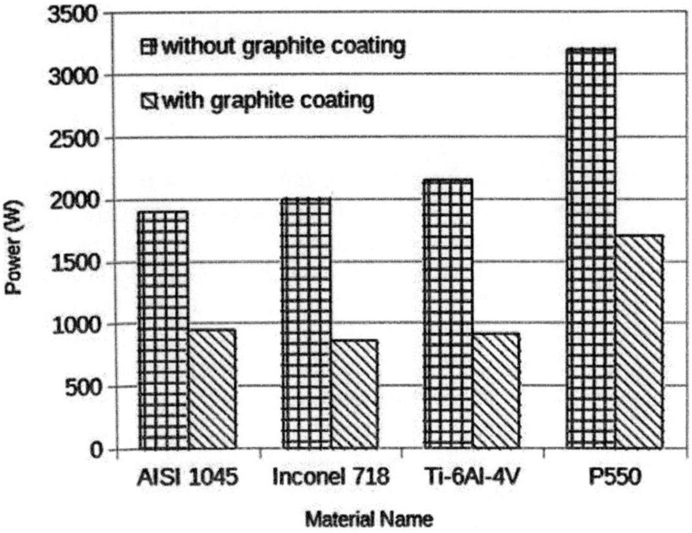

Fig. 6 Power needed for the materials Inconel 718, Ti–6Al–4 V, AISI 1045 and P550 to heat them to a temperature slightly below their melting point at a laser feed speed of 900 mm/min

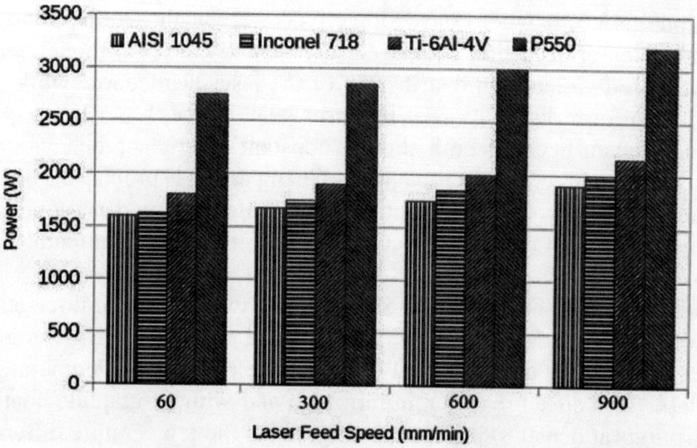

Fig. 7 Power versus laser feed speed plot for the materials Inconel 718, AISI 1045, Ti–6Al–4 V and P550

calculated at the surface of the workpiece, the material with higher thermal diffusivity (P550) requires a higher power to reach the necessary temperature difference at the surface-higher thermal diffusivity leads to quick penetration of heat from the surface into the workpiece. Because of its high absorptivity, graphite coating reduced the power requirement of each material approximately by 50%.

Then, the relationship between the laser feed speed and power requirement of each material was studied. Figure 7 shows the power versus laser feed speed plot for the four materials of interest. From Fig. 7, it is understood that power and laser feed speed have a linear proportionality relation. This can be explained based on the fact that with an increase in laser feed speed, the time available to heat the workpiece decreases, and hence, more power would be required to compensate for the lack of time available to heat the workpiece.

Another important observation from the model is that the distance (from the starting point of the laser beam) to reach a target temperature increases as the velocity increases. The reason for this can be explained by the time available for the heat to transfer in the direction of motion of the laser beam. Five equally spaced points were considered along the path of the laser-p1, p2, p3, p4 and p5. The laser beam was traced on the workpiece at three different speeds, viz. 0.005, 0.01 and 0.015 m/s. The target temperature difference was set as 1000. From Fig. 8a, the target temperature was reached at the first point p1 itself at a laser feed speed of 0.005 m/s. From Fig. 8b, the target temperature was reached at the second point p2 at a laser feed speed of 0.01 m/s. From Fig. 8c, the target temperature was reached at the 3rd point p3 at a laser feed speed of 0.015 m/s.

Another factor that was analysed is the grain size. From the literature, it was found that the effect of heating on the grain size which in turn affects the yield strength of the material is governed by the Hall–Petch equation. Hall–Petch equation is given

Fig. 8 Temperature versus time plot at different laser feed speeds (v)

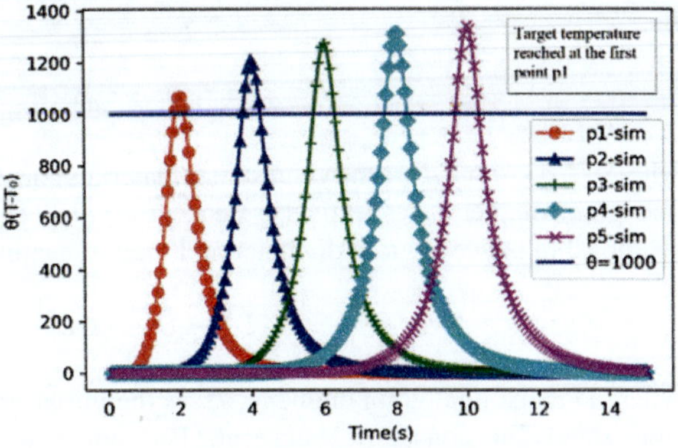

(a) v=0.005 m/s

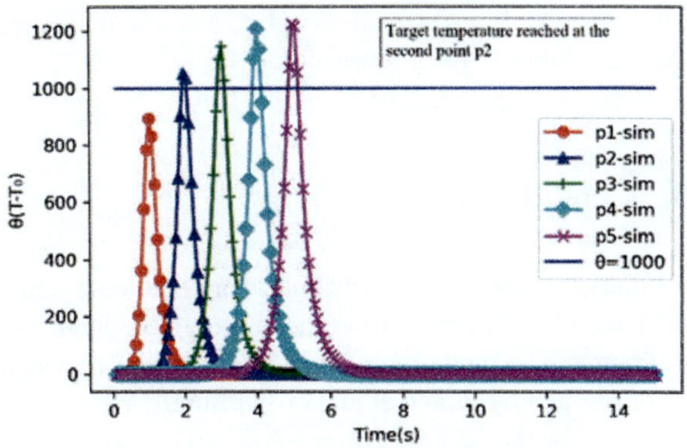

(b) v=0.01 m/s

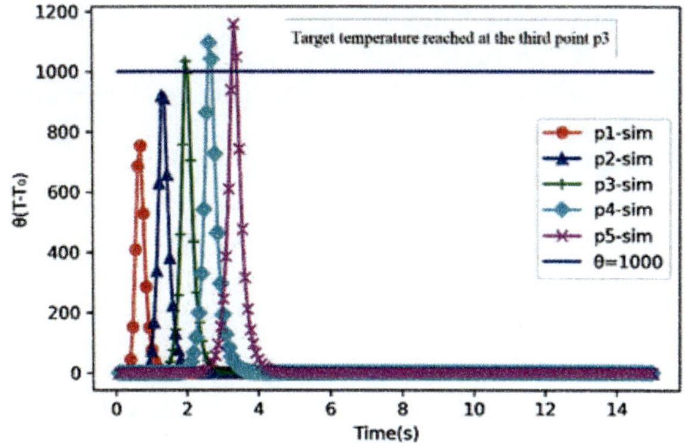

(c) v=0.015 m/s

by

$$\sigma_y = \sigma_0 + k/(d)^{1/2} \tag{14}$$

where σ_y, σ_0, k and d are the yield stress, materials constant, strengthening coefficient and grain diameter, respectively. Beck et al. [18] experimentally established the relationship between grain diameter and time of heating as below.

$$D^n - D_0^n = Ct \tag{15}$$

where D is the final grain diameter, D_0 is the initial grain diameter, t is the heating time, n and C are constants. Huda et al. [19] studied the grain size variation of Inconel 718 concerning temperature. It was found that there was no significant change in grain size when the material was heated at a constant temperature for 20 min to 2 h. Therefore, the grain size effect in LAM can be neglected since LAM involves heating the material only for a few seconds.

5 Conclusions

- Gaussian heat source distribution predicted the peak temperature of the workpiece during laser heating more accurately compared to a uniformly distributed heat source. The deviation of peak temperature between experiments and the model developed by Shang et al. was within 0.15%.
- Amongst the four materials—AISI 1045, Inconel 718, Ti-6Al-4 V and P550, P550 requires the highest power (above 3000 W) to reach a peak temperature that is slightly less than the melting point.
- The use of graphite coating reduces the power requirement by approximately 50% irrespective of the material.
- An increase in laser feed speed increases the power required to reach a particular peak temperature.
- Another important observation is that the distance (from the starting point of the laser beam) to reach a target temperature increases as the velocity increases.
- The effect of the grain size on the LAM process is found to be negligible as the time required for heating is very small.

Acknowledgements The research work was supported by the Science and Engineering Research Board (SERB) under the Start-up Research Grant (SRG) programme (Sanction Order No:

SRG/2019/000708). The authors are also grateful to the Department of Mechanical Engineering and the Central Facility for Materials and Manufacturing (CFMM), Indian Institute of Technology Palakkad, for providing the necessary software facilities for the work.

References

1. Shang Z, Liao Z, Sarasua JA, Billingham J, Axinte D (2019) On modelling of laser assisted machining: forward and inverse problems for heat placement control. Int J Mach Tools Manuf 138(November 2018):36–50
2. Rajagopal S, Plankenhorn DJ, Hill VL (1982) Machining aerospace alloys with the aid of a 15 kW laser. J Appl Metalwork 2(3):170–184
3. Anderson M, Patwa R, Shin YC (2006) Laser-assisted machining of Inconel 718 with an economic analysis. Int J Mach Tools Manuf 46(14):1879–1891
4. Shi B, Attia H, Vargas R, Tavakoli S (2008) Numerical and experimental investigation of laser-assisted machining of Inconel 718. Mach Sci Technol 12(4):498–513
5. Rashid RAR, Sun S, Wang G, Dargusch MS (2012) An investigation of cutting forces and cutting temperatures during laser-assisted machining of the Ti-6Cr-5Mo-5V-4Al beta titanium alloy. Int J Mach Tools Manuf 63:58–69
6. Germain G, Santo PD, Lebrun JL (2011) Comprehension of chip formation in laser assisted machining. Int J Mach Tools Manuf 51(3):230–238
7. Dandekar CR, Shin YC, Barnes J (2010) Machinability improvement of titanium alloy (Ti-6Al-4V) via LAM and hybrid machining. Int J Mach Tools Manuf 50(2):174–182
8. Navas VG, Arriola I, Gonzalo O, Leunda J (2013) Mechanisms involved in the improvement of Inconel 718 machinability by laser assisted machining (LAM). Int J Mach Tools Manuf 74:19–28
9. Kim DH, Lee CM (2014) A study of cutting force and preheating-temperature prediction for laser-assisted milling of Inconel 718 and AISI 1045 steel. Int J Heat Mass Transf 71:264–274
10. Woo WS, Lee CM (2015) A study of the machining characteristics of AISI 1045 steel and Inconel 718 with a cylindrical shape in laser-assisted milling. Appl Therm Eng 91:33–42
11. Ayed Y, Germain G, Salem WB, Hamdi H (2014) Experimental and numerical study of laser-assisted machining of Ti6Al4V titanium alloy. Finite Elem Anal Des 92:72–79
12. Parida AK, Maity K (2017) Effect of nose radius on forces and process parameters in hot machining of Inconel 718 using finite element analysis. Eng Sci Technol Int J 20(2):687–693
13. Xu D, Liao Z, Axinte D, Sarasua JA, M'Saoubi R, Wretland A (2020) Investigation of surface integrity in laser-assisted machining of nickel-based superalloy. Mater Des 194:108851
14. Parida AK (2019) Analysis of chip geometry in hot machining of inconel 718 alloy. Iran J Sci Technol Trans Mech Eng 43:155–164
15. Oh NS, Woo WS, Lee CM (2018) A study on the machining characteristics and energy efficiency of Ti-6Al-4V in laser-assisted trochoidal milling. Int J Prec Eng Manuf Green Technol 5(1):37–45
16. Woo HG, Cho HS (1999) Three-dimensional temperature distribution in laser surface hardening processes. Proc Inst Mech Eng Part B: J Eng Manuf 213(7):695–712
17. Boo KS, Cho HS (1990) Transient temperature distribution in arc welding of finite thickness plates. Proc Inst Mech Eng Part B: J Eng Manuf 204(3):175–183
18. Beck PA, Kremer JC, Demer LJ, Holzworth ML (1948) Grain growth in high-purity aluminum and in an aluminum-magnesium alloy. Trans Am Inst Min Metall Eng 175:372–400
19. Huda Z, Zaharinie T, Metselaar HSC, Ibrahim S, Min GJ (2014) Kinetics of grain growth in 718 Ni-base super alloy. Arch Metall Mater 3

Development of Aluminium Alloy Thin-Walled Sections by Chemical Milling Process

B. V. Subba Rao, G. Gopi, Indumathi Vijey, J. John Rozario Jegaraj, P. Kiran, and B. Shivadayal Rao

Abstract Chemical milling has emerged as a non-conventional manufacturing process for realizing the thin-walled shells to reduce the weight of the hardware. Aluminium alloy (AA 2014) is widely used for aerospace and defence applications due to its high-strength/weight ratio. In this present study, various chemical milling masking agents, chemical milling compositions, process parameters, surface finish, and mechanical and microscopic studies are carried out for establishing the suitable process on aluminium alloys. These results are documented and concluded in this paper for the realization of thin-walled aluminium alloy shells. This process is not limited and it may be adapted to other types of hardware where the conventional manufacturing process is highly difficult.

Keywords Chemical milling · Thin-walled shells · Aluminium alloy

1 Introduction

The chemical milling process is involved in the removal of material from the components uniformly throughout the surface in presence of a suitable etching solution [1, 2]. This process is highly useful for achieving the precision dimensions on thin-walled sections where the components are highly difficult to manufacture by conventional manufacturing processes [1–5]. It is an economical process and will not create any

B. V. Subba Rao (✉) · G. Gopi · I. Vijey · J. John Rozario Jegaraj · P. Kiran · B. Shivadayal Rao
Defence Research and Development Laboratory, Kanchanbagh, Hyderabad 500058, India
e-mail: subbarao@drdl.drdo.in

G. Gopi
e-mail: ggopi@drdl.drdo.in

I. Vijey
e-mail: indumathi@drdl.drdo.in

J. John Rozario Jegaraj
e-mail: johnmfrg@drdl.drdo.in

P. Kiran
e-mail: pkiran@drdl.drdo.in

U. S. Dixit et al. (eds.), *Advances in Forming, Machining and Automation*, Lecture Notes in Mechanical Engineering, https://doi.org/10.1007/978-981-19-3866-5_19

distortions in the components. This process can be carried out from room temperature to high temperature up to 100 °C also depending on material removal rate. A huge no. of components also can be processed simultaneously depending on the etching solution bath capacity.

Aluminium alloys are widely in the aerospace industry and defence application due to their high strength to weight ratio. Various airframe sections are manufactured with aluminium alloys due to the above advantage. However, achieving the airframe section thickness in terms of 0.2–0.5 mm is highly difficult to achieve by conventional manufacturing processes [6]. The above critical and precise airframe dimensions can be achieved by the chemical milling process uniformly and without inducing any external stresses [7–11]. An alkaline solution consisting of surfactant and alkaline sulphides is generally employed for this chemical milling process at a certain process temperature for aluminium alloys.

Various chemical etching maskants and chemical etching solutions are available in the open literature (1–5). However, the effectiveness of maskant and the effect of each chemical in the chemical etching solution were not much discussed in the literature. There is not much data available for the comparison of mechanical properties of mechanically milled and chemically etched samples (17). Hence, the aim is to study the various chemical etching maskants for their effectiveness, the effect of each chemical in the etching solution and mechanical properties comparison, etc.

2 Experimental

In the present study, aluminium alloy (AA 2014) is considered for the experimental study as this material is widely used in the fabrication of thin-walled airframe sections which are fabricated from forged form for defence and aerospace applications. The composition of aluminium alloy (AA 2014) is given in Table 1.

The chemical milling process consists of various steps, i.e. 1. degreasing, 2. application of masking agent wherever milling is to be avoided, 3. chemical milling process, 4. dimensional inspection, 5. removal of masking agent.

The masking agent is to be selected based on the adhesive property and reduces the edge effect during the milling process. Various masking agents are available commercially, and three different types of masking agents are considered for the present study. Various chemical milling compositions are available in literature like ferric chloride, i.e. acid and alkali-based etching solutions [4]. However, alkali-based milling solutions produce a higher milling rate, and hence, alkali-based milling solution is selected for our present study. Sodium hydroxide is the simple, cost-effective,

Table 1 Chemical composition of aluminium alloy (AA 2014)

Element	Cu	Si	Fe	Mn	Mg	Zn	Ti	V	Al
Composition	3.9–5.0	0.9–1.2	0.5–0.7	0.4–1.2	0.2–0.8	0.25	0.15	0.15	*Remaining*

and easily available chemical chosen for alkali-based chemical milling composition. Along with sodium hydroxide, sodium sulphide and sodium gluconate are selected for retaining a better surface finish as per the literature. Chemical milling rate and surface finish are studied concerning various process parameters like chemical composition, solution temperature, time duration, etc., for effective chemical milling of aluminium alloys without compromising surface finish after the chemical milling process [12–16]. In addition to the above properties, the mechanical properties [17, 18] also have been studied to understand the mechanical behaviour after the chemical milling process as this process reduces the surface stresses during the process [17]. These properties are compared with specimens processed with the mechanical milling process.

3 Results and Discussion

3.1 Masking Agent

Three types of masking agents supplied from different manufacturers like M/s R. M. lacquer (red masking agent, acrylic–based), Southfield paints (green masking agent, vinyl–based), Shailtex Lubricants Pvt. Ltd. (Beige masking agent, Neoprene latex–based) have been studied for effective adhesion and controlling the edge effect during the chemical milling process. All these masking agents were applied on thin-walled shells and dried overnight for curing. All three masking agents exhibit a similar adhesion with aluminium alloy before the chemical milling process. However, the lacquers which are supplied from M/s R. M. lacquer and Southfield paints form a thin layer on components and could not control the edge effect during the chemical milling process. The significant areas are also affected by using these lacquers. However, the masking agent supplied from M/s Shailtex Lubricant Pvt. Ltd (Shailmask–522) is highly effective in terms of adhesion and controlling the edge effect during the process. It is found that the removal of the shailmask–522 masking agent is easier physically without the use of any chemical which is highly recommended for limiting the use of various health hazardous solvents (Fig. 1).

3.2 Chemical Composition

Sodium hydroxide is the main constituent for the removal of aluminium material by forming sodium aluminate precipitation which dissolves in water during the chemical milling process. It is observed that chemical milling was carried out in alkali etching solution of aluminium alloy is bumpy and nodular especially with copper-based aluminium alloys. For reducing the bumpy and nodular surface effect, sodium gluconate is selected for creating the wetting interface during the process as per the

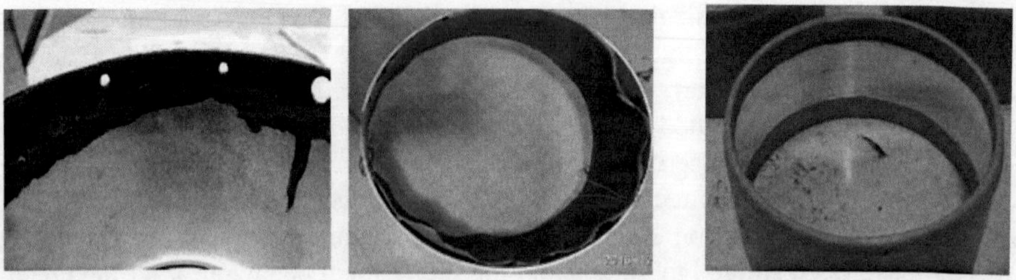

R.M. Lacquer Southfield masking agent, Shailtex lubricants

Fig. 1 Acrylic based masking agent, Vinyl based masking agent, Neoprene based masking agent

literature. Active sulphur containing salt like sodium sulphide/potassium sulphide is required which can precipitate the alloying elements like copper, zinc, magnesium, etc., and dissolve into the solution. This will help in further improving the surface finish of the component after chemical milling. Hence, sodium sulphide is selected for the above purpose which is simple, cheap, and easily available for the product ionization process.

By considering the all above chemicals, trials are conducted by varying the chemical composition, time, and temperature constant to understand the effect of milling rate and surface finish.

3.2.1 Sodium Hydroxide Concentration

Chemical milling rate is measured concerning sodium hydroxide concentration at the temperature of 300 °C for one hour process by keeping sodium sulphide at 5% and sodium gluconate at 10% composition in the chemical milling solution. The fresh solution is prepared for every composition and maintained at the solution temperature constant at 300 °C for studying the effect of sodium hydroxide concentration. Based on the study Fig. 2, it is observed that as the sodium hydroxide concentration increases, until 100 g/L, the milling rate increases and after that, there is a slight decrease in the milling rate concerning sodium hydroxide concentration. This may be due to the solution attaining a saturation level above 100 g/L sodium hydroxide concentration. Hence, 100 g/L concentration is chosen for the present study for making the process more economical way and reducing the chemical effect on working people.

3.2.2 Sodium Sulphide Concentration

Sodium sulphide is added to the chemical milling solution to reduce the bumpy and nodular deposition on the milling surface due to other alloying elements like copper and zinc. This study is conducted by keeping sodium hydroxide at 100 g/L and

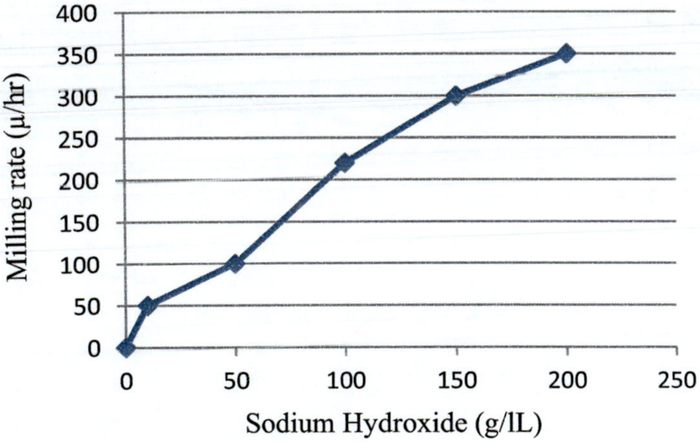

Fig. 2 Effect of sodium hydroxide concentration on milling rate

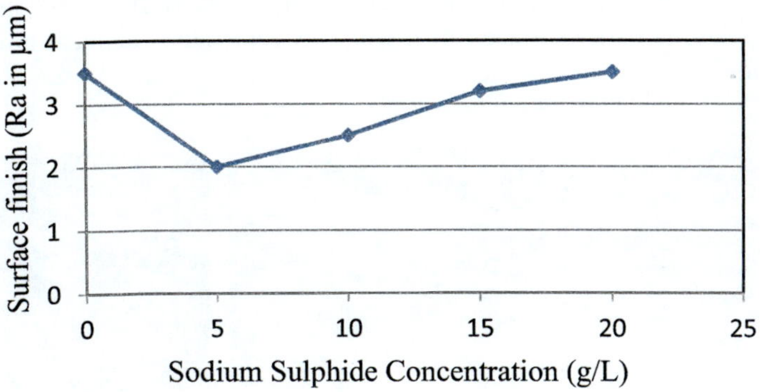

Fig. 3 Effect of sodium sulphide concentration on surface finish

sodium gluconate at10g/L at 30 °C temperature. Based on study Fig. 3, the effect of sodium sulphide up to 5 g/L is giving a better surface finish and roughness start to increase with sodium sulphide content above 5 g/L. This may be due to the excess presence of sulphide content to precipitate the copper element in the aluminium alloy which result in increasing the surface roughness.

3.2.3 Sodium Gluconate Concentration

Sodium gluconate is used as a wetting agent for reducing the wetting angle and improving the surface finish after the chemical milling process. Based on the study Fig. 4, the surface finish is improved until sodium gluconate concentration up to

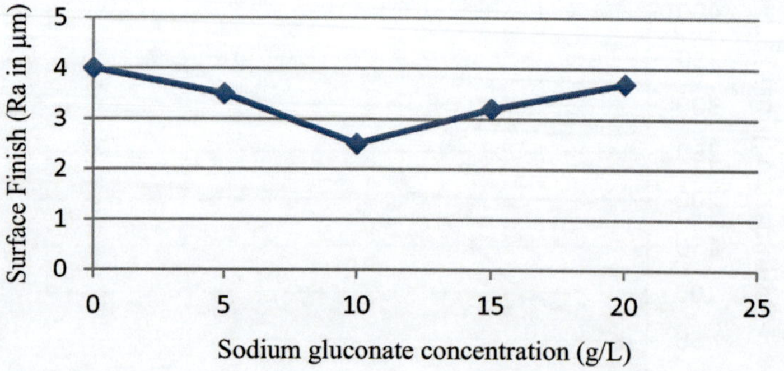

Fig. 4 Effect of sodium gluconate on surface finish

10 g/L. Once the concentration is more than 10 g/L, the surface roughness started to increase and it may be due to the over–saturation of the solution above 10 g/L.

Based on the above various chemicals effect, the composition with 100 g sodium hydroxide, 5 g of sodium sulphide, and 10 g of sodium gluconate in one litre of DM water is giving the better surface finish with a considerable chemical milling rate (Figs. 5 and 6).

Fig. 5 Surface microstructure after mechanical milling process at various magnifications

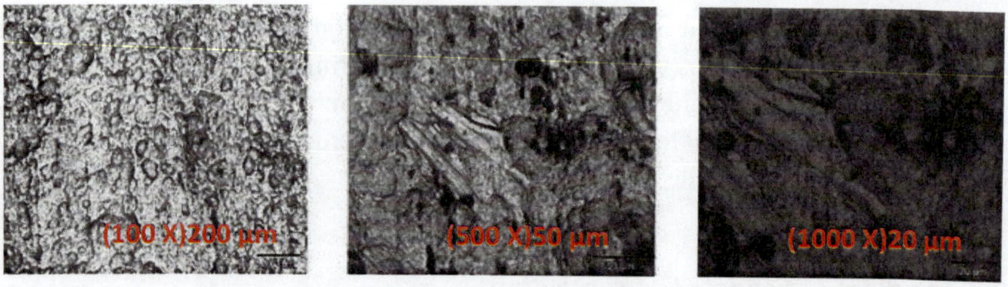

Fig. 6 Surface microstructure after chemical milling process at various magnifications

3.3 *Microstructure Analysis*

Based on the above optical microscopy analysis, it is understood that the mechanical milling process is giving a better surface finish and texture as it removes all elements in the material irrespective of other alloying elements. In contrast to the above, some of the alloying elements like copper and zinc are left out on the surface after the chemical milling process as the material removal rate varies with element nature. Hence, it is evident of precipitation of black precipitations are formed which are due to the formation of copper and zinc precipitates with free sulphur resulting in the formation of bumpy and nodular deposition.

3.4 *Mechanical Properties*

Sub-size tensile test specimens are prepared from mechanically milled and chemically milled test plates for 500 μm. Mechanical properties are compared for both mechanical and chemical milling test specimens which are subjected to chemical milling composition. It is observed that the mechanical properties of mechanically milled samples are slightly high due to the presence of compressive surface stresses generated during the milling process which helps for higher mechanical properties. However, material removal during the chemical milling process by chemical reaction is very slow and will remove atom by atom from the surface without the application of any external forces. These surface stresses are either compressive or tensile. However, in most cases, compressive stresses are generated during the mechanical milling process due to the application of compressive cutting tool forces which are completely absent in the chemical milling process (17–18). Due to this, it is observed that slight reduction in the mechanical properties which are true mechanical properties in absence of the surface compressive stresses generated during the mechanical milling process (Fig. 7).

4 Conclusions

- Aluminium alloy thin-walled shells can be realized by a simple and cost-effective chemical milling process.
- Visual appearance of various stages of a chemical milling operation is seen in Fig. 8 for reference.
- Various chemical milling maskants have been studied and compared. Neoprene-based masking agent has provided better adhesion and reduced the edge effect during the chemical milling process.

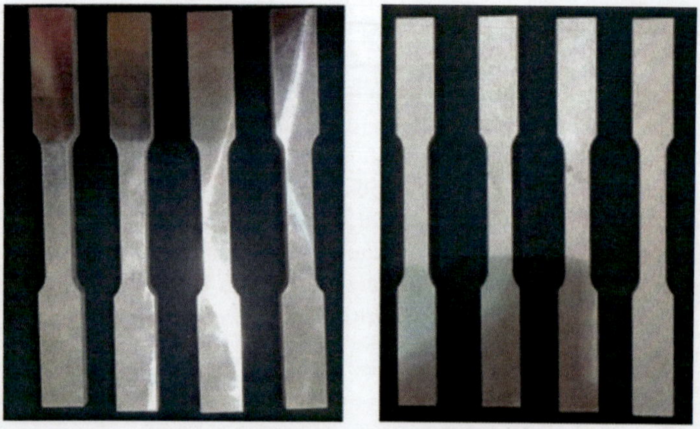

Fig. 7 Mechanical milled and chemical milled tensile test specimens

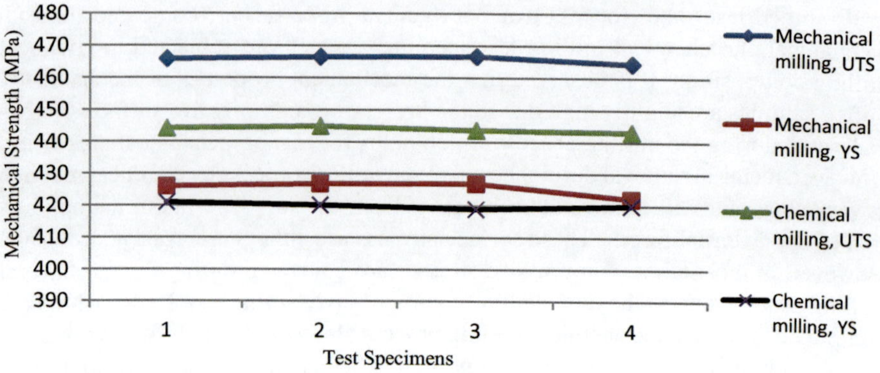

Fig. 8 Mechanical milling and chemical milling comparison

- By adding sodium salt of gluconate and active sulphide containing chemicals improve the surface finish by reducing the wetting angle at the milling interface and by precipitating the other alloying elements like copper and zinc.
- Chemical milling rate is independent of no. of components. Hence, the production rate can be increased by increasing the chemical milling tank capacity and more components can be loaded simultaneously.
- There is not much degradation of surface finish after the chemical milling process
- There is a slight reduction in the mechanical properties which are true mechanical properties of the material after the chemical milling process.
- Anodizing may be carryout without any difficulty after the chemical milling process.

(A) (B) (C)

Fig. 9 Various stages of chemical milling process a before milling, b during milling, c after milling and removal of masking agent

- Since this process is involved in the preparation of chemical milling solution and start using by monitoring the process parameters and milling rate, hence no machinery and capital investment are required.
- Chemical milling rate will depend on solution temperature, solution composition, time, and aluminium content which is already dissolved in the solution during the chemical milling process (Fig. 9).

Acknowledgements The authors express their sincere thanks to the Product Development Division of the Directorate of Engineering, DRDL, Hyderabad for their process planning support.

The authors also express their sincere thanks to the Alignment and Inspection Centre (AIC) of the Directorate of Engineering, DRDL, Hyderabad for non–destructive wall thickness measurement during the milling process.

References

1. Paul DeGarmo E, Black JT, Kohsern RA (1997) Materials and processes in manufacturing, 8th edn. Prentice–Hall, Englewood Cliffs
2. Langworthy M (1994) Chemical milling nontraditional machining process machining handbook. ASM
3. Chemical machining, metals handbook, vol 3, Machining, 8th edn. American Society for Metals, Metals Park, pp 240–248
4. Brinksmeier E, Lucca DA, Walter A (2004) Chemical aspects of machining processes. CIRP Ann Manuf Technol 53(2):685–699

5. Yankee HW (1979) Manufacturing processes. Chemical Milling Prentice–Hall, Englewood Cliffs, p 22
6. Dini JW (1984) Fundamentals of chemical machining. Am Mach Spec Rep 768:128
7. US Patent No.2795490: Process for etching aluminium alloy surfaces
8. NASA Document on chemical milling (Brief 66– 10312, Year 1966)
9. US Patent No. 3356550: chemical milling of aluminium alloys
10. US Patent No. 4588474: chemical milling processes and etchants therefor
11. US Patent No. 3464870: aluminium polishing process
12. Çakır O, Temel H, Kiyak M (2005) Chemical etching of Cu–ETP copper. J Mater Process Technol 162–163:275–279
13. Çakır O (8002) Chemical etching of aluminium. J Mater Process Technol 199(1–3):337–340
14. Fadaei Tehrani A, Imanian E (2004) A new etchant for the chemical machining of St304. J Mater Process Technol 149:404–408
15. Al-Ethari HA, Alsultani KF, Dakhil N (2003) Variables affecting the chemical machining of stainless steel 420. Int J Eng Innov Technol (IJEIT) 3(6):810–216
16. Ting HT, Abou–El–Hossein KA, Chuo HB (2009) Review of micro machining of ceramics by etching. Tans Non Ferrous Met Soc China 19:1–19
17. Book chapter on chemical milling by Eugene M. Langworthy
18. Effect of Milling speed and feed on surface residual stress of 7050–T7451 aluminium alloy

Effect of Magnetizing Parameters on Surface Integrity During Dry and Cryogenic Grinding of AISI D2 Tool Steel

Ashwani Sharma, Abhimanyu Chaudhari, Mohd Zaheer Khan Yusufzai, and Meghanshu Vashista

Abstract An experimental study was carried out to compare the role of process parameters, such as magnetizing frequency, magnetic field intensity, and depth of cut on the ground surface of AISI D2 tool steel under different grinding modes, i.e., dry and cryogenic conditions. The Barkhausen noise (BN) response from the ground surface of the workpiece was measured in terms of root mean square (RMS) and Peak. It was noted that variation in magnetizing frequency and depth of cut (DOC) inversely affect the RMS and Peak value of the BN signal. Also, variation in magnetic field intensity (MFI) directly affects the BN response parameters. The magnetizing frequency, MFI, and DOC were ranged between 10 to 40 Hz, 200 to 500 Oe, and 10 to 40 μm, respectively. The result revealed a better BN response of the cryogenic ground sample over the dry ground sample owing to fewer obstacles or hurdles for movement of the magnetic domain wall. A linear correlation between process parameters and BN responses could be obtained.

Keywords AISI D2 tool steel · Barkhausen noise analysis · Grinding

1 Introduction

In general, the metallurgical sector requires a technique that can characterize material accurately, quickly, and easily without causing any damage to the material being evaluated. The Barkhausen noise (BN) method allows several benefits like

---

A. Sharma (✉) · A. Chaudhari · M. Z. K. Yusufzai · M. Vashista
Department of Mechanical Engineering, Indian Institute of Technology (BHU), Varanasi 221005, India
e-mail: ashwanisharma.rs.mec18@itbhu.ac.in

A. Chaudhari
e-mail: abhimanyuchaudhari.rs.mec18@itbhu.ac.in

M. Z. K. Yusufzai
e-mail: mzkhan.mec@iitbhu.ac.in

M. Vashista
e-mail: mvashista.mec@iitbhu.ac.in

U. S. Dixit et al. (eds.), *Advances in Forming, Machining and Automation*, Lecture Notes in Mechanical Engineering, https://doi.org/10.1007/978-981-19-3866-5_20

the higher intensity of insertion, quick measurement, and compact machine over different methods. It is an economical non-destructive technique (NDT) that can analyze samples of all shapes and sizes under various environmental conditions. When a ferromagnetic substance is magnetized by an external magnetic field, BN signals are produced owing to the movement of magnetic domain walls. The movement of magnetic domain walls is hindered due to the defects like inclusion or dislocation and grain boundaries. The magnetic domain walls movement continued till it reached the stability point, and the pick-up coil received this movement in the form of a response signal.

The Barkhausen noise is used to evaluate the surface integrity of intricate profiles such as gears and camshafts [1]. Therefore, as compared to other NDT methods, BN can evaluate the surface integrity of a ground component in [2]. The BN signal penetration depth depends upon the applied magnetic field frequency, and the skin depth from the material to be evaluated depends upon the frequencies of low pass filter and high pass filter. These two analyzing parameters are required to be controlled for the assessment of surface integrity from different depth regions. Surface integrity involves surface oxidation, tensile residual stress, change in microhardness and microstructure, and surface roughness. Aside from that, surface integrity has a substantial impact on service life, as poor surface integrity can lead to product failure before the expected service life. Therefore, the highest intensity of insertion by which BN signal is characterized relies on the depth of the magnetic field, magnetic frequency, medium's permeability, length separating the poles of the magnetic yoke, and BN signal's responsiveness. To obtain data of different insertion depths for BN parameters, the magnetizing frequency and magnetic voltage are critical parameters [3]. An increment in the frequency of the magnetic field can improve the extent of measurement [4]. The applied magnetic voltage over the magnetic coil influences the insertion intensity [5]. An influence of change in magnetic pulse on BN magnitude while assessing the time and space distribution of magnetic flux density was investigated by Augustyniak et al. [6]. Another study found variations in magnetic characteristics of BN signals upon distinct frequencies during evaluating surface decarburization of steels at varying depths [7]. Yusufzai et al. [8] reported variation in BN signal parameters in terms of RMS and number of pulses at various input parameters such as MFI and magnetizing frequency. Microhardness and microstructure outcomes correlated with both BN signals. RMS increases, and the number of pulses decreases as the magnetizing frequency rises at a constant 1000 Oe MFI, while the RMS and number of pulses rise when the magnetic field strength rises. Shrivastava et al. [9] conducted grinding tests on AISI D2 tool steel with an alumina wheel under dry and flood conditions. They analyzed the ground sample by the BN technique. It was found a linear relationship between response and process BN parameters. They also found decreasing trend of BN parameters (RMS and Peak) with an increase in DOC.

The precision surface grinding is usually a form and finishes material removal method in the manufacturing of various machine components, which needs significant surface finish as well as fine tolerances. The excessive temperature affects the surface integrity characteristics of workpiece materials and also affects the abrasive wheel, which depreciates the grinding productivity. To subdue the aforementioned limitation, cutting fluid is necessary to remove the heat from the machining region and adjacent area of the active cutting edge of grinding wheel grains. Moreover, the cutting fluid removes grinding chips as well as provides adequate lubrication. Besides, traditional cooling is insufficient at a larger grinding wheel velocity to seize a grinding region that begins on the interface connecting the workpiece and abrasive wheel because of the hydrodynamic boundary layer formed throughout the periphery of the grinding wheel [10]. Hence, it is difficult for the coolant to enter in cutting region to restrict the heat at a greater depth of cut, which results in a thermal burn, surface, and sub-surface microcracks occur and also affect surface integrity [11]. Furthermore, heat generated in the grinding zone is high, due to which sometimes coolant gets heated and dissipated instantly, which harms the operator's health and contaminates the surroundings [12]. Consequently, eco-friendly cutting fluid and lubricant, known as cryogenic cooling, perform a critical role in industrial applications. Therefore, it is sustainable and efficiently constrains the cutting region temperature [13].

In precision engineering components fabrication industries, better surface integrity of machine components is required because the surface integrity considerably influences the component's service duration. Currently, various methods like X-ray diffraction, energy dispersive X-ray analysis, and light optical microscope are used by researchers and engineers to understand the surface integrity parameters of the ground workpiece. These methods are tedious, expensive, and time-consuming [14]. Therefore, various fabrication manufacturers get a novel machine to probe the specimen during online production recognized as the BN technique. After the critical literature survey, it was found the BN method is frequently utilized in the machining of ferromagnetic materials. However, there is very little work present on the ground surface investigation of hard-to-cut steel. Therefore, the present study evaluated AISI D2 tool steel on various BN and grinding parameters.

2 Experimentation Procedure

The AISI D2 tool steel samples of rectangular shape with $80^L \times 15^W \times 8^T$ mm dimensions were prepared for grinding operation in plunge surface grinding mode using HMT 455 H grinder machine (Fig. 1). Many pilot experiments were carried out on tool steel samples to decide the range of input parameters. The samples were ground at a depth of 10–40 microns with a table speed of 8 m/min. Figure 2 shows the thermal damage on the ground sample at low and high DOC under various grinding modes. All experiment was completed as per the block diagram as presented in Fig. 3. Table 1 shows the details of the grinding parameters.

The reason for selecting the AISI D2 tool steel material is that it is more suitable for the application of punch, stamping, and injection dies because of its high wear and corrosion resistance. The chemical composition of the work material is illustrated in Table 2.

Fig. 1 Experimental setup of cryogenic grinding

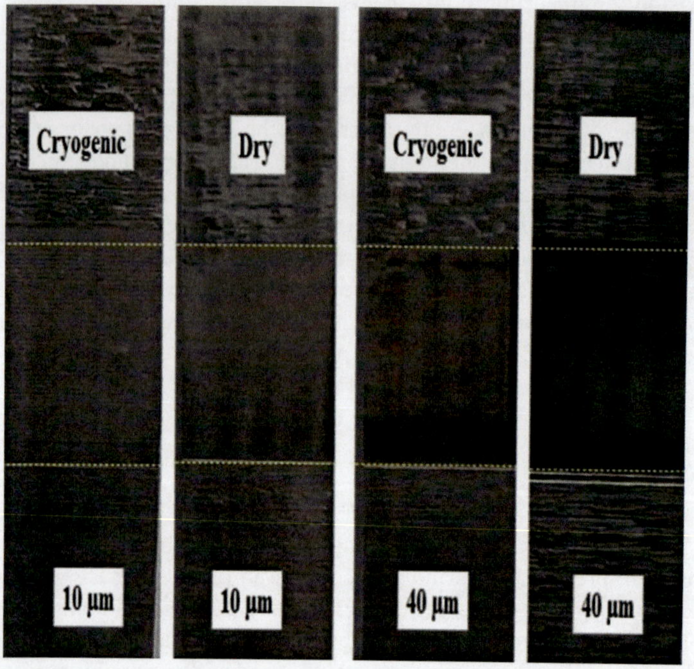

Fig. 2 Actual photographs of ground samples at different depths of cut

The Barkhausen noise technique is used to analyze the surface integrity of work material. The BN response parameters (i.e., RMS and Peak) were measured using the commercially available BN analyzer supplied by Technofour Magstar, India (Fig. 4). A cuboid shape type magnetic probe with two magnetizing yokes and a pick-up coil

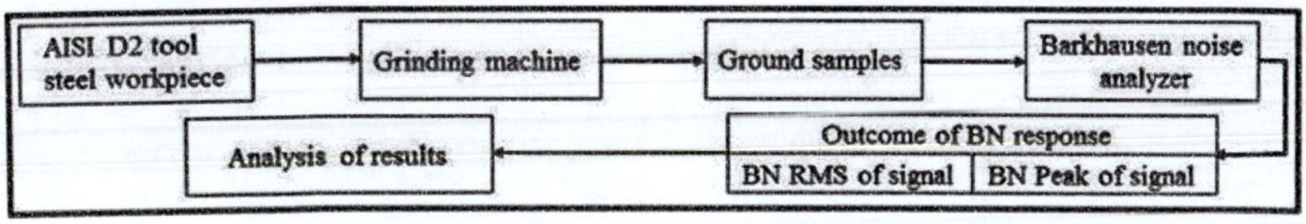

Fig. 3 Block diagram of the experimentation

Table 1 Grinding parameters

Parameters	Setting
Grinding mode	Plunge grinding
Grinding mode	1. Dry grinding 2. Cryogenic grinding
Grinding wheel	Alumina (AA60K5V6)
Work material	AISI D2 tool steel
Wheel speed	39.42 m/s
Table speed	8 m/min
Downfeed	10–40 μm, step of 10
Dresser	Single-point diamond

Table 2 Chemical composition of the AISI D2 tool steel (wt. %)

Elements	Fe	C	Cr	Si	Mn	Mo	V
Weight %	Balance	1.55	11.80	0.30	0.40	0.90	0.80

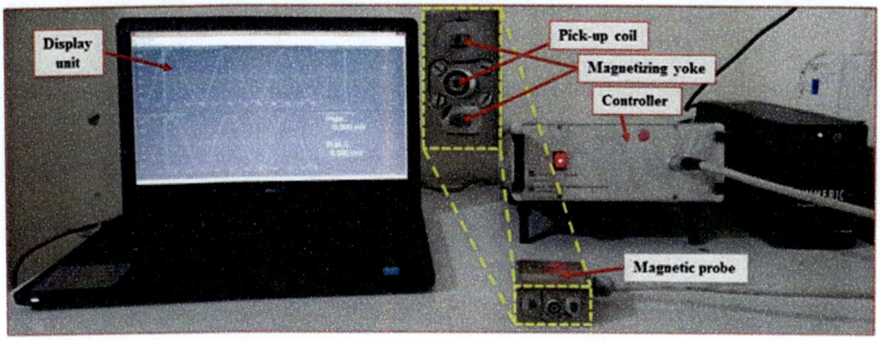

Fig. 4 Barkhausen noise analyzer

at the center was used to create a magnetic field and take the BN signal. The ground surface of the sample was cleaned before the measurement of BN response parameters obtained by BN signal. For BN measurement, the location was selected at the central area of samples. The RMS and Peak value of BN signal were chosen for analysis in the present study. The average value of five observations for each sample (analyzed

Table 3 Parameters for BN analysis

Magnetizing frequency	Magnetic field intensity	Filter frequency
10, 20, 25, 30, 40 Hz	200, 300, 400, 450, 500 Oe	100–300 Hz

by Magstar software) was taken for analysis. Table 3 represents the details of BN parameters used for analysis.

3 Results and Discussion

The Barkhausen noise analyzer is a quick, eco-friendly, and non-destructive method for analyzing ground samples. The BN analysis presents major advantages compared to the destructive method.

3.1 *Results*

During BN analysis of tool steel material, it was observed that RMS and Peak value decrease with an increase in magnetizing frequency (f) and DOC. The depth penetration of signal in BN analysis depends on the applied MFI, applied magnetizing frequency, and frequency range of band-pass-filter. Consequently, the depth penetration can be risen by using higher MFI, low magnetizing frequency, and selecting the proper low pass filter and high pass filter frequency.

The analysis from the outcome of BN parameters with magnetizing frequency, MFI, and DOC of AISI D2 tool steel material was done and plotted the graph with the help of Tables 4, 5, 6, and 7.

In the present research work, Fig. 5 shows the variation in BN response parameters with input parameters. According to Fig. 5a, RMS value of BN signal decreases with an increase of magnetizing frequency, and Fig. 5b also represents the decreasing

Table 4 Outcome values of BN parameters with variation in magnetizing frequency

Magnetizing frequency (Hz)	Ground sample (dry) at DOC: 40 μm, MFI: 450 Oe		Ground sample (cryogenic) at DOC: 40 μm, MFI: 450 Oe	
	RMS values (mV)	Peak values (mV)	RMS values (mV)	Peak values (mV)
10	0.035	0.332	0.041	0.391
20	0.032	0.259	0.038	0.375
30	0.03	0.239	0.037	0.298
40	0.029	0.229	0.035	0.278

Table 5 Outcome values of BN parameters with variation in depth of cut

DOC (μm)	Ground sample (dry) at *f*: 10 Hz, MFI: 450 Oe		Ground sample (cryogenic) at *f*: 10 Hz, MFI: 450 Oe	
	RMS values (mV)	Peak values (mV)	RMS values (mV)	Peak values (mV)
10	0.046	0.464	0.056	0.552
20	0.045	0.396	0.052	0.464
30	0.041	0.38	0.046	0.415
40	0.035	0.332	0.041	0.391

Table 6 Outcome values of BN parameters with variation in magnetic field intensity

MFI (Oe)	Ground sample (dry) at *f*: 25 Hz, DOC: 40 μm		Ground sample (cryogenic) at *f*: 25 Hz, DOC: 40 μm	
	RMS values (mV)	Peak values (mV)	RMS values (mV)	Peak values (mV)
200	0.01	0.063	0.011	0.083
300	0.013	0.093	0.025	0.225
400	0.025	0.19	0.044	0.4
500	0.042	0.352	0.061	0.483

Table 7 Variation in BN RMS and peak with change in depth of cut under dry and cryogenic grinding

DOC (μm)	Ground sample (dry) at *f*: 25 Hz, MFI: 500 Oe		Ground sample (cryogenic) at *f*: 25 Hz, MFI: 500 Oe	
	RMS values (mV)	Peak values (mV)	RMS values (mV)	Peak values (mV)
10	0.055	0.508	0.069	0.596
20	0.05	0.41	0.06	0.532
30	0.043	0.391	0.057	0.483
40	0.042	0.352	0.058	0.483

trend of Peak value with respect to the magnetizing frequency in increasing value. It clearly indicates that BN parameters for both the ground samples decrease with an increase in the magnetization frequency as well as the value of BN parameters were higher in the case of cryogenic grinding in comparison with the dry grinding.

From Fig. 5c, it is noticed that with the increase in DOC of the ground sample, there is a reduction in BN (RMS), and the same trend is seen from Fig. 5d between (Peak) and DOC.

The effect of variation in MFI and DOC on BN response parameters of different ground samples under dry and cryogenic grinding modes was observed at constant magnetizing frequency 25 Hz for MFI input parameter and constant MFI 500 Oe for DOC grinding input parameter (Fig. 6). RMS and Peak values were found in increasing trends concerning the increasing trend of MFI, as shown in Fig. 6a and b.

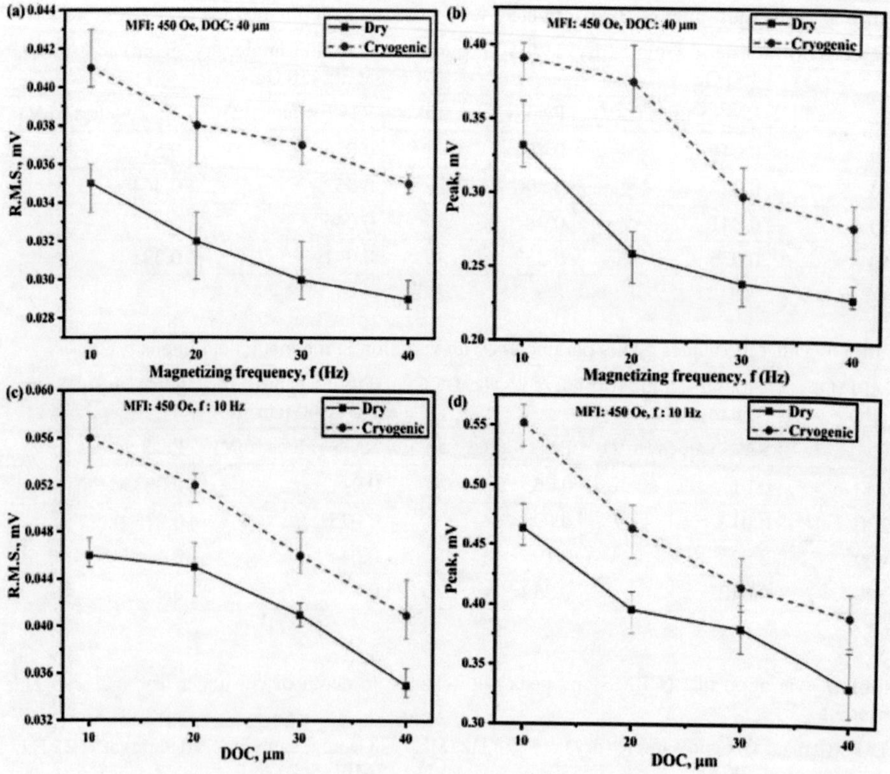

Fig. 5 Variation in BN parameters **a** RMS and **b** Peak with magnetization frequency, **c** RMS and **d** Peak with the depth of cut of ground sample

In Fig. 6c and d, the variation of RMS and Peak parameters with the DOC was seen oppositely against the RMS and Peak parameters with the MFI.

3.2 Discussion

Barkhausen noise parameters, i.e., RMS and Peak of AISI D2 tool steel ground sample with different DOC under two grinding modes, viz. dry and cryogenic, depict BN analysis affected by magnetizing frequency and MFI. In this present work, BN parameters were also influenced by the DOC. The ground sample was magnetized by applying the external magnetic field using magnetizing yoke of the magnetic probe. In the direction of the applied magnetic field, the magnetic domains align. Further, this magnetic field penetrated the ground sample with the influence of magnetizing frequency. Higher magnetizing frequency indicates less penetration depth to the ground sample. It reduces the coverage range of the penetrating magnetic field,

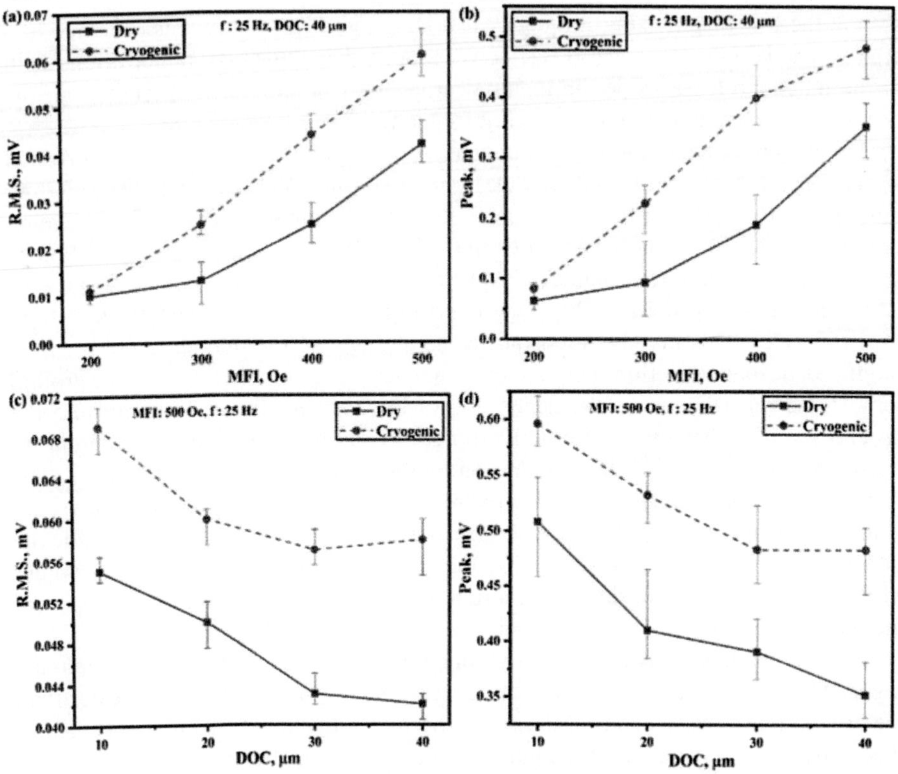

Fig. 6 Variation in BN parameters **a** RMS and **b** Peak with magnetic field intensity, **c** RMS and **d** Peak with the depth of cut of ground sample

resulting in the movement of more limited magnetic domain walls. A magnetic domain is an area inside the ground sample, i.e., defined by a magnetic polarity boundary. Motions of magnetic domains interior the material depended on the number of voltage pulses. Therefore, less movement of the magnetic domain meant less number of voltage pulses, resulting in a low value of RMS and Peak BN parameters. In the case of domain wall movement, a pinning site is formed that produces the resistance. According to Sharma et al. [15], grain boundaries are accountable for this resistance or obstacle to the movement of the domain wall. After that, they found poor results of BN response.

Magnetizing frequency, MFI, and DOC were in a wide range (i.e., 10–40 Hz, 200–500 Oe, and 10–40 μm, respectively) to perceive the effect of low magnetizing frequency, MFI and DOC, and high magnetizing frequency, MFI, and DOC on the movement of magnetic domains. In the ground sample, the BN RMS and BN Peak decreased (0.046–0.035 mV and 0.464–0.332 mV in dry, 0.056–0.041 mV, and 0.552–0.391 mV in cryogenic mode) with the increase in magnetizing frequency (10–40 Hz), as shown in Fig. 5a and b. Here, a higher value of BN parameters was seen at the low magnetizing frequency (10 Hz), and a low value was obtained at a high magnetizing frequency (40 Hz). It was mainly due to the high penetration depth of the magnetic field inside the material at a low magnetizing frequency and low penetration depth at a high magnetizing frequency.

The outcomes of the response parameters, such as RMS and Peak values, increased with an increment in MFI (200–500 Oe). The RMS and Peak values in dry grinding mode ranged from 0.01 to 0.042 mV and 0.063 to 0.352 mV, respectively. In cryogenic grinding mode, BN values ranging from 0.011 to 0.061 mV in RMS and 0.083 to 0.483 mV in Peak were obtained. From these results, it was noted less value of RMS and Peak at low MFI (200 Oe) and a high value of RMS and Peak at high MFI (500 Oe). This may be a reason for the strength of the magnetic field, which produced the magnetization energy to move the magnetic domain wall into the ground sample. Lower MFI means the generation of lower magnetization energy into the ground sample. Usually, a minimum strength of the magnetic field is essential to cause motion of the magnetic domains and allows them to cross the obstacles. The differences in values of BN parameters were noticed significantly less or close to zero at low MFI (Fig. 6a and b) because of more obstacles and less field energy inside the material. The increased field energy by the strength of the magnetic field at higher MFI provided the high accelerated motion to magnetic domain walls to jump over the obstacles/hurdles.

On the other hand, BN results were found higher at the low DOC and lower at the higher DOC. It was also remarkable to observe the cryogenic plotline greater than the dry plotline (Figs. 5 and 6). Because of thermal damage over the nascent surface, this restricts the movement of the magnetic domain wall. High DOC indicates high thermal damage, as shown in Fig. 2. Kumar et al. [16] reported that the root means square (RMS) and Peak value of the BN signal are inversely affected by DOC in the analysis of ground steel. They were also obtained dissimilarities trend in BN (RMS) with DOC. Sharma et al. [15] observed that BN profile changes from low to high in BN signal with increased MFI at different frequencies. They stated that better BN response parameters regarding higher RMS and Peak were received in cryogenic grinding mode than dry condition.

4 Conclusions

The study of BN results presented the following conclusions.

1. The BN response outcomes, namely RMS and the Peak value of the BN signal, gets affected by the magnetic variables and grinding process parameters. These response outcomes have lower values in dry grinding than cryogenic grinding due to less movement of the magnetic domain wall into the ground sample.
2. The penetration depth of the magnetic field into the sample is affected by the frequency of the applied magnetic field. Increased magnetizing frequency reduces the penetration depth of the magnetic field, which reduces the overall magnetic field's coverage area.

3. BN outcomes, i.e., RMS and Peak decrease with an increment in magnetizing frequency at constant MFI 450 Oe and DOC of 40 μm. In the cryogenic grinding, ~17 and 18%, higher values of RMS and Peak were observed at 10 Hz frequency, compared to dry grinding.
4. The grinding process parameter, i.e., DOC significantly affects the magnetic responses because of the large heat accumulated in the nascent surface at higher DOC, which further restricts the magnetic domain wall displacement.

References

1. Moorthy V, Shaw BA, Mountford P, Hopkins P (2005) Magnetic Barkhausen emission technique for evaluation of residual stress alteration by grinding in case-carburized En36 steel. Acta Mater 53(19):4997–5006
2. Vashista M, Ghosh S, Paul S (2009) Application of micromagnetic technique in surface grinding for assessment of surface integrity. Mater Manuf Process 24(4):488–496
3. Awale A, Shrivastava AK, Chaudhari A, Vashista M, Yusufzai MZK (2021) Micro-magnetic characterization of ground AISI D2 tool steel using hysteresis loop technique. Int J Mater Prod Technol 62(1–3):180–198
4. Makar JM, Tanner BK (2000) The effect of plastic deformation and residual stress on the permeability and magnetostriction of steels. J Magn Magn Mater 222(3):291–304
5. Harrison IS (2005) Detecting white layer in hard turned components using non-destructive methods. Dissertation, Georgia Institute of Technology
6. Augustyniak M, Augustyniak B, Piotrowski L, Chmielewski M, Sablik MJ (2005) Evaluation of time and space distribution of magnetic flux density in a steel plate magnetized by a C-Core. In: Proceedings of the of 5th international conference on Barkhausen noise and magnetic testing, pp 191–198
7. Mayos M, Segalini S, Putignani M (1987) Electromagnetic nondestructive evaluation of surface decarburization on steels: feasibility and possible applications. In: Review of progress in quantitative nondestructive evaluation. Springer, Boston, MA, pp 1691–1699
8. Yusufzai MZK, Raja AR, Gupta SK, Vashista M (2021) Barkhausen noise analysis of friction stir processed steel plate. Indian J Eng Mater Sci 27(3):670–676
9. Shrivastava AK, Sharma A, Awale AS, Yusufzai MZK, Vashista M (2021) Assessment of grinding burn of AISI D2 tool steel using Barkhausen noise technique. J Inst Eng (India) Ser C 1–12
10. Awale AS, Vashista M, Yusufzai MZK (2020) Multi-objective optimization of MQL mist parameters for eco-friendly grinding. J Manuf Process 56:75–86
11. Yin GQ, Gong YD, Li YW, Cheng J (2018) Research on force and temperature characteristics of novel point grinding wheels. J Mech Sci Technol 32(8):3817–3834
12. Chaudhari A, Awale AS, Chakrabarti AK (2019) Surface integrity characterization of austenitic, martensitic, and ferritic stainless steel under different grinding processes. Mater Res Express 6(11):1165c9
13. Hong SY (2001) Economical and ecological cryogenic machining. J Manuf Sci Eng 123:331–338
14. Lasaosa A, Gurruchaga K, Arizti F, Martinez-De-Guerenu A (2017) Induction hardened layer characterization and grinding burn detection by magnetic Barkhausen noise analysis. J Nondestr Eval 36(2):27

15. Sharma A, Chaudhari A, Awale AS, Yusufzai MZK, Vashista M (2021) Effect of grinding environments on magnetic response of AISI D2 tool steel. Russ J Nondestr Test 57(3):212–221
16. Kumar S, Yadav M, Agrawal P, Zaheer Khan M, Vashista M (2011) Assessment of microhardness profile in grinding using Barkhausen noise technique at various analysis parameters. Int Sch Res Not

Effect of Peak Current on Material Removal Rate During EDM of Ti–6Al–4V Using Cold Treated Brass Electrode

B. K. Tharian, Munna Kumar, P. B. Dhanish, and R. Manu

Abstract Ti–6Al–4V or Titanium Grade 5 alloy is the workhorse material of many vital industries, including aerospace, automobile, biomedical, and die industries. The titanium alloy is not preferred to machines using conventional techniques because of the poor machinability due to its hardness and poor thermal conductivity. Electric discharge machining can cut any material irrespective of its hardness and is one of the potential machining operations for machining titanium alloys. Apart from its ability to cut any material, the EDM process's material removal rate (MRR) is very low. The current research focuses on exploring the effect of peak current on MRR using cold treated brass electrode compared with a non-treated brass electrode. MRR increases with an increase in peak current, and it drops at higher peak current values. Cold treatment of brass electrodes has a positive effect on the MRR when compared with the non-treated electrode.

Keywords Cold treatment · MRR · Peak current

1 Introduction

Electric discharge machining (EDM) process plays a vital role in several industries like aerospace, automotive, biomedical, and die industries owing to its ability to machine any material irrespective of its hardness [1]. Material removal in EDM is based on spark erosion. In this process, the workpiece and the tool will act as separate

B. K. Tharian (✉) · M. Kumar · P. B. Dhanish · R. Manu
Department of Mechanical Engineering, National Institute of Technology Calicut, Calicut, Kerala 673601, India
e-mail: bibinktharian@gmail.com

M. Kumar
e-mail: munna_m200529me@nitc.ac.in

P. B. Dhanish
e-mail: dhanish@nitc.ac.in

R. Manu
e-mail: manu@nitc.ac.in

U. S. Dixit et al. (eds.), *Advances in Forming, Machining and Automation*, Lecture Notes in Mechanical Engineering, https://doi.org/10.1007/978-981-19-3866-5_21

electrodes, and dielectric fluid acts as a medium between them. The plasma of a very high temperature occurs when the supply is fed between the electrodes [2]. Thus, the material at the ends of plasma gets removed by melting and vaporization. One of the significant limitations of this process is its slow pace, i.e., the lesser value of material removal rate [3]. Researchers are working on several methods to improve the EDM performance by improvising the tool, workpiece, and other machining conditions like the rotary tool, ultrasonic-assisted magnetic field-assisted, dry, cryo-treated, powder mixed, and a lot more.

Lightweight, high-temperature application, and bio-compatibility are factors favoring the usage of titanium alloy in a wide range of aerospace and biomedical industries. Titanium Grade 5 alloy (Ti–6Al–4V) is one of the most widely used titanium alloys [2]. EDM is one of the most promising techniques for machining titanium alloy. One of the challenges while machining titanium alloys using EDM is the poor thermal expansion and thermal conductivity, which demands more energy for material removal [3]. Cryogenic treatments can be applied to enhance the properties of metals [4]. Several researchers [5–11] worked on the sub-zero treatment of EDM electrodes to improve the EDM performance in terms of MRR, TWR, and surface integrity. Sub-zero treatments can be divided into cold treated (−50 to −80 °C), shallow treated (−80 to −160 °C), and deep cryo-treated (−160 to −196 °C) based on the temperature to which they are cooled.

Sundaram et al. [5] experimentally studied the EDM performance based on a cryogenically treated tool. Copper tool electrode was cold treated at −101 °C and deep cryogenic treated at −184 °C and observed that the effect of deep cryogenic treatment was more significant than cold treatment on MRR of Be-Cu material but negligible effect on TWR of both treatments. Abdulkareem et al. [6] evaluated the effect of cooling of the copper electrode on electrode wear (EW) and surface roughness (SR) during the die-sinking of electrical discharge machining on titanium alloy (Ti–6Al–4V) using cryo-cooled copper electrode. EW decreased by 27% due to cryogenic cooling of the electrode compared to the normal electrode. Peak current was the most significant parameter for EW than pulse-on time. In this work, the authors employed cryogenic cooling during the machining by modifying the electrode with provisions for liquid nitrogen to pass through the electrode. Gill et al. [7] investigated the machinability of Ti6246 alloy in electric discharge drilling with the aid of a cryogenically treated electrode. He observed that MRR, TWR, and SR improved by 8.5, 34.78, and 9.01%, respectively. The deep cryogenic treatment is recommended for a long duration of EDD and where high precision is required.

Hui et al. [8] explored the effect of cryogenically cooled electrodes on discharge gap and discharge characteristics during electrical discharge machining of Ti–6Al–4V with distilled water as the dielectric medium. As per their observations, equivalent heat transfer and specific heat capacity increased due to the effect of cryogenic treatment. Due to that, EWR decreased, discharge gap, and energy density of the discharge channel reduced due to this corner size machining improved. Sanjeev et al. [9] investigated the effect of cryogenic treatment on both materials, i.e., the tool and the workpiece. Three grades of titanium alloys were cryogenically treated and machined with cryogenically treated electrodes during powder mixed electric discharge machining (PMEDM). The author observed that peak current was the most significant factor for TWR. A significant effect of cryogenic treatment on the tool was observed but on the workpiece, and it is not significant. In another research, the same authors [10] studied the effect of deep cryogenic treatment on TWR during electric discharge machining of Ti–5Al–2.5Sn titanium alloy. Peak current was the most significant factor affecting the TWR of the Cu-W electrode. TWR of deep cryogenic treated electrodes improved by 15.86% compared to non-treated electrodes. Goyal et al. [11] studied the cryogenically treated and non-treated AISI D2 tool steel as a workpiece with cryogenic cooled, cryogenic treated, and the non-treated copper electrode on EDM. MRR is increased by 18%, and TWR and surface roughness improved by 26 and 11%, respectively.

Manoj et al. [12] studied the MRR, TWR, white layer thickness, and SR of the Ti–6Al–4V with the copper electrode on EDM. The author observed that peak current imposes a positive effect on MRR, TWR, Ra, and thickness of the white layer. EDM machined surface detects the formation of TiC, and surface crack density increases by increasing peak current. Tahir et al. [13] explored the effect of cold treatment on brass wire electrodes. Cold treatment is carried out at −70 °C for 24 h for refining the microstructure of the brass wire. Cold treated brass wires at low energy conditions reduce the thickness of the recast layer and wire material infusion when compared with non-treated brass wire. Fatih et al. [14] used the deep and shallow cryogenic treatment on Ti–6Al–4V for improving the machinability of Ti–6Al–4V alloy wire electrodischarge machining based on MRR and SR. The author observed that MRR increases by 5% and SR decreases by 24%, and that aging treatment lowers both machinability and electrical conductivity of the material. In this study, both treatments, i.e., shallow and deep, increase the MRR; however, the deep treatment made more improvement.

Naveen et al. [15] investigated the MRR and SR of Ti–6Al–4V with the cryo-treated and coated copper electrode. They observed that MRR increased by 30.85% with the cryo-treated copper electrode, and TWR and SR decreased. Jagtar et al. [16] employed 3D printing to fabricate complex electrodes with cooling channels which reduced EWR by 42.84%. This work exposed the possibilities of 3D printing that can provide cryogenic cooling channels which will reduce the temperature of the electrode by the cooling effect on the electrode, and this cooling effect will further reduce the wear of the electrodes which is one of the challenges faced during the EDM operation with highly conductive electrodes.

In the present research work, the effect of peak current is analyzed while performing EDM on Ti–6Al–4V using cold treated brass electrodes, and the same has been compared with the non-treated brass electrode.

2 Materials and Methods

Experiments were conducted on Electronica xpert1 EDM die-sinking machine using cold treated and non-treated brass electrodes. Table 1 shows the chemical composition of the Ti–6Al–4V used for experimentation.

Cold treatment of brass electrodes was performed at −80 °C for 24 h in a sub-zero treatment unit. To avoid the thermal shock by exposing to sub-zero temperatures, electrodes were gradually brought to −80 °C at a temperature rate of 1.67 °C per minute. Each run has been conducted for a machining time of 10 min. Experiments were varied in 5 levels of peak current (I_p), and other experimental conditions were summarized and listed in Table 2, and the experimental setup is shown in Fig. 1, and the specimen after machining is shown in Fig. 2.

The weight loss method [17] has been used to calculate material removal rate using

$$\text{MRR} = (\text{mass before machining} - \text{mass after machining})/(\text{density} \times \text{time}) \quad (1)$$

Weight measurements are done using analytical weigh balance equipment (Shimadzu AUX220). As the dielectric fluid is EDM oil, the workpiece and tool were cleaned and dried before weighing. To get rid of the carbon layer formed on the tool, the tool was polished when required using a 3 disc polishing machine (Model BP-3TV).

Table 1 Chemical composition of Titanium Grade 5 alloy (Ti–6Al–4V)

Element	Al	V	Fe	Cu	Ti
Weight (%)	6.05	4.25	0.25	1.5	87.95

Table 2 Experimental conditions

Workpiece	Ti–6Al–4V plates of 6 mm thickness
Tool	Non-treated and cold treated brass electrode 6 mm dia
Machining time	10 min
Dielectric fluid	EDM oil
Flushing	Side flushing
Polarity	Straight
Levels of peak current	9, 15, 21, 27, 33 A
Pulse-on time	200 μs
Gap voltage	40 V
Duty cycle	8

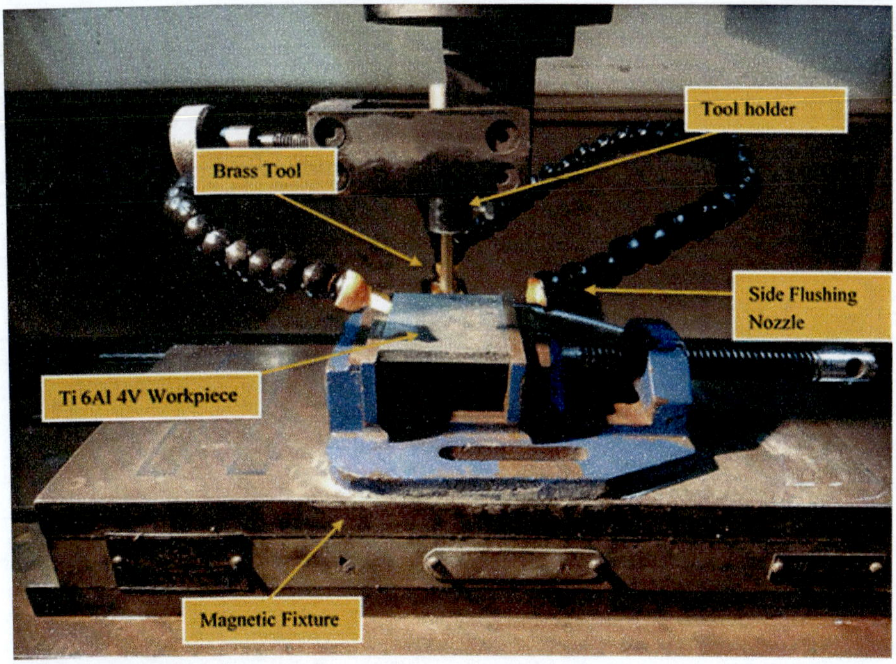

Fig. 1 Experimental setup

3 Results and Discussion

Experiments were conducted on five levels of peak current and the results are tabulated in Table 3. In the case of both the electrodes, the MRR value increases with the peak current value, a similar observation was made in previous works [12].

In Fig. 3, the highest value of MRR is obtained at a peak current value of 27 A, after that MRR drops. This could be the effect of a very high peak current value, which leads to the poor cooling performance by the dielectric due to overuse of dielectric consumption and thereby reducing the MRR. So, the outermost layer will be solidified back into the surface affecting the material removal. This problem of recast layer and surface integrity can be solved to an extent by using appropriate machining strategies [18]. Several researchers contributed to this changing trend of MRR by observing a layer of TiC at the outer [7]. When comparing the MRR using non-treated and cold treated brass electrodes, it is clear that the cold treatment has a significant positive effect on the MRR (refer to Table 3). Sub-zero treatments improve the conductivity of the electrodes [6], which increases MRR [11] due to the efficient transfer of discharge energy.

Magnification and profile of the surfaces that correspond to maximum MRR (Experiment No. 4, i.e., 27 A) were taken using Bruker Alicona InfiniteFocus G5. While analyzing the profiles and images, it is clear that higher currents reduce the

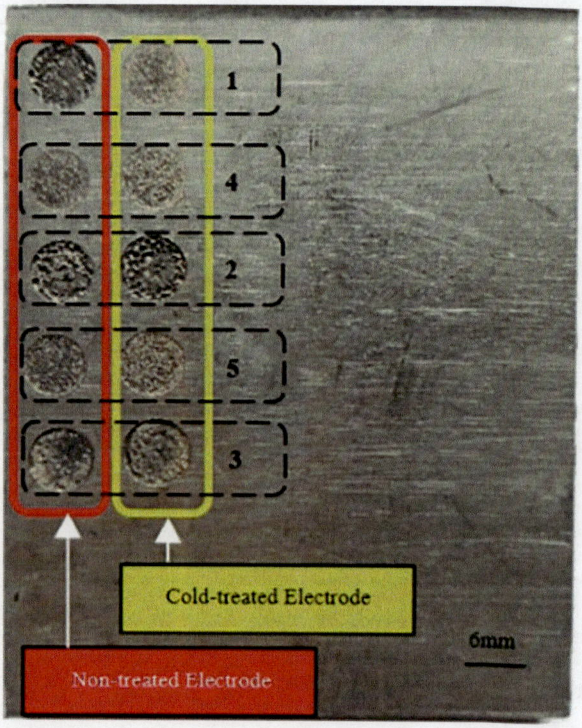

Fig. 2 EDM surfaces using non-treated and cold treated brass electrodes

Table 3 Experimental results

					MRR (mm^3/min)		
Expt no.	Peak current (A)	Voltage (V)	Pulse-on time (μs)	Duty cycle	Non-treated	Cold treated	Percentage increase
1	9	40	200	8	0.115	0.124	7.843
2	15	40	200	8	0.239	0.287	20.08
3	21	40	200	8	0.368	0.47	27.61
4	27	40	200	8	0.463	0.625	35.12
5	33	40	200	8	0.406	0.504	23.89

surface quality by creating deeper craters. Higher peak currents cause a large amount of discharge energy, which in turn produces deeper craters on the surface.

Compared to the surfaces produced by non-treated electrodes, the surface produced by cold treated electrodes has an inferior surface finish. These observations were made by comparing the areal surface roughness values (Fig. 4). The areal surface roughness of the surface produced by cold treated and non-treated electrodes was 8.072 and 6.066 μm, respectively.

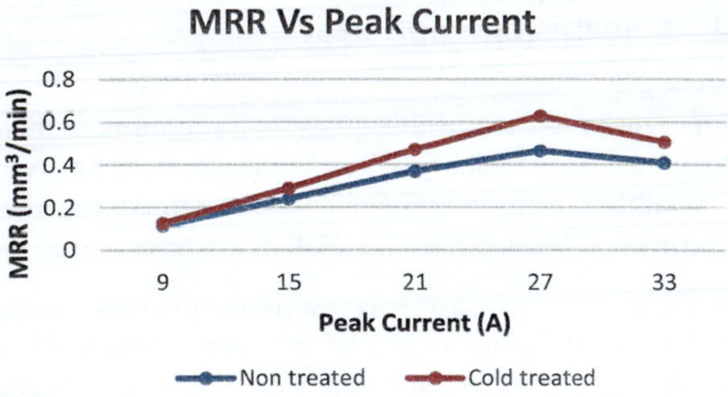

Fig. 3 MRR versus peak current for cold treated and non-treated electrodes

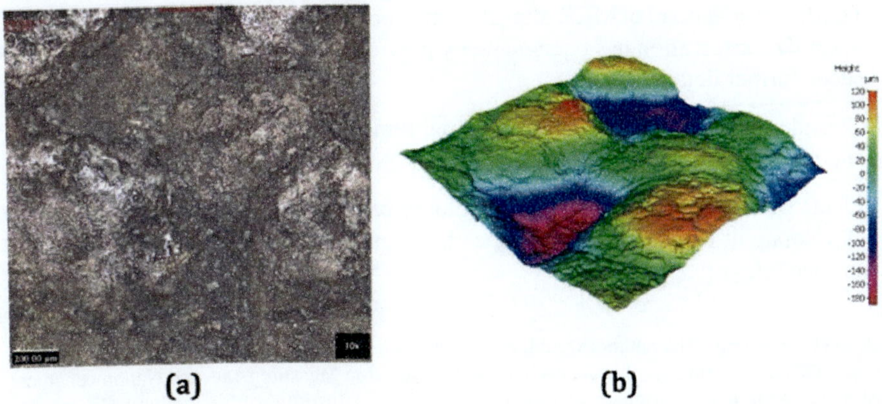

Fig. 4 Magnified image (10×) and profile of EDM surfaces using non-treated brass electrodes at 27 A peak current

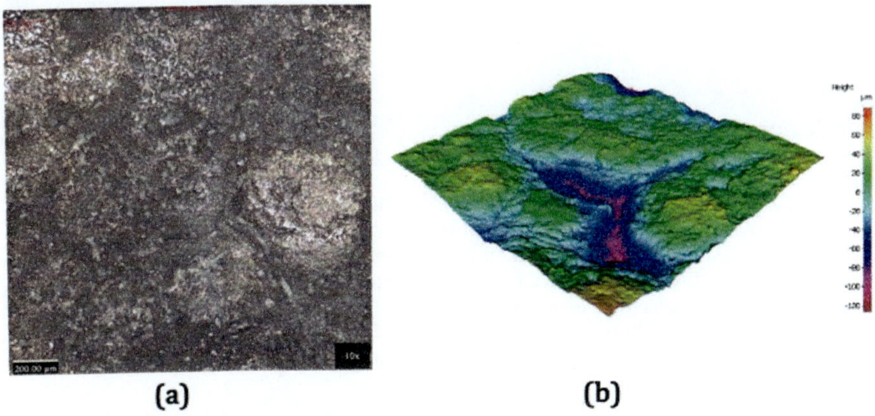

Fig. 5 Magnified image (10×) and profile of EDM surfaces using cold treated brass electrodes at 27 A peak current

4 Conclusion and Future Scope

The work reported in this study is a preliminary study to show the effect of cold treatment on enhancing MRR. In this study, the effect of peak current on MRR was investigated using cold treated brass electrodes and the same was compared with the non-treated brass electrodes. Following observations were made.

- Cold treated electrodes can significantly improve the material removal rate of the EDM process up to 35.12% (corresponding to 27 A peak current).
- Debris accumulation at the tool electrode gap can cause abnormal discharges which in turn causes the sudden drop in MRR.
- The decreasing trend of MRR after 27 A may be due to the carbide layer formation. A similar observation has been made by previous researchers also. However, this needs further detailed study.

Considering the improvement of material removal rate, this work can be further extended in the following directions.

- Electrodes fabricated by powder metallurgy can be used for further studies.
- Development of machining strategies that can improve MRR without affecting surface integrity.

Acknowledgements The authors would like to acknowledge: • Department of Science and Technology (DST), Government of India and Centre for Precision Measurements and Nanomechanical Testing, Department of Mechanical Engineering, National Institute of Technology Calicut, for providing the facility purchased under the scheme 'Fund for Improvement of Science and Technology' (FIST—No. SR/FST/ETI-388/2015) during the period of the research work. • Cryo-Lab was sponsored by ISRO in Amal Jyothi College of Engineering for providing the facility for cold treatment.

References

1. Pandey AK, Dubey AK (2012) Simultaneous optimization of multiple quality characteristics in laser cutting of titanium alloy sheet. Opt Laser Technol 44(6):1858–1865
2. Pramanik A, Basak A (2018) Sustainability in wire electrical discharge machining of titanium alloy: understanding wire rupture. J Clean Prod 198:472–479
3. Lin G et al (2012) Electrical discharge machining of Ti_6Al_4V with a bundled electrode. Int J Mach Tool Manufact 53(1):100–106
4. Reitz W, Pendray J (2001) Cryoprocessing of materials: a review of current status. Mater Manuf Process 16(6):829–840
5. Sundaram MM, Yildiz Y, Rajurkar KP (2016) Experimental study of the effect of cryogenic treatment on the performance of electro discharge machining, pp 1–8
6. Abdulkareem S, Ali Khan A, Konneh M (2010) Cooling effect on electrode and process parameters in EDM. Mater Manuf Process 25:462–466
7. Gill SS, Singh J (2010) Effect of deep cryogenic treatment on machinability of titanium alloy (Ti-6246) in electric discharge drilling. Mater Manuf Process 25:378–385

8. Hui Z et al (2016) Effect of cryogenic cooling of tool electrode on machining titanium alloy (Ti–6Al–4V) during EDM. Mater Manuf Process 31:475–482
9. Kumar S et al (2017) Modeling the tool wear rate in powder mixed electro-discharge machining of titanium alloys using dimensional analysis of cryogenically treated electrodes and workpiece. Proc Inst Mech Eng Part E J Process Mech Eng 231:271–282
10. Kumar S et al (2017) Effect of cryogenically treated copper-tungsten electrode on tool wear rate during electro-discharge machining of Ti-5Al-2.5Sn alloy. Wear 386–387:223–229
11. Goyal R, Singh S, Kumar H (2018) Performance evaluation of cryogenically assisted electric discharge machining (CEDM) process. Mater Manuf Process 33:433–443
12. Kumar M, Datta S, Kumar R (2019) Electro-discharge machining performance of Ti–6Al–4V alloy: studies on parametric effect and phenomenon of electrode wear. Arab J Sci Eng 44:1553–1568
13. Tahir W et al (2019) Surface morphology evaluation of hardened HSLA steel using cryogenic-treated brass wire in WEDM process. Int J Adv Manuf Technol 104:4445–4455
14. Çakir FH, Çelik ON (2020) Tribological properties of cryo-treated and aged Ti6Al4V alloy. Trans Indian Inst Met 73:799–809
15. Naveen Anthuvan R, Krishnaraj V (2020) Effect of coated and treated electrodes on Micro-EDM characteristics of Ti-6Al-4V. J Brazilian Soc Mech Sci Eng 42:1–16
16. Singh J, Singh G, Pandey PM (2021) Electric discharge machining using rapid manufactured complex shape copper electrode with cryogenic cooling channel. Proc Inst Mech Eng Part B J Eng Manuf 235:173–185
17. Jabbaripour B et al (2012) Investigating the effects of EDM parameters on surface integrity, MRR and TWR in machining of Ti–6Al–4V. Mach Sci Technol 16:419–444
18. Sun Y, Gong Y (2017) Experimental study on the microelectrodes fabrication using low speed wire electrical discharge turning (LS-WEDT) combined with multiple cutting strategy. J Mater Process Technol 250:121–131

Emulating Chatter with Process Damping in Turning Using a Hardware-in-the-Loop Simulator

Govind N. Sahu, Pulkit Jain, Mohit Law, and Pankaj Wahi

Abstract Improved chatter vibration-free cutting performance at low speed occurs due to process damping. An additional cutting force component that arises due to interference of the vibrating tool with the cut surface is thought to be responsible for this damping. Experimentally identifying and isolating the mechanisms causing process damping is difficult due to its complex relationship with tool wear. This paper hence proposes the use of a hardware-in-the-loop (HiL) simulator to investigate and emulate the stability of turning with process damping. The HiL simulator has a hardware layer consisting of a flexure representing a flexible workpiece and an actuator that emulates the virtual cutting forces computed in the software layer. Controlled experiments on the HiL simulator agree with model predictions and confirm an increase in the chatter vibration-free cutting performance at low speeds in the presence of process damping. These results can instruct investigating other nonlinearities co-occurring with process damping to develop solutions to mitigate them.

Keywords Turning · Process damping · Chatter · Hardware-in-the-loop simulator

1 Introduction

Self-excited regenerative-type chatter vibrations fundamentally limit the cutting performance capability of machine tool systems. Chatter occurs due to the interaction of the cutting process with the machine tool dynamics, and when it occurs, it increases tool wear and may damage components of the machine tool. Models for chatter are hence instrumental in prescribing ways to avoid it [1]. Models prescribe cutting at parameters that lie below the boundaries of stability. Sometimes, at low cutting speeds, experimental observations depart from model predictions that suggest a lower boundary than observed in practice [2]. The gain in stability limit at lower cutting speeds is usually attributed to the friction between the tool flank face and freshly cut surface, which adds process damping to the system [2]. The thrust force

G. N. Sahu (✉) · P. Jain · M. Law · P. Wahi
Department of Mechanical Engineering, Indian Institute of Technology Kanpur, Kanpur 208016, India
e-mail: govindsahu15@gmail.com

U. S. Dixit et al. (eds.), *Advances in Forming, Machining and Automation*, Lecture Notes in Mechanical Engineering, https://doi.org/10.1007/978-981-19-3866-5_22

at the tool flank face acts against the relative vibration velocity of the tool-workpiece and purportedly induces damping in the cutting process [3]. Process damping is thought to depend on the amplitude and wavelength of vibrations and their complex relationship with vibration frequencies, cutting speeds, and tool wear [2–4].

Since process damping results in a preferential increase in the stability boundary at low speeds, different models have been proposed to characterize the main mechanisms responsible for the improvement in cutting performance. Wallace and Andrew [3] helped reveal that process damping occurs due to the distribution of cutting forces over the contact zone between flank face and freshly cut surface. A detailed qualitative theoretical and experimental proof of stability with process damping was established by Sisson and Kegg [2]. An experimental technique was proposed in [4] to identify the dynamic cutting force coefficients responsible for process damping effects. Excellent summaries of a detailed theoretical and experimental investigation on process damping in turning and milling processes can be found in [4, 5].

Although there exist models that characterize process damping and though some of these models can guide experimentation, the efficacy of most of the existing models is prevaricated by the complex relationship between the amplitude and wavelength of vibrations, vibration frequencies, cutting speeds, and tool wear. Furthermore, since chatter increases tool wear and since this relationship is also not well defined, experiments to confirm the increase in the stability boundary at low cutting speeds in the presence of process damping can sometimes be challenging to perform. In this light, this paper offers an alternate and controlled way of investigating the effects of process damping in cutting processes. We suggest the use of a hardware-in-the-loop (HiL) simulator that has hardware and software layers in which sensors and actuators are used to emulate the physics of the cutting process in the presence of process damping. Since cutting is only emulated on the HiL simulator, vagaries of uncertainties due to tool wear are avoided, and the platform allows for controlled investigations of complex cutting behavior.

HiL simulators have been successfully used previously to study the stability of turning [6–12] and milling processes [13, 14]. HiL simulators have also been used to test different active damping control strategies to mitigate machine tool vibrations [6, 8, 11–13]. Though the stability of the cutting processes and measures to control vibrations have been successfully emulated on HiL simulators, emulating chatter in turning with process damping has not been reported in the literature. As such, the emulations presented in this paper are new and are our modest claim to a contribution to state of the art. Emulating such a phenomenon can be instructive for the investigations of process damping co-occurring with nonlinear cutting force characteristics in other cutting processes and in developing active damping solutions to mitigate them.

The rest of the paper is structured as follows: At first, the mechanical model of the turning process with process damping is presented in Sect. 2. This section also discusses the solution for stability using the semi-discretization method [15]. A brief description of the HiL simulator for the turning process is described in Sect. 3. In Sect. 4, a discussion on validation of the HiL simulator for linear cutting force characteristics while neglecting process damping effect is presented. The results of emulated stability with process damping in turning are discussed in Sect. 5, followed by the main conclusions of the present work.

2 Mechanical Model of Turning with Process Damping

The mechanical model of a turning process with process damping is shown in Fig. 1. We assume the tool is rigid, and the workpiece is flexible, and that it can be approximated as a single degree of freedom system. The flank face of the tool interferes with the cut surface and contributing to the process damping phenomenon is also shown in Fig. 1.

From Fig. 1, the resulting governing equation of motion representing turning process dynamics can be written as:

$$\ddot{x}(t) + 2\zeta\omega_n\dot{x}(t) + \omega_n^2 x(t) = \frac{F_f(t) + F_{pd}(t)}{m}, \tag{1}$$

Wherein, m, ζ, and ω_n are the mass, damping ratio, and natural frequency of the workpiece. $F_f(t)$ is the cutting force in feed direction which is assumed to be linear in the present case, i.e., $F_f(t) = K_f b h(t)$, in which K_f is the empirically identified cutting force coefficient, b is the depth of cut, and $h(t) = h_0 + x(t-\tau) - x(t)$ is the total chip thickness. h_0 is the mean chip thickness; $x(t)$ and $x(t-\tau)$ are the vibration amplitudes corresponding to the present and the previous revolutions, respectively. $\tau = 60/N$ is the spindle period, also known as regenerative delay, and N is the spindle speed in rpm. During machining, the tool/workpiece gets excited due to cutting forces, and relative vibrations between tool and workpiece may become large, leading to an instantaneous loss of contact of the tool with the workpiece [16]. Due to

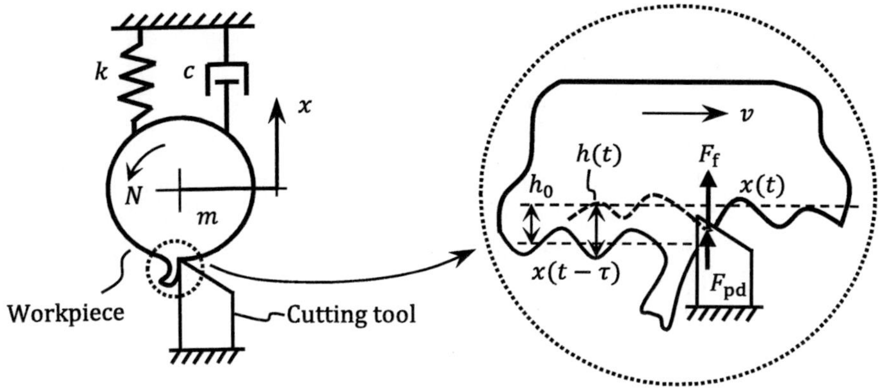

Fig. 1 Mechanical model of turning process with process damping; the inset shows the variation in chip thickness due to regenerative effect and the interference of the tool flank and machined surface

the tool jumping out of cut phenomenon, the regenerative effects do not only depend on just the previous revolution but also multiple previous revolutions vibration data, and the expression for the total chip thickness will hence become $h(t) = x_{\min} - x(t)$, wherein $x_{\min}$ is the smallest value of $\{h_0 + x(t-\tau), 2h_0 + x(t-2\tau), \ldots\}$ taken over several previous revolutions.

F_{pd} in Eq. (1) is the process damping force. For analysis herein, we assume it takes the form suggested in [4], i.e., $F_{\text{pd}}(t) = -Cb\dot{x}(t)/v$, in which C is the velocity-dependent cutting force coefficient and v is the cutting speed. The process damping force is vibration velocity proportional, and since this velocity is related to the frequency (ω), wavelength (λ), and cutting velocity ($v = \frac{\omega}{2\pi} \times \lambda$), the process damping force also depends on these parameters. Furthermore, since the cutting speed is a function of the diameter (d) of the workpiece being cut ($v = \pi \text{d}N/60$), process damping forces increase with a reduction in the spindle speed and workpiece diameter, which ultimately increases the stability limit at low spindle speed. Moreover, since these forces oppose the motion of the cutting tool, it is to be subtracted from the cutting force resulting from chip formation.

On substituting the expressions of $F_{\text{f}}(t)$ and $F_{\text{pd}}(t)$ in Eq. (1), the governing equation of motion forms a delay differential equation (DDE). The DDE with process damping can be solved for stability using the Nyquist frequency domain [4] or the time domain [5] or using the semi-discretization method (SDM) [15].

We use the well-established SDM due to its robustness in handling nonlinearities in the cutting process. For solving Eq. (1) using SDM, it is converted into state-space form as:

$$\dot{\mathbf{x}}(t) = \mathbf{A}(t)\mathbf{x}(t) + \mathbf{B}(t)\mathbf{x}(t-\tau), \tag{2}$$

wherein

$$\mathbf{A}(t) = \begin{bmatrix} 0 & 1 \\ -(\omega_{\text{n}}^2 + \frac{K_{\text{f}}b}{m}) & -(2\zeta\omega_{\text{n}} + \frac{Cb}{mv}) \end{bmatrix}, \tag{3}$$

$$\mathbf{B}(t) = \begin{bmatrix} 0 & 0 \\ \frac{K_{\text{f}}b}{m} & 0 \end{bmatrix}. \tag{4}$$

In SDM, the stability of the cutting process is obtained by determining the Eigenvalue of the transition matrix from Eq. (2) [15]. The limiting stability condition is established when the modulus of Eigenvalue is unity. Using this condition, a theoretical stability chart is constructed by scanning different values of depth of cut for the spindle speed range of interest. The stability chart without process damping can be constructed by simply putting $C = 0$ in Eq. (2).

3 Hardware-in-the-Loop Simulator

The hardware-in-the-loop (HiL) simulator consists of a hardware layer and a software layer, as shown in Fig. 2. The HiL simulator used herein has been built and reported in our previous work [11, 12], wherein the characteristics of the key components of the HiL simulator have been detailed.

The hardware layer contains a flexure that represents a flexible workpiece and a shaker that emulates the cutting force onto the flexure. The shaker has a force capacity of 100 N, which is connected to the flexure via a stinger. A load cell and an accelerometer are mounted on the flexure to monitor input force and output response, respectively. The hardware layer also includes a cRIO-9036 controller with an onboard FPGA module and plugged-in analog to digital converter (ADC) and digital-to-analog converter (DAC). The software layer has been programmed in NI-LabVIEW 2018, which includes data acquisition, filtering, numerical integration, calculation of regenerative cutting forces, and process damping forces. It also

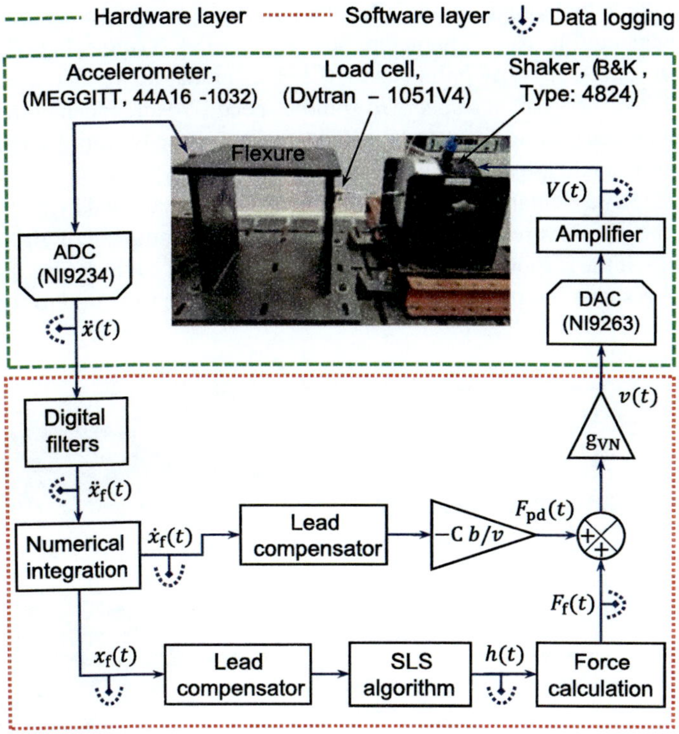

Fig. 2 Block diagram of the hardware-in-the-loop simulator for turning process with process damping

includes a surface location storage (SLS) algorithm that incorporates tool jumping out of cut and multiple regenerative effects [12].

The total delay in the mechatronic HiL simulator is estimated based on the method suggested in [11] and compensated using a phase-lead compensator in the software layer, see Fig. 2. The modal parameters of the flexure coupled with the shaker are as follows: $m = 12.98$ kg, $c = 119.4$ Ns/m, $k = 7.15$ N$\times 10^6$/m, and $f_n = 118.1$ Hz. The calculated total cutting force, i.e., the sum of regenerative cutting force and process damping force, is converted into voltage by multiplying it with a voltage-to-force gain of the shaker ($g_{VN} = 0.038$ V/N) and then transferred it to the shaker through its power amplifier.

4 Validation of HiL Simulator for a Linear Force Model

The HiL simulator is validated for the orthogonal turning process without the process damping effect ($C = 0$). We assume a cold-rolled AISI1045 steel workpiece and carbide grooving tool (edge width—2.4 mm, rake angle—0°, clearance angle—7°) with linear cutting force characteristics, K_f =1384 N/mm^2 [4]. For finding experimental critical stability points using the HiL simulator, we increase the depth of cut (b) in steps of 5 μm at a specified spindle speed (N) and mean chip thickness (h_0). The measured cutting force and displacement response for representative cases of stable and unstable are shown in Fig. 3.

Initially, for a particular combination of speed, depth of cut, and mean chip thickness (N, b, h_0), the flexure is perturbed due to the static component of virtual cutting force ($K_f b h_0$); if the response starts decaying with time, the cut is deemed stable, see Fig. 3. If the response starts to grow with time and whenever the first tool out of cut

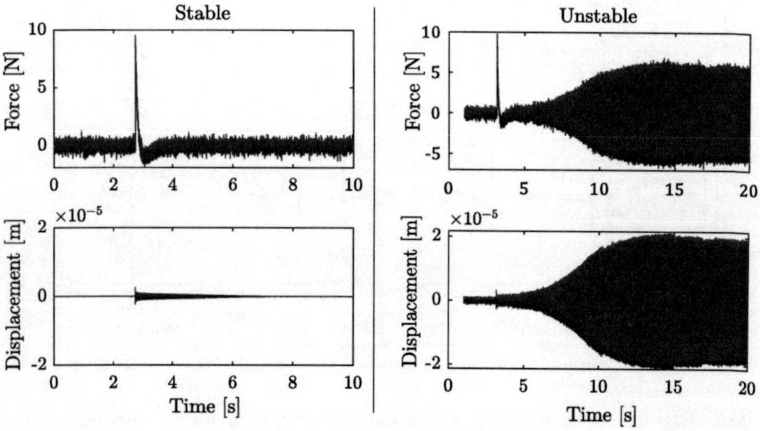

Fig. 3 Theoretical and experimental time domain response for stable ($N = 2600$ rpm and $b = 0.05$ mm) and unstable case ($N = 2600$ rpm and $b = 0.15$ mm)

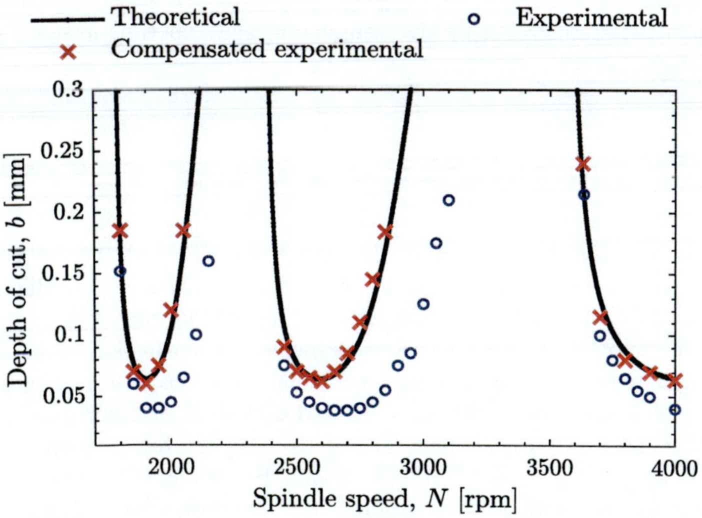

Fig. 4 Theoretical and experimental stability results with and without compensation of delay

is detected, the cut is considered unstable, and corresponding N, b, and oscillation frequency are recorded for that case. In this way, experiments are conducted for the spindle speed range of interest, see Fig. 4.

It is clear from Fig. 4 that experimentally obtained stability behavior is different from the model prediction. Such differences are also observed in earlier reported work [8, 11] and are mainly attributed to delay in the mechatronic HiL simulator. Hence, the delay is systematically identified by the method suggested in [11] and is found to be ~ 0.8 ms. Since the delay in the HiL simulator is equivalent to the negative phase, a phase-lead compensator is designed and compensated suitably. The transfer function of the phase-lead compensator is given below:

$$C(s) = \frac{8.07 \times 10^8\left[\left(\frac{s}{254.2}\right)^2 + 0.7\left(\frac{s}{254.2}\right) + 1\right]}{(s+2)(s+1457.5)(s+5314.1)} \tag{5}$$

Equation (5) is implemented in the software layer, see Fig. 2, and we repeated the chatter experiments with the compensator. The experimental procedure for finding stability points with a compensator is the same as those discussed without a compensator. The experimentally identified stability behavior with the compensator is overlaid with the model prediction, as shown in Fig. 4.

As is evident from Fig. 4, the experimentally observed stability behavior on the HiL simulator with delay compensated system matches up well with the theoretically predicted stability boundaries, whereas for the case of the delay not being compensated, the experiments diverge from prediction. Having validated the HiL simulator for the orthogonal turning process with linear force model, process damping phenomenon is introduced in the HiL simulator, and experiments with those are discussed next.

5 Emulating Stability with Process Damping

All analysis presented hereafter is for the case of the HiL with the delay compensated. Experiments on the HiL simulator are performed with process damping and for assumed linear cutting force characteristics. For the tool and workpiece material under consideration, the process damping coefficient is taken to be $C = 10^6$ N/m, and the workpiece diameter was assumed to be 35 mm. If the tool geometry and/or workpiece material change, the coefficients (K_{f} and C) will change accordingly. Vibration velocities necessary to calculate the process damping forces were obtained by numerically integrating the accelerometer signal. Experimentally obtained stability and theoretical results with process damping effect are shown in Fig. 5. Since process damping is generally a low-speed phenomenon, results shown in Fig. 5 are limited to the low-speed range of 500—1200 rpm.

As is evident from Fig. 5, the stability boundary at lower speeds is higher than at relatively higher speeds, and experiments on the HiL simulator can closely capture modeled behavior. These findings are consistent with earlier reported cutting experimental results [2, 4] and demonstrate how the HiL simulator can also prove effective in investigating process damping-related stability characteristics. Performing real cutting experiments at low spindle speed requires high torque at a given power rating of a spindle motor. Hence, investigation of process damping with different tool geometries and workpiece materials, workpiece diameter, and nonlinear force

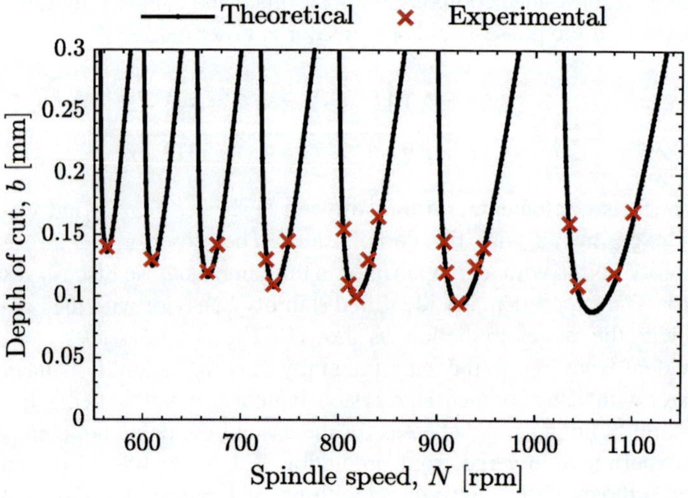

Fig. 5 Theoretical and emulated experimental stability behavior with process damping

models on the real machine is challenging and not safe. These can easily be investigated on the present HiL simulator. Though present work focuses on the emulation of regenerative chatter with process damping on the HiL simulator, abovementioned other investigations will be part of future studies.

6 Conclusions

The stability of the orthogonal turning process with the process damping phenomenon prevalent in the real cutting process is experimentally emulated using a validated hardware-in-the-loop simulator. We show that experimentally emulated stability behavior is in good agreement with the model predictions. Results reveal that stability limits with process damping at lower spindle speeds are higher than at higher spindle speeds, and that these observations are consistent with the other earlier reported experimental works. Since the HiL simulator is non-destructive, it offers a cost-effective, repeatable, safe, and pedagogical tool for controlled experiments. Moreover, since it allows for experiments while not suffering from the vagaries of uncertainties due to tool wear, it facilitates further analysis to characterize the complex process damping phenomenon. Though emulations herein discussed stability in turning in the presence of process damping for the case of a linear force model and a defined process damping coefficient and workpiece diameter, other similar investigations to understand the role of the damping coefficient, workpiece diameter, and other force model types are possible and can be as easily emulated on the HiL simulator.

Acknowledgements This work was supported by the Government of India's Impacting Research Innovation and Technology (IMPRINT) initiative through project number IMPRINT 5509.

References

1. Altintas Y, Weck M (2004) Chatter stability of metal cutting and grinding. CIRP Ann Manuf Technol 53:619–642
2. Sisson TR, Kegg RL (1969) An explanation of low-speed chatter effects. J Eng Ind 91:951–958
3. Wallace PW, Andrew C (1965) Machining forces: some effects of tool vibration. J Mech Eng Sci 7:152–162
4. Eynian M (2010) Chatter stability of turning and milling with process damping. PhD Thesis, The University of British Columbia
5. Saleh K (2013) Modeling and analysis of chatter mitigation strategies in milling. PhD Thesis, The University of Sheffield
6. Ganguli A, Deraemaeker A, Horodinca M, Preumont A (2005) Active damping of chatter in machine tools—demonstration with a "hardware in the loop" simulator. J Syst Control Eng 219:359–369
7. Matsubara A, Tsujimoto S, Kono D (2015) Evaluation of dynamic stiffness of machine tool spindle by non-contact excitation tests. CIRP Ann Manuf Technol 64:365–368

8. Mancisidor I, Beudaert X, Etxebarria A, Barcena R, Munoa J, Jugo J (2015) Hardware-in-the-loop simulator for stability study in orthogonal cutting. Control Eng Pract 44:31–44
9. Stepan G, Beri B, Miklos A, Wohlfart R, Bachrathy D, Porempovics G et al (2019) On stability of emulated turning processes in HIL environment. CIRP Ann Manuf Technol 4–7
10. Beri B, Miklos A, Takacs D, Stepan G (2020) Nonlinearities of hardware-in-the-loop environment affecting turning process emulation. Int J Mach Tools Manuf 03611
11. Sahu GN, Vashisht S, Wahi P, Law M (2020) Validation of a hardware-in-the-loop simulator for investigating and actively damping regenerative chatter in orthogonal cutting. CIRP J Manuf Sci Technol 29:115–129
12. Sahu GN, Jain P, Wahi P, Law M (2021) Emulating bistabilities in turning to devise gain tuning strategies to actively damp them using a hardware-in-the-loop simulator. CIRP J Manuf Sci Technol 32:120–131
13. Ganguli A, Deraemaeker A, Romanescu I, Horodinca M, Preumont A (2006) Simulation and active control of chatter in milling via a mechatronic simulator. J Vib Control 12:817–848
14. Miklos A, Takacs D, Toth A, Wohlfart R, Porempovics G, Molnar TG et al (2017) The development of high-speed virtual milling test. In: Proceedings of ASME 2017 Dynamic Systems and Control Conference DSCC2017, 11–13 Oct 2017, Tysons, Virginia, USA, pp 1–10
15. Insperger T, Stépán G (2004) Updated semi-discretization method for periodic delay-differential equations with discrete delay. Int J Numer Methods Eng 141:117–141
16. Tlusty J, Ismail F (1984) Basic non-linearity in machining chatter. CIRP Ann Manuf Technol 30:299–304

Experimental and Statistical Analysis of Process Parameters on Micro-milling of Ti–6Al–4V Alloy

Mohan Kumar, Ankit Jain, Shashank Shukla, and Vivek Bajpai

Abstract This research paper focuses on the statistical analysis of cutting parameters: axial depth, radial depth, and feed per tooth on the top burr formation in the down milling of Ti–6Al–4V alloy. Analysis of variance (ANOVA) has been used to find the main parameters influencing the top burr formation in down milling. It was observed that axial depth of cut (contribution 60.07%) remains the most significant factor followed by the radial depth of cut (27.27%) and feed per tooth (5.90%) influencing top burr width.

Keywords Top burr · Down milling · ANOVA

1 Introduction

The major problem that mechanical machining, either macro- or micro-scale, i.e., milling, turning, drilling, is facing the burr formation during the machining process. Burrs deteriorate the accuracy and precision of the machined parts. Deburring is the most known burr-removal operation after machining. However, it is only limited to macro-machined components and cannot be used for micromachined parts or features because it deteriorates the dimensional accuracy. Further, the developed setup of deburring for micromachining is very complex and expensive. Various approaches have been utilized in the recent past to make deburring economical to the maximum extent. Researchers worked on different optimization techniques for machining and tool parameter optimization for burr minimization. However, the best approach for burr minimization would be first to understand the burr formation mechanism. It is challenging and complicated to comprehend the burr formation mechanism theoretically due to the implication of higher-order three-dimensional elastic and plastic deformation of material [1]. Burr formation has been classified based on its formation mechanism [2], shapes [3], and locations [4]. Among all the burrs (mainly entry burr, exit burr, and top burr) formed in micro-milling operation, top burr formed

M. Kumar · A. Jain (✉) · S. Shukla · V. Bajpai
Department of Mechanical Engineering, Indian Institute of Technology (ISM), Dhanbad, India
e-mail: mechankit92@gmail.com

U. S. Dixit et al. (eds.), *Advances in Forming, Machining and Automation*, Lecture Notes in Mechanical Engineering, https://doi.org/10.1007/978-981-19-3866-5_23

in down milling is more dangerous for safety, surface quality, and productivity of product [5]. Fecova et al. [6] found that those chips with a volumetric coefficient lie between 40 and 50 were better for machining; however, they got a higher volumetric coefficient in down milling concerning milling. It was found the top burr formed in micro-milling is more significant due to the size effect [7]. Filiz identified that the size of the top burr formed on the down milling side is much larger than on the up milling side. The formation of the top burr starts at the lower undeformed chip thickness region when the tool begins to rotate. The cutting edge radius and uncut chip thickness are comparable in micro-milling operation, and the uncut chip thickness size influences the cutting responses [8]. Hence, furthermore, study requires optimizations of cutting parameters to reduce the burr (mainly down milling) to the maximum extent. This paper presents an analysis to identify the most predominant factor that affects the top burr formation in the down milling operation of Ti6Al4V alloy.

2 Material and Methods

The experiment was conducted on an indigenously developed micro-milling setup which is shown in Fig. 1. Three-axis micro-milling setup has stacked *XY* and *Z* stages assembled on a granite frame. The stages have an accuracy of 2.5 μm with a positional resolution of 0.3125 μm. An end mill tool (AlTiN coating) having a diameter of five hundred microns and tool edge radius of 4.46 μm is used to do an experiment on Ti6Al4V specimen having dimension 25 × 25 × 5 (mm). Fixed process parameters (Table 1) for the Taguchi's L9 orthogonal array are decided based on literature [9].

Fig. 1 In lab developed high-speed micro-milling setup

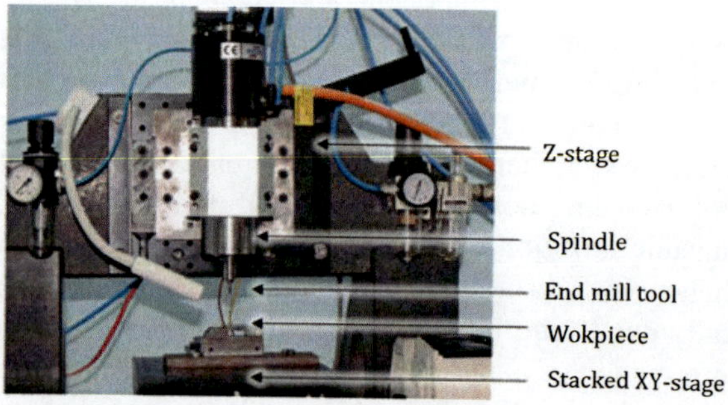

Table 1 Taguchi's L9 orthogonal arrays

Ex. No.	Radial *X/Y*-axis (mm)	Axial *Z*-axis (μm)	Feed per tooth (μm)
1	0.5	0.015	6
2	0.5	0.03	8
3	0.5	0.045	10
4	0.375	0.015	8
5	0.375	0.03	10
6	0.375	0.045	6
7	0.25	0.015	10
8	0.25	0.03	6
9	0.25	0.045	8

2.1 Design of Experiment (DOE)

The experiments are designed and planned to empower the statistical approach via Taguchi's method. An L9 array-based design of experiments was executed as shown in Table 1. Each experiment was repeated twice to get more precise results. The L9 array-based design of experiments minimizes the number of tests to half than a full factorial design, which saves resources.

3 Results and Discussion

3.1 Surface Morphology

The morphological analysis of the top burr formation at different experimental process parameters has been analyzed through the SEM image analysis as shown in Fig. 2. It has been observed to be a top burr formation at the down milling operation.

3.2 Statistical Analysis

After the completion of the experiment procedure, the obtained result was statically analyzed with the help of ANOVA variance analysis [10]. ANOVA is used to break-down the total dependent variable into the component of variance and help in the evaluation of their significance. It was executed by calculating the sum of squares (SSs) for every parameter using following equation

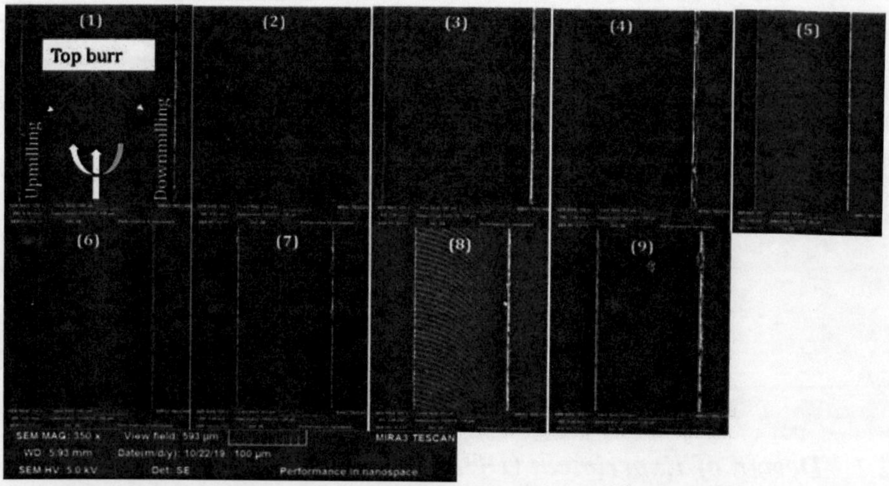

Fig. 2 SEM images showing top burr width at different experimental parameters as shown in Table 1

$$ss = \sum_{I=1}^{n} (Y_i - Y')\tag{1}$$

Variance analysis has been performed as it depicts the clearance of the significant and insignificant variable in calculating the response parameter concerning the axial depth of cut, radial depth of cut, and feed per tooth. The percentage contribution can be determined by the equation

$$\%\text{CR} = \frac{\text{MSS} - \text{SS}_{\text{res}}}{\text{SS}_{\text{total}}} \times 100\tag{2}$$

Table 2 shows the detailed processed parameter analysis and other statistics of R^2, which validate the model because the R^2 value comes closer to 1 to confirm the model analysis.

It has been observed that burr forms in down milling side is more concerning up milling. This paper focuses on the analysis of process parameters on the top burr width of the down milling side. Table 2 presents the ANOVA with CRs of each machining parameter. It has been observed that axial depth of cut, radial depth of cut, and feed per tooth have contribution ratios of 60.07, 27.27, and 5.90%, respectively. As it can also be observed from the main effects plot that it exhibits similar trend, i.e., as feed per tooth, the axial and radial depth of cut increasing top burr width increasing as shown in Fig. 3. The contact area between the tool and the workpiece increased as the axial depth of cut increased, subsequently the extrusion and scratching processes expanded. The cutting force increases, and hence, the maximum effective strain is gradually increased. This results in more material deformation, and hence, larger burr formed. With the increase in radial depth of cut, the cutting flute has to remove

Table 2 ANOVA analysis for top burr formation

Source	F	Adj.SS	Adj.MS	F-Value	P-Value	Significance	CR (%)
Axial depth (μm)	1	24.149	24.148	72.12	0.000*	Significant	60.07
Radial depth(μm)	1	11.145	11.145	33.28	0.002*	Significant	27.27
Feed per tooth (μm/tooth)	1	2.6767	2.675	7.99	0.037*	Significant	5.90
Error	5	1.674	0.334				
Total	8	39.664					

$R^2 = 95.78\%$ R^2 (adj) $= 93.24\%$
* Significant at 95% confidence level

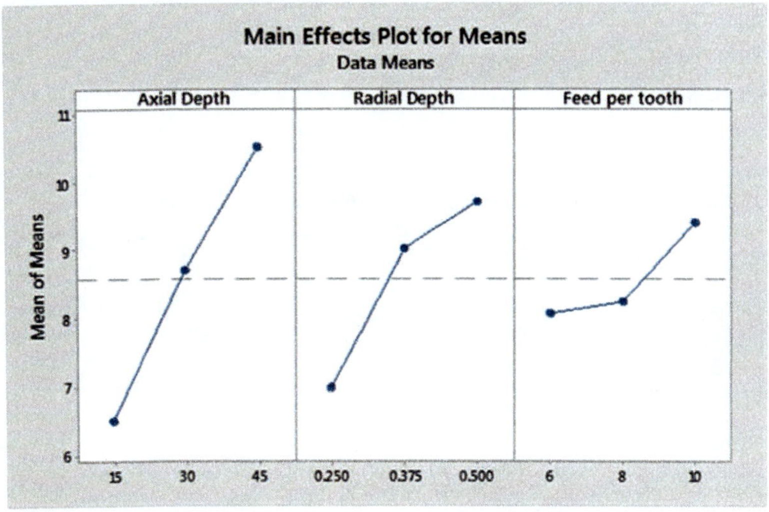

Fig. 3 Mean effect plot for burr width related to machining parameters

more material, thereby large deformation occurs, and hence, large burr is formed. As feed per tooth increases, the shearing phenomenon predominates over plowing in micro-milling, which results in rubbing and bulging, and hence, more burr produced [11].

Figures 4 and 5 show the residual and normal probability plot for the burr formation at different cutting conditions. It is known that residual plot plays an important role in the deviation to show the error in the experimental analysis with the experiment order; the experiment number 1 shows the minimum error having machining parameter of axial 0.5 μm, radial 0.015 μm, and feed of 6 μm, which may be considered to the near-optimum parameter for the burr formation. Now, from the normal probability plot, it has been observed that the observation parameter is closer to the mean line,

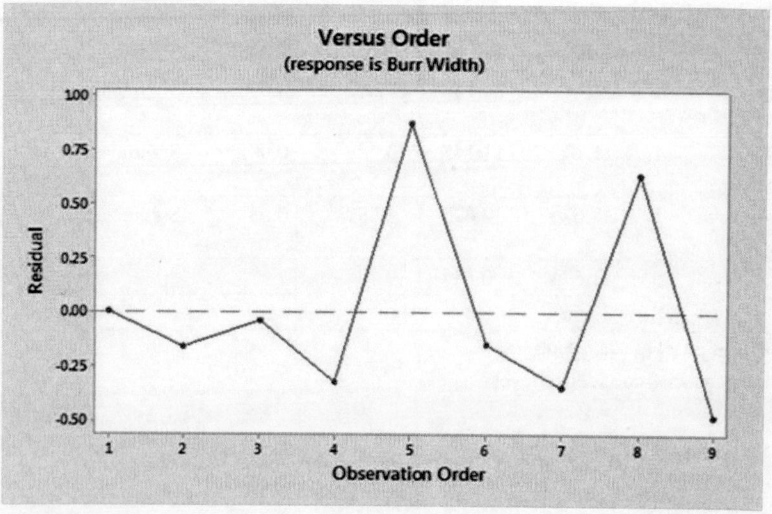

Fig. 4 Residual plot

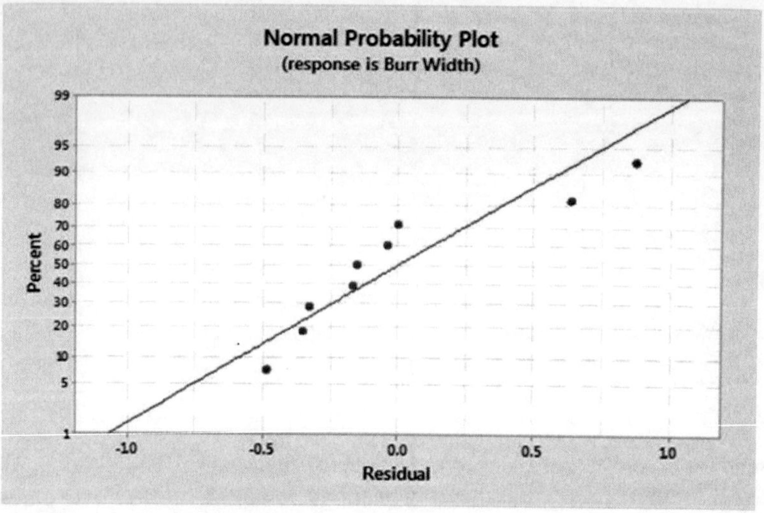

Fig. 5 Normal probability plot

which confirms the validation of the model with a regression coefficient of 0.95 approx. as shown in Table 2.

Figure 6 shows the contour plot between the axial depth and radial depth concerning the burr width formation. It has been observed that as the axial depth and radial depth increases burr with are showing in the increasing trend, this might be due to an increase in chip load with the increase of radial depth of cut. Higher

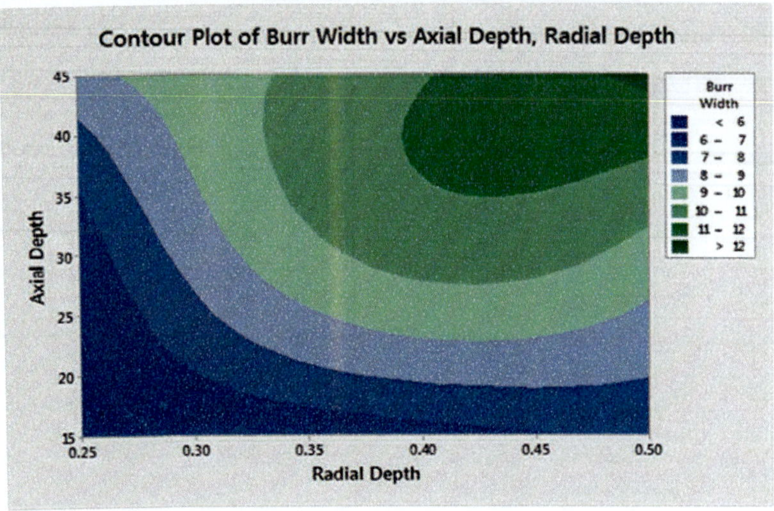

Fig. 6 Contour plot of burr width versus axial and radial depth

chip load leads to larger plastic deformation of material at a high radial depth of cut. This leads to more chips and burrs formation. Figure 7 shows the variation of burr width with axial depth and feed per tooth; it is observed that at lower feed per tooth with an increase in the axial depth burr width decreases, but at the certain axial depth of approximately 25 μm burr width increases slowly up to the 7–8 μm, as the axial feed increasing above 25 μm, burr width suddenly increases up to the 12 μm. Now,

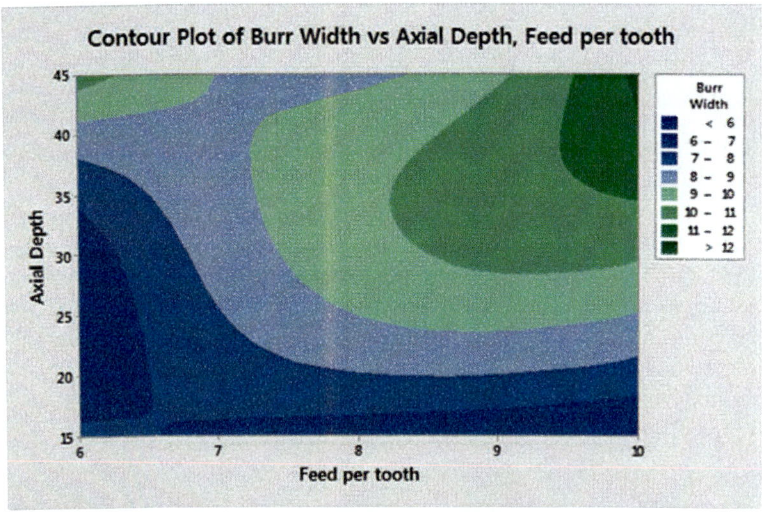

Fig. 7 Contour plot of burr width versus axial and feed per tooth

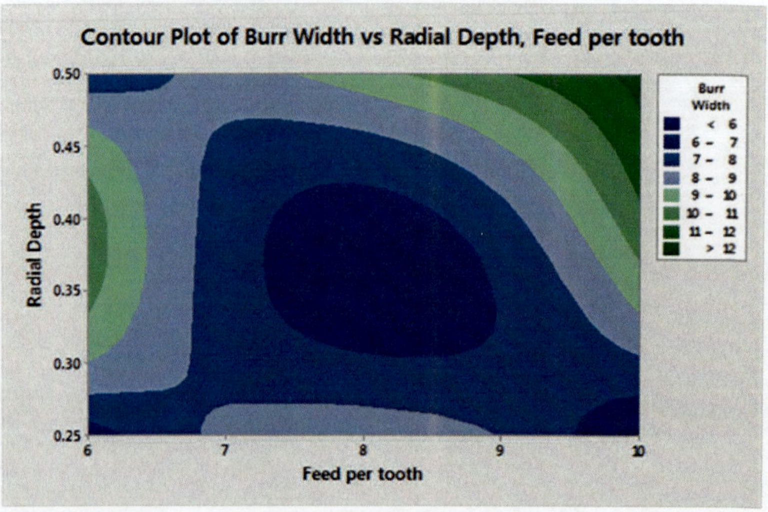

Fig. 8 Contour plot of burr width versus radial depth and feed per tooth

Fig. 8 shows the variation of burr width with the radial depth and feed per tooth; it is found to be burr with a minimum at a radial depth of 0.35 µm and 8 feed per tooth; however, in this case, radial depth plays a significant role in burr formation, due to at constant high feed at initial condition 10, minimum burr formed but at the radial, the depth increases above 0.35 µm, burr width found to be increases.

4 Conclusions

From the above discussion, the following conclusion has been obtained:

1. The experiment has been successfully carried out with top burr formation in a down milling operation.
2. Axial depth of cut comes out to be the most influencing parameter which affects top burr width with the contribution of 60.07% followed by the radial depth of cut (27.27%) and feed per tooth (5.90%), respectively.
3. Top burr width increases with an increase in the axial depth and radial depth.
4. Feed per tooth shows the less impact parameter as compared to the axial depth and radial depth.
5. This analysis helps in the selection of appropriate parameters for optimizations to reduce top burr size to a maximum extent.

References

1. Nakayama K, Arai M (1987) Burr formation in metal cutting. CIRP Ann Manuf Technol 36(1):33–36
2. Aurich J, Dornfeld D, Arrazola P, Franke V, Leitz L, Min S (2009) Burrs-analysis, control and removal. CIRP Ann Manuf Technol 58(2):519–542
3. Hashimura M, Hassamontr J, Dornfeld D (1999) Effect of in-plane exit angle and rake angles on burr height and thickness in face milling operation. J Manuf Sci Eng 121(1):13
4. Wang G, Zhang C (2004) Mechanism of burr formation in milling. KEM 278(81):259–260
5. Kumar M, Bajpai V (2019) Experimental investigation of top burr formation in high-speed micro-milling of Ti6Al4V alloy. Proc Inst Mech Eng Part BJ Eng Manuf 1–9
6. Fecova V, Michalik P, Zajac J, Mihok J, Berthoty J (2014) The chip in the up and down milling process. In Adv Mater Res 856:379–383
7. Schueler G, Engmann J, Marx T, Haberland R, Aurich J (2010) Burr formation and surface characteristics in micro-end milling of titanium alloys. In: Burrs—analysis control and removal, pp 129–138
8. Rehman G, Husain Imran Jaffery S, Khan M, Ali L, Khan A, Ikramullah Butt S (2018) Analysis of burr formation in low speed micro-milling of titanium alloy (Ti-6Al-4V). Mech Sci 9(2):231–243
9. Jaffery S, Khan M, Ali L, Mativenga P (2016) Statistical analysis of process parameters in micromachining of Ti-6Al-4V alloy. Proc Inst Mech Eng Part B J Eng Manuf 230(6):1017–1034
10. Slack N (1994) A review of 'managing for total quality: from deming to taguchi and SPC'. Int J Prod Res 32(7):1754 (Logothetis N (ed), Prentice Hall, 1992, [pp. xv + 447] Price: £24·95)
11. Chen L, Deng D, Pi G, Huang X, Zhou W (2020) Burr formation and surface roughness characteristics in micro-milling of microchannels. Int J Adv Manuf Technol 111(5–6):1277–1290

Experimental Investigation on the Cutting Performance of Textured Inserts in Dry Turning of Ti–6Al–4V

Amal S. Siju and Sachin D. Waigaonkar

Abstract The application of textured tools has been emerged as a promising research front to attain better tribological properties in machining. Despite the benefits of microstructured tools being established, manufacturers are yet to implement the same on turning inserts. This study aims to investigate the performance of textured inserts under dry machining of Ti–6Al–4V. Cutting tests were performed using innovative dual textured uncoated carbide inserts using six different machining lengths. The cutting forces were monitored, and the worn inserts were collected for analysis using a scanning electron microscope. A detailed comparison emphasizing the variations in cutting friction and tool wear was made for a convectional non-textured insert. The effect of rake surface textures on the cutting tool temperature was also investigated using a thermal camera. Based on the obtained results, a few key conclusions are made on overwhelming studies on tool texturing and real-life applications of textured inserts.

Keywords Ti–6Al–4V · Tool wear · Cutting tool temperature · Textured inserts

1 Introduction

Machining is one of the most reliable metal-removal processes for attaining high surface finish and accuracy. However, over the past decade, titanium (Ti) alloys have been extensively used for various parts and equipment in applications such as aerospace and bioimplants. Some of the parts to be fabricated often demand dry or near-dry processes [1] due to ecological necessities and/or considering its application. On the other hand, machining these materials demands new strategies in-process guidance and tool design [2] due to adverse effects on tool life in dry cutting conditions. It is known that tool life is one of the most affected economic considerations in dry machining. A considerable amount of research has been done

A. S. Siju (✉) · S. D. Waigaonkar
Department of Mechanical Engineering, BITS Pilani, K. K. Birla Goa Campus, Goa 403726, India
e-mail: amals@goa.bits-pilani.ac.in

U. S. Dixit et al. (eds.), *Advances in Forming, Machining and Automation*, Lecture Notes in Mechanical Engineering, https://doi.org/10.1007/978-981-19-3866-5_24

in this regard, and many methods have been successfully implemented to increase the tool life [3].

Recently, texturing of tools in conventional machining, as well as in drilling and milling operations, has gotten significant attention. Texturing of cutting tools shows the potential to deliver improved tribological properties such as friction, wear, and lubrication. In recent years, micro-structuring of cutting tools rake faces has been on the research front. Various researchers around the globe have successfully investigated the better tribological conditions exhibited by the textured inserts [4, 5]. Researchers have shown that rake face textures on a cutting tool insert can considerably reduce the friction along with the tool-chip interface and tool wear under various machining conditions [6, 7]. For example, using an experimental analysis, Obikawa et al. [8] stated that micro-structures help to improve the cutting conditions, and they are more effective when the contact length of tool and chip is large, and deeper notches are to be preferred. Another investigation using microstructured cutting tools showed a superior behavior for adhesion and tribology for both dry and wet conditions [6]. They also point out that effects are strongest using lubrication. This understanding corresponds with the findings of [9], where they observed significant improvement in tribological aspects of cutting. When the textured inserts were used in MQL [6] and with solid lubricants [4], they showed a further increase in the effectiveness of textured tools. Apart from improved tribological aspects in cutting, textured tools have shown a reduction in shear angle [10], cutting temperature, etc. [11]. In many studies, the microstructure approach was also subject to FEM simulations to examine their effects, and the results show good agreement with experimental results [12].

Despite many foregoing and ongoing researches on the benefits of textured tools [13, 14], manufacturers are yet to implement the same in their tool design. One of the prime reasons attributing to this is the cutting conditions and specific tool-workpiece combinations to be used. It is known the that cutting parameters such as feed, cutting speed, and depth of cut will significantly influence the performance of textured turning inserts [15]. The geometrical features of the textures also influence their performance and adequacy [16]. Therefore, the effects of textured inserts are questionable when they are not used in specific cutting conditions. Apart from this, another essential aspect missing in the current research involving the performance of textured inserts is the effect of machining time. A comprehensive analysis of existing literature in this area shows that this aspect is to be analyzed more closely to comment on textured tools' practical application. So, in the present study, textured inserts' performance was analyzed when machined under different machining lengths. The textured inserts are exclusively tested in the machining of titanium (Ti) alloy in dry conditions.

2 Experimental Details

The Ti 5th-grade alloy (Ti–6Al–4V) was used as the workpiece material in the present study. The hollow 120-mm-long-cylindrical workpiece had a 50 mm outside diameter and a 30 mm internal diameter. The chemical composition and the mechanical properties of this alloy can be found from [15]. The turning experiments were conducted on a conventional three-jaw lathe. To evaluate the performance of textured tools, the variation in cutting forces was monitored. The cutting force measurement was carried out using a Kistler 9747 multi-component piezoelectric dynamometer mounted on the tool post, which was wired to an amplifier and data accusation system. The data acquisition and filtering were carried out using a four-channel charge amplifier type 5070A provided by Kistler. The software tool DynoWare, provided by Kistler™, was used to monitor and analyze the collected data. The arrangements for mounting the dynamometer are made by removing the tool post, as shown in the experimental setup in Fig. 1.

A conventional triangular-shaped uncoated inserts supplied by Sandvik, India, with code name TCMW 16 T3 H13A were used for performing machining operations, whose major and relevant properties are given in Table 1. This insert has got the static geometry of 0° for the rake angle, 7° for the clearance angle, and 90° for the major (or side) cutting edge/relief angle (when fixed on the tool holder). A

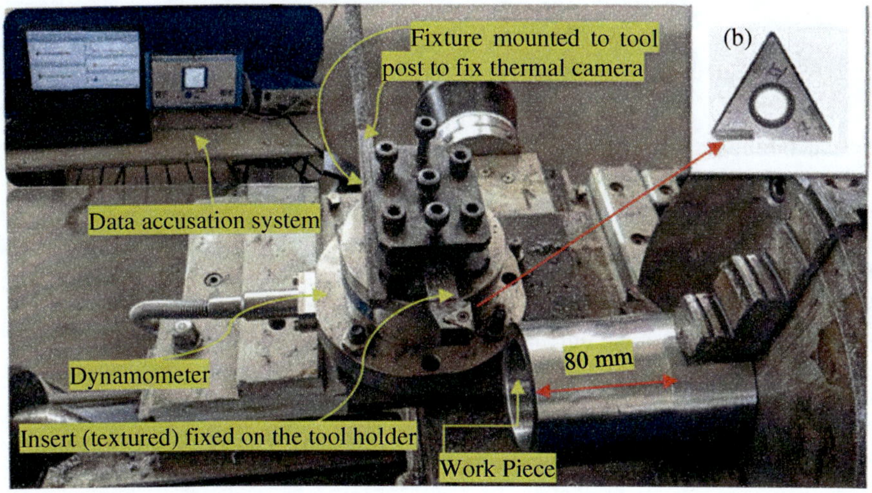

Fig. 1 Schematic of machining setup showing, **a** Tool dynamometer arrangement, **b** actual photograph of the insert with textures engraved

Table 1 Chemical composition of the carbide inserts

Elements	WC	CO	VC + Cr_3C_2
Composition in wt.%	88.4–90%	9.5–10.5%	0.5–1.1%

commercially available tool holder STGC 1616 H 16 (the holder provides a side relief angle of 0°), supplied by Sandvik India, was used for performing the orthogonal cutting operations. The turning experiments were conducted using a dual rake surface textured insert (DTI). The selected texture geometry was based on our earlier results on the effect of textural geometry on the enhancement of cutting performance [17]. The design and fabrication details of textured inserts can be found in our recent publications [17, 18].

All machining experiments were carried out in dry conditions, i.e., no cutting fluid was used. For all experiments, 0.2 mm/rev feed and a depth of cut of 1 mm were used. In addition, all the cutting experiments were performed using a cutting speed of 92 m/min. These cutting parameters are opted based on earlier findings [15, 17] and the current work in the pipeline. To validate the effects of textured inserts, both textured and conventional inserts (CI) were used to perform cutting experiments by varying the cutting lengths (machining length).

During the cutting, the cutting temperature was monitored at a point on the tool surface away from the chip flow direction. The temperature was monitored using a FLIR infrared thermal camera mounted on a fixture arrangement. The reported temperature is the average of temperature two repeated measurements repeated toward the end of the total cutting length. Each set of experiments was repeated twice using fresh cutting edge. After each experiment, the inserts were collected for characterization purposes using a scanning electron microscope (SEM).

3 Results and Discussion

The machining experiments were conducted for different machining lengths (for the longest being roughly 3 min) without a break using selected cutting conditions. The comparison was made based on matching length. Since each cut was 80 mm, multiple cuts were done using the same insert (viz., 80 mm, 160 mm ..., etc.) to attain a predetermined machining length (each cut was taking roughly 40 s). While machining, the changes in cutting speed due to a decrease in workpiece diameter were neglected. Then, the comparisons were made between CI and DTI using the observed reduction in cutting forces and friction. From the measured values of cyclic forces, the moving average (for the entire machining duration in the respective set) of measured variation in cyclic forces was used to calculate the main cutting force (F_c) and thrust force (F_t). Finally, the coefficient of friction (COF, μ) was calculated using Eq. (1)

$$\mu = F_c / F_t \tag{1}$$

The observed differences in the cutting forces and friction are presented as 'average reduction %'. The reasons for the obtained results are addressed for the underlying mechanisms and are discussed below.

3.1 Textured Inserts Effects on Cutting Forces and Friction

Figure 2 depicts the distribution of reduction % in cutting forces and coefficient of friction for textured cutting tools compared to those of the untextured tools versus various machining lengths. Initially, a reduction in cutting force was observed for textured tools. But interestingly, the percentage reduction in F_c reached a maximum when machined 160 mm (7.2%) and 240 mm (8.5%). Thereafter, the textured inserts exhibited an increase in cutting force, i.e., considerably lowered the reduction %. After approximately after 3 min of machining, textured tools had no advantage on the main cutting force, and eventually, textured inserts became ineffective. When machined about 560 mm continuously, it is noted that the DTI resulted in about 4% higher cutting force compared to cutting performed using CI.

A similar kind of trend in reduction in F_t can be seen from Fig. 2. However, the maximum reduction in F_t with the use of DTI was noted to be 23% when compared to CI. This was observed when DTI was used for machining 240 mm. Roughly, about 400 mm of machining using DTI, the thrust forces started increasing, and the results were not in favor of textured inserts. Figure 2 also indicates that, initially, cutting with DTI showed a 12% reduction in COF. Then, when machined 240 mm, it reached the maximum, which was about 19%. After that, the reduction % of COF drops to −2%, and thereafter, it reaches the lowest at 560 mm (about −4%). These observations are in good agreement with the results discussed related to F_c and F_t.

The reason behind these variations can be attributed following reasons. After machining almost 400 mm, it is possible that the relatively weak texture walls to worn out or the workpiece adhesion due to chip debris which are generated because of derivative cutting filled the textures. These aspects are discussed in [19, 20] and verified in Sect. 3.2. Further, it also appears that the thin-walled texture structures are prone to damage under the chip flow. In addition, texturing may help diffusion wear and crater wear progress much faster as compared to conventional inserts. The above factors lead to reduced effectiveness of the textured inserts.

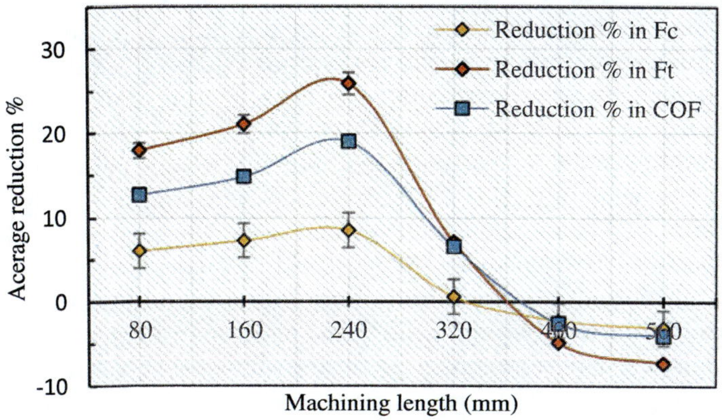

Fig. 2 Reduction % in F_c, F_t, and COF for textured inserts versus machining length

So, it is evident that it is difficult to attain the maximum efficacy of textured inserts for the entire machining duration even it is used in a specific machining regime where it is most effective. On the other hand, a significant reduction in the cutting force and friction was exhibited by DTI when the machining time was short. Obviously, a reduction in cutting forces and friction will also influence the shear angle, chip thickness, cutting energy, etc. However, based on the above observations, it should be assumed that these variations are also transitory. The variation in chip morphology and tool wear land in similar machining regimes can be found in [17, 18]. Also, a general trend is reported in many published articles [21].

3.2 *Textured Inserts Effects on Cutting Tool Temperature*

When machining Ti alloys, the cutting temperature on the rake surface of the tool usually reaches as high as 900 °C at the cutting speed used in this study [22]. This temperature measurement is difficult using a thermal camera as the peak temperature happens somewhere near 0.3–0.7 mm from the cutting edge [23], where the chip is usually in contact with the rake face. So, for reference, the cutting temperature was measured on the tool rake surface, very close to the cutting zone at a fixed point (the point was marked 1 mm from the main cutting edge and 2 mm from the nose, focused using the laser pointer of the thermal camera). On the other hand, the temperature on the rake surface depends on many parameters such as cutting tool substrate, coating material and cutting environment. In addition, the flow of heat from the primary zone into the tool, the flow of heat into the tool from the secondary zone, etc., also contribute to the heat accumulation on the rake face. The above factors further depend on parameters such as thermal conductivities and diffusivity of the tool and workpiece and other cutting parameters. So, the heat measured at a fixed point on the rake face should be comparable most of the time in the same cutting regime.

Figure 3a, b depicts the example of the temperature measured on the insert rake face using the thermal camera. As seen, a decrease in cutting temperature for DTI was observed, indicating that textured inserts help to reduce the cutting zone temperature. However, it is interesting to note that that hot spot occurred near the secondary cutting edges, i.e., the texture walls, when examining the images using FLIR tools™ software for image analysis.

The average measured temperature after each cut is plotted against machining length and is depicted in Fig. 4. As seen, the tool temperature initially is higher for DTI insert, i.e., when machined for 80 mm, when compared to CI. Then, DTI shows a reduction in the temperature. About 10 and 18% reductions can be observed, respectively, at 160 and 240 mm of machining. Thereafter, no significant reduction in tool temperature was observed for DTI. The initial increase in tool temperature can be linked to the derivative cutting. Derivative cutting temporarily increases the tool temperature, but as time progresses, the rate of derivative cutting reduces due to the micro-welding of chip debris in the microtextures. Eventually, the advantage of textures fades off, and they do not help to reduce the friction. In contrast, the increase

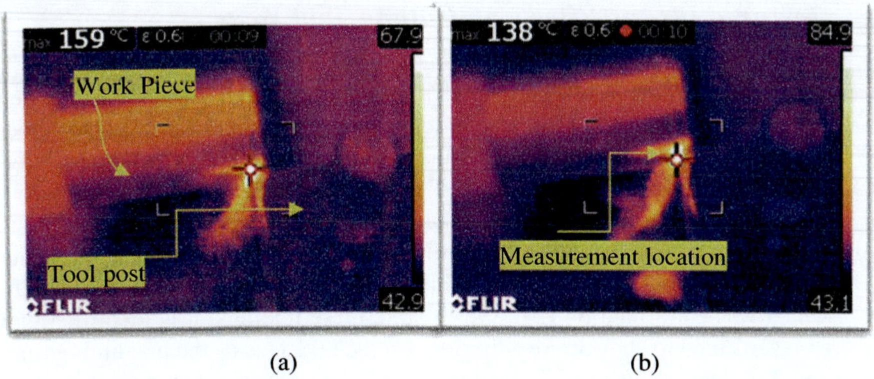

Fig. 3 Thermal camera image showing cutting insert temperature after 170 mm of machining using **a** CI and **b** DTI

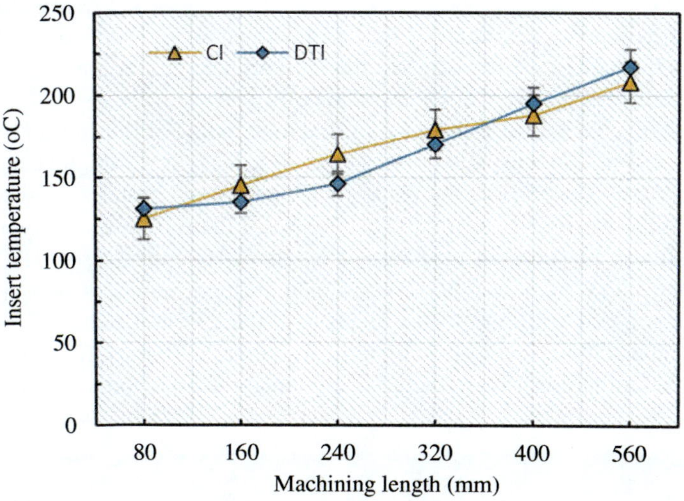

Fig. 4 Variation in tool temperature versus machining length

in cutting temperature can be attributed to crater wear, notch wear, and thermal cracks on the cutting edge. One may assume that due to texturing, the cobalt binder on the inserts is not intact, and the diffusion of cobalt into the insert at high temperature may make the flowing chip carry away the relatively fragile tungsten carbide. This could weaken the cutting edge leading to premature failure of tool insert.

3.3 Textured Inserts Effects on Cutting Tool Wear

In the previous section, it was found that the texture effects diminished when machined for larger lengths. It is also evident that these changes will also affect the tool wear. Therefore, herein, the rake surface and the flank surface of the textured tools were examined using SEM. From SEM micrographs presented in Fig. 5a, b, severe adhesions of chip debris on the rake surfaces can be seen. These adhesions were gradually increasing as the cutting time increased. On the other hand, some earlier studies [6] show that the textured inserts reduce adhesion by approximately 15%, which was confirmed by [24, 25]. Despite these findings, it is identified that textured inserts tend to show more adhesions on the rake face as the machining time increases. This can be linked to derivative cutting, the derivative cutting generates a great amount of heat [19, 26]. So, the trapped chips get welded inside the textures due to this temporary increase in cutting temperature. The chip welding is more in dry cutting of Ti–6Al–4V since there is no cutting fluid, and the Ti–6Al–4V is a poor conductor of heat. In addition, these adhesions are usually not chipped off as the cutting progresses because of the extra surface area generated for work-tool adhesion on the rake face due to the textures. It should be further noted that the work adhesions on the surface fill the textures and nullify their effects. Furthermore, these adhesions may induce diffusion wear and lead to chipping of the cutting edge.

The crater wear is inconspicuous due to the presence of textures and adhesions on the rake face. So, in each case, the flank wear is used to characterize tool wear. The flank face of the inserts was examined using SEM. Figure 6 shows the typical flank face of DTI and CI insert after machining a length of 160 mm. As seen, the primary wear mechanisms were adhesion, abrasion, and attrition (adhesion with wear). We noted that compared to CI inserts, the intensity of flank wear is reduced for textured inserts at low machining time (machining length up to 320 mm). It is also possible that at higher cutting lengths, an increase in temperature may induce diffusion wear

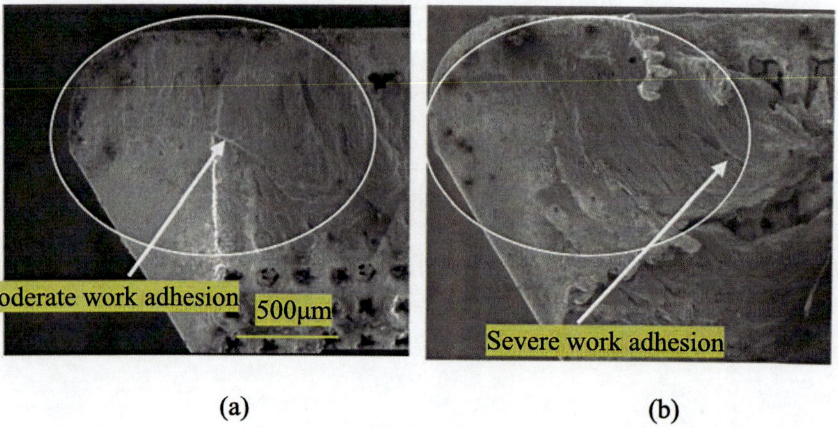

Fig. 5 Rake face of DTIs after **a** 160 mm and **b** after 320 mm of machining length

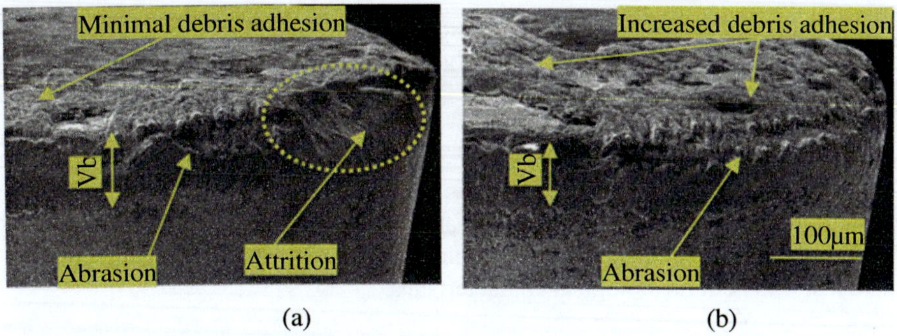

Fig. 6 Flank face of **a** CI, **b** DTI after 160 mm of machining

to progress, especially when the cobalt binder is disturbed for textured inserts, as discussed earlier. However, this mechanism is not investigated in this study.

For further analysis, the maximum worn width at the flank face of the tool, i.e., flank wear (Vb_{max}), was recorded using *ImageJ* software from each insert. Flank wear evolution was tracked by plotting Vb_{max} versus cutting length. The progressive tool wear graphs for both DTI and CI are presented in Fig. 7. It is interesting to note that an initial wear rate region, then a wear stabilization region, and finally an acceleration phase where the wear rate increases rapidly were seen for both inserts. However, initially after the first cut, the results show an approximately 6% reduction in Vb_{max} for DTI. Further, increase in machining length increases the reduction percentage to 18% and then to 22% after 240 mm of machining. After that, the effects were not significant, and the Vb_{max} increased considerably for DTI when compared to CI. The agreements between tool wear and cutting forces and friction plotted earlier are

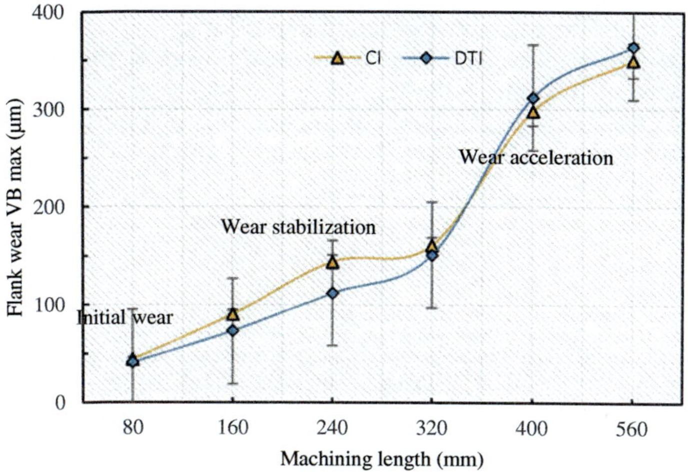

Fig. 7 Progression of flank wear (Vb_{max}) versus machining length

correct. As mentioned earlier, the damage to the textures and increase in crater wear due to accumulation of temperature could be related to the observed increase in flank wear for DTI.

It is evident that the effect of the textured inserts depends on the machining regime [6], texture geometrical parameters [18], workpiece and tool combinations, and other cutting parameters. However, based on the results obtained in this study, one may conclude that the effects of tool textures are inadequate in improving tribological aspects of cutting when machined for a prolonged period, even in a matching regime where it shows better tribological characteristics.

4 Conclusions

The present study investigates the effectiveness of textured inserts from a practical point of view. To achieve this objective, machining experiments were carried out using different machining lengths, and effects in reducing cutting forces and friction were monitored. Based on the analysis carried out, the following findings were obtained.

- Textured inserts show a reduction in cutting forces and cutting friction, which is limited to a specific cutting condition and further for specific machining time/length
- Dual textured inserts show an average reduction of 8.5, 23, and 19%, respectively for F_c, F_t, and COF, while the tool temperature showed about 18% reduction when machined 240 mm.
- In general, textured inserts show a reduction in flank wear, but it progressively decreases with machining time. The cutting tool wear could be split into three regions. Though this is as expected for conventional non textured inserts, surprisingly, the same was observed for textured inserts.
- The average flank wear was initially less than (about 6–22%) conventional non-textured inserts. But, as the machining time increased, it was significantly impacted. After machining 560 mm, the wear rate progress for textured inserts and the use of textured inserts did not show any advantage.

References

1. Prasad CLVRSV, Ramana SV, Pavani PL, Ramji K (2013) Experimental investigations for the prediction of wear zones on the rake face of tungsten carbide inserts under dry machining conditions. Procedia CIRP. 8:528–533. https://doi.org/10.1016/j.procir.2013.06.145
2. Palakudtewar RK, Gaikwad SV (2014) Dry machining of superalloys: difficulties and remedies 3:277–282
3. Pimenov DY, Mia M, Gupta MK, Machado AR, Tomaz ÍV, Sarikaya M, Wojciechowski S, Mikolajczyk T, Kaplonek W (2021) Improvement of machinability of Ti and its alloys using

cooling-lubrication techniques: a review and future prospect. J Market Res 11:719–753. https://doi.org/10.1016/j.jmrt.2021.01.031
4. Orra K, Choudhury SK (2018) Tribological aspects of various geometrically shaped micro textures on cutting insert to improve tool life in hard turning process. J Manuf Process 31:502–513. https://doi.org/10.1016/j.jmapro.2017.12.005
5. Patel KV, Jarosz K, Özel T (2021) Physics-based simulations of chip flow over micro-textured cutting tool in orthogonal cutting of alloy steel. J Manuf Mater Proc 5:65. https://doi.org/10.3390/jmmp5030065
6. Singh R, Dureja JS, Dogra M, Gupta MK, Mia M, Song Q (2020) Wear behavior of textured tools under graphene-assisted minimum quantity lubrication system in machining Ti–6Al–4V alloy. Tribol Int 145:106183. https://doi.org/10.1016/j.triboint.2020.106183
7. Li C, Qiu X, Yu Z, Li S, Li P, Niu Q, Kurniawan R, Ko TJ (2021) Novel environmentally friendly manufacturing method for micro-textured cutting tools. Int J Precis Eng Manuf Green Technol 8:193–204. https://doi.org/10.1007/s40684-020-00256-w
8. Obikawa T, Kamio A, Takaoka H, Osada A (2011) Micro-texture at the coated tool face for high-performance cutting. Int J Mach Tools Manuf 51:966–972. https://doi.org/10.1016/j.ijmachtools.2011.08.013
9. Gajrani KK, Sankar MR (2017) ScienceDirect state of the art on micro to nanotextured cutting tools. Mater Today Proc 4:3776–3785. https://doi.org/10.1016/j.matpr.2017.02.274
10. Soni H, Mashinini PM (2021) An analysis on tool-chip interaction during dry machining of SS316 using textured carbide tools. Arab J Sci Eng 46:7611–7621. https://doi.org/10.1007/s13369-021-05499-6
11. Tamil N, Zeman P, Hoier P, Beno T, Klement U (2019) Investigation of micro-textured cutting tools used for face turning of alloy 718 with high-pressure cooling. J Manuf Process 37:606–616. https://doi.org/10.1016/j.jmapro.2018.12.023
12. Parida AK, Rao PV, Ghosh S (2020) Performance of textured tool in turning of Ti–6Al–4V alloy: numerical analysis and experimental validation. J Braz Soc Mech Sci Eng 42:1–14. https://doi.org/10.1007/s40430-020-02333-2
13. Elias JV et al (2021) Tool texturing for micro-turning applications—an approach using mechanical micro indentation. Mater Manuf Proces 36:84–93. https://doi.org/10.1080/10426914.2020.1813899
14. Sugihara T, Kobayashi R, Enomoto T (2021) Direct observations of tribological behavior in cutting with textured cutting tools. Int J Mach Tools Manuf 103726. https://doi.org/10.1016/j.ijmachtools.2021.103726
15. Siju AS, Waigaonkar SD (2021) Effect of rake surface textures spacing and width on the cutting performance of inserts in dry machining of titanium alloy. 23:136–159
16. Law M, Karthik R, Sharma S, Ramkumar J (2020) Finish turning of hardened bearing steel using textured PcBN tools. J Manuf Process 60:144–161. https://doi.org/10.1016/j.jmapro.2020.10.051
17. Siju AS, Gajrani KK, Joshi SS (2021) Dual textured carbide tools for dry machining of titanium alloys. Int J Refract Metals Hard Mater 94:105403. https://doi.org/10.1016/j.ijrmhm.2020.105403
18. Siju AS, Waigaonkar SD. Effect of rake surface textures spacing and width on the cutting performance of inserts in dry machining of titanium alloy
19. Kishawy HA, Salem A, Hegab H, Hosseini A, Balazinski M (2021) Micro-textured cutting tools: phenomenological analysis and design recommendations. CIRP Ann 00:1–4. https://doi.org/10.1016/j.cirp.2021.04.081
20. Duan R, Deng J, Ai X, Liu Y, Chen H (2017) Experimental assessment of derivative cutting of micro-textured tools in dry cutting of medium carbon steels. 3531–3540. https://doi.org/10.1007/s00170-017-0360-8
21. Machado AR, da Silva LRR, de Souza FCR, Davis R, Pereira LC, Sales WF, de Rossi W, Ezugwu EO (2021) State of the art of tool texturing in machining. J Mater Process Technol 293:117096. https://doi.org/10.1016/j.jmatprotec.2021.117096

22. Mane S, Karagadde S, Joshi SS (2019) Study of cutting edge radius effect on the cutting forces and temperature during machining of Ti6Al4V. 309–316. https://doi.org/10.1007/978-981-32-9072-3_26
23. Mane S, Ramchandani J, Marla D, Joshi SS (2019) Experimental investigation of oil mist assisted cooling on the orthogonal cutting of Ti6Al4V. Procedia Manuf 34:369–378. https://doi.org/10.1016/j.promfg.2019.06.180
24. Kumar CS, Patel SK (2018) Effect of WEDM surface texturing on Al_2O_3/TiCN composite ceramic tools in dry cutting of hardened steel. Ceram Int 44:2510–2523. https://doi.org/10.1016/j.ceramint.2017.10.236
25. Durairaj S, Guo J, Aramcharoen A, Castagne S (2018) An experimental study into the effect of micro-textures on the performance of the cutting tool. Int J Adv Manuf Technol 98:1011–1030. https://doi.org/10.1007/s00170-018-2309-y
26. Duan R, Deng J, Lei S, Ge D, Liu Y, Li X (2019) Effect of derivative cutting on machining performance of micro-textured tools. J Manuf Process 45:544–556. https://doi.org/10.1016/j.jmapro.2019.07.037

Experimental Investigations on Tool Path Strategies in Ball-End Milling

S. Niven Matthew and M. Kanthababu

Abstract The selection of the optimal tool path strategy and appropriate cutting conditions, especially in the machining aspects of free form surfaces are important to achieve higher productivity. In this work, optimal tool path selection for machining concave surfaces is carried out to achieve minimum surface roughness (Ra), cutting forces and machining time in ball-end milling of aluminium alloy 6061 (Al 6061). Various tool path strategies including 3D offset, raster, concentric, radial zig-zag, zig, and concentric zig-zag are employed. The cutting parameters considered were feed rate, spindle speed and step over. Design of experiments (DoE) along with the statistical analysis of variance (ANOVA) and mean plots are used to determine the significant parameters and their levels for each tool path strategy. Conflicting results were obtained from the ANOVA and the mean plot analyses, to achieve optimum levels that led to lower surface roughness, cutting forces and machining time. It is observed that lower values of surface roughness and cutting forces are obtained with the low level of feed rate and step over in all six tool path strategies. However, high feed rate and step over the lead to decreased machining time. Hence, grey relational analysis is carried out considering multiple performance characteristics such as minimum surface roughness, cutting forces and machining time, and optimal combinations of the milling process parameters for each tool path strategies are identified.

Keywords Ball-end milling · Tool path strategy · Aluminium alloy 6061 · Surface roughness · Cutting forces · Machining time · Grey relation analysis

1 Introduction

High precision parts with free-form surfaces are required in several applications. Ball-end tools are widely used for machining free form surfaces in CNC milling for various applications including in the manufacturing of moulds and dies. Machining of free

S. Niven Matthew · M. Kanthababu (✉)
Department of Manufacturing Engineering, College of Engineering Guindy, Anna University, Chennai 600025, India
e-mail: kb@annauniv.edu

U. S. Dixit et al. (eds.), *Advances in Forming, Machining and Automation*, Lecture Notes in Mechanical Engineering, https://doi.org/10.1007/978-981-19-3866-5_25

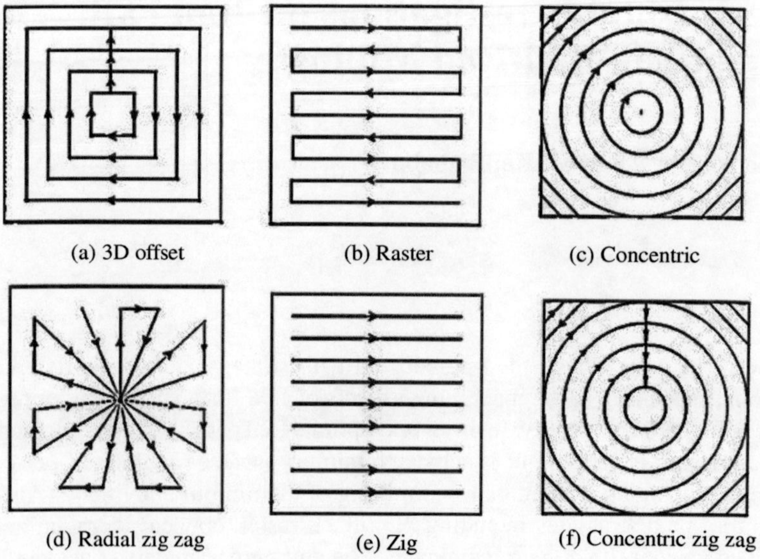

Fig. 1 Schematic of different tool path strategies

form surfaces is generally achieved by a given machining tool path strategy. Proper selection of tool path strategy and appropriate combinations of process parameters are crucial for accomplishing desired machined surfaces considering surface roughness, cutting forces, machining time, temperature, tool life and workpiece surface integrity [1, 2].

The different tool path strategies commonly used in CNC milling are 3D offset, raster, concentric, radial zig-zag, zig and concentric zig zag. The schematic of different tool path strategies used for machining concave surfaces is shown in Fig. 1. In the 3D offset milling strategy, the tool path starts at the periphery of the inner surface of the workpiece to be machined or the tool may start at the centre of the workpiece and then proceed outwards. In the raster tool path, the tool movement is parallel in the *XY* plane and tracks the surface in the *Z*-direction. The concentric tool path is also called the spiral tool path, wherein the tool moves in a circular path from a given focal point by having constant contact between the tool and workpiece. In the case of the radial path, the tool moves to the central point, thereafter it retracts back to the original position and then moves to the centre, whereas in the radial zig-zag tool path strategy, the tool moves continuously in a zig-zag manner. In the zig-zag tool path stratergy, the tool moves in circular form and continuously moves in the zig-zag manner. In the concentric zig-zag-tool path strategy, the tool moves in circular form and continuously moves in the zig-zag form [3–6].

Adriano et al. [7] have investigated different tool path strategies while milling a mould cavity during finishing operation and observed that the tool path strategies have a great influence on the surface roughness and machining time. Researchers have also studied the influence of different tool path strategies on surface roughness [8]. Few researchers have studied the effect of cutting forces on different tool orientations [9–11]. However, only a few have investigated the influence of cutting forces in different tool path strategies, especially while machining free form surfaces [12–16]. Therefore, proper evaluation of tool path strategies considering minimum surface roughness, minimum cutting forces, and minimum machining time is needed in ball-end milling of concave surfaces. The objective of this study is to analyse different machining tool path strategies such as 3D offset, raster, concentric, radial zig-zag, zig and concentric zig-zag in a 3axis CNC milling while machining free form surface (concave surface). The influence of milling process parameters on surface roughness, machining time and cutting forces on the tool path strategies are investigated and the most significant parameters that affect the surface roughness, cutting forces and machining time are identified in each tool path strategy using analysis of variance (ANOVA) and mean plot analysis.

2 Experimental Details

Machining has been carried out using HURCO 3axis CNC milling centre of model VM10 (Figs. 2 and 3). This machine has a maximum spindle speed of 10,000 rpm, a maximum feed rate of 24 m/min. The machining is carried out in a dry condition. The workpiece material used in the work is Al 6061-T6, which is widely used in many industrial applications particularly in architectural, structural, and motor vehicles

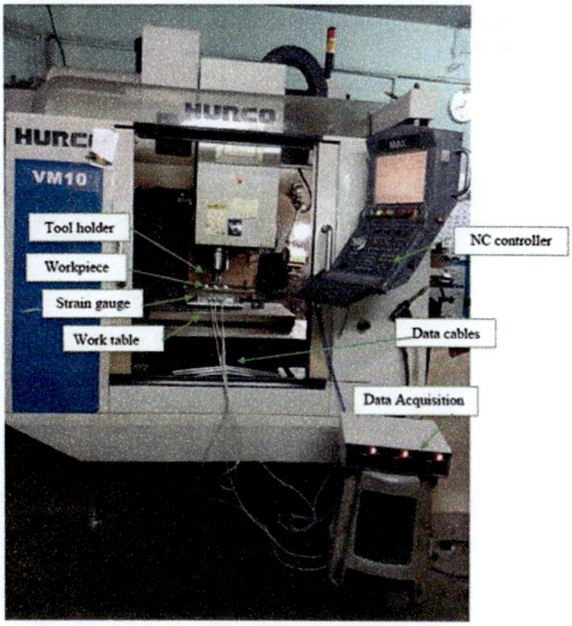

Fig. 2 Photograph of the experimental setup

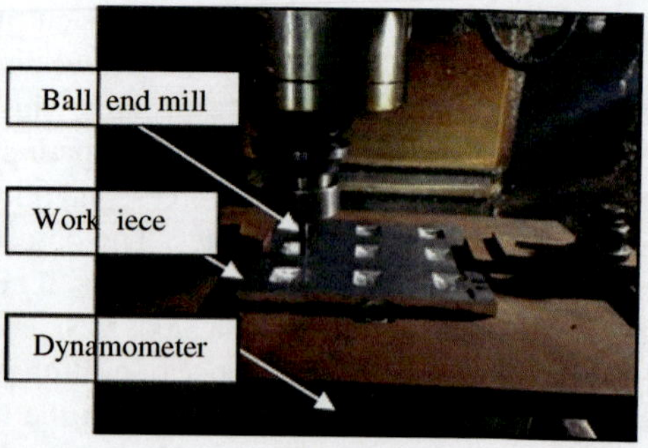

Fig. 3 Closer view of the machining zone

due to its properties such as high corrosive resistance, and strength. The chemical composition is of Al 6061 which is 1.2% Mg, 0.8% Si, 0.7% Fe, 0.4% Cu, 0.5% Mn, 0.35% Cr and 0.25% Zn. About 0.15% Ti and Al balance, which is measured using optical emission spectroscopy. The mechanical properties of the Al 6061 are shown in Table 1. During each experimental trial, the workpiece is machined for an area of 20 mm × 20 mm and a depth of 4 mm (Fig. 4). The cutting tool used in this study is a ball-end mill made up of solid carbide material of 6 mm diameter and 30° helix angle with four flutes.

Cutting force measurements Fx, Fy and Fz are carried out using a 3axis strain gauge dynamometer. The surface roughness is measured using Surfcorder (SE3500),

Table 1 Mechanical properties of Al 6061

Thermal conductivity (W/m K)	Tensile strength (MPa)	Hardness (HB)	Density (g/cm^3)	Max temperature in work (°C)
167	276	107	2.7	652

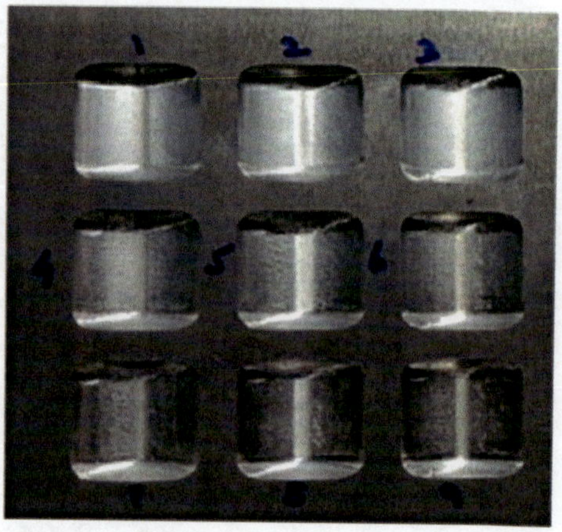

Fig. 4 Typical machined surface using zig tool path strategy

Table 2 Machining parameter and their levels

Machining parameter	Unit	Level		
		1	2	3
Feed rate	mm/min	900	1200	1500
Spindle speed	rpm	3000	3500	4000
Step over	mm	0.06	0.12	0.18

contact type surface roughness tester with phase-corrected Gaussian filter, traverse the length of 5.6 mm and cut-off length of 0.8 mm, while the machining time is recorded from the beginning to the end of the cut, using a standard digital time clock.

The DoE for each tool path strategy is carried out according to L9 orthogonal array (three factors with three levels each). The cutting parameters considered are feed rate, spindle speed and step over (Table 2). Accordingly, nine experimental trials were performed with six different tool path strategies such as 3D offset, raster, concentric, radial zig-zag, zig and concentric zig zag. The tool path strategies used in this investigation are simulated on NX–CAM software before machining.

3 Result and Discussion

The following section deals with the analysis of surface roughness, machining time and cutting forces.

3.1 Analysis of Surface Roughness

The surface roughness obtained during experimental trial runs with each tool path strategy is given in Table 3. The mean effect plots of surface roughness are shown in Fig. 5, and the ANOVA Table of surface roughness is given in Table 4.

From the main effect plots shown in Fig. 5, it is observed that the minimum surface roughness is achieved with low feed rate and low step over in all six different tool path strategies. The surface roughness is also found to be minimum at higher spindle speed in all six tool path strategies excluding the zig-tool path strategy (Fig. 5). In the case of the zig-tool path strategy, the surface roughness is found to be lower at low spindle speed (Fig. 5e). The surface roughness values increase with the increase in the feed rate which may be due to the larger cutting area [17]. In the case of step over, it is observed, when the step over increases, the surface roughness is also found to be increasing. This may be due to the increase in the distance between the tool path (step over), which results in higher surface roughness values. Therefore, low

Table 3 Experimental results of surface roughness

Ex. No.	Input parameter			Surface roughness [*Ra*] (µm)					
				Tool-path strategy					
	Feed (mm/min)	Spindle speed (rpm)	Step over (mm)	3D Offset	Raster	Concentric	Radial zig zag	Zig	Concentric zig zag
1	900	3000	0.06	0.616	1.093	0.917	0.593	0.783	1.590
2	900	3500	0.12	0.833	1.670	1.200	0.787	1.667	1.497
3	900	4000	0.18	0.913	2.050	1.140	0.730	1.610	1.260
4	1200	3000	0.12	0.960	1.643	1.157	0.800	0.973	1.540
5	1200	3500	0.18	1.350	2.513	1.370	1.000	1.850	1.710
6	1200	4000	0.06	0.766	1.243	0.977	0.693	1.420	1.243
7	1500	3000	0.18	1.680	2.227	1.337	0.927	1.957	1.703
8	1500	3500	0.06	1.110	1.190	1.070	1.057	1.657	1.373
9	1500	4000	0.12	1.470	1.147	0.990	0.804	1.683	1.706

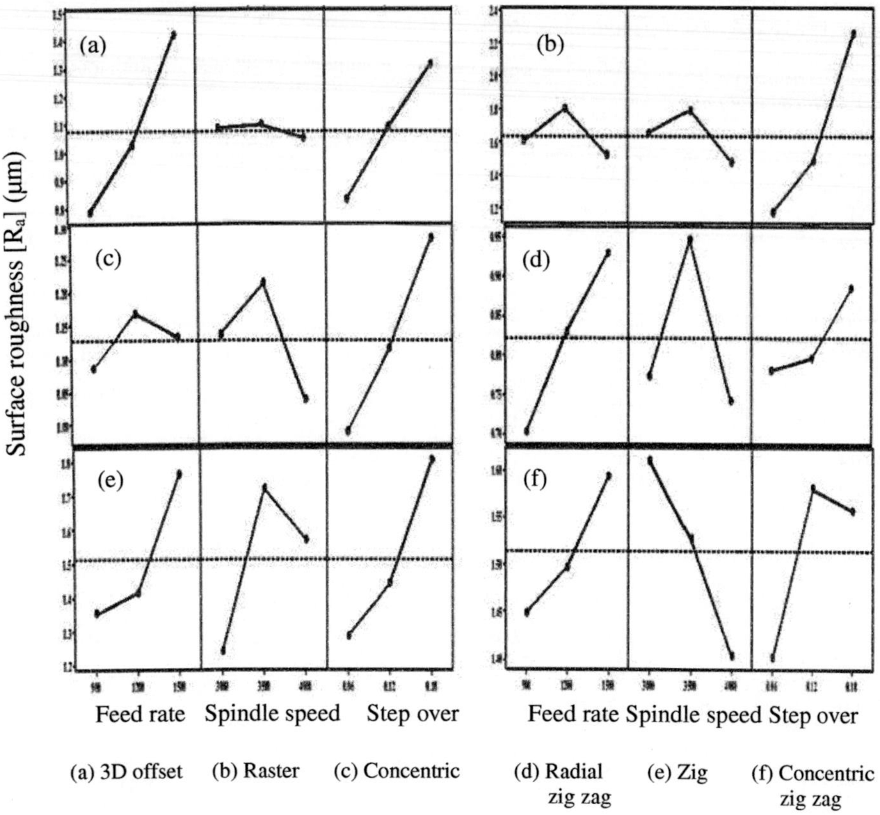

Fig. 5 Mean effect plot of surface roughness

feed rate, low step over and higher speed are recommended to achieve lower surface roughness.

From ANOVA Table 4a, it is observed that in the case of 3D offset tool path strategy, the feed rate and step over are found to be significant, while the spindle speed is found to be insignificant. From Table 4b and c, it is observed that the stepover is found to be significant, whereas the feed rate and spindle speed are found to be insignificant in raster and concentric tool path strategies. It is also observed that the feed rate and the spindle speed are found to be significant in the radial zig-zag tool path strategy (Table 4d). From Table 4e and f, it is observed that in the case of zig and concentric zig-zag tool path strategies, none of the machining process parameters are significant as the *P*-value is greater than 0.05 in these two strategies.

Table 4 ANOVA of surface roughness (*R*a)

Source	DF	Adj SS	Adj MS	*F*-value	*P*-value	Remarks
(a) 3D offset tool path						
Feed rate	2	0.612	0.306	17.37	0.054	Significant
Spindle speed	2	0.003	0.001	0.11	0.904	Insignificant
Step over	2	0.351	0.175	9.96	0.091	Significant
Error	2	0.035	0.017	–	–	–
Total	8	1.003	–	–	–	–
(b) Raster tool path						
Feed rate	2	0.122	0.061	4.74	0.174	Insignificant
Spindle speed	2	0.145	0.072	5.64	0.151	Insignificant
Step over	2	1.883	0.941	72.86	0.014	Significant
Error	2	0.025	0.012	–	–	–
Total	8	2.178	–	–	–	–
(c) Concentric tool path						
Feed rate	2	0.010	0.005	1.07	0.483	Insignificant
Spindle speed	2	0.047	0.023	4.99	0.167	Insignificant
Step over	2	0.130	0.065	13.70	0.068	Significant
Error	2	0.009	0.004	–	–	–
Total	8	0.198	–	–	–	–
(d) Radial zig-zag tool path						
Feed rate	2	0.077	0.038	10.11	0.090	Significant
Spindle speed	2	0.073	0.036	9.68	0.094	Significant
Step over	2	0.019	0.009	2.50	0.286	Insignificant
Error	2	0.007	0.003	–	–	–
Total	8	0.177	–	–	–	–
(e) Zig tool path						
Feed rate	2	0.297	0.148	2.26	0.307	Insignificant
Spindle speed	2	0.371	0.185	2.83	0.261	Insignificant
Step over	2	0.426	0.213	3.24	0.236	Insignificant
Error	2	0.131	0.065	–	–	–
Total	8	1.226	–	–	–	–
(f) Concentric zig-zag tool path						
Feed rate	2	0.032	0.016	0.27	0.787	Insignificant
Spindle speed	2	0.065	0.032	0.55	0.647	Insignificant
Step over	2	0.056	0.028	0.47	0.679	Insignificant
Error	2	0.120	0.060	–	–	–
Total	8	0.275	–	–	–	–

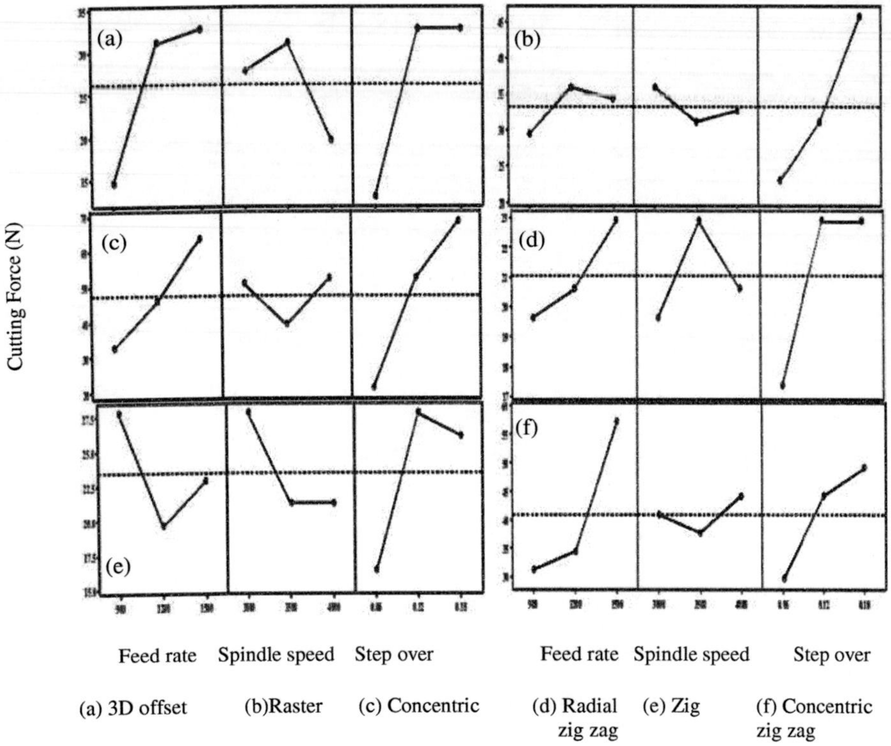

Fig. 4 Mean effect plot of cutting force

3.2 Analysis of Cutting Forces

The cutting forces obtained from the different machining conditions in each experimental trial run are given in Table 5. The mean effect plots for cutting forces in different tool path strategies are shown in Fig. 6. From Fig. 6, it is observed that the minimum cutting force is achieved with low feed rate and low step over in all the six tool path strategies except the 3D offset tool path strategy. From Fig. 6, it is also observed that the increase in feed rate results in an increase in the cutting forces. In the case of step over, it is also observed that the increase in the step over results in the increase in the cutting forces. In the case of spindle speed, it is observed that mostly medium spindle speed results in lower cutting forces. Cutting forces also increase with the increase in the step over, which may be due to the increase in the size of the uncut chip thickness in all six tool path strategies [1, 2, 18].

The results of the ANOVA of cutting forces are shown in Table 6. From ANOVA Table 6a, e, it is observed that all the process parameters such as feed rate, spindle speed and step over are found to be significant in the case of 3D offset and zig tool path strategies. From Table 6b, d, it is also observed that the stepover is found to be significant, whereas the feed rate and spindle speed are found to be insignificant in

Table 5 Experimental results of cutting forces

Ex. No.	Input parameter			Cutting force in Z-direction Fz (max) (N)					
				Tool path strategies					
	Feed (mm/min)	Spindle speed (rpm)	Step over (mm)	3D offset	Raster	Concentric	Radial Zig zag	Zig	Concentric zig zag
1	900	3000	0.06	4.905	19.62	14.72	14.72	24.53	19.62
2	900	3500	0.12	24.53	24.53	29.43	24.53	29.43	34.34
3	900	4000	0.18	14.72	44.15	53.95	19.62	29.43	39.24
4	1200	3000	0.12	39.24	39.24	49.05	19.62	29.43	34.34
5	1200	3500	0.18	44.15	44.15	63.77	24.53	19.62	39.24
6	1200	4000	0.06	9.81	24.53	24.53	17.65	9.81	29.43
7	1500	3000	0.18	39.24	49.05	88.29	24.53	29.43	68.67
8	1500	3500	0.06	24.53	24.53	24.53	19.62	14.72	39.24
9	1500	4000	0.12	34.34	29.43	78.48	24.53	24.53	63.77

Table 6 ANOVA of cutting force

Source	DF	Adj SS	Adj MS	*F*-value	*P*-value	Remarks
(a) 3D offset tool path						
Feed rate	2	593.41	296.703	36.96	0.026	Significant
Spindle speed	2	208.53	104.264	12.99	0.071	Significant
Step over	2	770.02	385.010	47.96	0.020	Significant
Error	2	16.06	8.028	–	–	–
Total	8	1588.01	–	–	–	–
(b) Raster tool path						
Feed rate	2	69.49	34.75	1.86	0.350	Insignificant
Spindle speed	2	37.35	18.67	1.00	0.501	Insignificant
Step over	2	807.34	403.67	21.56	0.044	Significant
Error	2	37.45	18.72	–	–	–
Total	8	951.62	–	–	–	–
(c) Concentric tool path						
Feed rate	2	1459.4	729.72	10.11	0.090	Insignificant
Spindle speed	2	304.8	152.39	2.11	0.321	Insignificant
Step over	2	3480.3	1740.14	24.10	0.040	Significant
Error	2	144.4	72.20	–	–	–
Total	8	5388.9	–	–	–	–
(d) Radial zig zag tool path						
Feed rate	2	16.91	8.453	1.61	0.383	Insignificant
Spindle speed	2	16.91	8.453	1.61	0.383	Insignificant
Step over	2	61.90	30.951	5.90	0.045	Significant
Error	2	10.49	5.245	–	–	–
Total	8	106.20	–	–	–	–
(e) Zig tool path						
Feed rate	2	101.615	50.808	19.01	0.050	Significant
Spindle speed	2	85.543	42.772	16.00	0.059	Significant
Step over	2	229.799	114.900	42.98	0.023	Significant
Error	2	5.346	2.673	–	–	–
Total	8	422.304	–	–	–	–
(f) Concentric zig-zag tool path						
Feed rate	2	1218.99	609.50	19.00	0.050	Significant
Spindle speed	2	64.16	32.08	1.00	0.500	Insignificant
Step over	2	625.53	312.77	9.75	0.093	Insignificant
Error	2	64.16	32.08	–	–	–
Total	8	1972.84	–	–	–	–

the case of the raster, concentric and radial zig-zag tool path strategies. It is observed in Table 6f that feed rate is significant for the concentric zig-zag tool path strategy.

3.3 Analysis of Machining Time

The machining time recorded during different combinations of machining conditions in each experimental trial run is given in Table 7. The mean effect plots of machining time for different tool path strategies at three levels are shown in Fig. 7. The mean effect plots indicate that the machining time decreases with the increase in the feed rate and step over in all the six tool path strategies. It is observed that the increase in the feed rate and step over leads to decreased machining time due to an increase in tool movement and reduced tool path length [2]. The spindle speed does not affect the machining time, and it is found to be an insignificant factor. Similar trends were observed in all six tool path strategies.

Table 8 shows the ANOVA results of machining time obtained at various tool path strategies. From Table 8, it is found that step over is the most significant factor in all six tool path strategies. However, the feed rate is also found to be significant along with step over in raster tool path strategy. The spindle speed is found to be an insignificant factor in all the strategies studied in this analysis.

From the above ANOVA and mean plot analyses, the study observed that lower values of surface roughness and cutting forces are obtained with a lower level of feed rate and step over in all six tool path strategies. However, higher feed rate and step over the lead to decreased machining time. Therefore, conflicting results were obtained to achieve optimum levels that lead to low surface roughness, machining time and cutting force. Therefore, there is a need for a multi-objective optimization technique to achieve lower cutting force, lower surface roughness (Ra) and lower machining time to improve productivity. Hence, grey relation analysis has been carried out and presented below.

4 Grey Relation Analysis

The grey relation analysis converts the multi-objective optimization problem into a single objective problem [19, 20]. In grey relation analysis, the experimental results of surface roughness, machining time, and cutting force are first normalized in the range from zero to one. Thereafter, the grey relational coefficient is calculated with the help of Eqs. (1)–(3), from the normalized experimental data, which represents the relevance between the desired and actual experimental results.

In this study, to achieve lower roughness with minimum machining time and minimum cutting forces, 'lower the better' condition is chosen and expressed as,

Table 7 Experimental results of machining time

Input parameter				Machining time (min)					
				Tool path strategies					
Ex. No.	Feed (mm/min)	Spindle speed (rpm)	Step over (mm)	3D offset	Raster	Concentric	Radial zig zag	Zig	Concentric zig zag
1	900	3000	0.06	6.46	8.40	10.13	11.20	11.24	4.42
2	900	3500	0.12	3.25	4.35	4.50	6.25	5.47	2.27
3	900	4000	0.18	2.06	3.09	3.20	3.56	3.56	1.5
4	1200	3000	0.12	2.49	3.49	4.37	4.44	5.26	2.14
5	1200	3500	0.18	1.48	2.39	3.05	3.15	3.41	1.43
6	1200	4000	0.06	5.08	7.14	8.40	9.00	10.45	4.18
7	1500	3000	0.18	1.24	2.24	3.05	2.53	3.34	1.43
8	1500	3500	0.06	4.25	6.25	8.44	8.00	10.10	4.11
9	1500	4000	0.12	2.24	3.23	4.26	4.10	5.15	2.13

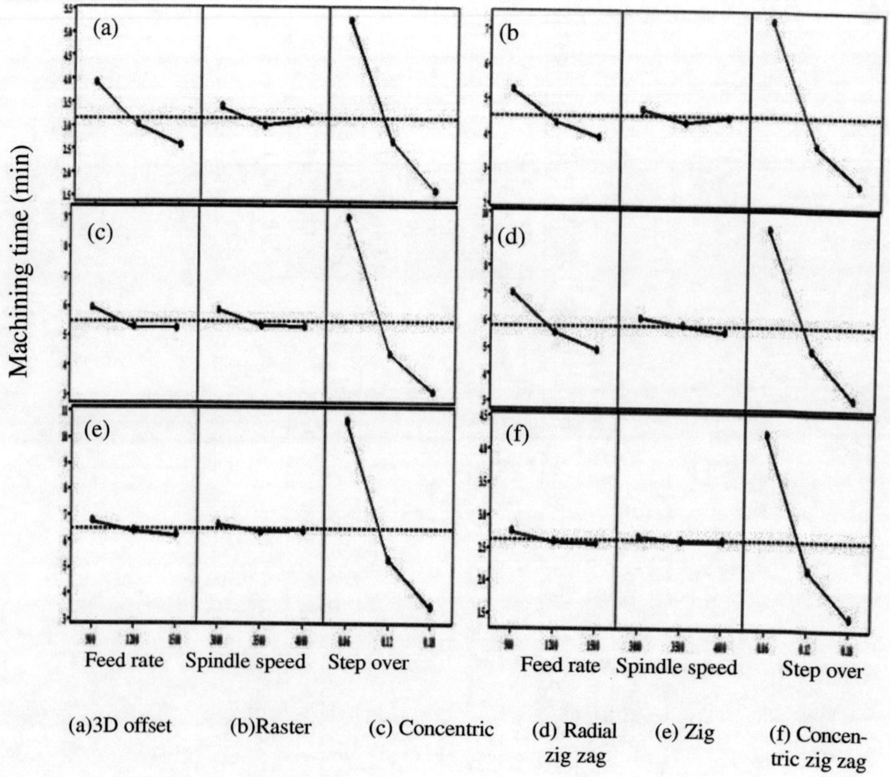

Fig. 7 Mean effect plot of machining time

$$x_i(k) = \frac{\max y_i(k) - y_i(k)}{\max y_i(k) - \min y_i(k)} \tag{1}$$

where $x_i(k)$ is the value after linear normalization, $y_i(k)$ is the experimental re-response, max $y_i(k)$ is the largest value of $y_i(k)$ for the kth response and min $y_i(k)$ is the smallest value of $y_i(k)$ for the kth response. The grey relation coefficient ξ_i is calculated from the below equation,

$$\xi_i(k) = \frac{\Delta \min + \zeta \Delta \max}{\Delta o_i(k) + \zeta \Delta \max} \tag{2}$$

where $\Delta o_i(k)$ represents the deviation of absolute value between $y_0(k)$ and $y_i(k)$, Δmin and Δmax are the smallest and largest value of $\Delta o_i(k)$, respectively. ζ is the distinguishing coefficient, and the value of ζ is considered as 0.5.

After averaging the grey relation coefficient, the grey relation grade γ_i is observed from the below equation,

Table 8 ANOVA of machining time

Source	DF	Adj SS	Adj MS	*F*-value	*P*-value	Remarks
(a) 3D offset tool path						
Feed rate	2	2.829	1.414	8.87	0.101	Insignificant
Spindle speed	2	0.253	0.126	0.79	0.557	Insignificant
Step over	2	21.384	10.692	67.07	0.015	Significant
Error	2	0.318	0.159	–	–	–
Total	8	24.785	–	–	–	–
(b) Raster tool path						
Feed rate	2	2.957	1.478	11.52	0.080	Significant
Spindle speed	2	0.218	0.109	0.85	0.540	Insignificant
Step over	2	36.011	18.005	140.29	0.007	Significant
Error	2	0.256	0.128	–	–	–
Total	8	39.444	–	–	–	–
(c) Concentric tool path						
Feed rate	2	0.930	0.465	1.96	0.338	Insignificant
Spindle speed	2	0.589	0.294	1.24	0.446	Insignificant
Step over	2	57.604	28.802	121.46	0.008	Significant
Error	2	0.474	0.237	–	–	–
Total	8	59.598	–	–	–	–
(d) Radial zig-zag tool path						
Feed rate	2	7.120	3.560	6.66	0.131	Insignificant
Spindle speed	2	0.380	0.190	0.36	0.738	Insignificant
Step over	2	63.345	31.672	59.26	0.017	Significant
Error	2	1.068	0.534	–	–	–
Total	8	71.915	–	–	–	–
(e) Zig tool path						
Feed rate	2	0.491	0.245	3.75	0.211	Insignificant
Spindle speed	2	0.137	0.068	1.04	0.489	Insignificant
Step over	2	82.838	41.419	630.96	0.002	Significant
Error	2	0.131	0.065	–	–	–

(continued)

$$\gamma_i = \frac{1}{n}\sum_{k=1}^{n} \xi_i(k) \tag{3}$$

where n represents the number of experimental responses.

The results of the grey relation coefficient and the grey relation grade from the experimental data of different six tool path strategies are given in Table 9. The

Table 8 (continued)

Source	DF	Adj SS	Adj MS	*F*-value	*P*-value	Remarks
Total	8	83.598	–	–	–	–
(f) Concentric zig-zag tool path						
Feed rate	2	0.052	0.026	5.89	0.145	Insignificant
Spindle speed	2	0.007	0.003	0.81	0.552	Insignificant
Step over	2	12.504	6.252	1410.32	0.001	Significant
Error	2	0.008	0.004	–	–	–
Total	8	12.573	–	–	–	–

maximum value of grey relation grade in Table 9 indicates the closure value of optimum process parameters. From Table 9, it is observed that the maximum value of grey relation grade for 3D offset, raster, concentric and radial zig-zag path strategies is found in experiment number one with input process parameter's combinations of 1000 mm/min of feed rate, 3000 rpm of spindle speed and 0.06 mm of step over. These combinations are considered to be the best multiple performance characteristics among the nine experimental results as per the grey relation analysis. In the case of zig tool path strategy, it is found at the experiment number five, with combinations of input process parameters (i.e.) 1250 mm/min of feed rate, 3500 rpm of spindle speed and 0.18 mm of step over resulted in the best multiple performance characteristics, while in the case of concentric zig-zag tool path, it is found at the experiment number three, with combinations of input process parameters (i.e.) 900 mm/min of feed rate, 4000 rpm of spindle speed and 0.18 mm of step over resulted in best multiple performance characteristics.

5 Conclusion

The effect of different tool path strategies such as 3D offset, raster, concentric, radial zig-zag, zig and concentric zig-zag while machining curved surfaces were investigated during CNC ball-end milling of Al 6061. The effect of input milling process parameters such as feed rate, step over and spindle speed are studied on the responses such as surface roughness, machining time and cutting force in each strategy.

Conflicting results were obtained from ANOVA and mean plot analyses to achieve optimum levels that lead to low surface roughness, machining time and cutting force. It is observed that lower values of surface roughness and cutting forces are obtained with a lower level of feed rate and step over in all six tool path strategies in this work. However, higher feed rate and step over the lead to decreased machining time. Therefore, a multi-objective optimization technique using grey relation analysis has been carried out to achieve lower cutting force, lower surface roughness and lower machining time to improve productivity.

Table 9 Grey relation coefficient and grey relation grade

Ex. No.	Grey relation coefficient			Grey relation grade	Rank
	Surface roughness	Machining time	Cutting force		
(a) 3D offset tool path strategy					
1	**1.000**	**0.333**	**1.000**	**0.778**	**1**
2	0.710	0.565	0.500	0.592	4
3	0.642	0.761	0.667	0.690	2
4	0.607	0.676	0.364	0.549	7
5	0.420	0.916	0.333	0.556	6
6	0.780	0.405	0.800	0.662	3
7	0.333	1.000	0.364	0.566	5
8	0.519	0.464	0.500	0.494	9
9	0.384	0.723	0.400	0.502	8
(b) Raster tool path strategy					
1	**1.000**	**0.333**	**1.000**	**0.778**	**1**
2	0.552	0.593	0.750	0.632	5
3	0.426	0.784	0.375	0.528	9
4	0.563	0.711	0.429	0.568	7
5	0.333	0.954	0.375	0.554	8
6	0.826	0.386	0.750	0.654	4
7	0.385	1.000	0.333	0.573	6
8	0.880	0.434	0.750	0.688	3
9	0.929	0.757	0.600	0.762	2
(c) Concentric tool path strategy					
1	**1.000**	**0.333**	**1.000**	**0.778**	**1**
2	0.445	0.709	0.714	0.623	4
3	0.504	0.959	0.484	0.649	3
4	0.486	0.728	0.517	0.577	8
5	0.333	1.000	0.429	0.587	7
6	0.791	0.398	0.790	0.659	2
7	0.350	1.000	0.333	0.561	9
8	0.597	0.396	0.790	0.594	6
9	0.756	0.745	0.366	0.622	5
(d) Radial zig-zag tool path strategy					
1	**1.000**	**0.333**	**1.000**	**0.778**	**1**
2	0.545	0.538	0.333	0.472	8
3	0.629	0.808	0.500	0.646	2
4	0.528	0.694	0.500	0.574	5

(continued)

Table 9 (continued)

Ex. No.	Grey relation coefficient			Grey relation grade	Rank
	Surface roughness	Machining time	Cutting force		
5	0.363	0.875	0.333	0.524	7
6	0.699	0.401	0.626	0.575	4
7	0.410	1.000	0.333	0.581	3
8	0.333	0.442	0.500	0.425	9
9	0.524	0.734	0.333	0.530	6
(e) Zig tool path strategy					
1	1.000	0.333	0.400	0.578	4
2	0.399	0.650	0.333	0.461	9
3	0.415	0.947	0.333	0.565	5
4	0.755	0.673	0.333	0.587	3
5	**0.355**	**0.983**	**0.500**	**0.612**	**1**
6	0.480	0.357	1.000	0.612	2
7	0.333	1.000	0.333	0.556	6
8	0.402	0.369	0.666	0.479	8
9	0.395	0.686	0.400	0.493	7
(f) Concentric zig-zag tool path strategy					
1	0.402	0.333	1.000	0.579	6
2	0.479	0.640	0.625	0.581	4
3	**0.932**	**0.955**	**0.556**	**0.814**	**1**
4	0.440	0.678	0.625	0.581	5
5	0.333	1.000	0.556	0.630	3
6	1.000	0.352	0.714	0.689	2
7	0.337	1.000	0.333	0.557	7
8	0.642	0.358	0.556	0.519	8
9	0.335	0.681	0.357	0.458	9

The maximum value of grey relation grade in Table 9, indicated in bold values is the closure value of optimum process parameters

From the grey relation analysis, it is found that the combinations of input parameters such as 1000 mm/min of feed rate, 3000 rpm of spindle speed, and 0.06 mm of step over are found as optimal parameters for 3D offset, raster, concentric and radial zig-zag tool path strategies. In the case of the zig tool path, the combinations of 1250 mm/min of feed rate, 3500 rpm of spindle speed and 0.18 mm of step over are found as optimal parameters, and in the concentric zig-zag tool path strategy, 900 mm/min of feed rate, 4000 rpm of spindle speed and 0.18 mm of step over are found as optimal combinations of parameters.

References

1. Shajari S, Sadeghi MH, Hassanpour H, Jabbaripour B (2012) Influence of machining strategies on surface roughness in ball-end milling of inclined surfaces. Adv Mater Res 488:836–840
2. Shajari S, Sadeghi MH, Hassanpour H (2014) The influence of tool path strategies on cutting force and surface texture during ball-end milling of low curvature convex surfaces. Sci World J Article ID 374526
3. Ikua BW, Tanaka H, Obata F, Sakamoto S, Kishi T, Ishii T (2002) Prediction of cutting forces and machining error in ball-end milling of curved surfaces-II Experimental verification. Precis Eng 26(1):69–82
4. Ali RA, Mia M, Khan AM, Chen W, Gupta MK, Pruncu CI (2019) Multi-response optimization of face milling performance considering tool path strategies in the machining of Al-2024. Materials 12(7):1013
5. Camposeco-Negrete C (2013) Optimization of cutting parameters for minimizing energy consumption in turning of AISI 6061 T6 using Taguchi methodology and ANOVA. J Clean Prod 53:195–203
6. Monreal M, Rodriguez CA (2003) Influence of tool path strategy on the cycle time of high-speed milling. Comput Aided Des 35(4):395–401
7. de Souza AF, Machado A, Beckert SF, Diniz AE (2014) Evaluating the roughness according to the tool path strategy when milling free form surfaces for mold application. Procedia CIRP 14:188–193
8. Toh CK (2004) Surface topography analysis in high-speed finish milling inclined hardened steel. Precis Eng 28(4):386–398
9. Kim GM, Cho PJ, Chu CN (2000) Cutting force prediction of sculptured surface ball-end milling using Z-map. Int J Mach Tools Manuf 40(2):277–291
10. Chu CN, Kim SY, Lee JM, Kim BH (1997) Feed-rate optimization of ball-end milling considering local shape features. CIRP Ann 46(1):433–436
11. Ng EG, Lee DW, Dewes RC, Aspinwall DK (2000) Experimental evaluation of cutter orientation when ball nose end milling Inconel 718. J Manuf Process 2(2):108–115
12. Guzel BU, Lazoglu I (2004) An enhanced force model for sculptured surface machining. Mach Sci Technol 8(3):431–448
13. Budak E, Ozturk E, Tunc LT (2009) Modeling and simulation of 5-axis milling processes. CIRP Ann 58(1):347–350
14. Huang Y, Liang SY (2003) Cutting forces modeling considering the effect of tool thermal property-application to CBN hard turning. Int J Mach Tools Manuf 43(3):307–315
15. Zhang L, Zheng L (2004) Prediction of cutting forces in milling of circular corner profiles. Int J Mach Tools Manuf 44(2–3):225–235
16. Zheng L, Li Y, Liang SY (1998) A generalized model of milling forces. Int J Adv Manuf Technol 14(3):160–171
17. Mali RA, Aiswaresh R, Gupta TVK (2020) The influence of tool-path strategies and cutting parameters on cutting forces, tool wear and surface quality in finish milling of Aluminum 7075 curved surface. Int J Adv Manuf Technol 108:589–601
18. Tangjitsitcharoen S, Thesniyom P, Ratanakuakangwan S (2017) Prediction of surface roughness in ball-end milling process by utilizing dynamic cutting force ratio. J Intell Manuf 28(1):13–21
19. Lin CL (2004) Use of the Taguchi method and grey relational analysis to optimize turning operations with multiple performance characteristics. Mater Manuf Processes 19(2):209–220
20. Angappan P, Thangiah S, Subbarayan S (2017) Taguchi-based grey relational analysis for modeling and optimizing machining parameters through the dry turning of Incoloy 800H. J Mech Sci Technol 31(9):4159–4165

Experimental Study on Surface Integrity of Single-Crystal Nickel-Based Superalloy Under Various Machining Processes

Srinivasa Rao Nandam, A. Venugopal Rao, Amol A. Gokhale, and Suhas S. Joshi

Abstract Nickel-based superalloys have the ability to retain superior mechanical and chemical properties at elevated temperatures. Therefore, these alloys become an ideal material choice for use in the hot sections of gas turbine engines. Advanced cast, single-crystal castings offer higher creep strength and thermal fatigue resistance than the columnar grained and polycrystalline castings. The castings undergo certain precision machining operations to obtain required fitment geometries during the engine assemblies. However, the nickel-based superalloys are considered as difficult-to-cut materials. High cutting forces and heat are generated while machining these alloys, which results in premature failure of cutting tools and poor integrity of the machined surfaces. The surface integrity aspects are of prime concern in the aerospace applications. This paper presents the results of the vital surface integrity aspects (i.e., surface topography and surface metallurgy) of a 2nd-generation single-crystal nickel-based superalloy under various precision machining process, namely milling, grinding WEDM and AWJM, which induces certain thermo-mechanical loads. It is observed that the ground surface has a highest surface finish (0.2 μm Ra) by the cutting and polishing action of multiple cutting points under higher cutting velocity. WEDM surface has lowest machining effect zone (10 μm) though it has highest surface roughness, misorientation angle, and degree of work hardening on the machined surface. Recrystallization and surface cracks were not observed at the machined surfaces under the subjected machining conditions.

Keywords Single-crystal nickel-based superalloy · Machining process · Surface integrity · Surface topography · Surface metallurgy

---

S. R. Nandam (✉) · A. Venugopal Rao
Defence Metallurgical Research Laboratory, DRDO, Hyderabad 500058, India
e-mail: srinivas-nandam.dmrl@gov.in

S. R. Nandam · A. A. Gokhale · S. S. Joshi
Department of Mechanical Engineering, IIT Bombay, Mumbai 400076, India

U. S. Dixit et al. (eds.), *Advances in Forming, Machining and Automation*, Lecture Notes in Mechanical Engineering, https://doi.org/10.1007/978-981-19-3866-5_26

1 Introduction

Nickel-based superalloys have an exceptional combination of high-temperature strength, toughness, and resistance to degradation in corrosive and oxidizing environments. Therefore, these alloys become an ideal material choice for use in the hot sections of gas turbine engines for aerospace and power generation applications [1]. A noteworthy feature of nickel-base alloys is that they can be used for load-bearing applications at temperatures up to 80% of melting temperature (1000 °C) of the alloy [2].

The major phases present in nickel-based superalloys are γ and γ'. The continuous matrix, γ is a face-centered cubic (FCC) nickel-based austenitic phase that contains a high percentage of high strength and refractory elements. This matrix has high mechanical properties by the solid solution strengthening and high modulus. The primary strengthening phase, γ' is $Ni_3(Al/Ti)$ coherently precipitating phase with an ordered L_{12} (FCC) crystal structure. Interestingly, the flow stress of the γ' increases with increasing temperature up to 650 °C due to Kear Wilsdorf lock [3]. In addition, γ' is quite ductile and thus imparts strength to the matrix without lowering the fracture toughness.

Nickel-based superalloys can be cast into components of complex shapes and/or internal configurations with metallurgically controlled uniform microstructure. Vacuum investment casting process is employed to produce cost-effective components with a high degree of dimensional accuracy, surface finish, and structural integrity [4]. Initially, the gas turbine blades were made as polycrystal (equi-axed) castings; later, development of the columnar-grained castings resulted in some significant improvements in creep strength and thermal fatigue resistance. Advanced cast, single-crystal (SC) nickel-based superalloy castings offer further improvement of high-temperature capabilities and resulting in superior turbine engine performance and durability [5]. The SC castings thus produced by the vacuum investment casting process undergo various machining operations to obtain fitment geometries for the engine assemblies. The nickel-based superalloys are considered as hard and difficult-to-cut materials due to high strength, hot hardness, fracture toughness and work hardening, and low thermal conductivity [6]. Higher cutting forces and heat are generated while machining these alloys, which result in premature failure of cutting tools and poor integrity of the machined surfaces [7].

The surface integrity (SI) of a machined component is the sum of all the elements that describe the conditions existing on or at the surface. The term was coined initially by Field and Kahles in the year 1964 [8]. The formation of a white layer at the machined surface is attributed to severe plastic deformation, grain refinement, phase transformation under rapid heating and quenching [9–12]. The SI aspects of the component are of prime concern in aerospace applications as the failures are primarily caused due to fatigue, creep, and stress corrosion, etc. [13].

Although there have been numerous studies on different aspects of cutting process over the years on superalloys, the properly published literature on SI aspects in the machining of SC nickel-based superalloys is not available. The SC superalloys

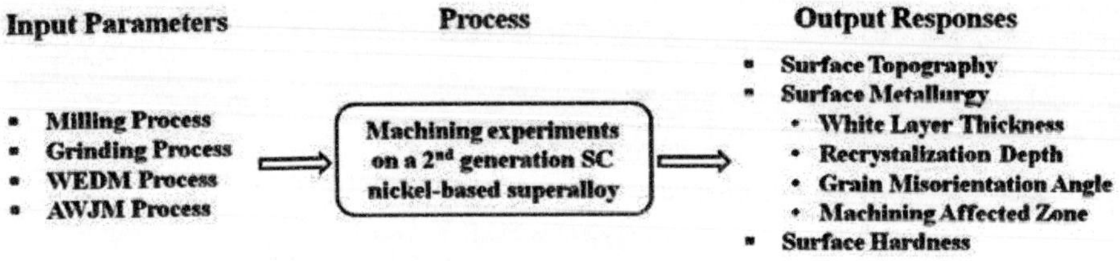

Fig. 1 Overall theme of the experimental investigation

are highly sensitive to thermo-mechanical loads. Hence, a systematic experimental study has been carried on a commercial second-generation single-crystal nickel-based superalloy, CMSX-4 superalloy by using various precision conventional and unconventional machining process, such as vertical milling, surface grinding, wire electric discharge machining (WEDM), and abrasives water jet machining (AWJM), respectively, which induces certain thermo-mechanical loads.

This paper focuses on the surface topography and surface metallurgy of the machined surfaces. The overall theme of the experimental study with input parameters and output responses is presented in Fig. 1.

2 Experimental Setup

2.1 Work Material

The single-crystals of the superalloy were supplied by Cannon Muskegon Corporation, USA, in the form of round rods with axis parallel to [001] direction. The nominal composition of CMSX-4 superalloy by weight percent is shown in Table 1.

This alloy maintains its ultra-high strength even at high temperatures by solid solution strengthening and precipitation hardening through the balanced alloy chemistry [14]. Rhenium element in this alloy mainly partitions into the γ matrix, which retards coarsening of γ' strengthening phase and increases γ/γ' misfit. Therefore, this alloy exhibits good resistance to excessive γ' growth during service. The physical and mechanical properties of CMSX-4 alloy from the supplier technical data are shown in Table 2.

Table 1 Nominal chemical composition of CMSX-4 alloy in wt.%

Cr	Co	Mo	Ti	Al	W	C	Ta	Re	Hf	Ni
6.5	9.6	0.6	1.0	5.6	6.4	0.05	6.5	3.0	0.1	Balance

Table 2 Physical and mechanical properties of CMSX-4 alloy

Property	Unit	Value
Density @ RT	kg/m^3	8700
Thermal conductivity @ RT	W/m K	8
Melting temperature	°C	1380
0.2% proof yield strength @ 760 °C	MPa	980
Ultimate tensile strength @ 760 °C	MPa	1200
Elongation @ 760 °C	%	10
Reduction of area @ 760 °C	%	20

2.2 *Specimen Preparation*

Solid casting blocks were cut by a precision WEDM process from the fir tree side of SC turbine blade castings, which were manufactured with <001> grain growth direction from the prefix bottom seeding and directional solidification technique through vacuum investment casting process. The test specimens were of length of 47 mm, width of 30 mm, and thickness of 12 mm. The micrographs of transverse to the grain growth direction of the base metal in as-cast condition are shown in Fig. 2. The optical micrographs reveal that the casting has four-fold dendrite core morphology and eutectic phases in the inter-dendrite region. These eutectic phases can be seen as bright lamellar regions in the inter-dendrite region. The high magnification SEM images (Fig. 2b) of the dendritic core show the uniform distribution of γ' phase embedded in the γ matrix.

Fig. 2 Micrographs of transverse to the grain growth direction of the CMSX-4 alloy in as-cast condition **a** Optical under 50× magnification and **b** SEM under 50,000× magnification

2.3 Machining Experiments

It is known that the milling and grinding process is conventional machining process, wherein the cutting action carries out by the combination of thermal and mechanical loads. The WEDM process is an unconventional machining process, wherein the cutting action carries out by the thermal erosion. The AWJM process is also an unconventional machining process, wherein the cutting action is carried out by the mechanical erosion from the high kinetic energy of water and abrasives. The thermo-mechanical loads in qualitative and specific energy in quantitative manner of the subjected machining process are shown in Fig. 3.

Milling Process: A knee-type conventional vertical milling machine (*FN2V model from M/s HMT, India*) of spindle motor power of 5.5 kW and rotational speed of 1800 per min was used. The sample was firmly fixed on the table, and face milling operations were performed with the solid carbide insert number TPN 2204 PDR 3040 (*M/s Sandvik Coromant, Pune)* fitted into a 100-mm-diameter cutter of 8 flutes. The flood coolant with a stable emulsion of 1:20 water ratio was employed with cutting speed of 50 m/min, table feed of 16 mm/min, and depth of cut of 0.2 mm.

Grinding Process: Grinding experiments were performed by a conventional horizontal spindle-type surface grinding machine (*PRAGA 452P model from M/s HMT, India*). The power rating of the spindle drive was 1.5 kW, and the maximum speed was 2800 rpm. A conventional tool room-type (white color) alumina-abrasive wheel of specification A60K5V8 (*M/s Carborundum Universal Limited, Chennai*) was used. The wheel has an outer diameter of 200 mm, inner diameter of 31.75 mm, and width of 20 mm. The table speed of 1.7 m/min, wheel speed of 1540 m/min, and depth of cut of 0.03 mm were used along with the soluble cutting oil, Servocut S grade with 3% concentration.

WEDM Process: The cutting experiments were conducted with 0.25-mm-diameter hard brass wire by a submerged-type precision WEDM machine (*CNC Ultracut S2 model from M/s Electronica Machine Tools, Pune*). Gap voltage of 36 V, current of

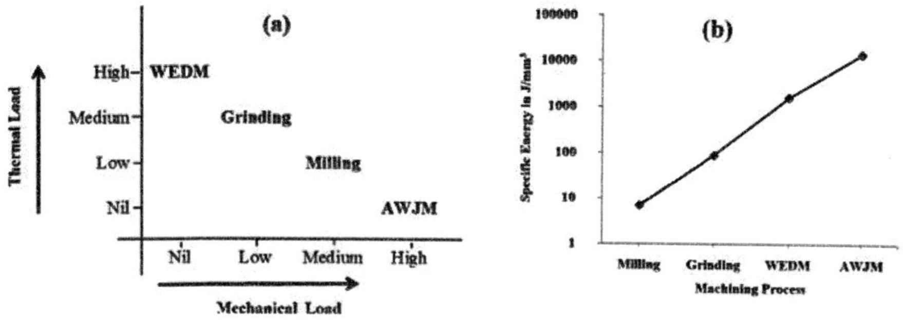

Fig. 3 Precision machining processes **a** thermo-mechanical loads and **b** semi-log plot of specific energies

1.4 A, time ON of 10 μs, cutting speed of 0.5 mm/min, and wire tension of 11 N were used along with the de-ionized water with total dissolved salts of below 3 ppm as dielectric.

AWJM Process: The cutting experiments were conducted with 80 mesh size garnet abrasives and 0.7 mm diameter sappier orifice nozzle by a water jet machining center (*55,100 model from M/s Omax, USA*). The parameters of surface quality of 5 number (i.e., high), water jet pressure of 255 MPa, abrasive flow rate of 340 g/min, traverse speed of 35 mm/min, and stand-off distance of 3 mm were used during the experiments.

2.4 Characterization Techniques

The surface roughness measurements were carried out on the machined surface by using a contact-type instrument (*Form Talysurf Intra model from M/s Taylor Hobson, UK*) with a 2 μm conisphere-shaped diamond stylus at 3 different locations along the machining direction. A cross-section of 12 mm × 12 mm with 10 mm thick was extracted from the machined surfaces of each machining process by using a precision WEDM. These geometries were mounted into the 1-inch size Bakelite molds and polished using a series of silicon carbide emery papers and diamond paste. The cleaned surfaces were etched by a commercial Kalling's 2 regent. The micro-structural analysis was performed by an optical microscope (*DM 4000 M model from M/s Leica Microsystems GmbH, Germany*) and a scanning electron microscope (*EVO-18 SEM model from M/s Carl Zeiss, Germany*). The white-layer thickness was measured on the digital micrographs by using an image processing software (*Image J*). The recrystallization, crystal orientation, and mis-orientation angle studies were carried by using the electron backscatter diffraction (EBSD) technique (*Gemini SEM 300 model from Carl Zeiss GmbH, Germany*). The hardness on the machined surfaces was measured by using a micro-Vickers hardness instrument (*MMT XZ model from M/s Matsuzawa, Japan*) with diamond pyramid indenter under 0.5 kg load and dwell time of 10 s.

3 Results and Discussions

Optical images of the cross-section of the machined surfaces in Fig. 4 show that the machined surfaces have not consisted of micro-cracks.

The surface machined by WEDM has highest surface irregularities, and AWJM has the highest waviness.

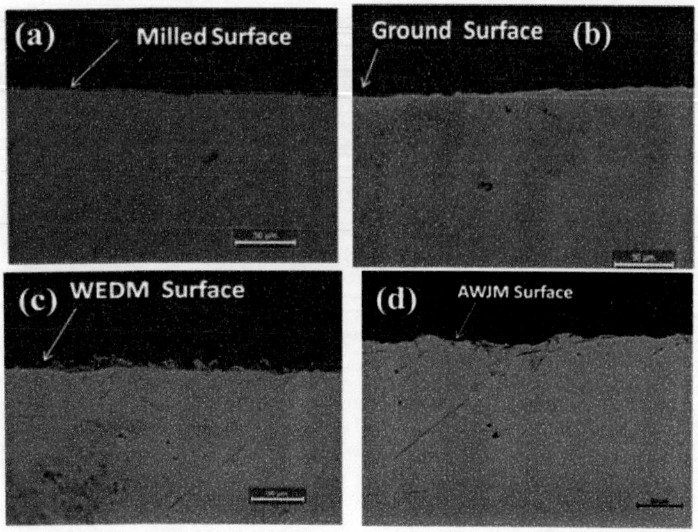

Fig. 4 Optical images (200X) of the machined surfaces of **a** milled, **b** ground, **c** WEDM and **d** AWJM process

3.1 Surface Roughness

The results of the average surface roughness are reported in Fig. 5. It is observed that the ground surface has the highest surface finish due to the cutting and polishing action of abrasives. The WEDM surface has the lowest surface finish due to higher surface irregularities by the discrete spark erosions and deposition of debris.

The milled surface has a better surface finish due to the engagement of multiple cutting inserts.

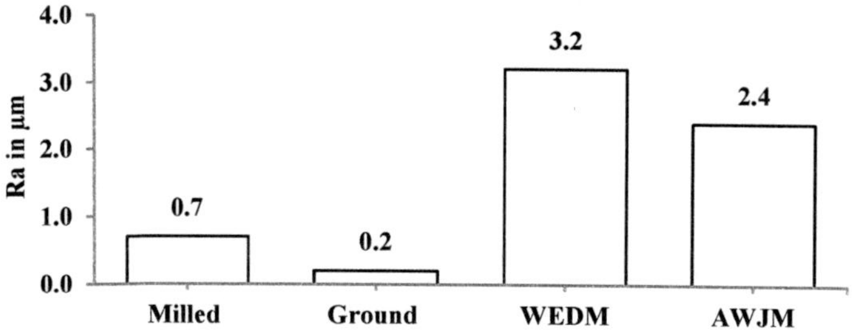

Fig. 5 Average surface roughness of the machined surfaces under subjected conditions

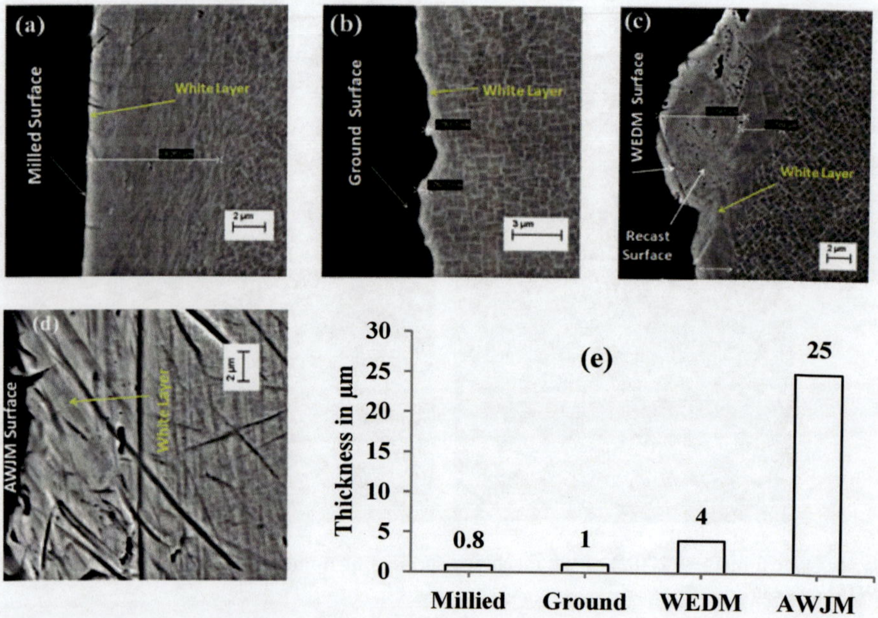

Fig. 6 SEM images (10,000×) of machined surfaces of **a** milled, **b** ground, **c** WEDM, **d** AWJM, and **e** results of white-layer thickness

3.2 White-Layer Thickness

The SEM images of the cross-section of the machined surfaces are shown in Fig. 6.

It can be seen that the machined surface by AWJM has the highest white-layer thickness due to higher mechanical loads while cutting, whereas the milled surface has the lowest value due to lower specific energy.

3.3 Recrystallization Depth

The results of the recrystallization studies on the machined surfaces by EBSD technique are shown in Fig. 7.

It is observed that none of the machining process could generate recrystallization defects under applied working conditions; however, a significant deformation (shown in red color at the location of 3rd column in the figure) was observed for the milling process due to multidirectional cutting forces.

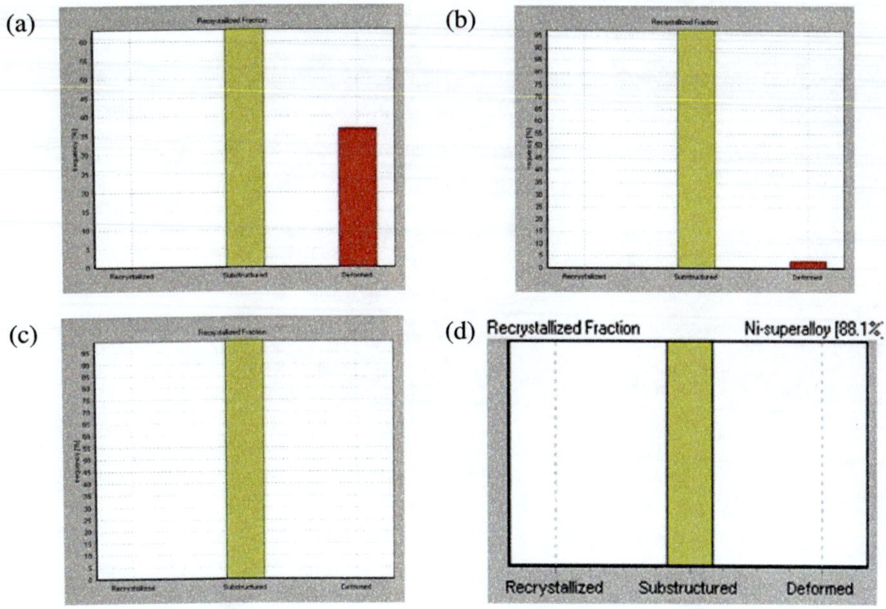

Fig. 7 Recrystallization results of **a** milled, **b** ground, **c** WEDM and **d** AWJM surfaces

3.4 Crystal Orientation Direction

The machined surfaces were further analyzed by EBSD technique for crystal orientation mapping. The results along with inverse pole figure (IPF) color key are shown in Fig. 8.

It was observed that the machined surface has the crystal orientation equal to <001> direction; it was the same as that of the base metal casting.

3.5 Misorientation Angle

The EBSD results of the distribution of grain boundary misorientation angles are shown in Fig. 9. It was observed that the surfaces under all machining process have a higher fraction of misorientation angle in the range from 0 to 5° only. AWJM surface has negligible distribution beyond 0 to 5° range, whereas the WEDM surface has a small number of misorientation angles beyond 5° and up to 15°.

The spread of WEDM surface could be due to the high temperature exposures by the discrete electric sparks. Formation of the crack-free surface was attributed to low misorientation angles.

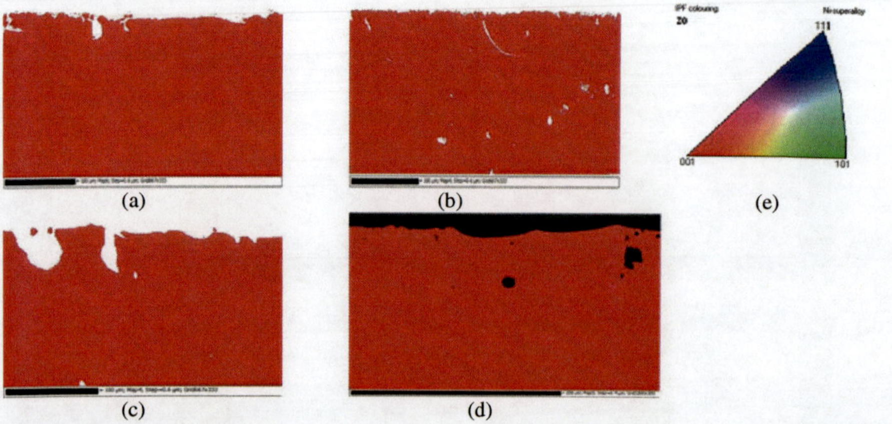

Fig. 8 Crystal orientation results of **a** milled, **b** ground, **c** WEDM and **d** AWJM surfaces of CMSX-4 superalloy, and **e** IPF key

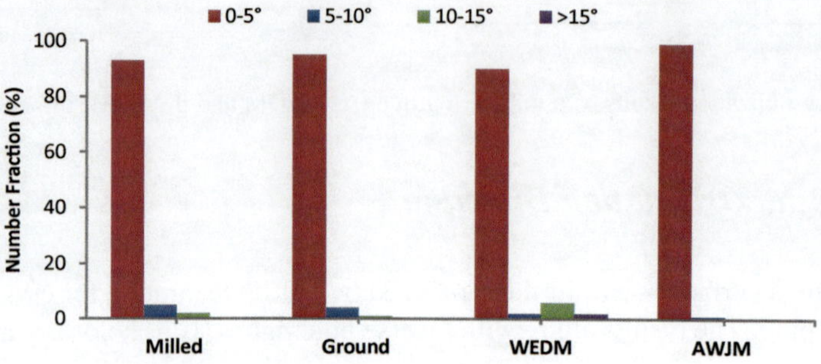

Fig. 9 Misorientation angles of the grain boundaries of milled, ground, WEDM and AWJM surfaces of CMSX-4 superalloy

3.6 *Machining Affected Zone*

The deformation images of the machined surface by EBSD technique are shown in Fig. 10. It can be seen that the milled surface has the highest deformation zone up to the maximum depth of 100 μm from the surface, whereas the WEDM has the lowest deformation depth of 10 μm.

The ground surface has the lower deformation depth than the milled surface. The ground surface images reveal the existence of slip lines below the machined surface. This could be due to localized thermo-mechanical loads.

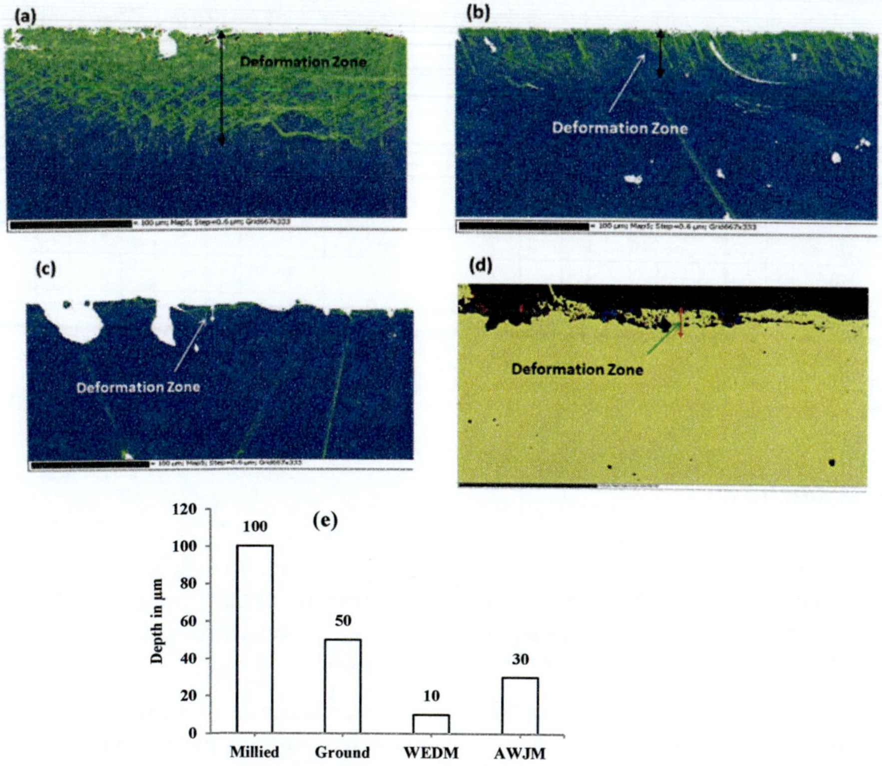

Fig. 10 Machining affected zone of CMSX-4 superalloy under **a** milling, **b** grinding, **c** WEDM and **d** AWJM process

3.7 Surface Hardness

The average surface hardness values of the machined surfaces are shown in Fig. 11a.

The degree of surface hardening (DSH) in percentage was calculated by using the below equation:

$$DSH = \frac{(MSH - BSH)}{BSH} \times 100 \quad (1)$$

where ***MSH*** is the surface hardness of the machined surface and ***BSH*** is the surface hardness of the base metal in VHN. The results of DSH are shown in Fig. 11b. It can be seen that the machined surfaces have less than 10% DSH.

The WEDM has a higher DSH due to highest thermal loads during the cutting. The AWJM has a lower DSH though it has highest specific energy. This could be due to the direction of dominant cutting force in parallel to the machined surface.

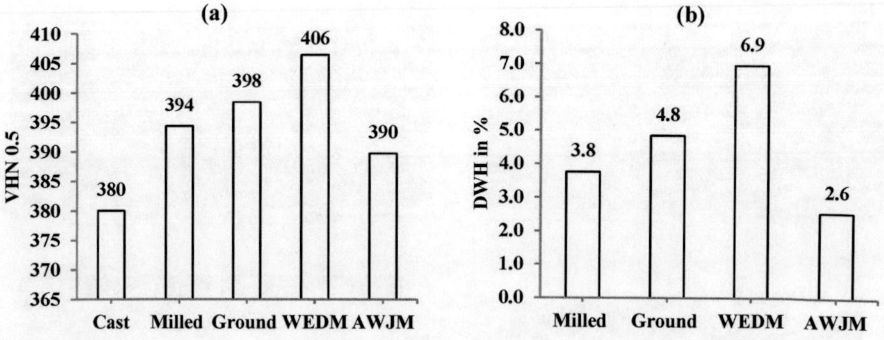

Fig. 11 Surface hardness of the machined surfaces of CMSX-4 alloy under various machining process **a** average Vickers hardness and **b** DSH

4 Conclusions

Machining experiments were conducted on a 2nd-generation single-crystal nickel-based superalloy, CMSX-4 alloy in as-cast condition by milling, grinding, WEDM and AWJM process, which induces certain thermo-mechanical loads of various order. The following conclusions are drawn from the experimental study.

- The ground surface has achieved the best surface roughness of 0.2 μm Ra due to the cutting and polishing actions by the abrasives, whereas the WEDM surface has achieved the poor surface roughness of 3.2 μm Ra due to the deposition of discrete micro-debris on the machined surface. The machined surface of AWJM has highest waviness due to the divergence of the water jet.
- The specific energy is attributed to the formation of white-layer thickness. The AWJM surface has the highest value due to highest specific energy (mechanical load), and the milled surface has the lowest due to least specific energy.
- Though the milled process has lower specific energy, the milled surface has the highest machining affected zone with high plastic deformation depth due to the multidirectional action of cutting forces during cutting. The AWJM surface has the least amount of machining affected zone due to alignment of dominant cutting force in parallel direction to the machined surface.
- The thermal loads are attributed to the formation of misorientation angles. WEDM surface has a wide range of misorientation angles up to 15° by the high thermal loads, whereas AWJM has the least values of the angles.
- The thermal loads are also attributed to the hardening of the machined surfaces The WEDM surface has the highest amount of DSH due to the formation of a recast layer under rapid heating and cooling.
- Surface cracks, changes in the crystal orientation, and recrystallization were not observed on any of the machined surfaces under the machining conditions.

The chosen machining parameters, therefore, appear appropriate for machining the CMSX-4 superalloy, which obtained a higher integrity on the machined surfaces.

Acknowledgements The authors are grateful to the *Defense R&D Organization* for the financial support. The authors are thankful to the *Director, DMRL,* officers, and staff of *Armor Design and Development, Directional Solidification, Electron Microscopy, Mechanical Engineering and Structure and Failure Analysis Groups* of DMRL for their support. The authors are also grateful to the faculty and students of MMMF Lab of IIT Bombay for their help.

References

1. Pollock TM, Tin S (2006) Nickel based superalloys for advanced turbine engines: chemistry, microstructure, and properties. J Propuls Power 22(2):361–374
2. Reed RC (2006) The superalloys—fundamentals and applications. Cambridge University Press
3. Onyszko A, Kubiak K, Sieniawski J (2009) Turbine blades of the single crystal nickel based CMSX-6 superalloy. J Achiev Mater Manuf Eng 32(1):66–69
4. Spikesley GE (1979) Investment casting. Mater Eng Appl 1:328–334
5. Frasier DJ, Whetstone JR, Harris K, Erickson GL, Schwer RE (1990) Process and alloy optimization for CMSX-4 superalloy single crystal airfoils. In: Presented at COST 501/505 Conference, Belgium, pp 1281–1300
6. Arunachalam R, Mannan MA (2000) Machinability of nickle-based high temperature alloys. J Mach Sci Technol 4(1):127–168
7. Coelho RT, Silva LR, Braghini Jr A, Bezerra AA (2004) Some effects of cutting edge preparation and geometric modifications when turning Inconel 718 at high cutting speeds. J Mater Proces Technol 148(1):147–153
8. Bellows G, Tishler N (1970) Introduction to surface integrity, Pamplet 1, TM70-974, GE, Aircraft Engine Group, Cincinnati, USA
9. Arunachalam RM, Mannan MA, Spowage C (2004) Surface integrity when machining age hardened IN 718 with coated cutting tools. Int J Mach Tools Manuf 1481–1491
10. Saoubi RM, Outeiro JC, Chandrasekaran H, Dillon OW Jr, Jawahir IS (2008) A review of surface integrity in machining and its impact on functional performance and life of machined products. Int J Sustain Manuf 1(1/2):203–236
11. Pawade RS, Joshi SS, Brahmankar PK (2008) Effect of machining parameters and cutting edge geometry on surface integrity of high-speed turned IN718. Int J Mach Tools Manuf 48:15–28
12. Thakur A, Gangopadhyay S (2016) State-of-the-art in surface integrity in machining of nickel-based superalloys. Int J Mach Tools Manuf 100:25–54
13. Carter TJ (2005) Common failures in gas turbine blades. Eng Fail Anal 12:237–247
14. Harris K, Wahl JB (2004) Improved single crystal superalloys. CMSX-4 and CMSX-486 Superalloys 45–52

Finite Element Analysis of Drilling on LPBF-Produced Maraging Steel

Mohit Godara, Jino Joshy, Basil Kuriachen, K. P. Somashekar, and Jose Mathew

Abstract AM processes are highly demanded nowadays because of their design freedom, but with that, it is also important to meet the required properties for an AM-produced part. Compared to conventionally produced parts, the anisotropy in additively manufactured parts is one of the major concerns nowadays. The surface roughness of as produced part is high, which leads to an increase in the residual stresses in the body. Holes produced by laser powder bed fusion (LPBF) are mostly covered with unmelted powder particles and also geometrical irregularities due to step effects. This work presents a FEM-based drilling process simulation on LPBF-produced maraging steel which is widely accepted for aerospace as well as in tooling applications due to its mechanical properties like ultra-high strength, good weldability, and high toughness. Simulation of temperature distribution, von Mises stress, and shear stress distribution is performed by ANSYS Academic 2020 R1. On comparing the results of different build directions with conventionally produced parts, it was found that the maximum cutting temperature in conventionally produced parts is about 8% and 6% higher than the horizontally and vertically produced parts. Whereas, the maximum average von Mises stress generated in the horizontally produced part is about 2% higher than the conventionally produced part and 3.5% higher than the vertically produced part. To understand the generation and straining of chips, the shear stress distribution is also discussed.

Keywords Additive manufacturing · Selective laser melting · Maraging steel · Drilling · Simulation

---

M. Godara · J. Joshy · B. Kuriachen (✉) · J. Mathew
Advanced Manufacturing Center, Department of Mechanical Engineering, National Institute of Technology Calicut, Kattangal 673601, India
e-mail: bk@nitc.ac.in

M. Godara
e-mail: godaramohit10@gmail.com

J. Joshy
e-mail: jino.arimboor@gmail.com

K. P. Somashekar
Department of Technical Education, Bangalore, India

U. S. Dixit et al. (eds.), *Advances in Forming, Machining and Automation*, Lecture Notes in Mechanical Engineering, https://doi.org/10.1007/978-981-19-3866-5_27

1 Introduction

According to ASTM, additive manufacturing (AM) is a process in which materials such as plastic or metal are deposited layer by layer to produce a 3D model. This process helps us to produce complex geometries which are not easy to produce using conventional methods [1]. Many AM techniques are present for producing metal parts; some of these are wire based, and some are based on powder. The most popular wire and powder-based AM techniques are shaped metal deposition, selective laser sintering, selective laser melting, and electron beam melting. The basic concepts of all the AM processes are the same, i.e., first, the information of part geometry is delivered by a 3D model, and the same is sliced into layers of defined thickness [2]. Other than design freedom, AM process has material and resource efficiency and also environment friendliness. Due to production flexibility applications of AM fabricated parts increase day by day, especially in automotive, medical, aerospace, biological systems, etc. [3].

Laser powder bed fusion (LPBF) is an AM process under the powder bed fusion technique. In this process, complete melting and solidification of metal powder are done layer by layer with the help of laser power. This process involves two main steps which are deposition of powder and exposing the laser beam by a scanning system. Each layer is formed by scanning a laser spot as per part drawing, after melting the powder particles fuses together, and get solidified. This process continues till the formation of the complete part [4].

Kumar et al. [5] have studied the microstructures and mechanical properties of LPBF-produced maraging steel, and found that the ductility of LPBF-produced maraging steel is much lower than the conventionally produced part. Liverani et al. [6] have discussed that even after aging treatment ductility and impact toughness were found very low for LPBF-produced part as compared to conventionally manufactured part. Zhang et al. [7] have already discussed that the surface quality of LPBF-produced part is very poor and the main reason behind these defects. Many researchers already studied the effects of conventional processes like turning, milling, mechanical polishing, electropolishing, and laser polishing to reduce the surface roughness. To improve the surface characteristics of LPBF-produced holes such as surface roughness, hole quality, reduction in form error, etc. with the above mentioned processes is not possible. So, one possible way is by conventional drilling on LPBF-produced parts as many researchers studied the effects of conventional and non-conventional material removal processes [8, 9] Another reason to use the drilling process here is that according to Hassanin et al. [10] most of the hole produced by LPBF process for cooling channels are found partially or fully blocked due to presence of unmelted powder particles.

Before machining, it is very important to know the effect of the manufacturing process, i.e., build direction of an LPBF-produced material. In the present study, it has been discussed in the terms of temperature generation, von Mises stress, and shear stress distribution. And as all these variables are very difficult to obtain through experimental techniques, a 3D simulation is effective and can provide accurate results.

Many simulations based on milling are already conducted [11–13] and discussed, but very limited studies are conducted on drilling. As per the authors' knowledge, no drilling simulation is performed on LPBF-produced maraging steel, and the effect of different build directions on the cutting process is very crucial. As discussed above, it is also important to compare the simulations results with conventional produced maraging steel; here, for this study, properties of material produced by forging have been used.

2 Material Properties and Chip Formation Criterion

For this study, maraging steel (18-Ni-300) is taken under consideration due to its wide applications in aerospace, aircraft industry, and tooling applications due to its ultra-high strength, high toughness, low coefficient of thermal expansion, good weldability, and dimensional stability during heat treatment [14]. This steel utilizes both age hardening and martensitic microstructure to improve mechanical properties. The major applications of maraging steel are rocket engines, conformal cooling channels in injection molding, and gear components of helicopters. Nowadays, demand for LPBF-produced maraging steel is more due to its less production cost and lower weight. The chemical composition of the maraging steel powder is reported in Table 1.

The effect of build direction has a substantial effect on both static and dynamic properties. Building orientation determines the direction of defects, size of the defect, and distribution of defect in the LPBF-produced part. Mainly, it affects fatigue strength, cracks growth rate, yield strength, and ultimate strength significantly. Lack of fusion is more in perpendicular direction compared to building direction, and this is also true for porosity as more porosities were found in vertical build direction than horizontal [15]. But, after heat treatment, the mechanical properties of LPBF-produced part tend to increase and show similar results as conventionally produced part. Mechanical properties of LPBF-produced maraging steel and conventional produced maraging steel [14, 16] are shown in Table 2.

In the present study, a finite element (FE) drilling model is proposed for the prediction of cutting temperature distribution, shear stress, and von Mises stress distribution in different build directions of LPBF-produced maraging steel, and a $20 \times 20 \times 10\text{mm}^3$ cuboid is designed for both horizontal and vertical directions as well as for conventionally produced maraging steel same sample. Different properties as mentioned above in Table 2 are used in this study. To drill holes, an 8-mm-diameter cobalt-coated tungsten carbide (WC) drill bit with a point angle of 120° and helix

Table 1 Chemical composition (in wt%) of maraging steel (18-Ni-300)

Fe	Ni	Co	Mo	Ti	Cr	Cu	Mn	C
Balance	18.660	9.10	5.10	0.754	0.069	0.612	0.086	0.015

Table 2 Summary of mechanical properties of LPBF-produced and conventionally produced maraging steel

	Conventional	Horizontal	Vertical
Ultimate tensile strength (MPa)	1158–2030	1147	1035
Yield strength (MPa)	1056–2100	976	794
Elastic modulus (GPa)	190	185	189
Poisson's ratio	0.3	0.27	0.27
Density (kg/m^3)	8200	8100	8000
Thermal conductivity (W/m.°C)	20–25	14.2–28.6	14.2–28.6

angle of 30° has been used, and mechanical properties of tool material [17] are shown in Table 3. To assemble the drill bit and workpiece material, coincident mating is applied by using 3D CAD software Solidworks 2020 as shown in Fig. 1.

As drilling is a nonlinear process, explicit dynamics model has been chosen for analysis. To perform this simulation work, Ansys workbench software has been used. The simulation of the drilling operation is shown in Fig. 2 obtained from Ansys. The minimum element size in the contact region is 0.5 mm in fine mesh, and the maximum element size is 2.5 mm for coarse meshing. The drilling parameters are mentioned in Table 4. Here, to reduce the simulation time, element size is kept large in areas other than the contact region. For this current model, mass scaling and time scaling have been imposed, and due to these techniques, it is possible to obtain accurate results.

Mainly, two types of approaches are used for simulating the chip in drilling or milling processes. They are (a) damage initiation criterion or (b) modified material

Table 3 Material composition and mechanical properties of drill bit material

Density (kg/m^3)	14,950
Ultimate tensile strength (MPa)	1516.84
Young's modulus (GPa)	634.317–641.21
Poisson's ratio	0.26

Fig. 1 Drill bit design used in drilling simulation

Fig. 2 FEM model of drilling

Table 4 Simulation parameters and boundary conditions of drilling analysis

Number of elements	48,764
Number of nodes	49,738
Total number of cycles	16,581
Simulation end time	4.616 s
Cutting speed	25.13 m/min
Feed rate	0.16 mm/rev
Depth	10 mm

model with material softening approach with temperature increase. As temperature increases, the effect of flow softening and strain hardening effect increases. The most popular model which considers the effects of strain, strain rate, and temperature is Johnson–Cook material constitutive model [18], and damage criteria have been used in this proposed model.

$$\sigma_{JC} = \left[A + B\epsilon^{n}\right]\left[1 + Cln\left(\frac{\dot{\epsilon}}{\dot{\epsilon}_{o}}\right)\right]\left[1 - \left(\frac{T - T_{r}}{T_{m-T_{r}}}\right)^{m}\right] \tag{1}$$

The expression for the Johnson–Cook model described by Eq. (1) based on average flow stress and all the required parameters are given in Table 5

Here, ε is equivalent plastic strain; $\dot{\epsilon}$ and $\dot{\epsilon_{o}}$ are equivalent and reference plastic strain rate; T is the temperature of the material cutting zone; T_{m}andT_{r} are melting and reference temperature of the material; n is strain hardening constant; m is the thermal softening index; A, B, C are initial yield stress, strain hardening parameter, and strain rate hardening parameter.

Table 5 Johnson–Cook strength model parameters for maraging steel [18]

Johnson–Cook model	A (MPa)	B (MPa)	C	n	m
Value	1000	3200	0.03	0.01021	1.03

2.1 Modeling Assumptions

1. Drill tool used for modeling is considered as an isothermal rigid body.
2. Our workpiece (mentioned as the base) is fixed from all directions and is provided rotational motion and movement in the z-direction as remote displacement.
3. Contact between tool and workpiece is assumed as frictionless as very limited thermal properties are available for LPBF-produced maraging steel.

3 Drilling Simulation of LPBF-Produced Maraging Steel

Finite element simulation of drilling process for 18-Ni-300, maraging steel was performed in different built directions (horizontal and vertical build directions) using the same set of cutting parameters, and a comparative study with conventionally produced maraging steel is also performed. Here, the tool is assumed as a rigid body and workpiece as a flexible body whose material is viscoplastic material, which reduced the computation effort. As it is very difficult to measure the stresses during the process practically, so, comparison of von Mises stresses has been simulated. Von Mises stress also calculates the fracture strength and yield criterion of material as once its value exceeded the average flow stress, yielding in material takes place. All three variables considered in this study for simulations are difficult to measure or analyze practically. Analysis of shear stress is very important to understand as it influences the energy requirement in any machining process.

The material properties of the horizontally and vertically produced part have been added through additive manufacturing suite in Ansys workbench, in which z-direction is considered as build direction, i.e., horizontal considers as 0° and vertical property provided on the same suite as 90° build direction. The major problem that occurs with AM-produced material is a large number of defects, and due to this, there is an anisotropy in the material properties. In the current study, no other separate porosity model is used; all changes in obtained results are due to changes in mechanical properties only. LPBF material properties are also changing with the powder size, layer thickness, and laser power. The mechanical properties considered in Table 2 are for 40 μm layer thickness and 400 W laser power supply [16].

4 Result and Discussion

The comparative study of cutting temperature distribution is shown in Fig. 3. It is showing that temperature increases as the drill approached the base and reached the maximum, and after that, it shows constant as it requires more simulation time for stable results which will increase computation efforts. Figure 4 is showing the

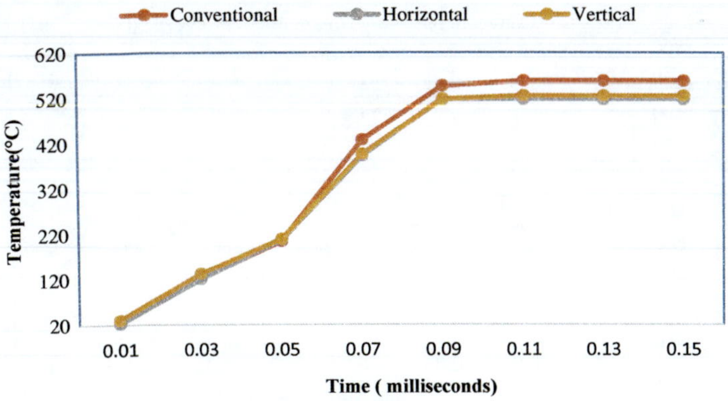

Fig. 3 Change in temperature with machining time

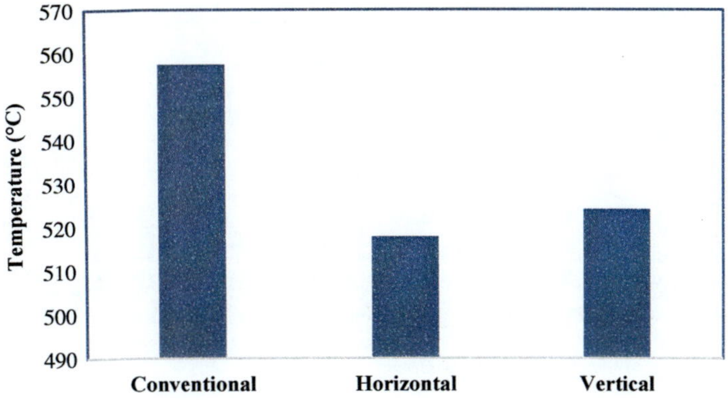

Fig. 4 Comparison of maximum temperature

maximum temperature achieved in the drilling process, and the maximum temperature generated is for the conventionally produced (~557 °C). When the cutting temperature generation is more which indicates the hard contact and high friction between tool and workpiece. The main reason for high friction is high value of hardness of the specimen. So, it is clear that the conventional part is harder than the LPBF-produced part, and in comparison of different building directions, the temperature distribution for a sample produced in both directions is almost the same. This generation of higher temperature can also lead to more tool wear in drilling of the conventionally produced part than LPBF-produced part.

The comparison of average von Mises stress is shown in Fig. 5, and it is maximum at the interaction of workpiece and drill. The comparison of maximum average von Mises stress is shown in Fig. 6. The maximum average von Mises stress was found in 0° build part which is 1653 MPa, and this is too high compared to the value of yield strength at room temperature. This may be due to variation of properties with

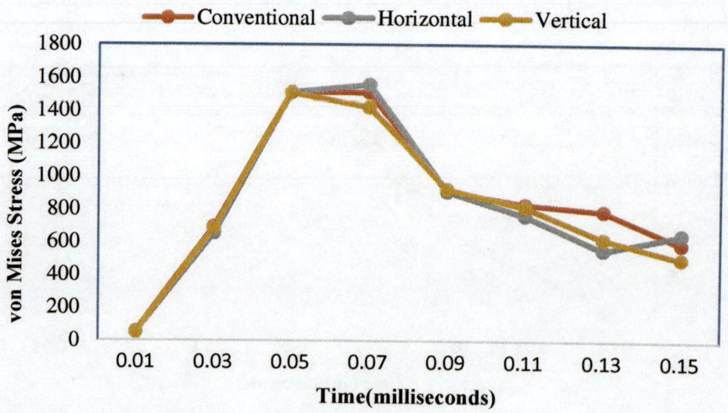

Fig. 5 Change in von Mises stress with cutting time

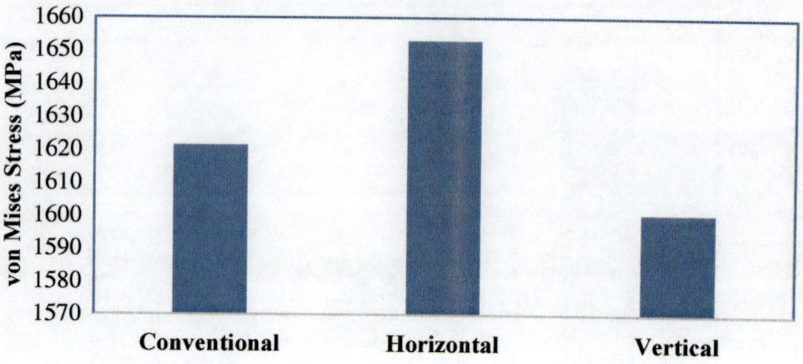

Fig. 6 Comparison of average von Mises stress

temperature as with the increase in temperature the properties like yield and ultimate strength will decrease. Shen et al. [15] reported that yield stress of horizontally produced part is more than vertically produced part and thus necessitates the current study to understand the variation of stress in different directions.

Similarly, Fig. 7 shows the change in the shear stress with the time, and it is also maximum at the interaction of drill bit tip and workpiece. This is very important for the generation of chips, higher the value of shear stress means the heating and straining of the chip increases. As for this work, a two-flute drill bit has been used, so both flutes were responsible for shear stress generation, and one was positive, and the other was considered as negative for a particular sample the distribution of shear stress in the sample is shown in Fig. 8. The reason behind such distribution is that surfaces that will be in contact with flutes will experience more shear stress. The comparative study of shear stress among different samples is shown in Fig. 9, and it was found that shear stress was maximum in the vertically build workpiece.

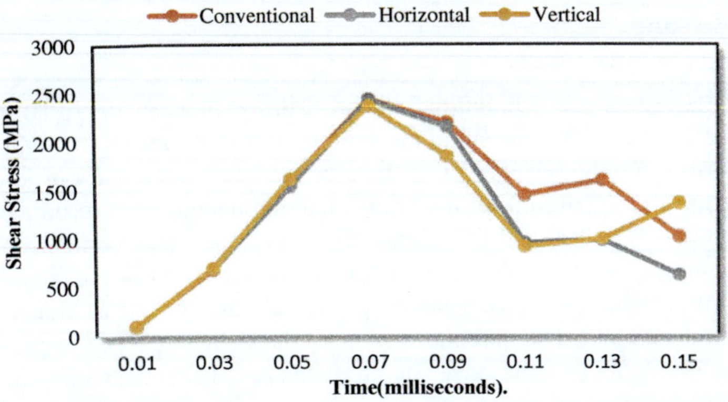

Fig. 7 Comparison of shear stress

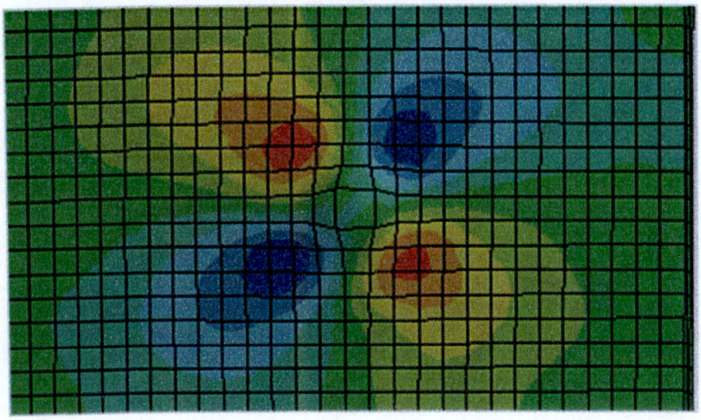

Fig. 8 Shear stress distribution

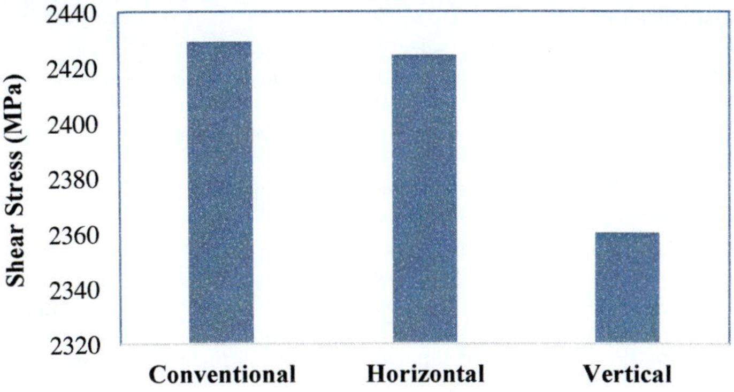

Fig. 9 Comparison of max. shear stress

5 Conclusion

The paper described the simulation of the drilling process of LPBF-produced maraging steel (18-Ni-300) and simulated temperature distribution, von Mises stress, and shear stress, and the results were compared with conventionally produced part, and found that the maximum temperature of the conventionally produced part is 40 °C and 33 °C, i.e., 8% and 6% higher than horizontal and vertically produced part. Whereas, the average value of von Mises stress of the horizontally produced part is 2% more than the conventionally produced part and 3.5% higher than the vertically produced part. The maximum value of the shear stress of the conventional part is about 3% higher than the vertically produced part. From the simulation results, it is clear that defects play an important role in material properties; here, in LPBF-produced part due to lots of anisotropy in the mechanical properties in different build directions, lots of variation occurred in results also.

Acknowledgements The authors would like to thank the Department of Science and Technology, Govt. of India for supporting this work under the scheme FIST (No. SR/FST/ETI-388/2015).

References

1. ASTM (2021) https://www.astm.org/Standards/additive-manufacturing-technology-standards.html. Last accessed 15 May 2021
2. Becker TH, Dimitrov D (2016) The achievable mechanical properties of SLM produced maraging steel 300 components. Rapid Prototyping J 22:487–494
3. Simufact (2021) https://www.simufact.com/additive-manufacturing.html. Last accessed 18 June 2021
4. William E (2014) Frazier: metal additive manufacturing: a review. J Mater Eng Perform 23:1917–1928
5. Kumar P et al (2020) Fatigue strength of additively manufactured 316L austenitic stainless steel. Acta Mater 199:225–239
6. Liverani E et al (2017) Effect of selective laser melting (SLM) process parameters on microstructure and mechanical properties of 316L austenitic stainless steel. J Mater Proc Technol 249:255–263
7. Zhang B, Li Y, Bai Q (2017) Defect formation mechanisms in selective laser melting: a review. Chin J Mech Eng (Engl Edition) 30:515–527
8. Fortunato A et al (2018) Milling of maraging steel components produced by selective laser melting. Int J Adv Manuf Technol 94:1895–1902
9. Han S et al (2020) Effect of abrasive flow machining (AFM) finish of selective laser melting (SLM) internal channels on fatigue performance. J Manuf Process 59:248–257
10. Hassanin H et al (2018) Tailoring selective laser melting process for titanium drug-delivering implants with releasing micro-channels. Addit Manuf 20:144–155
11. Uçak N et al (2019) Finite element simulations of cutting force, torque, and temperature in drilling of Inconel 718. Procedia CIRP 82:47–52
12. Karabulut Y, Kaynak Y (2020) Drilling process and resulting surface properties of Inconel 718 alloy fabricated by selective laser melting additive manufacturing. Procedia CIRP 87:355–359
13. Gao X et al (2011) Simulation of stainless steel drilling mechanism based on Deform-3D. Adv Mater Res 160–162:1685–1690

14. Matmatch (2021) https://matmatch.com/learn/material/maraging-steel. Last accessed 13 June 2021
15. Shen L-C et al (2020) Effects of build direction on the mechanical properties of a martensitic stainless steel fabricated by selective laser melting. Materials 13:1–18
16. RENISHAW (2021) https://www.renishaw.com. Last accessed 25 July 2021
17. AZO materials (2021) https://www.azom.com. Last accessed 26 July 2021
18. Jakus AE et al (2012) Dynamic deformation and fragmentation response of maraging steel linear cellular alloy. AIP Conf Proc 1426:1363

Investigation on Micro Electric Discharge Machining of Polymer Nanocomposites Modified by Graphene Nanoplatelet

Rahul Vishwakarma, Rajesh Kumar Verma, and Kishore Debnath

Abstract In this present work, investigations have been made to evaluate the feasibility of micro-electrical discharge machining. The micro-sized hole is highly required in die-making applications, bioelectronics setups, optical, and aircraft components that are not achievable by the conventional machining process. The development of the micro-sized blind holes is performed in carbon fiber reinforced polymer (CFRP) composites modified by graphene nanoplatelets (GNPs) by using the micro-electrical discharge machining (μEDM) process. The input machining control factors were voltage (60, 80, and 100 V), pulse on time (25, 35, and 45 micro sec.), and weight percentage of GNP (0.25, 1.00, and1.75). The machining process was analyzed to control the material removal rate (MRR). An efficient metaheuristics Jaya algorithm was applied with varying constraints to optimize the MRR value. A field electron scanning electron microscopy (FESEM) was used to examine the quality of the machined hole and surface damages generated during the machining operation. The microscopy results show that the addition of GNP has a considerable impact on nanocomposite micromachining to improve the MRR.

Keywords CFRP · Graphene · μEDM · Material removal rate · Scanning electron microscopy

1 Introduction

Nowadays, carbon fiber reinforced polymers (CFRPs) are replacing traditional materials in the fabrication of lightweight mechanical applications, such as spacecraft, automobile, civil, and marine constructions. [1–4]. CFRP has superior mechanical and physical properties. However, the further addition of carbon nanomaterials

R. Vishwakarma · R. K. Verma (✉)
Department of Mechanical Engineering, Madan Mohan Malaviya University of Technology, Gorakhpur 273010, India
e-mail: rkvme@mmmut.ac.in

K. Debnath
Department of Mechanical Engineering, National Institute of Technology, Meghalaya 793003, India

U. S. Dixit et al. (eds.), *Advances in Forming, Machining and Automation*, Lecture Notes in Mechanical Engineering, https://doi.org/10.1007/978-981-19-3866-5_28

(CNMs) could boost the electrical, thermal, and interlaminar properties of laminates for a broader scope of applications [5, 6]. The conventional machining of polymer laminates generates lots of defects on the machined portion, and it is also difficult to produce a complex product. So, micromachining is highly recommended to form a sophisticated component used in biomedical, aerospace, and engineering component.

The electrical discharge machining (EDM)/micro-electrical discharge machining (μEDM) method is an alternative method for machining polymer composites. Machining of polymer in μEDM is conceivable in two ways. The first technique applies a skinny layer of conductive tape on the workpiece, which has to be machined. This method is known as assisting electrode (AE) method [7]. Another method is the reinforcing of a conductive nanoparticle in the polymer composite. The insertion of conductive filler raises the overall electrical properties of the material. This allows the composites to be readily machined by EDM/μEDM process. Fan et al. [8] conducted experiments on CFRP/epoxy nanocomposite materials with the addition of 1% GNP nanofiller material and found that electrical conductivity increased by 14%. Kostagiannakopoulou et al. [5] investigated that thermal properties and electrical conductivity were improved 176% and 48% correspondingly with the accumulation of 15% (weight concentration) of GNP nanofiller into the epoxy matrix. Kumar et al. [9] experimented with micro-electrical discharge drilling while making micro-sized holes in CFRP laminates with a rotating tungsten carbide tool of dia. 120 μm. It was concluded that better hole quality is achieved at lower discharge energy. Priti et al. [10] perform the micro electrochemical discharge machining (μECDM) process on carbon fiber laminates. The authors have studied the effect of different input parameters optimizing the MRR and hole enlargement by the Entropy-VIKOR method has optimized the MRR and overcut. The authors concluded that the Entropy-VIKOR method gives the finest value with a minimum calculation. Moreover, the authors also examine the damaged surface of the machined hole with the optical micrograph. Sahu et al. [11] have carried electro discharge machining (EDM) on titanium alloy with the copper-tungsten-boron carbide tool. The authors have optimized the input parameters of machining with the VIKOR method to maximize the material removal rate and minimize the material removal rate the tool deterioration as well as surface roughness. Furthermore, they stated that currently plays a significant role while machining. Additionally, the algorithm easily applies to the process parameter with lesser computational effort. Until et al. [12] studied the machining behavior of a carbon fiber epoxy-based composite and discovered degradation in machining quality, uneven cutting of fibers, and composite tearing during the electrical discharge machining process.

From the literature survey, it was found that there has been limited research performed on the CFRP nanocomposite by μEDM machining. However, no work has been introduced to "Jaya" in μEDM of CFRP/GNPs epoxy nanocomposite. This effort has been made to overcome the drawbacks of existing optimization modules.

2 Development of CFRP/GNPs Epoxy Nanocomposite

CFRP/GNPs epoxy nanocomposites were formed with the GNP, carbon fiber, epoxy, and hardener. Carbon fabric of 250GSM, plain woven and bidirectional, was used. During the fabrication process, Lapox L-12 resin and Lapox K-6 hardener were used. The samples are prepared with three different weight concentrations of GNP, i.e., 0.25%, 1.0%, and 1.75%. The initially different weight percentages of GNP were properly mixed with acetone separately and repeated the sonication process in the bath sonicator. The mixture of epoxy, acetone, and GNPs was heated on a heating plate for evaporation of acetone. As a result, GNP was appropriately dispersed in epoxy resin. The hardener was added 10:1 ratio and mixed properly by hand in a cooled mixture. This dispersed solution was equally scattered over the carbon fabric pile with the help of a brush and roller by using the hand layup process. This process was repeated up to the 12 layers of carbon ply with stacking to get the 1.5 mm thickness of the final laminates. The wet prepared sample was placed under a hydraulic press to synthesize the laminate. After the 24 h sample was cured. The sample was removed and cut in the required dimension. All the procedures are mentioned in Fig. 1; Table 1.

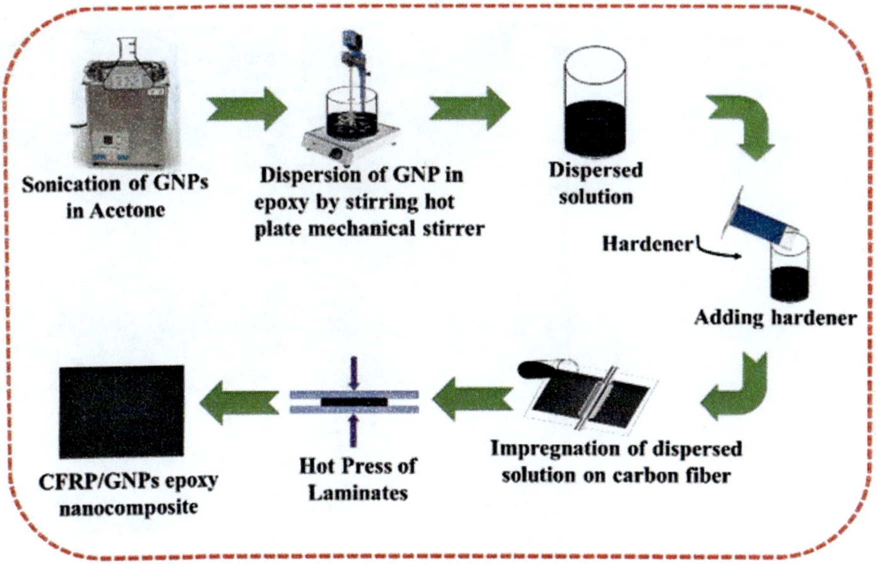

Fig. 1 Development of sample

Table 1 Electrical conductivity of the different sample

S. No	Reinforcing material matrix	Electrical conductivity (S/m)
1	CFRP/GNPs-0%	22.5
2	CFRP/GNPs-0.25%	49.87
3	CFRP/GNPs-1.00%	108
4	CFRP/GNPs-1.75%	127.84

3 Methodology and Experimentation

Micro electrical discharge machining was performed on the "HYPER 15" made by the Sinergy nano system. It has equipped with an RC (resistance–capacitance)– type pulse generator. The setup's spindle was powered by a DC motor with a speed range of 300-3000 rpm. In this setup, hydrocarbon-based dielectric fluid is used for this machining process. The machining was done on CFRP/GNPs epoxy nanocomposite with tungsten carbide tool electrode of 960 μm diameter to make a blind hole up to 0.5 mm depth. The machining responses were examined in terms of the MRR of the blind hole as shown in Figs. 2 and 3; Table 2.

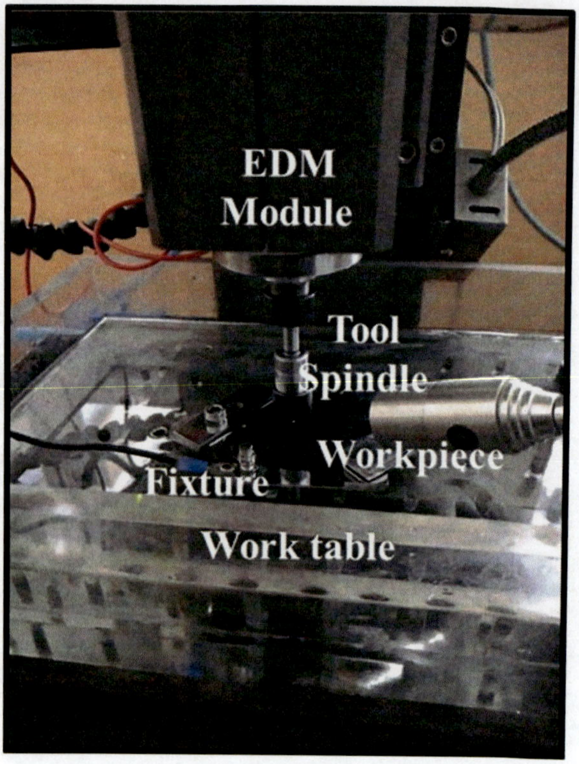

Fig. 2 Machining setup of μEDM

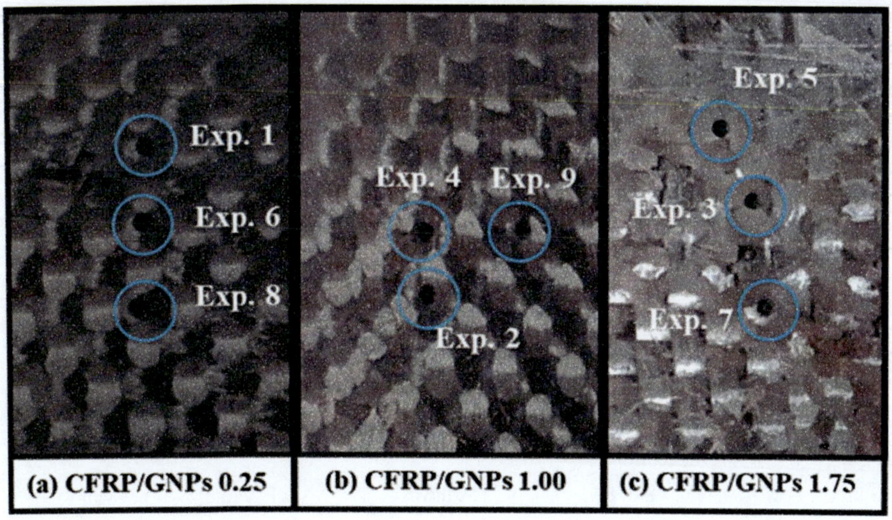

Fig. 3 Machined sample

Table 2 Process parameter and their levels

Symbol	Parameter	Level 1	Level 2	Level 3
V	Voltage	60 V	80 V	100 V
P	Pulse on time	25 μs	35 μs	45 μs
W	Weight % of GNP	0.25	1.00	1.75

4 Result and Discussion

The machining performance of the CFRP/GNPs epoxy composite was examined during the μEDM process. Jaya algorithm was projected to achieve the optimal values of MRR. Table 3 is summarized L9 orthogonal array (OA), corresponding MRR response values.

4.1 Analysis of Variance

The effect of process parameter to the response, variance analysis (ANOVA) was performed with a 95% confidence level ($\alpha = 0.05$), as shown in Table 4. The ANOVA result shows that the voltage (with P-value for MRR $0.000 < 0.05$ significant) was the most significant parameter. The adequacy of the model was the acceptable performance of the established model in R-sq and R-sq-adjustable in Table 4. The effect of voltage, pulse on time, and weight concentration was shown in Fig. 4

Table 3 L9 orthogonal array and value of MRR

Exp. No.	V	P	W	MRR (μm^3/sec)
1	60	25	0.25	98,625
2	60	35	1.00	121,270
3	60	45	1.75	137,596
4	80	25	1.00	129,454
5	80	35	1.75	151,247
6	80	45	0.25	139,070
7	100	25	1.75	191,024
8	100	35	0.25	148,045
9	100	45	1.00	178,754

Table 4 ANOVA for machining

Source		Adj SS	Adj MS	F-Value	*P*-Value
Regression		493,827,114	1,646,090,380	36.23	0.001
V		328,217,837	3,282,178,371	72.23	0.000
P		970,714,082	970,714,082	21.36	0.006
W		685,378,688	685,378,688	15.08	0.012
Error		227,188,607	45,437,721		
S		R-sq	R-sq(adj)	R-sq(pred)	
6740.75		95.60%	92.96%	83.74%	

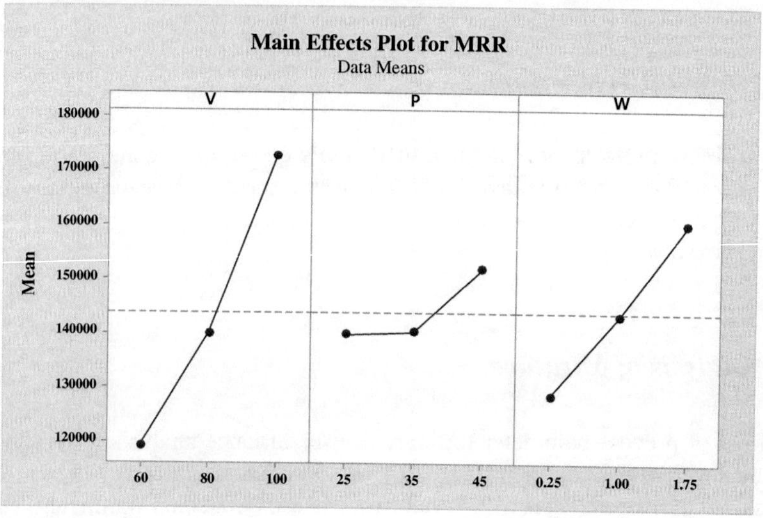

Fig. 4 Main effect plot of MRR

$$\text{Regression Equation for MRR} = -10647 + 1169\,\text{V} + 1272\,\text{P} + 14250\,\text{W} \quad (1)$$

4.2 Jaya Algorithm

The majority of optimization algorithms are probabilistic and require common regulating factors such as sample size, no. of population, and so on. Furthermore, some algorithms necessitate their own set of control settings. The right tuning of the parameters is a critical component that influences the efficiency of the algorithm. The Jaya method generates ‘s’ initial solutions at random between the upper and lower boundaries of the process variables. Following that, Eq. 2 is used to update each variable of each solution stochastically. The best solution has the most fitness, whereas the worst solution has the lowest fitness [13].

$$Y'_{s,t,u} = Y_{s,t,u} + p_{1,s,u}\left(Y_{s,\text{best},u} - \left|Y_{s,t,u}\right|\right) - p_{2,s,u}\left(Y_{s,\text{worst},u} - \left|Y_{s,t,u}\right|\right) \quad (2)$$

where $Y_{s,t,u}$ is the value of the variable ‘s’ for the best candidate, $Y_{s,\text{worst},u}$ is the value of the variable s for the worst candidate, and $Y'_{s,t,u}$ is the updated value of $Y_{s,t,u}$ in every iteration. $p_{1,s,u}$ and $p_{2,s,u}$ are the two random numbers for the s_{th} variable during the nth iteration (Fig. 5).

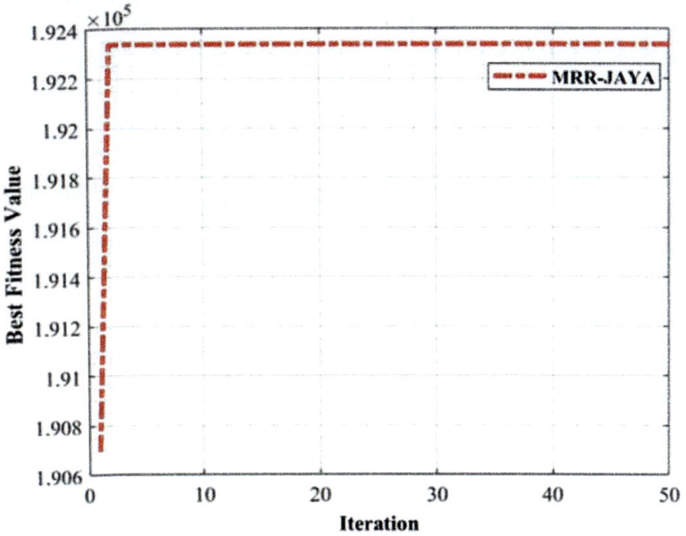

Fig. 5 Iteration of Jaya algorithm

Table 5 confirmatory test of machining

	Optimal machining setting			% Improvement		
Response	Taguchi L9 OA (V-3, P-1, W-3)	Jaya (V-3, P-3, W-3)	Confirmatory test (V-3, P-3, W-3)	Jaya to Taguchi	Confirmatory to Taguchi	Confirmatory to Jaya
MORE	191,024.00	192,337.74	208,745.88	0.68774%	8.4%	7.86%

4.3 *Confirmatory Test*

With the optimal parametric setting (V3, P3, and W3), the confirmatory test has been performed and validated with the machining outcomes of the optimal setting obtained by the Taguchi-based Jaya algorithm (Table 5). The confirmatory result indicates that the performance of MRR enhanced 8.4% concerning Taguchi and 7.86% concerning the Jaya algorithm. Also, the hole quality has been improved, which is shown in Fig. 7.

5 Surface Analysis of Drilled Hole

The FESEM analysis of micro-drilled hole generated through the μEDM process is shown in Fig. 6. During the μED machining, very high intense heat is generated in between the tool electrode and workpiece. Due to insufficient flushing, the eroded debris particles are suddenly quenched and adhere to the surface of the drilled hole [14]. This can be easily seen in Fig. 6. The solidified debris from the drilled hole was minimized with the effect of efficient flushing of the molten debris during the machining process. The spalling effect of a drilled hole affects the drilled hole's periphery and is caused by the rapid speed of the tool electrode. For a smooth drilled surface, the tool rotation of the electrode should be optimal. The matrix is burned from the excess heat. The fibers are breakage from the side surface of the drilled hole. The FESEM image of the confirmatory test is also given in Fig. 7, which shows the better MRR and hole quality of the machined area.

6 Conclusion

In this study, micromachining has been performed to produce a micro-drilled hole in CFRP/GNPs hybrid composite material. From the obtained investigations, the following conclusions can be drawn.

- Formation of micro holes in CFRP/GNPs epoxy composite is possible with the μED machining approach by reinforcing conductive nanoparticles.

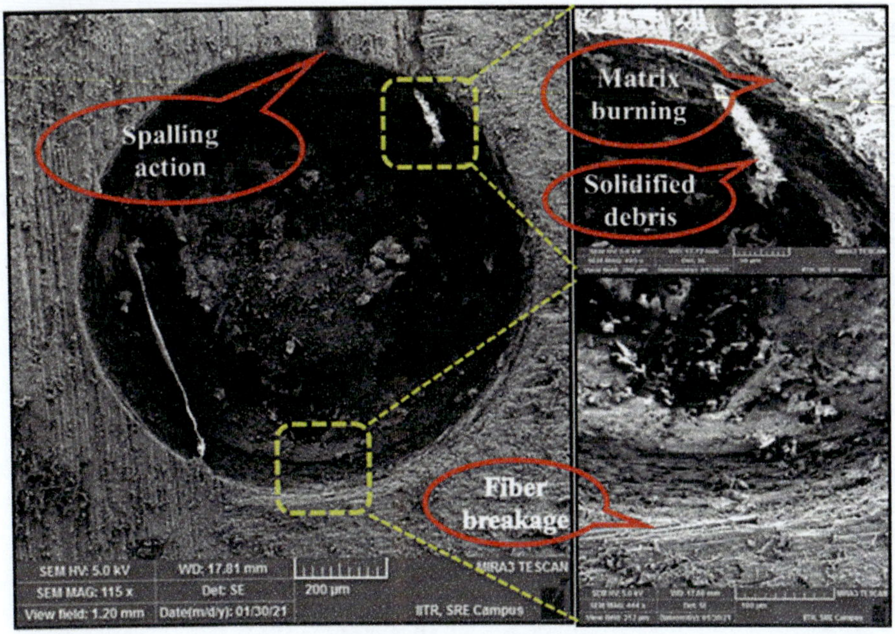

Fig. 6 FESEM analysis of drilled hole

Fig. 7 FESEM analysis of confirmatory drilled hole

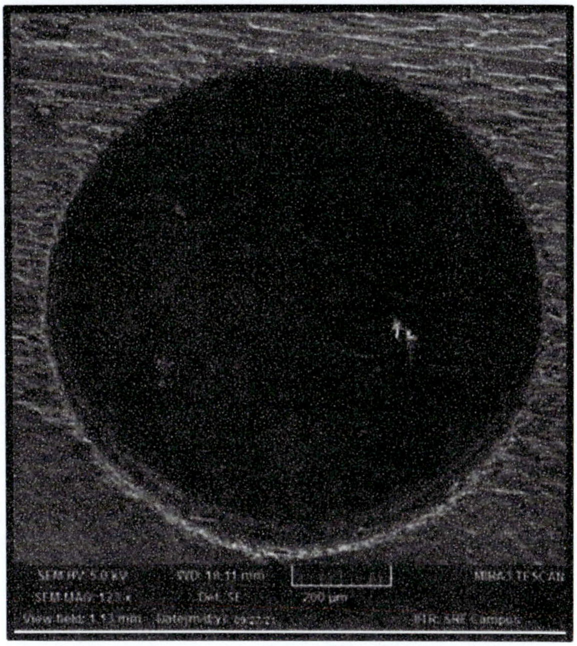

- It has been seen that the optimal setting for the higher MRR at level 3 for voltage, pulse duration, and amount of GNP %, which is 100 V, 45 μs, and 1.75 GNP%, respectively.
- Jaya algorithm was effectively applied to optimize the MRR, and it can be recognized that the confirmatory result (MRR) was improved by 0.68774% over OA.
- The improvement of the confirmatory test over Taguchi and the confirmatory test to Jaya algorithm was 8.4 percent and 7.86 percent, respectively.
- Pultrusion of fibers, spalling of fibers, and burning of a matrix are visible in the high-resolution microscopic image of the drilled surface due to intense heat generation

References

1. Anand A, Bhanuprakash L, Varghese S (2020) Study of flexural strength of CFRP composites with functionalized graphene oxide. Mater Today Proc 22:1932–1938. https://doi.org/10.1016/j.matpr.2020.03.093
2. Phonthammachai N, Li X, Wong S, Chia H, Tjiu WW, He C (2011) Fabrication of CFRP from high-performance clay/epoxy nanocomposite: preparation conditions, thermal-mechanical properties, and interlaminar fracture characteristics. Compos Part A Appl Sci Manuf 42(8):881–887. https://doi.org/10.1016/j.compositesa.2011.02.014
3. Li Y, Zhang H, Huang Z, Bilotti E, Peijs T (2017) Graphite nanoplatelet modified epoxy resin for carbon fibre reinforced plastics with enhanced properties. J Nanomater 2017:5194872. https://doi.org/10.1155/2017/5194872
4. Quan D, Alderliesten R, Dransfeld C, Murphy N, Ivanković A, Benedictus R (2020) Enhancing the fracture toughness of carbon fiber/epoxy composites by interleaving hybrid meltable/non-meltable thermoplastic veils. Compos Struct 252:112699. https://doi.org/10.1016/j.compstruct.2020.112699
5. Jang JU, Park HC, Lee HS, Khil MS, Kim SY (2018) Electrically and thermally conductive carbon fibre fabric reinforced polymer composites based on nano carbons and an in-situ polymerizable cyclic oligoester. Sci Rep 8(1):7659. https://doi.org/10.1038/s41598-018-25965-w
6. Kandare E et al (2015) Improving the through-thickness thermal and electrical conductivity of carbon fiber/epoxy laminates by exploiting the synergy between graphene and silver nano-inclusions. Compos Part A Appl Sci Manuf 69:72–82. https://doi.org/10.1016/j.compositesa.2014.10.024
7. Annebushan Singh M, Kumar Sarma D (2018) Parametric and subsurface analysis of MWCNT alumina composites in WEDM process. Ceram Int 44(2):2186–2197. https://doi.org/10.1016/j.ceramint.2017.10.174
8. Fan W, Li JL, Zheng YY (2015) Improved thermo-oxidative stability of three-dimensional and four-directional braided carbon fiber/epoxy hierarchical composites using graphene-reinforced gradient interface layer. Polym Test 44:177–185. https://doi.org/10.1016/j.polymertesting.2015.04.010
9. Kumar R, Kumar A, Singh I (2018) Electric discharge drilling micro-holesoles in CFRP laminates. J Mater Proc Technol 259:150–158. https://doi.org/10.1016/j.jmatprotec.2018.04.031
10. Priti MS, Singh S (2020) Micro-Machining of CFRP composite using electrochemical discharge machining and process optimization by Entropy-VIKOR method. Mater Today Proc 44(part 1):260–265. https://doi.org/10.1016/j.matpr.2020.09.463

11. Mohanty PP, Sahoo SK, Sahu AK (2016) Electro Discharge machining of Ti-alloy (Ti6Al4V) and 316L stainless steel and optimization of process parameters by grey relational analysis (GRA) method. Adv 3D Print Addit Manuf Technol 22:65–78. https://doi.org/10.1007/978-981-10-0812-2
12. Antil P, Singh S, Manna A (2020) Experimental investigation during electrochemical discharge machining (ECDM) of hybrid polymer matrix composites. Trans Mech Eng 44:813–824. https://doi.org/10.1007/s40997-019-00280-5
13. Venkata Rao R (2016) Jaya: a simple and new optimization algorithm for solving constrained and unconstrained optimization problems. Int J Ind Eng Comput 7(1):19–34. https://doi.org/10.5267/j.ijiec.2015.8.004
14. Dutta H, Debnath K, Sarma DK (2019) A study of material removal and surface characteristics in micro-electrical discharge machining of carbon fiber-reinforced plastics. Polym Compos 40(10):4033–4041. https://doi.org/10.1002/pc.25264

Material Removal Rate and Tool Life Improvement Studies on Machining of Fused Silica Ceramic Radome Using Diamond Tip Brazed (DTB) Tool

J. Vimal Kumar, K. Theenathayalan, J. John Rozario Jegaraj, and M. Ravi Sankar

Abstract Ceramic radomes are made of fused silica and are widely used for critical aerospace applications. In this paper, an attempt has been made to improve the tool life while machining fused silica ceramic radome. Various methodologies such as dry machining, wet machining, green ceramic machining (GCM) and machining using poly crystalline diamond (PCD) inserts/diamond tip brazed (DTB) tool have been evolved for machining of ceramic radomes. From the literature, it is found that wet machining with the application of demineralized water in limited quantity proved to be beneficial in terms of improved tools life and better machinability. Wear of PCD tools was found to be much slower in the case of wet machining. It is also found that green ceramic machining can be adopted for machining of ceramic radome with higher raw material stock. In this work, an attempt has been made to study the use of the DTB tool for the machining of fused silica ceramic radome. The study revealed that the DTB tool is beneficial in terms of cost-effectiveness, tool life, material removal rate (MRR) and machining time over PCD inserts.

Keywords Ceramic radome machining · Green ceramic machining (GCM) · PCD inserts · Diamond tip brazed (DTB) tool

1 Introduction

A radome, an acronym coined from radar dome, is a cover or structure placed over an antenna that protects the antenna from its physical environment. Ideally, the radome

J. V. Kumar (✉)
Research Centre Imarat, DRDO, Hyderabad, India
e-mail: vimalkumarjain@rcilab.in

K. Theenathayalan · J. J. R. Jegaraj
Defence Research and Development Laboratory, DRDO, Hyderabad, India
e-mail: johnmfrg@drdl.drdo.in

M. R. Sankar
Department of Mechanical Engineering, IIT, Tirupati, India
e-mail: evmrs@iittp.ac.in

U. S. Dixit et al. (eds.), *Advances in Forming, Machining and Automation*, Lecture Notes in Mechanical Engineering, https://doi.org/10.1007/978-981-19-3866-5_29

is a radar dome that is radio frequency (RF) transparent so that it does not degrade the electrical performance of the enclosed antenna in any way [1]. The usage of ceramics in aerospace and missile applications is significantly increasing day by day. Ceramics is one of the optimum materials for missile radomes due to its superior electromagnetic and dielectric constant properties.

The radome is used in the front section of the missile for the transmission of electro-magnetic waves to find the target. The radome is made of ceramic materials because high-speed Mach vehicles during flight at high altitudes are subjected to severe aerothermal and mechanical loading. The surfaces of high-speed missiles are subjected to aerodynamic heating, mechanical stresses, and erosion. Consequently, radomes for high-speed missiles must have both good high-temperature strength and suitable dielectric properties within the entire temperature range at which the missiles operate.

Silica (SiO_2) has proven useful for making high-temperature radomes. However, there exists a continuing need for radome materials having greater high-temperature strength and erosion resistance together with good dielectric properties. The main problem with the use of ceramics is that machining is highly expensive. This prohibits the replacement of metal parts with ceramic parts in many industries in which machined parts are widely used, such as the automotive, aerospace and semiconductor industries [2].

As per literature, the traditional technology to machine ceramics is grinding using making very small chips produced by the cutting edges of abrasive particles. The grinding process removes ceramics materials with low productivity. With the ever-increasing number of ceramics materials in the marketplace, there is a pressing need to improve traditional methods to machine ceramic materials for cost reduction and quality assurance to achieve the full potential of ceramics [3]. To explore the possibility of using cutting fluids as an efficient means to achieve satisfactory machining performance [4].

Jain et al. [5] have experimented with the wet environment, and it is evident from results that the application of distilled water is having an advantage over the dry environment in the case of fused silica machining. Significant tool life improvement (about three times) has been observed in the case of wet machining (with distilled water) compared to dry machining for all the investigated machining conditions. This may be attributed to the reduction of tangential cutting force in the case of machining with distilled water. Moreover, the application of distilled water may have a favourable tribological effect on fused silica.

Jain et al. [6] carried out a machinability study of green and sintered fused silica samples in respect of material removal rate and consumption of diamond cutting insert. A new methodology was proposed for improving the machinability of ceramics. It is concluded that the proposed sequence of machining yielded three times improvement in MRR and 43% improvement in tool life.

From the literature survey, it is found that wet machining with an application of demineralized water in limited quantity proved to be beneficial in terms of improved tools life and better machinability. Wear of PCD tools was found to be much slower

Table 1 Physical, mechanical, thermal and electrical properties of fused silica

Properties	Value
Density	2.2 gm/cc
Elastic modulus	73 GPa
Shear modulus	31 GPa
Hardness	600 kg/mm^2
Maximum use temperature	1100 °C
Thermal conductivity	1.38 W/m–K
Coefficient of thermal expansion	0.55×10^{-06}/°C
Dielectric strength	30 kV/mm
Dielectric constant	@ 1 MHz

in case of wet machining, and it is also found that green ceramic machining can be adopted for machining of ceramic radome with higher raw material stock.

In this paper, an attempt has been made to improve the tool life while machining fused silica ceramics using diamond tipped brazed tool. The study revealed that the DTB tool is beneficial in terms of cost-effectiveness, tool life, MRR and machining time over PCD inserts.

2 Material

The radome is made of fused silica (SiO_2) ceramic material. It is a non-crystalline form of silicon dioxide. It is a highly cross-linked three-dimensional structure that gives high use temperature and low thermal expansion coefficient [7]. The properties of fused silica is given in Table 1.

3 Component Description

Ceramic radome is a long Ogive shape with one closed hemispherical end and another end is open to keep the antenna system. Its overall size is the inner diameter of 170 mm with wall thickness 5.0 mm and 470 mm total length. As this ceramic radome is moulded, it requires finish machining to final sizes. Machining requires maintaining the total length and machining of external and internal profiles. Figure 1 shows the CAD model of ceramic radome.

Fig. 1 CAD model of ceramic radome

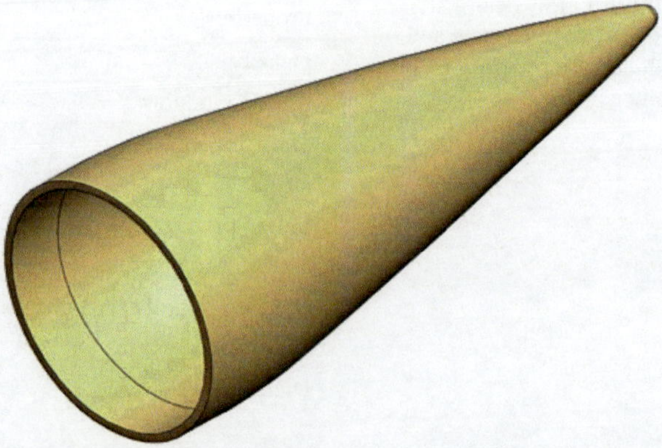

4 Process Sequence for Ceramic Radome

Machining of ceramics is of great concern due to its poor machinability. It is characterized by the following parameters. Metallic machining is characterized by ductile failure by shearing action, and metal machining parameters are well defined and the use of coolant is possible in almost all metals. In ceramic materials, machining is characterized by brittle failure, leading to uncontrolled chip removal; the material is removed in the form of powder [4].

- Due to radome geometry
  - Holding of Ogive shape radome with sufficient clamping force
  - Requirement of long boring bar for internal turning
- Due to material (ionic or covalent bond)
  - High hardness
  - High brittleness
  - Heat insulator
- Due to process
  - Proper selection of cutting tool
  - Establishment of cutting parameter
  - Proper selection of coolant.

The flow process chart established for machining of radome is shown in Fig. 2

5 NDT on Sintered Ceramic Radome

Radiography testing is carried out on ceramic radome to assess the presence of cracks or any defects, as the presence of crack is the main cause of breaking of radome. No cracks and voids are the acceptance criteria of the radome [8].

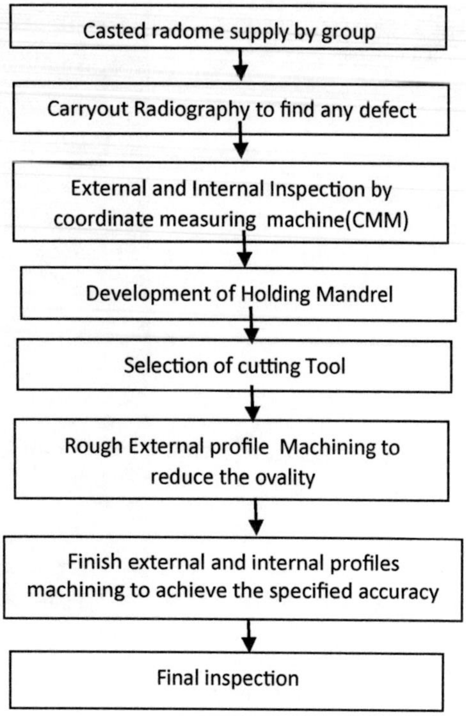

Fig. 2 Established process sequence for machining of the radome

6 Dimensional Inspection Using Coordinate Measuring Machine (CMM)

Inspection of casted ceramic radome is carried out using CMM at four locations (0, 90, 180 and 270 deg) at the interval of 5 mm before machining (Fig. 3). This helps in finding out the ovality and other geometrical and dimensional variations on cast radome. After inspection, these dimensions were taken into consideration for the design and development of machining holding fixtures (Fig. 4).

7 Development of Holding Fixtures

Holding fixtures are required to carry out the external profile machining of the casted ceramic radome. The holding fixture is made of nylon material, and it is having a critical contoured profile. The major challenge is to match exactly with casted ceramic radome. The machining of the holding fixture (Fig. 4) is completed and matched with casted radome for external profile machining.

Fig. 3 CMM inspection of casted radome

8 Experimentation

Two samples of the size of 50 mm diameter and 100 mm length as shown in Fig. 5 are considered for the experimentation. Two types of cutting tool inserts (PCD insert and diamond tip brazed (DTB)) are chosen for the study, as shown in Figs. 6 and 7. The turning operation has been carried out to reduce diameter 50 to 40 mm using a diamond insert. Material removal rate (MRR) and tool life are estimated. The

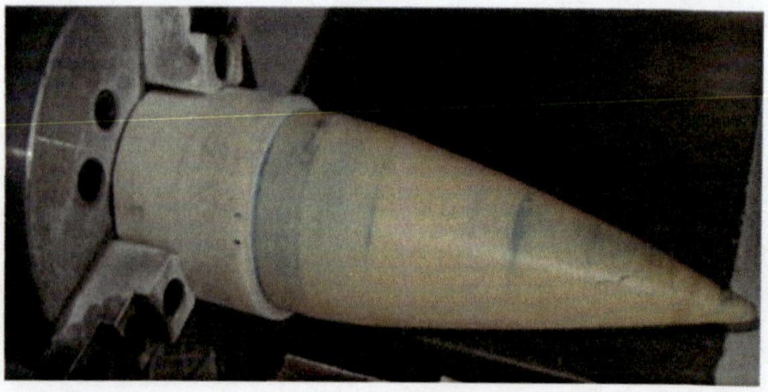

Fig. 4 Holding fixture for external machining of radome

Fig. 5 Fused silica sample

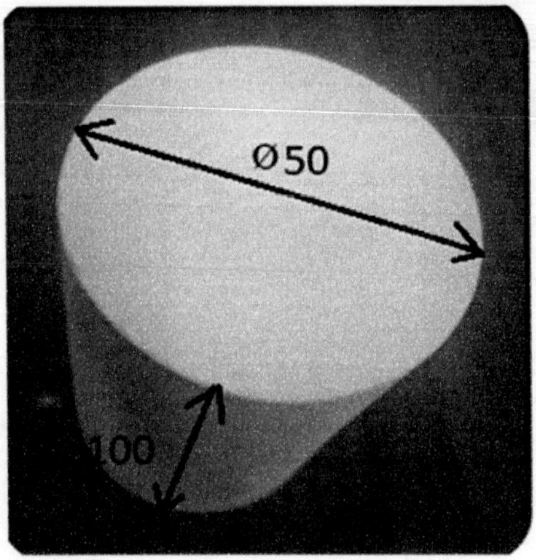

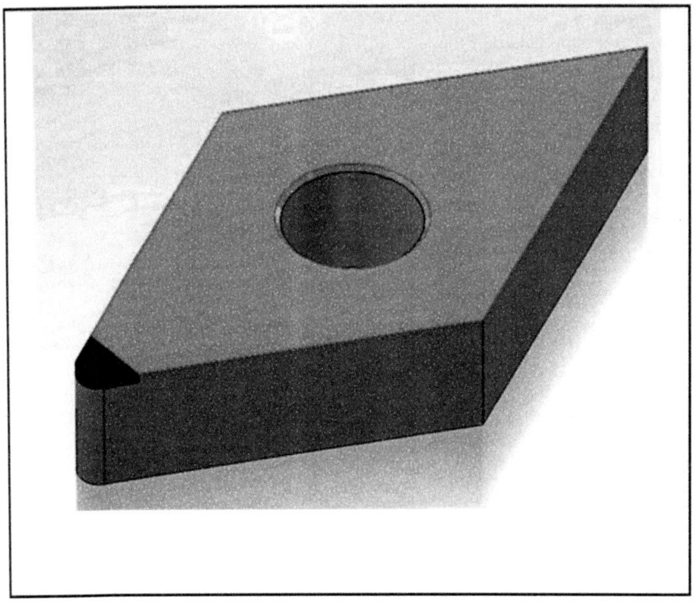

Fig. 6 PCD insert

first sample has been machined using PCD insert, and the second sample has been machined with DTB tool.

Machining Parameters

The cutting parameters, established in the previous study, have been used for this study also, with slight alterations [6]. The parameters used for both the tools are

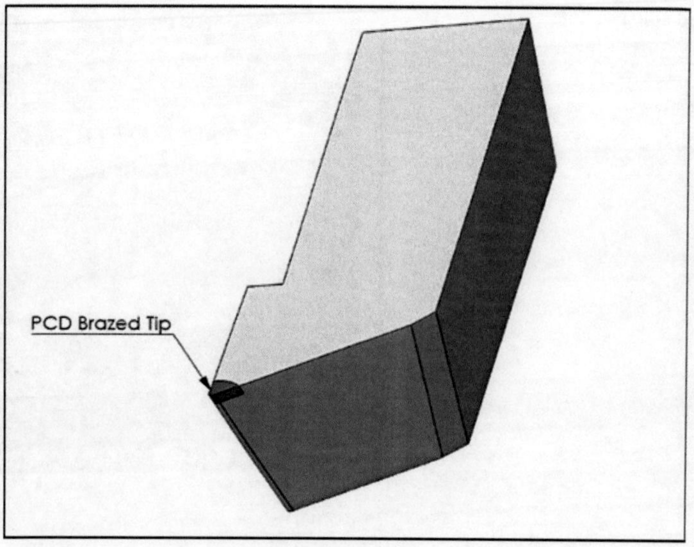

Fig. 7 Diamond tip brazed (DTB) tool

Table 2 Machining parameters

Parameter	PCD inserts	DTB tool
Cutting speed, Vc (m/min)	50	50
Feed, f (mm/rev)	0.06	0.08
Depth of cut, d (mm)	0.3	0.4

listed in Table 2. All the experiments have been conducted on CNC lathe machine. Demineralized water as a coolant has been used while machining sintered samples.

9 Results and Discussion

As it can be seen from Table 3, using the DTB tool results in significant improvement in both MRR and tool life. There is an improvement of 103.2% in material removal rate while machining fused silica radome under wet machining conditions. Also, 242.9% improvement has been observed in tool life in the DBT tool when compared to PCD insert. Tool life in terms of time taken for flank wear was not measured. Instead, the total material removed before the first failure of the tool is considered as a measure of tool life in this study.

Following advantages are envisaged with the use of DTB tool.

- Consumptions of PCD inserts have been found more and the same is overcome with the use of the DTB tool

Table 3 MRR and tool life calculations

Parameters	PCD tool	DBT tool
MRR		
Speed, N (rpm)	70	80
Cutting speed, Vc (mm/min)	10,990	12,560
Feed, f (mm/rev)	0.06	0.08
Depth of cut, d (mm)	0.3	0.4
Material removal rate, MRR (mm^3/min)	197.82	401.92
% Improvement in MRR	–	**103.2%**
Tool life		
Starting diameter, D_o (mm)	50	50
Final diameter, before changing insert, D_f (mm)	49.2	47.2
Length of cut, L (mm)	85	85
Total material removed before tool life (mm^3)	5295.29	18,159.8
% Improvement in tool life (In terms of total material removed before tool life)	–	**242.9%**

- Requirement of frequent change of PCD insert is eliminated, leading to improved machining time
- The cost-effectiveness of the DTB tool is far better when compared to PCD inserts.

Hence, it is felt by the authors that the use of wet machining and application of DTB tool along with green ceramic machining concept is the best choice for machining of fused silica ceramic radome in respect of better material removal rate, tool life, machining time and cost-effectiveness.

10 Conclusion

- Fused silica radome (shown in Fig. 8) using the DTB tool is successfully machined with the proper selection of cutting parameters.
- Machined ceramic radome is supplied to the user. He tested and achieved a good result.
- Adopting the process improvement like using DTB tools instead of PCD inserts result in an improvement 103.2% in material removal rate (MRR) and enhanced 242.9% in the tool life.
- The use of diamond tip brazed (DTB) tools further resulted in improvement in machining time and reduction of machining cost instead of using PCD Inserts.

Fig. 8 Machined fused silica radome

Acknowledgements The authors wish to express their sincere thanks and gratitude to Sri. U. Raja Babu, OS and Director, RCI, DRDO for providing permission and facilities to carry out the research work. The authors also thank Sri B. R. K. Reddy, OS and Technology Director, DRFT and Associate Director, RCI and his team for providing casted radome, and Sri K. Venkat Reddy Scientist 'F', Head, Radiography Section, DRDL and other team members for their continuous support and encouragement to carry out this work.

References

1. Dennis JK (2009) Analysis of radome enclosed antennas, 2nd edn. Artech House Publishers, Norwood
2. Groover MP (1996) Fundamental of modern manufacturing; material, processes and systems, 2nd edn. Prentice Hall, New Jersey
3. Marinescu IM (2007) Handbook of advanced ceramics machining, 1st edn. CRC Press, Bocaraton
4. Gates RS, Hsu SM (1999) Chemo-mechanical machining of ceramics, 1st edn. NIST Publications, Gaithersburg
5. Jain VK, Podder B, Saikumar S, Yadav DR (2012) Dimond turning of ceramic radome for aerospace application. In: 4th international and 25_{th} AIMTDR, vol.1, Jadavpur University, Kolkata, pp 277–280
6. Jain VK, Theenathayalan K (2017) Experimental studies on improvement of material removal rate and cutting tool life while machining Gel-cast fused silica ceramics. In: 10th international conference on precision, meso, micro and nano engineering COPEN-10. IIT Madras, Chennai, pp 887–890
7. Ruth EW (1986) ASM handbook materials characterization, American society for metals, 9th edn.(10) ASM international, USA
8. Harris JN, Welsh EA (1972) Acceptance criteria for slip-cast fused silica radome, Technical Report No.3, Engineering experiment station, Georgia

Optimization of CNC Green Milling Process Parameters: An Integrated MCDM Approach

Sandeep Kumar and Abhishek Singh

Abstract In the present industrial scenario, green machining is must be required to improve the beneficial responses with minimum consumption of energy and emission of pollutants. In this research work, multi-objective optimization of green milling parameters on machining of H_{21} steel using TOPSIS-MOORA-CRITIC method and work considered three process parameters (CS, FR and DC), and three response parameters (MRR, SR and SE). The material used for the milling cutter is high-speed steel. Initially, experimentations are performed using Taguchi's L_{27}-OA and multi-response optimizations are done with TOPSIS-CRITIC and MOORA-CRITIC methods. Additionally, comparative analyses are carried out between single response and multi-response optimization techniques and the result shows optimum combinations of milling parameters using Taguchi's analysis of mean are CS 1200 rpm, FR 0.35 mm/tooth, and DC 0.6 mm for material removal rate; cutting speed 1200 rpm, feed rate 0.15 mm/tooth and depth of cut 0.4 mm for surface roughness; cutting speed 400 rpm, rate feed 0.15 mm/ tooth and depth of cut 0.2 mm for specific energy. The optimal parametric combination of process parameters obtained using TOPSIS-CRITIC as well as MOORA-CRITIC method is speed 1200 rpm, feed 0.35 mm/tooth and depth of cut 0.6 mm. The results of ANOVA for MRR revealed that the cutting speed is the most significant parameter and the speed and depth of cut have the highest % contribution about 37.25% and 27.44%, respectively. The results of ANOVA for surface roughness revealed that the speed is the most significant parameter and speed and interaction of speed and speed have the highest % contribution about 63.87% and 20.99%, respectively. The results of ANOVA for specific energy revealed that the feed is the most significant parameter and feed and speed have the highest % contribution about 41.31% and 35.29%, respectively.

Keywords Milling · Green manufacturing · Optimization · TOPSIS · MOORA · CRITIC · ANOVA

S. Kumar (✉) · A. Singh
Department of Mechanical Engineering, NIT, Patna, India
e-mail: sandeepk.ph21.me@nitp.ac.in

U. S. Dixit et al. (eds.), *Advances in Forming, Machining and Automation*, Lecture Notes in Mechanical Engineering, https://doi.org/10.1007/978-981-19-3866-5_30

1 Introduction

H_{21} steel is hot work tool steel that exhibits excellent red hardness and strength. It is largely used in tool, die and mould industries. The machining of $H_{21}$is difficult because of its high hardness. In the industrial globalization era, manufacturers are required to find the techniques to enhance the quality of products and raise productivity economically. Hence, as a basic machining process, milling is generally used in industries because of its ability to produce excellent surface quality with high material removal rate [1, 2]. In the milling process, surface quality is an important machining parameter to evaluate the eminence of the machined surface of the product which largely depends on surface characteristics such as wear resistance, friction and fatigue. It depends on machining conditions and process variables [3]. In the machining process, the surface quality of product and productivity are important factors of consideration, but it conflicts with each other as productivity increase the quality of product decrease. Hence, it is necessary to optimize the surface quality and productivity simultaneously [4]. Presently, manufacturers are aware of the consumption of energy and its environmental impacts. To produce a quality product at a high production rate in minimum energy consumption because energy consumption is proportional to pollution generated by industry [5].

CNC machining process is consumed approximately about 50–70% of the total energy utilized by the manufacturing sector and produced many pollutants [6]. It is also produced fine and powder chip particles with dust released into the air. As result, serious, health and environmental problems are occurring when these poisonous fine particles enter the body of operators and nearby people through the intake, breathing, and skin contact. Hence, the high productivity and quality of products with minimum consumption of energy is still a challenging task for the industry. Therefore, the manufacturing industry needs to implement green manufacturing (GM) to decrease the consumption of energy and disposal of pollutants. GM is contributing to all industrial conditions to enhance efficiency, reduce environmental impacts and health issues [7, 8].

The milling process parameters largely influence the quality of products, material removal rate, energy consumption and amount of waste generation (solid or gaseous). Therefore, the selection of optimal machining process parameters to improve the response variable such as productivity, efficiency and quality of products in minimum consumption of energy is a challenging task. Hence, milling process parameters optimization is a major task to attain green milling. Thus, in past many researchers have been attempt to change the process conditions and used much optimization technique to obtain the desire response and attain the green milling such as, investigation of effects of input parameters (DC, FR, CS, concavity and axial relief angle) on surface roughness in slot milling of AL2014-T6 using response surface methodology (RSM) [9]; optimization of process variables (CS, FR, DC and width of cut) on response parameters cutting energy (CE), MRR and SR in milling of medium carbon steel (C45) using RSM and grey relation analysis [10]; optimization of input parameters on surface roughness in end milling of AISI-6062 using genetic algorithm (GA) [11]; optimization of input parameters (CS, FR and rake angle) on SR in end milling of Ti6Al4V using ANN-GA [12]; optimization of process variables (CS, FR, DC) on cutting force during end milling of Al2014-T6 using GA-RSM [13]; optimization and effects analysis of parameters (CS, FR, DC) on SR and machining time in face milling of different material such as copper, brass, aluminium and mild steel using PSO [14]; optimization of process variables on SR and MRR in face milling of stainless (austenitic) steel using RSM-desirability function [15]; optimization of process parameters on SR in end milling of inconel-718 using ANN-GA [16]; optimization of process parameters (CS, FR, axial and radial depth) on SR and cutting force in milling of 7050 aluminium alloy using RSM and TLBO algorithm [17]; optimization of process variable on SR in micro end milling of AISI-1040 using RSM [18]; optimization of input parameters on SR in micro milling of inconel-718 using SVM [19]; optimization of input parameters on MRR and SR during end milling of Al-6061using ANN-RSM [20]; optimization of input parameters on SR and TWR in milling of TC-17 alloy using Taguchi's-GRA [21]; Modelling and optimization of process variables on chatter vibration during milling of Al-7075T6 using FEM [22]; optimization of input parameters on MRR, SR and dimensional deviation during end milling of EN-31 steel using AHP-VIKOR-WASPAS [23]. Thus, in the past, insufficient literature are reported on multi-response optimization of green milling input parameters such as CS, FR and DC on output parameters such MRR, SR and specific energy on H21 hot work tool steel. The aim of the present work is the optimization of green milling parameters on H21steel using TOPSIS-GRA with the CRITIC method, and the work considers three input parameters CS, FR and DC, and three output parameters as MRR, SR and SE.

2 Materials and Methods

2.1 Work Materials

In the present study, H_{21} hot work tool steel of dimension 30 × 30 × 15 mm is used for experimentation. The chemical composition and properties of H21 [23] steel are shown in Tables 1 and 2, respectively. The material used for the milling cutter is high-speed steel.

Table 1 Chemical composition of H_{21} steel

C	Mn	Si	Cr	Ni	W	V	Cu	P	S
0.30	0.30	0.20	3.60	0.3	8.5	0.40	0.25	0.03	0.03

Table 2 Properties of H_{21} steel

Density (g/cm^3)	Poisson's ratio	Specific heat (cal/g °C)	Hardness (Rockwell)	Melting point (°C)	Modulus of elasticity (GPa)
8.19	0.27–0.30	0.110	40–55	1432	190–210

Table 3 Process parameters and their levels

Parameters	Symbols	Units	Level-1	Level-2	Level-3
Cutting speed (A)	CS	rpm	400	800	1200
Feed rate (B)	FR	mm/tooth	0.15	0.25	0.35
Depth of cut (C)	DC	mm	0.2	0.4	0.6

2.2 *Design of Experiment*

Taguchi's method of design of experiment (DOE) is a useful statistical tool for experimental design and determination of the response of milling input parameters on performance parameters. In Taguchi's method of DOE, orthogonal array (OA) is used to reduce the number of experimental trials, cost, and time of experiments [24, 25]. In the present work, DOE is performed using L27-OA and three milling input process parameters considered for the study are varied at three levels as illustrated in Table 3, and other parameters are considered constant.

2.3 *Experimental Setup and Response Measurement*

In this research work, the end milling process is carried out using DMC 635 V DMG ECOLINE, Deckel Maho Germany and Siemens 810D, 3axis, CNC vertical milling machine according to L_{27}-OA and recorded the output responses of twenty-seven experimental conditions with input milling process parameters. At each parametric setting three trails are performed for better accuracy in response variables. Finally, investigate the response characteristics of experimental data to determine the optimal milling input milling process parameters (Fig. 1).

The material removal rate is calculated as the difference of weights of the workpiece before and after the experiments and each parametric setting are performed in three experimental trials. The average of three experimental trials is considered as a response for analysis. To estimate the consumption of energy in milling machining, the electrical power drawn from the power grid is recorded at the main switch of the CNC machine for every 0.1 s on the LabView interface. Thus, the average power is calculated and multiple with cycle time to determine the total energy required for material removal from the workpiece. The specific energy is the quantity of energy consumed for the removal of 1mm^3 material during milling operation, and specific energy values for each experimental run according to L_{27}-OA are calculated and

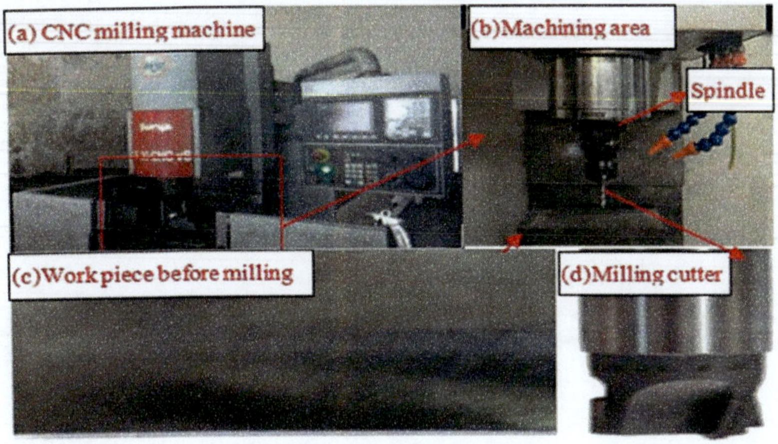

Fig. 1 Experimental setup of CNC milling machine

considered as a response for analysis. The SR of the milled surface is determined using Mitutoyo surftest SJ-301. Three readings are recorded at each parametric and their average is considered as a response for analysis. All results of experimental trials are illustrated in Table 4.

2.4 Criteria Importance Through Criteria Inter-Correlation (CRITIC)

CRITIC technique is a multi-criteria decision-making technique to calculate the weights of responses in multi-response optimization with conflicting responses. The manufacturers are largely concerned about economic machining (high productivity and quality) and assigning equal importance to all responses may lead to dissatisfaction among customers and industrialists [26]. Hence, in this work CRITIC method was used to estimate the weights of response MRR, SR and SE and estimated values of the weight of 0.283, 0.303 and 0.414, respectively.

2.5 Technique for Order Preference by Similarity to Ideal Solution (TOPSIS)

TOPSIS is an MCDM method used to determine the most suitable alternative among finite sets in multi-objective optimization problems. In this method select the criteria or preference value from alternative which is close to the ideal best positive solution

Table 4 L_{27}orthogonal array and responses

Sl. No	Cutting speed (rpm)	Feed rate (mm/tooth)	Depth of cut (mm)	MRR (gm/min)	SR (μm)	Specific energy (J/mm^3)	Preference value	Assessment value
1	400	0.15	0.2	3.6	4.3	24.44	0.000	−1.420
2	400	0.15	0.4	6.7	3.6	27.02	0.120	−1.519
3	400	0.15	0.6	9.9	4.3	28.33	0.219	−1.657
4	400	0.25	0.2	5.7	5.2	32.41	0.037	−2.386
5	400	0.25	0.4	10.9	4.4	34.5	0.190	−2.365
6	400	0.25	0.6	15.1	6.1	37.23	0.274	−2.850
7	400	0.35	0.2	8.2	6.3	38.66	0.073	−3.379
8	400	0.35	0.4	16.0	6.4	34.87	0.325	−2.546
9	400	0.35	0.6	23.9	5.1	51.07	0.345	−4.355
10	800	0.15	0.2	7.8	1.8	30.1	0.174	−1.654
11	800	0.15	0.4	15.1	1.5	32.28	0.359	−1.600
12	800	0.15	0.6	22.3	1.4	36.25	0.491	−1.635
13	800	0.25	0.2	12.6	1.3	44.35	0.212	−3.452
14	800	0.25	0.4	23.8	1.7	43.49	0.431	−2.616
15	800	0.25	0.6	32.9	2.1	51.16	0.514	−3.104
16	800	0.35	0.2	18.7	2.1	44.44	0.316	−3.179
17	800	0.35	0.4	36.8	2.7	45.02	0.655	−1.567
18	800	0.35	0.6	55.1	2.7	48.73	0.876	0.726
19	1200	0.15	0.2	11.9	1.9	37.41	0.225	−2.447
20	1200	0.15	0.4	23.4	1.9	39.5	0.469	−2.039
21	1200	0.15	0.6	34.8	1.8	41.23	0.677	−1.133

(continued)

Table 4 (continued)

Sl. No	Cutting speed (rpm)	Feed rate (mm/tooth)	Depth of cut (mm)	MRR (gm/min)	SR (μm)	Specific energy (J/mm^3)	Preference value	Assessment value
22	1200	0.25	0.2	19.6	2.0	42.66	0.351	−2.820
23	1200	0.25	0.4	38.6	2.0	44.87	0.690	−1.247
24	1200	0.25	0.6	55.7	1.7	59.07	0.752	−1.179
25	1200	0.35	0.2	29.1	2.4	51.1	0.449	−3.529
26	1200	0.35	0.4	54.7	1.7	54.28	0.797	−0.351
27	1200	0.35	0.6	81.3	1.8	60.67	1.000	4.599

and extreme from the ideal negative best solution. In the TOPSIS method, multiple-attribute optimization problems with conflicting responses are converted to the single or composite problems by preference index value and rank the preference values in decreasing order [27]. The higher preference value among all alternatives is selected as an optimal solution, and calculated preference values are illustrated in Table 4.

2.6 *Multi-Objective Optimization Based on Ratio Analysis (MOORA)*

MOORA is an MCDM method used to estimate the optimal parametric setting of machining variables among all alternatives with conflicting responses as beneficial (maximization) and no beneficial (minimization) criteria. In this method, multi-objective problems are converted into a single objective problem as assessment values and rank of each alternative [28]. The alternatives of the highest assessment value are selected as an optimal solution, and calculated assessment values are illustrated in Table 4.

2.7 *Analysis of Variance (ANOVA)*

ANOVA is a statistical technique applied to estimate the significant effects and relative importance of process parameters on response characteristics. It is also used to determine the statistical significance (P-value) and influence of process parameters (F-value) [29–31]. The results of the analysis of variance of response parameters, MRR, SR and SR are represented in Tables 5, 6, 8, respectively.

Table 5 Means of preference and assessment values

Response parameters	Preference value			Assessment value		
Process parameters	CS	FR	DC	CS	FR	DC
Level-1	0.1759	0.3038	0.2041	−2.497	−1.678	−2.696
Level-2	0.4476	0.3834	0.4484	−2.009	−2.447	−1.761
Level-3	0.6011	0.5373	0.5720	−1.127	−1.509	−1.176
Delta	0.4252	0.2336	0.3679	1.370	0.938	1.520
Rank	1	3	2	2	3	1

Table 6 Results of ANOVA for material removal rate

Sources	DoF	Square SS	Adj.SS	Adj.MS	F_{Val}	P_{Val}	Percentage contribution (%)
Regression	9	9184.87	9184.87	1020.54	252.07	0.000	99.26
CS	1	3447.27	48.69	48.69	12.03	0.003	37.25
FR	1	1969.83	73.97	73.97	18.27	0.001	21.29
DC	1	2539.47	48.74	48.74	12.04	0.003	27.44
CS*CS	1	0.02	0.02	0.02	0.01	0.942	0.00
FR*FR	1	16.12	16.12	16.12	3.98	0.062	0.17
DC*DC	1	0.27	0.27	0.27	0.07	0.800	0.00
CS*FR	1	375.20	375.20	375.20	92.67	0.000	4.05
CS*DC	1	530.67	530.67	530.67	131.07	0.000	5.73
FR*DC	1	306.03	306.03	306.03	75.59	0.000	3.31
Error	17	68.83	68.83	4.05			0.74
Total	26	9253.70	9253.70				100.00

3 Results and Discussion

In the present research work, influence of input parameters on three response parameters is investigated according to L_{27}- OA and observed experimental outputs are revealed in Table 4. The mean of response parameters at different levels of process parameters is used to determine optimal parametric combination as higher is better for beneficial and lower is better for non-beneficial responses. The multi-response optimization is performed on preference values and assessment values using the TOPSIS-MOORA-CRITIC method. Hence, in this study, optimal levels of process parameters for material removal rate, determined from Fig. 2, are CS 1200 ppm (A_3), FR 0.35 mm/ tooth (B_3) and DC 0.6 mm (C_3), i.e. A_3, B_3, C_3.

The optimal levels of input parameters for minimization of SR determined from Fig. 3 are CS 1200 ppm (A_3), FR 0.15 mm/tooth (B_1) and DC 0.4 mm (C_2), i.e. A_3, B_1 and C_2.

The optimal levels of input parameters for minimization of specific energy determined from Fig. 4 are CS 400 rpm (A_1), FR 0.15 mm/tooth (B_1) and DC 0.2 mm (C_1), i.e. A_1, B_1 and C_1.

In this study, multi-objective optimization is performed using the TOPSIS-MOORA-CRITIC method. The experimental trail 27 has highest preference value as well as assessment value and hence parametric combination of experiment 27 is optimal parametric combination of parameters for multi-objective optimization are CS 1200 rpm (A_3),FR 0.35 mm/tooth (B_3) and DC 0.6 mm(C_3), i.e. A_3,B_3 and C_3.

Table 5 demonstrates the mean of preference and assessment value of each parametric level, and Fig. 5 illustrates the optimum levels of preference value are CS 1200 rpm (A_3), FR 0.35 mm/tooth (B_3) and DC 0.6 mm (C_3), i.e. A_3, B3, C3. Moreover, the most influential milling parameter is calculated by rank. It is a difference

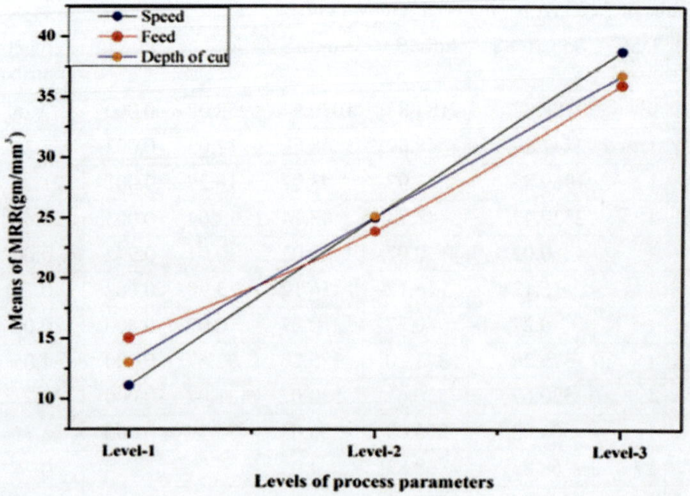

Fig. 2 Main effect plots of means for material removal rate

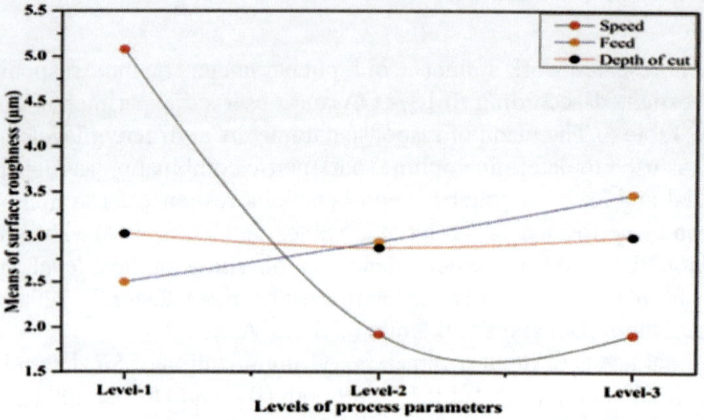

Fig. 3 Main effect plots of means for surface roughness

of maximum and minimum mean preference values as shown in Table 5, and results show cutting speed (CS) is an extremely influential parameter on the preference value of multi-response characteristics followed by the depth of cut (DC) and feed rate (FR) according to their rank.

Fig. 6 illustrates the optimum levels of assessment value are CS 1200 rpm (A_3), FR 0.35 mm/tooth (B_3) and DC 0.6 mm (C_3), i.e. A3, B3 and C3. The depth of cut is the most influential parameter on the assessment value of multi-response characteristics followed by cutting speed and feed rate according to their rank.

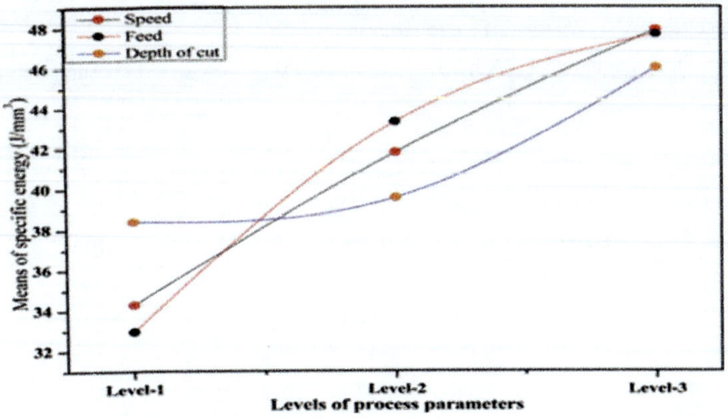

Fig. 4 Main effect plots of means for specific energy

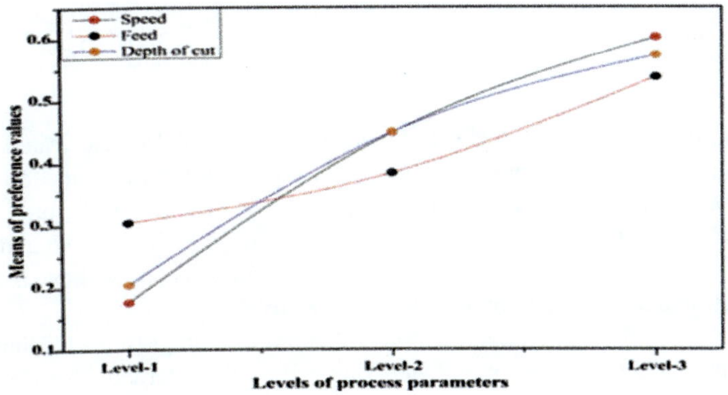

Fig. 5 Main effect plots of means for preference value

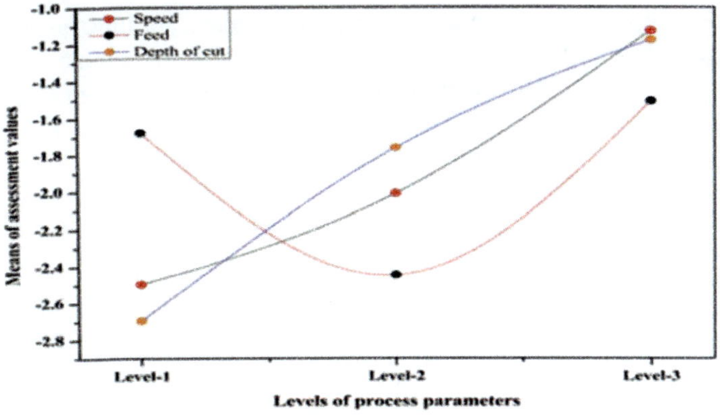

Fig. 6 Main effect plots of means for assessment values

Table 7 Results of ANOVA for surface roughness

Sources	DoF	Square SS	Adj.SS	Adj.MS	F_{Val}	P_{Val}	Percentage contribution (%)
Regression	9	66.7136	66.7136	7.4126	31.96	0.000	94.42
CS	1	45.1250	12.3703	12.3703	53.34	0.000	63.87
FR	1	4.2050	0.3398	0.3398	1.47	0.243	5.95
DC	1	0.0050	0.0268	0.0268	0.12	0.738	0.01
CS*CS	1	14.8313	14.8313	14.8313	63.95	0.000	20.99
FR*FR	1	0.0091	0.0091	0.0091	0.04	0.846	0.01
DC*DC	1	0.1157	0.1157	0.1157	0.50	0.489	0.16
CS*FR	1	2.3408	2.3408	2.3408	10.09	0.006	3.31
CS*DC	1	0.0408	0.0408	0.0408	0.18	0.680	0.06
FR*DC	1	0.0408	0.0408	0.0408	0.18	0.680	0.06
Error	17	3.9427	3.9427	0.2319			5.58
Total	26	70.6563	70.6563				100.00

The results of analysis of variance for material removal rate illustrated in Table 5 represents milling parameter CS is the most significant parameters followed by DC, FR, the interaction of CS and DC, the interaction of CS and FR, interaction of FR and DC. Also, the interaction of CS and CS, FR and FR and DC and DC are insignificant because its p_{values} are larger than 0.05. The CS and DC have the highest % contribution about 37.25% and 27.44%, respectively.

The results of ANOVA for SR illustrated in Table 7 that represents milling parameter CS is the most significant parameter followed by the interaction of CS and CS, the interaction of FR and CS and the rest of all parameters are insignificant because its p_{values} are larger than 0.05. The CS and interaction have the highest % contribution about 63.87% and 20.99%, respectively.

The results of analysis of variance for specific energy are illustrated in Table 8 that represents milling parameter. FR is the most significant parameters following the interaction of FR and FR, and the rest of all parameters are insignificant because their p_{values} are larger than 0.05. The FR and CS have the highest % contribution about 41.31% and 35.29%, respectively.

4 Conclusions

The present work analyzed the performance of green milling process parameters on H_{21} steel for MRR, SR and SE, and multi-objective optimization is performed using an integrated MCDM approach. The main conclusions are as follows:

- The optimal parametric setting of milling process parameters obtained using analysis of mean are: CS 1200 rpm, FR 0.35 mm/tooth and DC 0.6 mm for MRR;

Table 8 Results of ANOVA for specific energy

Sources	DoF	Square SS	Adj.SS	Adj.MS	F_{Val}	P_{Val}	% Con
Regression	9	2179.63	2179.63	242.182	23.75	0.000	92.63%
CS	1	830.42	15.47	15.467	1.52	0.235	35.29%
FR	1	972.11	80.08	80.078	7.85	0.012	41.31%
DC	1	258.17	32.82	32.818	3.22	0.091	10.97%
CS*CS	1	2.81	2.81	2.811	0.28	0.606	0.12%
FR*FR	1	54.16	54.16	54.160	5.31	0.034	2.30%
DC*DC	1	42.05	42.05	42.047	4.12	0.058	1.79%
CS*FR	1	0.80	0.80	0.801	0.08	0.783	0.03%
CS*DC	1	6.28	6.28	6.279	0.62	0.443	0.27%
FR*DC	1	12.83	12.83	12.834	1.26	0.278	0.55%
Error	17	173.35	173.35	10.197			7.37%
Total	26	2352.99	2352.99				100.00%

CS 1200 rpm, FR 0.15 mm/tooth and DC 0.4 mm for SR; CS 400 rpm, FR 0.15 mm/tooth and DC 0.2 mm for SE.

- The result of experimental trial 27 has the highest rank and optimal response using TOPSIS-CRITIC as well as the MOORA-CRITIC method. The optimal levels of process parameters are CS 1200 rpm, FR 0.35 mm/tooth and DC 0.6 mm.
- The optimal parametric combination milling parameters obtained using analysis of mean of preference value as well as assessment value is CS 1200 rpm, FR 0.35 mm/tooth and DC 0.6 mm.
- The results of ANOVA for MRR revealed that CS is the most significant milling process parameter. The CS and DC have the highest % contribution about 37.25% and 27.44%, respectively.
- The results of ANOVA for SR revealed that CS is the most significant milling process parameter. The CS and interaction of CS and CS have the highest % contribution about 63.87% and 20.99%, respectively.
- The results of ANOVA for SE illustrate that FR is the most significant parameters milling process parameter. The FR and CS have the highest % contribution about 41.31% and 35.29%, respectively.
- The results of ANOVA for SE revealed the process parameters that have a larger % of contribution are not significant in comparison with the parameter that has a low % of contribution but is significant.

References

1. Koshy P, Dewes RC, Aspinwall DK (2002) High speed end milling of hardened AISI D2 tool steel (~58 HRC). J Mater Proc Technol 127(2):266–273
2. Lee TS, Lin YJ (2000) A 3D predictive cutting-force model for end milling of parts having sculptured surfaces. Int J Adv Manuf Technol 16(11):773–783
3. Ghosh G, Mandal P, Mondal SC (2019) Modeling and optimization of surface roughness in keyway milling using ANN, genetic algorithm, and particle swarm optimization. Int J Adv Manuf Technol 100:1223–1242
4. Moshat S, Datta S, Bandyopadhyay A, Pal PK (2010) Optimization of CNC end milling process parameters using PCA-based Taguchi method. Int J Eng Sci Technol 2(1):92–102
5. Negrete CC, Nájera JDDC (2019) Optimization of energy consumption and surface roughness in slot milling of AISI 6061 T6 using the response surface method. Int J Adv Manuf Technol 103:4063–4069
6. Kumar S, Jagadish, Singh AK, Kumar N (2020) Multi-objective optimization of CNC drilling parameters on HcHcr steel using Taguchi's based utility concept and GRA-PCA methods. Int J Adv Eng Sci Technol Res (IJAESTR) (special issue ICAME-2020):55–63
7. Sheng P, Srinivasan M (1995) Multi-objective process planning in environmentally conscious manufacturing: a feature-based approach. Ann CIRP 44(1):433–437
8. Li XX, Li WD, He FZ (2018) A multi-granularity NC program optimization approach for energy efficient machining. Adv Eng Softw 115:75–86
9. Wang MY, Chang HY (2004) Experimental study of surface roughness in slot end milling AL2014-T6. Int J Mach Tools Manuf 44:51–57
10. Yan J, Li L (2013) Multi-objective optimization of milling parameters - the tradeoffs between energy, production rate and cutting quality. J Clean Prod 52:462–471
11. Brezocnik M, Kovacic M, Ficko M (2004) Prediction of surface roughness with genetic programming. J Mater Proc Tech 28–36
12. Zain AM, Haron H, Sharif S (2012) Integrated ANN–GA for estimating the minimum value for machining performance. Int J Prod Res 50(1):191–213
13. Dikshit MK, Puri AB, Maity A (2014) Analysis of cutting forces and optimization of cutting parameters in high speed ball-end milling using response surface methodology and genetic algorithm. Procedia Mater Sci 5:1623–1632
14. Raja SB, Baskar N (2012) Application of particle swarm optimization technique for achieving desired milled surface roughness in minimum machining time. Expert Syst Appl 39(5):5982–5989
15. Selaimia AA, Yallese MA, Bensouilah H, Meddour I, Khattabi R, Mabrouki T (2017) Modeling and optimization in dry face milling of X2CrNi18-9 austenitic stainless steel using RMS and desirability approach. Meas J Int Meas Confed 107:53–67
16. Ozcelik B, Oktem H, Kurtaran H (2005) Optimum surface roughness in end milling Inconel 718 by coupling neural network and genetic algorithm. Int J Adv Manuf Technol 27(3–4):234–241
17. Lil B, Tian X, Zhang M (2020) Modeling and multi-objective optimization of cutting parameters in the high-speed milling using RSM and improved TLBO algorithm. Int J Adv Manuf Technol 111:2323–2335
18. zcelik B, Bayramoglu M (2016) The statistical modeling of surface roughness in high-speed flat end milling. Int J Mach Tools Manuf 46(12):1395–1402
19. Lu XH, Hu XC, Wang H, Wang H, Si LK, Liu YY, Gao L (2016) Research on the prediction model of micro-milling surface roughness of Inconel718 based on SVM. Ind Lubr Tribol 68(2):206–211
20. Shaik JH, Srinivas J (2017) Optimal selection of operating parameters in end milling of Al-6061 work materials using multi-objective approach. Mech Adv Mater Modern Processes 3(5)
21. Wang Z, Li L (2021) Optimization of process parameters for surface roughness and tool wear in milling TC17 alloy using Taguchi with grey relational analysis. Adv Mech Eng 13(2):1–8
22. Jafarzadeh E, Movahhedy MR, Khodaygan S, Ghorbani M (2018) Prediction of machining chatter in milling based on dynamic FEM simulations of chip formation. Adv Manuf 6:334–344

23. Kumar GVA, Reddy DVV, Nagaraju N (2018) Multi-objective optimization of end milling process parameters in machining of en 31 steel: application of AHP embedded with VIKOR and WASPAS methods. I-manag J Mech Eng 8(4):39–46
24. Jagadish, Kumar S, Soni DL (2021) Performance analysis and optimization of different electrode materials and dielectric fluids on machining of high carbon high chromium steel in electrical discharge machining. Proc Nalt Acad Sci India Sect A Phys Sci
25. Kumar S, Singh I (2016) The influence of process parameters on cutting speed of WEDM using Taguchi's technique. Int J Emerg Technol (Special Issue on ICRIET-2016) 7(2):332–337
26. Patel Gowdru Chandrashekarappa M, Kumar S, J Pimenov DY, Giasin K (2021) Experimental analysis and optimization of EDM parameters on HcHcr steel in context with different electrodes and dielectric fluids using hybrid Taguchi-based PCA-utility and CRITIC-utility approaches. Metals 11(419)
27. Slebi-Acevedo CJ, Pascual-Muñoz P, Lastra-González P, Castro-Fresno D (2019) Multi-response optimization of porous asphalt mixtures reinforced with aramid and polyolefin fibers employing the CRITIC-TOPSIS based on Taguchi methodology. Materials 12(22):3789
28. Rao RV, Kalyankar VD (2014) Optimization of modern machining processes using advanced optimization techniques : a review. Int J Adv Manuf Technol 73(5–8):1159–1188
29. Kumar S, Ramola IC (2016) Analysis of performance parameter for high carbon high chromium steel on die sinking EDM using Taguchi technique. GGGI J Eng Technol 5(1)
30. Kumar S, Singh I (2016) Analysis of influence of EDM process parameters on tool wear rate using Taguchi technique. CPIE-2016 NIT Jalandhar
31. Kumar S, Jagadish, Ray AK (2021) Multi-objective optimization of CNC drilling parameters on machining of HcHcr Steel Using Taguchi's and Grey relational analysis. In: Advances in thermal engineering, manufacturing and production management. Springer

Optimization of Process Parameters for Improved Surface Finish in Heat Pipe-Assisted Turning of AISI 1040 Steel

M. Parthiban, D. J. Hiran Gabriel, and B. Vaishnavi

Abstract The purpose of this experiment is to find the best settings for getting a satisfactory surface finish while turning AISI 1040 steel with a heat pipe and a coated tungsten carbide insert tool. Using a heat pipe in the machining zone, the heat from the tool chip interface was removed during the turning operation. To establish the best turning settings for a superior surface polish, the Taguchi approach was applied. In Taguchi L9 orthogonal array, parameters like feed speed and depth of cut are considered input parameters and examined. Experiments were conducted with input parameters distributed as on the L9 orthogonal array. The effects of the input parameters on the output parameters of surface finish were analyzed using signal to noise ratio (S/N ratio), and Analysis of Variance (ANOVA) in MINITAB 18 and optimum values of the parameters were obtained. For verification of the experiment, a linear regression equation was established to predict the theoretical values of surface finish. The experiment was repeated with the optimized parameters, and surface roughness was calculated. The obtained values theoretically and experimentally did not have a considerable deviation in the values hence proving the correctness of the conducted experiment.

Keywords Heat pipe-assisted machining · Green manufacturing · Cutting fluid elimination · Sustainable manufacturing

1 Introduction

Turning is the most frequent method for processing circular bars in the industry. To achieve an excellent surface finish in turning operations, selecting the best process parameters is vital. Surface roughness is a metric for assessing the quality of a

M. Parthiban (✉) · D. J. H. Gabriel · B. Vaishnavi
Department of Mechanical Engineering, PSG College of Technology, Coimbatore, Tamil Nadu 641004, India
e-mail: mpn.mech@psgtech.ac.in

B. Vaishnavi
e-mail: bvi.mech@psgtech.ac.in

U. S. Dixit et al. (eds.), *Advances in Forming, Machining and Automation*, Lecture Notes in Mechanical Engineering, https://doi.org/10.1007/978-981-19-3866-5_31

machined component [1]. To achieve such surface roughness, the parametric optimization of the control variables is a vital process. Only a suitable selection of optimal machining settings could achieve the productivity goals of increasing production rate, improving product quality, lowering production costs, and minimizing environmental impact at the same time [2]. The assessment of the best combination of cutting parameters in any machining process necessitates a complete understanding of the process and has long intrigued the interest of researchers and professional engineers. The Taguchi technique is the most efficient method for deciding the proper parameters for improving surface quality. Surface quality is critical in the automotive, aerospace, and bearing sectors since these surfaces must resist intense loading contact during motion [3].

Nowadays, it is also essential in making the machining environment eco-friendly. Increasing pollution levels and restrictions from legislative bodies prohibit the usage of cutting fluids during the machining process. As a response, researchers have made several attempts in the past to assess the feasibility of addressing some of these issues using various approaches, such as the examples given below [4–7].

- Optimization of the cutting parameters
- Microstructure inclusions in the cutting tool
- Changing the cutting tool's geometry
- Changing the cutting mechanism, and
- Incorporation of cooling mechanism to the tool holder.

Heat pipes are heat conducting devices that help transmit heat between two solid contacts using phase transition and thermal conductivity concepts. Heat pipes are passive heat carriers because they do not need an active source of energy to function. In addition, heat pipes do not require any moving or rotating parts to function, which is another benefit [8]. Heat pipes are used primarily when a large amount of heat must be transported, yet there is only limited space available for heat transfer. Heat pipes were once utilized in space applications, but they are now widely utilized to remove heat from microprocessors in mobile phones and laptops. In the manufacturing sector, heat pipe finds its application in metal cutting, die-casting, and injection molding [9]. Heat pipes have been effectively integrated into metal cutting equipment by several researchers. Heat pipes have been installed in the tool holder to reject metal working heat from the inserts to the environment without using an enormous volume of cutting fluid [10, 11].

Debnath et al. [9] investigated the milling of AISI 1040 with heat pipe-aided cooling. For experimental purposes, an electrolytic copper ring-shaped heat pipe using distilled water as the working fluid was designed and manufactured. Using EDM, a custom provision was established in the cutting tool allowing the heat pipe to be included within the tool. Experiments were carried out, and a comparison between dry machining and heat pipe-assisted machining was conducted. The investigations revealed a drop in cutting temperature and lowered cutting forces in the heat pipe-assisted machining process. In addition, the machined components were of higher quality than those produced by dry machining.

Anton et al. [7] used custom-made U-drills with integrated heat pipes to experiment with milling cast iron blocks. The heat pipe was positioned axially on the drill. The findings revealed that the heat pipe helped decrease the temperature at the tool-work contact. A noticeable increase in tool life was seen as a result of the reduced cutting temperature and force. Compared to dry drilling, the quality of the holes produced using heat pipe-assisted U-drill had a better surface quality.

During the hard turning of AISI 4340, Shokrani et al. [12] investigated the performance of heat pipes with little fluid application of coolant. It was discovered that using heat pipes to evacuate heat from the metal cutting zone lowered the cutting temperature. In addition, tool wear and cutting forces were also reduced by 15% and 2.9% during heat pipe-aided machining.

Heat pipe-assisted machining is identical to dry machining; the main difference is that heat pipes are included within the cutting tool. The integrated heat pipe utilizes thermal conduction to evacuate heat from the cutting insert, lowering the cutting temperature during the machining operation. In addition, dry chips were produced through heat pipe-assisted machining, making them easy to recycle. The use of cutting fluid is also eliminated throughout the procedure. As a result, the cost of purchasing and maintaining cutting fluid is eradicated. In addition, there are no toxic residues released into the environment, assuring the safety of both the environment and the operators [6, 7].

It is clear from the literature that heat pipe-assisted machining is a beneficial approach. There are, however, far less attempts to optimize the influencing variables. As a result, the major goal of this study is to investigate heat pipe-aided machining of AISI 1040 steel and uses Taguchi's approach and ANOVA to find the best process factors that impact surface roughness.

2 Materials and Method

2.1 Material

The material selected for the investigation of the heat pipe-assisted turning operation is AISI 1040. The elemental composition of AISI 1040 is listed in Table 1.

Table 1 Elements in AISI 1040 and its weight percentage

Element	Content (%)
Iron, Fe	98.6–99
Manganese, Mn	0.60–0.90
Carbon, C	0.370–0.440
Sulfur, S	≤0.050
Phosphorous, P	≤0.040

Table 2 Structural specifications of the heat pipe

Structural parameters	Measurement (mm)
Evaporator length	10
Adiabatic section length	5
Condenser length	85
The outer diameter of the container	6.1
Container wall thickness	1
Fin width	10
Fin thickness	0.5
Fin length	85

The material was a choice of interest as the material finds its application in the manufacturing of automobile parts such as couplings, crankshafts, and body parts. The workpiece procured for the investigation was 65 mm in diameter and 350 mm in length.

2.2 Heat Pipe Fabrication

The heat pipe is a device that efficiently transports heat across two interfaces by combining the theories of phase transition and thermal conductivity. A hollow cylindrical-shaped heat pipe is fabricated using a 1 mm thick electrolytic copper sheet. The structural specifications of the heat pipe are given in Table 2.

A porous wick structure on the inside walls of the heat pipe facilitates the heat transfer. The wick structure is necessary for capillary forces to flow the working fluid in the heat pipe. Since the heat pipe's working temperature is below 300 °C, distilled water is employed as the working fluid. By utilizing the EDM process, the condenser part of the heat pipe is implanted into the tool. During the hole-making process, it is made sure that the condenser section of the heat pipe is in contact with the insert. The adiabatic section of the heat pipe acts as a transfer region where no heat is exchanged with atmospheric air but the working fluid is carried to the condenser section.

The fins are welded to the heat pipe to increase the area of contact of the condenser section with the atmosphere air since no active cooling is provided for the cooling of the heat pipe. The fabricated heat pipe is shown in Fig. 1.

2.3 Experimental Design

The Taguchi approach was applied to organize the experimental runs by varying the feed, speed, and depth of cut input parameters. The experiment was arranged using an L9 orthogonal array table. The following Table 3 gives the three variables and

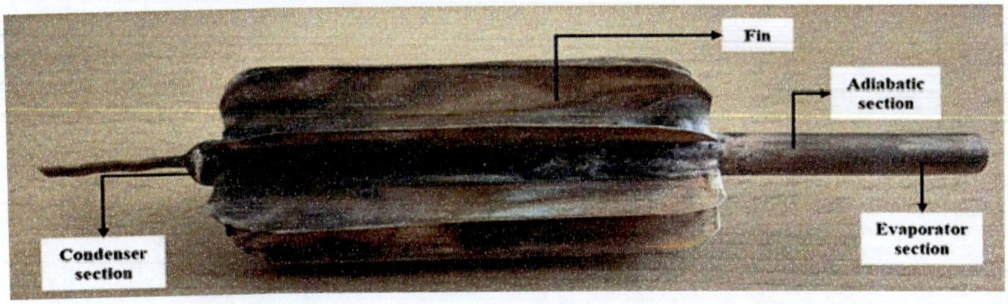

Fig. 1 Fabricated heat pipe

Table 3 Parameters and levels

Parameters	Level 1	Level 2	Level 3	Unit
Cutting speed	80	160	240	m/min
Feed rate	0.10	0.15	0.20	mm/rev
Depth of cut	0.25	0.50	0.75	mm

Table 4 Experimental design

S. No	Cutting speed (m/s)	Feed rate (mm/rev)	Depth of cut (mm)
1	80	0.10	0.25
2	80	0.15	0.50
3	80	0.20	0.75
4	160	0.10	0.50
5	160	0.15	0.75
6	160	0.20	0.25
7	240	0.10	0.75
8	240	0.15	0.25
9	240	0.20	0.50

the three levels of variation of each factor. Table 4 gives the L9 orthogonal array obtained from MINITAB 18 using Taguchi's design of experiments.

2.4 Experimental Setup and Measuring Responses

Turning operations were carried out on a conventional lathe machine Kirloskar Turn master-35. Tungsten carbide Al–Ti–N-coated inserts of specification SNMG 120,408 were used for the experiment. To hold the insert, suitable insert holder PSBNR2525M12 was used in the investigation. The experimental setup is shown in Fig. 2. A K-Type thermocouple is used to measure temperature at the meal working

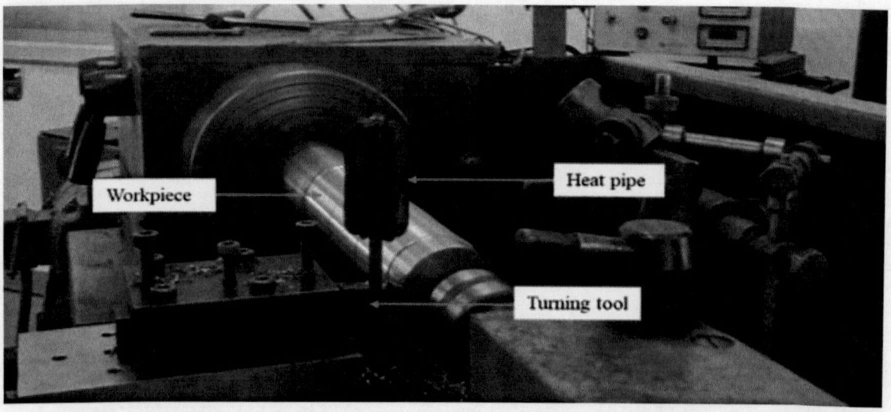

Fig. 2 Experimental setup

zone.

Surface roughness was measured using the surface roughness tester Mahr TR100 surface roughness tester of type Mahr Surf GD 25. Surface roughness was typically measured at four distinct locations across the machined surface, with the probe movement set at 12.5 mm. The average of all the recorded values was used to guide any subsequent manipulations.

3 Results and Discussion

3.1 Comparison of Cutting Temperature and Surface Roughness

The foremost aim of the present evaluation is to determine the optimum turning process parameters for obtaining lower surface roughness during the turning process of AISI 1040 assisted with heat pipe cooling.

The experimental results of the turning process were tabulated in Table 5.

3.2 Determination of Ideal Turning Parameters for Surface Roughness

The foremost objective of the presented investigation is to determine the optimum turning process variables for obtaining lower surface roughness on the turning process of AISI 1040 assisted with heat pipe cooling. According to Taguchi, the goal of parameter design should be to identify the best amounts of the control factors, such

Table 5 Experimental responses

Run	Surface roughness (μm)		Cutting temperature (°C)	
	Dry machining	Heat pipe-assisted machining	Dry machining	Heat pipe-assisted machining
1	0.9038	0.7032	143.53	129.18
2	0.8512	0.7612	147.32	132.59
3	0.9471	0.8388	136.25	122.63
4	1.1143	0.7188	129.62	116.66
5	1.2768	0.8768	138.51	114.96
6	0.9815	0.7917	141.23	117.22
7	0.754	0.6781	117.26	97.33
8	0.7948	0.548	126.87	105.30
9	0.962	0.542	132.65	92.86

that the response variable is resistant to noise-induced variability. Thus, according to Taguchi, there were only three possible objectives for any experiment in optimization.

- Obtain a minimized response (smaller is better)
- Obtain a maximized response (larger is better)
- Achieve the anticipated target value as a response.

Taguchi defined the signal to noise (S/N) ratio, which evaluates the fluctuation in response data, to achieve each of these objectives. The S/N ratio is maximized to enhance the quality characteristic while minimizing the influence of noise components. If the performance measure (y) is replicated n times for each experiment in the specified orthogonal array, the S/N ratio was determined calculated as follows:

i. For minimized response, it is calculated as

$$Y = -10log_{10} + \left[\frac{1}{n}\sum y_i^2\right]db \tag{1}$$

ii. For maximized response, it is calculated as

$$Y = -10log_{10} + \left[\frac{1}{n}\sum y_i^{-2}\right]^{-2} db \tag{2}$$

iii. To achieve the desired target, it is calculated as

$$Y = -10loglog\left(\frac{s^2}{y^2}\right) \tag{3}$$

The presented investigation was aimed to optimize the process parameter for better surface finish. It is always desired that lower values for surface roughness were preferred. "Smaller is better" quality characteristic is used to evaluate the optimum turning process parameters that affect the surface roughness. Table 6 gives the rank of the turning process parameters for surface roughness generated using MINITAB 18 software based on the delta value. Cutting speed, followed by the depth of cut and feed rate, significantly impacts surface roughness, as seen by the S/N ratio. As a result, the S/N ratio plot and mean plot were used to determine the best process parameters for the turning operation, as shown in Figs. 3 and 4. Cutting speed of 240 m/min, feed rate of 0.10 mm/rev, and cut depth of 0.50 mm was the best process parameters for achieving the lowest surface roughness.

The optimal parameters are listed in Table 7 achieving preferable surface roughness.

Table 6 S/N ratio response table for surface roughness

Parameter	Level 1	Level 2	Level 3	Delta	Rank
Cutting speed	2.318	2.013	4.674	2.661	1
Feed rate	3.135	2.912	2.959	0.223	3
Depth of cut	3.437	3.519	2.049	1.47	2

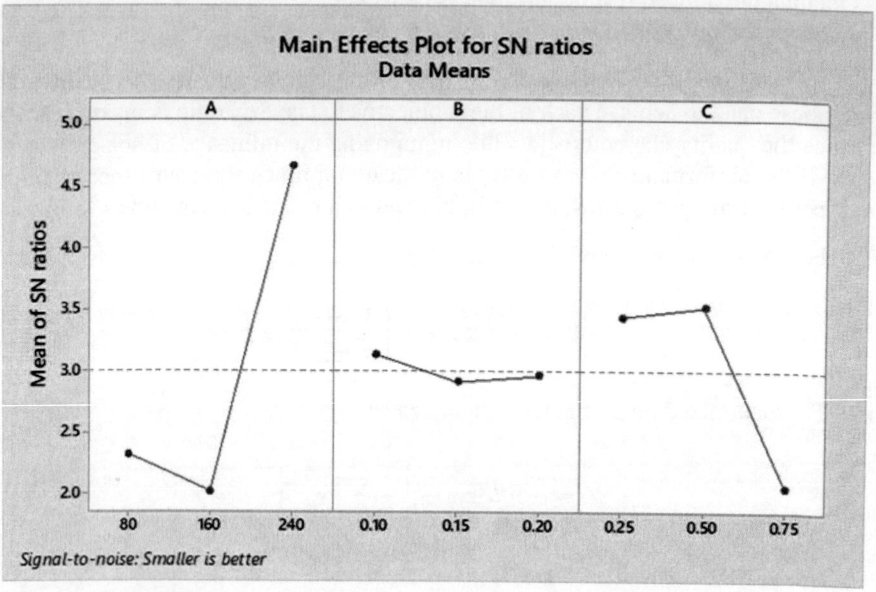

Fig. 3 Main effects plot for S/N ratio for surface roughness

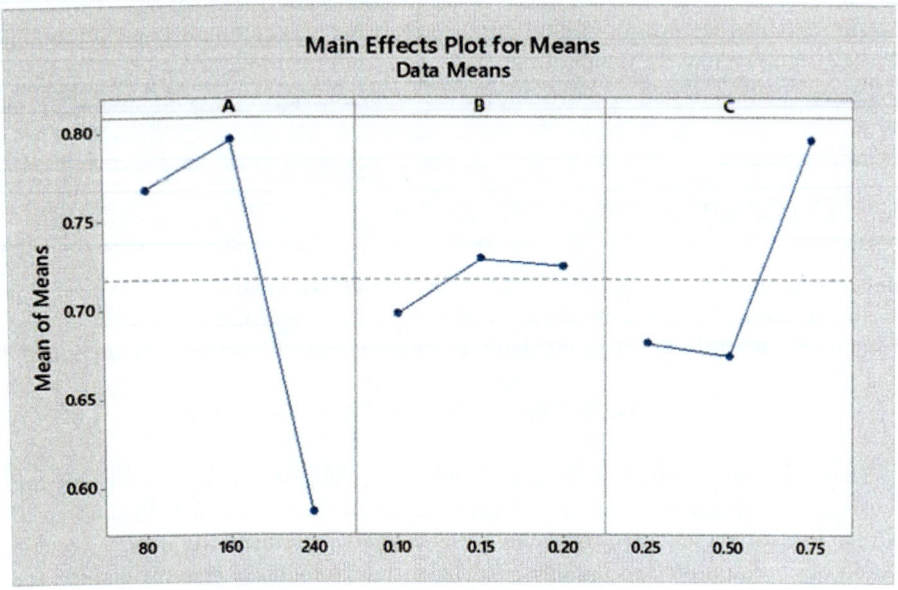

Fig. 4 Main effects plot for means for surface roughness

Table 7 Optimal turning parameters

S. No	Variable	Optimum level
1	Cutting speed (V_c)	240 m/min
2	Feed rate (F_r)	0.10 mm/rev
3	Depth of cut (D)	0.50 mm

3.3 Analysis of Variance (ANOVA)

The relevance of each parameter in the investigation is determined using ANOVA test. The result of ANOVA attained from MINITAB 18 software for surface roughness is given in Table 8.

Table 8 ANOVA table for surface roughness

Source	DF	Seq. SS	Adj. SS	Adj. MS	Contribution (%)
Cutting speed	2	0.077294	0.077294	0.038647	70.59
Feed rate	2	0.001722	0.001722	0.000861	6.57
Depth of cut	2	0.027787	0.027787	0.013894	20.38
Error	2	0.00269	0.00269	0.001345	2.46
Total	8	0.109493			100.00

The obtained results show that cutting speed has a significant contribution of 70.59%, followed by the depth of cut at 20.38%, and feed rate at 6.57%. The error has a contribution of just 2.46% to the surface roughness, which is less than 15%.

3.4 *Confirmation Test*

Once the optimal process parameters have been identified, a confirmation experiment is run with them. A linear regression model is used to compute the anticipated surface roughness (SR). MINITAB 18 yielded the linear regression equation shown below.

$$\mathrm{SR} = 0.743 - 0.001132\,\mathrm{Vc} + 0.268\,\mathrm{Fr} + 0.228\,\mathrm{D} \tag{4}$$

The optimized parameters obtained from the signal to noise ratio function were substituted in the above equation to obtain a theoretical value of 0.5551 micron. For confirmation, the surface roughness was derived experimentally using a combination of optimum process parameters. The surface roughness results obtained for optimum conditions were 0.5540 micron. Theoretically, predicted value using the linear equation was very close to the experimental value.

4 Conclusion

Turning experiments were conducted on AISI 1040 assisted with heat pipe for cooling purposes. The Taguchi method was used to investigate the impact of turning process variables under the investigation. The following are some of the essential conclusions derived from the findings: Cutting speed of 240 m/min, feed rate of 0.1 mm/rev, and cut depth of 0.50 mm was determined to be the optimum values for the variables to obtain good quality of surface finish as a response. Cutting speed had a more significant influence (70.59%) on surface roughness than continuing depth of cut (20.38%) by and feed rate (6.57%), according to an ANOVA study.

References

1. Kalpakjian S (2001) Manufacturing engineering and technology, Pearson Education India
2. Sharma VS, Dogra M, Suri N (2009) Cooling techniques for improved productivity in turning. Int J Mach Tools Manuf 49:435–453
3. Chan C, Lee W, Wang H (2013) Enhancement of surface finish using water-miscible nano-cutting fluid in ultra-precision turning. Int J Mach Tools Manuf 73:62–70
4. Arulraj GA, Wins KLD, Raj A (2014) Artificial neural network assisted sensor fusion model for predicting surface roughness during hard turning of H13 steel with minimal cutting fluid application. Proc Mater Sci 5:2338–2346

5. Liang L, Liu X, Li X-Q, Li Y-Y (2015) Wear mechanisms of WC–10Ni 3 Al carbide tool in dry turning of Ti6Al4V. Int J Refract Metal Hard Mater 48:272–285
6. Adler D, Hii W-S, Michalek D, Sutherland J (2006) Examining the role of cutting fluids in machining and efforts to address associated environmental/health concerns. Mach Sci Technol 10:23–58
7. Anton S, Andreas S, Friedrich B (2015) Heat dissipation in turning operations by means of internal cooling. Proc Eng 100:1116–1123
8. Das SR, Dhupal D, Kumar A (2015) Experimental investigation into machinability of hardened AISI 4140 steel using TiN coated ceramic tool. Measurement 62:108–126
9. Debnath S, Reddy MM, Yi QS (2016) Influence of cutting fluid conditions and cutting parameters on surface roughness and tool wear in turning process using Taguchi method. Measurement 78:111–119
10. Ghosh S, Rao PV (2015) Application of sustainable techniques in metal cutting for enhanced machinability: a review. J Clean Prod 100:17–34
11. Robinson Gnanadurai R, Varadarajan AS (2016) Investigation on the effect of cooling of the tool using heat pipe during hard turning with minimal fluid application. Eng Sci Technol Int J 19(3):1190–1198
12. Shokrani A, Dhokia V, Newman ST (2012) Environmentally conscious machining of difficult-tomachine materials with regard to cutting fluids. Int J Mach Tools Manuf 57:83–101

Particulate Damping Effects on Tribological Properties During Hard Boring

G. Lawrance, P. Sam Paul, D. S. Ebenezer Jacob Dhas, D. Raskin Benny, and J. Hari Vignesh

Abstract In the boring process, the productivity in the manufacturing sector is affected mainly due to its overhanging length. In this study, an attempt is made to enhance the tribological properties using a particulate damper. A particulate damper is offered using particles inside the tool which dissipate energy through friction and inelastic collisions. A 27 run Taguchi technique was conducted in a cutting experiment with varying particle size, particle material, and particle fillings and followed by an analysis of variance on tribological properties. The results were compared to conventional boring tools and found the particle damper can enhance the tribological properties effectively.

Keywords Particulate damper · Boring method · Tribological properties · Taguchi technique · Analysis of variance

1 Introduction

The primary goal of machining operations is to achieve a superior material removal rate within a short period while retaining acceptable quality [1]. To expand the pre-drilled holes, a boring method is utilized. Tool vibration is a major problem since it has an impact on the product's quality owing to the thin proportion of the boring instrument [2]. It occurs due to the interaction between the waves produced on the workpiece surface by the earlier cut and the cutting forces [3]. Using a passive vibration mechanism in the boring method increases damping ability, according to Daghini et al. [1]. Ramesh et al. [4] recognized inactive damping methods as a budget-saving and are a widely utilized approach for vibration reduction, and projected an ANN model in a variety of machining situations. Transversal particulate damping was also proposed by Popplewell and Smergicil [5]. The particulate damping approach is based on the impact damper, which involves placing small weights into a crater in a pulsating assembly. Within the hollow walls of the exciting assembly, the elements

G. Lawrance (✉) · P. S. Paul · D. S. E. J. Dhas · D. R. Benny · J. H. Vignesh
Department of Mechanical Engineering, Karunya Institute of Technology Sciences, Coimbatore 641 114, India
e-mail: lawrancedevaraj@gmail.com

U. S. Dixit et al. (eds.), *Advances in Forming, Machining and Automation*, Lecture Notes in Mechanical Engineering, https://doi.org/10.1007/978-981-19-3866-5_32

transfer energy and slide brushing [6]. Senthilkumar et al. [7] investigated the influence of lead and copper elements in particulate dampers and discovered that the surface roughness was decreased. Devaraj et al. [8] investigated the influence of zinc, aluminum, copper, and silicon particulates with different masses on vibration and surface irregularity. Khatake and Nitnaware [9] used an inactive impediment method to reduce vibration and surface roughness in drilling bars with various overhang lengths. Olson [10] investigated particulate restraining on a cantilever beam and used energy–displacement relationships to compute the force between particulates. The effect of a particle damper on tribological properties during the boring of AISI4340 steel with a 45 HRC was investigated in this study. The tribological properties consist of surface roughness and tool wear. The boring tool's cavity is filled with small balls, the diameter of which is lesser than the cavity diameter. As the boring instrument vibrates, the particle inside dissipates energy by colliding with other particles and the cavity wall, resulting in successful damping. A serious cutting experiment was conducted on varying sphere size, sphere material, and filling of spheres in the boring cavity which is followed by a signal-to-noise ratio and analysis of variance.

2 Materials and Methods

A boring instrument with the design S25T PCLNR12F3 was utilized in this research. The workpiece is made of AISI 4340 steel (45 HRC), with an outer diameter of 80 mm, a diameter of 40 mm, and a length of 100 mm [11]. A boring tool with a particulate damper is shown in Figs. 1 and 2. Electrical discharge machining is used to bore the boring tool, which has a diameter of 10 mm and a length of 200 mm. The particulates were placed in the drilling cavity and secured with a screw.

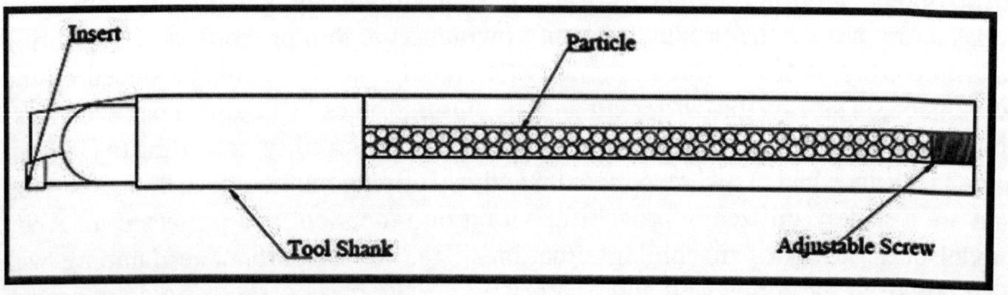

Fig. 1 Boring bar with particulate damper

Fig. 2 Photograph of the boring bar with particulate damper

3 Experimental Method

Kirloskar turn master-35 lathe was used for 27 runs; Mitutoyo—SJ 210 was employed to assess surface coarseness, and toolmakers compound microscope was utilized to examine the tool wear. To improve machining parameters, a conventional group of particulate damper parameters was developed, including particulate filling in the boring cavity (T) (tightly filled, loosely filled, and partially filled), particulate dimensions (S) (2 mm, 4 mm, and 6 mm), and particulate material (R) (chromium steel, mild steel, and stainless steel). Table 1 illustrates a particulate damper with three levels of variable parameters. The feed rate was set to 0.05 mm/min, the machining speed to 110 m/min, and the cut depth to 0.6 mm. Machining trials took two minutes in a dry machining environment.

Table 1 Impact particulate damper designed with three-level parameters

S. No	Parameters	Level 1	Level 2	Level 3
1	Particulate filling	T_1 (tightly filled)	T_2 (loosely filled)	T_3 (partially filled)
2	Particulate material	R_1 (chromium steel)	R_2 (mild steel)	R_3 (stainless steel)
3	Particulate dimensions	S_1 (2 mm)	S_2 (4 mm)	S_3 (6 mm)

3.1 Design of Experiments

To choose the levels of machining parameters, analyze the data, and forecast the best machining parameters, Taguchi parameter design (L_{27}) is used. The Taguchi technique makes use of a feature criterion (S/N ratio) that may be determined employing equality (1).

$$\eta - 10\log(\text{M.S.D.}) \tag{1}$$

The mean square deviation (M.S.D.) is used to match the outputs in the current investigation as a minima quality feature, according to Yang and Tarng [12].

$$\text{M.S.D} = \frac{1}{M}\sum_{i=1}^{m} S_i^2 \tag{2}$$

The following equation is used to validate and forecast the best quality characteristic.

$$\eta' = \eta_m + \sum_{i=1}^{0} (\eta_i - \eta_m) \tag{3}$$

The influence of contributing factors on the machinability of the boring method is determined using ANOVA. ANOVA is often used to determine the important machining factors that impact the material's machinability. The total sum of squared deviations is employed to rank the machining parameter and is calculated using the equality below.

$$\text{SS}_T = \sum_{i=1}^{n} (\eta_i - \eta_m)^2 \tag{4}$$

4 Result and Discussion

For best results, the particulate filling on the boring cavity should be kept at Level 3 (partially filled), particulate dimensions at Level 1 (2 mm), and particulate material at Level 3 (stainless steel), as shown in Figs. 3 and 4. Table 2 shows tribological property experimental results and signal-to-noise ratios. Minitab software was used to calculate the signal-to-noise ratio for surface roughness and tool wear. Figures 3 and 4 indicate that particulate filling on boring cavities should be retained at Level 3 (partially filled), particulate dimensions at Level 1 (two millimeters), and particulate material at Level 3 (Stainless steel) for optimum results.

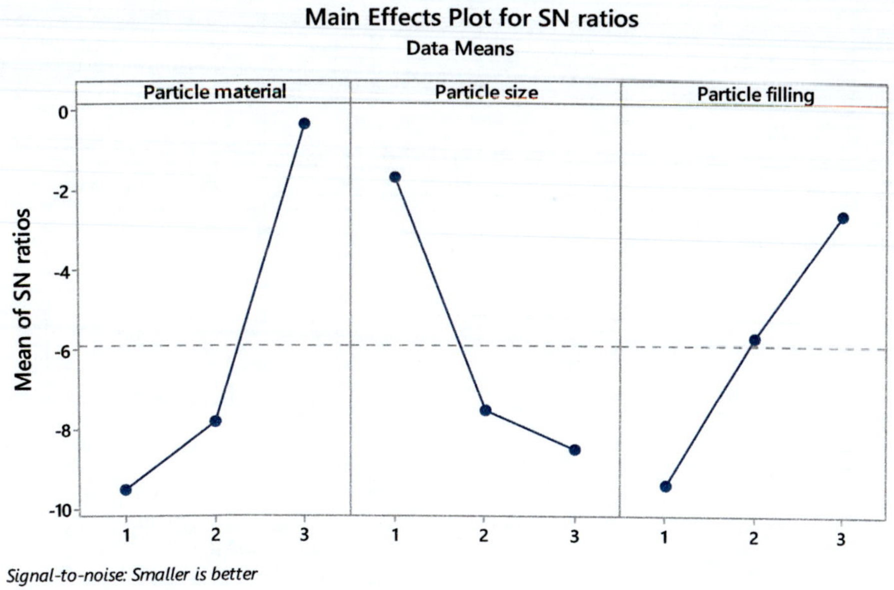

Fig. 3 Surface roughness—Mean S/N ratios

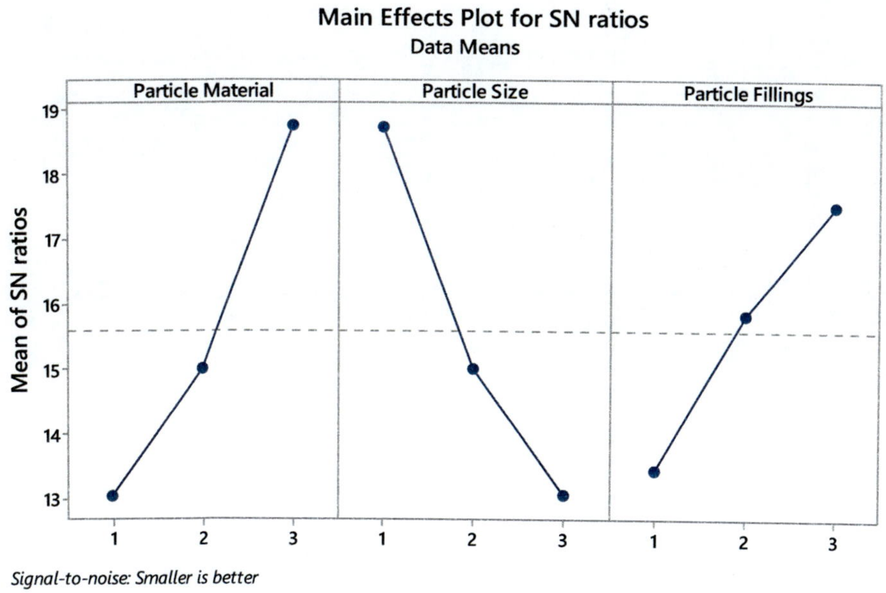

Fig. 4 Tool wear—Mean S/N ratios

Table 2 Results from experiments and the S/N ratios

Standard order	Factor			Surface roughness (μm)	Tool wear (mm)	S/N ratio for surface roughness (dB)	S/N ratio for tool wear (dB)
	Particulate material	Particulate dimensions	Particulate filling				
1	R_1	S_1	T_1	2.83	0.218	−9.04	13.23
2	R_1	S_1	T_2	1.985	0.149	−5.96	16.54
3	R_1	S_1	T_3	1.846	0.141	−5.32	17.02
4	R_1	S_2	T_1	3.7	0.2855	−11.36	10.89
5	R_1	S_2	T_2	3.245	0.2465	−10.22	12.16
6	R_1	S_2	T_3	3.184	0.1985	−10.06	14.04
7	R_1	S_3	T_1	4.353	0.306	−12.78	10.29
8	R_1	S_3	T_2	3.224	0.287	−10.17	10.84
9	R_1	S_3	T_3	3.519	0.244	−10.93	12.25
10	R_2	S_1	T_1	2.317	0.1635	−7.30	15.73
11	R_2	S_1	T_2	1.568	0.1065	−3.91	19.45
12	R_2	S_1	T_3	1.423	0.1075	−3.06	19.37
13	R_2	S_2	T_1	3.311	0.254	−10.40	11.90
14	R_2	S_2	T_2	2.691	0.192	−8.60	14.33
15	R_2	S_2	T_3	2.672	0.137	−8.54	17.27
16	R_2	S_3	T_1	3.972	0.3135	−11.98	10.08
17	R_2	S_3	T_2	2.783	0.2265	−8.89	12.90
18	R_2	S_3	T_3	2.449	0.2005	−7.78	13.96
19	R_3	S_1	T_1	1.574	0.117	−3.94	18.64
20	R_3	S_1	T_2	0.637	0.089	3.92	21.01

(continued)

Table 2 (continued)

Standard order	Factor			Surface roughness (μm)	Tool wear (mm)	S/N ratio for surface roughness (dB)	S/N ratio for tool wear (dB)
	Particulate material	Particulate dimensions	Particulate filling				
21	R_3	S_1	T_3	0.105	0.042	19.58	27.54
22	R_3	S_2	T_1	2.969	0.1655	−9.45	15.62
23	R_3	S_2	T_2	1.668	0.1165	−4.44	18.67
24	R_3	S_2	T_3	0.549	0.0995	5.21	20.04
25	R_3	S_3	T_1	2.624	0.188	−8.38	14.52
26	R_3	S_3	T_2	1.437	0.149	−3.15	16.54
27	R_3	S_3	T_3	1.343	0.153	−2.56	16.31

The fractional influences of particulate material, particulate dimensions, and particulate filling in particulate dampers for surface roughness are 43.32%, 28%, and 23%, respectively, as shown in Table 3. When associated with particulate dimensions and filling, it is apparent that particulate material has the most important role in decreasing surface roughness.

Table 4 shows that the proportional influence of particulate material, particulate dimensions, and particulate filling in particulate dampers for tool wear is 38.55%, 36.72%, and 20.26%, respectively. The error rate for surface roughness is 5.68%, and tool wear is 4.45%, being quite little, according to ANOVA analysis. According to Table 5, comparison experiments for improved parameters with and without particulate damper revealed that surface roughness decreased by 96%, and tool wear subsided by 86%.

Table 3 Analysis of variance—surface roughness

Source	DoF	Seq SS	Adj SS	Adj MS	F	P	Contribution%
Particulate material	2	13.0433	13.0433	6.5216	76.17	0.000	43.32
Particulate dimensions	2	8.4260	8.4260	4.2130	49.20	0.000	28
Filling of particulates	2	6.9218	6.9218	3.4609	40.12	0.000	23
Error	20	1.7125	1.7125	0.0856			5.68
Total	26	30.1036					

Table 4 Analysis of variance—tool wear

Source	DoF	Seq SS	Adj SS	Adj MS	F	P	Contribution%
Particulate material	2	0.051568	0.051568	0.025784	86.49	0.000	38.55
Particulate dimensions	2	0.049126	0.049126	0.024563	82.40	0.000	36.72
Filling of particulates	2	0.027114	0.027114	0.013557	45.48	0.000	20.26
Error	20	0.005962	0.005962	0.000298			4.45
Total	26	0.133769					

Table 5 Comparison experiments

S. No	Tribological properties		With damper	Without damper	%Reduction
1		Surface roughness (μm)	0.105	3.2	96
2		Tool wear (mm)	0.042	0.31	86

As compared to other passive dampers, the particulate damper has a greater influence on damping. The first is friction between particulates, which is influenced by the coefficient of friction, and the second is particulates impact force, which is influenced by the density and hardness of the constituent particulates. ANOVA analysis of Tables 3 and 4 revealed that particulate material is the most important characteristic, followed by particulate dimensions and particulate fillings. Particulates made of mild steel, chromium steel, and stainless steel are reflected in this study. The hardness and density of the particulates' substance have an impact on their damping properties [13]. The absorption of vibratory energy by losses that arise during the impact of granular particulates that travel freely inside the boundary of a cavity in a boring tool is known as particulate damping. If the particulate's hardness and density are high, the vibrating structure's energy is dissipated more efficiently by particulate contacts, and hence, surface roughness is minimized during the boring method. When tool friction is reduced, the bouncing of the tool in and out of the workpiece is reduced, and irregularities on the surface are reduced, resulting in a smoother surface finish [14].

ANOVA analysis of Tables 3 and 4 revealed that particulate dimension is the most important feature after particulate material. Particulate dimensions of 2 mm, 4 mm, and 6 mm are investigated in this study. Smaller spheres have higher energy degeneracy owing to rubbing, resulting in lower surface roughness and tool wear, as seen in Figs. 3 and 4 [15]. When the particulates' diameter is greater than half the hole, energy is dissipated further due to collision, resulting in more surface roughness and tool wear. The maximum hole gap of about 8 mm is obtained, while the 2 mm particulates were used in the experiments. As the mass of the particulates is smaller and the distance between the particulates and the cavity wall is smaller, the damping effect of the particulates increases, increasing the impact momentum transfer.

The particulate damper is a chamber filled with particulates that disperse energy by friction and impact. It is also necessary to consider the number of particulates enclosed within the cavity. ANOVA analysis of Tables 3 and 4 revealed that particulate fillings are the least important factor when compared to particulate dimensions and particulate material. Particulate fills such as firmly filled (100%), loosely filled (50%), and partly filled (80%) are taken into account in this study. Figures 3 and 4 illustrate that when particulates are tightly filled, there is no energy degeneracy between them, whereas the minute they are loosely filled, there is more energy degeneracy between them, resulting in a higher vibration. The energy degeneracy from one particulate to the subsequent is exact when the particulates are partially compressed, which enhances the tribological characteristics.

5 Conclusions

The effect of a particulate damper on tribological properties during boring of hardened AISI 4340 steel is studied. The experimental results, S/N ratio, and ANOVA analysis show the effect of particle damper on machining parameters. From the present investigation, we concluded:

1. A particulate damper in a boring tool with stainless steel is partially filled with smaller particles to accomplish enhance tribological properties.
2. When compared to the conventional boring bar, particulate dampers are more effective.

References

1. Daghini L et al (2009) Design, implementation and analysis of composite material dampers for turning operations. World Acad Sci Eng Technol 53:613–620
2. Lawrance G et al (2019) Suppression of tool vibration in boring process: a review. J Inst Eng (India): Ser C 100:1053–1069
3. Lawrance G et al (2017) Attenuation of vibration in boring tool using spring controlled impact damper. Int J Interact Des Manuf (IJIDeM) 11:903–915
4. Ramesh K et al (2013) Investigation of chatter stability in boring tool and tool wear prediction using neural network. Int J Mater Prod Technol 46:47–70
5. Popplewell N, Semercigil SE (1989) Performance of the bean bag impact damper for a sinusoidal external force. J Sound Vib 133:193–223
6. Fowler BL et al (2000) Effectiveness and predictability of particle damping. In: Smart structures and materials 2000: Damping and isolation 3989:356–367
7. Senthilkumar M et al (2011) A case study of vibration control in a boring bar using particle damping. Int J Sci Technol Eng 3:177–184
8. Devaraj S et al (2014) Surface quality enrichment using fine particle impact damper in boring operations. Int J Res Eng Technol 3:531–535
9. Khatake P, Nitnaware PT (2013) Vibration mitigation using passive damper in machining. Int J Mod Eng Res 3:3649–3652
10. Olson SE (2003) An analytical particle damping model. J Sound Vib 264:1155–1166
11. Lawrance G et al (2019) Influence of magnetorheological elastomer on tool vibration and cutting performance during boring of hardened AISI4340 steel. J Mech Sci Technol 33:1555–2156
12. Yang WH, Tang YS (1998) Design optimization of cutting parameters for turning operations based on the Taguchi method. J Mater Proc Technol 84:122–129
13. Sathishkumar BK et al (2014) Impact of particle damping parameters on surface roughness of bored surface. Arab J Sci Eng 39:7327–7334
14. Sam Paul P et al (2018) Effectiveness of particle and mass impact damping on tool vibration during hard turning process. Proc Inst Mech Eng Part B J Eng Manuf 232:776–786
15. Diniz AE et al (2019) Evaluating the use of a new type of impact damper for internal turning tool bar in deep holes. Int J Adv Manuf Technol 101:1375–1390

Theoretical and Experimental Study on Forces in Ball End Magnetorheological Finishing Process

Z. Alam, D. A. Khan, F. Iqbal, and S. Jha

Abstract Ball end magnetorheological finishing (BEMRF) is a newly developed nanofinishing process employing a magnetically stiffened abrasive ball to finish the surfaces. Precise control over the stiffness makes this process capable of finishing the material up to nanometer level. It relies on magnetorheological polishing (MRP) fluid which is a composition of micron-sized particles of abrasive and ferromagnetic substance dispersed in a fluid medium. Knowledge of forces involved in finishing is essential for understanding the process mechanism. In the BEMRF process, normal and shear forces constitute the finishing forces. In this study, theoretical models to predict both these forces are proposed. The finishing forces are modeled considering the magnetic flux density variation in the working gap. An experimental study on the finishing forces is also carried out based on the following three process parameters: current, working gap, and tool speed. The experiments are performed on a three-axis CNC BEMRF machine with mild steel as the workpiece material. During experimentation, the selected process parameters are varied over a certain range and a dynamometer is used for recording the normal and shear forces acting on the workpiece surface during the finishing process in real time. For lower finishing current values, the results show good agreement between the modeled theoretical values and experimentally recorded data.

Keywords Magnetorheological · Nanofinishing · Finishing force · Polishing fluid

Z. Alam (✉)
Department of Mechanical Engineering, IIT (ISM), Dhanbad, India
e-mail: zafar@iitism.ac.in

D. A. Khan
Department of Mechanical Engineering, NIT, Hamirpur, India
e-mail: dilshad@nith.ac.in

F. Iqbal
Institute for Integrated Micro and Nano Systems, University of Edinburgh, Edinburgh, UK
e-mail: faiz.iqbal@ed.ac.uk

S. Jha
Department of Mechanical Engineering, IIT, Delhi, India
e-mail: suniljha@mech.iitd.ac.in

U. S. Dixit et al. (eds.), *Advances in Forming, Machining and Automation*, Lecture Notes in Mechanical Engineering, https://doi.org/10.1007/978-981-19-3866-5_33

1 Introduction

Nanofinishing processes, among all types of finishing processes, are crucial due to the requirement of precise controlling of finishing parameters to control the finishing forces. The magnetorheological finishing process is one of the nanofinishing processes in which the finishing forces can be controlled precisely to achieve a very high level of surface finish. Polishing mediums exhibiting magnetic behavior [1] are utilized in this process which gets the form of the abrasive tool when magnetized by some magnetic field sources such as an electromagnet. Specifically, the magnetic abrasive fluid utilized in magnetorheological finishing is termed magnetic abrasive polishing (MRP) fluid [2]. The MRP fluid has a different composition based on desired finishing level, physical and mechanical properties, and chemical affinity of the workpiece material to be finished. In general, MRP fluid is composed of constituting tiny particles having magnetic properties such as carbonyl iron particles (CIPs), granules of hard non-magnetic materials (abrasives), and a carrying substance such as oil or water.

Finishing of the lens, metallic mirrors, electronic interconnects, and medical implants are a few examples that require a high level of surface finishing for their proper and efficient functioning. Finishing processes help in enhancing the service life of components, especially under fluctuating loads by removing the stress concentrators such as indents and hairline cracks. Due to the peculiar features of magnetorheological finishing, it is gaining high attention from industries and research institutions involved in precision finishing. Almost all counties of the world including India are working extensively on achieving a high level of finishing using the concept of magnetorheological finishing. They have extended their research on the finishing of different types of materials, components of different configurations, and future possibilities of achieving a high level of finishing in comparatively less time.

Industrial needs of nanofinishing of different configurations of workpieces have evolved different variants of the magnetorheological finishing. One such process is ball end magnetorheological finishing (BEMRF) which was developed for the finishing of complex configurations, finishing off the selective desired area of the surface, and deep cavities in the workpieces [3]. In the published literature, it is found that BEMRF has exhibited its capability to finish different types of materials such as steel [4], copper [5, 6], silicon [7], and PLA [8].

2 Ball End Magnetorheological Finishing Process

Figure 1 illustrates the working principle of BEMRF and the associated finishing forces in BEMRF acting upon workpiece surface during finishing. In BEMRF, the magnetic pole is generated at the tip of the vertical spindle by energizing the electromagnet placed around the vertical spindle [9]. The stiffness of the MRP fluid between the spindle tip and the workpiece surface is decided by the magnetic flux

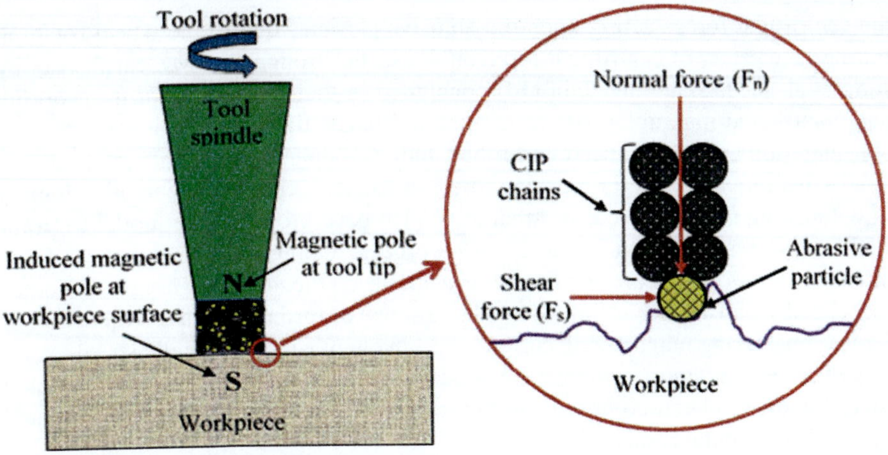

Fig. 1 Schematic of BEMRF process

density generated in the gap. In the case of the ferromagnetic workpiece, the spindle tip induces an opposite magnetic pole on the workpiece surface which enhances the magnetic flux density in the gap between the tooltip and workpiece surface. This is contrary to the finishing of the non-magnetic workpiece where the second pole is missing and the magnetic flux lines flair out at the workpiece surface. Due to the magnetic gradient between the spindle tip and the workpiece surface, the tiny magnetic particles (CIPs) attract toward the spindle tip in the region of higher magnetic flux density, and the abrasive particles cluster near the workpiece surface, i.e., at the region of lower magnetic flux density which ensures the availability of most of the abrasive content on the workpiece surface. The tiny magnetic particles (CIPs) work as a binding substance to grip the abrasive particles in between chains of magnetic particles. The stiffness of MRP fluid imparts two basic forces on abrasive particles in finishing, one is a normal force that comes from magnetic particles (CIPs), and the other is a shear force which is the result of the yield strength of the magnetized MRP fluid [10]. The yield strength of the MRP fluid inhibits the deformation of the MRP fluid under the action of rotation of the spindle while finishing.

The mechanism of material removal and the physics of the interaction of abrasive with the CIPs and workpiece surface can be understood by the nature and effect of finishing forces involved in the process. As the new variants of the magnetorheological finishing evolved, the researchers performed the analytical and simulation studies and have developed the mathematical model to analyze the effect of different process parameters, but in the case of BEMRF, there is a scarcity of such studies in broad-spectrum considering various aspects of the finishing. DeGroote et al. [11] reported that the drag forces increase linearly on increasing the abrasive concentration in the MRP fluid. Shorey et al. [12] reported that on increasing the abrasive concentration the drag force decreases which are contrary to findings of DeGroote et al. [11] in his studies. They also reported that these results are influenced by the centrifugal forces acting upon the abrasive particles due to the wheel rotation. Further, the effect of centrifugal force on the surface roughness was carried out by Jung et al. [13]. Spara and Jain [14] concluded in their research that the working gap contributed maximum toward normal and tangential forces followed by CIPs concentration and wheel speed. In further studies, Sidpara and Jain [15] have developed a theoretical model for finishing forces associated with the magnetorheological flow finishing (MRFF) process. Iqbal et al. [16] have proposed the model for transient roughness reduction in the BEMRF process taking normal and shear force into consideration. The present literature available on the modeling aspects of forces associated with BEMRF is limited and has a scope of further studies on modeling of forces considering magnetizing current, working gap, and tool spindle speed.

In the present study, a theoretical model for finishing forces, both normal and shear forces, has been proposed and the results of theoretical studies are validated by the experimental results.

3 Modeling of Forces

In the BEMRF process, normal and shear forces constitute the finishing forces. These forces are dependent on magnetic flux density and are responsible for providing the cutting action to the active abrasive particles, i.e., the bottom-most layer of the abrasives that touches the workpiece surface and participates in dislodging/smoothening the roughness peaks. The normal force provides the required thrust to the active abrasive particle to indent the workpiece surface, and the shear force gives the strength to the indented active abrasive to dislodge the roughness peaks. In this way, these forces simultaneously execute the finishing action.

3.1 Normal Force

As stated earlier, the normal force is a function of magnetic flux density in the working gap. It not only depends upon the magnitude but the gradient of flux density also influences the normal force. Since the magnetic flux density plays a major role in influencing the normal force, it becomes necessary to model the flux density first. For the BEMRF process, Alam and Jha [10] modeled the axial variation of flux density $B(z)$ in the working gap as

$$\vec{B}(\mathrm{z}) = \frac{\mu_0 \times m_{\mathrm{tool}}}{4\pi z^2} + \frac{\mu_0 \times m_{wp}}{4\pi (t - z)^2} \tag{1}$$

where μ_o is free space's absolute permeability, z is the distance from the tip of the tool measured in the axial direction, t is the working gap, and m_{tool} and m_{wp} are the pole strength of the tooltip and workpiece, respectively.

When the current to the electromagnet-based BEMRF tool is switched on, the electromagnet energizes and generates the magnetic field responsible for magnetizing the tool spindle as well as the MRP fluid. Upon magnetic excitation, the CIPs polarize magnetically and arrange themselves in chain form along the magnetic field lines. Assuming the chain structure near the workpiece surface to be a half body-centered cubic (BCC) where a single active abrasive is surrounded by four CIP, the normal force [17, 18] per abrasive F_{n_abr} is given by

$$F_{n\_\text{abr}}(z) = \frac{m\chi_{m\_\text{CIP}}}{\mu_o} B(z) \frac{dB(z)}{dz} \tag{2}$$

where m and χ_{m_CIP} are CIP's mass and magnetic susceptibility, respectively.

The diameter of the abrasive and CIP used here is assumed to be 19 and 18 μm, respectively. For this size of abrasive and CIP, the edge of the BCC unit cell (Fig. 2) is calculated as 20.785 μm. So for a circular tooltip of 10 mm diameter, a total number of active abrasives $N_{\text{act\_abr}}$ are (area of tooltip/area of a BCC unit cell) 181,872. Hence, the total normal force F_n during the BEMRF process is shown by

$$F_n = N_{\text{act\_abr}} \times F_{n\_\text{abr}} \tag{3}$$

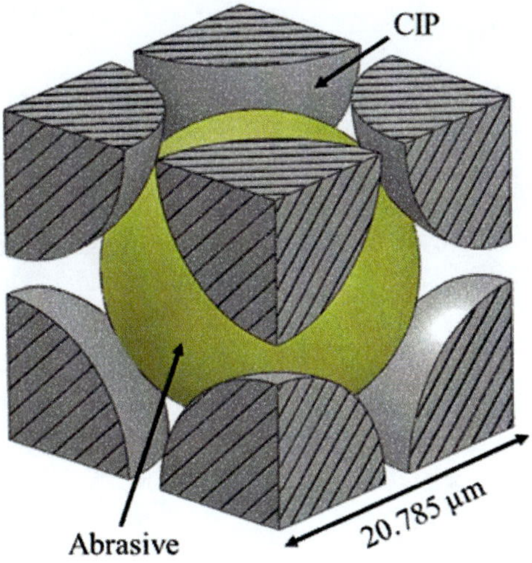

Fig. 2 BCC structure of energized MRP fluid with abrasive at center and CIP at corners

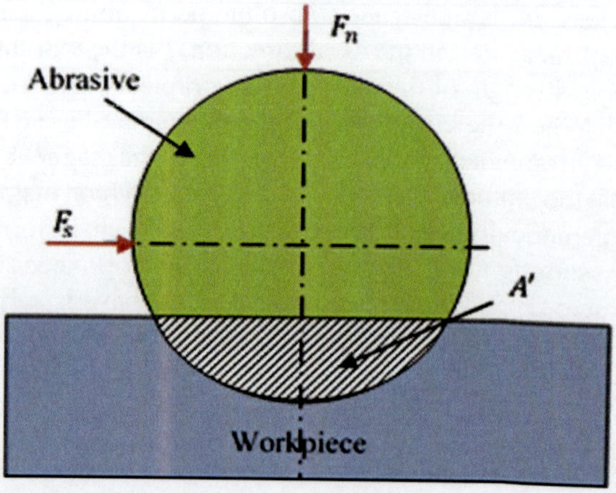

Fig. 3 Abrasive indenting the workpiece surface

3.2 *Shear Force*

The role of shear force in the BEMRF process is to give the required strength to the indented abrasive particle to remove/smoothen the roughness asperities. Two factors influence the shear force in the BEMRF process. One is the yield strength of the polishing fluid, and the other is the projected area of the non-indented part of the active abrasive particle as shown in Fig. 3. Therefore, the shear force applied by an active abrasive particle F_{s_abr} can be shown by

$$F_{s\_abr} = (A - A^{'})\tau_y \tag{4}$$

where A is the active abrasive particle's total projected area, τ_y is the polishing fluid's yield strength, and A' is the projected area of the indented part of the active abrasive particle.

The yield strength of the MRP is a function of magnetic flux density. Using a magneto-rheometer, the yield strength's variation at different values of magnetic flux density was experimentally found and a quadratic equation is fitted as shown by

$$\tau_y = -0.017B^2 + 50.182B - 2064 \tag{5}$$

where B is in mT and τ_y in Pascal. Hence, the total shear force F_s during the BEMRF process is given by

$$F_s = N_{act\_abr} \times F_{s\_abr} \tag{6}$$

4 Experimentation

Figure 4 shows an experimental setup in which the BEMRF tool is mounted in the vertical *Z*-axis of a precisely controlled 3-axis CNC machine. The MRP fluid is pumped through the hollow tool spindle, and when it reaches the tip of the tool, it is energized by the tool electromagnet. This energized fluid is then pressed against and rotated over the workpiece. Due to the interaction of the energized MRP fluid with the workpiece surface, finishing forces are generated which are captured using a dynamometer placed at the bottom of the workpiece fixture as shown in Fig. 4.

The recording of the dynamic forces is initiated even before the energized fluid makes a contact with the workpiece surface. It is continued till the tool assembly is

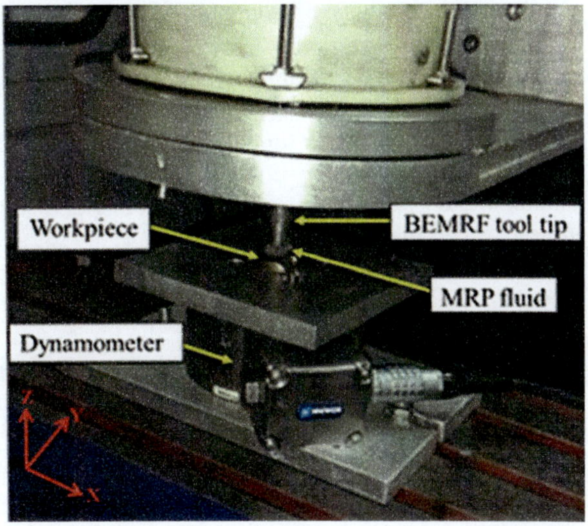

Fig. 4 BEMRF experimental setup with a dynamometer

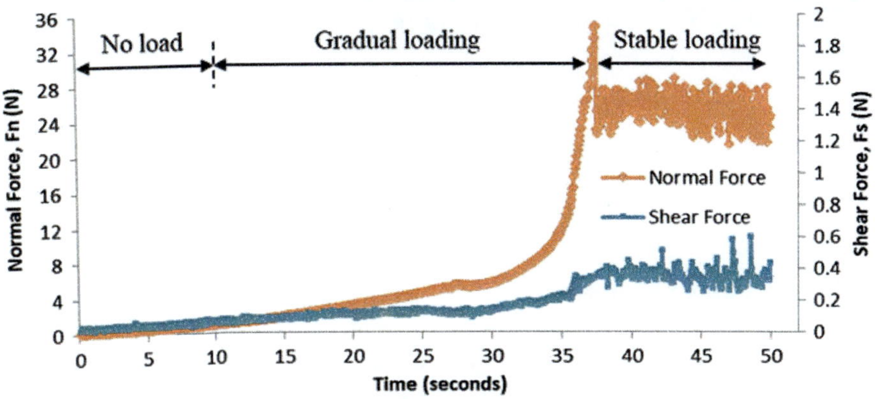

Fig. 5 Sample recording of normal and shear force in BEMRF process

Table 1 Selected parameters, their levels, and absolute values

Parameters	Levels				
	−1.682	−1	0	1	1.682
Current C (A)	2.3	3	4	5	5.7
Working gap G (mm)	0.73	0.8	0.9	1.0	1.07
Spindle speed S (rpm)	216	250	300	350	384

slide down the Z-axis and touches the workpiece surface before starting to rotate. The forces are maximum when the fluid makes contact with the workpiece surface and gets stabilized thereafter when the fluid starts rotating. A sample recording of the finishing forces is shown in Fig. 5 in which the different loading regions are highlighted. It should be noted that the forces in the stabilized region are considered for all the experiments.

The BEMRF process parameters selected for experimentation for the measurement of forces are magnetizing current (C), working gap (G), and tool spindle rotational speed (S). The MRP fluid used here consists of 20 vol% CIP (average particle size of Ø18 μm), 20 vol% SiC (average particle size of Ø19 μm), and 60 vol% carrier medium (heavy paraffin oil + grease). A circular disk of mild steel (Ø35 × 6 mm) is chosen as workpiece material.

The five levels of the selected process parameters and their absolute values are shown in Table 1. A central composite design technique was used for experimentation, and a total of 20 readings of both the forces were recorded. The response (F_n and F_s) as a function of the input variables (current, gap, and spindle speed) is given by quadratic equations [Eqs. (7) and (8)] having a very good coefficient of correlation ($R^2 = 0.98$ for normal force, F_n and $R^2 = 0.97$ for shear force, F_s). Within the selected range of the above process parameters, these equations can be used to obtain forces against any absolute values of the process parameters.

$$F_n = -48.1 + 9.82\,\mathrm{C} + 106.7\,\mathrm{G} + 0.095\,\mathrm{S} - 56.7\,\mathrm{G}^2 - 0.000183\,\mathrm{S}^2 - 7.39\,\mathrm{CG} \quad (7)$$

$$F_s = -7.13 + 0.39\,\mathrm{C} + 17.85\,\mathrm{G} + 0.002194\,\mathrm{S} - 0.0362\,\mathrm{C2} - 11.58\,\mathrm{G2} \quad (8)$$

5 Results and Discussion

The results of the theoretical model are compared with the experimental work to validate the proposed model. The value of magnetic flux density in the working gap used for theoretical modeling of both normal and shear force in this work is

Table 2 Comparison of theoretical and experimental value of normal force at 3 and 4 A magnetizing current

Sl. No.	Current (A)	Normal force per abrasive particle ($\times 10^{-4}$ N) [10]	F_n (N)		% error
			Theoret (Eq. 3)	Exp. (Eq. 7)	
1	3	1.368	24.87	21.44	13.80
2	4	1.981	36.02	23.99	33.38

Table 3 Comparison of the theoretical and experimental value of shear force at 3 and 4 A magnetizing current

Sl. No.	Current (A)	Shear force per abrasive particle ($\times 10^{-6}$ N) (Eqs. 1,4,5)	F_s (N)		% Error
			Theoret. (Eq. 6)	Expt. (Eq. 8)	
1	3	6.384	1.161	0.478	58.82
2	4	7.883	1.433	0.589	58.89

taken from Alam and Jha [10]. The theoretical and the experimental values of both the finishing forces are calculated at different values of magnetizing current and compiled in Tables 2 and 3 for normal force and shear force, respectively. In both cases, percent errors are calculated by

$$\%\text{Error} = ((\text{Theoret.} - \text{Expt.})/\text{Theoret.}) \times 100\% \tag{9}$$

From Table 2, it can be seen that the error's magnitude increases at higher values of magnetizing current. This is because of the widening difference between the value of theoretical and experimental data of normal force. In theoretical conditions, high values of magnetizing current lead to enhanced flux density in the working gap, and hence, magnetization of CIP increases. This further leads to an increase in the theoretical value of normal force. However, in actual/experimental conditions, the CIPs achieve their magnetic saturation value with an increase in the magnetic flux density. Therefore, further increase in flux density (by increasing current) does not increase the normal force, and hence, the error gets magnified.

In the case of shear force, it is observed, from Table 3, that the magnitude of error between the theoretical and experimental value is very large. This is because of the very high theoretical value obtained as compared to experimental results. This is attributed to the fact that the theoretical model for predicting shear force is based on magnetic flux density which is further accounted for only axial variation in the working gap (Eq. 1) and not accounted for radial variation. However, from Fig. 6 (measured experimentally by Alam and Jha [10]), it can be seen that there is a considerable amount of variation in flux density in the radial direction too from the center to the periphery of the tooltip. It is because of this that the experimental value of shear force, which depends upon the magnetic flux density in the working gap, is substantially lower than the theoretical value. If both axial and radial variations of

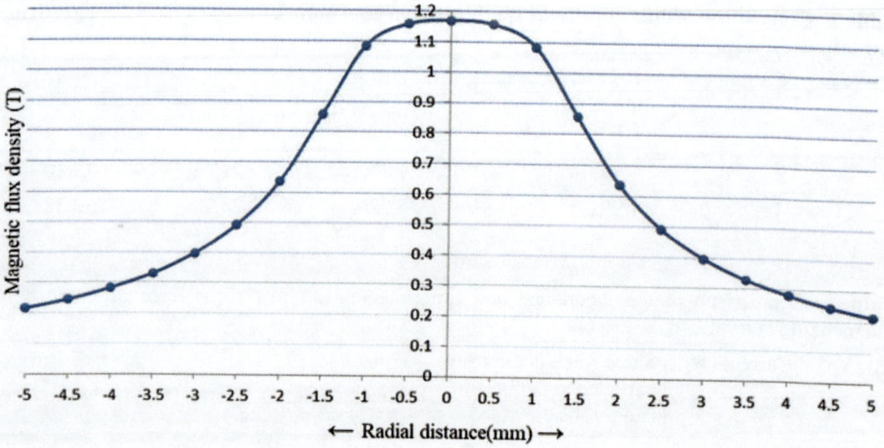

Fig. 6 Radial variation of magnetic flux density along the tooltip (at 4 A current) [10]

magnetic flux density in the working gap are simultaneously accounted for calculation of the shear force, the theoretical values will be closer to experimental values and the percentage in the error will be reduced.

6 Conclusion

The following can be concluded from this study on the forces involved in the BEMRF process:

- In the theoretical model for finishing forces, it is seen that both normal and a shear forces are directly influenced by the magnetic flux density in the working gap.
- By drawing a comparison to the experimental way of measurement of forces using dynamometers, it can be said that for finishing with lower current values the theoretical model can be used to accurately predict the forces. This method will help in eliminating the use of expensive dynamometers and yet get fairly accurate values of finishing forces.
- Comparing the theoretical and experimental results, the % error for shear force is very high. This is due to the assumption that in theoretical modeling of shear force only axial variation (not radially) in magnetic flux density is considered.

References

1. Huang J, Zhang JQ, Liu JN (2005) Effect of magnetic field on properties of MR fluids. Int J Mod Phys B 19(01n03):597–601

2. Sidpara A, Das M, Jain VK (2009) Rheological characterization of magnetorheological finishing fluid. Mater Manuf Processes 24(12):1467–1478
3. Alam Z, Iqbal F, Jha S (2015) Automated control of three-axis CNC ball end magneto-rheological finishing machine using PLC. Int J Autom Control 9(3):201–210
4. Alam Z, Iqbal F, Ganesan S, Jha S (2019) Nanofinishing of 3D surfaces by automated five-axis CNC ball end magnetorheological finishing machine using a customized controller. Int J Adv Manuf Technol 100(5–8):1031–1042
5. Khan DA, Jha S (2018) Synthesis of polishing fluid and novel approach for nanofinishing of copper using ball-end magnetorheological finishing process. Mater Manuf Process 33(11):1150–1159
6. Alam Z, Khan DA, Jha S (2019) MR fluid-based novel finishing process for nonplanar copper mirrors. Int J Adv Manuf Technol 101(1–4):995–1006
7. Saraswathamma K, Jha S, Rao PV (2015) Experimental investigation into ball end magnetorheological finishing of silicon. Precis Eng 42:218–223
8. Kumar A, Alam Z, Khan DA, Jha S (2019) Nanofinishing of FDM-fabricated components using ball end magnetorheological finishing process. Mater Manuf Process 34(2):232–242
9. Alam Z, Khan DA, Jha S (2018) A study on the effect of polishing fluid volume in ball end magnetorheological finishing process. Mater Manuf Process 33(11):1197–1204
10. Alam Z, Jha S (2017) Modeling of surface roughness in ball end magnetorheological finishing (BEMRF) process. Wear 374–375C:54–62
11. DeGroote JE, Marino AE, Wilson JP, Bishop AL, Lambropoulos JC, Jacobs SD (2007) Removal rate model for magnetorheological finishing of glass. Appl Opt 46(32):7927–7941
12. Shorey AB, Jacobs SD, Kordonski WI, Gans RF (2001) Experiments and observations regarding the mechanisms of glass removal in magnetorheological finishing. Appl Opt 40(1):20–33
13. Jung B, Jang KI, Min BK, Lee SJ, Seok J (2009) Magnetorheological finishing process for hard materials using sintered iron-CNT compound abrasives. Int J Mach Tools Manuf 49(5):407–418
14. Sidpara A, Jain VK (2011) Experimental investigations into forces during the magnetorheological fluid-based finishing process. Int J Mach Tools Manuf 51(4):358–362
15. Sidpara A, Jain VK (2012) Theoretical analysis of forces in magnetorheological fluid based finishing process. Int J Mech Sci 56(1):50–59
16. Iqbal F, Alam Z, Jha S (2020) Modelling of transient behavior of roughness reduction in ball end magnetorheological finishing process. Int J Abras Technol 10(3):170–192
17. Ida N (2004) Engineering electromagnetics. Springer, New York, (Chapter 8 and 9)
18. Stradling AW (1993) The physics of open-gradient dry magnetic separation. Int J Miner Process 39(1):1–8

Automation

A Machine Learning Approach for Prediction of Surface Roughness from the Images of Machined Components

A. Ramesh Babu

Abstract In this paper, a machine learning approach integrating with machine vision is proposed to predict the roughness of machined components. At first, specimens are prepared by using a CNC milling process, and then, their roughness values (R_a) are measured with a stylus instrument. With an in-house prepared machine vision setup, images are captured and analysed for their variation in the grey-level intensity patterns for surface roughness values. Images are scanned across the surfaces to generate the dataset of grey-level intensity profiles and developed the machine learning (ML) model by training the dataset. ML model is implemented using the Python programming language by utilizing the image processing, data science, and ML libraries. Finally, the model is validated by using a test dataset.

Keywords Surface roughness · Machine learning · Image processing · Computer vision · Machining

1 Introduction

It is very much essential to machine the components as per the specified dimensional and form accuracy, as well as surface finish to satisfy the functional performance and aesthetic requirements of mechanical components of final products. It is a well-known fact that smoother surfaces exhibit higher wear resistance, higher corrosion resistance, and higher fatigue strength. These properties are very much needed in many engineering and medical applications as well. However, in certain applications like surface coatings, rougher surfaces are desirable to improve the adhesive properties.

Typically, different machining and surface finishing processes produce surfaces with different roughness values (R_a) in the range of 0.05 μm (superfinishing) to 25 μm (shaping) [1]. Two techniques, namely contact stylus profilometry and optical profilometry, are in practice to measure the roughness. Contact profilometers use

A. R. Babu (✉)
Department of Mechanical Engineering, PSG College of Technology, Coimbatore 641004, India
e-mail: arb.mech@psgtech.ac.in

U. S. Dixit et al. (eds.), *Advances in Forming, Machining and Automation*, Lecture Notes in Mechanical Engineering, https://doi.org/10.1007/978-981-19-3866-5_34

diamond-tipped probes that physically move along the surface and surface irregularities are registered as amplified profiles of peaks and valleys. Whereas, the optical profilometers use light to measure the surface irregularities, and their operation is based on principles such as focus variation microscopy, confocal microscopy, and coherence scanning interferometry. Since several commercially available profilometers are desktop versions, machined components have to be taken to the measurement room to measure surface features.

With the advent of advanced and intelligent manufacturing practices, demand for in situ measurement of surface roughness is increasing rapidly. Because it provides real-time feedback to make suitable decisions to control the machining process or for online monitoring. In situ measurement refers to the measurements taken at the machine without disturbing the component's position on the machine table, however, the machining process is stopped during the measurement. Commercially available profilometers are generally desktop type and not suitable for in situ measurement.

Quinsat and Tournier [2] proposed a method for in situ measurement of surface roughness of mechanical components using chromatic confocal sensors. The sensor was attached to the spindle of the 5-axis machining centre, and roughness values were evaluated for milling and polishing operations. Similarly, Fu et al. [3] proposed an experimental setup using a chromatic confocal sensor and developed a software programme to calculate the roughness values. A robotic arm was employed to align the chromatic confocal sensor to the workpiece with a standoff distance of 6.5 mm. It was observed that measurement error while comparing with a standard stylus-based profilometer (Talysurf PGI 800) was less than 50 nm [3].

In another optical method based on the laser scattering principle, researchers attempted to establish the relationship between the surface roughness R_a value and laser scattered density pattern [4]. For this purpose, lapped, diamond-turned, and ground surfaces with R_a values in the range of 0.005–6 μm were considered. About 3–9% error was observed with the correlation obtained from laser scattering principle compared with stylus instrument [4]. The non-contact optical methods require very expensive optical/laser sensors as well as complex alignment and control systems. In this point of view, image processing techniques are a viable option for evaluating the surface features. Further, vision-based methods are easy to install, suitable for in situ measurement and automation.

Jeyapoovan and Murugan [5] created the database of the images for milled surfaces along with measured surface roughness values (R_a). When the surface roughness has to be evaluated for a new specimen, its image is compared with images in the database for the difference in image intensity pattern considering *Hamming distance* and *Euclidean distance* as the metrics [5]. Similarly, Kumar et al. [6] applied an image processing method to evaluate the surface roughness of components produced by an incremental forming process. In this work, images of formed components, whose surface roughness is in the range of 0.6–3.6 μm, are stored in a database and compared with images of components of the components to be tested. Typically, a comparison of pixel intensity values has been done using *Euclidean distance, Hamming distance,* and Wavelet-based methods [6].

Chiou et al. [7] proposed a vision-based system to monitor the surface roughness of milled components remotely. In this study, 4-axis CNC milling machine was considered and surface roughness values for the different cutting conditions (spindle speed 1000–10,000 rpm, feed 0.84–6.35 mm/s) are correlated to mean intensity values of images. A good agreement between the surface roughness values and the mean intensity of images was observed and noted as a linear relationship [7].

In contrast to above, Lee and Tarng [8] evaluated the surface roughness of turned components by considering the arithmetic average of the grey-level intensity of the images G_a, extracted along a straight line. The correlation was built based on the polynomial model by considering G_a and machining parameters including cutting speed, feed rate, and depth-of-cut. It was observed that error in the measurement is in the range of 0.2–12% while comparing the *Ga-based* correlation to surface roughness measured by contact type profilometer (R_a) [8]. Similarly, Rajneesh Kumar et al. [9] also considered the grey-level intensity of the images (G_a) to relate to surface roughness value (R_a) for machined specimens obtained from the grinding, milling, and shaping processes. A linear regression model was developed to evaluate R_a value as a function of G_a and cutting parameters, i.e. speed, feed, and depth-of-cut.

Further, the effect of magnification of images on correlation was studied and noted that magnification of the images is ineffective in the case of milling and shaping. However, it was found to be effective for the grinding process. It was observed that the developed regression model has the maximum error of 2%, 6.44%, and 8.2% for grinding, milling, and shaping processes, respectively [9].

Alessandro et al. [10] applied the convolutional neural network (CNN) method to classify images that represent the roughness values in the range of 0.2–2.0 μm produced with the EDM machining process. A dataset of 4400 images was used to train the network and observed that the error of the prediction model is around 10%. Since it is a classification method, several hundreds of images corresponding to discrete values of surface roughness are needed, which is cumbersome [10]. Similarly, Achmad et al. [11] built CNN models for turning, slot milling, and side milling processes considering the datasets of images of 41,680 for slot milling, 45,200 for side milling and 73,600 for turning process. The accuracy of the prediction of the model is in the range of 81–91%.

In the present work, a novel method is followed to generate the dataset of grey-level intensity profiles from the images of machined surfaces, and then, a machine learning model is proposed as follows.

2 Machine Vision Setup and Experimental Data

To develop the prediction model for evaluating the surface roughness, aluminium plates of 100 × 100 × 6 mm are taken and machined using a CNC vertical milling machine. By varying the spindle speed (in the range of 275–1050 rpm), federate (in the range of 120–280 mm/min) and depth-of-cut (constant 0.5 mm), specimens with different roughness values (R_a) are obtained. Stylus instrument (TIME-TR1100) is used to measure the surface roughness and the measured roughness values (R_a) are shown in Table 1.

Then, a machine vision setup is prepared, in-house, using a CCD camera (IDS USB 3.0, resolution of 2456 × 2054 pixels −5 MP), 25 mm focal length lens, co-axial light (100 × 100 mm size, make-VST) and a stand as shown in Fig. 1. The workpiece is placed on a flat surface and co-axial light is placed just above the workpiece to illuminate it. Lens is attached to the camera and positioned at a certain height from the co-axial light and adjusted the focus till the image is clear on IDS manager, an image capturing software. Figure 2a shows photographs of the specimens taken with a normal camera, and Fig. 2b, c, d shows images captured through IDS image capturing software for specimen numbers 1, 7, and 13, respectively.

The roughness profiles can be seen from the images and noted that peak and valley spots are dominant as the roughness values increase. Interpreting this intensity pattern and correlating it with roughness is a challenging task since it depends on the quality of the image, interpretation method, and efficiency of the model. In the present work, images are captured with a professional machine vision camera and

Table 1 Surface roughness values measured with a stylus instrument

Specimen No.	R_a (μm)	Specimen No.	R_a (μm)	Specimen No.	R_a (μm)
1	0.63	6	1.41	11	10.25
2	0.68	7	5.75	12	11.92
3	0.79	8	6.39	13	16.21
4	0.82	9	7.12		
5	0.93	10	7.65		

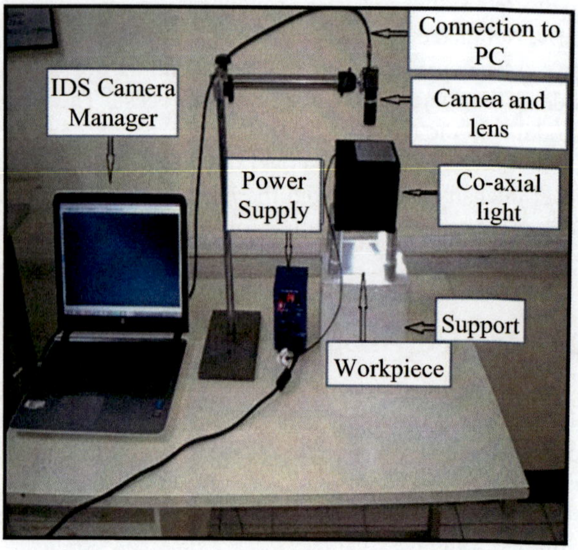

Fig. 1 Proposed machine vision setup

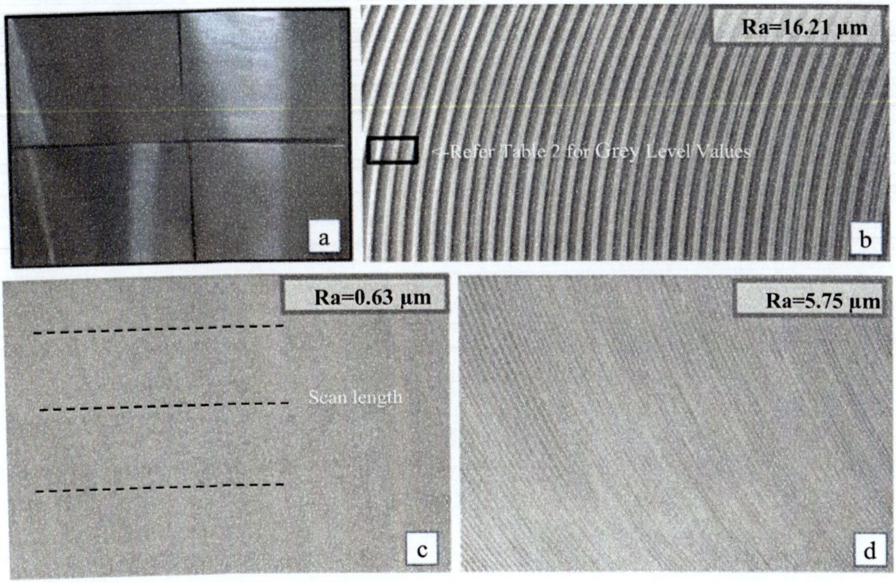

Fig. 2 Photographs of the specimens and images captured using the proposed setup

capturing software. Further, from Fig. 2, it is clear that the quality of the images is good and clear. The next step is the extraction of the digitation data and relating it to roughness texture is an important task as follows.

3 Extraction and Analysis of Digital Data from Image

Digital information extracted from the images, i.e. grey-level values are analysed to understand the relation between the surface texture and corresponding grey-level values. This analysis is carried out using Python programming executed through *Jupyter* notebook. Python provides several built-in libraries needed for image processing, data science, machine learning, mathematical functions, and plotting.

Images captured through the proposed machine vision setup are stored in a local drive of the PC and processed through the Python programming language. At first, coloured images are converted into grey images and their grey-level values are extracted. Grey-level values for a specific window (6 rows and 10 columns) for the image (Fig. 2b) are shown in Table 2. It is well known that grey-level values are in the range of 0–255, and 0 represents black and 255 represents white. Variation in these grey-level values represents the texture of the image, particularly surface roughness in the present case. Since surface roughness is measured by the stylus instrument by moving the probe parallel to the edge on the surface, variation in grey-level values along a horizontal line is considered for this study.

Table 2 Grey-level value of the image (Fig. 2b) at the specified window

	0	1	2	3	4	5	6	7	8	9
400	113	120	128	138	145	146	148	154	160	161
401	116	120	127	137	145	146	147	151	158	161
402	118	121	127	135	142	148	146	149	154	158
403	121	126	130	136	140	146	146	149	153	155
404	126	132	136	138	139	142	145	151	155	155
405	129	136	140	140	139	137	144	153	158	155

In this work, all the images are taken in the sizes of 450 × 649 except the last image (specimen no. 13) which is 542 × 1307 in size. To analyse the variation in grey-level values, intensity values along the scanned lines, as shown on the image (Fig. 2b), are considered. Variation in grey-level values along a horizontal scanned line at position (10, 10) with a length of 400 pixels for images (Fig. 2 b, c, d) is shown in Fig. 3.

From Fig. 3, it is observed that variation in the grey-level values looks similar to a typical roughness profile generated by stylus instruments. Hence, it is more appropriate to consider grey-level values along a straight line, while developing the correlation between the roughness values and grey-level pattern. In this work, the arithmetic average of the grey-level values (G_a) is calculated as given in the equation below.

$$G_{\mathrm{a}} = \left(\sum (|y_1 - y_m| + |y_2 - y_m| + \ldots + |y_n - y_m|)\right. \tag{1}$$

$$y_m = (y_1 + y_2 + \ldots + y_n)/n \tag{2}$$

where

n Number of pixels
y_i Grey-level intensity value for ith pixel
y_m Mean of grey-level intensity values

It can be seen from Figs. 3 and 4 that G_a values have a linear relationship with surface roughness. As the roughness increases, G_a value also increases. In this work, a machine learning approach is proposed to find this relationship as discussed below.

4 Preparation of Dataset and Machine Learning Model

Machine learning (ML) is one of the important components in data science and the subset of artificial intelligence (AI). Machine learning algorithms generate the models through the training of the labelled or unlabelled data to predict or classify

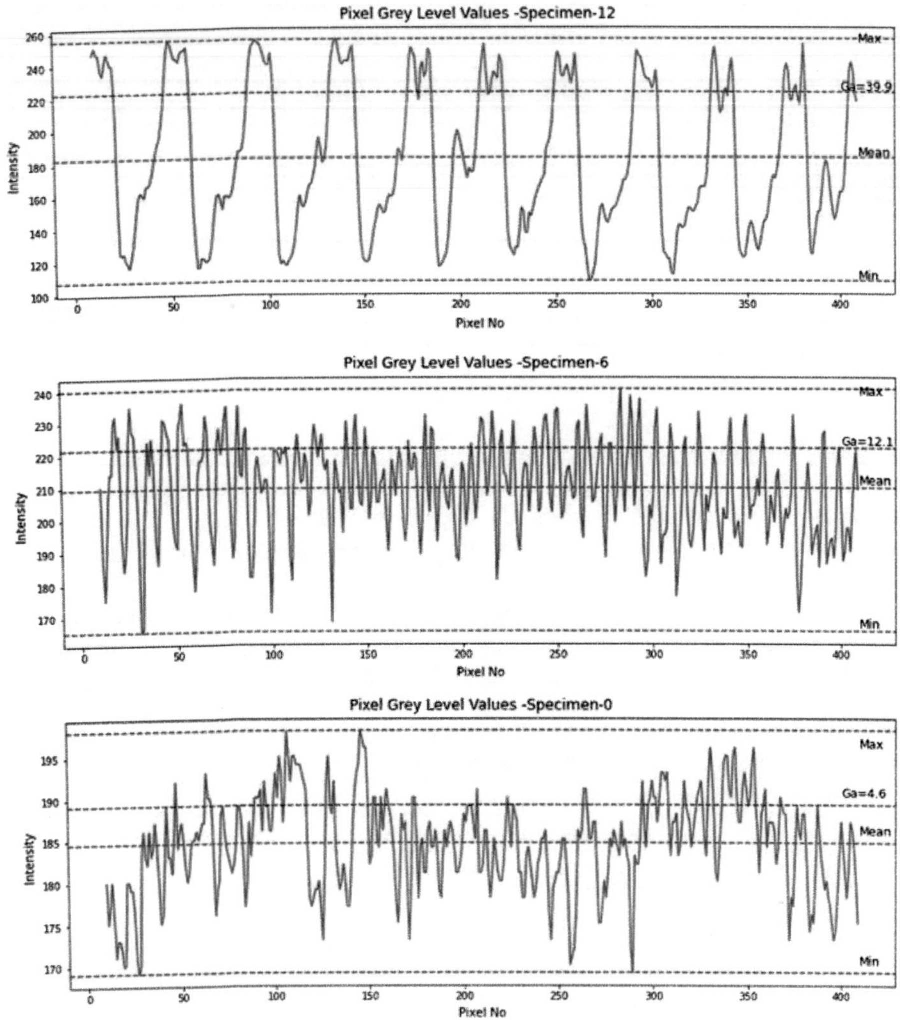

Fig. 3 Variation of grey-level values along a straight line for specimens 1, 7, and 13 (index of specimens starts from 0)

the data. In the present work, a prediction model is developed to predict the surface roughness of the specimen from the given image corresponding to a specimen. Since the images of the machined components are captured in a non-contact manner, this method is more suitable for in situ measurement which is a very important feature in advanced machining centres.

A large amount of data is needed for ML algorithms to attain the higher accuracy of the model. In this work, a novel approach is followed to generate a large dataset by scanning along the several horizontal lines on the images of machined surfaces (shown as dotted lines in Fig. 2). The purpose of the scanning process is

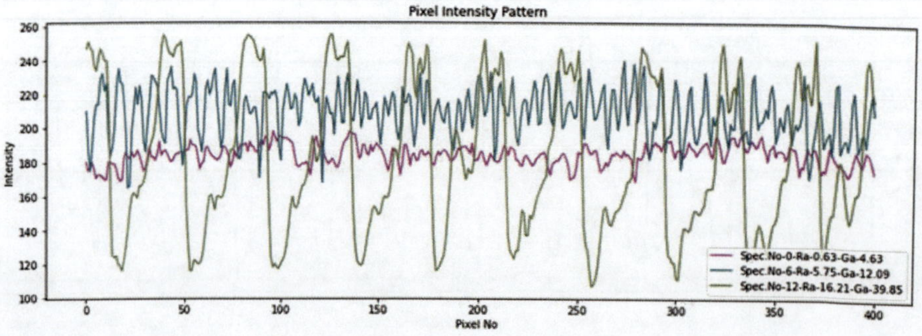

Fig. 4 Variation in pixel intensity patterns among the specimens 1, 7, and 13

to extract the grey-level values along a straight line across the workpiece. Then, the arithmetic average of the grey-level values (G_a) for each of the scanned lines is calculated as expressed in equation (1). Grey-level values for typical scanned lines and corresponding G_a values and measured R_a values are shown in the dataset of Table 3.

Each row in Table 3 corresponds to a scanned line considered on the image of the surface. In this case, 3 scanned lines are considered on each image and since there are 13 specimens, a total of 39 rows can be seen in Table 3. The last two columns

Table 3 Structure of the dataset for the prediction model

	Grey-level intensity values along scanned lines								Specimen No.	R_a	G_a
	0	1	2	3	4	…	398	399			
0	180	175	177	180	176	…	180	175	0	0.63	4.63
1	183	177	185	193	192	…	195	195	0	0.63	4.04
2	203	201	202	202	202	…	190	191	0	0.63	3.62
3	198	183	194	201	198	…	195	188	1	0.68	4.43
4	188	195	195	190	190	…	198	205	1	0.68	4.04
5	193	188	187	187	197	…	203	200	1	0.68	4.65
6	199	203	202	200	197	…	182	183	2	0.79	4.56
7	201	199	198	197	194	…	185	177	2	0.79	4.53
8	199	198	198	199	199	…	179	181	2	0.79	5.15
…	…	…	…	…	…	…	…	…	.	…	…
33	174	160	146	129	110	…	117	114	11	11.92	25.66
34	182	171	147	129	122	…	109	113	11	11.92	23.75
35	133	152	153	148	147	…	140	134	11	11.92	12.61
36	135	134	129	123	119	…	164	163	12	16.21	26.44
37	175	191	214	204	175	…	136	122	12	16.21	29.34
38	98	101	106	122	131	…	158	170	12	16.21	27.59

```
For S = 1 to number of specimens
    Read the Image S
    Convert S into Grey Image
        L= Length of a scanned line (input)
        For k = 1 to number of scanned lines (along y axis)
        Get the grey level values along the L
        Calculate G_a as per the equation (1)
        Add grey level values, G_a and R_a to dataset
Spilt the data into training set and testing set
Train and fit the model using ML regression model
Validate the model
Get the coefficients
Predict the values
```

Fig. 5 Algorithm for the proposed model

show roughness values (R_a) and the arithmetic average of grey-level values (G_a), respectively. Elements in columns 0 to till the specimen number indicate the grey-level values of the scanned line. For example, (180, 175, 177, 180, 176, ...180, 175), (183, 177, 185, 193, 192, ...195, 195) and (203, 201, 202, 202, 202, ...190, 191) indicate the grey-level values for the 1st, 2nd, and 3rd scanned lines of length 400 pixels on specimen number 0. Size of the dataset and its values depend on the number of scanned lines, and its position and length. The dataset is programmatically generated and will be the input for the ML prediction model. Then, the model is trained with the ML linear regression algorithm. The major steps in the proposed algorithm are shown in Fig. 5.

With the proposed algorithm, as described in Fig. 5, it is found that there is good agreement between surface roughness and grey-level intensity patterns of images corresponding to the surfaces, as discussed below.

5 Results and Discussion

To develop the prediction model, labelled data of R_a and G_a is considered. In this case, 80% of data is used to train the model and the remaining 20% is used to the test model. Table 4 shows typical data consisting of original G_a and R_a values, and G_a and R_a values considered for training and testing.

Since G_a is linearly related to R_a, a simple linear regression technique is applied while training the ML model. The relation between the G_a and R_a is shown in Fig. 6, as the scattered plot along with the regression model evolved through the ML approach. From Fig. 6, it can be seen that R_a values are linearly related to G_a values. However, there is variation in G_a values of scanned lines for the same specimens.

Table 4 Data considered for training and testing the machine learning (ML) algorithm

Sl. No.	G_a	R_a	X_train (G_a)	y_train (R_a)	X_test (G_a)	y_test (R_a)
0	4.63	0.63	4.44	0.82	4.04	0.68
1	4.04	0.63	13.89	7.65	15.01	7.65
2	3.62	0.63	12.09	5.75	14.07	7.65
3	4.43	0.68	7.71	1.41	25.66	11.92
4	4.04	0.68	3.62	0.63	23.75	11.92
5	4.65	0.68	27.59	16.21	13.97	7.12
6	4.56	0.79	12.68	5.75	4.30	0.82
7	4.53	0.79	26.44	16.21	14.32	6.39
8	5.15	0.79	8.25	1.41	NaN	NaN
9	3.9	0.82	12.61	11.92	NaN	NaN
…	…	…	…	…	…	…
37	29.34	16.21	NaN	NaN	NaN	NaN
38	27.59	16.21	NaN	NaN	NaN	NaN

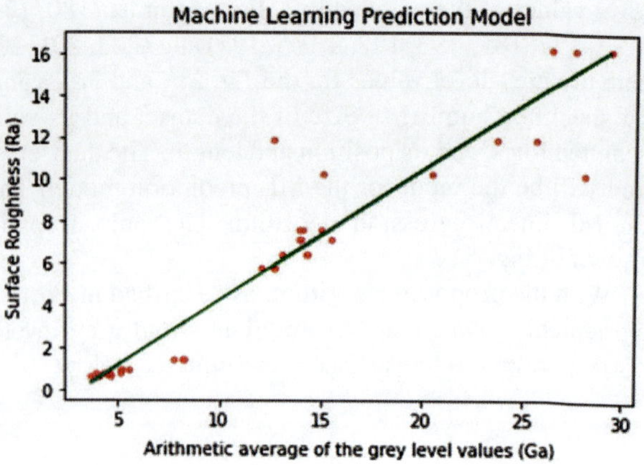

Fig. 6 Prediction model with the proposed ML model

$$R_a = -1.93 + 0.62 \times G_a \tag{3}$$

Correlation arrived through the ML model along with coefficients is shown in Eq. (3). The model is validated with test data. The predicted values for the test data are shown in Table 5. It can be noted that predicted values are closer to actual values and variation is in the range of 0.1–0.6 μm.

Table 5 Predicted surface roughness values with the proposed ML approach

Test No.	Actual (R_a) (μm)	Predicted (R_a) (μm)
1	0.68	0.56
2	7.65	7.33
3	7.65	6.75
4	11.92	13.91
5	11.92	12.73
6	7.12	6.69
7	0.82	0.72
8	6.39	6.91

6 Conclusions

The proposed machine learning approach shows a good agreement of surface roughness with the grey-level intensity pattern of the images corresponding to the machined surfaces. It is observed that three is only a slight error in the range of 0.1–0.6 μm for predicted and actual values. This is due to the certain variations in grey-level intensity patterns among the scanned lines for a given specimen. This variation occurs due to several factors including lay direction and certain black spots/patches on the image. There is a good scope for further study in this direction to build a more robust model. Further, a greater number of experiments and images may improve the performance of the model. Similarly, several uncertainty parameters in measurement, preparation of specimens, and capturing the images have an impact on the quality of the prediction model.

Applying the machine learning prediction model with the generation of a large set of data is the novelty of the present work. Even though there are certain correlations available in ligature, based on the arithmetic average of grey-level values with surface roughness, they are limited to a single scanned value per specimen and thus cannot guarantee the precision in prediction.

The proposed machine vision setup consists of a professional industrial camera, lens, and lighting system, found to be suitable for non-contact prediction of surface roughness. However, one can explore a different combination of image capturing systems.

References

1. Whitehouse DJ (2004) Surfaces and their measurement. In: 1st (ed), Butterworth-Heinemann
2. Quinsat Y, Tournier C (2012) In situ non-contact measurements of surface roughness. Precis Eng 36:97–103
3. Fu S, Kor WS, Cheng F, Seah LK (2020) In-situ measurement of surface roughness using chromatic confocal sensor. Proc CIRP 94:780–784

4. Tay CJ, Wang SH, Quan C, Shang HM (2003) In situ surface roughness measurement using a laser scattering method. Optics Commun 218:1–10
5. Jeyapoovan T, Murugan M (2013) Surface roughness classification using image processing. Measurement 46:2065–2072
6. Gandla PK, Inturi V, Kurra S, Radhika S (2020) Evaluation of surface roughness in incremental forming using image processing-based methods. Measurement 164
7. Chiou RY, Kwon Y, Tseng TL, Mauk M (2015) Experimental study of high speed CNC machining quality by noncontact surface roughness monitoring. Int J Mech Eng Robotics Res 4:282–286
8. Lee BY, Tarng YS (2001) Surface roughness inspection by computer vision in turning operations. Int J Mach Tools Manuf 41:1251–1263
9. Kumar R, Kulashekar P, Dhanasekar B, Ramamoorthy B (2005) Application of digital image magnification for surface roughness evaluation using machine vision. Int J Mach Tools Manuf 45:228–234
10. Giusti A, Dotta M, Maradia U, Boccadoro M, Gambardella LM, Nasciutia A (2020) Image-based measurement of material roughness using machine learning techniques. Proc CIRP 95(10)
11. Rifai AP, Aoyama H, Tho NH, Dawal SZM, Masruroh NA (2020) Evaluation of turned and milled surfaces roughness using convolutional neural network. Measurement 161

A Review of Smart Condition Monitoring System for Gearbox

Manvir Singh Lamba, Amandeep Singh, and J. Ramkumar

Abstract Gears are the most critical and commonly used machine elements. They are used for a variety of applications throughout various industries like energy, aerospace, transportation, mining, agriculture, manufacturing, etc. The failure of gear may result in cataclysmic shutdowns causing significant production, economic losses and even human casualties. Fault diagnosis is an important component of condition-based maintenance and it has gained much attention for the safe operation of the gearboxes. But in many applications, fault diagnosis is not accurate. Sensor-based condition monitoring systems help in enhancing the quality of measurement data. A lot of research has been done in developing machine learning algorithms and models to extract information of fault status from the sensor data. Data acquisition of the meaningful sensor data from the gearbox is an open challenge. This makes the estimation of the remaining useful time of gearbox more challenging. This paper presents a literature review of the different IoT-based techniques used to acquire health status data from the gearbox.

Keywords Condition monitoring · Gearbox

Abbreviations

s_i	The ith member of dataset
N	The number of points in the dataset
CF	Crest factor

M. Singh Lamba (✉)
Department of Mechanical Engineering, Thapar Institute of Engineering and Technology, Patiala, India
e-mail: mlamba_be19@thapar.edu

A. Singh · J. Ramkumar
Department of Mechanical Engineering, Indian Institute of Technology Kanpur, Kanpur, India
e-mail: adsingh@iitk.ac.in

J. Ramkumar
e-mail: jrkumar@iitk.ac.in

U. S. Dixit et al. (eds.), *Advances in Forming, Machining and Automation*, Lecture Notes in Mechanical Engineering, https://doi.org/10.1007/978-981-19-3866-5_35

$S_{peak-peak}$	The peak-to-peak value of the signal
RMS	The root mean square value of the vibration signal.
EO	Energy operator
$\triangle \overline{x}$	The mean value of signal $\triangle x$
$\triangle x_i$	$s_{i+1}^2 + s_i^2$
Kurt	Kurtosis
ER	Energy ratio
$\sigma(d)$	The standard deviation of the difference signal
$\sigma(r)$	Standard deviation of the regular signal
d_i	The ith point of the differential signal in the time record
r_i	The ith point in the time record of the residual signal
r_{ij}	Is the ith point in the jth time record of the residual signal
J	The current time record
I	The data point number per reading
M	The current time record in the run ensemble
SLF	Sideband level factors
s_{std}	The standard deviation of the time signal average
Si	The amplitude of the ith sideband around fundamental gear meshing frequency
SBI	Sideband index
k	The number of sidebands
Smax i	The ith maximum linear amplitude of sideband
E	The envelop of the band passed signal
$\overline{E}$	The mean value of the enveloped signal
$\lvert\overline{s}(t)\rvert$	The envelope of the analytic signal
s(t)	Is an input analog signal
$\overline{s}(t)$	The Hilbert transform of the input signal

1 Introduction

In the gearbox transmission system, different rotational speeds and specific torques between electric and laid machines are to be maintained. Monitoring and assessment of gearbox transmission systems are crucial to avoid production losses, downtime and financial losses and human efforts. For example, in 2006, 'Zhuoying 4' rocket's propulsion system was damaged due to a gear failure [1]. Continuous monitoring of gear health can avoid sudden industry shutdowns.

Some of the factors causing faults in gears are design, inadequate lubrication, misalignment, fluctuating loads, etc. (Fig. 1).

In the literature, different techniques have been used to monitor gear transmission systems such as acoustic emissions, oil debris, vibration-based analysis, time–frequency signal analysis, motor current signal analysis and infrared thermography

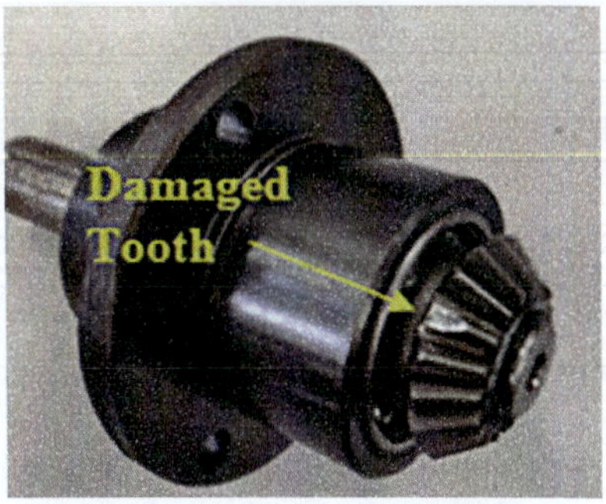

Fig. 1 Pinion having a damaged tooth [2]

[3]. Gears are the main part of the gearbox transmission system. They are responsible for transmitting the circular motion through the contact of cogwheels inside of them. Degradation of gears and reduction in mechanical properties is caused by the friction between them. The most common and initial wear in gears is the appearance of uniform gear [4] (Fig. 2).

Studies have shown that tooth breakage, cracks in the root, pitting and tooth surface damage are some of the faults found in gears [5, 6]. Studies indicate that 65% of gearbox damage is due to faults in the gears [7]. The most common defects in gear could be classified into the following [2, 8]:

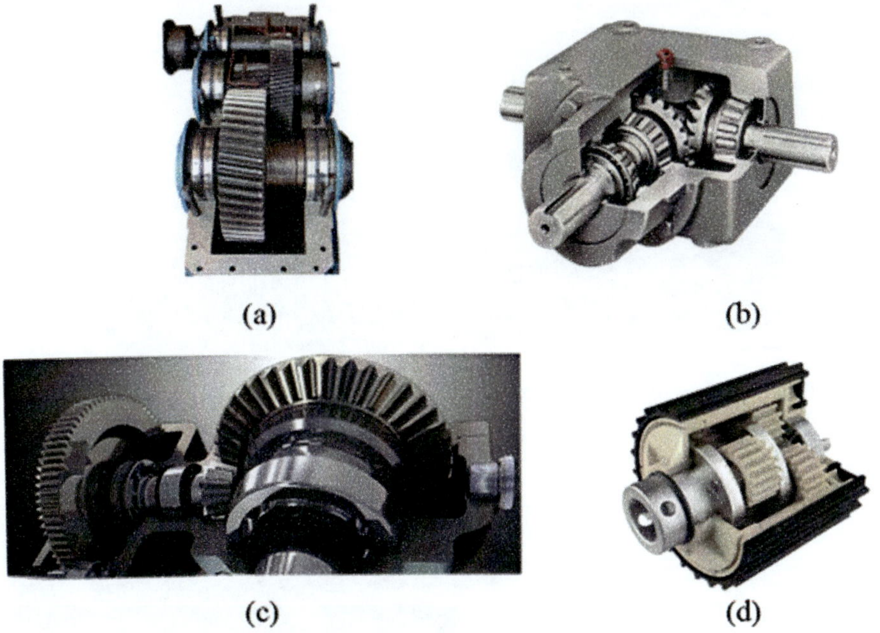

Fig. 2 Various types of industrial gearbox: **a** helical, **b** bevel, **c** worm and worm wheel, **d** planetary [2]

1. Fatigue: Fatigue is the result of cyclic loading. It includes crack initiation, cracks propagation and fracture. It is also known as high-cycle fatigue if the yield strength of the material is more than the cycle stress.
2. Contact fatigue: A frequent cyclic stress originates at the cracks on a surface commonly causing macro and micro pitting (Fig. 3).
3. Wear: Removal of material from the surface of the gear due to electrical, chemical or mechanical actions (Fig. 4).
4. Scuffing: When the surface film breaks down due to high temperature sliding between two surfaces (Fig. 5)

Maintenance strategies like run-to-break maintenance and time-based preventive maintenance are not useful in today's world. A higher rate of reliability is required for today's applications. The cost of preventive maintenance has increased

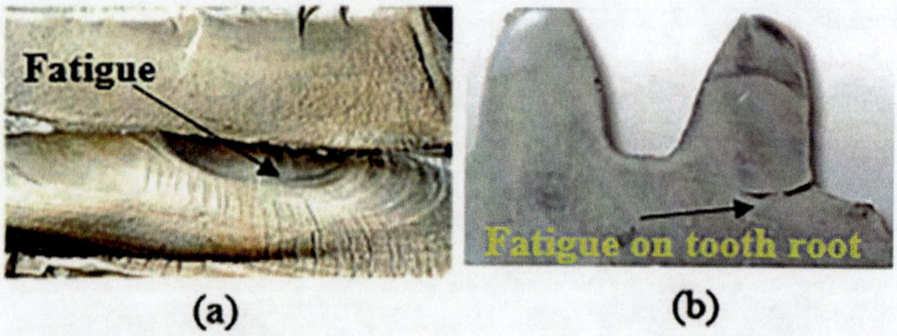

Fig. 3 Fatigue on gear tooth: **a** surface, **b** root filler [2]

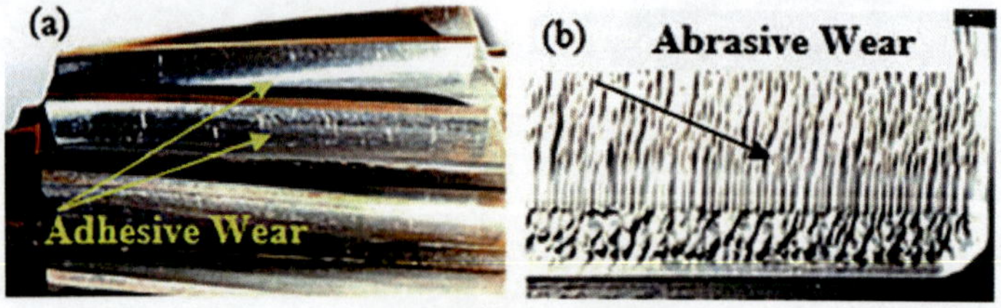

Fig. 4 Different tropes of wear: **a** adhesive, **b** abrasive [2]

Fig. 5 Scuffing on a gear tooth surface [2]

and thus conditioning monitoring is preferred in the industry. Conditioning monitoring consists of three steps, data acquisition, data processing and preparing data decision models. Data acquisition deals with obtaining relevant data related to system health. Data processing deals with signal processing and makes the data ready to be processed by the machine learning models. The last step is to create and test the decision-making model [9]. Researchers use condition monitoring systems to study the detection of gear defects and finding the root cause of the defect [10]. Designers can use this study to make enhancements in gear designs to improve performance and minimize failures. Condition monitoring systems deal with various types of input data, for instance, vibration, acoustic emission, temperature, oil debris analysis, etc. Systems based on vibration analysis, acoustic emission and oil debris are the most common and are very well established in the industry [11].

Advancements in IoT have led to innovative digital solutions to solve problems faced in traditional industries. The use of IoT in condition monitoring systems leads to predictive maintenance which is the most important use case of Industry 4.0. The geometric complexity of the gearbox and high-speed rotation of gears in the gearbox makes the acquisition of health data more challenging. This study presents the review of the techniques used by the researchers to acquire the meaning of full health status data of the gearbox. Researchers have developed many machine learning algorithms and models from the sensor data to study gearbox health. Hence, this is not included in this study.

2 Conditional Indicators

Gearbox condition indicators deal with the data distribution. Traditional CIs deal with data distribution. The main differences between these CIs are in the signal from which the computations are made. Raw, residual and difference signals are used to determine the conditional indicators. Some of the indicators used are as follows [12–14] (Table 1).

2.1 RMS

RMS value computed of a vibration signal describes the amount of energy in a signal. Using RMS, we can observe the effect of a fluctuating signal. RMS was originally used to identify the heat produced from a resistor when it was subjected to an alternating current. From the formula of RMS, it is clear that any isolated peaks in the signal do not affect the value of RMS. Hence, this parameter is not sensitive to tooth failure in its early stage rather it is sensitive to changes in load and speed. As the gearbox wears out, the vibration signal increases cause the RMS to increase. For a sine wave, the RMS value is around 0.707 times the amplitude. RMS is also known as the standard deviation of the signal average [13, 15–19].

Table 1 List of conditional indicators with their formula

‘	Formula	Fault	References
RMS, delta RMS	$\text{RMS} = \sqrt{\frac{1}{N}\sum_{i=1}^{N} s_i^2}$	Depicts the overall condition of a gearbox. Its value increases as tooth failure progresses. Can identify pitting damage	[13, 15–19]
Crest factor	$\text{CF} = \frac{S_{\text{peak-peak}}}{\text{RMS}}$	This parameter can indicate very small surface damages at an early state	[15, 16]
Energy operator	$\text{EO} = \frac{N^2 \sum_{i=1}^{N} (\Delta x_i - \Delta \overline{x})^2}{\left(\sum_{i=1}^{N} (\Delta x_i - \Delta \overline{x})^2\right)^2}$	High values in EO indicate serve pitting or scuffing	[14, 20, 21]
Kurtosis	$\text{kurt} = \frac{N \sum_{i=1}^{N} (s_i - \overline{s})^2}{\left(\sum_{i=1}^{N} (s_i - \overline{s})^2\right)^2}$	This parameter describes the peaks and the flats in the signal data. So, defects like gear wear and breakage are identified	[15, 16, 18, 22]
Energy ratio	$ER = \frac{\sigma(d)}{\sigma(r)}$	This parameter efficiently indicates heavy wear like when more than one tooth is damaged	[14, 23]
FM4	$FM4 = \frac{N \sum_{i=1}^{N} (d_i - \overline{d})^2}{\left(\sum_{i=1}^{N} (d_i - \overline{d})^2\right)^2}$	This parameter indicates defects in gear meshing and gear tooth, like wear, scuffing, pitting and tooth bending due to root crack	[15, 17, 24, 25]
NA4	$\text{NA4} = \frac{N \sum_{i=1}^{N} (r_i - \overline{r})^4}{\frac{1}{M} \sum_{j=i}^{M} (\sum_{i=1}^{N} (r_{ij} - \overline{r_j})^2)^2}$	This parameter identifies an increase in gear damage or when damage progresses from one tooth to another	[17, 21, 26, 27]
Sideband level factors	$\text{SLF} = \frac{\sum_{i=1}^{N} s_{\text{igearmesh}+i}}{s_{\text{std}}}$	This parameter is used to detect gear misalignment	[13, 14, 28, 29]
Sideband index	$SBI = \frac{1}{k} \sum_{i=1}^{k} S_{\max i}$	This parameter is used to obtain the pinion quality	[28–30]
NB4	$\text{NB4} = \frac{N \sum_{i=1}^{N} (E_i - \overline{E})^4}{\frac{1}{M} \sum_{j=i}^{M} (\sum_{i=1}^{N} (E_{ij} - \overline{E_j})^2)^2}$ $E = \lvert \overline{s}(t) \rvert = \sqrt{s^2(t) + \overline{s}^2(t)}$	This parameter identifies the transient load cause by a few damaged gear teeth	[14, 21, 31]

$$\mathrm{RMS} = \sqrt{\frac{1}{N}\sum\nolimits_{i=1}^{N} s_i^2} \tag{1}$$

2.2 Crest

An advantage of using crest factor over RMS is that it can detect a defect in its early stage. When a small defect occurs, a few high-amplitude peaks are observed in the signal. This causes the crest factor to increase significantly as the denominator would decrease due to fewer peaks and the numerator would increase due to high-amplitude peaks [13, 15, 16].

$$\mathrm{CF} = \frac{S_{\mathrm{peak-peak}}}{\mathrm{RMS}} \tag{2}$$

2.3 Energy Operator

The energy operator is a nonparametric method to detect defects. This method is based on calculating the energy of the oscillations in a signal. Energy operator is a simple algorithm and is filter-free, i.e., there is no need for initial information. One of the most prominent drawbacks of EO is its inability to work with noise. The advantage of using it provides increased signal-to-interference ratio and increased amplitude modulation [14, 20, 21].

$$\mathrm{EO} = \frac{N^2 \sum_{i=1}^{N}(\Delta\, x_i - \Delta\, \overline{x})^2}{\left(\sum_{i=1}^{N}(\Delta\, x_i - \Delta\, \overline{x})^2\right)^2} \tag{3}$$

2.4 Kurtosis

The kurtosis is the normalized fourth moment of a signal. The number and amplitude of peaks in a signal are statistically measured with this parameter. If a defect occurs, a peak is observed in the vibration signal causing an increase in kurtosis. For an ideal Gaussian distribution, the kurtosis is three. While calculating this parameter three is subtracted, so kurtosis for an ideal gearbox will be zero [15, 16, 18, 22].

$$\text{kurt} = \frac{N\sum_{i=1}^{N}(s_i - \overline{s})^2}{\left(\sum_{i=1}^{N}(s_i - \overline{s})^2\right)^2} \tag{4}$$

2.5 Energy Ratio

The primary principle behind this parameter is that when a defect occurs, energy is transferred from the regular meshing component to the remainder of the signal. It compares the amount of energy in the difference signal with the amount of energy in the normal components signal. This parameter can efficiently identify heavy wear, i.e., when multiple gear teeth are being damaged [13, 14, 23].

$$\text{ER} = \frac{\sigma(d)}{\sigma(r)} \tag{5}$$

2.6 FM4

It is one of the most popular vibration diagnostic parameters. This parameter measures how flat or peaked the signal is. In the frequency domain, the shaft and meshing frequencies, their harmonics, and the first-order sidebands are removed to generate a difference signal. If the vibration signal from the gearbox is a difference signal with a Gaussian distribution, then FM4 assumes the gearbox is in good condition. On the other hand, a gearbox with multiple defective teeth will produce a difference signal with major peaks, and the kurtosis will decrease and the distribution will flatten out [15, 17, 24, 25].

$$\text{FM4} = \frac{N\sum_{i=1}^{N}\left(d_i - \overline{d}\right)^2}{\left(\sum_{i=1}^{N}\left(d_i - \overline{d}\right)^2\right)^2} \tag{6}$$

2.7 NA4

The NA4 parameter was created to address the flaws in the FM4 parameter. FM4 became less susceptible as the number and intensity of defects increased in the gearbox. The difference signal is used to compute FM4, while the residual signal is used to calculate NA4. Another modification is that the NA4 option now includes trends. NA4 is equal to the ratio of the data record's kurtosis divided by the square of the average variance [17, 21, 26, 27].

$$\mathrm{NA4} = \frac{N\sum_{i=1}^{N}(r_i - \overline{r})^4}{\frac{1}{M}\sum_{j=i}^{M}(\sum_{i=1}^{N}\left(r_{ij} - \overline{r_j}\right)^2)^2} \quad (7)$$

2.8 Sideband Level Factors

SLF stands for sideband level factor and is an indicator of a single tooth or gear shaft degradation. This parameter is based on the concept that a tooth defect will generate amplitude modulation of the vibration signal. This parameter is zero for an ideal healthy gearbox [13, 14, 28, 29].

$$\mathrm{SLF} = \frac{\sum_{i=1}^{N} \mathrm{si}_{\mathrm{gearmesh}+i}}{s_{\mathrm{std}}} \quad (8)$$

2.9 Sideband Index

The average amplitude of the fundamental gear mesh frequency's sidebands is defined as the sideband index (SBI). This parameter is good for detecting local gear faults This CI is driven by a rise in the amplitude of the fundamental gear meshing frequency sidebands, which indicates a local failure [28–30].

$$\mathrm{SBI} = \frac{1}{k}\sum_{i=1}^{k} S_{\max i} \quad (9)$$

2.10 NB4

The NA4 parameter serves as the foundation for the NB4 parameter. The residual signal is used to calculate NA4, whereas the envelope signal is used to calculate NB4.The gear meshing frequency is used to bandpass filter a raw vibration signal. An equation known as Hilbert transform is used to generate an analytic signal after the undesirable parts of the signal have been removed [14, 21, 31].

$$NB4 = \frac{N \sum_{i=1}^{N} \left(E_i - \overline{E}\right)^4}{\frac{1}{M} \sum_{j=i}^{M} (\sum_{i=1}^{N} \left(E_{ij} - \overline{E_j}\right)^2)^2} \quad (10)$$

$$E = |\overline{s}(t)| = \sqrt{s^2(t) + \overline{s}^2(t)} \quad (11)$$

3 Smart Conditioning Monitoring

In general, a fault is related to the measurable properties of a gearbox through physical effects. The physical effects can be influenced by disturbances. The measurable properties are the measurands that are acquired by the sensors. In general, a physical effect will influence more than one measurand, and a measurand will depend on more than one physical effect. The actual measurement of the measurands by the sensors is subjected to a measurement error. The goal is to minimize the measurement error as the measurement error and the model uncertainty can cause errors in the estimation of the fault status. The basic requirement for choosing measurands are signal–noise ratio (SNR) should be high and the measurands should be determinable and robust toward disturbances [32].

Analysis of vibration signals is an appropriate technique for smart monitoring of gearboxes. Any change in the gearbox condition will lead to an effect on the vibration signal. That means defects on gear will change the amplitude and phase of the gear's vibrations signal [33].

Traditional data acquisition systems use the acquisition card to measure the data and data is sent into the computer for processing. Advances in sensor technologies have led to smart monitoring technologies. These technologies are small in size, portable, can start quickly and use wireless communication. With smart monitoring technologies, researchers benefit from high-speed microcontrollers and enhance the quality of data acquisition [34].

A typical setup consists of three main parts: sensors, a signal conditioner with an A/D interface and a computer. Several sensors of different types can be strategically attached to the gearbox body. The signals from the sensor grid should be compatible with the monitoring application, to achieve such signal conditioning and data acquisition hardware is a must. Lastly, the signal is processed in the computer from which the health of the gearbox is obtained. Here, various machine learning algorithms and models can be used on the signal data to predict the fault in the gearbox before it occurs (Fig. 6).

When a fault occurs in a gearbox, some disturbances are observed. These disturbances are measured by sensors. The data from the sensors is then passed on to the conditional monitoring system as shown in Fig. 7. While designing this kind of system, the designer should consider what kind of distances are to be measured, the position of the sensors to measure the disturbances and develop the conditional monitoring model which will accurately show the health of the gearbox [32].

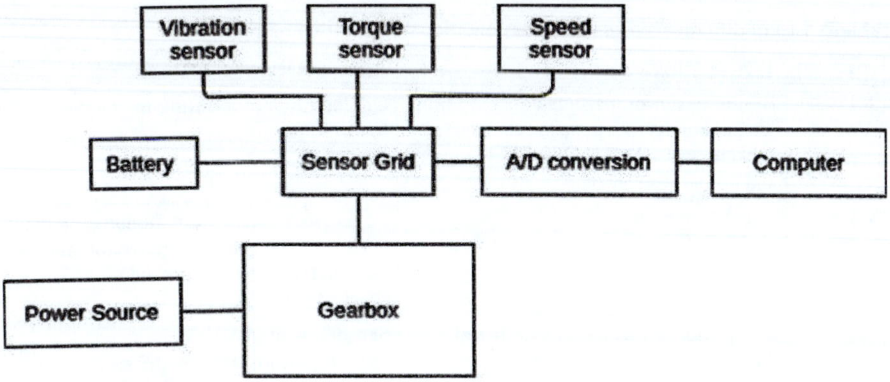

Fig. 6 Diagram showing the normal setup of sensor-based monitoring of a gearbox

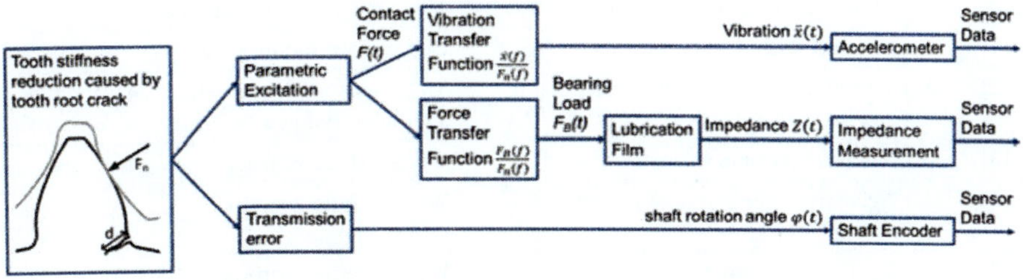

Fig. 7 Relationship between a tooth root crack and measurands [32]

Choosing the right sensors for smart condition monitoring has three criteria. They are as follows [32]:

1 Choosing the measurand (vibration, oil analysis, inverter current, etc.).
2 Choosing the sensor position that yields the best data quality.
3 Choosing the sensor principle and the actual sensor model for example choosing between piezoelectric and MEMS accelerometers.

The location of the sensor is critical for data quality. Integration of sensors close to the location of interest will reduce uncertainties and enhance the quality [35]. The table below displays the list of sensors used by different researchers to determine the health of the gearbox (Table 2).

Studies indicate that most of the researchers have used accelerometers to measure the vibration from the gearbox. Most of them have placed sensors on top of the gearbox. This approach is not very accurate as the signal does contain a lot of noise. Some studies have placed multiple sensors at different locations to solve this problem. The position does make a lot of difference to obtain the actual result. In only a few studies, researchers have tried to incorporate sensors as part of the gearbox. There are not many references found in the literature. This approach is complex but can

Table 2 List of different sensors used for conditional monitoring of a gear

Reference	Type of sensors used	Application
[36]	Humidity sensor, particle concentration sensor, conductivity sensor, dielectric constant sensor, viscosity sensor, quality and properties sensor	Oil condition monitoring in gearboxes for wind turbines
[37]	Accelerometers	12 accelerometers were mounted on the outside of the gearbox, generator, and the main bearing to measure the vibration acceleration
[38]	Self-powered wireless sensor based on the ADXL345 accelerometer	Sensors were installed on the wind turbine gearbox at three different positions
[39]	Built-in position sensors, motor encoder and linear scale	Feed-axis gearbox condition monitoring and fault detection
[40]	Motor encoder and linear scale	Mechanical drive components
[41]	Infrared sensor	Gearbox oil monitoring
[42]	Vibration accelerometer, microphone and acoustic emission sensor	Multi-sensor fused model in automobile and industrial gearboxes
[43]	Abrasive particle sensor	Wind turbine gearbox
[44]	Accelerometer and acoustic emission sensor	Wind turbine gearbox
[45]	Radial accelerometer	Wind turbine gearbox
[46]	Sound sensors	Feed-axis gearbox
[47]	Three-way acceleration sensor	Planetary gearbox
[48]	3 accelerometer and an acoustic emission sensor	
[49]	Accelerometer and acoustic emission sensor	1 stage gearbox with two gears
[50]	Infrared thermal sensor	A three-way gearbox with straight-cut bevel gears,
[51]	Accelerometer, temperature sensor, speed sensor and an acoustic emission sensor	Industrial gearboxes
[52]	2 torque, speed sensors and acceleration sensors	Two parallel spur gear stages and one planetary gear stage
[53]	Accelerometer sensor, speed sensor	Wind turbine gearbox
[54]	Piezoelectric energy harvester and an accelerometer	Wind turbine gearbox
[55]	Vibration sensor	Wind turbine gearbox
[56]	Strain sensor	Sun gear, planetary gear and ring gear
[57]	Ultrasound sensors, acoustic emission and vibration sensors	Wind turbine gearbox

(continued)

Table 2 (continued)

Reference	Type of sensors used	Application
[58]	Acoustic emission, vibration, speed and torque sensors	Helical gearbox
[59]	Piezoelectric vibration transducer	Gearbox in agricultural machines
[60]	Acoustic emission, vibration, speed and torque sensors	Turbojet engine gearbox test rig
[61]	Vibration and torque sensors	Gearbox in industrial machines
[62]	Vibration sensor	Gearbox in automobiles and industrial machines
[63]	Developed a built-in crack detection sensor	Developed an experimental rig
[64]	acceleration sensor and high-speed camera	Developed an experimental rig
[65]	Wireless accelerometer and hall effect sensor	Developed testbed
[66]	Accelerometer (DYTRAN 3097A2)	Vibration monitoring on a lathe
[67]	Tachometer, non-contact accelerometer (NI 9233)	Rotor-bearing test rig

produce accurate results as there are no external factors affecting the data quality. Some of the opportunities and challenges are as follows:

- Challenges like contact measurement, localized information, noise contamination and high computation costs make conditional monitoring of a gearbox complex and cost ineffective.
- With the single sensor used on top of the gearbox, the signal is very noisy. This might not give accurate results about all of the gearbox faults.
- Not all of the faults can be detected early, as early vibration signals are not good enough for fetching meaningful information.
- Studies show that the position of sensors is critical. Optimization location is challenging.
- Researchers have used multiple types of sensors. This helps to get information about different faults but optimization of sensor fusion remains an open challenge.
- Manufacturing a gearbox with all the required sensors included can produce accurate results. But this is too complex and provides opportunities for researchers to build smart gearboxes.

4 Summary

In this study, we have listed down the different conditional indicators used by researchers. Various sensors like vibration, torque, speed and acoustic emission sensors have been used by researchers to monitor gearbox health. The methods where sensors were used externally were not able to provide accurate results to identify defects in a gearbox. Opportunities and challenges are discussed in the study. Researchers can use the opportunity to develop a product like a smart gearbox which can send the quality data to machine learning algorithms for predicting the defect before it occurs and accurately determine the remaining useful life of the gearbox.

References

1. Elasha F, Greaves M, Mba D, Fang D (2017) A comparative study of the effectiveness of vibration and acoustic emission in diagnosing a defective bearing in a planetry gearbox. Appl Acoust 115(1):181–195
2. Kumar A, Gandhi CP, Yuqing Z, Kumar R, Xiang J, Latest developments in gear defect diagnosis and prognosis: a review. Measurement 158:107735. https://doi.org/10.1016/j.measurement.2020.107735
3. Resendiz-Ochoa E, Saucedo-Dorantes JJ, Benitez-Rangel JP, Osornio-Rios RA, Morales-Hernandez LA (2020) Novel methodology for condition monitoring of gear wear using supervised learning and infrared thermography. Appl Sci 10(2):506
4. Liu X, Yang Y, Zhang J (2016) Investigation on coupling effects between surface wear and dynamics in a spur gear system. Tribol Int 101:383–394
5. Zhang R, Gu F, Mansaf H, Wang T, Bal AD (2017) Gear wear monitoring by modulation signal bispectrum based on motor current signal analysis. Mech Syst Signal Process 94:202–213
6. Dempsey PJ (2003) Integration oil debris and vibration measurement for intelligent machine health monitoring. PhD. Thesis, The University of Toledo, Toledo, OH, USA
7. Allianz Versicherungs-AG (1978) Handbook of loss prevention, Springer-Verlag, Berlin
8. Michel AN, Miller RK (1983) Gears and their vibration, Marcel Dekker, Inc
9. Martin KF (1994) A review by discussion of condition monitoring and fault-diagnosis in machine tools. Int J Machi Tools Manuf 34:527–551
10. Miljkovic D (2016) Brief review of vibration-based machine condition monitoring. Hrvatska Elektroprivreda, Zagreb, Croatia,
11. Wang P, Lee CM (2019) Fault diagnosis of a helical gearbox based on an adaptive empirical wavelet transform in combination with a spectral subtraction method. Appl Sci 9:1696
12. Verucchi C, Bossio J, Bossio G, Acosta G (2016) Misalignment detection in induction motors with flexible coupling by means of estimated torque analysis and MCSA. Mech Syst Signal Process 80:570–581
13. Večeř P, Kreidl M, Šmíd R (2005) Condition indicators for gearbox condition monitoring systems. Acta Polytech 45(6)
14. Zhu J, Nostrand T, Spiegel C, Morton B (2014) Survey of condition indicators for condition monitoring systems. PHM_CONF, 6(1)
15. Decker HJ (2002) Crack detection for aerospace quality spur gears. NASA TM-2002–211492, ARL-TR-2682, NASA and the US Army Research Laboratory
16. Bechhoefer E, Qu Y, Zhu J, He D (2013) Signal processing techniques to improve an acoustic emission sensor. Annual Conf Prognost Health Manag Soc 4:1–8
17. Li JC, Limmer JD (2000) Model-based condition index for tracking gear wear and fatigue damage. Wear 241:26–32
18. Wu S, Zuo MJ, Parey A (2008) Simulation of spur gear dynamics and estimation of fault growth. J Sound Vib 317:608–624
19. Ma H, Pang X, Feng R, Song R, Wen B (2015) Fault features analysis of cracked gear considering the effects of the extended tooth contact. Eng Fail Anal 48:105–120

20. Ma J (1995) Energy operator and other demodulation approaches to gear defect detection. In: Proceedings of the 49th meeting of the Society for Machinery Failure Prevention Technology (MFPT), Virginia Beach, VA, pp 127–140
21. Zakrajsek JJ, Lewicki DG (1996) Detecting gear tooth fatigue cracks in advance of complete fracture. NASA M-107145, ARL TR-970, NASA and the US army aviation systems command
22. Wang W (2001) Early detection of gear tooth cracking using the resonance demodulation technique. Mech Syst Signal Process 15:887–903
23. Zhang XH, Kang JS, Zhao JS, Cao DC (2013) Features for fault diagnosis and prognosis of gearbox. Chem Eng Trans 33:1027–1032
24. Decker HJ, Lewicki DG (2003) Spiral bevel pinion crack detection in a helicopter gearbox. In: Proceedings of the American Helicopter Society 59th annual forum, Phoenix, AZ, pp 1222–1232
25. Shen CH, Wen J, Arunyanart P, Choy FK (2011) Vibration signature analysis and parameter extractions on damages in gears and rolling element bearings. Int Schol Res Netw. ISRN Mech Eng 402928:10
26. Zakrajsek JJ, Townsend DP, Decker HJ (1993) An analysis of gear fault detection methods as applied to pitting fatigue failure data. NASA TM-105950, AVSCOM TR-92-C-035, NASA and the US army aviation systems command, January
27. Dempsey PJ, Zakrajsek JJ (2001) Minimizing load effects on NA4 gear vibration diagnostic parameter. In: 55th Meeting sponsored by the society for machinery failure prevention technology, Virginia Beach, Virginia, 2–5, April
28. Antolick LJ, Branning JS, Wade DR, Dempsey PJ (2010) Evaluation of gear condition indicator performance on rotorcraft fleet. In: 66th American Helicopter Society
29. Zhang XH, Kang JS, Bechhoefer E, Zhao J (2014) A new feature extraction method for gear fault diagnosis and prognosis. Eksploatacja i Niezawodnosc—Mainten Reliab 16(2):295–300
30. Szczepanik A (1989) Time synchronous averaging of ball mill vibration. Mech Syst Signal Process 3:99–107
31. Zakrajsek JJ, Handschuh RF, Decker HJ (1994) Application of fault detection techniques to spiral bevel gear fatigue data. In: Proceedings of 48th meeting of the Society for Mechanical Failure Prevention Technology, Wakefield, ME, April
32. Martin G, Vogel S, Schirra T, Vorwerk-Handing G, Kirchner E (2018) Methodical evaluation of sensor positions for condition monitoring of gears
33. Wang W, Mc FP (1995) Decomposition of gear motion signals and its application to gearbox diagnostics. J Vib Acoust 117:363
34. Hızarcı B, Ümütlü R, Ozturk H, Kiral Z (2016) Smart condition monitoring of worm gearbox
35. Martin G, Schork S, Vogel S, Kirchner E (2018) Potential through mechatronic machine elements. Konstruktion 70(01–02):71–75
36. Salameh JP, Cauet S, Etien E, Sakout A, Rambault L (2018) Gearbox condition monitoring in wind turbines: a review. Mech Syst Signal Process. 111:251–264. ISSN 0888–3270. https://doi.org/10.1016/j.ymssp.2018.03.052
37. Zhang Z, Verma A, Kusiak A (2012) Fault analysis and condition monitoring of the wind turbine gearbox. IEEE Trans Energy Convers 27(2):526–535. https://doi.org/10.1109/TEC.2012.2189887
38. Lu L, He Y, Ruan Y, Yuan W (2021) Wind turbine planetary gearbox condition monitoring method based on wireless sensor and deep learning approach. In IEEE transactions on instrumentation and measurement, vol 70, pp 1–16. Art no. 3503016. https://doi.org/10.1109/TIM.2020.3028402.
39. Zhou Y, Runa T, Mei X, Jiang G, Sun N (2011) Feed-axis gearbox condition monitoring using built-in position sensors and EEMD method. Robot Comput Integr Manuf 27(4):785–793. ISSN 0736–5845. https://doi.org/10.1016/j.rcim.2010.12.001. https://www.sciencedirect.com/science/article/pii/S0736584510001778
40. Plapper V (2001) Sensorless machine tool condition monitoring based on open NCs. In: Proceedings of the 2001 IEEE international conference on robotics and automation, Seoul Korea, pp 3104–108

41. Rauscher M, Tremmel A, Schardt M, Koch A (2017) A non-dispersive infrared sensor for online condition monitoring of gearbox oil. Sensors 17(2):399
42. Praveen Kumar T, Saimurugan M, Hariharan RB, Siddharth S, Ramachandran KI (2019) A multi-sensor information fusion for fault diagnosis of a gearbox utilizing discrete wavelet features. Meas Sci Technol 30(8):11. https://doi.org/10.1088/1361-6501/ab0737
43. Zhang L, Yang Q (2020) Investigation of the design and fault prediction method for an abrasive particle sensor used in wind turbine gearbox. Energies 13(2):365
44. Soua S, Van Lieshout P, Perera A, Gan TH, Bridge B (2013) Determination of the combined vibrational and acoustic emission signature of a wind turbine gearbox and generator shaft in service as a pre-requisite for effective condition monitoring. Renew Energ 51:175–181, ISSN 0960–1481. https://doi.org/10.1016/j.renene.2012.07.004
45. Vamsi I, Sabareesh GR, Penumakala PK (2019) Comparison of condition monitoring techniques in assessing fault severity for a wind turbine gearbox under non-stationary loading. Mech Syst Signal Process 124:1–20, ISSN 0888–3270. https://doi.org/10.1016/j.ymssp.2019.01.038
46. Vanraj, Dhami SS, Pabla Guoping Ding BS (Reviewing Editor) (2017) Optimization of sound sensor placement for condition monitoring of fixed-axis gearbox. Cogent Eng 4:1. https://doi.org/10.1080/23311916.2017.1345673
47. Wu Z, Zhang Q, Cheng L, Tan S (2019) A new method of two-stage planetary gearbox fault detection based on multi-sensor information fusion. Appl Sci 9(24):5443
48. Alkhadafe H, Al-Habaibeh A, Daihzong S, Lotfi A Optimising sensor location for an enhanced gearbox condition monitoring system. In: 25th International congress on condition monitoring and diagnostic engineering, Huddersfield, UK
49. Loutas TH, Sotiriades G, Kalaitzoglou I, Kostopoulos V (2009) Condition monitoring of a single-stage gearbox with artificially induced gear cracks utilizing on-line vibration and acoustic emission measurements. Appl Acoust 70(9):1148–1159, ISSN 0003-682X. https://doi.org/10.1016/j.apacoust.2009.04.007
50. Li Y, Gu JX, Zhen D, Xu M, Ball A (2019) An evaluation of gearbox condition monitoring using infrared thermal images applied with convolutional neural networks. Sensors 19(9):2205
51. Adeyemi O, Onsy A, Sherrington I (2020) Towards the development of a tribotronic gearbox. In: Ball A, Gelman L, Rao B (eds) Advances in asset management and condition monitoring. smart innovation, systems and technologies, vol 166. Springer, Cham. https://doi.org/10.1007/978-3-030-57745-2_102
52. Pan Y, Hong R, Chen J, Singh J, Jia X (2019) Performance degradation assessment of a wind turbine gearbox based on multi-sensor data fusion. Mech Mach Theory 137:509–526, ISSN 0094-114X. https://doi.org/10.1016/j.mechmachtheory.2019.03.036
53. Luo H et al (2014) Effective and accurate approaches for wind turbine gearbox condition monitoring. Wind Energy 17(5):715–728
54. Lu SL, He Y, Wang T, Shi T, Li B (2019) Self-powered wireless sensor for fault diagnosis of wind turbine planetary gearbox. IEEE Access 7:87382–87395. https://doi.org/10.1109/ACCESS.2019 2925426
55. Garcia MC, Sanz-Bobi MA, Pico JD (2006) SIMAP: Intelligent system for predictive maintenance: application to the health condition monitoring of a windturbine gearbox, Comput Indus 57(6):552–568, ISSN 0166–3615. https://doi.org/10.1016/j.compind.2006.02.011
56. Yoon J, He D, Van Hecke B (2015) On the use of a single piezoelectric strain sensor for wind turbine planetary gearbox fault diagnosis. IEEE Trans Industr Electron 62(10):6585–6593. https://doi.org/10.1109/TIE.2015.2442216
57. Howard TP (2016) Development of a novel bearing concept for improved wind turbine gearbox reliability. PhD Thesis, University of Sheffield, Sheffield, UK
58. Alkhadafe H, Al-Habaibeh A, Lotfi A, (2016) Condition monitoring of helical gears using automated selection of features and sensors. Measurement 93:164–177, ISSN 0263–2241. https://doi.org/10.1016/j.measurement.2016.07.011
59. Andreas B, Karl-Ludwig K Development of a sensor- and cloud-based condition monitoring system for the detection of gear damage. https://www.ndt.net/article/dgzfp2020/papers/v4.pdf

60. Rezaei A, Dadouche A (2012) Development of a turbojet engine gearbox test rig for prognostics and health management. Mech Syst Signal Process. https://doi.org/10.1016/j.ymssp.2012.05.013
61. Razavi Far R et al. (2019) Information fusion and semi-supervised deep learning scheme for diagnosing gear faults in induction machine systems. IEEE Trans Indus Electr 66(8):6331–6342. https://doi.org/10.1109/TIE.2018.2873546
62. Suresh S, Naidu VPS (2021) Gearbox health condition monitoring using DWT features. In: Proceedings of the 6th national symposium on rotor dynamics, Bangalore, India, 2–3 July 2019; Rao JS, Arun Kumar V, Jana S (eds); Springer, Singapore, vol 329, pp 361–374
63. Iba D, Futagawa S, Miura N, Iizuka T, Masuda A, Sone A, Moriwaki I (2019) Development of smart gear system by conductive-ink print (impedance variation of a gear sensor with loads and data transmission from an antenna) 8. https://doi.org/10.1117/12.2515339
64. Iba D et al. (2019) Vibration analysis of a meshing gear pair by neural network (visualization of meshing vibration and detection of a crack at tooth root by VGG16 with transfer learning. In: Smart structures and NDE for energy systems and industry 4.0, (Presented at the Smart Structures and NDE for Energy Systems and Industry 4.0 International Society for Optics and Photonics), vol. 10973, 109730Y, Accessed 8. 8. 2021
65. Kwan C, Ayhan B, Yin J, Liu X, Ballal P, Athamneh A, Ramani A, Lee W, Lewis F (2009) Real-time system condition monitoring using wireless sensors 1–8. https://doi.org/10.1109/AERO.2009.4839681
66. Mouleeswaran S, Vikram M, Pradeep B (2015) Vibration monitoring for defect diagnosis on a machine tool: a comprehensive case study. Int J Acoust Vibr 20:4–9
67. kumar MS, kumar SS (2014) Condition monitoring of rotating machinery through vibration analysis. J Sci Indus Res 73

Application of Artificial Intelligence for Failure Prediction of Engine Through Condition Monitoring Technique

Suvendu Mohanty and Swarup Paul

Abstract Engine failure prediction, to date, has become more challenging for adequately diagnosing and assigning appropriate maintenance decision-making processes. This paper investigates the health of an engine through experimental observation using an artificial neural network (ANN). Lubricating oil analysis has been performed for diagnosing quantitative analysis, i.e. wear particle concentration (WPC), severity index (SI), wear severity index (WSI), and percentage of large particles (PLP). An ANN model using a nonlinear autoregressive with exogenous input (NARX) architecture has been employed to predict quantitative outputs. Finally, a data-driven approach by applying an artificial neural network to understand the system degradation from accumulated condition monitoring data is studied. Topology 3–18–4 from NARX (ANN) was optimal in developing a predictive failure model with regression coefficients (0.9985–0.9999), having an error autocorrelation factor bounded within 95% confidence limit and lowered MSE and MAPE values as 0.00093 and 3.56. The application of neural networks is increasingly attractive and seems to be the right choice for a data-driven diagnostic approach. In addition, the outcomes from the ANN data are validated with the experimental set so that the strength of the model is reflected and a pattern of failure from the historical monitoring of the operating engines is predicted.

Keywords Condition monitoring · Failure · Diagnosis · NARX · ANN

1 Introduction

Failure of an engine is not desirable to provide continuous power production, a vital energy source for automobiles and industrial sectors. In 2018, Khan et al. conducted failure assessments on the exhaust valves of heavy-duty natural gas engines overheated. He observed that failure occurred much earlier than the expected life span [1]. In 2009, Hirani investigated that maintenance contributes significantly to the

S. Mohanty (✉) · S. Paul
Production Engineering Department, National Institute of Technology, Agartala, Tripura, India
e-mail: suvendu.nita@gmail.com

U. S. Dixit et al. (eds.), *Advances in Forming, Machining and Automation*, Lecture Notes in Mechanical Engineering, https://doi.org/10.1007/978-981-19-3866-5_36

life of the technology capital assets. It describes a combination of all administrative, technical, and sporting stuff to operate maintenance and other physical layouts to re-establish appropriate working conditions [2]. In 2009, Levi and Eliaz investigated the goal of condition monitoring is to carry out maintenance activity optimally concerning time, cost, and quality. Failure analysis of a Wankel engine detected a high rise in WPC quantity. The spectroscopic analysis found the presence of a ferrous and non-ferrous particle, which shows a shift in WPC level from fatigue to rubbing wear mode [3]. In 2013, Biswas et al. suggested even for a better diagnostic step, both vibrations and ferrographic techniques. Oil analysis focuses on identifying the material deterioration level early [4]. In 2012, Prajapati et al. reviewed that condition monitoring is not only on identifying and assessing device faults but also on investigating, predicting, and monitoring faults compared to breakdowns and preventive maintenance [5]. In 2002, suggested that predictive maintenance is routine monitoring of the existing condition of rotating equipment, operational productivity, and various parameters providing evidence to assess the optimum time between measures to minimize costs and to decrease the number of indefinite delays [6]. In 2016, Asgari et al. investigated a new innovative approach using the NARX black box study to capture the turbine system's dynamic behaviour [7]. In 2016, Kiakojoori and Khorasani implemented an algorithm like Elman neural network, when combined with the Bayesian model, has effectively shown a better prediction in the degradation of turbine blades [8]. In 2001, Lu et al. applied engine diagnostic with the AANN model has resulted in a high reduction of noise level and increased fault tolerance [9]. In 1999, Laghari and Boujarwah suggested ANN application has become increasingly important not only in failure diagnostic but also in predicting different wear modes by using a developed neural classifier system through pattern recognition techniques [10]. In 2010, Huang et al. investigated new soft computing tools like GA-based neuro-fuzzy networks combined with a neural net for better fault prediction [11]. In 2017, Li et al. applied specific algorithms like SVM. These machine learning features can be adopted with condition monitoring data to predict fault at more advanced stages before failure [12].

Therefore, it is seen that the researchers have experimented with different aspects of engines using various tribological investigations of engines. But the unconventional fuel lubricating oil analysis with optimization technique has not been highlighted significantly by the researchers. This present work focuses on combining condition monitoring as a diagnostic approach with an ANN-developed model based on NARX architecture. After testing over many lubricant oil samples, an ANN diagnostic system has been designed at an optimal parameter setting to predict fault at an early stage. The predictive model developed from wear quantitative datasets can monitor the engine performance, help prevent maintenance, and warn of any maintenance that requests replacement or repair.

2 Experimental Procedures

2.1 *Sample Collection*

The sample collection for the present case study is done from a four-stroke, twelve-cylinder vertical CNG engine from the oil and natural gas industry. The working hours of the engine range from 17–18 h per day with an interval of 1–1.5 h following 6–7 h of steady running. The collection of lubricant samples from the CNG engine was done under normal operating conditions. The thermal properties of oil range at an approx. limit of 30000–32000 F, and speed of engine maximum up to 1100 RPM. The capacity of the lubricating chamber was 350 L with filter replacement within 800 h. Seven CNG engine lubricating oil samples were taken, keeping a periodical interval of 800 h from the start. Tetrachloroethylene (C_2Cl_4) was used as a cleaning solvent, and for preventing contamination, dry containers were used for safely preserving the lubricating oil samples. To retain the wear particles uniformly, the oil collection was done immediately after switching off the engine.

Quantitative Ferrographic technique. Quantitative ferrography helps analyse the nature, magnitude, and trend of growth in wear rate by the particle size distribution of wear debris, as shown in Fig. 1a, b. This characterizes and distinguishes different wear situations. Oil samples, along with solvent tetrachloroethylene (C_2Cl_4), are shaken in a test tube to reduce the oil's viscosity. This is made to flow through a precipitator tube under symphonic action. A magnet is placed beneath the glass tube. The magnetic attraction arrests the ferrous particles. Larger particles (5 μm) are deposited at the entry, while smaller particles (1–2 μm) are arranged away from the entry. The magnetic force is proportional to the particle volume, whereas the viscous force resisting motion is proportional to the particle area. The motion downward through the glass tube is proportional to the effective particle diameter. Two light beams pass through the precipitator tube. The first beam is located at the tube entry vicinity where large (L) particles are deposited, and the second beam crosses the tube where the smaller (S) particles are deposited. The number of particles deposited is measured by the attenuation of light from a light source, the light passing being detected by the photoelectric transducer.

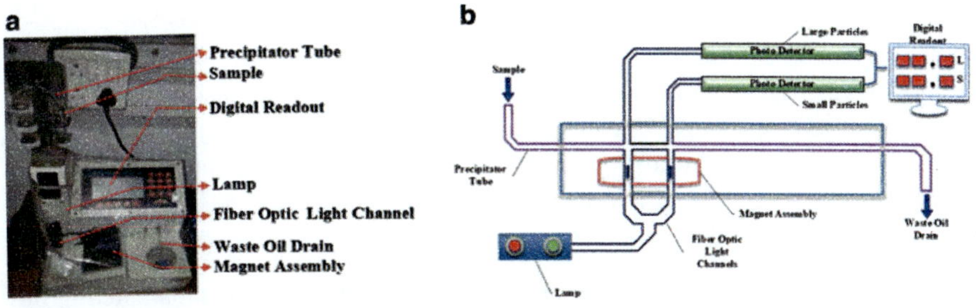

Fig. 1 **a** DR-V ferrograph and **b** schematic diagram of DR-V ferrograph

3 Nonlinear Autoregressive with Exogenous Input (NARX) Model

Statistical algorithms are more successful for prognostic technologies because of the versatility in data exploration and accuracy in complex systems. The fact that they can integrate advanced and contemporary strategies by constructing integrative diagnostic tools across a wide range of data series makes them particularly important. Neural networks are functional implementations of mathematical and theoretical models to simulate complex nonlinear dynamic structures applied to engineering tasks. Multi-step long-term dynamic modelling predictions are appropriate for dynamic system prognostic algorithms because they can be calculated faster and more comfortably than different prognosis methods. Therefore, recursive neural networks are commonly used as the most common diagnostic technique based on data, and many researchers in multiple domains have shown their merits by incorporating different approaches. One such technique for modelling multi-step prediction is NARX. NARX model is described as a time series recurrent neural system which can examine the condition-based maintenance of the engine. The diagnostic model based upon NARX was designed to learn from the current monitoring data and forecast wear conditions inside the system.

3.1 *Identification of Input and Output Data Collection*

The NARX model relies on several input variables and output variables. In the present study, the input model is the engine rpm, the engine oil temperature, and the engine's running hours, and they are used to predict four outputs: WPC, WSI, SI, and PLP. The data is normalized before computing and is calibrated to a range (-1 to $+1$) by leveraging equation, where Q_{k} is the normalized P_{k} estimate and $P_{\mathrm{k,\,max}}$, $P_{\mathrm{k,\,min}}$ are the extreme and lowest values.

$$Q_{\mathrm{k}} = -1 + \left\{ 2.0 \times \left(\frac{P_{\mathrm{k}} - P_{\mathrm{k,max}}}{P_{\mathrm{k,max}} - P_{\mathrm{k,min}}} \right) \right\}$$

Analysis of Dataset. The literature review demonstrates that the NARX basis for the design can be precisely used to train and test various sample data parts. To create the model, 75% of the input data was hand-picked as the training component and a delay line of ten time steps. The cross-recognition and validation of neural datasets have been correlated with a supplementary dataset. To prevent mistrust, selecting 15% from the cross-recognition dataset is a pivotal reason to address the potential overfitting of the neural network.

Determination of activation function. The transfer function is implemented in the NARX model, specifically designed to change the number of neurons in the output model. To design suitable neural circuits, weight and bias are used, and activation functions create a nonlinear connection between neurons. The hyperbolic tangent sigmoid activation function below is preferred among several functions because it is differentiable, continuous, and nonlinear to predict improved outcomes.

$$F(X) = \frac{2}{\left(1 + e^{-2x}\right) - 1}$$

Determination of training algorithm **.** The study aimed to perform a feed-forward backpropagation with the Levenberg–Marquardt training algorithm and mitigation learning function. The mean square error is the objective of the ANN dataset to estimate data accurately. The parameter for optimizing the TRAINLM function of Levenberg–Marquardt keeps weight and bias factors updated. To correctly compute the ANN model, recurrence was accomplished by varying the configuration of hidden neurons from 7 to 30.

Determination of mathematical variables. The statistical assessment uses various sets of statistical criteria to estimate the performance sets outcomes. Statistical parameters are discussed in Table 1, with their limitations mentioned.

where n, a_i, b_i, and $\overline{a}$ = the total data, actual value, predicted value, and actual measured mean value, respectively.

Table 1 Various mathematical variables

Statistical measurements	Formula	Accuracy
R (correlation coefficient)	$R = \sqrt{1 - \left\{\frac{\sum_{i=1}^{n}(a_i - b_i)^2}{\sum_{i=1}^{n} b_i^2}\right\}}$	> 0.9
R^2	$R^2 = 1 - \left\{\frac{\sum_{i=1}^{n}(a_i - b_i)^2}{\sum_{i=1}^{n} b_i^2}\right\}$	–
Mean square error (MSE)	$\text{MSE} = \frac{1}{N}\left[\sum_{i=1}^{n}(a_i - b)^2\right]$	< 0.001
Mean absolute percentage error (MAPE)	$\text{MAPE} = \left[\sum_{i=1}^{n}\left\lvert\frac{a_i - b_i}{a_i}\right\rvert\right] \times \frac{100\%}{n}$	< 5%

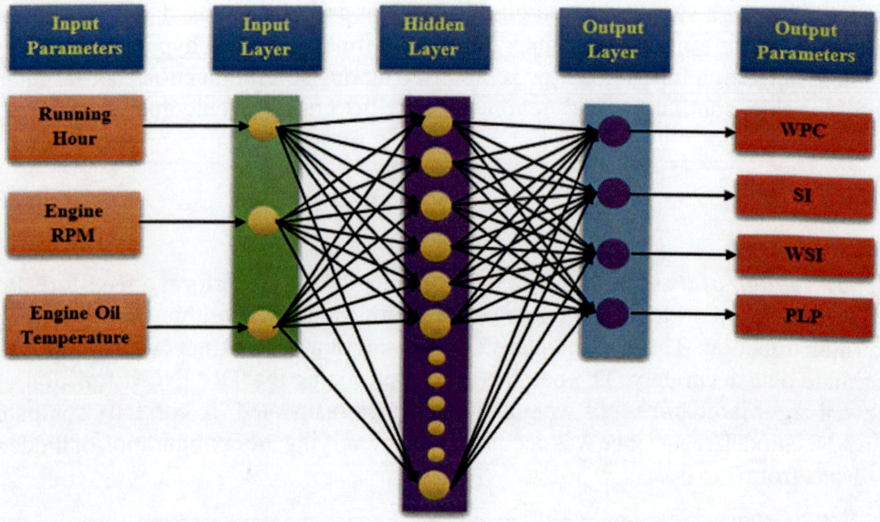

Fig. 2 ANN model configuration

4 Results and Discussion

4.1 Modelling of ANN (NARX Architecture)

The study developed the model with a marginal number of experimental process parameters then compared the outcome. The ANN model was constructed using NARX architecture which resembles a time series neural prediction system. In this study, there are three inputs concerning the engine parameters. Fig. 2. shows the framework model. There are four sets of output (WPC, SI, WSI, and PLP), and each set contains 40 data. Each output does not have the same unit; hence, the results cannot be linked. That is why the information datasets are adjusted or normalized to build it analogously.

Next, a multi-layer perceptron model is formed with NARX architecture having feed-forward error backpropagation and tapped delay lines to define the hidden relationship between input and output. The data is trained by matching the engine parameter with the quantitative parameters obtained from the experimental analysis.

Optimization of neuron topology. The NARX model is iterated a large number of times by changing the number of neurons in the hidden layer to produce an optimized failure prediction. The topology (3–18–4) is deemed best for predicting input–output parameters from the previous table. The adaptive network with 18 neurons produces maximum R and minimum MSE and MAPE values. The performance plot for achieving a minimum gradient of error 1e−10 has been attained with an epoch level of 2500 iterations and a learning rate of 0.001. The total average value of R is 0.97803, and 0.98906 for training, 0.94653 for validation, and 0.96363 for testing are shown in Fig. 3. The MSE and MAPE values acquired are 0.00093 and 3.56, respectively. A statistical error autocorrelation measurement has been achieved within the 95% confidence limit shown in Fig. 4. After the optimum topology is reached, the defined framework is further evaluated to determine its feasibility in a realistic way to define the structure. This model's reliability by statistical analysis measures makes it feasible to be a powerful identification method for diagnostic purposes and for detecting the pattern of failure of the engine.

NARX model validation. The objective of designing the NARX failure predictive model is to verify the output responses with the experimental outcomes. The quantitative parameters for each unit are validated with NARX outcomes. A regression plot between the model and experimental outcomes is done to minimize the error, as shown in Figs. 6, 8, 10, and 12, making the NARX model a robust identification tool. The comparison between the experimental and NARX outcomes for the units is given in Figs. 5, 7, 9, and 11.

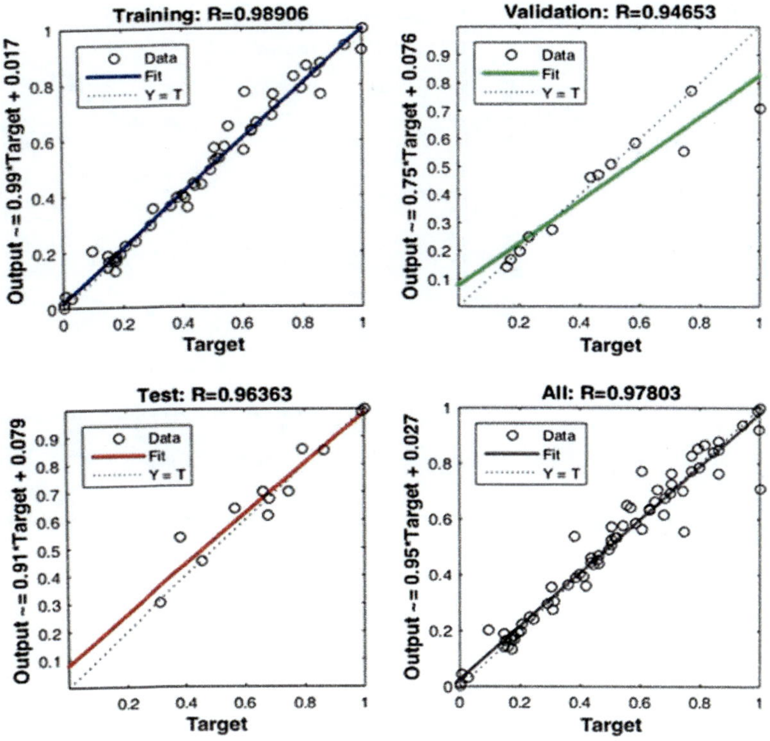

Fig. 3 Overall, R values for the optimized model

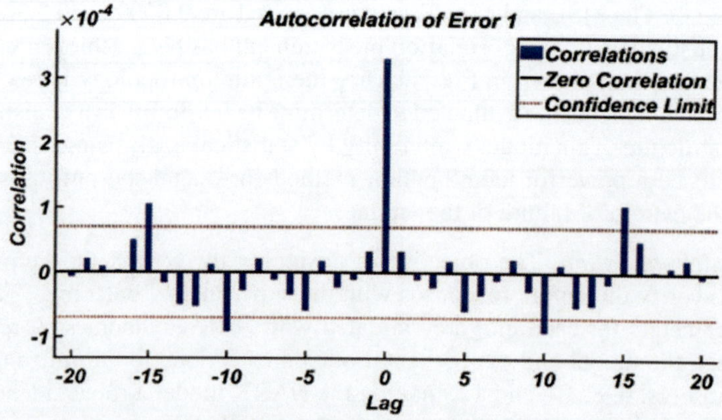

Fig. 4 Correlation measurement for the optimized model

Fig. 5 Experimental WPC versus ANN WPC

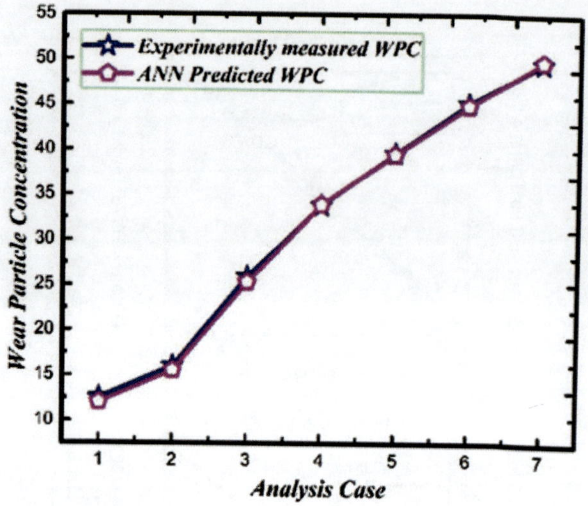

Fig. 6 Regression value of WPC

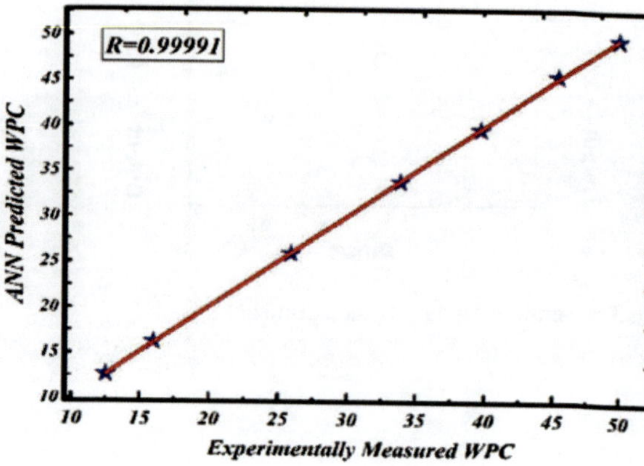

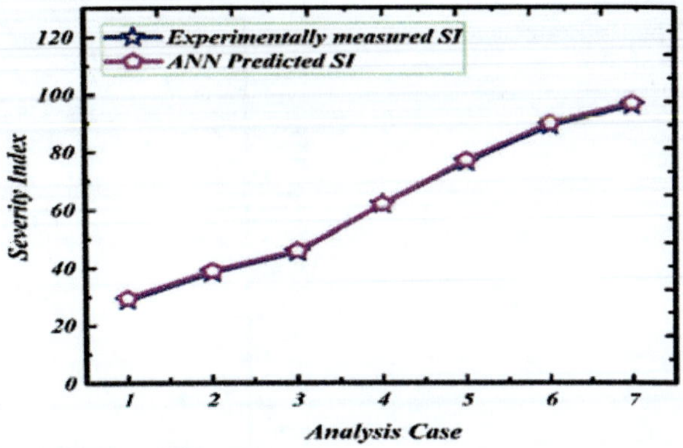

Fig. 7 Experimental SI versus ANN SI

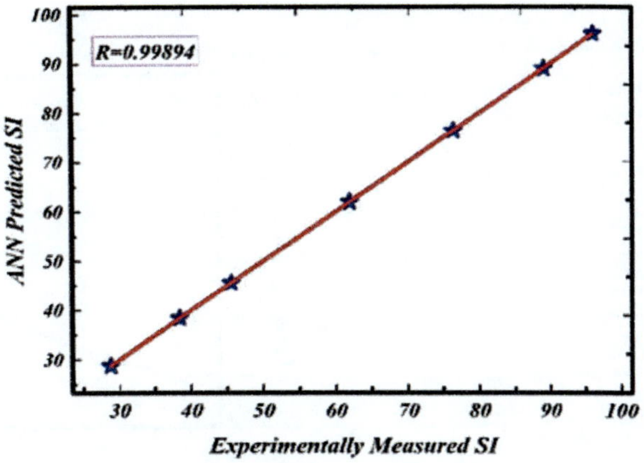

Fig. 8 Regression value of SI

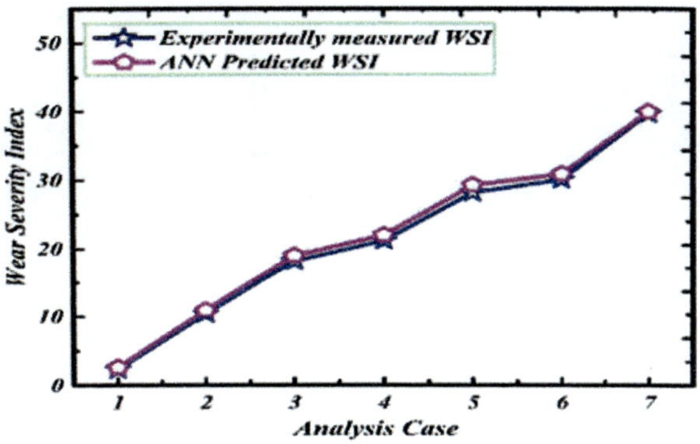

Fig. 9 Experimental WSI versus ANN WSI

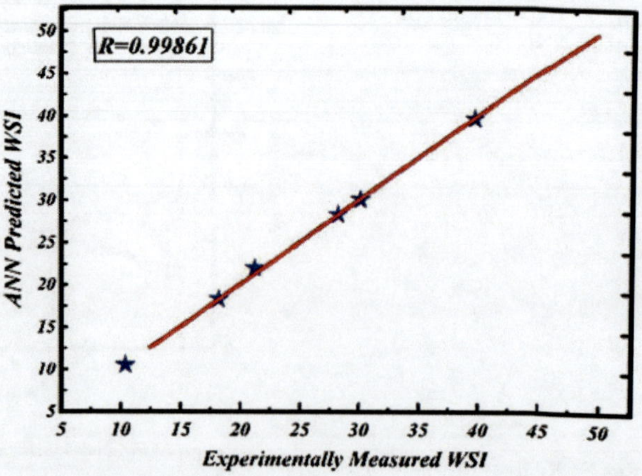

Fig. 10 Regression value of WSI

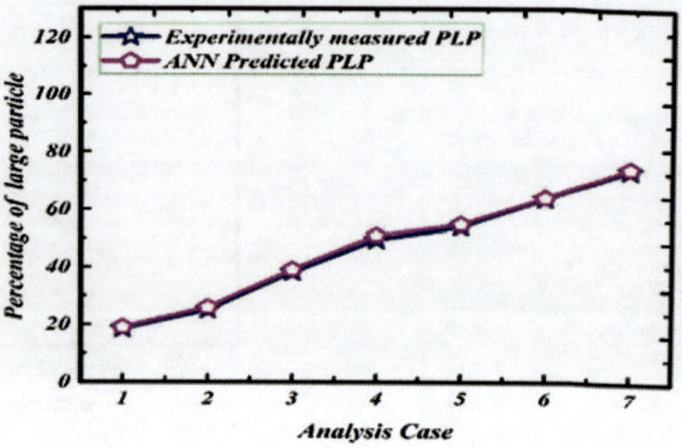

Fig. 11 Experimental PLP versus ANN PLP

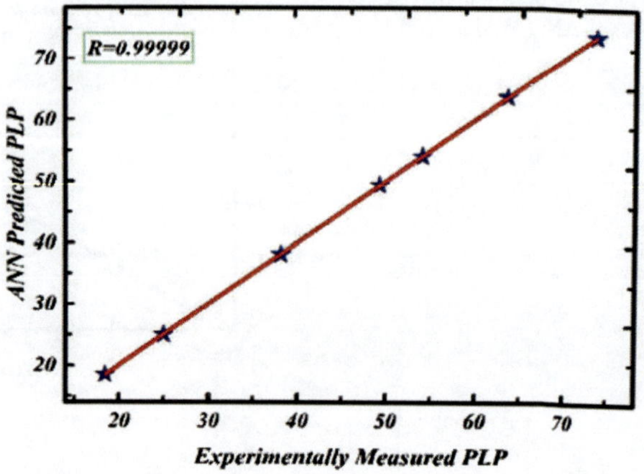

Fig. 12 Regression value of PLP

5 Conclusions

Based on the experimental and soft computational work, the following conclusions are given below:

1. Quantitative ferrographic analysis has been conducted successfully to monitor the health and understand the cause of failure inside the engine.
2. Topology 3–18–4 from NARX (ANN) was optimal in developing a predictive failure model from the experimental datasets.
3. The NARX model validation reflects the high value of regression coefficients (0.9985–0.9999), having an error autocorrelation factor bounded within the 95% confidence limit. Statistical measures with lower MSE and MAPE values as 0.00093 and 3.56, respectively, drive the neural model to accurately capture the failure trend and increase fault tolerance capacity to classify as an ANN-based failure predictive model.

References

1. Khan MI, Khan MA, Shakoor A (2018) A failure analysis of the exhaust valve from a heavy-duty natural gas engine. Eng Fail Anal 85:77–88
2. Hirani H (2009) Root cause failure analysis of outer ring fracture of four-row cylindrical roller bearing. Tribol Transac 52(2):180–190
3. Levi O, Eliaz N (2009) Failure analysis and condition monitoring of an open-loop oil system using ferrography. Tribol Lett 36:17–29
4. Biswas RK, Majumdar MC, Basu SK (2013) Vibration and oil analysis by ferrography for condition monitoring. J Inst Eng (India) Series C 94:267–274
5. Prajapati A, Bechtel J, Ganesan S (2012) Condition-based maintenance: a survey. J Quality Mainten Eng
6. Mobley RK (2002) An introduction to predictive maintenance. Elsevier
7. Asgari H, Chen X, Morini M, Pinelli M, Sainudiin R, Spina PR, Venturini M (2016) NARX models for simulation of the start-up operation of a single-shaft gas turbine. Appl Therm Eng 93:368–376
8. Kiakojoori S, Khorasani K (2016) Dynamic neural networks for gas turbine engine degradation prediction, health monitoring and prognosis. Neural Comput Appl 27(8):2157–2192
9. Lu PJ, Zhang MC, Hsu TC, Zhang J (2001) An evaluation of engine faults diagnostics using artificial neural networks. J Eng Gas Turbines Power 123(2):340–346
10. Laghari MS, Boujarwah A (1999) Wear particle texture classification using artificial neural networks. Int J Pattern Recognit Artif Intell 13(03):415–428
11. Huang ZY, Yu ZQ, Li ZX, Geng YC (2010) A fault diagnosis method of rolling bearing through wear particle and vibration analyses. Appl Mech Mater 26:676–681
12. Li Q, Zhao T, Zhang L, Sun W, Zhao X (2017) Ferrography wear particles image recognition based on extreme learning machine. J Electr Comput Eng

Application of MOORA Method in a Multi-Criteria Decision-Making Problem of an Automobile Parts Manufacturing Company

Rishav Raj Singh, S. R. Maity, and Divya Zindani

Abstract In the past decades, supplier selection emerged as a crucial process for any organization. Only making maximum profits and low-cost production are not only the main goal of an organization in these years. In this study, supplier selection has been done for an automobile parts manufacturing company from north India, with the implementation of multi-objective optimization based on ration analysis MOORA method on the eight different main criteria. The main criteria are Quality (*Q*), Cost (*C*), Transportation and delivery time (*TD*), flexibility and production capacity (*FPC*), payment term (*PT*), packaging of the product (*PP*), minimum order quantity (*MOQ*) and customer relations (*CR*). The weight of individual criteria and sub-criteria are calculated with the help of the Entropy weight method. After implementing the MOORA method, the ranking of the suppliers has been marked in which supplier 4 (0.0919) is best among all and supplier 1 (−0.3272) came last in the rank marking. The ranking showed that a supplier needs to be approachable at all front of the criteria which benefit industry, supplier, and customer too. Limitation and future scopes are also suggested.

Keywords Supplier selection · MOORA · Entropy · Ranking

1 Introduction

To satisfy the need of the customers, good quality products are mandatory, for good quality products appropriate suppliers and their selection for raw materials are a must.

R. R. Singh · S. R. Maity (✉)
Department of Mechanical Engineering, National Institute of Technology Silchar, Silchar 788010, India
e-mail: saikat.jumtech@gmail.com

R. R. Singh
e-mail: rishav_rs@mech.nits.ac.in

D. Zindani (✉)
Department of Mechanical Engineering, Sri Sivasubramaniya Nadar (SSN) College of Engineering, Kalavakkam 603110, India
e-mail: saikat.jumtech@gmail.com

U. S. Dixit et al. (eds.), *Advances in Forming, Machining and Automation*, Lecture Notes in Mechanical Engineering, https://doi.org/10.1007/978-981-19-3866-5_37

Supplier selection is a procedure of fetching the finest suppliers with upright price and quality at the proper time and quantity [1]. Correct suppliers lead to appropriate raw materials which move for good quality of products and good quality product complete the needs of customers. In the process for selection of suppliers, based on a performance score and then ranking their order gives the idea to selection accordingly. The performance of the overall supply chain is thoroughly dependent on the supplier selections and it is a fundamental issue too [2]. The selection of suppliers can be of different parameters which complete the objectives. The method of the selection procedure for suppliers can be done in many ways but ranking the suppliers is the important aspect for it.

The multiple-criteria decision-making (*MCDM*) process, in which decision-makers (*DMs*) assess various criteria, further chose the most suitable possibilities [3]. Simplicity, simple mathematical form, and a large number of inputs for various criteria are some of the advantages of this method. Supplier relation management (*SRM*) implies all processes, contains the best suppliers' selection, best settlement pricing, delivery, and the demand–supply correct ratio [4]. In recent times, supplier selection and quality of products are the necessary criteria for the growth of companies. These criteria significantly affect the overall performance of any organization along with customer satisfaction. Customer selection is depending on the quality of the product, the quality of the product depends on the selection of suppliers for raw materials. In the other dimension of supplier selection in the Indian automobile industry, the industrial environmental image was also considered for green supplier selection. The researcher studied, investigated, and ranked the environmental criteria with the help of the analytical hierarchy process (*AHP*) while considering the Indian automobile industry [5].

The MOORA method was first introduced by researcher Brauers to solve complex and contradictory decision-making problems. The researcher used the MOORA approach for the ranking of the maintenance contractors with taking the view of the clients and contractors [6]. The MOORA method involves fewer calculations and extensiveness which may evaluate large numbers of criteria [7]. This method is a reliable source for resolving multi-objective optimization problems, for permanent up-gradation of the process [8]. The best part of the MOORA method required less time for athematic calculations and the least criteria for evaluation [9].

The weights of the separate criteria, sub-criteria were learned by employing the entropy weight method. This method is broadly studied and applied in research work. The entropy weight method minimizes the human factors on the weight calculation which enhances the objectivity of the work [10].

2 Methodology

The objective of this study was to rank the suppliers with the help of the MOORA and entropy method. In this work, to calculate the weights of criteria, sub-criteria

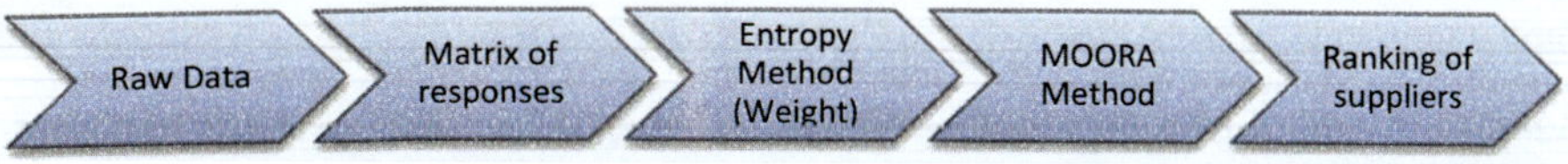

Fig. 1 Flow chart of the methodology

Entropy method, the predefined procedure is used then MOORA was applied to rank the suppliers. The flow chart of the methodology is shown in Fig. 1.

To approach the supplier selection on various criteria, A company that has 50+ employees, selected in North India situated at State Punjab. This company was established in the Year 2004 to manufacture tractor parts, PDC Drills for multiple companies according to their requirements. A company manufacture parts for multiple companies, they have a wide range of suppliers for raw materials. As per the information collected through the visit, with the procurement department, the company doesn't follow any kind of proper mechanism to select the suppliers for the products. To select the suppliers, they consider the price of the raw materials, quality, and availability of the raw materials.

Our visits, helped us to understand the expectation of the company from the supplier and vice versa as well as to complete a pilot survey for the questionnaire too. The pilot survey leads to modifying the questionnaire with more parameters and criteria. The questionnaire is divided into 8 different parameters (A–H), further sub-divided ($A1$, $A2$...$H1$, $H2$) according to the questions and suppliers' names as $S1$...$S5$. After the detailed discussions with a company as well as the supplier's questionnaire was divided into 8 criteria which are quality, cost, transportation and delivery time, flexibility and production capacity, payment term, packaging of the product, minimum order quantity, and customer relations. All the criteria have a different set of questions in the questionnaire. To collect the response, this questionnaire provides a 5-point Likert scale (1–5) facility in which 1 is least and 5 is highest. After the extensive research review, the following is the list of criteria and sub-criteria on basis of supplier's ranking has been done and the same has been tabulated in Table 1.

2.1 *Entropy Weight Method*

The Entropy weight method decides the weight of individual response without any consideration of the decision of the researcher. The basic concept of this method is considered in such a manner that top value is more helpful than smaller value. In this paper, the following steps are used to calculate the weight of the criteria. The weight calculation procedure is as follows [14]:

Table 1 List of criteria and sub-criteria used in this work [12–14]

Main criteria	Sub criteria
Quality (Q)	Inspection of product on quality ($A1$)
	No of defected product ($A2$)
	Performance of the product ($A3$)
	Quality/experience ($A4$)
Cost (C)	Holding cost ($B1$)
	Reduce in cost due to late ($B2$)
	Any discount offers on purchasing quantity ($B3$)
Transportation and delivery time (TD)	JIT delivery ($C1$)
	Speed of delivery/supply ($C2$)
	Transportation cost ($C3$)
	Delivery time flexibility ($C4$)
Flexibility and production capacity (FPC)	Capacity utilization ($D1$)
	A variety of products can be produced by the supplier ($D2$)
	Change in operation methods ($D3$)
	Respond to changes of quantity ($D4$)
Payment term (PT)	On-time payment ($E1$)
	Depending on payment conditions ($E2$)
Packaging of the product (PP)	Development in the packaging ($F1$)
	Modification to new products ($F2$)
Minimum order quantity (MOQ)	Holding stock on required quantity ($G1$)
	Supplier's adaption on requested quantity ($G2$)
Customer relations (H)	On the basis of contact ($H1$)
	On the basis of relationship ($H2$)

Step I: Normalization of the order of decision matrix to obtain the data outcomes P_{ij}.

$$P_{ij} = \frac{x_{ij}}{\sum_{i=1}^{m} x_{ij}} \tag{1}$$

Step II: Finding of the entropy from the data collected using the following equation:

$$E_j = -k\sum_{i=1}^{m} P_{ij}\ln P_{ij} \tag{2}$$

where, $k = 1/\ln(m)$.

Step III: Determining the different criteria weight:

$$W_j = \frac{1 - E_j}{\sum_{j=1}^{n}\left(1 - E_j\right)} \tag{3}$$

2.2 MOORA Method

The next step according to Fig. 1. Flow chart of the methodology. MOORA method is to be implemented to mark the rank of the suppliers. The procedures follow:

Step I: Normalize the Decision Matrix: based on the selection problem, the alternatives and attributes values in the decision matrix is to be normalized by the equation

$$x_{ij}^{*} = \frac{x_{ij}}{\left[\sum_{i=1}^{m} x_{ij}^{2}\right]^{1/2}} \quad \text{(where } j = 1, 2 \ldots n) \tag{4}$$

Step II: Estimation of assessment values (y_i)

$$y_i = \sum_{j=1}^{g} w_j x_{ij}^{*} - \sum_{j=g+1}^{n} w_j x_{ij}^{*} \quad \text{(where } j = 1, 2 \ldots n) \tag{5}$$

3 Results and Discussion

In this section, calculation of the values according to procedures of both the methods i.e., entropy weight method and MOORA were calculated. First, the weight of the criteria with the help of entropy weight method, which is tabulated in Table 2, and then ranking of the suppliers by MOORA method, which is tabulated in Table 4.

Table 2 Weight (W_j) values through entropy method

Sub criteria	W_j
A1	0.0031
A2	0.0430
A3	0.0088
A4	0.0051
B1	0.1218
B2	0.1158
B3	0.0088
C1	0.0152
C2	0.0051
C3	0.0305
C4	0.1218
D1	0.0050
D2	0.1032
D3	0.0514
D4	0.0836
E1	0.0031
E2	0.1158
F1	0.0305
F2	0.0691
G1	0.0247
G2	0.0053
H1	0.0152
H2	0.0144

3.1 Entropy Method

Following the above 2.1 procedures of the calculating weight of the criteria, sub-criteria, Table 2 shows the values of weights. These values were calculated according to the procedure mentioned in the above section.

Now, after finding the weights of all the criteria, the second section is about the ranking of the suppliers through the MOORA method was performed. As per 2.2, the step I is a normalization of the decision matrix, step II is the Estimation of assessment values and ranking the suppliers. The numerical values of both the steps are tabulated in Tables 3 and 4 respectively.

Step I: Normalize Decision Matrix, x_{ij}*:

Step II: Estimation of assessment values (y_i): ranking of the suppliers will follow after assessment values.

As per the final calculation, the weight of the criteria which is illustrated in Table 2 with the help of the entropy weight method, and the weights were given

Table 3 Normalization of decision matrix

Sub criteria	X_{ij}^*
$A1$	10.7703
$A2$	3.7417
$A3$	7.6811
$A4$	8.5440
$B1$	5.5678
$B2$	3.6056
$B3$	7.6811
$C1$	10.0000
$C2$	8.5440
$C3$	6.4807
$C4$	5.5678
$D1$	10.3441
$D2$	7.4162
$D3$	3.3166
$D4$	6.7823
$E1$	10.7703
$E2$	3.6056
$F1$	6.4807
$F2$	7.1414
$G1$	7.3485
$G2$	9.8995
$H1$	10.0000
$H2$	10.4403

Table 4 Ranking of suppliers

Supplier's name	y_i	Rank
S1	−0.3272	5
S2	0.0149	3
S3	0.0900	2
S4	0.0919	1
S5	0.0139	4

priority according to the weight values. The five suppliers of ABC company were ranked through the MOORA method as S4 (0.0919), S3 (0.900), S2 (0.0149), S5 (0.0139), S1 (−0.03272). After the evaluation, supplier 4 (S4) ranked high on the chosen criteria assessed.

4 Conclusions

In this research work, the MOORA method along with the entropy weight method was used to work out the multi-criteria optimization problem in supplier selection. There are many other methods to solve multi-criteria decision-making problems and each, method has its uses and limitations. In this problem, suppliers are ranked with the help of the MOORA method and this study can further extend to various other criteria like reputation of suppliers and warranty as well as sub-criteria accordingly. A comparative study can be done by taking various methods for the same criteria and sub-criteria in different areas. And at last, the best choice for the ABC company to select supplier S4, which is the best among all the suppliers participated in this supplier selection problem.

Acknowledgements The authors express their sincere thanks to *the automobile parts manufacturing company* for providing necessary data and inputs and *National Institute of Technology, Silchar (Assam).*

References

1. Ayhan MB, Kilic HS (2015) A two-stage approach for supplier selection problem in multi-item/multi-supplier environment with quantity discounts. Comput Ind Eng 85:1–12
2. Sanayei A, Mousavi SF, Yazdankhah A (2010) Group decision-making process for supplier selection with VIKOR under fuzzy environment. Expert Syst Appl 37:24–30
3. Wu Q, Zhou L, Chen Y, Chen H (2019) An integrated approach to green supplier selection based on the interval type-2 fuzzy best-worst and extended VIKOR methods. Inf Sci 502:394–417
4. Chopra S, Meindl P (2007) Supply chain management: strategy, planning, and operation, 3rd edn. Pearson Prentice Hall, Upper Saddle River, New Jersey
5. Mathiyazhagan K, Sudhakar S, Bhalotia A (2017) Modeling the criteria for selection of suppliers towards green aspect: a case in Indian automobile industry. Opsearch 55(1):65–84
6. Brauers WKM (2004) Optimization methods for a stakeholder society: a revolution in economic thinking by multi-objective optimization. Kluer Academic Publishers, Boston
7. Chakraborty S (2011) Applications of the MOORA method for decision making in a manufacturing environment. Int J Adv Manuf Technol 54(9–12):1155–1166
8. Chaturvedi V (2014) Parametric optimization of ECM process parameters by MOORA method. IJREAS 4(10):114–117
9. Tansel İç Y (2020) A multi-objective credit evaluation model using MOORA method and goal programming. Arab J Sci Eng 45(3):2035–2048
10. Zhi-Hong Z, Yi Y, Jing-Nan S (2006) Entropy method for determination of the weight of evaluating indicators in fuzzy synthetic evaluation for water quality assessment. J Environ Sci 18:1020–1023
11. Taherdoost H, Brard A (2019) Analyzing the process of supplier selection criteria and methods. Proc Manuf 32:1024–1034

12. Guarnieri P, Trojan F (2019) Decision making on supplier selection based on social, ethical, and environmental criteria: a study in the textile industry. Resour Conserv Recycl 141:347–361
13. Yang CM, Chen KS, Hsu TH, Hsieh HC (2019) Supplier selection and performance evaluation for high-voltage power film capacitors in a fuzzy environment. Appl Sci 9(23):5253
14. Khan Z, Sharma T, George J, Badoniya P (2019) Entropy weight based multi-objective optimization based on ratio analysis (MOORA) model for supplier selection in supply chain management. Int J Sci Eng Res 10:1545–1560

Comprehensive Survey and a Proposal for Blockchain-Enabled Supply Chain Management

G. R. Karpagam, L. Pavithra, and G. Swetha

Abstract Supply chain management (SCM) involves the whole process starting from supplying the raw material to the manufacturer, to the delivery of the final product. But, tracking each step of the process is becoming difficult taking into account its authenticity, quality, traceability and accountability. There is no proper record for the stakeholders to verify these parameters. Blockchain being one of the most popular and innovative technologies, can solve the issue by storing all this information in a decentralized, immutable and traceable manner. This allows all the stakeholders of the system to be up-to-date ensuring privacy. This paper proposes a blockchain-enabled SCM for the secure trading of militant weapons and goods for governments across the globe. The work also provides a comprehensive literature survey of papers in the same domain and a brief introduction and working of blockchain technology to point future research towards the same direction.

Keywords Blockchain · Supply chain management · Decentralization · Distributed ledger · Immutability

1 Introduction

Cyber-physical systems (CPS) promise solutions in manufacturing, logistics and supply chain management by combining physical objects with integrated computing facilities. CPS employs blockchain to achieve a distributed trust model by eliminating a third-party. Each block bundles an array of transaction records and their cryptographic chain links. Supply chain management (SCM) involves participants

G. R. Karpagam · L. Pavithra (✉) · G. Swetha
Department of Computer Science and Engineering, PSG College of Technology, Coimbatore 641 004, India
e-mail: 19z231@psgtech.ac.in

G. R. Karpagam
e-mail: grk.cse@psgtech.ac.in

G. Swetha
e-mail: 19z252@psgtech.ac.in

U. S. Dixit et al. (eds.), *Advances in Forming, Machining and Automation*, Lecture Notes in Mechanical Engineering, https://doi.org/10.1007/978-981-19-3866-5_38

like raw material suppliers, manufacturers, distributors, retailers and customers. It is the process of transporting the goods from manufacturer to customer.

2 Problem Statement

Have you ever thought about how important it is to deliver quality military weapons securely? It requires an extreme level of security to transport these weapons from manufacturer to rightful owner. Imagine a military force purchasing some weapons like firearms from manufacturers. On its delivery, it is found that there are some mechanical malfunctions in the firearm due to counterfeit products. On examining, the manufacturer and raw material supplier blamed each other, claiming that the error was not on their side and now the examiner tries to look into the records, and what if the manufacturer knocks down all those records? Can you imagine how dreadful it is if counterfeiting or grey market trade happens in military weapons or the spare parts of some crucial weapons? What would happen if this military weapon does not reach a rightful or intended owner?

Our objective is to use blockchain technology in supply chain management. Blockchain is a decentralized, immutable and distributed ledger technology to store data in a secured way. It stores all the data from the manufacture of raw materials to the final product where no one would be able to tamper with the data. It allows us to trace information and is transparent to all the stakeholders of the system. A private blockchain is one where only authorized manufacturers and suppliers will be able to take part under the supervision of central authority and can be used for militant purposes where information should be confidential. It prevents the replacement of original products with counterfeit products as all the activities are recorded in the blockchain. In a decentralized network, there is no single point of failure. If there are any issues with the quality of the product or delay in delivery, customers will be able to track those issues with the help of records. Hence, blockchain plays a crucial role in supply chain management.

3 Related Work

3.1 Comprehensive Literature Survey

This section describes the existing work in the area of blockchain in the following dimensions, namely concepts, consensus algorithms, currency, applications and challenges (see Table 1).

Table 1 Comprehensive survey

S. No.	Purpose/goal	Journal/conference paper	Remarks
Supply chain management ([1–5])			
1	Detailed analysis and categorization of various definitions of SCM to impart a clear understanding for the readers	Journal	Novices to professionals alike, the paper does a great job in defining the basics of SCM thus allowing researchers to have a strong foundation to do future work on SCM
2	Understand, explore and study the evolution of SCM and discuss the future of SCM	Article	Everything to be known about the evolution of SCM is depicted with the help of timeline diagrams, making it easy for fellow researchers. The paper answers various questions related to SCM in a detailed yet straightforward manner
3	Proposal of the solution to the issue of the fragile rural supply chain management system	Conference paper	The proposed solution employs the usage of cellphones and paper watermarking systems, efficient and cost-effective, which can be of great use for improving rural SCM
4	Develop and review environmental supply chain strategies	Research article	Outlines the role of environment in supply chain strategy development by conducting appropriate interviews and data collection
5	Identify the major problems with healthcare and SCM. Discuss solutions for the same	Journal	Detailed study of issues relating to patient safety, communication, waiting times and integration is provided along with appropriate examples
Blockchain and its applications ([2, 6–11])			
6	Blockchain-based access control management to maintain health records	Article	The proposed solution ensures data privacy and simultaneously hands over a great degree of authority to the user regarding control over data Issues of network traffic and reliability of pointers are a concern yet to be discussed

(continued)

Table 1 (continued)

S. No.	Purpose/goal	Journal/conference paper	Remarks
7	To summarize blockchain-based applications	Conference paper	The paper provides a timely summary for individuals and organizations interested in blockchains
8	A study of blockchain-based intelligent transportation system (ITS)	Conference paper	With case studies to support the model, the paper details the working of the conceptual model in a way which provides to be useful for future research in this domain
9	To employ blockchain in aiding 'smartization' of urban cities	Article	The concept of smart cities is new and including blockchain for incorporating information technology in urban cities is a unique approach
10	Present the applications of blockchain in marketing	Conference paper	Various preexisting research related to blockchain and marketing is explored along with some proposed applications proving to help widen the future scope of the model
11	Establish an understanding of blockchain for tourism. Research on a conceptual model for implementing the same	Journal	The paper demonstrates how blockchain can revolutionize the tourism industry by providing related preexisting research on the same
12	Provide an in-depth analysis of other papers that discuss blockchain and healthcare. Discuss papers that associate blockchain with other parallel technologies	Article	Erikson Julio et al. make it effortless for fellow researchers by clearly tabulating a comparison of the main features found in papers, jotting the positives and the negatives. Alongside the comparison, a brief working, application in healthcare and challenges faced by blockchain is neatly surveyed
Supply chain management enabled with blockchain ([12–17])			
13	Propose a blockchain-based framework for supply chain quality management	Conference paper	The paper contributes a comprehensive system architecture for SCQM. The work details are presented in a very organized way, magnifying every aspect of the framework

(continued)

Table 1 (continued)

S. No.	Purpose/goal	Journal/conference paper	Remarks
14	Study of solutions, issues and predictions for employing blockchain in solving real-world issues	Journal	The areas that the paper touches on are data exchange, contracts and supply chain management about healthcare With appropriate examples, the paper presents the projected use of blockchain in healthcare
15	Discuss blockchain SCM intersections so far and identify five use case clusters of blockchain technology in SCM	Article	The paper provides a systematic literature review of papers concerning SCM and blockchain technology. It also analyses the present-day trends of this intersection with the help of datasets
16	Provide a study on the current status of blockchain-enabled SCM. Detail an example of how it can be used in a global supply chain network	Article	The paper describes clearly how blockchain can be of advantage in terms of transparency and tracking for SCM and proposes a detailed solution with an example, making it easier for readers to grasp the concept
17	Aim to solve the issues in the food supply chain with the help of this technology. Propose a solution for the same	Conference paper	The potential of this technology is realized in the food supply chain domain. The paper also introduces a food supply chain model with HACCP and states hazard controls for the food supply chain
18	Enunciate the usability, merits, and demerits of blockchain and RFID technology for agri-food supply chain traceability systems and propose a model for the same	Conference paper	With a description of RFID and blockchain technology for Agri SCM, the paper provides a conceptual framework implementing the same. Along with the advantages of the proposed system, the challenges and risks are also presented, making the paper a panacea for future research

4 Background

The comprehensive survey describes the existing work in three dimensions, namely supply chain management, blockchain, and its applications, and supply chain management enabled with blockchain. Papers listed under SCM deal with how SCM is done in several fields and explore solutions to the existing problems in regards to SCM. These solutions deploy technologies that do not involve blockchain and its properties. The second dimension, namely blockchain and its applications provides remarks on papers that employ blockchain solutions to real-world problems exclusive of SCM. This has been provided to know how the properties of blockchain often help solve problems that cannot be incorporated wholly by other technologies. The third dimension is the amalgamation of the previous two and outlines remarks about papers that employ blockchain to solve problems in SCM. The literature survey of these papers provided here can provide future researchers to gain a basic understanding of how blockchain can be integrated with other domains and help efficiently solve problems in the same. Table 2 shows SCM with and without blockchain.

Blockchain 1.0—Cryptocurrency. The primary focus and use case of blockchain 1.0 was cryptocurrency. The current monetary systems work on centralized authority, having its demerits. Blockchain technology was initially proposed by Nakamoto [18] as an electronic cash system. While bitcoin is an application, blockchain is the underlying technology. Satoshi Nakamoto's white paper described it as a peer-to-peer electronic cash system. Bitcoin ruled out the client–server system by introducing the P2P system, allowing two parties to trust the immutability of their blocks rather

Table 2 SCM with and without blockchain

SCM without blockchain	SCM with blockchain
If there occur any mistakes in shipments, payments, or counterfeiting, it is difficult to trace and sometimes cannot even find the source	It is easy to trace the error because blockchain is a distributed ledger, it maintains the record of all activities and it is distributed amongst all authorized participants of the network
Not all participants have access to the record	Records are encrypted and distributed; thus, all authorized participants have their copy of the blockchain, thus they can review all transactions and activities, establishing a trustworthy environment
Data is mutable	Data is immutable. In the blockchain, each block is connected to the previous block through a hash function. If one tries to tamper with the block, it will be apparent to the network as its hash code changes and the hash code of all the blocks connected to it will also change. Thus, a person has to mine all the blocks connected to it and it is nearly impossible
It involves lots of physical documents and scrutinizing process at every entry and exit point	Reduces paperwork

than to trust a third party. Blockchain employs proof of work, which involves a lot of electricity to be spent by miners to mine a block. Blockchain 1.0 limited the technology to the financial world. **Blockchain 2.0— Ethereum** When developers and other blockchain enthusiasts got to know about the untapped potential of blockchain technology, a domino of tasks were put into action to extend this technology to other fields. The result of this is the second generation Blockchain 2.0, Ethereum. Ethereum is a decentralized programmable blockchain. It comprises smart contracts. They are scripts that run above the blockchain. The limitation that existed with Ethereum is its scalability issues. The transaction speed of Ethereum is a meagre 7–15 transactions per second. The load is increasing substantially day by day, slowing the network even more decreasing the transaction throughput. **Blockchain 3.0** The inefficiencies and hiccups of the previous generation led to the evolution of Blockchain 3.0. Blockchain 3.0 was designed keeping in mind the scalability issues. Interoperability allows communication between blockchains, thus increasing network speed. While IOTA, Nano, Cardano and others claim to be Blockchain 3.0, critics often see them as post-blockchains. Whether we are moving towards the next generation of still improvising Blockchain 3.0 is uncertain. While one portion of enthusiasts claim that Blockchain 4.0 happens when AI meets blockchain, the other portion claim that Blockchain 4.0 would focus on features like scalability, efficiency, energy conservation and so on. Some significant projects which are on the brink to move to the next generation are DeepChainBrain and SingularityNet. If proved revolutionary, they can drag the entire blockchain industry to the next generation.

The following are the terms and terminologies for the reader to understand the rest of the paper: **Consensus**. In a distributed, dynamically changing network, an agreement has to be reached by all participant nodes on a decision. This mechanism of making sure that all participant nodes agree on the current status of the blockchain and making sure that genuine, agreed-upon transactions are only present in the blocks is called a consensus mechanism. Proof of work and proof of state are some common consensus mechanisms. **Decentralization** A system is decentralized if its participants can connect either directly (peer-to-peer) or through a network of nodes that are not centrally managed. In a decentralized blockchain network, no one has to know or trust anyone else. As there is no trusted central authority governing the network, hence no single point of failure. **Distributed ledger** A distributed ledger is a type of database that is shared amongst all the participants of a decentralized network. The distributed ledger records the transactions of the participants, such as the exchange of data or assets. There is a no central authority or third-party mediator to govern the network. Transaction data is distributed to all the nodes in the network. Each node has an updated copy of all data. **Proof of Stake** Unlike PoW, this hands responsibility to users. Instead of spending energy to create blocks, based on a node's stake, it is allowed to create a block. The other nodes can validate blocks and can get rewards. If they validate a malicious block, their stake is taken away. **Proof of Work** Proof of work (PoW) is a consensus algorithm that works by proving to others that computational effort was spent which qualifies them to add a new block to the network. Bitcoin, Litecoin, Monero and other cryptocurrencies use this mechanism. Fig. 1 depicts when do we need to use blockchain.

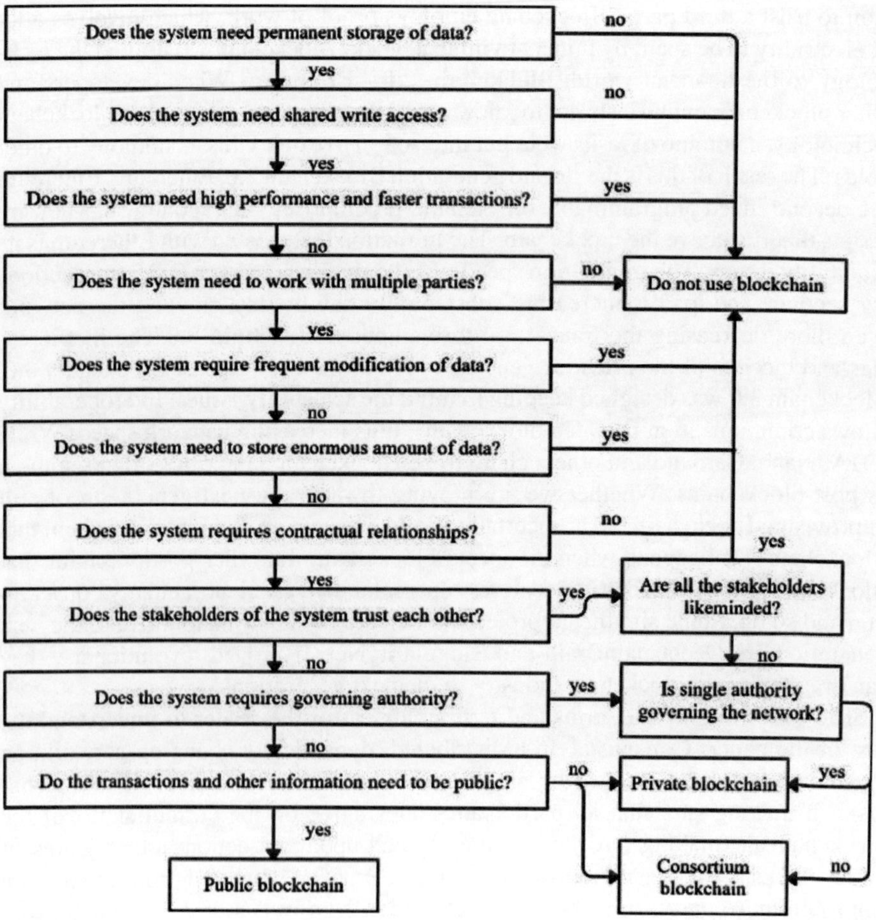

Fig. 1 When do we need to use blockchain

The following are the types of blockchain: *Permissionless blockchain.* In a permissionless blockchain, anyone can join the network. There is no controlling authority to govern the network; hence, participants do not require any permission to join the network. *Permissioned blockchain* in a permissioned blockchain, the participants need permission to join the network and make transactions. In this blockchain, users cannot easily join and view the transactions. These are mostly preferred by centralized organizations. *Public blockchain* A public blockchain is a permissionless blockchain in which anyone can join the network to read and submit transactions. In this blockchain, transactions are visible to all the participants in the network. *Private blockchain* A private blockchain is permissioned blockchain in which organizations control who is allowed to join the network. These are mostly used in organizations in which only authorized members are allowed to participate in the network. *Consortium blockchain* In a private blockchain, only a single organization governs the network, whereas in consortium blockchain groups of an organization come together to govern the network. They can decide upon who can read and write the transactions.

4.1 Methodology

The following steps say how does blockchain work.

Step 1: Transaction enters the blockchain network (see Fig. 2).
Step 2: Once the transaction enters the network, all the nodes verify and validate the transaction (see Fig. 3).
Step 3: This transaction will get into the transaction pool. The transaction pool is the place where all the unconfirmed transactions reside (see Fig. 4).
Step 4: Miners take those transactions from the translation pool (normally the ones which generate a high transaction fee) and try to solve the mathematical puzzle. The miner who solves the puzzle will create a block and add the transaction to the block.
Step 5: Block gets added to the blockchain.

Fig. 2 Sender initiates the transaction

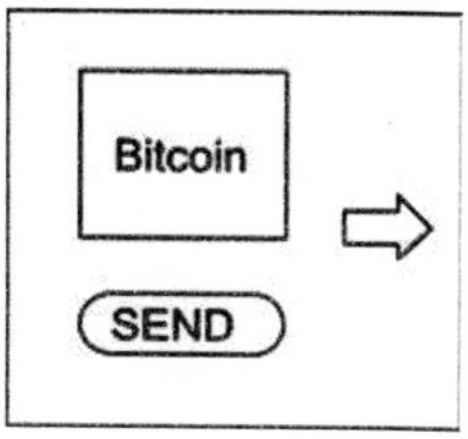

Fig. 3 Nodes verifying the transaction

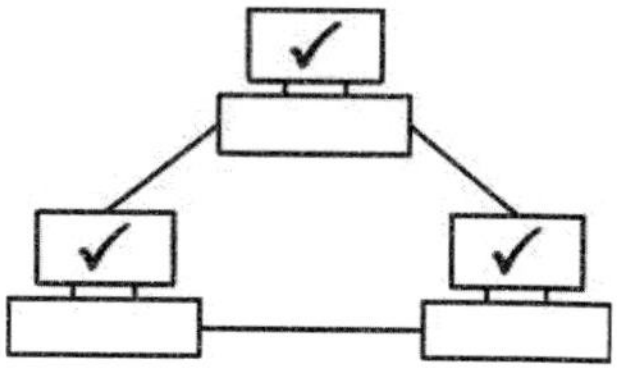

Fig. 4 Transaction pool

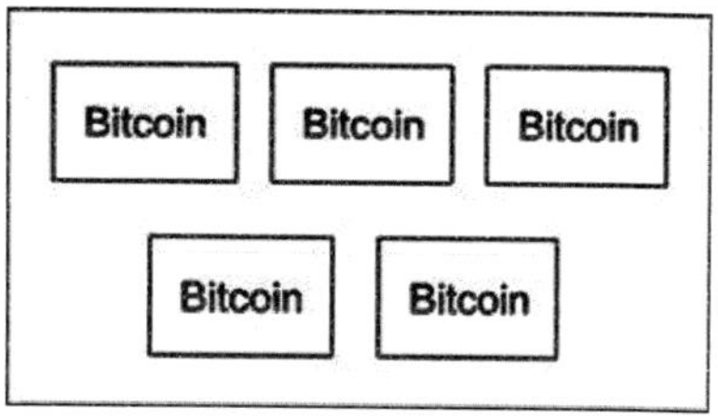

What would happen if more than one miner was able to solve the puzzle and in this case whose block will get added to the blockchain? Here comes the longest chain rule. Based upon the speed by which a block is created, i.e. miner, who is having a better processor will create a block faster than others. So, other miners (there will be a lot of miners who are trying to create new blocks) will keep on adding their block on top of that block. This builds a long chain of blocks. This longest chain will get added to the blockchain (see Fig. 5).
Step 6: Transaction has been completed (see Fig. 6).

Once the block was created and added to the blockchain, the person who initiated the transaction will get one confirmation. As more blocks get added to the blockchain, the number of confirmations increases. Usually, it is suggested to wait until six confirmations to assume it as fully confirmed without any chances of cancellation.

Fig. 5 Longest chain rule

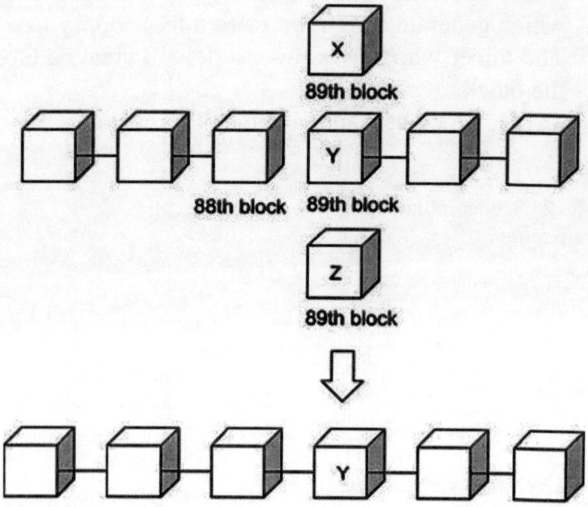

Fig. 6 Transaction is successful

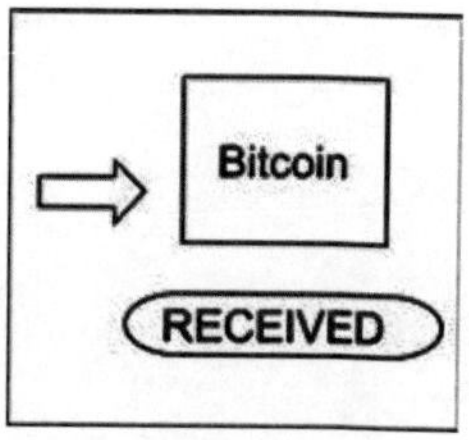

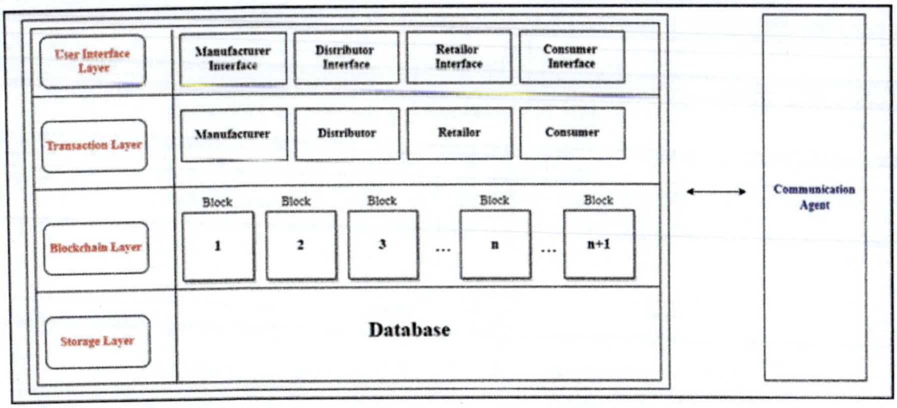

Fig. 7 Conceptual architecture

5 Conceptual Architecture

The conceptual architecture portrays the end-to-end working of the blockchain-enabled supply chain management system. It provides an avenue for solutions in the system through the various layers and their inherent components. This section describes the architecture and its working (see Fig. 7).

User Interface Layer deals with the actual user and is responsible for implementing the various user functionalities like displaying required data, the input of new data, updating of existing data, and more. In **the Transaction Layer**, the various transactions performed by the different tiers of users are visible here. In **Blockchain Layer,** the data captured is stored on the blockchain network in the form of blocks. The data is hashed using a hashing algorithm, encrypted with a private key, and then stored on the blocks. In **Storage Layer**, no central storage is recommended since blockchain is decentralized, hence stored in nodes/computers/systems across the network. The nodes have a copy of the blockchain which in turn contains the transactions.

6 Sequence of Actions

1. Manufacturers, distributors and customers register themselves and create an account with a blockchain application.
2. Manufacturer requests access to the blockchain application to append the details of their products, and the application verifies the authenticity of the manufacturer and validates the access. Now, the manufacturer sells their products to various distributors.

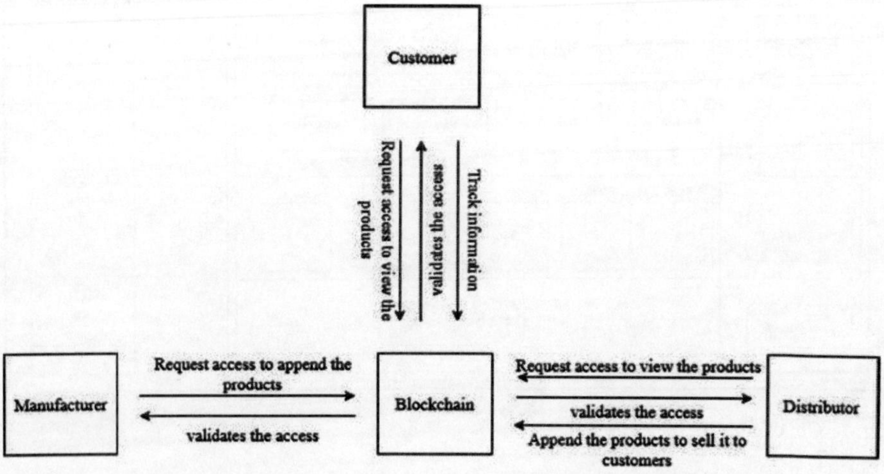

Fig. 8 Sequence of actions for blockchain-enabled supply chain management

3. Distributor requests access to the blockchain application to view the details of the manufacturer and their products. Once the blockchain validates the access, the distributor makes the transaction and buys the products from the manufacturer. Now, the distributor appends the product details in the blockchain to sell it to various customers.
4. Customer requests access to the blockchain application to view the product details and it displays back the products. Now the customer makes the transaction and buys the products from a distributor, and he will be able to track all the information about the status of the product in the blockchain (see Fig. 8).

The tools used for prototype implementation include Ethereum, MetaMask, Truffle [19] , Ganache and web3.js [20]. Ethereum is an open decentralized software platform that allows smart contracts and distributed applications (DApps) to be built and run without any interference from a third party. MetaMask acts as an Ethereum browser, Truffle is a development environment and Ganache uses a user-friendly GUI to view the blocks, transactions and related information. The web3.js is a collection of libraries that allow you to interact with a local or remote Ethereum node, using an HTTP or IPC connection. The web3 JavaScript library interacts with the Ethereum blockchain. It can retrieve user accounts, send transactions and interact with smart contracts.

7 Limitations

About 51% attack happens when more than 50% of the mining nodes are controlled by a group of miners, in which case they can control other miners from creating new blocks and can also be able to prevent the transactions from getting confirmed. This group of miners enjoys the monopoly of mining new blocks and getting rewards. Once the group of miners' controls half of the network, there will be two blockchains, the one controlled by the attackers and the second one is the genuine blockchain where other miners across the world are trying to create new blocks. As the group of miners will have more computational power, they will be able to mine more blocks and build a long blockchain.

This 51% attack is not possible in a private blockchain, as it is controlled by a central authority and only authorized people can be a participant in the network. The attack is hypothetical and it is nearly impossible to control half of the network as it needs tremendous computing power and a huge amount of investment. It also requires an enormous amount of electricity. One of the solutions to this attack is to penalize them by rejecting their blocks if they are creating more than three blocks by not broadcasting them to the network.

8 Discussion: Challenges and Opportunities

The inherent properties of blockchain: immutability, transparency, reliability and traceability make it ideal for solving problems existing with our current technologies. This makes blockchain technology an indispensable asset for the present and future generations. With research continuing on blockchain technology, it has proven to be useful beyond the financial domain. It is no longer limited to cryptocurrencies and its potential is now being comprehended in solving real-life problems. It continues to lend hands in revolutionizing healthcare, military applications, e-commerce, data privatization and more. While the opportunities are endless, our present-day advancements are currently not enough to overcome some challenges existing within this technology, the major challenge being scalability. With its importance and future scope being understood, more and more load is being placed on the network, making the transaction throughput very low compared to our existing technologies. Other significant challenges are interoperability, security and privacy issues, lack of skilled developers, environmental concerns and more. Challenges that strike us during the development of new technologies often provide a learning edge that can prove to be useful and may even lead to innovations. Hope the challenges faced by blockchain technology led to a deeper understanding of the technology and help us in transforming our future.

References

1. Handfield R, Sroufe R, Walton S (2005) Integrating environmental management and supply chain strategies. Bus Strateg Environ 14:1–19. https://doi.org/10.1002/bse.422
2. Linn L, Koo MB (2016) Blockchain for health data and its potential use in health IT and health care related research. Office of the National Coordinator for Health Information. https://www.healthit.gov/sites/default/files/11-74-ablockchainforhealthcare.pdf. Accessed 5 August 2020
3. Mentzer J, Dewitt W, Keebler J, Min S, Nix N, Smith C, Zacharia Z (2001) Defining supply chain management. J Bus Logistics 22. https://doi.org/10.1002/j.2158-1592.2001.tb00001.x
4. Parkhi S (2015) A study of evolution and future of supply chain management. AIMS Int J Manage 9:95–106
5. Sharma A, Subramanian L, Brewer E (2008) Secure rural supply chain management using low cost paper watermarking. https://doi.org/10.1145/1397705.1397711
6. Aguiar EJD et al (2020) A survey of blockchain-based strategies for healthcare. ACM Comput Surv (CSUR) 53:1–27
7. Antoniadis I, Kontsas S, Konstantinos S (2019) Blockchain applications in marketing
8. Chen W, Xu Z, Shi S, Zhao Y, Zhao J (2018) A survey of blockchain applications in different domains 17–21. https://doi.org/10.1145/3301403.3301407
9. Treiblmaier H, Rejeb A, Strebinger A (2020) Blockchain as a driver for smart city development: application fields and a comprehensive research agenda. Smart Cities 3:853–872. https://doi.org/10.3390/smartcities3030044
10. Tyan I, Yagüe M, Guevara-Plaza A (2021) Blockchain adoption in tourism: grounded theory-based conceptual model. ARA J Tourism Res/Revista de Investigación Turística 68–89
11. Yuan Y, Wang F (2016) Towards blockchain-based intelligent transportation systems. In: 2016 IEEE 19th international conference on intelligent transportation systems (ITSC), pp 2663–2668. https://doi.org/10.1109/ITSC.2016.7795984
12. Abeyratne SA, Monfared R (2016) Blockchain ready manufacturing supply chain using distributed ledger. Int J Res Eng Technol 05:1–10
13. Blossey G et al (2019) Blockchain technology in supply chain management: an application perspective. In: HICSS
14. Chen S et al (2017) A blockchain-based supply chain quality management framework. In: 2017 IEEE 14th international conference on e-business engineering (ICEBE), pp 172–176
15. Gaynor M et al (2020) Adoption of blockchain in health care. J Med Internet Res 22(9):e17423. https://doi.org/10.2196/17423
16. Tian F (2016) An agri-food supply chain traceability system for China based on RFID and blockchain technology. In: 2016 13th international conference on service systems and service management (ICSSSM), pp 1–6. https://doi.org/10.1109/ICSSSM.2016.7538424
17. Tian F (2017) A supply chain traceability system for food safety based on HACCP, blockchain and internet of things. In: 2017 international conference on service systems and service management, pp 1–6. https://doi.org/10.1109/ICSSSM.2017.7996119
18. Nakamoto S (2008) Bitcoin: a peer-to-peer electronic cash system. https://bitcoin.org/bitcoin.pdf
19. Truffle and metamask documentation. https://www.trufflesuite.com/docs/truffle/getting-started/truffle-with-metamask. Last accessed 29 June 2021
20. Web3js documentation. https://web3js.readthedocs.io/en/v1.3.0/. Last accessed 29 June 2021
21. Ganache documentation. https://www.trufflesuite.com/docs/ganache/overview. Last accessed 29 June 2021
22. Geth documentation. https://geth.ethereum.org/docs/. Last accessed 29 June 2021

Design and Development of Multipurpose Robot with Rocker-Bogie Suspension and Manipulator

A. Mohan, A. Vidyadharan, M. Mouvlieswaran, T. Nimalathith, and R. Naga Dhatshana

Abstract There are places where humans cannot go and inspect physically, as it will lead to serious problems. For example, the area with natural calamity, radioactive hazards, terrorist occupation, or any unlawful acts, humans cannot go directly without prior experience or knowledge. Thus, there is a need for security mobile robots that can inspect surveillance and rescue operations. This paper proposes a design of a mobile manipulator with a Rocker-Bogie mechanism inspired by NASA's Mars rover sojourner. It has the capability to move on different types of terrains, it utilizes Rocker-Bogie mechanism with six wheels each fixed with individual 1.58 Nm DC motors based on total tractive torque calculation for locomotion which ensures the robot moves on edges, 45-degree slope, potholes, and bumps on its path. The dynamic simulation was carried out to check the structure of the locomotion part. The maximum stress from the analysis was 56 MPa, while the maximum allowable stress is 165 MPa. The manipulator has 4 degrees of freedom in which maximum and minimum end effector's position are verified with help of forwarding kinematics. The manipulator helps to pick different objects and it can also be used in open doors as it detects obstacles in its path and route accordingly.

Keywords Multipurpose robot · Rocker-Bogie mechanism · Manipulator · Dynamic simulation · Forward kinematics · Locomotion

1 Introduction

The mobile robot is a type of robot, which is flexible and not stationary, and service robot applications are non-industrial. The existing papers focus on developing only a certain part of the mobile robot-like locomotion, manipulator improvement, communication, robot vision, energy management, etc. This designing a mobile robot with all the mentioned characteristics of a robot. The need for service robots in places like terrorist occupation and natural calamities is very important, and hence, the goal of

A. Mohan (✉) · A. Vidyadharan · M. Mouvlieswaran · T. Nimalathith · R. Naga Dhatshana
Department of Mechanical Engineering, PSG College of Technology, Coimbatore 641004, India
e-mail: mohan3mmm@gmail.com

U. S. Dixit et al. (eds.), *Advances in Forming, Machining and Automation*, Lecture Notes in Mechanical Engineering, https://doi.org/10.1007/978-981-19-3866-5_39

the paper is to design and develop reliable and cost-efficient robots that can serve both surveillances even in an open area with the help of a Rocker-Bogie mechanism and mobile manipulation like delivering and inspecting with the help of image processing.

2 Literature Survey

Kumar et al. [1] proposed a robot that was used to inspect and patrol the area specified in a location. The robot mails the image of a trespasser through an android application for further processing. Yusoffa et al. [2] proposed a mobile robot mounted with a gripper and LEDs and it was manually operated with a joystick. The robot used a radio frequency module for communication. Joshi et al. [3] explained the inclusion of video streams in the robot. It was controlled via a Web browser for navigation. Raspberry Pi NoIR Camera and ultrasonic distance sensors were used to avoid obstacles. Raspberry Pi 3 Model B microcontroller was operated via Wi-Fi.

Xu et al. [4] developed a security patrol inspection robot in a defined area. The robot used a Web-based monitoring platform and included obstacle avoidance. The robot also incorporated a GPS module for location and was built to have an autonomous operation. Hasan et al. [5] proposal are about the design of an unlimited distance range robot for operation and software used to control Visual Basic. Visual streaming was included with a Vmcap, a VNC technology. Toha and Zainol [6] designed a mobile robot with intelligent inclined motion control. The Rocker-Bogie mechanism is used to locomote in rough terrain in calamity areas.

Yang et al. [7] proposed a mobile robot that is made with improved precision of the 3D reconstruction. A TOF depth camera with a low resolution was used. Binocular stereo vision with high resolution was used but not ideal for low-texture scenes. The purpose of this study is to design and develop a wireless-based manually operated cost-efficient mobile robot for rescue, surveillance, and inspection. And have the functionality of image processing in security aspects with the capability to move in different conditions like slopes, edges, and obstacles.

3 Specifications of the Robot

The technical specifications of the robot parts are shown in Tables 1 and 2. A manipulator is used to collect samples and can be used to remove obstacles in the path. The main advantage of the Rocker-Bogie mechanism is it can be used in different types of terrains like slope, staircase, and off-road. The differential step will balance the robot during the movement on different level surfaces. Arduino Uno is the microprocessor used for motion control along with the Bluetooth module. And Raspberry Pi 3 B microprocessor is used for image processing applications.

Table 1 Specification of locomotion

Parameter	Value
Type of locomotion	Rocker-Bogie mechanism
Speed	58 rpm
No. of wheels	6
No. of drivers	6 DC servo motor
Velocity	0.3 m/s
Acceleration	0.6 m/s^2
The torque required per wheel	0.42 Nm
Mass	3 kg
Max climbing angle	45°

Table 2 Specification of manipulator

Parameter	Value
Type-based on configuration	Revolute coordinate or articulated coordinate robots
DOF	4
Payload	0.2 kg
No. of links	2
Link 1 length	250 mm
Link 2 length	250 mm
Total reach	500 mm
End effector	Gripper
Gripper type	Servo-electric
Gripper opening width	70 mm
Distance from ground	250 mm

The 3D model of this project was developed using design software as shown in Fig. 1 based on the requirements of the objectives. Each part has a specified function in this robot. It includes a manipulator, Rocker-Bogie mechanism, and electronic components required for its locomotion and processing.

4 Design Analysis of Locomotion

The Rocker-Bogie mechanism is used for suspension on rough terrains, which is widely used in space rovers to overcome the planet's crests and trough. This paper focused on the impact of the same mechanism in service-oriented applications, especially on rescue operations. The robot will have to undergo the same rough terrains or even worse in a rescue operation. In this paper, a dynamic analysis and a real-time

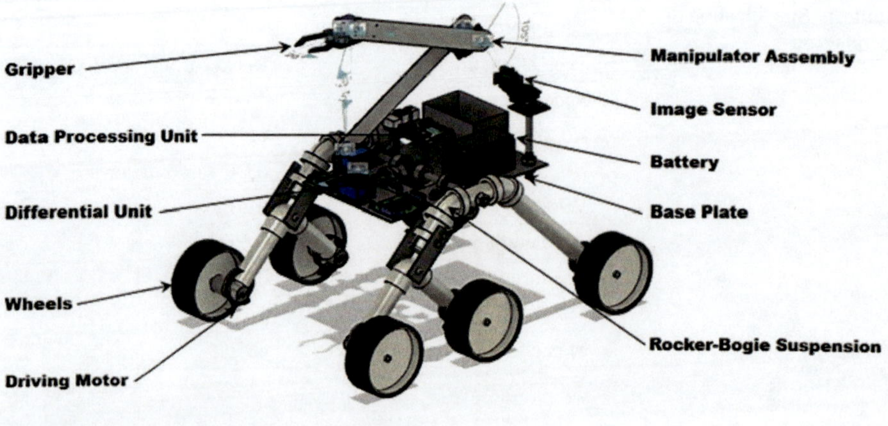

Fig. 1 Final CAD model of robot

experiment on a wave path and debris path to justify the use of the Rocker-Bogie mechanism in service applications have been done.

4.1 *Dynamic Analysis on Links of Rocker-Bogie Mechanism*

Initially, the speed of the robot was defined as 0.3 m/s, based on the application. Using the MSC Adams View Software, the deformation and stress-induced on the structure have been derived during the action of movement on the inclined surface. In this design, out of the 8 rotational joints available, 6 rotational joints are for the 6 wheels and 2 rotational joints for the bogie mechanism. The base plate is fixed with the right and left side wheelbase through 2 rigid joints.

Figure 2 shows the maximum and minimum value of the deformation due to the inclined surface. The blue color on the frame represents the value of deformation that occurs on the frame during the climbing. From the analysis, it was found that there is only a minimum deformation which is 0.35 mm at the time of 1.23 s.

Figure 3 shows the deformation due to the off-road surface. Here, a minimum

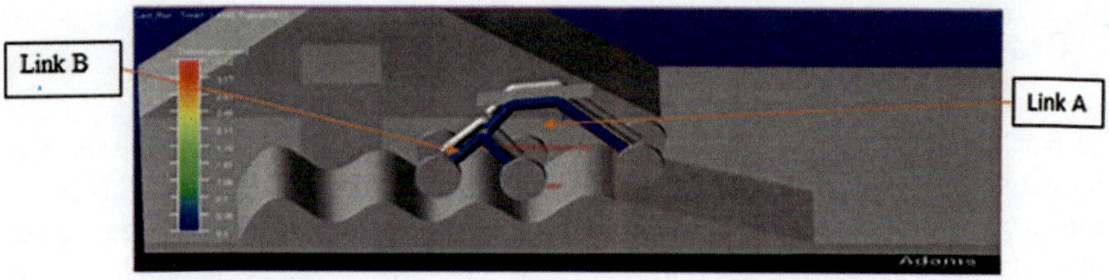

Fig. 2 Deformation due to an inclined surface

Fig. 3 Deformation due to off-road surface

deformation of 0.54 mm occurs at 3.1 s which is high compared to the inclined surface. But both deformation values are low when compared to the structure dimensions.

Figure 4 shows the stress-induced on the frame during the dynamic analysis simulation. The legend on the left side of the simulation shows the maximum and minimum values of stress-induced on the frame. The maximum allowable stress for aluminum is 165 MPa. The maximum stress value obtained through the dynamic analysis on link A is 40.29 MPa, which defines that the design is safe. The factor of safety obtained is 4. And the maximum stress value obtained through the dynamic analysis on link B is 14.83 MPa, which defines that the design is safe. The factor of safety for this design is 11.

The links A and B are the only meshed parts made of a flexible body (in analysis) to study its behavior. The wind resistance is not considered in dynamic simulation. Table 3 shows the details of elements considered for the analysis. The "solid" element type is selected since the meshed model does not have a beam or plate. The "tetrahedral" element shape is selected because the CAD model has a complex geometry. "Linear"

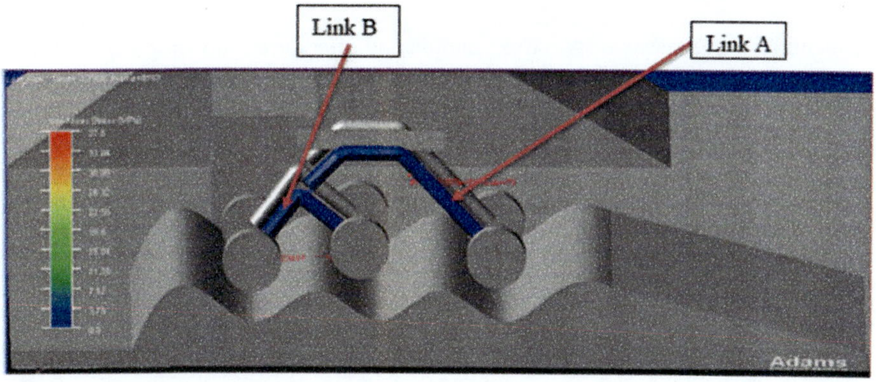

Fig. 4 Stress due to the inclined surface

Table 3 Details of element chosen for analysis

Element parameters	Selected parameters
Element type	Solid
Element shape	Tetrahedral
Element order	Linear

Fig. 5 Stress due to off-road surface

element order ("first-order") was selected to reduce the analysis time and complexity instead of using the "quadratic" element.

The maximum stress value due to the off-road surface obtained through the dynamic analysis on link A is 25.49 MPa (Fig. 5), which defines the design as safe. The factor of safety is 6. The maximum stress value obtained through the dynamic analysis on link B is 26.93 MPa, which defines that the design is safe. The factor of safety of this design is 6.

5 Design Analysis of Manipulator

The manipulator consists of two links and an end effector as shown in Fig. 6. From its base, the joints are arranged as follows: twisting pair and 3 rotary pairs powered by servo motors. Each link has a length of 250 mm, width of 30 mm, and thickness of 2 mm. The material for the link is aluminum, as it is low in weight.

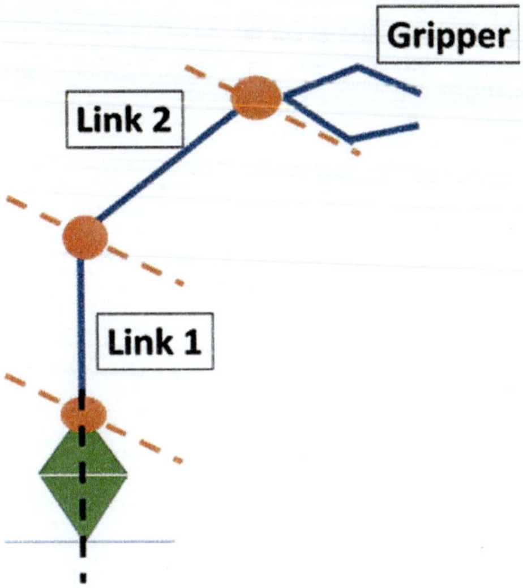

Fig. 6 Manipulator representation

5.1 Forward Kinematic Analysis

The forward kinematics is a vital step in designing a manipulator. The goal is to calculate the end effector pose from the position of other joints. Hence, the manipulator is manually operated with the help of a forward kinematics approach as it is most suitable to find the relative position and orientation of the link from base to end effector. Here, Denavit-Hartenberg (D-H) convection is utilized to do forward kinematics. Roboanalyzer software is being used to do the forward kinematics for the designed robot as shown in Figs. 7 and 8. The purpose of the Roboanalyzer software is to calculate the D-H matrix for the given rotation of joints and to obtain the end effector position. This helped to determine the maximum and minimum positions of the end effector.

[010 0.29351 0 −10 000 100 0.0873 0.05580] is the maximum position of end effector.

[−0.422 0.906 0 0.0766 0 −0.4222610 −100 −0.0873 0.031940] is the minimum position of end effector.

The servo motors are named A, B, C, and D from the base, respectively. The diagram of the extended manipulator with force and perpendicular distance is represented in Fig. 9. The total torque (T_T) required by a servo should be an addition of torque required to lift the load (T_L) and the torque required to move the link from its rest position (T_I). The load torque is calculated using the formula T_L = (Force in N) * (Perpendicular distance in m). Inertial torque is calculated using the formula T_I = (Moment of inertia in kg m^2) * (Angular acceleration in rad/sec^2).

The total torque required for servo A and B is 1.5 Nm (15 kg cm). The total torque required for servo C is 0.85 Nm (approx. 8.5 kg cm). The total torque required for servo D is 0.1 Nm (approx. 1 kg cm). The static analysis was performed on the

Fig. 7 Initial position of the robot in Roboanalyzer software

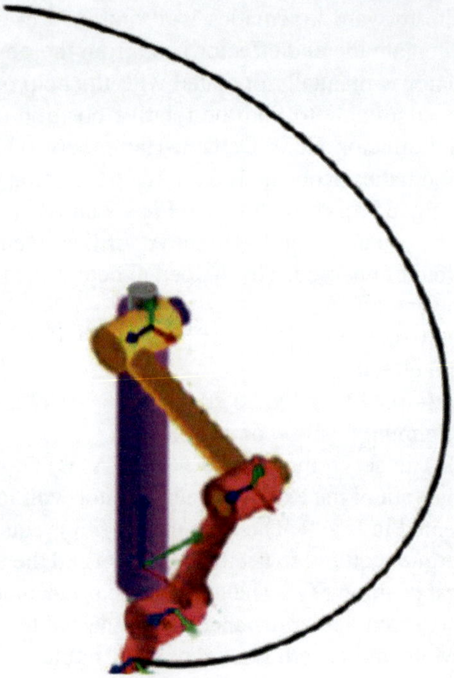

Fig. 8 Movement of joints done to obtain D-H parameters

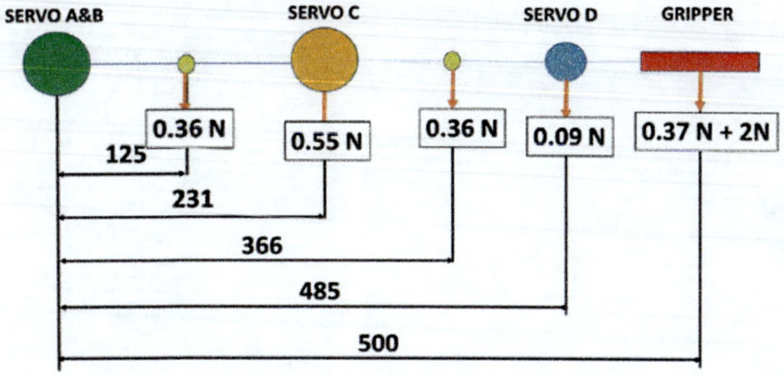

Fig. 9 Diagram of the extended manipulator with force and perpendicular distance

manipulator link to check its deformation. The link is constrained in 6 DOF at one end, and a force of 5 N (500 g) is applied at the other end, while the actual payload is 200 g. It was observed that the maximum deflection of the link is 0.089 mm, which is negligible.

6 Prototype Development

The fabricated prototype has reduced features from the actual model. The aluminum material used for the structure of the robot was replaced by PVC and wood. Manipulator DOF was reduced to 3. The robot model designed does not include differential assembly, pan, and tilt assembly. The camera is placed in front of the setup for image capturing in its path. Figure 10 shows the picture of the assembled robot.

The connection flowchart (Fig. 11) shows the communication of the robot with the other modules. The manipulator is controlled via Bluetooth with HC-05 Bluetooth module via an app created using MIT App inventor. The Bluetooth module relates to Arduino Uno and is then connected with 3 L298N motor drivers. Each motor driver is connected to 2 motors. A Pi camera used for vision relates to Raspberry Pi 3 B+ which visually streams the video to the laptop via a Wi-Fi connection. YOLO algorithm is used for image classification purposes, which can be used for the identification of objects.

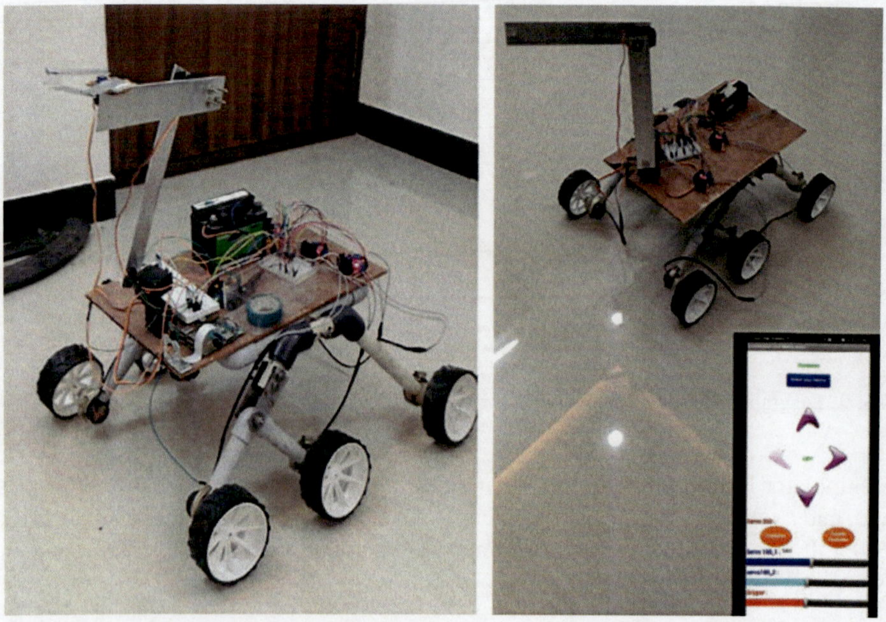

Fig. 10 Final assembly of the locomotion setup

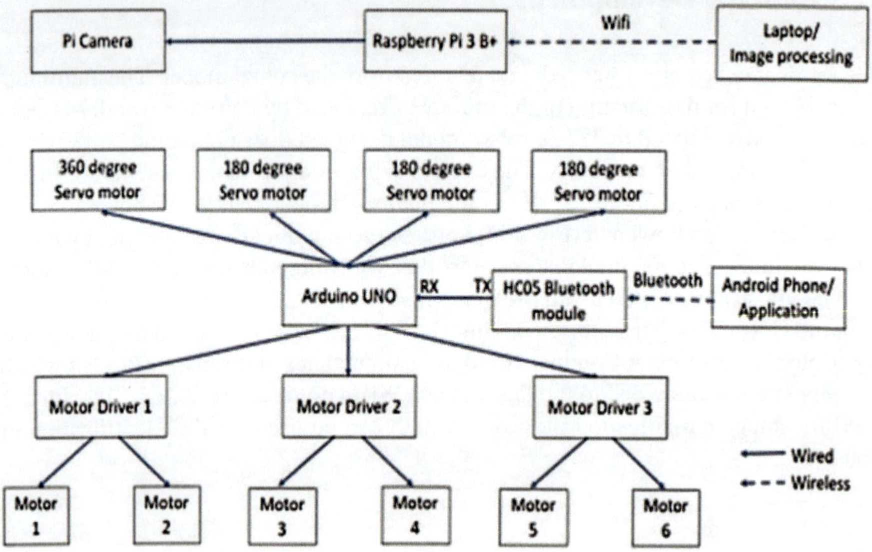

Fig. 11 Connection flowchart

7 Results and Discussion

The results show that a robot has been built in the view of service aspects with a Rocker-Bogie mechanism. The prototype was fabricated and tested successfully as shown in Figs. 12 and 13. From the dynamic analysis, it was confirmed that the structure of the robot can withstand stress due to movement on rough terrains. The maximum allowable stress of the material is significantly less than the stress-induced on the structure to validate its factor of safety. The maximum allowable stress for

Fig. 12 Manipulator crossing edges

Fig. 13 Manipulator climbing in slope

aluminum is 165 MPa. The maximum stress value obtained through the dynamic analysis on the Rocker-Bogie link is 40.29 MPa, which defines that the design is safe. Thus, the proposed manipulator exhibits a factor of safety of 4.

The robot was able to cross door edges and climb the slope of 35° (Figs. 12 and 13). This robot was tested in a rough collapsed environment with scattered bricks, which is equivalent to debris, and it was able to successfully cross through the area (Fig. 14). Object detection is done by image capturing and processing as shown in Fig. 15.

Fig. 14 Robot moving in debris

Fig. 15 Testing runs for object detection

The image processing is tested in an indoor environment and was able to identify the objects (Fig. 15). The image sensor can be able to discriminate the object detection based on the programming given as input.

8 Conclusions

Initially, to find a solution to the problem identified, designing a mobile robot in a CAD model was done. After the validation of the design, classification of the design parts as manipulator and locomotion was done. The specifications of the robot were decided and accordingly performed the kinematic analysis and deformation analysis for the manipulator followed by dynamic analysis for locomotion. After the analysis, the manipulator was fabricated and assembled, respectively. Programming of the Arduino Uno for teleoperation, Raspberry Pi for visual streaming and image processing was executed. The built manipulator was experimented with in different terrains by enabling image processing for visual streaming. All the analyzes were implemented using the Rocker-Bogie mechanism.

The performance appeared good in the analysis as well as experimental conditions and thus it would be well suited for service applications. The image processing ability for the robot was introduced for making the robot autonomous in future, which will be used for autonomous navigation and human identification in calamity areas.

References

1. Kumar CN, Ramesh B, Shivakumar G, Manjunath JR (2014) Android based autonomous intelligent robot for border security. Int J Innov Sci Eng Technol 1:544–548
2. Yusoff MAK, Samin RE, Ibrahim BSK (2012) Wireless mobile robotic arm. Proc Eng 41:1072–1078
3. Joshi SA, Tondarkar A, Solanke K, Jagtap R (2018) Surveillance robot for military application. Int J Eng Comput Sci 7(5):23939–23944
4. Xu F, Li Z, Yuan K (2007) The design and implementation of an autonomous campus patrol robot. In: IEEE international conference on robotics and biomimetics (ROBIO), pp 250–255
5. Hasan SM, Rasid SMR, Mallik A, Rokunuzzaman M (2018) Development of a wireless surveillance robot for controlling from long distance
6. Toha SF, Zainol Z (2005) System modelling of rocker-bogie mechanism for disaster relief. Proc Comput Sci 76:243–249
7. Yang Y, Meng X, Gao M (2017) Vision system of mobile robot combining binocular and depth cameras. J Sens

Development of Industry 4.0 Curriculum Based on Industry-Academia Collaboration and Testbed Demonstrator Concept

Ganesh Kumar Nithyanandam, Javier Munguia, Muruthanayagam Marimuthu, and Rudramoorthy Rangasamy

Abstract The relevance of Industry 4.0 has been widely discussed at a governmental level, from academic literature, and at specialists' industrial events, yet the specifics of what Industry 4.0 entails and how to introduce students to this discipline remain unclear. As a joint research collaboration, PSG College (Coimbatore, India) and Newcastle University (UK) embarked on a "Royal Academy"—a funded project aimed at developing strategies for Industry 4.0 curriculum development, embedded in higher education. The approach involved: (1) benchmarking best practices across other institutions regarding Industry 4.0 initiatives, (2) identifying enabling technologies that support the development of Industry 4.0 at a local level, (3) developing student-led projects at the intersection of various digital manufacturing technologies, and (4) applying these projects in an industry-relevant context. The results show a high level of engagement of students in the applied technology project, and the potential to use such pilot projects as prototype "testbeds" that can potentially attract industrial interest for a more widespread take of these technologies.

Keywords Industry 4.0 · Industry-academia collaboration · Augmented reality · 3D printer · Internet of Things

G. K. Nithyanandam (✉)
PSG College of Technology, Coimbatore, India
e-mail: gkn.mech@psgtech.ac.in

J. Munguia
Newcastle University, Newcastle, UK
e-mail: javier.munguia@newcastle.ac.uk

M. Marimuthu
PSG Industrial Institute, Coimbatore, India

R. Rangasamy
PSG CARE, Coimbatore, India

U. S. Dixit et al. (eds.), *Advances in Forming, Machining and Automation*, Lecture Notes in Mechanical Engineering, https://doi.org/10.1007/978-981-19-3866-5_40

1 Introduction

The World Economic Forum (WEF) 2016 [1] suggested that technical education directly affects a country's productivity by developing human capital. Thus, they will execute the existing tasks more quickly to today's industry requirements. For this, the workforce's educational level needs to be improved to nurture the entrepreneurship thought process [2, 3] in young students' minds. Cappelli [4] reported that the industry tends to hire professionals who have entrepreneurial abilities. This is because entrepreneurs are creatives who would solve business problems more quickly than traditional managers or the workforce [5]. In addition, entrepreneurs give more ideas to improve their productivity which benefits the industry at large. Therefore, the educational system must cultivate young minds with entrepreneurship thought processes from the first year of engineering school.

A preliminary study was conducted to investigate the industry needs. A set of questionnaires was sent to more than 300 industries requesting what types of ICT technology are required so that they could improve their day-to-day productivity. The majority of the industries responded that they need low-cost Industry 4.0 solutions for survival. Currently, these technologies are available in the market, but they are proprietary to large corporations.

Every industry is unique in the way they operate even though the operations are the same. Therefore, the available Industry 4.0 self-products need to be customized as per the plant floor requirements. Currently, the training cost for these proprietary solutions is very expensive and it may be difficult for MSME (micro, small, and medium enterprises) to implement these technologies at affordable prices. Mettl [6] reported that on average an industry spends around 25% in learning and development (L&D) costs, which could be approximately Rs. 41,838 per employee per annum whereas according to the American Society for Training and Development (ASTD) and Harvard Business Publishing (HBP), the US spends US$331 per employee per annum [7]. One solution to solve this problem is to offer industry-needed technologies as a curriculum.

An attempt is made to develop the curriculum in the field of Industry 4.0. A curriculum is created by several methods: (a) experts dictating the courses required to form a curriculum; (b) accreditation committee defining a set of rules and regulations to create curriculum; (c) brainstorm with industry and create industry-ready curriculum (industry-academia collaboration). This paper focuses on developing the industry-ready Industry 4.0 curriculum so that the students will be ready for today's industry requirements. Similarly, the industries could hire these industry-ready students and put them to work on day one without providing any type of training.

The authors received a joint research industry-academia collaboration project to develop an Industry 4.0 curriculum. In this, Indian engineering academia (PSG College of Technology) and a UK engineering academia (Newcastle University) collaborate with an Industry partner (PSG Industrial Institute) in India. Royal Academy of Engineering sponsored this one-year project.

2 Literature Review

A curriculum is a collection of lessons, assessments, and other academic content that is taught in a school, university, program, or class by a teacher [8]. Teachers use the curriculum to describe what they need to teach in their classroom. Many educational experts [8, 9] complained that the engineering curriculum in India is outdated and it needs to be reversed. On the other hand, industries complained that they are unable to find qualified skilled workers [10]. If they can find qualified workers, they need to train them which is an additional cost [7]. The only solution to solve this problem is to create an industry-ready curriculum. A popular two-wheeler industry—TVS motors created a polytechnic college called CPAT-TVS [11] to educate the students with their industry needs. Later, they hire qualified skilled workers from their institution. The University of Toledo [12] created a Co-op program with Dana Corporation where the students exercise the day-to-day activities of Dana Corporation in their class activities. This led the students to know the current practices of the industry. With this model, Dana Corporation hires the students who have cleared the Co-op course at the University of Toledo.

Before starting the project, authors visited in-person to several Universities and industries in India and abroad to understand how Industry 4.0 technologies were practiced. National Institute of Technology (NIT) at Tiruchirappalli has a Center of excellence in manufacturing laboratory in alliance with Siemens, where Siemens showcase their products for research purposes [13]. This laboratory is created to foster industry-academia collaboration in research and development. Similarly, Newcastle University has a Mind sphere Innovation Network (MINe) [14] laboratory where Siemens showcases their cloud-based Internet of Things platform which interfaces with the existing cyber-physical research assets and provides necessary data available in real-time to improve teaching and research collaborations between academics and industry partners.

Indian Institute of Technology (IIT) at Kharagpur created the smart wireless applications and networking (SWAN) group [15] where fundamental research activities toward wireless technology and industrial Internet of Things are carried out. The focus of their research activities is a simulation, analytical, and other experimental methodologies of their research activities. Similarly, the Indian Institute of Technology (IIT) at Madras created IITM Research Park [16, 17] to build a knowledge and innovation ecosystem so that fostering of partnerships and assisting new ventures are created.

The above research groups created their laboratories as workshop models and their research is not focused on any one industry problem. On the other hand, the Barton research group [18] created a smart manufacturing laboratory at the University of Michigan with a cloud-based digital manufacturing framework. It has a Production as a Service (PaaS) concept to connect the end users who need small batch customized manufacturing, with manufacturers that have existing underutilized resources. Similarly, Wayne State University formed the Digital/SMART Manufacturing Demonstration Center (D/SDC) as part of the Michigan State Digital Acceleration (SDA) initiative. The focus of this center is to develop the next generation of digital manufacturing platforms based on real-time manufacturing and design data [19, 20]. The above two research group focuses on providing smart manufacturing solutions to automobile industries.

Similarly, Binghamton University has Smart Electronics Manufacturing Laboratory (SEMLab) [21] where precision assembly processes of electronics manufacturing are studied using data science and artificial intelligence principles. The main focuses of their research are to develop algorithms and integrate with their six collaborative partners. The current practices and futuristic technology related to 3D printing are observed in the TCT show [22], whereas IMHX show provided the current and futuristic intelligent warehouse solutions [23]. In the coming sections, the implementation of each Industry 4.0 technology will be explained.

3 Methodology

The objective of this paper is to create an industry-ready Industry 4.0 curriculum. Figure 2 shows the methodology adopted to create the curriculum. First, an industry partner who needs Industry 4.0 solutions were selected. To begin with, a popular product was selected. Currently, the selected industry manufactures and assembles the product manually. They use less than 3% digital technology in their process. The industry faces a high number of rejections, scarcity of available skilled workers, high employee turnover, high machine idle time, low productivity, huge inventory, etc. To solve the problem, a detailed process map was derived with time study and motion study (Fig. 1).

The selected case study industry comes under MSME. Therefore, it is very difficult for them to purchase proprietary solutions. So, they were seeking low-cost Industry 4.0 solutions.

An attempt was made to transform this traditional factory into low-cost Industry 4.0 solutions by introducing the Testbed demonstrator concept for, the first time in India for pump assembly. This testbed demonstrator acts as a digital twin of the pump assembly. The details of this testbed demonstrator will be explained later.

Several low-cost Industry 4.0 solutions were introduced in this testbed demonstrator. These solutions were integrated with an Odoo ERP package. Reverse engineering was applied to gather what skills are required to implement these technologies. From brainstorming sessions, the knowledge gained in implementing these technologies was categorized into nine modules. A detailed syllabus of each module was developed. Then a framework was derived on how these modules will be trained to students and industries.

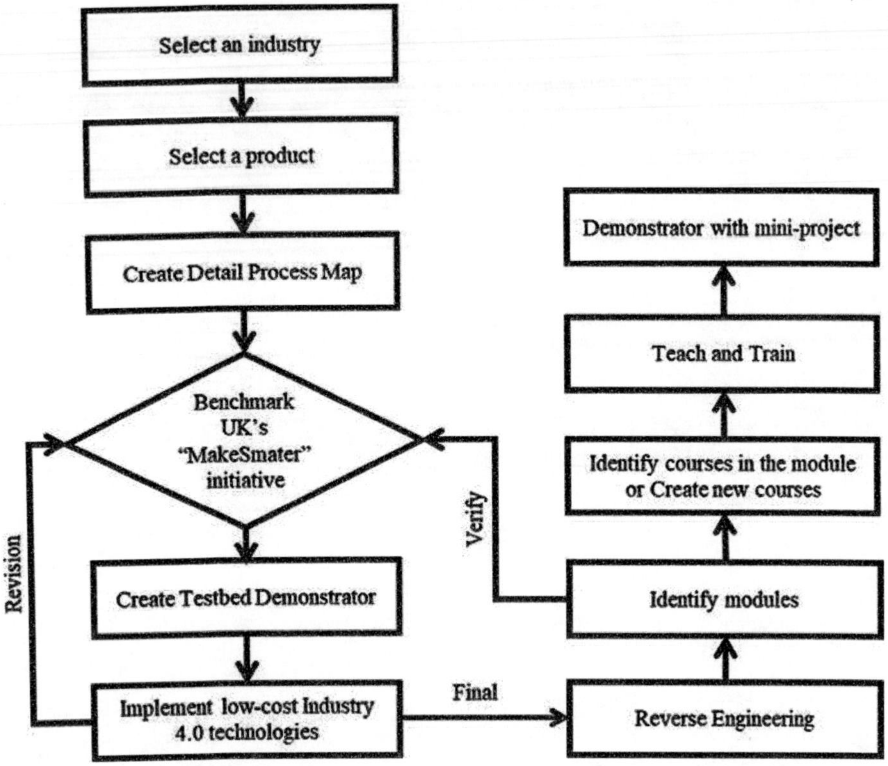

Fig. 1 A framework was adopted to create an Industry 4.0 curriculum

4 Testbed Demonstrator

The selected industry partner is a pioneer in manufacturing machine tools and pumps. The industry operates its complete operations in almost two acre. For this study, eight stages five HP radial submersible pumps were selected. A detailed process map on how this pump is assembled was collected. It takes 64 steps to assemble this pump. The majority of the assembly operations were carried out in a single station.

Having these learning from various institutions, a testbed demonstrator concept was developed at Industry 4.0 Research Laboratory at PSG College of Technology replicating the complete pump assembly operations. The detailed process map was studied, and it was transformed into six stations with loading and unloading stations. Figures 2 and 3 show the schematic representation and physical testbed demonstrator, respectively. These six stations were designed in such a way that they are line balanced, single-piece flow conveyor systems. The size of this testbed demonstrator is 15 feet by 25 feet.

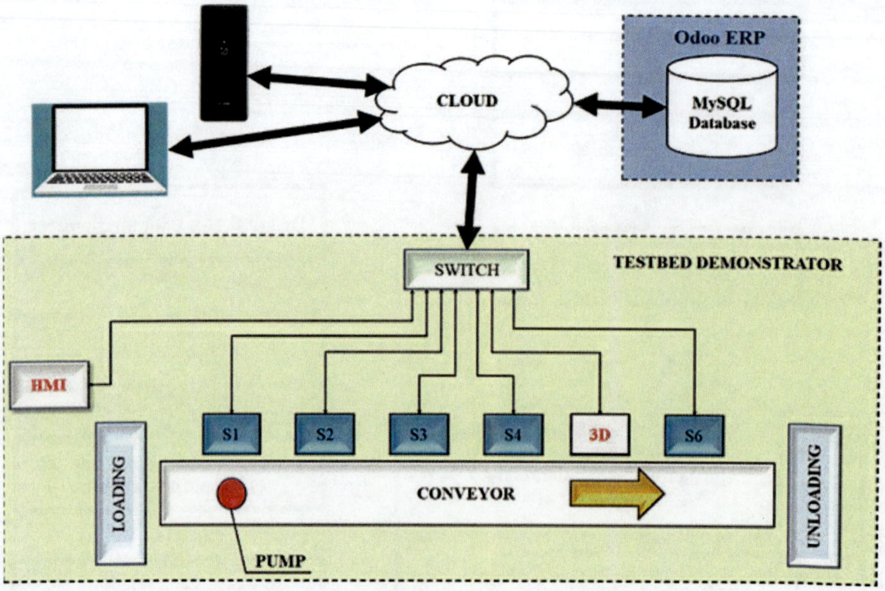

Fig. 2 Schematic representation of the testbed demonstrator for pump assembly

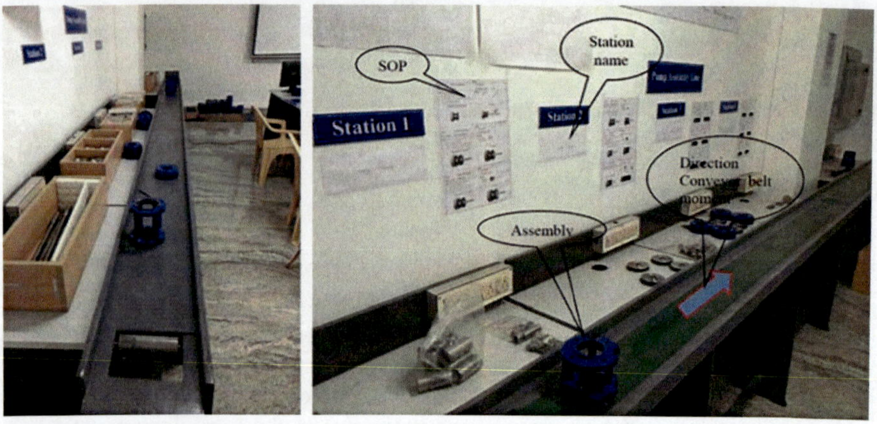

Fig. 3 Actual testbed demonstrator for pump assembly

In this testbed demonstrator, an attempt was made to implement several low-cost Industry 4.0 ICT solutions. These ICT solutions were introduced based on lessons learned from visiting various institutions within India and abroad. Some of the ICT solutions are discussed as follows:

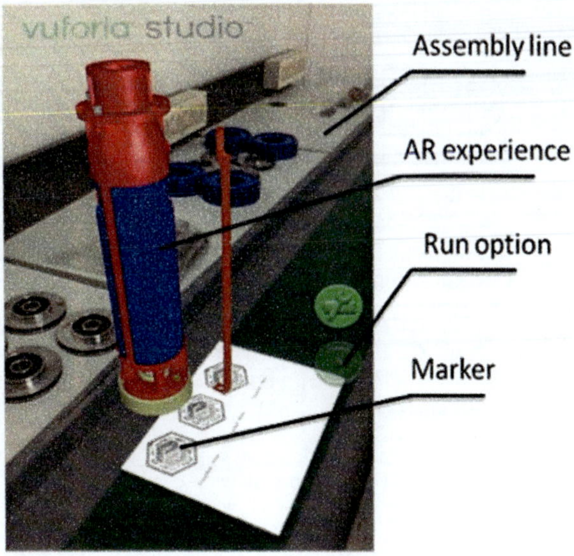

Fig. 4 Marked-based AR experience with the fully assembled system as shown by the AR app on a standard android smartphone

4.1 *Augmented Reality*

Currently, a pump is assembled manually at one station (workshop model) by a skilled worker. It is time-consuming and tedious process. The quality of the pump assembly is based on the skilled worker and his or her mood.

A mark-based augmented reality (AR) was developed for an android operating system using unity packages. The 3D images of the pump components overlay over the physical components. When a smartphone scans the marker, AR demonstrator how the pump needs to be assembled step-by-step. Figure 4 shows the demonstration of AR for the selected pump in the testbed demonstrator.

With the proposed technology, even unskilled workers can assembly the pump within a specified time. This technology was developed by a graduate student, and he received the best final year project award for that period. The details on how this technology was developed are illustrated in the earlier publication [24].

4.2 *Controlling 3D Printer Using Parametric Programming*

Currently, a pump is assembled manually at one station (workshop model) by a skilled worker. It is time-consuming and tedious process. The quality of the pump assembly is based on the skilled worker and his or her mood. The geometric dimensions of the pump's 3D subcomponent could be customized either from previous station data or remotely (mobile application) through parametric programming.

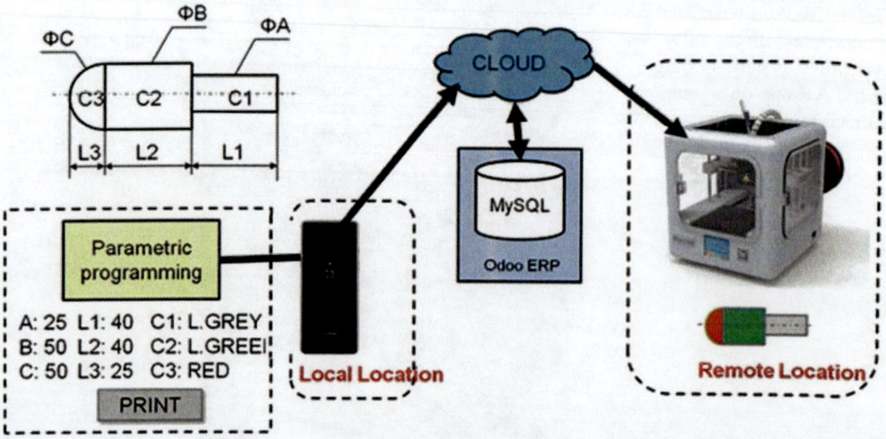

Fig. 5 Controlling 3D printer remotely

Currently, the case study industry does not use 3D printer technology in any operations. In the testbed demonstrator, a 3D printer is introduced at station 5 (S5) of the testbed demonstrator (see Fig. 2). Figure 5 shows the working principle of this technology. The 3D printer modifies its print-specific objects, and it functions based on parametric values. The parametric values are collected either from station 2 (S2) or remotely.

The 3D printer at S5 will start operating based on the demand from the previous station. In this case, S5 operates when the pump moves to station 2 (S2). By the time it reaches station 4 (S4), the subcomponent required for the pump assembly will be ready at S5. The part produced at S5 is then assembled at station 6 (S6) to complete the pump assembly as a whole.

This technology has override functionality that can be controlled from the mobile application. It gives the ability to monitor and control 3D printers autonomously. This technology was developed by three undergraduate students, and they are currently working on it.

4.3 Automatic Line Stoppage

Autonomous systems collect data, analyze it and make the corrective actions on their own. It provides intelligent feedback for the corrective actions. The proposed automatic line stoppage (ALS) technology, as shown in Fig. 6, is one where the operator needs to complete his or her assembly operations within the "start" (beginning of the operational sequence) and "end" (end of the operational sequence) imaginary lines. The "start" imaginary line is defined by a green marker whereas the "end" imaginary line is defined by a red marker. The yellow marker defines the 60% completion of the operational sequence. When it reaches the yellow marker, a sound warning buzzer is

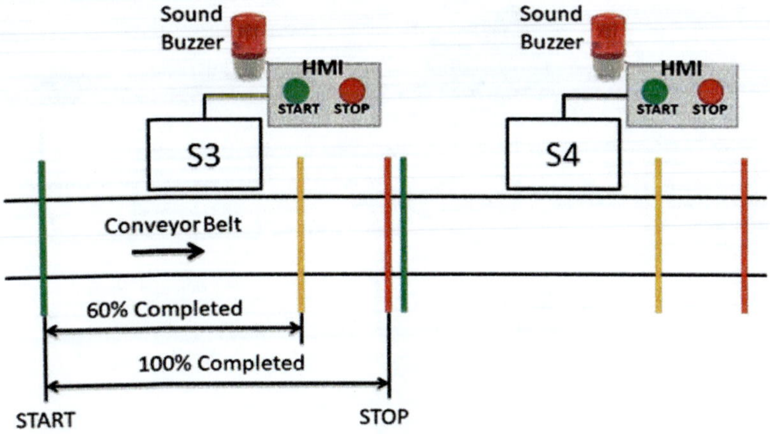

Fig. 6 Automatic line stoppage system for pump assembly

heard. If the operator did not complete the operations by the red marker, the conveyor system stops automatically. The IR sensors are also placed to notify the start, end, and 60% completion of the station's operation. It can be restarted by pressing the start button in the human–machine interface (HMI). The conveyor system could be stopped at any time by pressing the stop button in the HMI.

Currently, all the operations in the case study industry are manual. The proposed model is to automate some of the manual operations. This ALS technology is implemented in all the stations of the testbed demonstrator. This technology was developed by three undergraduate students using IR sensors, Arduino board, sound buzzer, and servomotor (to control the conveyor system).

4.4 eKanban

eKanban is a scheduling system that helps to order parts when the volumes of the parts in the bin are low. Typically, industries follow a two or three cards system to order required parts manually. The kanban system sends an electronic signal to the ERP system automatically to order the parts based on rules (when the volume of the parts in a bin reaches a certain quantity, take some action). Currently, there is no kanban system exists and they order parts manually in the selected case study industry.

A load sensor is placed underneath each bin and several rules are defined as a part of the algorithm. The working principle of this technology is explained in Fig. 7.

When 80% of the bin is exhausted, then the load sensor sends the signal to ERP to order that part automatically. If the ordered part is not filled in the bin within a specified time (say 15 min) and/or the bin is exhausted by 90%, then it gives a

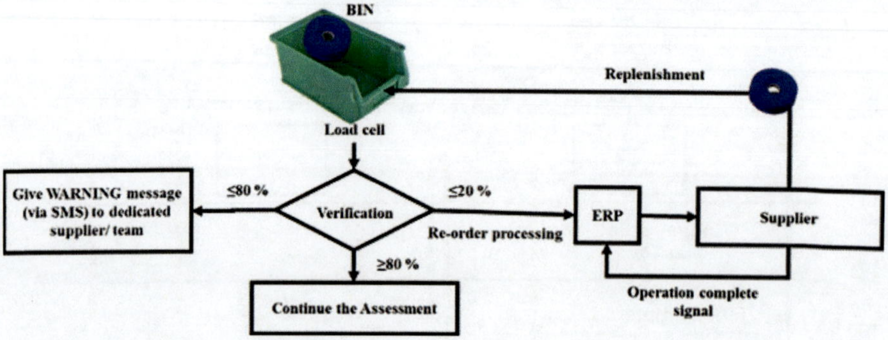

Fig. 7 eKanban system for pump assembly

warning SMS message to the registered person and a loud buzzer noise is heard. In this process, it is assumed that replenishment order scripts are in place in the ERP package to execute.

This technology is implemented on the testbed demonstrator. In other words, each station in the testbed demonstrator has the intelligence to reorder its parts by itself. This technology was developed by three undergraduate students.

4.5 Part Tracking and Traceability

Part tracking and traceability is one used to record the location of the assets on the plant floor. It helps to identify which part (subcomponent) is assembled in which product (pump). It helps to verify the history, location, and identification of parts on the assembly line. This technology helps the company with data on how much inventory it has in hand, where its inventories exist, the status of its inventories (damaged, returned, rejected, and on hold) and it helps to deter theft and loss. This system helps to track the location of the pump when a recall or any maintenance activity is required.

Currently, the parts are tracked and traced manually in the case study industry. It is time-consuming, and the occurrence of human errors is possible. This technology is implemented in each station of the testbed demonstrator. It was designed and developed by three undergraduate students using RFID, GSM, Raspberry Pi, Python, Andon, MySQL, and Microsoft Visual Studio C#.

4.6 Obsolescence Management System

The obsolete parts are the ones that are rejected due to design change, wrong parts being supplied, hazardous materials, environmental issues, high material cost, etc.

When these parts are assembled with or without the operator's knowledge, it could lead to an expensive proposition such as recall, bad image to the company, etc. Therefore, it is a good practice to implement an obsolescence management system to avoid such issues. In short, the obsolescence management system is nothing but tracking the parts which are no longer needed and helps to avoid obsolete parts into the system. This system is integrated with ERP to verify only approved parts.

Currently, this technology is not available in the case study industry. In the testbed demonstrator, a bar code scanner is in place in each station. The bar code label is scanned to verify whether the part (component) is obsolete or not before assembly. If it is identified, it makes a buzzer noise, and the data is recorded for the location of the obsolete parts. The management would take appropriate actions to remove the obsolete parts from the plant floor. This technology was designed and developed by three undergraduate students using bar code, Raspberry Pi, Andon, MySQL, and Python.

4.7 Human-Less Gantry Robot

In recent times, the scarcity of skilled laborers in the field of Industry 4.0 is high demand around the world. Even if they are available, they are very expensive. India has a more than 1.2 billion population. Implementing full automation may not be a good idea. So, a semi-automated gantry robot is designed and developed in station 2 of the testbed demonstrator. Here, "human-less" means the assembly line is operated with fewer humans (not eliminating humans). During the covid kind of situation, this technology complements the assembly line to meet the production when the industries run with limited employees (Fig. 8).

The gantry robot was proposed where repetitive actions exist. Figure 9 shows the three-axis gantry robot arm. Figure 10 shows the gantry robot arm with a vision

Fig. 8. 3-axis gantry robot arm

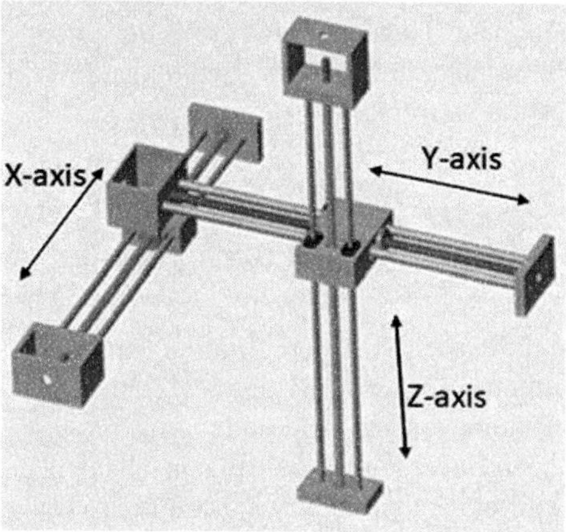

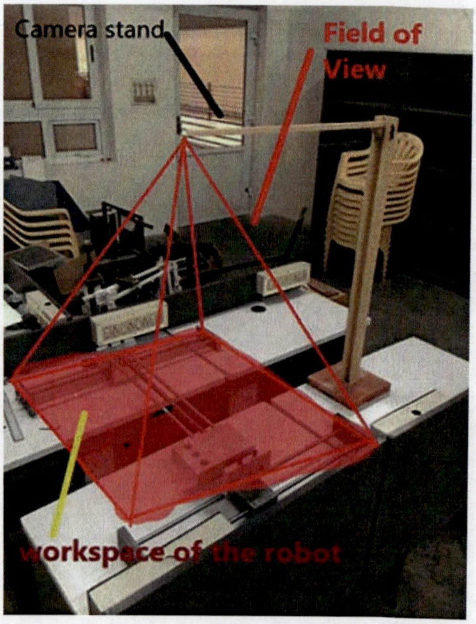

Fig. 9 Gantry robot with vision system

system. Each axis is connected to servomotors separately. At the end of the *Z*-axis, the end effector is installed. A vision system is mounted on the bottom of the end effector.

The vision system uses a machine-learning algorithm to detect the pump parts. Using a 3-axis gantry robotic arm, it picks the parts from the loading station and assembles them in the designated workplace. Currently, the pump is assembly by a skilled worker at one station and all the operations are manual.

With this technology, the complete pump assembly is assembled by a gantry robot using image processing technology with limited human intervention. The proof of concept of this technology is completed with the integration portion yet to be complete. This technology was developed by the graduate student using Raspberry Pi, Camera, Tensor flow, Python, and MySQL, and he received excellent comments from reviewers.

5 Industry 4.0 Curriculum

In the earlier work, a preliminary smart manufacturing framework was proposed but without cyber security and integration elements [25]. For this study, a new Industry 4.0 framework is proposed, which is shown in Fig. 10.

It is interesting to note that mechanical engineering domain knowledge is required to develop any Industry 4.0 solutions. To make the system user-friendly, easy to use, and very productive, electrical and electronics engineering and computer science

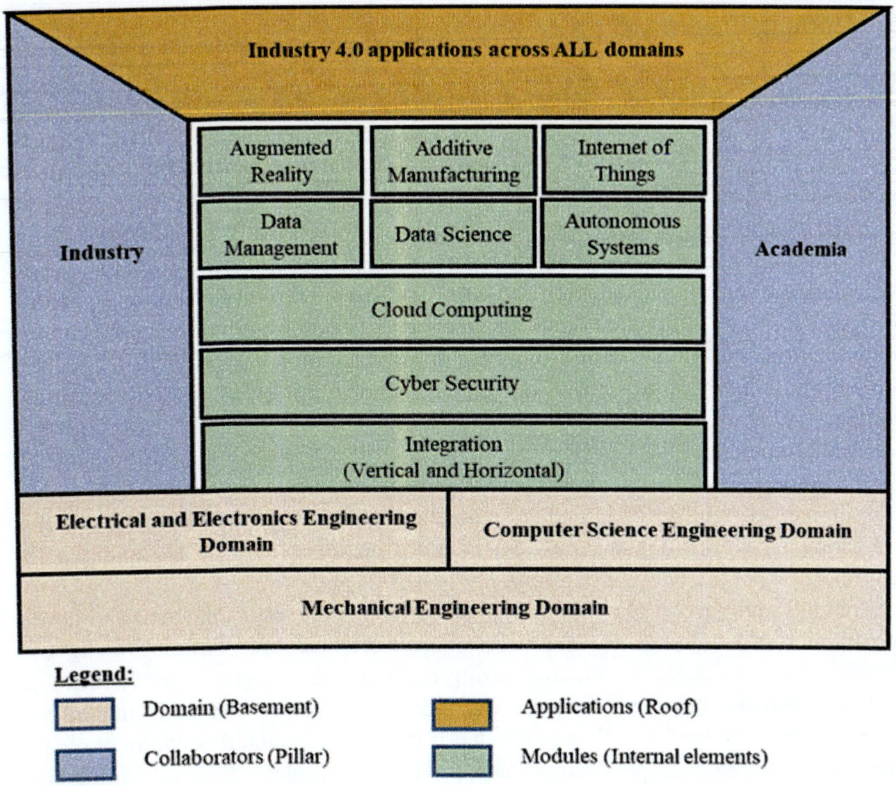

Fig. 10 Proposed Industry 4.0 curriculum

engineering domain knowledge are required. In other words, these two domains are like supporting elements in implementing Industry 4.0 solutions.

While developing these technologies, it was identified that the following engineering courses were used: engineering mechanics, engineering graphics, introduction to manufacturing engineering, design for manufacture and assembly, computer-aided design, basic electrical and electronics, programming skills, database management system, information technology for manufacturing applications, integrated product design and development, Internet of Things, PLC programming, and data science.

From this, the key technologies were derived from bird's eyes which are identified as augmented reality, additive manufacturing, Internet of Things, autonomous systems, data management, and data science. The data collected from these various systems needs to be stored in the cloud and it should be computed with high security. For these technologies to work effectively and efficiently, they should be integrated vertically and horizontally.

6 Conclusion

When each technology was implemented, different types of skill sets required for the students are observed as they could take their career path in the field of Industry 4.0. PSG took several methods to deliver the knowledge gained and they are listed based on end user (audience):

Industry and faculty audience: PSG CARE organized two months workshop on "Systems approach to pump design and manufacture". For this, the following courses were derived: selection and design of a pump, design validation of a pump, structural design of a pump, tolerance stack up and GD&T, process planning for a pump, manufacturing system design for pumps, quality assurance strategies, lean concepts for pump assembly, and sensors and IoT. Experts in each domain delivered their content.

Faculty audience: PSG CARE organized a series of workshops on Industry 4.0 such as Industry 4.0 for beginners, industrial Internet of Things using Thingworx, and augmented reality for beginners using Unity/Vuforia. In the coming months, the following workshops are planned: PLC for beginners and CAD modeling for beginners.

Students audience: The syllabus of the following courses is designed and developed: Internet of Things for engineers, object-oriented programming using C#, information technology for manufacturing applications, integrated product design and development, business analytics, etc. These courses are offered to undergraduate and graduate students. The integrated product design and development course were offered in an interdisciplinary manner where the students from all engineering domains were able to participate in the class. At the end of the course, the students demonstrated a multidisciplinary prototype model showcasing how mechanical, electrical and electronics and computer science be a part of the product design. This course was offered three times and was well accepted among students.

References

1. Woodruff J Factors affecting economic development and growth. https://smallbusiness.chron.com/factors-affecting-economic-development-growth-1517.html. Last accessed 25 June 2021
2. Grant C The contribution of education to economic growth, K4D knowledge, evidence, and learning for development. https://assets.publishing.service.gov.uk/media. Last accessed 25 June 2021
3. Chmura M Millions are entrepreneurs in the United States, Babson thought and action. https://entrepreneurship.babson.edu/millions-are-entrepreneurs-in-the-united-states. Last accessed 25 June 2021
4. Cappelli P Your approach to hiring is all wrong—outsourcing and algorithms won't get you the people you need. Harvard Business Review Magazine. https://hbr.org/2019/05/your-approach-to-hiring-is-all-wrong. Last accessed 25 June 2021

5. The Economic Times, Industry only 5% of adult Indians establish their own business: survey. https://economictimes.indiatimes.com/industry/miscellaneous/only-5-of-adult-indians-establish-own-business-survey/articleshow/63356372.cms?from=mdr. Last accessed 25 June 2021
6. Financial Express, Companies spend 25% of L&D budget on training sales teams, says Mettl's survey. https://www.financialexpress.com/industry/companies-spend-25-of-ld-budget-on-training-sales-teams-says-mettl-survey/1303493/. Last accessed 25 June 2021
7. Singh N Companies spent $331 per employee on training, development in 2011, Times of India. https://timesofindia.indiatimes.com/business/india-business/companies-spent-331-per-employee-on-training-development-in-2011/articleshow/22463818.cms. Last accessed 25 June 2021
8. Stauffer B (2020) What is a curriculum and how do you make one? Appl Educ Syst
9. Kaushika P Higher education syllabus outdated, Times of India. https://timesofindia.indiatimes.com/home/education/news/higher-education-syllabus-outdated/articleshow/2458330.cms. Last accessed 25 June 2021
10. Gunjan RK No jobs pending education loans: the collapse of the grand engineering dream. https://www.news18.com/news/india/. Last accessed 25 June 2021
11. CPAT-TVS. https://www.cpat-tvs.org/. Last accessed 25 June 2021
12. The University of Toledo, 20 years and 20,000 students—Co-op anniversary celebration. https://www.utoledo.edu/engineering/docs/. Last accessed 25 June 2021
13. NIT Siemens Center of Excellence in Manufacturing. http://siemenscoe.nitt.edu/index.php. Last accessed 12 Sept 2019
14. Newcastle University Siemens and Newcastle University launch UK's second mind-sphere lab. https://www.ncl.ac.uk/press/articles/archive/2018/05/mindsphere/. Last accessed 18 August 2020
15. Misra S (2018) Smart wireless applications and networking. Research group. http://cse.iitkgp.ac.in/~smisra/swan/. Last accessed 18 August 2020
16. IITM Research Park. https://respark.iitm.ac.in/. Last accessed 14 Sept 2019
17. IIT Madras. https://www.iitm.ac.in/research-park/iitm-research-park. Last accessed 14 Sept 2019
18. Barton Research Group, University of Michigan. https://brg.engin.umich.edu/research/smart-manufacturing/. Last accessed 26 June 2019
19. Wayne State University, Smart Manufacturing Demonstration Center. https://engineering.wayne.edu/news/wayne-state-welcomes-industry-partners-for-first-look-at-new-smart-manufacturing-demonstration-center-35640. Last accessed 26 June 2019
20. Wayne State University. https://engineering.wayne.edu/news/wayne-state-and-cisco-team-up-to-accelerate-digital-manufacturing-28568. Last accessed 26 June 2019
21. Smart Electronics Manufacturing Laboratory, Binghamton University. https://www.binghamton.edu/ieec/facilities/seml/index.html. Last accessed 05 July 2019
22. 3D printing and additive manufacturing intelligence. https://www.tctmagazine.com/tct-events/tct-3sixty-uk/new-3d-printing-technologies-tct-show/. Last accessed 11 August 2019
23. IMHX Homepage. https://www.imhx.net/en/home.html. Last accessed 11 August 2019
24. Nithyanandam GK, Kothandaraman PK, Munguia J, Marimuthu M (2020) Implementing marked-based augmented reality in discrete industry—a case study. In: IEEE-HYDCON
25. Nithyanandam GK, Sellappan SK, Ponnumuthu S (2018) A cyber-physical system improves the quality of machining in CNC milling machine—a case study. In: AIMTDR 2018 conference, advances in forming, machining and automation. Springer Singapore, pp 421–430

Development of IoT Enabled 3D Printed Smart Prosthetic Leg

K. Abhinav Rohan, Niyaz A. Sindhwani, R. Nikitha, D. Kondayya, and J. John Rozario Jegaraj

Abstract Prosthetics are used to help people with permanent disabilities and provide an opportunity to lead a normal and independent life. Prosthetics are customized, according to the use of the physical attributes of the individual. However, conventional prosthetics are heavier, and the manufacturing of them is expensive and time-consuming. To have customization of prosthetics and to have the light weight-optimized design of prosthetics, additive manufacturing techniques or 3D printing were introduced in prosthetics. Additive manufacturing in prosthetics provides a great advantage in customization and optimization of prosthetics design to each individual with reduced manufacturing lead time. Nowadays, many medical devices such as implants and prostheses are been manufactured using additive manufacturing techniques in a very short period. Recently, IoT has also been a major effective technology in the medical field in terms of medication and research of the human body. IoT technology helps to convert prosthetics into smart prosthetics. IoT is used to detect the changes in and out of the human body and pass the data accordingly to take decisions. The main objectives of the work are (1) design of below-knee prosthesis according to amputee's residual leg, (2) analyzing the design using ANSYS by applying suitable loads, (3) fabricate the prosthetic leg using fused deposition modeling (FDM) technology and embed suitable sensors on to the prototype, and (4) integration of sensors with a mobile application using IoT. The prototype was fabricated using FDM technology and the material utilized is ABS-M30 plastic. It can further be developed into a fully functional model which can be used in real-life conditions.

Keywords Prosthetic leg · Internet of Things (IoT) · Fused deposition modeling (FDM) · Sensors

---

K. Abhinav Rohan · N. A. Sindhwani · R. Nikitha · D. Kondayya (✉)
Sreenidhi Institute of Science and Technology, Hyderabad 501301, India
e-mail: dkondayya@sreenidhi.edu.in

J. John Rozario Jegaraj
Defence Research and Development Laboratory (DRDL), Hyderabad 500058, India
e-mail: johnmfrg@drdl.drdo.in

U. S. Dixit et al. (eds.), *Advances in Forming, Machining and Automation*, Lecture Notes in Mechanical Engineering, https://doi.org/10.1007/978-981-19-3866-5_41

1 Introduction

When a part of the body is lost, one might experience a deep sense of grievance. However, prosthetics gave great hope to amputees to regain some of their functionality. The science of creating artificial body components is called prosthetics. The artificial parts are called prostheses. Prosthetic implants are used to replace a missing part of the human body that might be lost due to some kind of trauma, a congenital disorder, or any accident. There can be multiple reasons. Prosthetic devices help to replace the missing part of the body. Limb prosthesis is gaining popularity. Advancements are taking place in the field of prosthetics which are helping amputees to recover and feel almost the same as they had their limbs functioning well. Only cosmetic benefits were achieved through prosthetics. With progression happening at a rapid pace in the field of research and technology, the present-day prosthetics can provide amputees not only support to restore gait movement but also improve the functionality of the missing part [1]. There are two types of a prosthesis for the lower limb, namely the above-knee prosthesis and the below-knee prosthesis.

Recent research in both material sciences and technology has resulted in significant advancements in prosthetic limbs. It is a pathway to fulfill an amputee's tantalizing glimpse of activities that can be achieved with almost normal gait movement and functionality with the use of bionic limbs. Mostly, the prosthetics are manufactured using engineering plastics or composites such as carbon fiber or fiberglass. In fused deposition modeling, the prosthetic leg is manufactured by depositing the suitable material layer by layer onto the work base [2]. Most prosthetic legs vary in complexity based on their functions and the type of feet required. Prosthetic legs that consist of a foot-and-ankle mechanism may include the following:

- **Simple model**: This model has a fixed foot-and-ankle position. A pivot might also exist to allow up-and-down movement.
- **Complex model**: It is used for walking on uneven roads, and thus it provides multi-axle movement.
- **High-performance model**: This is a higher-grade prosthetic leg. It allows a person to participate in high-intensity activities such as jumping, running, or jogging.

In this work, a simple model of a prosthetic leg is considered by taking suitable dimensions of a 60 kg person. The prosthetic leg was designed and analyzed using ANSYS software. Then a prototype of the prosthetic leg is 3D printed using a fusion deposition modeling (FDM) 3D printer. The sensors are used to extract the required data from the prosthetic leg, and the data is sent to the mobile device through an IoT system developed. As the material used in the fabrication of prosthetic leg is ABS-M30 plastic, the cost of fabrication including the 3D printing is less compared to other prosthetic legs fabricated with other expensive materials. As the developed prosthetic leg is just a prototype the cost analysis cannot be acquired as it depends on the amount of material required to produce the fully functional real-size model.

2 Work Methodology

By considering suitable dimensions, the prosthetic leg is designed using solidworks CAD modeling software. The analysis of the prosthetic leg design is performed using ANSYS software by considering acrylonitrile butadiene styrene (ABS) plastic as a material. The loading conditions are considered according to the literature review and field study that have been made. The load applied on the prosthetic leg model is 900 N which is approximately equal to 90 kg which includes the factor of safety for a 60 kg person [3] (Fig. 1).

After the finalization of the design meets the requirements, the CAD data is converted into standard tessellation language (STL) file format. The updated file format of the CAD data of components is then processed using Insight 3D software for 3D printing [4]. The slicing, tool path generation, and support generation are carried out in the Insight 3D software. The prosthetic leg is then 3D printed in a FDM 3D printer. The material used for 3D printing is ABS M-30 and the 3D printer used is STRATASYS 400 MC. After the 3D printed prosthetic leg is obtained, sensors are integrated with the prototype. The suitable sensors for the prosthetic leg are selected by considering the loading scenarios [5]. The selected sensors are ultrasonic sensors for obstacle detection and force resistive sensors for load detection. These sensors are further integrated with necessary electronics and bluetooth devices for suitable IoT interfaces. The integration for IoT is done with the 3D printed leg successfully, and the data of ultrasonic and load sensors is obtained using a mobile application through a Bluetooth module.

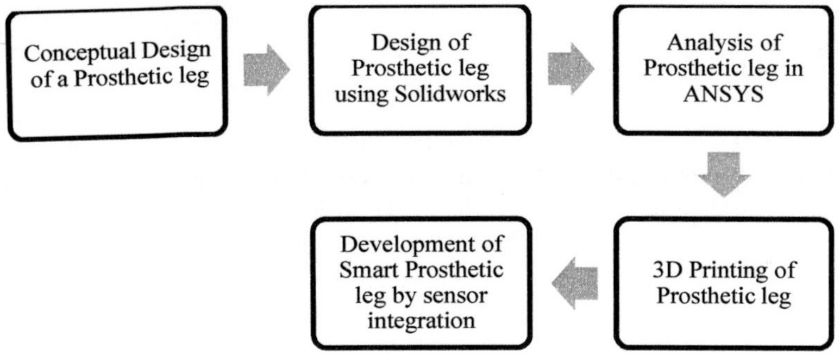

Fig. 1 The flow diagram of the work

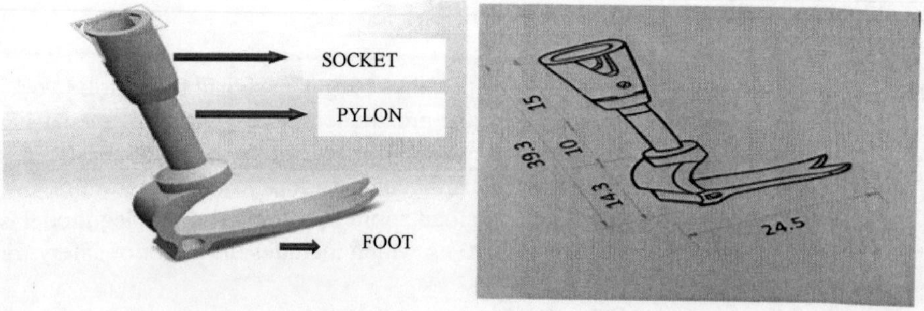

Fig. 2 Assembly view and dimensions of prosthetic leg design

3 Design of Prosthetic Leg

3.1 Design Consideration

The design of the prosthetic leg was carried out considering the weight of the person as 60 kg and the dimension as shown in Fig. 2. The conceptual design of the prosthetic leg is acquired from the ESIC medical hospital orthopedic wing, Hyderabad. This assembly consists of three parts—socket, pylon, and foot. The prosthetic leg is modeled in solidworks software in three parts. The total height of the prosthetic leg is around 393 mm (39.3 cm). All the dimensions mentioned in Fig. 2 are in centimeters.

3.2 Analysis

The analysis of the modeled below-knee prosthetic leg is performed using ANSYS software by considering the load conditions. The meshing considered for the prosthetic leg is tetrahedral meshing and the load considered is 60 kg. A multiplicative factor of 1.5 (factor of safety) is taken into consideration and multiplied with the 60 kg load [6]. Therefore, the total loading on the prosthetic leg is 900 N and is approximately equal to 90 kg. The analysis of the foot is performed by considering a vertical load of 900 N. The total deformation analysis and the von Mises stress analysis are performed on the foot part being a critical part. In this case, ABS plastic is considered as material for the analysis of the foot to have a lightweight foot for amputees. The results obtained after the analysis of the foot are well within the permissible limits. Then, the total assembly of the prosthetic leg is analyzed. The loading condition considered is that load acting on the top of the prosthetic leg at the socket. The analyzes performed are: Total deformation analysis and von Mises stress analysis. The analysis results of the foot and the assembly of the prosthetic leg are tabulated in Table 1. The results obtained upon the analysis are considered well

Table 1 Analysis results

S. No.	Analysis	Result (Max)
1	Total deformation analysis on foot	0.50568 mm
2	Von Mises stress analysis on foot	15.569 MPa
3	Total deformation analysis on prosthetic leg	11.066 mm
4	Von Mises stress analysis on prosthetic leg	20.193 MPa

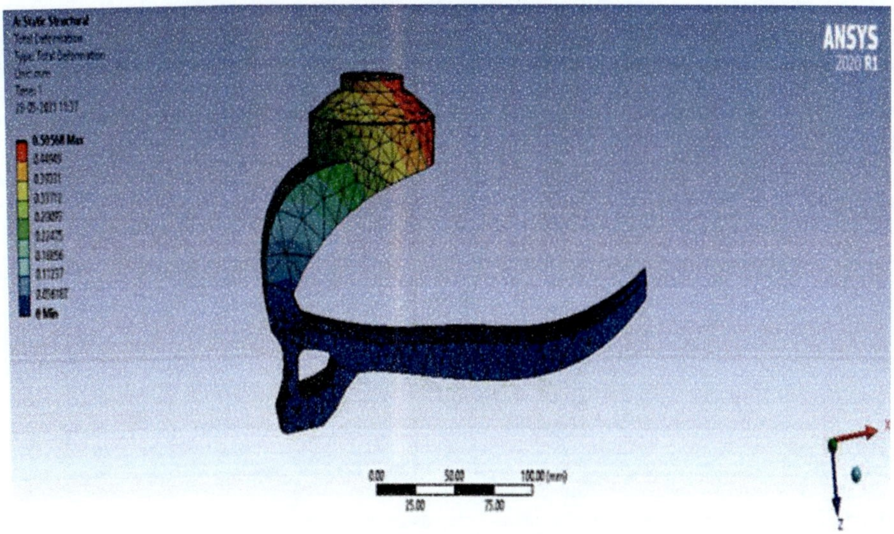

Fig. 3 Total deformation analysis of foot by applying a load of 900 N

within the permissible limits for ABS material. Hence, this prosthetic leg with ABS material is considered to be the finalized model for the fabrication (Figs. 3 and 4).

4 Fabrication

In this work, the three parts of the prosthetic leg considered for fabrication are socket, pylon, and foot. The dimensions of the parts are: Socket-height = 15 cm; pylon-height = 10 cm; foot-height = 14.3 cm; length = 24.5 cm; and width = 8.1 cm. The total height of the prosthetic leg assembly is 393 mm. The yield strength of ABS-M30 is 31 MPa, flexural strength is 58 MPa, and heat deflection value is 96 °C at 4.5 bar pressure, and the tensile modulus is 2200 MPa (Figs. 5 and 6).

Slicing operation is performed to all the parts individually in the Insight 3D modeling software. The tool path generation and support generation was carried out

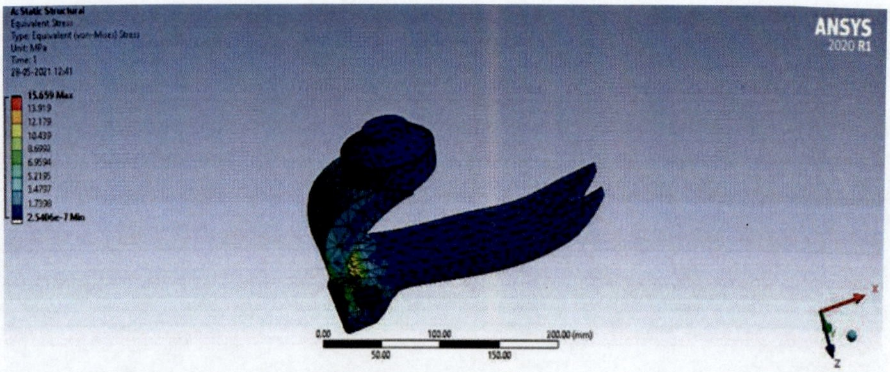

Fig. 4 von Mises stress analysis of the foot

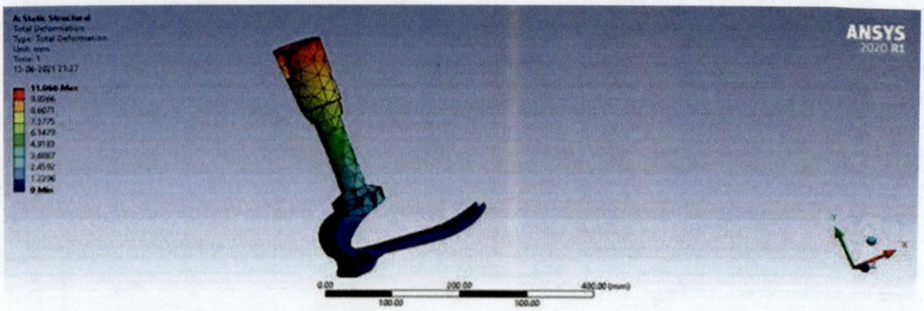

Fig. 5 Total deformation analysis of prosthetic leg assembly

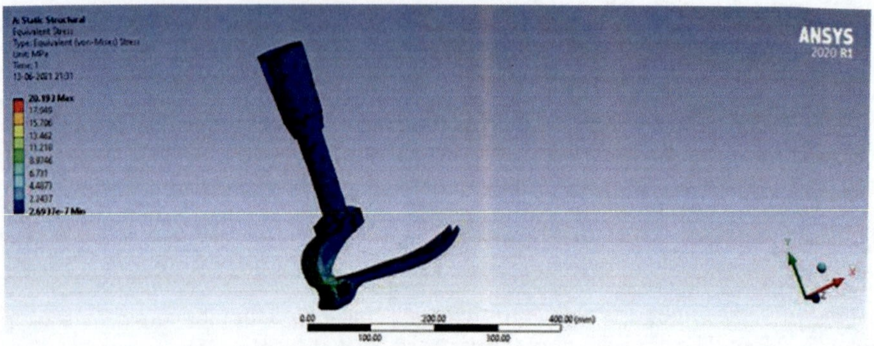

Fig. 6 von Mises stress analysis of the prosthetic leg assembly

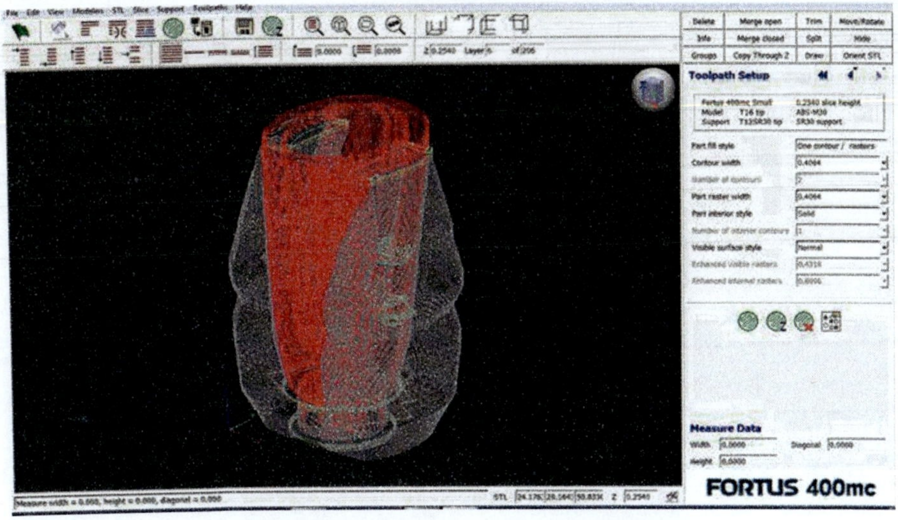

Fig. 7 Slicing and supports given to the socket

in the same software. Care is taken to orient the part in such a way that there is minimum support generation as shown in Fig. 7.

The 3D printing operation was carried out in a FDM 3D printer. The sliced files are fed into the 3D printer in the format of STL. The layer resolution input is 0.2 mm and the minimum wall thickness is 1.5 mm [7]. For the prototype, each component is scaled down to 1/3rd of its original length and printed (Fig. 8a). The time consumed for fabricating all the components of the prosthetic leg is 5 h 8 min. The material used for model and support is ABS plastic [8]. The supports provided for each component are removed after successful 3D printing and the unwanted materials on the surface of each component are removed by post-processing techniques [9]. Finally, all the three individual 3D printed components are assembled (Fig. 8b).

5 Sensor Integration

The sensors considered for the work are ultrasonic sensors and force resistive sensors and the electronic components used are a buzzer, bluetooth module (HC-05), and a breadboard [10]. The 3D printed part integrated with sensors is shown in Fig. 9. To control all the electronic components, a microcontroller (Arduino Uno) is used. A 5 V external power source is provided to Arduino Uno to execute operations. The Arduino Uno is connected to a computer through a USB cable and a code is written in Arduino IDE and then uploaded into the Arduino Uno. After uploading the code, the output of both the sensors, namely the ultrasonic sensor and the force resistor sensor are monitored in the serial monitor of Arduino IDE. The integration of sensors and

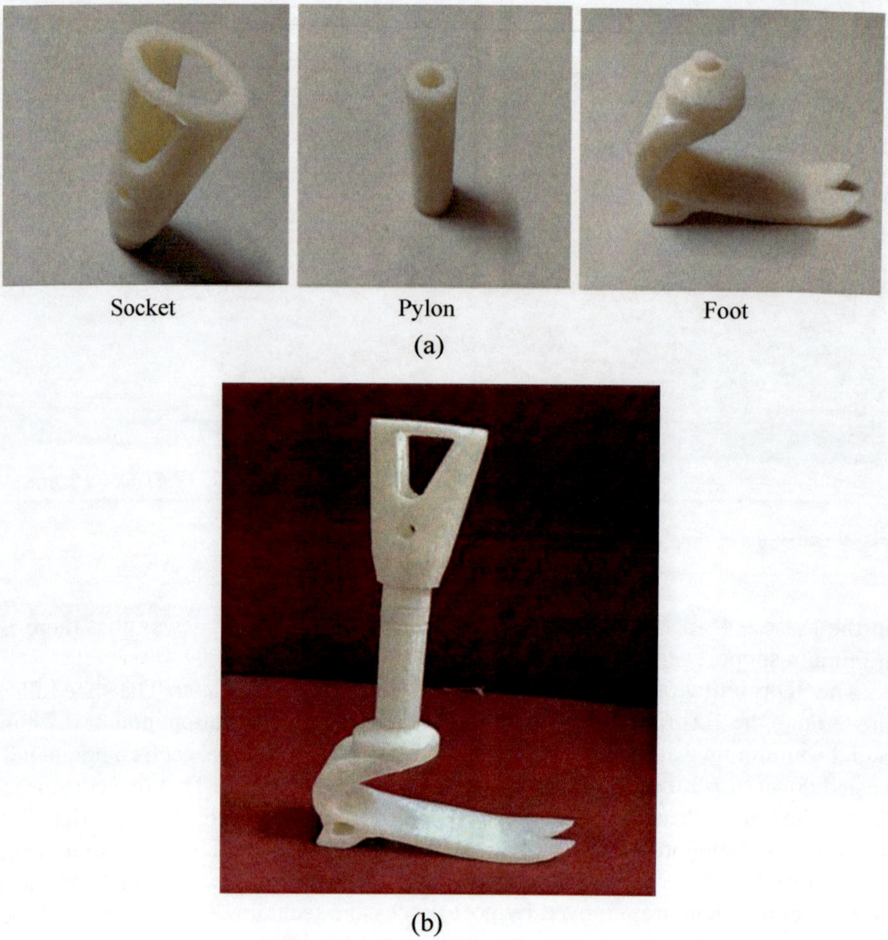

Fig. 8 **a** the fabricated individual components of the prosthetic leg, **b** the assembled model of 3D printed prosthetic leg

data acquisition systems associated with the design makes the proposed prosthetic leg more novel than the other available prosthetics.

6 Mobile Interface

A mobile application is developed on a platform known as modular creator. The mobile application format is Android Package (APK) [11]. The connectivity of the mobile application and the Arduino Uno is through bluetooth. The real-time sensor data is stored in the internal memory of the mobile device through the application

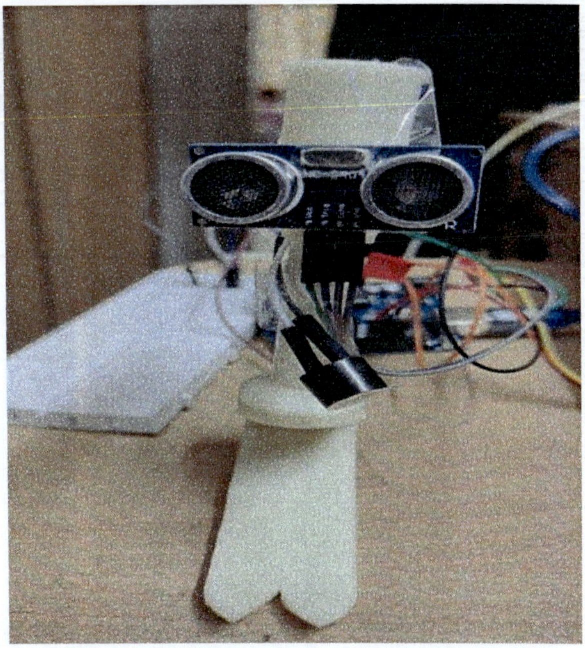

Fig. 9 Integration of IoT with the 3D printed prosthetic leg

in an excel file. Figure 10 shows the interface of the mobile application developed and the buttons that are created. The developed mobile application consists of three buttons, namely GetData, stop, and bluetooth select. Firstly, the mobile application is connected to HC-05 bluetooth using the bluetooth select button. As the successful pairing is achieved, the GetData button can be pressed to acquire the continuous data from the sensor to the mobile application. The stop button can be used to terminate the data flow from the sensor and GetData can be pressed at any moment to start the data flow again. The data acquired is simultaneously stored in the internal memory of a mobile device. Finally, the sensors are placed at the designated positions on the 3D printed prosthetic leg and thus integration of IoT with the prosthetic leg is achieved

7 Results and Discussions

The integration of the IoT and the 3D printed prosthetic leg is completed. When the load is applied on the prosthetic leg, the force resistive sensor detects the load and sends the data to the Arduino Uno. Further, the data is sent to the mobile application through a Bluetooth module. The mobile application stores the sensor data in the internal memory of the mobile in the form of an excel file as shown in Fig. 11. The excel file can be accessed at any time to monitor and acquire the required sensor data. The output of the ultrasonic sensor data is displayed on the serial monitor screen as shown in Fig. 12, and it shows the distance between the leg and obstacle and the

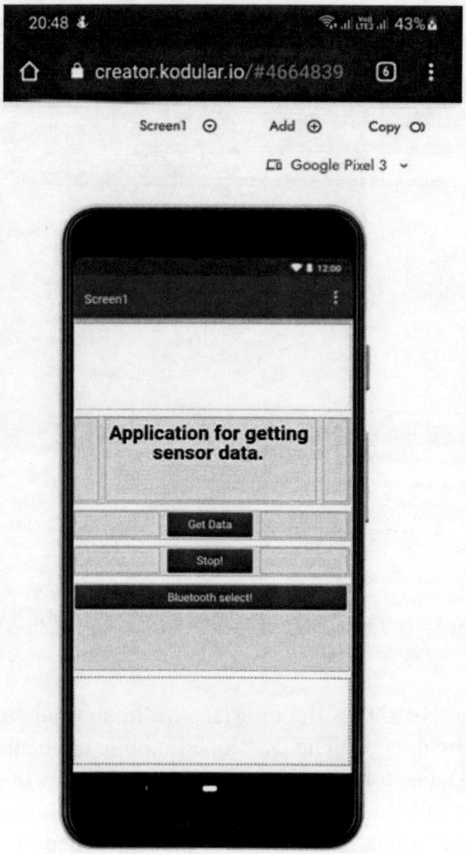

Fig. 10 The interface of the mobile application

force applied on it in Newtons (N).

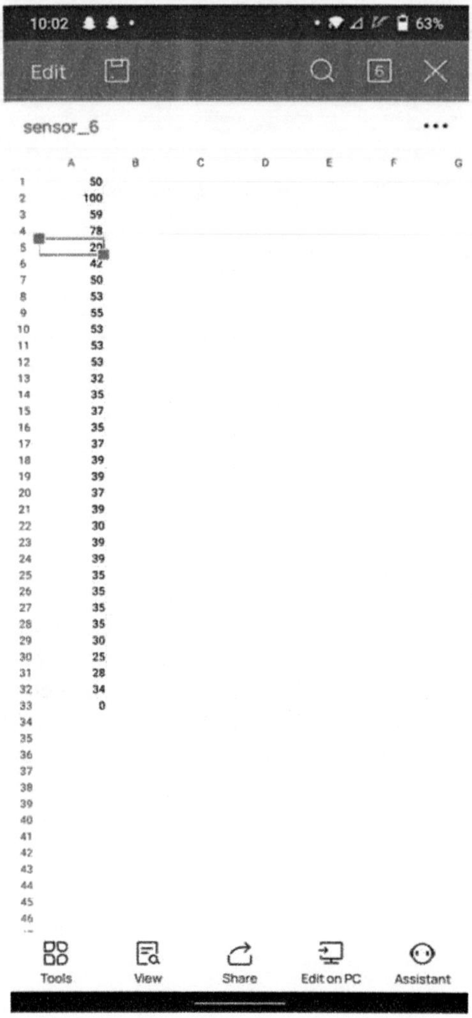

Fig. 11 Sensor data stored mobile interface storage (force is newton)

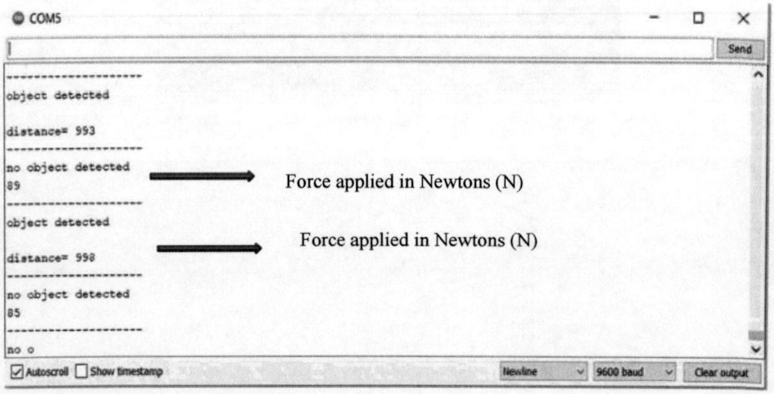

Fig. 12 Sensor data displayed on the serial monitor

8 Conclusions

The smart prosthetic leg is designed and developed with lightweight ABS material through the 3D printing process. The developed prosthetic leg was integrated with IoT sensors and successfully interfaced with mobile applications for quick monitoring of data. This prototype can be further developed and manufactured into a fully functional product by adding additional sensors and embedded electronics. The smart prosthetic leg can be further modified for different age groups, weights, and heights of people. The smart prosthetic leg will surely benefit the amputees in sensing the environmental conditions as well as provide more comfort.

Acknowledgements The authors express their sincere thanks to Director, *Defence Research and Development Laboratory (DRDL), Hyderabad and* Dr. Y. Ranjith Kumar, *Orthopedician, ESIC medical college and Hospital, Hyderabad* for providing necessary technical input and support.

References

1. Versluys R, Beyl P, Van Damme M, Desomer A, Van Ham R, Le-feber D (2009) Prosthetic feet: state-of-the-art review and the importance of mimicking human ankle-foot biomechanics. Disabil Rehabil Assistive Technol 4(2):65–75
2. Gibson I, Rosen D, Stucker B (2014) Additive manufacturing technologies, 3D printing, rapid prototyping, and direct digital manufacturing, 2nd edn. Springer, New York
3. Rochlitz B, Pammer D (2017) Design and analysis of 3D printable foot prosthesis. Periodica Polytech Mech Eng 61(4):282–287
4. Wan Fadzil WFA, Mazlan MA, Hanapiah FA, Abdullah AH (2019) 3D printed lower-limb socket for prosthetic legs. Int J Eng Res Manage 6(3):2349–2058
5. Biddiss E, Chau T (2007) Upper limb prosthesis user abandonment: a survey of the last 25 years. Prosthet Orthot Int 31(3):236–257

6. Bhandari S, Regina B (2014) 3D printing and its applications. Int J Comput Sci Inf Technol 2(2):378–380
7. Yap J, Renda G (2015) Low-cost 3D-printable prosthetic foot. In: Proceedings of the 3rd European conference on design4health. Sheffield UK
8. Farbman D, McCoy C (2016) Materials testing of 3D printed ABS and PLA samples to guide mechanical. In: Proceedings of the ASME 11th international manufacturing science and engineering conference
9. Pirjan A, Petrosanu D-M (2013) The impact of 3D printing technology on the society and economy. J Inf Syst Oper Manage 7(2):360–370
10. Islam T, Mukhopadhyay SC, Suryadevara N (2017) Smart sensors and internet of things: a postgraduate paper. IEEE Sens J 17(3):577–584
11. Kodular Homepage. https://www.kodular.io/. Last accessed 20 Sept 2021

Development of Quality Inspection System for an Impeller Using Convolutional Neural Network Model

B. Vaishnavi, D. J. Hiran Gabriel, V. Vijayanand, and S. Pradeep

Abstract This paper emphasizes the automation of quality inspection particularly using deep learning algorithm for visual inspection in identifying the surface defects of the impeller. The data collected from one of pump industry was processed using convolutional neural network-based binary classification model. The impeller test images were labeled and augmented and finally inspected with the help of convolutional neural network model. The inspected test images of the impeller have achieved an accuracy of 70.4% which were verified with receiver operating curve.

Keywords Deep learning · Quality inspection · Machine vision system · Automation · Python · Confusion matrix · ROC curve

1 Introduction

Manufacturing firms' long-term competitiveness depend on their ability to deliver fault less, high-quality products. Despite the great obstacles of growing product variation and complexities, as well as the requirement for low-cost production, quality inspection must be comprehensive and dependable [1]. As a result, high inspection volumes transform inspection operations as bottlenecks in the manufacturing process. And nowadays, the most difficult tasks in the manufacturing process are to ensure fast and accurate industrial inspection to guarantee the highest quality standards at a competitive price. An integrated solution for predictive model-based quality inspection in industrial manufacturing was highlighted in this paper, combining machine

B. Vaishnavi · D. J. Hiran Gabriel (✉) · V. Vijayanand
Department of Mechanical Engineering, PSG College of Technology, Coimbatore, Tamil Nadu 641004, India
e-mail: victorrajan1996@gmail.com

B. Vaishnavi
e-mail: bvi.mech@psgtech.ac.in

S. Pradeep
Department of Mechanical Engineering, Kumaraguru College of Technology, Coimbatore, Tamil Nadu 641004, India
e-mail: Pradeep2293.info@kct.ac.in

U. S. Dixit et al. (eds.), *Advances in Forming, Machining and Automation*, Lecture Notes in Mechanical Engineering, https://doi.org/10.1007/978-981-19-3866-5_42

learning techniques and computer vision technologies. It is one of the holistic approaches that includes target-oriented data collecting and processing, modeling and model deployment, and technological implementation.

To improve the accuracy and bring down the cost of an industrial visual inspection process, a deep learning application is paired with an optical camera. The rotating machinery division of PSG Industrial Institute followed manual visual inspection method for 8 stage submersible pump impellers. The occurrence of human error resulted in less quality assurance. Hence, this quality inspection system was automated by convolutional neural network-based binary classification model.

2 Literature Survey

An automated system was proposed for the ceramic tiles sector as part of the investigation to ensure visual inspection and quality control. A variety of approaches for identifying color and surface flaws in ceramic tiles were presented. The flaws were the emphasis of the industrial image processing approach, not the image processing methodology [1–3]. Javier et al. developed a method for profile subtraction in PCB examination that can help locate light-emitting diode (LED) chips on wafers in far less time. For picture segmentation, a fast template matching approach was employed, and blob analysis was performed to estimate the placements of possible chips. The technique was demonstrated to be considerably faster than the usual approaches for detecting LED chips in experimental studies on this system [4].

In the printing industry, Villalba-Diez et al. [5] show how a deep learning soft sensor application may be coupled with a high-resolution optical quality control camera to improve the accuracy and lower the cost of an industrial visual inspection process. Errors such as holes in the printing cylinder are unavoidable during the manufacture of gravure cylinders. In the investigations, a deep neural network (DNN) soft sensor that compares the scanned surface to the utilized engraving file and performs an autonomous quality control procedure by learning features from training data to improve defect detection performance and minimize quality inspection expenses. The developed DNN sensor achieved a fully automated classification accuracy rate of 98.4%.

Peres et al. [6] reduced the wastes and cost of repairs on a multistage assembly line from Volkswagen Auto Europa's body shop, particularly between the framing inspection stage and the body-in-white assembly evaluation, by comparing different ML techniques. They identified and conducted which ML technique performs well for their test data and reduced defects. Several machine learning classifiers were trained and tested on a range of metrics to predict geometric defects in a real automobile multistage assembly process. Between numerous human-operated assembly and pre-alignment stages on the line, two automated inspection stages are in operation. The findings showed that nonlinear models such as XGBoost and random forests are capable of modeling the complexity of such an environment and achieving a high true positive rate, as well as promising to improve existing quality control approaches by allowing defects and deviations to be addressed earlier and reducing scrap and repair costs.

Deshpande et al. [7] showed the application of one-shot recognition of the Siamese convolutional neural network on steel surfaces. In two methods, the vision-based technique adds to quality inspection automation. The first-fold is non-invasive, allowing the surface quality of steel after manufacturing to be assessed without inflicting any harm to the steel. The second aspect is that the created model just requires a little amount of labeled data to train new class pictures, making it easy to adapt to different applications. This combined use of deep learning and computer vision paves the door for a plethora of new industrial technologies. The network design utilized in this article was not optimum. Improved hyperparameter settings for the dataset are required for the model to perform better. In this case, just single-channel picture data of surface defects was employed to inspect steel surfaces. Sensor data with additional characteristics may be utilized to examine the texture of a surface in more depth.

Harshal Bhamare et al. [8] developed a system which measures the depth of tire treads using LabVIEW stereo vision and determined the correct depth in the tire's tread region of interest (ROI) using edge detection algorithm. Furthermore, this system allows the user to get information about the depth to off the tread through the Internet or a phone by connecting to a database via the Internet of Things. For safety concerns, the vehicle's tire tread depth could not be reduced below the limitation amount. If the tread depth is less than 1.6 mm, the tire should be changed. The tread depth was originally measured with a depth indicator while the vehicle was at rest. This type of assessment would be ineffective and reliant on humans. Image processing techniques were used to make the assessment of tire tread depth easier, with the goal of boosting efficiency and lowering labor intensity. To collect the tire's profile and quantify the groove depth, the system used an array of laser light sources and numerous image sensors arranged in a line.

Enciso-Aragón et al. [4] developed a fruit-oriented automatic quality inspection system. It was based on artificial intelligence algorithms using convolutional neural networks (CNN) and a fuzzy motor that evaluated a series of physical aspects of vital importance to determine the quality of a Persian lemon. The CNN was trained through transfer learning and obtained a 97.5% efficiency in the validation of characteristics. The methods proposed for the characterization process worked correctly, they faced problems in the identification of defects generated by the ambient lighting. The proposed CNN system consistently classified each of the lemons studied in the three proposed ranges according to the characteristics and rules established for the evaluation of these lemons, according to the predefined quality standards.

Huang et al. [8] developed a compact CNN-based model that not only achieves high performance on tiny defect inspection but they also worked on low-frequency central processing units (CPUs). A lightweight (LW) bottleneck and a decoder were included in the model. The bottleneck delivers rich features at a low computational cost thanks to a pyramid of lightweight kernels. The decoder is likewise made of

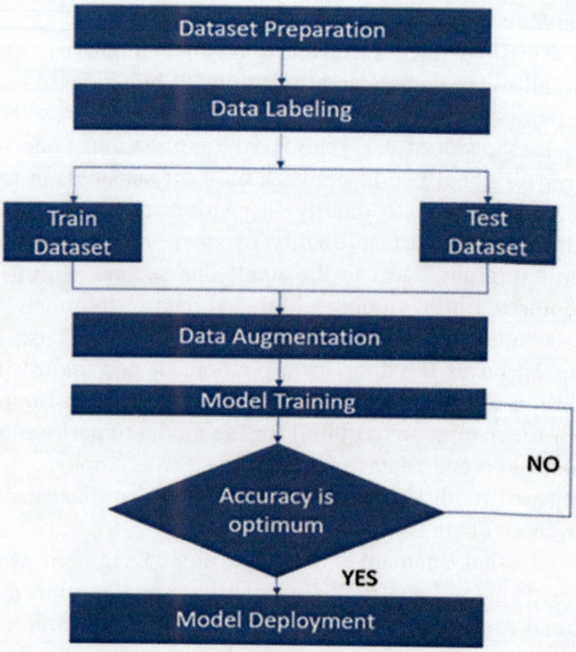

Fig. 1 Methodology for model deployment process

lightweight components, including a pooling layer and depth-wise separable convolution layers. The unnecessary weights and computations were substantially reduced thanks to these lightweight designs. Investigators used a variety of surface fault datasets to train the models. With an Intel i3-4010U CPU, the model successfully identified surface flaws in 30 ms. Results suggest that CNNs can be small and hardware-friendly for automated surface inspection applications (ASI).

Taweelerd et al. [9] in their investigation proposed an approach for products inspection of submersible pump impeller images by the vision system based on deep learning with convolutional neural network architecture for casting. The suggested method produced state-of-the-art results. With only 10 epochs and 1.51 min for model training, it was able to attain an average accuracy of 99.7%. When it comes to prediction, one picture takes 56.87 ms to forecast. It was concluded that CNN architecture can reach great accuracy, according to the trials. In the future, faulty items can be easily distinguished from non-defective ones, reducing human error in this sector [10] (Fig. 1).

3 Methodology

Initially dataset was prepared which includes images of the impeller having casting surface defects like blowholes, porosity, and cracks. The images of the defective impeller images and non-defective impeller images were taken and then stored in the folder as train dataset as shown in Figs. 2 and 3. Four hundred defective impeller images and two hundred non-defective images of the impeller were taken. After taking the images, it was processed with the help of data augmentation technique in the Spyder integrated development environment. Then CNN model for the binary classification of the impeller was developed, and test dataset was prepared

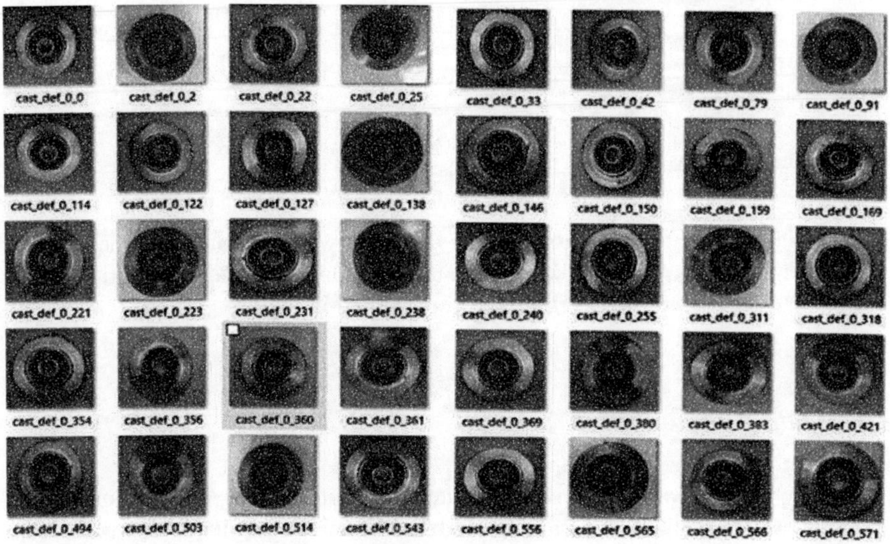

Fig. 2 Train dataset for defective impeller images

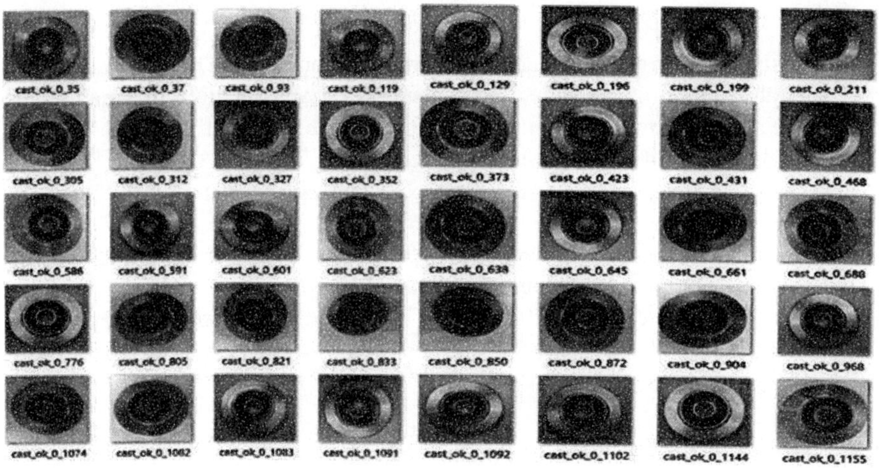

Fig. 3 Train dataset for non-defective impeller images

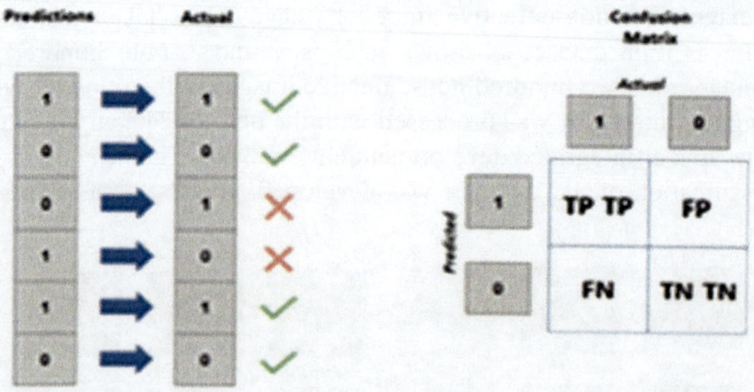

Fig. 4 Confusion matrix

and inspected. After this process, confusion matrix was generated and the receiver operating characteristics curve was generated for identifying the threshold.

The methodology of the work flow is represented in Fig. 1.

3.1 Data Augmentation

Data augmentation was done on the collected train dataset images for increasing the diversity of data available for training models, without having lot of new data. Data augmentation techniques such as cropping, padding, and horizontal flipping were done to train the neural networks. The schematic representation of confusion matrix is given in Fig. 4.

The below steps explain about the data augmentation process.

1. Load the image processing and deep learning libraries in the Python. image data generator and ski image were used to perform various functions like reading the image and performing computations with the pixel data of the image.
2. Load the image into the Python language
3. Reshape the image size according the model and apply various matrix operations on the loaded image.
4. Since image is a matrix of data, i.e., pixel values ranging from 0 to 255, using data augmentation we can multiply the available images.
5. The process was done for 20 times as it attained maximum accuracy for the model.

3.2 Model Training and Deployment

A confusion matrix was generated for the deployed model as given in Fig. 5. It describes the performance of a classification model. The performance measurement for machine learning classification problem where the output is of two classes was

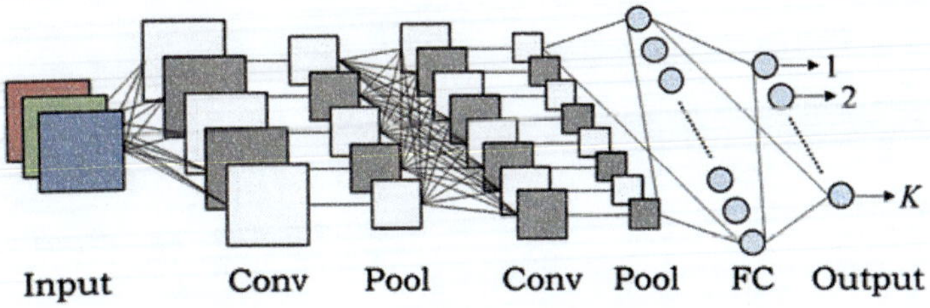

Fig. 5 Convolutional neural network architecture for the deployed model

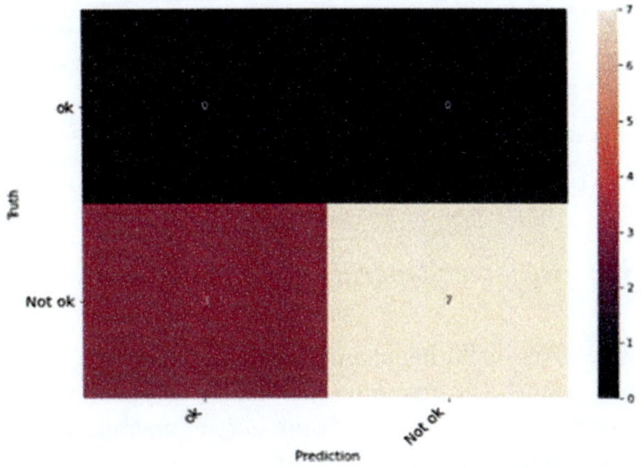

Fig. 6 Generated confusion matrix

calculated using the predefined functions in the Python library. With the help of confusion matrix recall, precision, specificity, accuracy, and most importantly AUC-ROC curve was generated. True Positive, False Positive, False Negative, True Negative, precision, and classification accuracy for the model was calculated. The below steps explain the algorithm in the confusion matrix and the generated confusion matrix is represented in Fig. 6.

Step 1: Load the obtained predictions values by the CNN model and actual values into the confusion matrix.

Step 2: Based on the given values, table will be generated showing all the values based on predicted value.

Step 3: The values obtained are TP, FP, FN, and TN

Step 4: The obtained true positive value will be helpful for validating the accuracy of the model.

Step 5: The ROC curve was used to avoid the creation of multiple confusion matrices.

Step 6: The ROC curve was generated for identifying the optimal threshold value. The threshold value specifies the maximum true positive rate which will be crucial for the accuracy of the model.

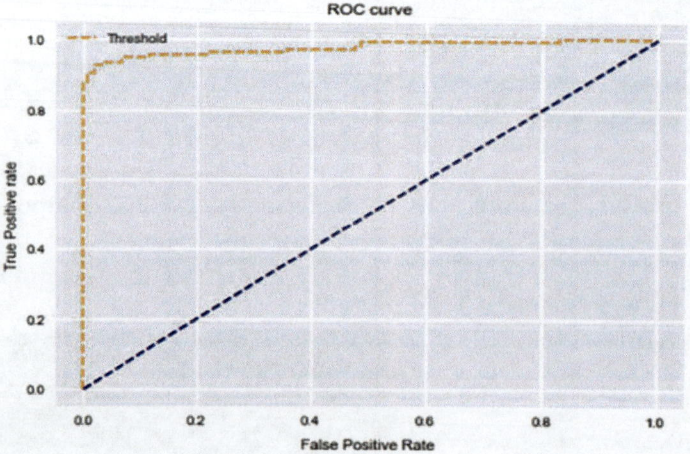

Fig. 7 Generated ROC curve

3.3 Receiver Operating Characteristic Curve

A receiver operating characteristic curve or ROC curve with false positive rate in *X* axis and true positive rate in *Y* axis was generated for identifying the threshold as represented in Fig. 7. After the curve generation, optimal threshold was identified. For a probability value of 0.5, the true positive values are high, and therefore, 0.5 is set as the threshold for the classification model.

4 Conclusion

The automation of quality inspection process using machine learning technique using CNN model has identified the defects with the help of deep learning architecture. The ROC curve was generated for identifying the threshold value by consolidating the various numbers of confusion matrix. The generated ROC curve showed a threshold of 0.5. After setting the obtained threshold, the model predicted the defects with an accuracy of 71.4%.

This results that the suggested algorithm outperforms the traditional inspection method by a significant margin with substantial speediness and correctness. The surface defects were identified with the help of eddy arrays. Further, the use of 3D reconstruction (point clouds) reduces the time and cost of the inspection. If computer vision and telocentric lens were combined for dimension inspections, the inspection time will also reduce and also it will achieve an accuracy of 0.00625 mm. The future work will be contributed to a complete automation of the quality inspection process which can be improvised by adding the geometrical measuring system and 3D contour reconstruction system.

Appendix

Program for data augmentation

```
from keras.preprocessing.image import ImageDataGenerator from skimage import io
    datagen = ImageDataGenerator(
    rotation_range = 45, #Random rotation between 0 and 45.
    width_shift_range = 0.2, #% shift.
    height_shift_range = 0.2,
    shear_range = 0.2,
    zoom_range = 0.2,
    horizontal_flip = True,
    fill_mode = ‘constant’, cval = 125) #Also try nearest, constant, reflect, wrap.
    x = io.imread(“Impeller.jpeg”) #Array with shape (256, 256, 3) x = x.reshape((1,)
+ x.shape) #Array with shape (1, 256, 256, 3).
    i = 0.
    for batch in datagen.flow(x, batch_size = 16,
    save_to_dir = ‘augmented’,
    save_prefix = ‘aug’,
    save_format = ‘png’):
    i + = 1
    if i > 20:
    break
```

Program for confusion matrix

```
from matplotlib import pyplot as plt
    from sklearn.metrics import confusion_matrix,
    classification_report
    import pandas as pd
    def print_confusion_matrix(confusion_matrix, class_names, figsize = (10,7),
fontsize = 14):
    df_cm = pd.DataFrame(
    confusion_matrix, index = class_names,

columns = class_names,

)
    fig = plt.figure(figsize = figsize)
    try:
    heatmap = sns.heatmap(df_cm, annot = True, fmt = “d”)
    except ValueError:
    raise ValueError(“Confusion matrix values must be
    integers.“)

heatmap.yaxis.set_ticklabels(heatmap.yaxis.get_ticklabels(), rotation = 0, ha =
‘right’, fontsize = fontsize)
```

heatmap.xaxis.set_ticklabels(heatmap.xaxis.get_ticklabels(), rotation = 45, ha = 'right', fontsize = fontsize)

plt.ylabel('Truth')

plt.xlabel('Prediction')

truth = ["ok", "ok", "ok", "ok",ok", "ok", "ok", "ok", "ok", "ok"] prediction = ["Not ok", "Not ok", "ok", "ok", "ok", "Not ok", "ok", "ok", "ok", "ok"]

cm = confusion_matrix(truth,prediction)

print_confusion_matrix(cm,["ok" "Not ok"])

print(classification_report(truth, prediction))

References

1. Gabriel DJH, Parthiban M (2020) A literature review on global challenges for third party logistics (TPL or 3PL) 9:27–30
2. Moganti M, Ercal F, Dagli CH, Tsunekawa S (1996) Automatic PCB inspection algorithms: a survey. Comput Vis Image Underst 63:287–313. https://doi.org/10.1006/cviu.1996.0020
3. Letichevsky AA, Letychevskyi OO, Skobelev VG, Volkov VA (2017) Cyber-physical systems. Cyber Syst Anal 53:821–834. https://doi.org/10.1007/s10559-017-9984-9
4. Enciso-Aragón CJ, Pachón-Suescún Cg, Jimenez-Moreno R (2018) Quality control system by means of CNN and fuzzy systems. Int J Appl Eng Res 13:12846–12853
5. Villalba-Diez J, Schmidt D, Gevers R, Ordieres-Meré J, Buchwitz M, Wellbrock W (2019) Deep learning for industrial computer vision quality control in the printing industry 4.0. Sensors (Switzerland) 19:1–23. https://doi.org/10.3390/s19183987.
6. Peres RS, Barata J, Leitao P, Garcia G (2019) Multistage quality control using machine learning in the automotive industry. IEEE Access 7:79908–79916. https://doi.org/10.1109/ACCESS.2019.2923405
7. Deshpande AM, Minai AA, Kumar M (2020) One-shot recognition of manufacturing defects in steel surfaces. 48:1064–1071. https://doi.org/10.1016/j.promfg.2020.05.146
8. Huang Y, Qiu C, Wang X, Wang S, Yuan K (2020) A compact convolutional neural network for surface defect inspection. Sensors (Switzerland). 20:1–19. https://doi.org/10.3390/s20071974
9. Taweelerd S, Chang CC, Tzou GY (2021) Vision system based on deep learning for product inspection in casting manufacturing: pump impeller images. J Phys Conf Ser 2020:012046. https://doi.org/10.1088/1742-6596/2020/1/012046
10. Tao F, Qi Q, Wang L, Nee AYC (2019) Digital twins and cyber–physical systems toward smart manufacturing and industry 4.0: correlation and comparison. Engineering 5:653–661. https://doi.org/10.1016/j.eng.2019.01.014

Developments in Fusion Deposition Modelling Process Using Industrial Robot

K. Venkateswarlu, A. Sri Harsha, G. Ravi Kiran Sastry, and CHR. Vikram Kumar

Abstract Fused deposition modelling (FDM) is an additive manufacturing process, where the material is deposited in horizontal layers, one on another to create a 3D object. In the conventional FDM process, the extruder head moves in two directions simultaneously to deposit one layer over a horizontal printing platform (heat bed) and the printing platform moves vertically downwards by a small amount (equal to layer thickness) to facilitate the deposition of the next layer. In conventional FDM machines, due to limited degrees of freedom, change in extruder orientation and printing on inclined planes are not possible. Recently, industrial robots are used to overcome these limitations using extruder as end effector. This paper reviewed various strategies adopted for conventions FDM and robotic FDM process to overcome various limitations. The major advantages of the FDM process with the robot are printing is possible on the inclined plane, multiple planes and curved surfaces without staircase effect.

Keywords Industrial robot · Additive manufacturing · Fused deposition modelling

1 Introduction

Manufacturing is the application of physical and chemical processes to alter the geometry, properties and appearance of a given starting material to make parts or

---

K. Venkateswarlu · G. Ravi Kiran Sastry
Department of Mechanical Engineering, NIT Andhra Pradesh, Tadepalligudem 534102, India
e-mail: grksastry@nitandhra.ac.in

A. Sri Harsha
Department of Mechanical Engineering, JNTUA, Ananthapuramu 515002, India

CHR. Vikram Kumar (✉)
Department of Mechanical Engineering, NBKR Institute of Science and Technology, SPSR, Vidyanagar, Nellore 524413, India
e-mail: cvikramkumar@gmail.com

U. S. Dixit et al. (eds.), *Advances in Forming, Machining and Automation*, Lecture Notes in Mechanical Engineering, https://doi.org/10.1007/978-981-19-3866-5_43

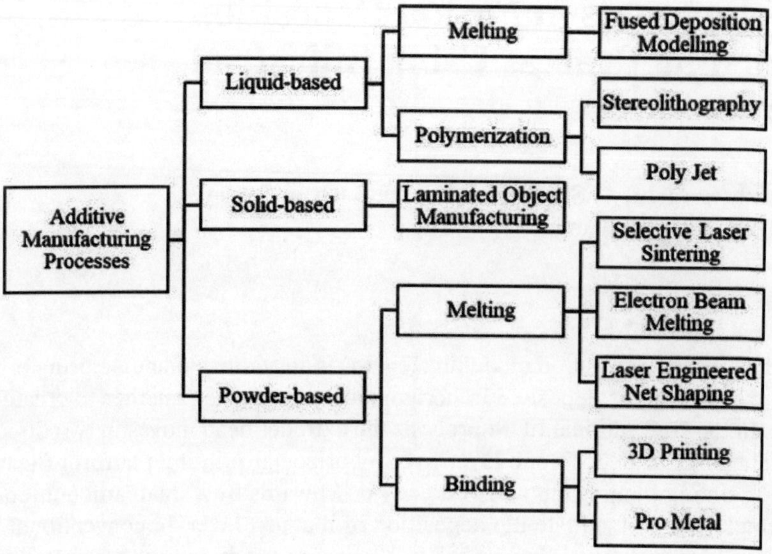

Fig. 1 Classification of additive manufacturing processes [4, 5]

products [1]. Depending on the geometry of the product, the manufacturing technology can be classified into three fundamental groups, they are subtractive manufacturing technology, formative manufacturing technology and additive manufacturing technology [2]. In additive manufacturing technology (AM), the desired geometry or shape of the object is obtained by the addition of material in contoured layers on top of each other. Additive manufacturing is also called layer technology since the material is deposited in layers, in which parts with complex shapes and geometries can be fabricated directly from 3D modelling computer-aided design (CAD) without using any special tooling. The AM process does not require detailed process planning [3]. The materials used for additive manufacturing are polymers, metals, ceramics and composites. Depending upon the type of material used additive manufacturing can be classified into three groups [4, 5] as shown in Fig. 1.

2 Fusion Deposition Modelling

Fused deposition modelling (FDM) is an additive manufacturing process where the fused material is deposited in horizontal layers one on another to create a 3D object. The digital model, i.e. CAD model is the input for this process. Software, hardware and materials work together to get the desired final product. Initially, this process is used for creating prototypes rather than end functional products. But, due to the availability of newly developed materials and higher precise machines, objects are produced with very close or the same mechanical and material properties as the end

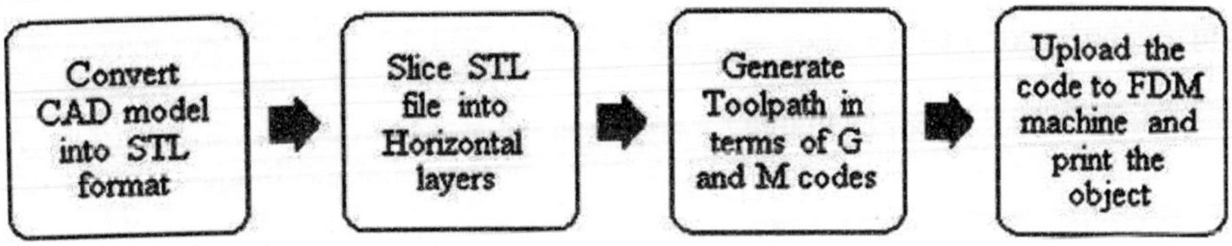

Fig. 2 Pre-processing steps in the FDM process [6]

functional products. In the conventional FDM process, the pre-processing steps to print an object are shown in Fig. 2. In the FDM process, first, the CAD model is generated by either scanning the existing object using reverse engineering principles [7] or the other way is to design a new CAD model.

Conventional FDM machine has a restricted degree of freedom (only 3). Because of the lower degree of freedom, it is only possible to print horizontal flat layers [8]. The staircase effect is one major limitation of the FDM process. This takes place because the material gets deposited in horizontal flat layers. This is quite noticeable along the curved surfaces of the object. This ultimately affects the part quality and yield lower detail of the object [3]. Many authors attempted to improve the capabilities of a conventional FDM machine. Tyberg et al. [9] introduced a new approach to slice an object adaptively to achieve the desired surface finish and to reduce the build time in the FDM process. Pandey et al. [10] presented a real-time slicing technique for the FDM process for the deposited layers on the edge profile. Thrimurthulu et al. [11] presented a methodology that uses a genetic algorithm to get an optimum orientation of the printed part for the FDM process for reducing surface roughness and build time. Adaptive slicing is considered to determine the optimum orientation of the object. Weiyin et al. [12] introduced an adaptive slicing algorithm based on the B-spline model to print objects with accurate dimensions and smooth surface finish. To minimize the build time, a selective hatching technique is presented in this study and implemented on kernel regions within the part. Chakraborty et al. [13] presented a tool path generation technique for the FDM extruder to deposit material in curved layers to eliminate the staircase effect. The presented technique is called a curve layer fused deposition modelling and it is implemented on parametric surfaces for testing. Jin et al. [14] proposed a new toolpath generation method for the FDM process to improve the quality and accuracy of the fabricated parts.

2.1 Influence of Various Parameters in the Conventional FDM Process

There are several parameters involved in the FDM process that will directly influence the mechanical properties, quality and printing time of the object. The outer boundary of each layer is called a shell. The other name of the shell is contour. Each layer can have more than one shell. Infill is a repetitive structure to fill up space inside the shell.

This is the internal structure of each layer. This is an invisible part that is covered with the outer layers. Figure 3 shows the working parameters within a layer. There are several infill patterns available for selection like concentric, grid, lines, triangle, zig-zag, etc. These infill patterns will directly influence the mechanical properties of the printed object. Figure 4 shows the different kinds of infill patterns used in the FDM process.

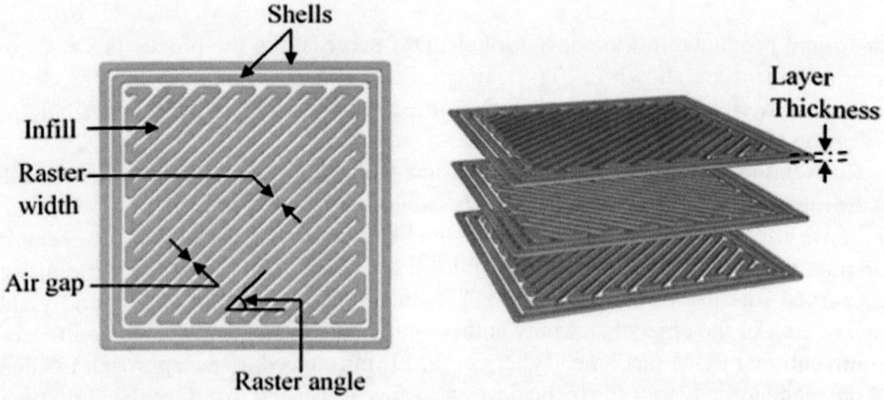

Fig. 3 Parameters within a Layer

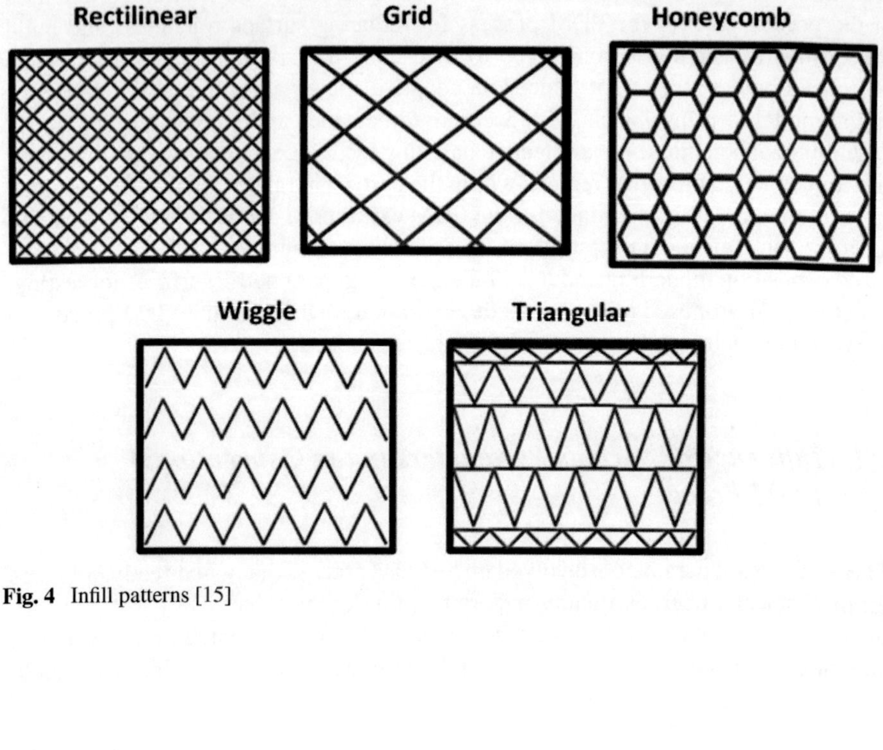

Fig. 4 Infill patterns [15]

Eric et al. [16] investigated the effect of process parameters on the tensile strength, the results showed that contour specimens exhibit high tensile strength and also indicated that strength of the specimens can be increased by adding adhesives. Kulkarni and Dutta [17] proposed two new deposition paths (spiral and modified raster) for better strength properties. Es-Said et al. [18] indicated that the tensile strength and impact resistance of the specimen with 0° raster angle is high compared to the other selected raster angles of 45°, 90°, 45°/−45° and 45°/0°. Anitha et al. [19] concluded that layer thickness has a significant effect on surface roughness. Ahn et al. [20] stated that the FDM printed specimens have 65–72% of the tensile strength and 80–90% of compressive strength than injection moulded ABS material. Lee et al. [21] determined the effect of layer thickness, raster width, raster angle and air gap on optimum elastic properties. Sun et al. [22] stated that the bed temperature and fabrication strategy affects the bonding strength of the deposited filaments. Zhang and Chou [23] addressed the issue of distortion due to residual stresses in the FDM process, it indicated that the printing speed and layer thickness are affecting the distortion. Sood et al. [24] investigated the effect of FDM process parameters on the shrinkage of the printed part made up of ABS material and indicated that shrinkage is significant along length and width. Sood et al. [25] studied the influence of FDM process parameters on tensile, bending and impact strengths and stated that with the increase in number of layers increases the strength. Fatimatuzahraa et al. [26] observed the effects of two types of raster orientation 0°/90° and 45°/−45° on tensile test, bending test, impact test and deflection test are conducted on the printed specimens. It indicated that specimen printed with 45°/−45° has high strength for bending, impact and deflection and tensile strength is with 0°/90° raster orientation. Bagsik et al. [27] analysed the influence of part orientation or build orientation and toolpath strategy on the mechanical properties of the FDM parts. Sood et al. [28] optimized the process parameters on the compressive strength using particle swarm optimization method, and stated low compressive strength is mainly due to anisotropy and distortion, and stated nonlinear complex relationship existing between the compressive strength and operating parameters. Croccolo et al. [29] developed analytical model between FDM process parameters and tensile strength and stiffness. The analytical model results are almost equal to the experimental results with a small margin of error, i.e. 4%.

2.2 *Multi-Nozzle FDM Process*

Jafari et al. [30] developed a freeform fabrication system to build a multiple number of ceramic actuators. This developed system is capable to print four types of materials within the shell (boundary) of a layer. Two subsystems were created in this work. One for positioning and the other for deposition. Khalil et al. [31] developed a multi-nozzle extruder system with four nozzles to deposit biopolymers in three-dimensional tissue scaffolds. Liu et al. [32] introduced a multi-nozzle extruder is used to extrude the polymers and showed that the multi-material scaffolds printed using the multi-nozzle extruder have enough strength and better hydrophilic nature. Espalin et al.

[33] modified two FDM machines to create a single 3D printing system that is capable of fabricating objects by depositing multiple materials within the layer. Liu et al. [34] developed a bioprinting platform to deposit multiple bioinks for fabricating complex parts within less time. Frutuoso [35] developed a new procedure to fabricate large parts in less time with the FDM process by using multiple printing extruders those work independently and simultaneously. The proposed approach is applied for four different components and indicated that the developed system is effective in reducing the build time of the large parts. Skylar et al. [36] developed a 3D printer to print voxelated matter with multi-nozzles. Cai and choi [37] presented a novel toolpath planning algorithm for additive manufacturing processes using multiple robotic actuators.

3 Hybrid Fused Deposition Modelling Process with Industrial Robot

In the conventional FDM process, the extruder moves in a horizontal plane due to limited degrees of freedom. An industrial robot is capable to move its end-effector precisely within the work volume, and it has more degree of freedom. In the hybrid fused deposition modelling process, the FDM extruder acts as the end-effector of the robot. Integrating the fused deposition modelling process with 6-axis industrial robotic arm is advantageous in the field of additive manufacturing. The manufacturing capabilities are enhanced by integrating the additive manufacturing process with other systems.

In the conventional FDM process, the extruder movements are controlled using G codes similar to CNC programming. To integrate the FDM process with the industrial robot the movements of the extruder are required to be controlled using a robot programming language. Ishak et al. [8] developed a 3D printer by attaching the 3D printing head to Motoman SV3X industrial robot's flange. Nonplanar toolpaths were developed for printing the components with refined geometries. Zhang et al. [38] developed a robotic printing system to deposit material on curved surfaces. In this work, new software was developed to generate the toolpath by converting the traditional G-code into robotic programming commands that are used to move the robotic arm. Results indicated that parts can be fabricated with improved surface finish and improved mechanical properties using the developed system. Swanson et al. [39] developed a process to fabricate 3D objects with an FDM extruder using FANUC LR-MATE 200iB robotic arm. A translator is developed to translate *G*-code into robotic commands. Alsharhan et al. [40] presented a robotic additive manufacturing process system to deposit material on nonplanar layers to improve mechanical strength. As the material extruding head is attached to the robot, it indicated that reorienting along the 3D paths is possible with the proposed system. Brooks et al. [41] designed and developed a system that is capable of depositing biomaterial on curved surfaces with the assistance of a six-axis industrial robot.

The robotic arm has more workspace compared to the conventional FDM machine; hence it can fabricate components of large size. The industrial robot can move its end-effector in the entire workspace precisely for manipulation. Volker et al. [42] presented a new approach for large-scale 3D printing of lightweight mesh structures using an industrial robot for architectural applications. Barnett and Gosselin [43] developed a 3D printer for large-scale printing using a 6-DOF cable-suspended robot. In this work, a statue of 2 m height is printed using polyurethane foam. Felsch et al. [44] presented the use of an industrial robot to fabricate large parts using AM technologies. In this work, a robotic system was developed to print large parts that are weighing 25 kg using thermoplastic material. It indicated that the proposed system is flexible and economical for fabricating large-scale parts.

Different approaches are developed to overcome the difficulties of conventional FDM during the printing of overhang structures. Dai et al. [45] developed a new system to print the objects using an industrial robot with multi-axis motion to reduce the need for support material for overhanging structures. Curved toolpaths were generated to deposit the material inside the volume. It indicated that large overhanging structures can be printed with the developed system without using support structures. Oxman et al. [46] developed a "Freeform Printing" approach using a 6-axis robot to print without additional support structures. It indicated that this approach is focussed on biologically inspired design to minimize the printing time and eliminate wastage by printing without support material. Ishak et al. [47] developed multi-plane FDM with an industrial robot arm to avoid the printing of support material for overhang structure using different slicing approaches. The extruder head with a 6-DOF robot arm enabled the printing of layers in multiple planes, whereas conventional FDM is restricted to single plane layering (*x*–*y* plane). Ishak et al. [48] developed a new algorithm to print lattice structures using an industrial robot in multiple planes for eliminating the usage of the support structure. The results indicated that by using the robotic printing platform, the need for support structure is eliminated and multi-directional printing is possible. Onstein [49] developed path generation algorithms for the FDM process using a robotic arm based on the surfaces of the object's CAD model. It indicated that it is possible to deposit material on curved surfaces by minimizing the need for support structures for overhanging features. Castelli and Giberti [50] developed a setup for robotic arm printing to fabricate 3D objects. This setup is used to build material in different planes to avoid material wastage for printing extra support material for overhanging features. It indicated that with the developed robotic arm printing setup, it is possible to print material on existing components due to the additional degree of freedom of the robot.

The major advantage of integrated FDM with the industrial robot is the generation of nonlinear toolpath printing on curved planes. The industrial robot has higher degrees of freedom and can be used to carry out multi-directional and curved layer printing. By attaching the FDM extruder to the robotic arm as an end effector, it is possible to change its orientation depending upon the geometric features of the object. Zhao et al. [51] developed two nonplanar slicing approaches, namely curved surface slicing and cylinder surface slicing to eliminate the use of support structures and decrease the number of layers. Curved slices were generated for selective volumes of the objects, and nonplanar toolpaths are generated for printing the curved portion. These curved slices are printed using a robotic fused deposition modelling system. Shembekar et al. [52] have proposed a collision-free trajectory planning approach for printing 3D objects having multiple curvatures. The nonplanar material deposition is used to print those curvatures in contrast to the planar material deposition used in conventional 3D printers. It indicated that collision with the printing surface was avoided by properly controlling the trajectory parameters with respect to surface curvature. The proposed approach was implemented by using a 6 DOF robot arm. Complex 3D structures with various curvatures were successfully fabricated using the proposed collision-free trajectory and a satisfactory surface finish is accomplished. Bhatt et al. [53] proposed a robotic cell for multi-resolution AM (additive manufacturing with two different nozzle sizes). The proposed robotic cell consists of two 6 degrees of freedom (DOF) robot manipulators. Algorithms were developed for decomposing parts into multi-resolution layers that consist of both planar and nonplanar layers. Based on the proposed algorithms, collision-free trajectories were generated for these robot manipulators. Shembekar et al. [54] presented trajectory planning algorithms for a 6 DOF industrial robot to print material on nonplanar layers. Trajectory parameters were considered in such a way that the extruder avoids collision with previously deposited layers by satisfying the robot arm's joint constraints. It indicated that the nonplanar layers were printed without encountering collisions using the developed trajectory algorithms. Xie et al. [55] presented a B-spline based smoothing algorithm for the robotic FDM process that removes sharp corners along the printing path while maintaining the required height from the printing surface. This algorithm is designed to generate a smooth printing path by avoiding sharp corners. Sri Harsha and Vikram Kumar [56] demonstrated the dual extrusion FDM process using the industrial robot, in which both nozzles are is used to extrude the build material. The results indicated that upto 50% printing can be save by extruding same build material in both nozzles. The FDM extruder can be manipulated in different configurations within the workspace of the industrial robot. Instead of printing planar layers on a horizontal plane, it is possible to deposit material in multiple planes and multiple directions. The flexibility of changing the extruder's orientation using the industrial robot enables simultaneous printing over multiple planes. Sri Harsha and Vikram kumar [57] developed a nonplanar layers of the curved objects to eliminate the staircase effect. The demonstrated the printing of aircraft wing using nonlinear tool path developed for curved layers. Bhatt et al. [58] addressed the advantages of using a reorienting deposition (build) platform to eliminate the use of support structures while printing objects with overhanging features. A slicing algorithm is developed in this work to print thin shell objects using a reorienting deposition platform and robotic extrusion system. The build platform has 3 degrees of freedom. Only 3 degrees of freedom of the robotic arm is utilized in this work. Toolpath generating algorithm is also developed for the material depositing extruder (tool) that is attached to the robotic arm. Few thin shell objects were printed using the proposed system. The results indicated that the build time is reduced drastically as there is no need for printing the support structure. This also reduces the cost as there is no need for printing extra material for the support structure. It indicated that the surface finish is also improved as the proposed system is capable to print nonplanar layers.

4 Conclusion

Recent literature indicated the FDM has a wide scope of applications for lightweight components in aerospace, automotive and medical industries due to its simplicity in use and capability to produce complex structures. A conventional FDM machine has three degrees of freedom, i.e. three translation motions. Various strategies developed in conventional FDM to print the curved layer without staircase effect are discussed in this paper. Due to this limited degree of freedom, conventional FDM is suitable to print only on a flat horizontal plane only. Integrating robotic and FDM technologies will be so advantageous in the field of manufacturing. Because of the higher degrees of freedom, the robot arm can move freely in a huge workspace compared to the conventional fused deposition modelling machines. Staircase effect can be eliminated by printing the curved surfaces of the object by changing the orientation of the extruder which is not possible in a conventional FDM machine.

References

1. Groover MP (2019) Fundamentals of modern manufacturing: materials, processes and systems. 7th Edn. Wiley, ISBN: 978-1-119-4521-7
2. Andreas G, Jan-Steffen H (2016) Additive manufacturing 3D printing for prototyping and manufacturing. Hanser, ISBN: 978-1-56990-582-1
3. Gibson I, Rosen DW, Stucker B (2020) Additive manufacturing technologies. 3rd edn. Springer Verlag, ISBN: 978-3030561260
4. Kruth JP (1991) Material incress manufacturing by rapid prototyping techniques. CIRP Ann 40(2):603–614. https://doi.org/10.1016/S0007-8506(07)61136-6
5. Wong KV, Hernandez A (2012) A review of additive manufacturing. Int Scholarly Res Notices 2012, Article ID 208760. https://doi.org/10.5402/2012/208760
6. Chua Ck, Leong KF (2010) Rapid prototyping: 3D printing and additive manufacturing principles and applications", 5th Edn. World Scientific, ISBN: 978-0000987570
7. Gibson I (2002) Software solutions for rapid prototyping. 1st edn. Wiley, ISBN: 978-1860583605
8. Ishak I, Fisher J, Larochelle P (2015) Robot arm platform for rapid prototyping: concept. In: Proceedings of the Florida conference on recent advances in robotics (FCRAR)
9. Tyberg J, Bøhn JH (1999) FDM systems and local adaptive slicing. Mater Des 20(2–3):77–82. https://doi.org/10.1016/S0261-3069(99)00012-6
10. Pandey PM, Reddy NV, Dhande SG (2003) Real time adaptive slicing for fused deposition modelling. Int J Mach Tools Manuf 43(1):61–71. https://doi.org/10.1016/S0890-6955(02)00164-5

11. Thrimurthulu K, Pandey PM, Venkata Reddy N (2004) Optimum part deposition orientation in fused deposition modeling. Int J Mach Tools Manuf 44(6):585–594. https://doi.org/10.1016/j.ijmachtools.2003.12.004
12. Weiyin M, Wing Chung B, Peiren H (2004) NURBS-based adaptive slicing for efficient rapid prototyping. Comput Aided Des 36(13):1309–1325. https://doi.org/10.1016/j.cad.2004.02.001
13. Chakraborty D, Aneesh Reddy B, Roy Choudhury A (2008) Extruder path generation for curved layer fused deposition modeling. Comput Aided Des 40(2):235–243. https://doi.org/10.1016/j.cad.2007.10.014
14. Jin Y, He Y, Xue Gh (2015) A parallel-based path generation method for fused deposition modelling. Int J Adv Manuf Technol 77:927–937. https://doi.org/10.1007/s00170-014-6530-z
15. Areir M () Development of 3D printed flexible supercapacitors: design, manufacturing, and testing. Doctor of Philosophy thesis submitted to Brunel University London. http://bura.brunel.ac.uk/handle/2438/16659
16. Eric F, Martin K, Unny M (1996) Mechanical and dimensional characteristics of fused deposition modeling build styles. Int Solid Freeform Fab Symp 419–442
17. Kulkarni P, Dutta D (1999) Deposition strategies and resulting part stiffnesses in fused deposition modeling. ASME J Manuf Sci Eng 121(1):93–103. https://doi.org/10.1115/1.2830582
18. Es-Said OS, Foyos J, Noorani R, Mendelson M, Marloth R, Pregger BA (2000) Effect of layer orientation on mechanical properties of rapid prototyped samples. Mater Manuf Process 15(1):107–122. https://doi.org/10.1080/10426910008912976
19. Anitha R, Arunachalam S, Radhakrishnan P (2001) Critical parameters influencing the quality of prototypes in fused deposition modelling. J Mater Process Technol 118(1–3):385–388. https://doi.org/10.1016/S0924-0136(01)00980-3
20. Ahn S, Montero M, Odell D, Roundy S, Wright PK (2002) Anisotropic material properties of fused deposition modeling ABS. Rapid Prototyping J 8(4):248–257. https://doi.org/10.1108/13552540210441166
21. Lee BH, Abdullah J, Khan ZA (2005) Optimization of rapid prototyping parameters for production of flexible ABS object. J Mater Process Technol 169(1):54–61. https://doi.org/10.1016/j.jmatprotec.2005.02.259
22. Sun Q, Rizvi GM, Bellehumeur CT, Gu P (2008) Effect of processing conditions on the bonding quality of FDM polymer filaments. Rapid Prototyping J 14(2):72–80. https://doi.org/10.1108/13552540810862028
23. Zhang Y, Chou K (2008) A parametric study of part distortions in fused deposition modelling using three-dimensional finite element analysis. J Eng Manuf 222(8):959–68. https://doi.org/10.1243/09544054JEM990
24. Sood AK, Ohdar RK, Mahapatra SS (2009) Improving dimensional accuracy of fused deposition modelling processed part using grey Taguchi method. Mater Des 30(10):4243–4252. https://doi.org/10.1016/j.matdes.2009.04.030
25. Sood AK, Ohdar RK, Mahapatra SS (2010) Parametric appraisal of mechanical property of fused deposition modelling processed parts. Mater Des 31(1):287–295. https://doi.org/10.1016/j.matdes.2009.06.016
26. Fatimatuzahraa AW, Farahaina B, Yusoff W (2011) The effect of employing different raster orientations on the mechanical properties and microstructure of fused deposition modeling parts. In: IEEE symposium on business, engineering and industrial applications (ISBEIA), pp 22–27. https://doi.org/10.1109/ISBEIA.2011.6088811
27. Bagsik A, Schöppner V (2011) Mechanical properties of fused deposition modeling parts manufactured with ULTEM 9085. In: Proceedings of 69th annual technical conference of the society of plastics engineers 1–5
28. Sood AK, Ohdar RK, Mahapatra SS (2012) Experimental investigation and empirical modelling of FDM process for compressive strength improvement. J Adv Res 3(1):81–90. https://doi.org/10.1016/j.jare.2011.05.001
29. Croccolo D, Agostinis M, Olmi G (2013) Experimental characterization and analytical modelling of the mechanical behaviour of fused deposition processed parts made of ABS-M30. Comput Mater Sci 506–518. https://doi.org/10.1016/J.COMMATSCI.2013.06.041

30. Jafari MA, Han W, Mohammadi F, Safari A, Danforth SC, Langrana N (2000) A novel system for fused deposition of advanced multiple ceramics. Rapid Prototyping J 6(3):161–175. https://doi.org/10.1108/13552540010337047
31. Khalil S, Nam J, Sun W (2005) Multi-nozzle deposition for construction of 3D biopolymer tissue scaffolds. Rapid Prototyping J 11(1):9–17. https://doi.org/10.1108/13552540510573347
32. Liu L, Xiong Z, Yan Y, Zhang R, Wang X, Jin L (2009) Multinozzle low-temperature deposition system for construction of gradient tissue engineering scaffolds. J Biomed Mater Res Part B Appl Biomater 88(1):254–263. https://doi.org/10.1002/jbm.b.31176
33. Espalin D, Alberto Ramirez J, Medina F, Wicker R (2014) Multi-material, multi-technology FDM: exploring build process variations. Rapid Prototyping J 20(3):236–244. https://doi.org/10.1108/RPJ-12-2012-0112
34. Liu W, Shrike Zhang Y, Heinrich MA, De Ferrari F, Lin Jang H, Mahwish Bakht S, Moisés Alvarez M, Yang J, Chen Li Y, Trujillo-de Santiago G, Miri AK, Zhu K, Khoshakhlagh P, Prakash G, Cheng H, Guan X, Zhong Z, Ju J, Harry Zhu G, Jin X, Ryon Shin S, Remzi Dokmeci M, Khademhosseini A (2016) Rapid continuous multimaterial extrusion bioprinting. Adv Mater 29(3):1–8. https://doi.org/10.1002/adma.201604630
35. Frutuoso N (2017) Tool-path generation for a multiple independent print head system for fused deposition modelling. Master thesis submitted to University of Lisbon
36. Skylar-Scott MA, Mueller J, Visser CW (2019) Voxelated soft matter via multimaterial multinozzle 3D printing. Nature 575:330–335. https://doi.org/10.1038/s41586-019-1736-8
37. Cai Y, Choi SH (2019) Deposition group-based toolpath planning for additive manufacturing with multiple robotic actuators. Proc Manuf 34:584–593. https://doi.org/10.1016/j.promfg.2019.06.223
38. Zhang GQ, Mondesir W, Martinez C, Li X, Fuhlbrigge TA, Bheda H (2015) Robotic additive manufacturing along curved surface—a step towards free-form fabrication. In: IEEE international conference on robotics and biomimetics (ROBIO), pp 721–726. https://doi.org/10.1109/ROBIO.2015.7418854
39. Swanson M, Spurgeon W, Vass T, Danielewicz M (2016) 3D printing robotic arm, project report: submitted to the faculty of Worcester Polytechnic Institute. Massachusetts. http://eps.novia.fi/assets/Sidor/2/1545/Final-Report-3D-Printing-with-Robotic-Arm.pdf
40. Alsharhan AT, Centea T, Gupta SK (2017) Enhancing mechanical properties of thin-walled structures using non-planar extrusion based additive manufacturing. In: Proceedings of the ASME 2017 12th international manufacturing science and engineering conference. https://doi.org/10.1115/MSEC2017-2978
41. Brooks BJ, Arif KM, Dirven S (2017) Robot-assisted 3D printing of biopolymer thin shells. Int J Adv Manuf Technol 89:957–968. https://doi.org/10.1007/s00170-016-9134-y
42. Volker H, Willmann J, Thoma A, Piskorec L, Hack N, Gramazio F, Kohler M (2015) Iridescence print: robotically printed lightweight mesh structures. 3D Printing and Addit Manuf 3(2) 117–122. https://doi.org/10.1089/3dp.2015.0018
43. Barnett E, Gosselin C (2015) Large-scale 3D printing with a cable-suspended robot. Addit Manuf 7:27–44. https://doi.org/10.1016/j.addma.2015.05.001
44. Felsch T, Klaeger U, Steuer J, Schmidt L, Schilling M (2017) Robotic system for additive manufacturing of large and complex parts. In: 22nd IEEE international conference on emerging technologies and factory automation (ETFA) 1–4. https://doi.org/10.1109/ETFA.2017.8247739
45. Dai C, Wang CL, Wu C, Lefebvre S, Fang G, Liu YJ (2018) Support-free volume printing by multi-axis motion. ACM Trans Graphs 37(4). https://doi.org/10.1145/3197517.3201342
46. Oxman N, Laucks J, Kayser M, Tsai E, Firstenberg E (2013) Freeform 3D printing: towards a sustainable approach to additive manufacturing. Mater Manuf Processes 479–483. https://doi.org/10.1201/B15002-93
47. Ishak I, Fisher J, Larochelle P (2016) Robot arm platform for additive manufacturing: multi-plane printing. In: Proceedings of the 29th Florida conference on recent advances in robotics. https://doi.org/10.1115/DETC2016-59438

48. Ishak I, Moffett MB, Larochelle P (2018) An algorithm for generating 3D lattice structures suitable for printing on a multi-plane FDM printing platform. In: Proceedings of the ASME 2018 international design engineering technical conferences and computers and information in engineering conference. https://doi.org/10.1115/DETC2018-85459
49. Onstein IF (2018) An additive manufacturing path generation method based on CAD models for robot manipulators. Mater thesis submitted to Norwegian University of Science and Technology. http://hdl.handle.net/11250/2561062
50. Castelli K, Giberti H (2019) A preliminary 6 dofs robot based setup for fused deposition modelling. Adv Ital Mech Sci Mech Mach Sci 68. https://doi.org/10.1007/978-3-030-03320-0_27
51. Zhao G, Ma G, Feng J (2018) Nonplanar slicing and path generation methods for robotic additive manufacturing. Int J Adv Manuf Technol 96:3149–3159. https://doi.org/10.1007/s00170-018-1772-9
52. Shembekar AV, Yoon YJ, Kanyuck A, Gupta SK (2018) Trajectory planning for conformal 3D printing using non-planar layers. In: Proceedings of the ASME 2018 international design engineering technical conferences and computers and information in engineering conference, vol 1A. https://doi.org/10.1115/DETC2018-85975
53. Bhatt PM, Kabir AM, Malhan RK, Shah B, Shembekar AV, Yoon YJ, Gupta SK (2019) A robotic cell for multi-resolution additive manufacturing. In: Proceedings of the international conference on robotics and automation (ICRA). https://doi.org/10.1109/ICRA.2019.8793730
54. Shembekar AV, Yoon YJ, Kanyuck A, Gupta SK (2019) Generating robot trajectories for conformal three-dimensional printing using nonplanar layers. ASME J Comput Inform Sci Eng 19(3). https://doi.org/10.1115/1.4043013
55. Xie F, Chen L, Li Z, Tang K (2020) Path smoothing and feed rate planning for robotic curved layer additive manufacturing. Robot Comput Integr Manuf 65. https://doi.org/10.1016/j.rcim.2020.101967
56. Sri Harsha A and Vikram kumar CR (2020) Investigation on dual nozzle fused deposition modelling using industrial robot. Adv Mater Process Technol. Publication date 20 Dec 5
57. Sri Harsha A, Vikram Kumar CR (2021) Fused deposition modeling of an aircraft wing using industrial robot with non-linear tool path generation. Int J Eng 34(1):272–82
58. Bhatt PM, Malhan RK, Rajendran P, Gupta SK (2020) Building free-form thin shell parts using supportless extrusion-based additive manufacturing. Addit Manuf 32. https://doi.org/10.1016/j.addma.2019.101003

Dynamic Vibration Analysis of Shape Memory Alloy Wire-Coupled Wind Turbine Blades

Yuvaraja Mani and Jagadeesh Veeraragu

Abstract Monitoring vibrational characteristics of the domestic wind turbine blade system under dynamic conditions are a tedious problem. Mitigation of vibration in domestic wind turbine blades is a meticulous process due to the variation of wind speed. The research on the effect of an active control system has depicted a significant vibration reduction. Experimentation on the effect of SMA at dynamic conditions is yet to be elaborately studied. The effect of Shape Memory Alloy (SMA) on vibration control at dynamic conditions is not effectively studied. The influence of super-elastic and shape memory characteristics of SMA at dynamic conditions is evaluated through operational modal analysis. The dynamic model of the tip displacement of the blade, coupled with nacelle movement represented the need for vibration control in the blade system. The experimental dynamic analysis performed on the wind turbine blade system depicted the effect of centrifugal stiffening and the induced frequency shift in the wind turbine blade system. The flapwise mode of the blade was shifted 15% in the operating speed of the turbine. The SMA embedment showed a significant reduction in the vibrational amplitude of the blade. The coupling of SMA wires on the wind turbine blades under passive and active conditions depicted a 57 and 64% reduction in the amplitude of flapwise vibration of the system.

Keywords Wind turbine blade · Shape memory alloy · Dynamic analysis · Composites · Vibration control · Super-elastic · Shape memory effect

1 Introduction

Flapwise vibration of wind turbine blades is considered to be important to wind turbines, which when transferred to the turbine system can cause swaying of the system or failure of the system. These vibrations of the blade system are to be controlled as the system's integrity reduces as the blades rotate under specific wind speeds. To maintain the integrity of the system, the blades are not allowed to rotate at

Y. Mani · J. Veeraragu (✉)
Department of Mechanical Engineering, PSG College of Technology, Coimbatore, India
e-mail: 1607RM08@psgtech.ac.in

U. S. Dixit et al. (eds.), *Advances in Forming, Machining and Automation*, Lecture Notes in Mechanical Engineering, https://doi.org/10.1007/978-981-19-3866-5_44

these specific speeds, hence reducing the operating range of the wind turbine system. Various types of dampers have been applied to the blade to reduce the impact of the flapwise vibrations, but the super-elastic and shape memory effect of SMA on the wind turbine blades is not studied.

Flapwise vibration causes a significant drop in performance and the range of operation of the turbine system. Many kinds of research including [1, 2, 3] have studied the influence of flapwise vibrations of blades on the wind turbine system's performance through input shaping, mathematical model formulation and study on coupling effects. The experimental analysis of the effects of flapwise vibration is not yet studied in detail. The article [4] emphasises the effect of rotating speed on the natural frequencies of the wind turbine blades. The authors performed the control of such vibrations using Shape Memory Alloy actuation [5]. Though the article portrays the influence of SMA wires in vibration control, dynamic analysis on such a system is not briefed. The experimentation on the composite domestic wind turbine blade system at dynamic conditions is performed by coupling the coupled blade system to a variable speed DC motor. The analysis commences with the identification of the system's independent natural frequencies and studying them under dynamic conditions.

2 Dynamic Model of Blade System with Nacelle

The vibration in the domestic wind turbine blade of 500 W rated power is considered in this research. The vibrational characteristics of the wind turbine blade are evaluated under operating conditions. The maximum operating speed of the turbine is 300 rpm. The analytical model of the blade was developed based on the Euler–Bernoulli beam theory with rigid support. The model depicts the natural frequency of the blade and the deflection caused by the wind load. The wind flow over the blade induces lift and drag forces on the blade and the lift force induces torque on the rotor. The torque produced in the rotor is utilised to generate power. The domestic blades are subjected to torsional stiffening due to the wind flow-induced torque.

The tensile force acting on the blade due to the centrifugal motion of the blade is added to the model to depict the effect of rotational velocity on the natural frequency of the blade. The blade is evaluated for the flapwise direction as the flapwise mode is considered critical for the operation of the turbine. The deflection of the blade from the equilibrium position (x_b) is given by,

$$\rho_b A_g \frac{\partial^2 x_b}{\partial t^2} + \frac{\partial^2}{\partial x^2}\left(E_b I_g \frac{\partial^2 x_b}{\partial x^2}\right) - \frac{\partial}{\partial x}\left(T \frac{\partial x_b}{\partial t}\right) = f_w(x, t) \tag{1}$$

The cross-sectional area (A_g) and sectional modulus (Ig) of the blade at the centre of gravity are considered for evaluation. The centrifugal tensile force (T) acting on the blade depends on the rotational velocity of the blade (Ω), length of the blade (L), the density of the blade (ρ_b) and hub radius. The tensile force is integrated along the length of the blade as,

$$T(x) = \int_{x}^{l} \rho_b A_g \Omega^2 (r + x) \mathrm{d}x \tag{2}$$

Considering the blade to the rigidly mounted as a cantilever condition, the following boundary conditions applied on the blade.

$$\text{at } x = 0, \quad y(x) = 0, \quad \frac{dy(x)}{dx} = 0 \tag{3}$$

$$\text{at } x = L, \quad \frac{\mathrm{d}^2 y(x)}{\mathrm{d}x^2} = 0, \quad \frac{\mathrm{d}^3 y(x)}{\mathrm{d}x^3} = 0 \tag{4}$$

The application of geometric properties of the selected blade and Eqs. (3) and (4) in (1) provide 32.5, 57.8 and 112 Hz, like the first, second and third flapwise mode of vibration of the blade. The mode shape extracted from the model is shown in Fig. 1.

The vibrational characteristics of the blade under centrifugal stiffening are evaluated for the operating speeds of the turbine. The natural frequency of the blade increases with the rotational speed of the turbine due to the increase in the stiffness of the blade. The increase in the normalised natural frequency, dividing the obtained natural frequency by respective fundamental frequency, is shown in Fig. 2.

The centrifugal tension induced on the blade under rotation increases the flapwise natural frequency of the blade, as the stiffness of the blade increases. The blade's natural frequency initially at 32.5 Hz increased to 36 Hz at 100 rpm, 39 Hz at 200 rpm and 40 Hz at 300 rpm, respectively. The vibration of the blade is added with the nacelle movement due to the wind load. The vibration of the nacelle is evaluated using a Lagrangian equation, considering the nacelle and turbine blades. The total load (P) acting on the blade is evaluated by evaluating the kinetic (T_n) and potential (V_n) energy associated with the turbine system.

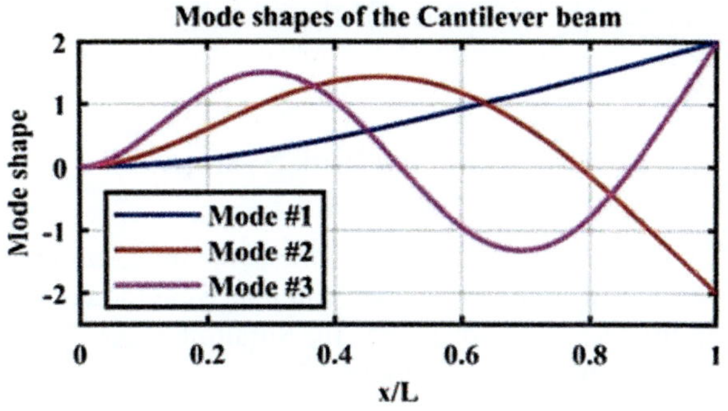

Fig. 1 Flapwise modes of turbine blade

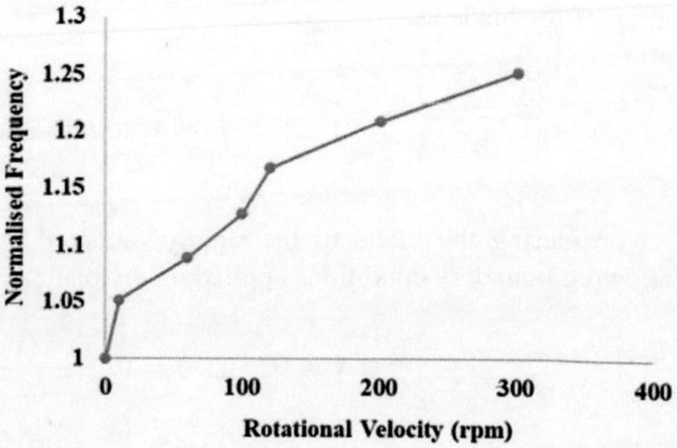

Fig. 2 Influence of rotational velocity on blade frequency

$$\frac{\mathrm{d}}{\mathrm{d}t}\frac{\partial T_n}{\partial x_i} - \frac{\partial T_n}{\partial x_i} + \frac{\partial V_n}{\partial x_i} = P_i \tag{5}$$

In Eq. (5), the kinetic and potential energy associated with the turbine system is given by,

$$T = \frac{1}{2}\dot{m}\sum_{i=1}^{3}\int_{0}^{l} v_{\mathrm{b}}^2 \mathrm{d}x + \frac{1}{2}M_n x_n \tag{6}$$

$$V = \frac{1}{2}EI\sum_{i=1}^{3}\frac{\partial^2 x_{\mathrm{b}}}{\partial x^2} + \frac{1}{2}K_n x_n^2 \tag{7}$$

$$P_1 = \frac{v_n^2 A}{3} + \frac{v_{n+l}^2 A}{3} + \frac{v_{n+l}^2 A}{3} \tag{8}$$

The energy of the turbine is based on the mass of the blade, absolute velocity (v_{b}) of the blade, rigidity modulus of the blade (EI), stiffness (K_n) of the nacelle and the displacement (x_n) of the nacelle.

3 Modelling the Turbine System

The wind load acting on the blade at the maximum wind speed is considered as the load acting on the system and laminar wind blowing across the turbine system. The turbine system is modelled in the Simulink environment of MATLAB. The modelled turbine system is shown in Fig. 3.

Under the wind load, the nacelle tends to vibrate and the displacement of the nacelle of the turbine under 100 rpm is shown in Fig. 4. The tip displacement of the blade coupled with the nacelle displacement gradually reduces and reaches near

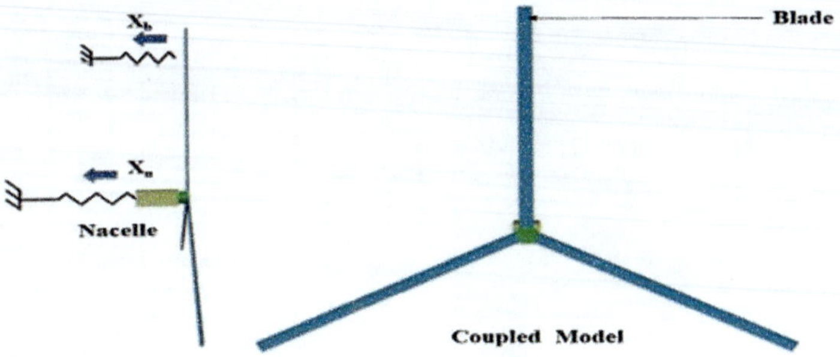

Fig. 3 Simulink dynamic model of the nacelle-blade system

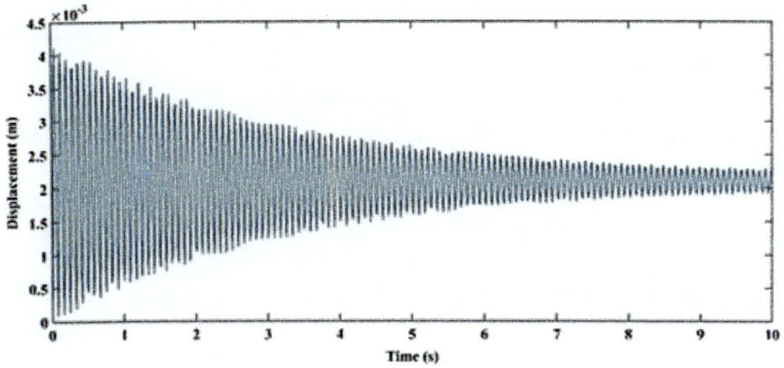

Fig. 4 Displacement of conventional blade system with nacelle movement

equilibrium by 8 s. This prolonged time to achieve equilibrium results in flutter vibration of the blade and leads to blade tower interaction. The experimental setup has to be developed to simulate the actual response of the blade under dynamic conditions.

The vibration characteristics of the blade have to be improved to reduce the flutter vibration and assist the blade to achieve equilibrium quicker. The SMA embedment on the blade has proven to improve the vibrational character of the structure. The improvement achieved by SMA embedment, under martensite and austenite characteristics of SMA, has to be evaluated under various operating conditions of the blade system.

4 Experimental Dynamic Modal Analysis on the Turbine System

The identification of the vibration characteristics of the blade system is initiated by isolating the blade system and the experimental setup. The coupled blade system

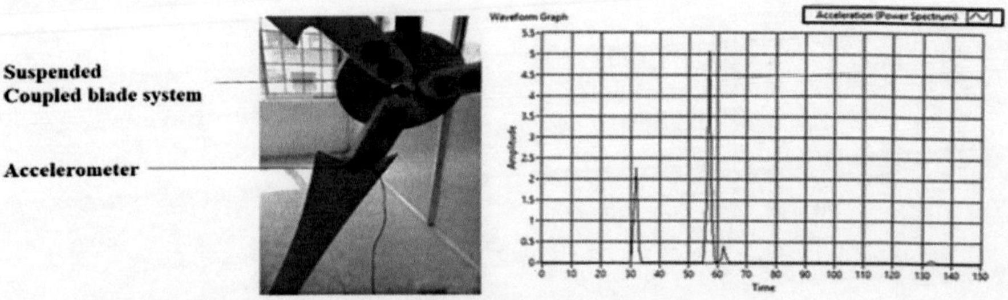

Fig. 5 Suspended blade system and modal response

is suspended and their natural frequencies are derived through modal analysis. A unidirectional accelerometer is used to obtain the out of rotational plane natural frequencies of the blade system.

The suspended system and frequency response function (FRF) of the system are depicted in Fig. 5. The FRF depicts the first and second flapwise bending frequencies of the blade system as 32 Hz and 58 Hz, respectively.

4.1 *Influence of Rotational Speed on Blade System*

The blade system is coupled to the motor and the vibrational characteristics are evaluated perpendicular to the rotational plane. Figure 6 shows the frequency response of the experimental system, where the frequencies 1 and 2 denote the blade system, similarly A and B denote the frequencies of the motor system. The contact stiffness achieved on coupling the blade system to the motor increases the natural frequency of the blade system.

The frequency of the blade and the motor system shifted due to the coupling effect portrayed by [3]. Table 1 denotes the shift in the frequency and the shift that happened due to the coupling effect. The first flapwise mode of the blade is critical for the behaviour of the turbine, and the targeted natural frequency for vibration control is 43 Hz.

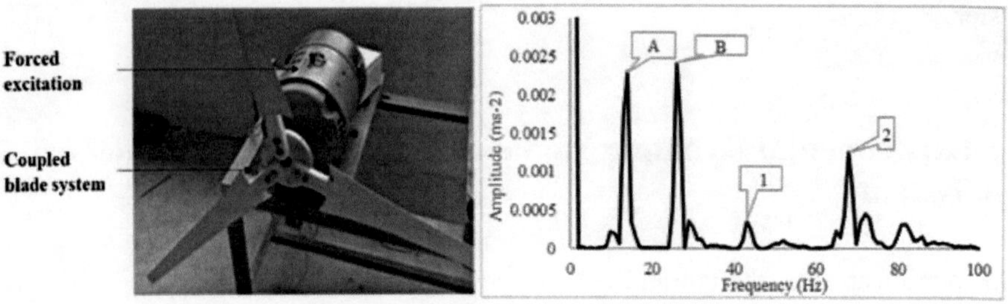

Fig. 6 Coupled experimental system and its frequency response

Table 1 Frequency shift achieved due to blade-nacelle coupling

Part	Targeted frequency (Hz)	Shifted frequency (Hz)
Motor	10, 25	14, 26
Blade	32, 58	43, 68

4.2 Dynamic Modal Analysis

The dynamic analysis is conducted to observe the vibration of the blade system under the different rotational speeds of the motor. The blade system is studied under the rotational speed of 100, 200 and 300 rpm. The experimental setup is shown in Fig. 7. Using the controller, the DC motor is provided necessary actuation, and the rotational speed is maintained.

The natural frequency of the system and its amplitude ratio are obtained from the frequency response performed on the blade system under variable rotational speed. The amplitude ratio thus obtained for different speeds is plotted and shown in Fig. 8.

From Fig. 8, the peak represents the influence of the blade and the motor under different rotor speeds. A and *B* regions represent the motor frequencies, similarly, 1 and 2 represent the blade frequencies. The targeted frequencies are shifted in the dynamic condition due to the centrifugal forces induced by the motor. In the idle

Fig. 7 Experimental setup for dynamic analysis

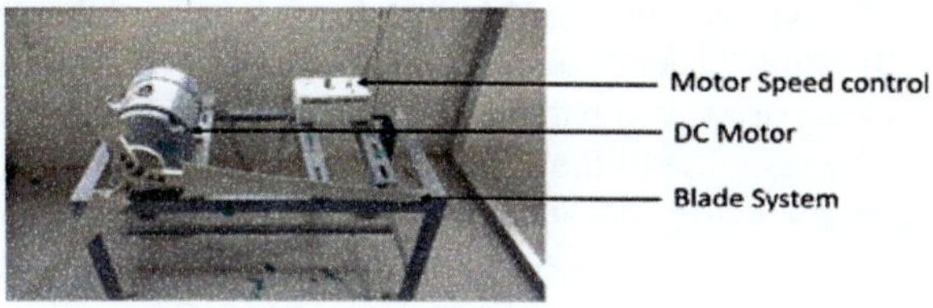

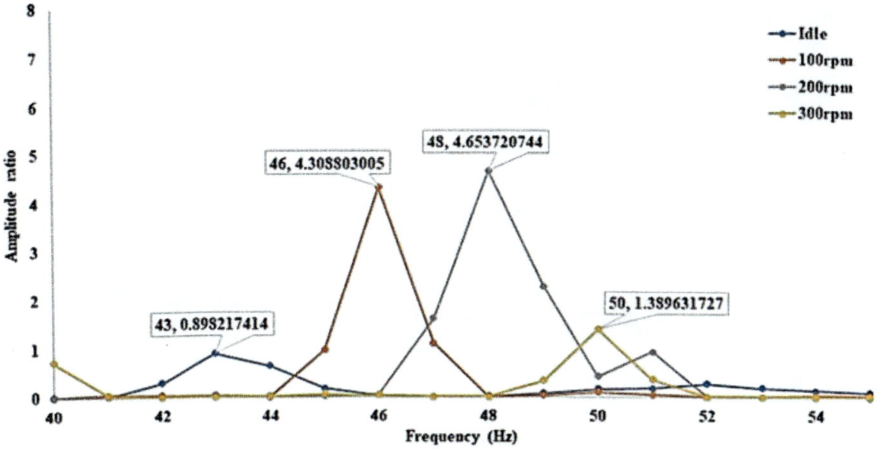

Fig. 8 Influence of rotational velocity on first flapwise frequency of the blade

condition, the motor had two frequencies and the blade had two frequencies. When the motor rotated at different speeds due to the centrifugal force, the amplitude and frequency increased. The rotational velocity is 200 rpm being critical with a high amplitude ratio. The vibration control method should be applied for the improvement in vibrational characteristics of the blade.

5 Shape Memory Alloy (SMA) in Vibration Control

The role of SMA in vibration control and adaptive stiffness behaviour is described by [6–8]. The incorporation of SMA in structures shows appreciable vibration control and increases the structural integrity of the system. The super-elastic behaviour of SMA contributes to the improved damping parameter of the structure, as portrayed by [5] and [9]. In this research, 0.4% volume fraction of SMA is embedded with each blade through high-temperature resistant masking tape as shown in Fig. 9. The selection of low volume fraction of SMA embedment relates to the high actuation force to weight ratio provided by SMA [10].

The Simulink model with the SMA embedment is evaluated with the 50% improvement in the damping ratio of the structure as depicted by a myriad of authors, namely [5, 10] and [11]. The response of the dynamic model under the influence of the martensite state of SMA. The displacement of the SMA embedded turbine system with coupled nacelle displacement is simulated for the same loading conditions, and the response is compared to the conventional blade system in Fig. 10.

The response of the system depicts the improvement in the response of the blade under the martensite state of SMA. The SMA embedded blade reaches a near equilibrium state within 4–5 s of the application of load. The reduction in the vibration is due to the super-elastic nature of the SMA. The advantages of shape memory effect and super-elasticity of SMA in austenite state can be utilised by providing actuation current to the SMA.

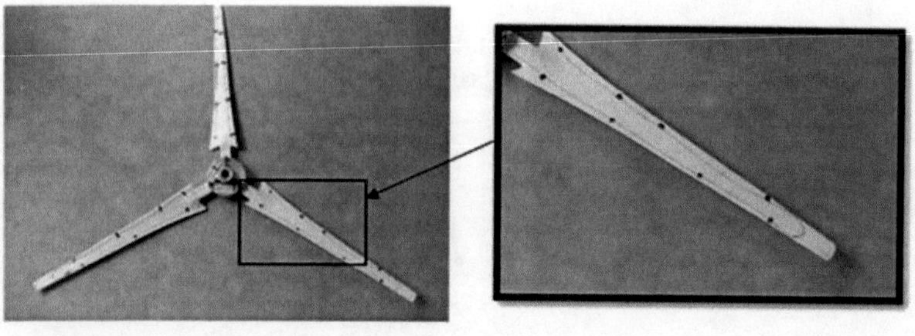

Fig. 9 Blade system coupled with 0.4% of SMA

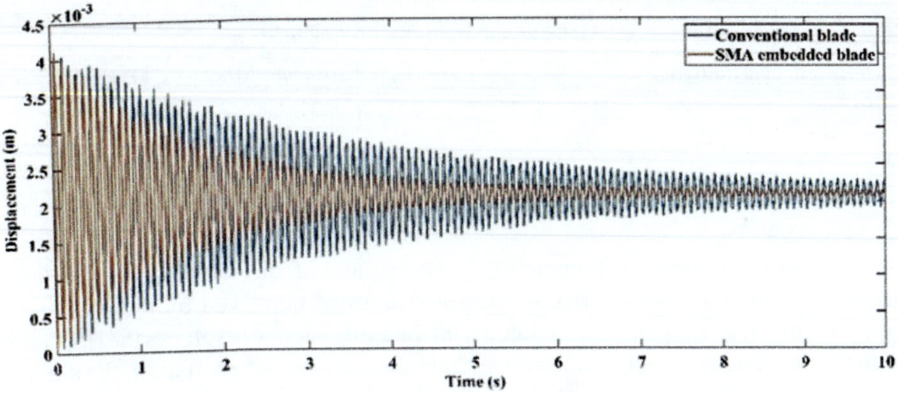

Fig. 10 Comparison of the response of SMA blade embedded blade and conventional system

5.1 Passive Vibration Control at Dynamic Conditions

The influence of SMA, in passive conditions, on the vibration characteristics of the blade are studied by performing the experimental analysis on the same rotational speeds. The vibrational characteristics of the blade system such as frequency and amplitude ratio of vibration are obtained for 100, 200 and 300 rpm. The experimental setup of the system under passive conditions.

Similarly, the SMA-coupled blade system is excited at different rotational speed, and their vibration characteristics are noted. The frequency response of the system under 100 rpm rotational speed is shown in Fig. 11. The frequency and amplitude ratio of the SMA-coupled system under different rotational speeds is tabulated in Table 2.

The experimental modal analysis of the blade system indicates the effect of centrifugal stiffening on the natural frequency of the blade. The flapwise natural frequency of the blade from 43 Hz increases to 46 Hz at 100 rpm, 48 Hz at 200 rpm and 50 Hz at 300 rpm, respectively. The experimental result is plotted against the analytical prediction in Fig. 11.

Fig. 11 Comparison of analytical prediction and experimental

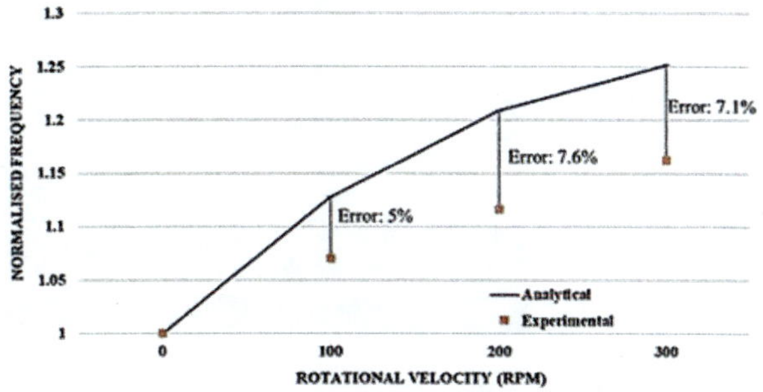

Table 2 Comparison of vibrational characteristics of blade system under different spccd

Motor speed (rpm)	Amplitude ratio	
	Without SMA	Passive
100	4.31	1.85
200	4.65	3.79
300	13.17	10.39

The error in the analytical prediction is a maximum of 7%, which could be eliminated with the usage of a correction factor. The trend portrayed by the analytical model for the frequency shift is validated with the experimental result. As the response of the blade system is being evaluated without the actual wind excitation, the amplitude ratio of the blade is evaluated for studying the severity of speed on the blade system. The amplitude ratio compares the amplitude of vibration of a dynamic system to the idle system's response.

5.2 *Active Vibration Control at Dynamic Conditions*

The SMA wires coupled to the wind turbine blade systemare actuated with an external power supply kit, which induces temperature difference on the SMA. The temperature increase is due to the high resistance offered by the SMA to the current. The induced temperature rise causes the SMA to convert from a martensitic to an austenitic state. The phase change induced causes the material to induce recovery stress onto the coupled structure. Thus, altering the stiffness of the structure. The influence of SMA under active conditions is studied with a constant current of 1 A, which results in the temperature of SMA reaching 60 °C. The experimental setup consisting of a variable power supply kit is shown in Fig. 12.

Similarly, the effect of active vibration control is studied under different rotational speeds such as 100, 200 and 300 rpm. The frequency response of the active control system is compared with the independent, without SMA and passive conditions in

Fig. 12 Experimental setup for active vibration control of the blade

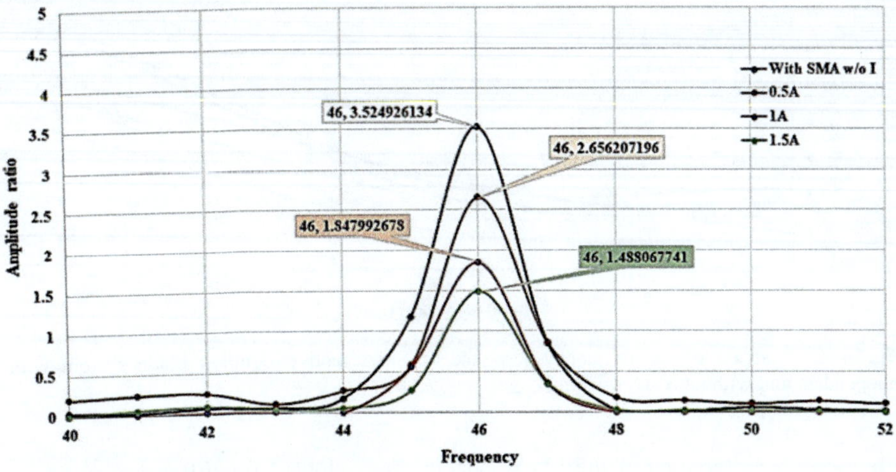

Fig. 13 Comparison of the frequency response of SMA (passive)-coupled blade system with independent and without SMA system

Table 3 Effectivity of vibration control methods on blade system

Motor speed (rpm)	Amplitude ratio			Percentage reduction	
	Without SMA	Passive	Active	Passive	Active
100	4.31	1.85	1.49	57.11	65.46
200	4.65	3.79	2.05	18.52	55.89
300	13.17	10.39	7.49	21.12	43.09

Fig. 13. The vibrational characteristics of the blade system under active control are compared with other conditions under different rotational speeds in Table 3.

The active vibration control of the blade system compared to passive control, in Figure 14, shows improvement in the dynamic response of the blade system. The shape memory effect and super-elasticity of SMA in austenite state have a drastic effect on vibration control of the blade system. The amplitude ratio obtained across various operating conditions of the blade is given in Table 3. The active vibration control on the turbine system results in a 65% reduction in amplitude ratio. A similar trend can be expected in the turbine's response to the wind excitation.

6 Conclusion

The analytical model of the blade system coupled with the nacelle movement depicted the requirement of a vibration control system for the blades. The model depicted improvement in the displacement response of the blade with SMA embedment. The

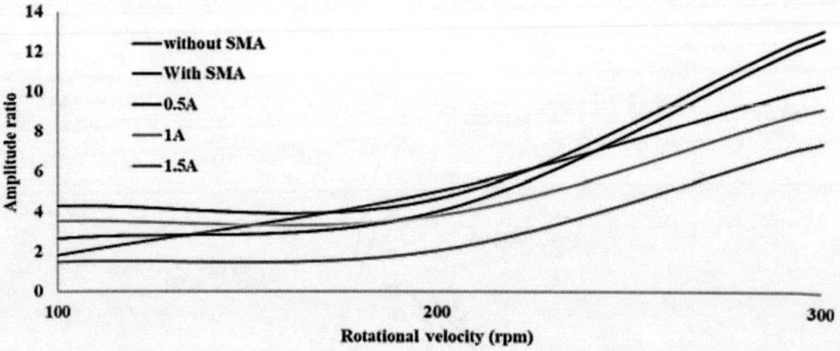

Fig. 14 Comparison of the frequency response of SMA (active)-coupled blade system with independent and without SMA system

vibration characteristics of the blade system are evaluated for different rotational speeds and the influence of centrifugal stiffening is depicted in the frequency shift. The super-elastic and shape memory effect characteristics of SMA are evaluated through the coupling of 0.4% volume fraction. The reduction in the amplitude ratio of vibration under passive conditions is induced by the increase in damping due to the super-elastic nature of SMA. The reduction can be visualised to a maximum of 57%, and the damping effect loses its potential with an increase in the centrifugal stiffening of the structure. The amplitude ratio reduction caused due to actuation of SMA under active conditions is due to the induction of recovery stress in the structure. The SMA wire has increased the stiffness of the structure thus allowing the structure to operate under a wider range of operations. The effect of SMA on embedment inside the wind turbine blade could be more effective.

References

1. Ju D, Sun Q (2014) Wind turbine blade flapwise vibration control through input shaping, IFAC Proc 19(3):5617–5622
2. Arrigan J, Pakrashi V, Basu B, Nagarajaiah S (2011) Control of flapwise vibrations in wind turbine blades using semi-active tuned mass dampers. Struct Control Heal Monit 18(8):840–851
3. Tang AY, Li XF, Wu JX, Lee KY (2015) Flapwise bending vibration of rotating tapered rayleigh cantilever beams. J Constr Steel Res 112:1–9
4. Özdemir Ö, Kaya MO (2006) Flapwise bending vibration analysis of a rotating tapered cantilever Bernoulli-Euler beam by differential transform method. J Sound Vib 289(2):413–420
5. Mouleeswaran SK, Mani Y, Keerthivasan P, Veeraragu J (2018) Vibration control of small horizontal axis wind turbine blade with shape memory alloy. Smart Struct Syst 21(3):257–262
6. Staino A, Basu B (2013) Dynamics and control of vibrations in wind turbines with variable rotor speed. Eng Struct 56(3):58–67
7. Gupta K, Sawhney S, Jain SK, Darpe AK (2003) Stiffness characteristics of fibre-reinforced composite shaft embedded with shape memory alloy wires. Def Sci J 53(2):167–173
8. Ni QQ, Xin Zhang R, Natsuki T, Iwamoto M (2007) Stiffness and vibration characteristics of SMA/ER3 composites with shape memory alloy short fibers. Compos Struct 79(4):501–507

9. Yuvaraja M, Senthil Kumar M (2012) Experimental studies on SMA spring-based dynamic vibration absorber for active vibration control. Eur J Sci Res 77(2):240–251
10. Mani Y, Veeraragu J, Sangameshwar S, Rangaswamy R (2020) Dynamic behavior of smart material embedded wind turbine blade under actuated condition. Wind Struct An Int J 30(2):211–217
11. Tak Lau K (2002) Vibration characteristics of SMA composite beams with different boundary conditions. Mater Des 23(8):741–749

Investigation on Actuation Performance of Shape Memory Alloy Actuator for Robotics Application

S. Jayachandran, Neelam Yashwanth, Hampa Faazil Hussain, and I. A. Palani

Abstract The shape memory alloy (SMA) actuator comprising SMA wire and kapton polyimide was fabricated. The actuation characteristics were analyzed using the open-loop programmable power supply. The maximum displacement at different parameters was analyzed. The actuator showed a two-way shape memory effect during analysis without the tedious process of training. Two different diameters of wire were used during the actuator fabrication. The composition analysis of two different diameter wires revealed that there is a slight variation in composition percentage. The differential scanning calorimetry (DSC) analysis showed the austenite transformation temperature for both the diameter wire (0.5 mm and 300 μm) lies in between 70–80 °C temperature range. The actuator of dimension 7×2 cm^2 showed a maximum displacement of 3 cm and 5×2 cm^2 reached a maximum displacement of 1.2 cm for 300 μm SMA wire.

Keywords NiTi · Shape memory alloy (SMA) · Differential scanning calorimetry (DSC)

1 Introduction

Shape memory materials possess a memory effect by which it transforms from martensite to austenite upon temperature stimuli [1–3]. Varying shapes by inducing temperature help in achieving high efficiency in MEMS applications [4]. Owing to its better mechanical and shape memory properties, NiTi is the most widely used and commercially available SMA [4]. The other reason for using NiTi is due to higher work output per unit volume of NiTi than other microactuation mechanisms. NiTi

S. Jayachandran · I. A. Palani (✉)
Mechatronics and Instrumentation Lab, Department of Mechanical Engineering, Indian Institute of Technology Indore, Madhya Pradesh, India
e-mail: palaniia@iiti.ac.in

N. Yashwanth · H. F. Hussain
Metallurgical Engineering and Materials Science, Indian Institute of Technology Indore, Madhya Pradesh, India

U. S. Dixit et al. (eds.), *Advances in Forming, Machining and Automation*, Lecture Notes in Mechanical Engineering, https://doi.org/10.1007/978-981-19-3866-5_45

SMA also exhibits pseudoelastic behavior which makes it useful for various MEMS applications [5–7]. The SMA alloys show excellent thermomechanical stability.

These unique features of SMA are highly useful in the field of robotics where the actuators have to be precise and powerful and also have the capability to perform multifunctional activities. SMA wires provide large forces during actuation due to the inherent nature to recover large recovery strain. The SMA wires provide only one-way actuation, which needs to be coupled into another element to produce out-of-plane deformation. One way of converting the one-dimensional deformation of SMA wire into out-of-plane deformation is through connecting or coupling with a rotating element. The main disadvantage may be the fatigue of the rotating element will decrease the performance of the actuator, and also many researchers have worked by embedding the soft polymeric matrix [6–9] to produce large continuous deformation.

In this work, the SMA wire is embedded along with kapton polyimide to produce a two-way shape memory effect without using mechanical joints or any soft polymeric matrix along with the SMA material. The actuation characteristics of the actuator were analyzed with an open-loop programmable power supply. The effect of varying the process parameters was also analyzed and corroborated with phase transformation temperature.

2 Experimental Details

The actuator consists of two components. The SMA wire is embedded inside the kapton polyimide flexible structure as shown in Fig. 1. Two different diameters of SMA wire were used for the actuator and a kapton polyimide of thickness 50 μm was used. The phase transformation temperature of the SMA wire was analyzed using differential scanning calorimetry (DSC). SMA wire of 30 mg was used for the analysis. The DSC data were recorded at a heating rate of 10 °C. The actuator performance of various configurations was analyzed using a programmable power supply as shown in Fig. 1. The maximum displacement of the actuator was measured in this open-loop system for different configurations. The actuator was fabricated in different configurations by changing dimensions, different strain percentages, and different wire diameters. The straining unit of the SMA wire is shown in Fig. 1. The gauge length was kept at 20 cm and the strain was made up to 8% in the wire.

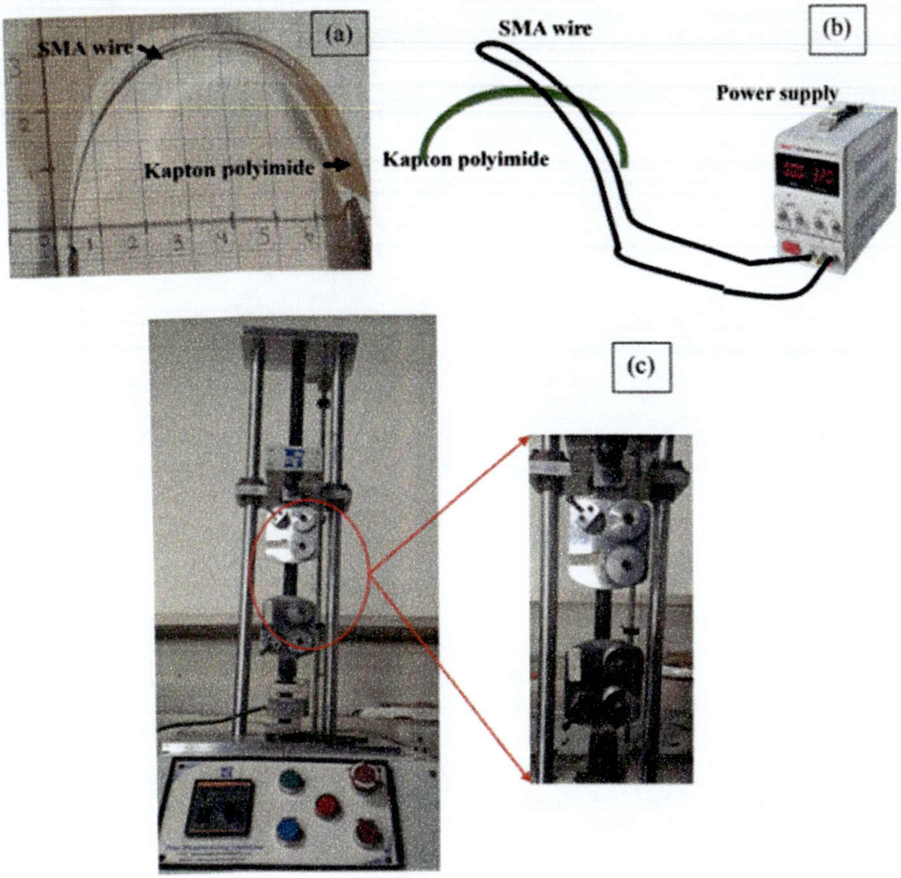

Fig. 1 Straining and fabrication of SMA composite structure

3 Results and Discussion

The actuator comprising SMA wire and kapton polyimide substrate will provide the two-way actuation without the tiresome process of training. During heating, the SMA wire tends to reach the austenite phase thus making the actuator reach straight shape as shown in Fig. 2. The change of shape during cooling where the actuator goes to a curved shape is due to the inherent elasticity of the kapton polyimide.

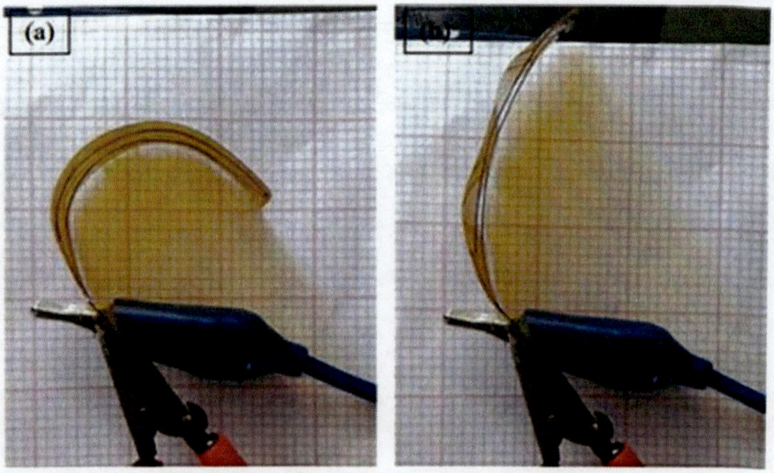

Fig. 2 Actuation **a** cooling and **b** heating

3.1 Compositional Analysis

The composition of the NiTi wire of both diameters was analyzed using energy dispersive spectroscopy. Small pieces of wire were mechanically cut for the characterization. There is a change or shift in the chemical composition. The SMA wire of 300 μm showed 51.6% of Ni and 48.4 wt% of Ti, whereas the wire of 0.5 mm diameter showed 51.2 wt% of Ni and 48.8 wt% of Ti. The composition analysis is shown in Fig. 3.

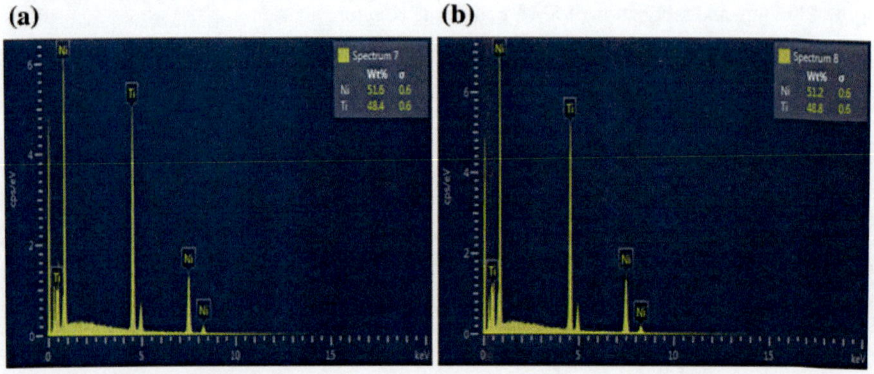

Fig. 3 Compositional analysis of **a** 300 um diameter wire and **b** 0.5 mm diameter wire

3.2 *Differential Scanning Calorimetry (DSC) Analysis*

Figure 4 shows the DSC curves for two different diameter SMA wire which is used in the actuator. The martensitic transformation is preceded by the *R*-phase in both the diameter of the wire (0.5 mm and 300 μm). There is a hysteresis of about 60–70 °C in both wire diameters. The austenite, martensite, and *R*-phase temperature are shown in Table 1. The transformation from martensite to *B*2 occurs almost in the same temperature range of 70–80 °C. The phase transformation temperature values are given in Table 1.

The broadening of peak during heating may be due to composition variation in which the Ni percentage is more than titanium [10]. The composition variation is evident from energy dispersive spectroscopy analysis.

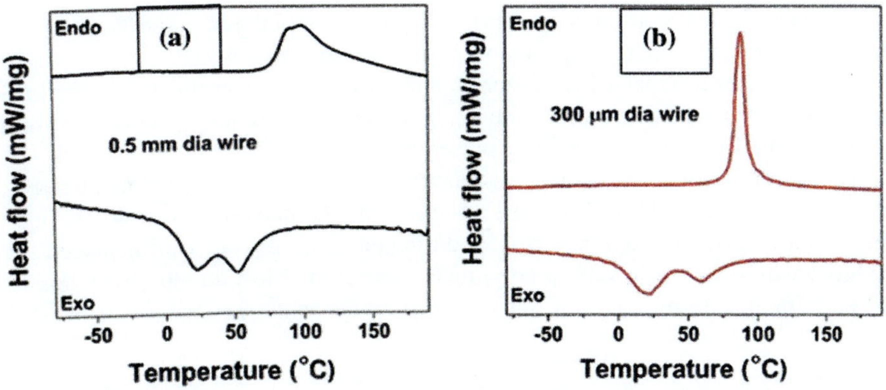

Fig. 4 DSC analysis: **a** 300 um diameter wire and **b** 0.5 mm diameter wire

Table 1 Transformation temperature

Transformation temperature	A_s	A_f	R_s	R_f	M_s	M_f
0.5 mm diameter wire	74	147	73	36	35	2
300 μm diameter wire	76	101	87	43	42	-3

3.3 Actuation Characteristics

3.3.1 Effect of actuator dimension

The actuation performance of the developed actuator was evaluated based on different SMA wire diameters. The SMA wire of two different diameters, namely 300 μm and 0.5 mm were considered. The wire diameter was tested by keeping other parameters constant. The current and voltage from the programmable power supply were 2 A and 4 V, respectively. The dimension of the actuator was taken as 5 × 2 and 7 × 2 cm^2 as shown in Fig. 5. Two different configurations were tested by placing the wire inside the kapton polyimide and the other by placing it outside the polyimide.

The maximum displacement of 3 cm was obtained with 300 μm wire diameter and 7 × 2 cm^2 actuator dimension. The same diameter wire with 5 × 2 cm^2 dimension deflected around 1.2 cm. The 0.5 mm diameter wire achieved a maximum displacement of 1 cm for 7 × 2 cm^2 actuator dimension and 0.50 cm for 5 × 2 cm^2 actuator dimension.

The increase in dimension increases the displacement of the actuator. The 300 μm diameter produces more actuation than 0.5 mm diameter wire as shown in Fig. 6. The smaller diameter dissipates heat faster than the higher diameter due to less surface area to volume of the SMA wire. The higher dimension actuator covers more area, which eventually results in higher displacement.

The placement of the SMA wire also plays a major part in the deflection/ displacement of the actuator. The SMA wire placed inside the kapton polyimide provides more force to move the kapton polyimide during the heating cycle. During placement of SMA wire outside the kapton polyimide, more heat is lost due to convection to the surroundings resulting in less displacement of the actuator.

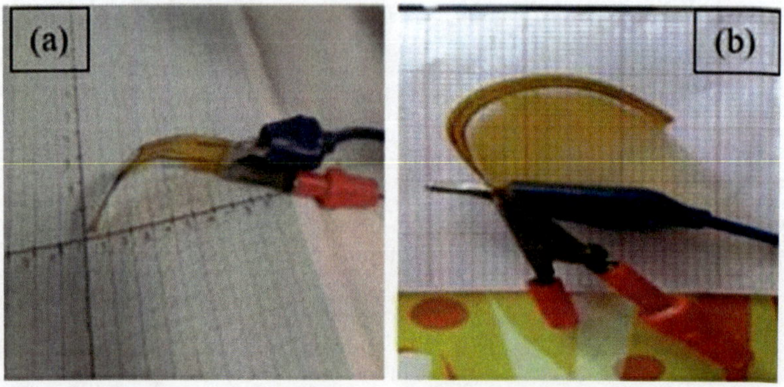

Fig. 5 Actuator of varying dimension: **a** 7 × 2 and **b** 5 × 2 cm^2

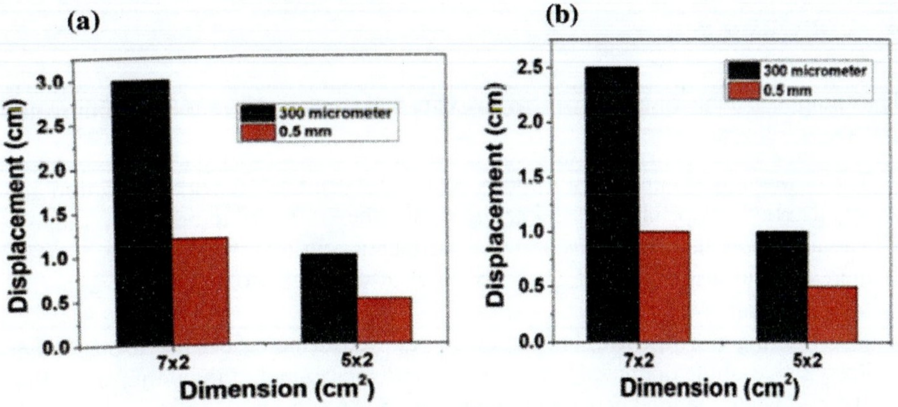

Fig. 6 Displacement **a** wire placed inside the polyimide and **b** wire placed outside the polyimide

3.3.2 Effect of Strain Percentage

The strain of the SMA wire was changed from 1 to 8% using a customized straining unit. The strained SMA wire is then placed along with kapton polyimide and actuation study was performed. The experiments were carried out for two different diameter wires for all the samples. The displacement for all the strain percentage is shown in Fig. 7. The SMA wire of both diameters showed maximum displacement at 8% strain.

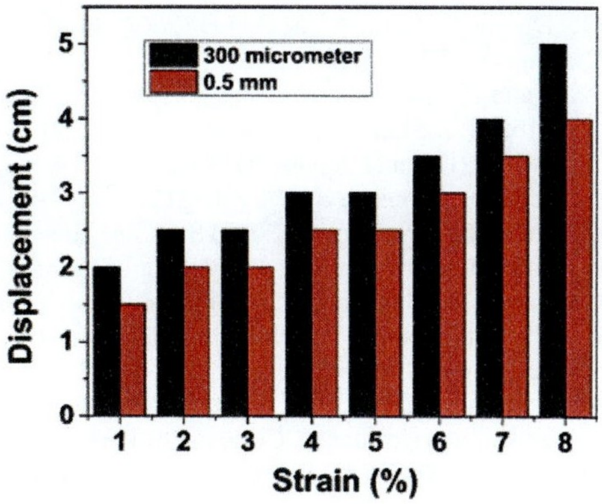

Fig. 7 Displacement against strain plot

4 Conclusions

The composite actuator comprising SMA wire and kapton polyimide was fabricated and analyzed.

- The differential scanning calorimetry results show the SMA wire of 0.5 mm and 300 μm both shape changes occurring during the 70–80 °C temperature range.
- The actuation characteristics showed the increase in dimension of the actuator increases the displacement. The maximum displacement of 3 cm was obtained for $7 \times 2\ cm^2$.
- The straining of wire at a higher percentage showed an increase in the displacement irrespective of the diameter of the wire. The displacement of the bimorph will be affected when the load acts on the fabricated SMA bimorph.
- The developed actuator can be useful in pick and place applications in the robotics field.

Acknowledgements The authors thank the National characterization facility of IIT Indore for their continuous support in providing composition analysis facility. We also thank CARS project (DYSL-SM/SMTECH/CARS-01) scheme under DRDO for providing financial support for carrying out the work.

References

1. Duerig T, Pelton A, Stöckel D (2002) An overview of nitinol medical applications. Mater Sci Eng A. 273–275:149–160
2. Kahn H, Huff MA, Heuer AH (1998) The TiNi shape-memory alloy and its applications for MEMS. J. Micromech Microeng 8:213–221
3. Wolf RH, Heuer AH (1995) TiNi (shape memory) films on silicon for MEMS applications. J Microelectromech Syst 4:206–212
4. Choudhary N, Kaur D (2016) Sensors and actuators a: physical shape memory alloy thin films and heterostructures for MEMS applications. A Rev Sens Actuators A Phys. 242:162–181
5. Seguin JL, Bendahan M, Isalgue A, Esteve-Cano V, Carchano H, Torra V (1999) Low temperature crystallised Ti-rich NiTi shape memory alloy films for microactuators. Sens Actuators A: Phys 74(1–3):65-9
6. Rodrigue H, Lee J-H, seop CY (2012) Application of SMA spring tendons for improved grasping performance. Composites Part B 5:1–5
7. Coyle S, Majidi C, Leduc P, Hsia KJ (2018) Bio-inspired soft robotics: material selection, actuation, and design. Extreme Mech lett 22:51–59
8. Seok S, Onal CD, Cho KJ, Wood RJ, Rus D, Kim S (2013) Meshworm: A peristaltic soft robot with antagonistic nickel titanium coil actuators. IEEE/ASME Trans Mechatron 18:1485–1497
9. Rodrigue H, Wang W, Kim DR, Ahn SH (2017) Curved shape memory alloy-based soft actuators and application to the soft gripper. Compos Struct 176:398–406
10. Petrini L, Migliavacca F (2011) Biomedical applications of shape memory alloys. J Metall 1:1–15

Investigation on Resting Orientation of Components Dropped from Different Heights

S. Udhayakumar, A. Mohan, A. Prabukarthi, and A. Megala

Abstract Part feeders are a critical part of an automated assembly system. Vibratory feeders which use vibration to feed material are the most used in industries. The feeders move the components to the required destination and drop them from a height. The feeders are expected to deliver the components at a specific rate and at a preferred orientation which changes with the component design and requirements. For designing a feeder system, certain parameters such as component size, configuration, the complexity of the component, and natural resting orientation of the component are to be known. The component's most probable natural resting orientation is to be determined theoretically and is to be known while designing a feeder. In this paper, the most probable natural resting orientation of two different components of the LEGO robot was analyzed using the theoretical stability method, and the results were compared with the experimental drop test results. The analysis is done based on the probability of occurrences of different orientations of the components, and a comparison is made between the theoretical stability method and the experimental drop test using the testing of the hypothesis. It was concluded that the experimental and theoretical results showed a good relation in the case of component-1, while the results from component-2 proved the theoretical conclusions to be insufficient to determine the most probable resting orientation of the component.

Keywords Resting orientation · Stability method · Drop test · Probability of occurrence · Testing of hypothesis

S. Udhayakumar · A. Mohan (✉) · A. Prabukarthi · A. Megala
Department of Mechanical Engineering, PSG College of Technology, Peelamedu, Coimbatore 641004, India
e-mail: mohan3mmm@gmail.com

S. Udhayakumar
e-mail: suk.mech@psgtech.ac.in

A. Prabukarthi
e-mail: apk.mech@psgtech.ac.in

U. S. Dixit et al. (eds.), *Advances in Forming, Machining and Automation*, Lecture Notes in Mechanical Engineering, https://doi.org/10.1007/978-981-19-3866-5_46

1 Introduction

As a process, material handling incorporates a wide range of manual, semi-automatic and automatic equipment, and systems. Part feeders constitute the automated material handling systems with the vibratory feeders being more preferred. Designing these feeders is a time-consuming process and requires knowledge of some critical aspects such as the part to be fed, the material of the feeder, the complexity of the part and the feeder desired orientation, and the natural resting orientations of the part. The preferred orientation of the part is necessary to adopt the gating mechanism to convert any orientation to the preferred orientation. The base of the gating system is formed after finding the natural orientation of the part. The most probable natural resting orientation of components with regular shapes and different sizes was determined using the theoretical stability method by Chua and Tay [1]. The determination of the most probable resting orientation of asymmetric sector-shaped components was done by Udhayakumar et al. [2]. An attempt was made to determine the most probable natural resting orientations of two complex components by stability method, and the results were compared with the experimental drop test results using testing of hypothesis.

2 Natural Resting Orientation

The natural resting orientation of a part is the orientation in which it rests when it is dropped from a height at which it possesses sufficient potential energy to change its orientation [3]. To determine the most probable natural resting orientation of a component, the probability of occurrence of each resting orientation of the component is to be determined by theoretical methods. Theoretically, the probability of occurrence of each resting orientation of the components can be determined using several methods such as the energy barrier method, critical solid angle method, centroid solid angle method, and stability method. In this paper, the theoretical determination of the probabilities of occurrence of each resting orientation of two complex components is done using the stability method.

3 Research Background

The stability method is to determine the probability of occurrence of orientations of parts with regular shapes by Chua and Tay [1]. The most probable natural resting orientations of asymmetric sector-shaped parts were determined by Udhayakumar et al. [2]. The stability method is used here to analyze the natural resting orientations of two complex parts of the LEGO robot and is compared with the drop test results.

The natural resting orientations of various components with different shapes were studied by researchers. The energy barrier method to analyze the natural resting orientation of simple-shaped parts like a square prism and rectangular prism was discussed by Boothroyd et al. [3]. The natural resting aspects of the cylindrical prismatic part and symmetrical T-shaped prism were analyzed by Lee et al. [4]. The resting orientations of the cylindrical prismatic part and symmetrical T-shaped prismatic part were analyzed using the centroid solid angle method and were validated by drop test methods. The natural resting behaviors of square prism and rectangular prism on the hard surface were analyzed using the hypothesis that the probability of the part coming to rest in a particular resting aspect on a hard surface is proportional to the difference between the centroid solid angle subtended by that resting aspect and the average of the critical solid angles of that aspect for coming to rest in its neighboring aspects and inversely proportional to the height of the center of gravity from that aspect, by Ngoi et al. [5]. The results obtained from the analyses of square and rectangular prisms had paved the way for extending the theory to analyze more complicated geometrical parts, such as T-shaped parts and L-shaped parts.

The natural resting orientations of three connectors of similar shapes, but different proportions were studied and analyzed by Juan and Rosario [6]. The probabilities of occurrences of different orientations of a connector were analyzed using centroid solid angle method and stability method and were validated by the drop test method. The methodology adopted in the development and implementation of the CAD system and the conceptual model of the feeder were discussed by Lo and Dick [7]. Functional precedence relation called part mating science, which is a description of how parts interact during assembly, was proposed by Nevins and Whitney [8].

The systematic approach to analyze geometrical and topological properties of parts designed in the CAD system and to select the orientation devices to feed the designed components was done by Ou-yang and Maul [9]. A case study is presented to illustrate the design methodology that encompasses selection, configuration, and parametric aspects of part feeder and that uses analytical, numerical, and design tools by Yeong and de Vries [10].

4 Components Understudy

The resting behavior of two different components used in the building of LEGO robots was studied to determine their respective static probability profiles. Figure 1 shows the two components understudy.

Different possible orientations of the component were considered, and the probability of occurrence of each of the orientations was determined by the stability method. Figures 2 and 3 show the different possible resting orientations of components 1 and 2, respectively.

Fig. 1 Components understudy

Fig. 2 Possible orientations of component-1

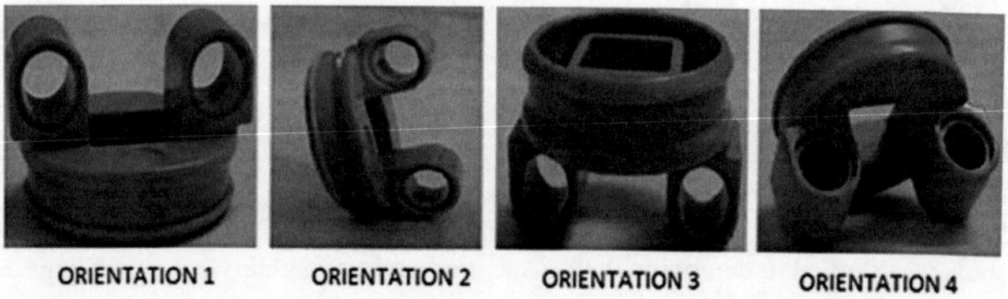

Fig. 3 Possible orientations of component-2

5 Stability Method to Determine the Natural Orientation

The stability method of determining the most probable natural resting orientation of components with regular shapes and different sizes was done by Chua and Tay [1].

Stability is the state of being able to keep in position. Based on logical analysis, the larger the contacting area, the more stable would be the component in that resting orientation. Also, the lower the center of gravity of the component from the base, the more stable would the component be in that orientation. Thus, stability is a function of the contact area and the distance of the center of gravity from the base.

Stability (S) is directly proportional to the contacting area (A) and inversely proportional to the distance of the center of gravity from the base (y) [1].

The generalized equation is given in Eq. (1).

$$P_i = N_i A_i / y_i / \sum N_i A_i / y_i \tag{1}$$

where

- P_i is the probability of occurrence of the orientation 'i'.
- N_i is the number of surfaces identical to and inclusive of the contacting surface 'i'.
- A_i is the area of contact of the orientation 'i', mm^2.
- y_i is the distance of the center of gravity from the base, mm.

5.1 *Determination of Probabilities*

The contact surface area and the distance of the center of gravity from the base were found using the software SOLIDWORKS. The component was designed to the specific dimensions, and the base or floor was fixed as a plane based on the desired orientation. The distance of the center of gravity from the base and the area of contact was determined using the software.

Tables 1 and 2 give the details about the center of gravity, the contacting surface area, and the probability of occurrence of each natural resting orientation of components-1 and 2, respectively. It was found that orientation 2 of component-1 had the highest probability of occurrence of 74%. It was also observed that orientations 1 and 4 of component-2 had the probabilities of occurrence of 36% and 30%, respectively, which was more than those of the other two orientations.

Table 1 Result of stability method for component-1

Orientation	Height of center of gravity y (mm)	Contact surface area A (mm^2)	Contact surface area/height (mm)	Probability of resting orientation
Orientation 1	7.09	6.87	0.97	0.17
Orientation 2	12.76	52	4.07	0.74
Orientation 3	14.41	6.87	0.48	0.09
			$\Sigma A/y = 5.52$	

Table 2 Result of stability method for component-2

Orientation	Height of center of gravity y (mm)	Contact surface area A (mm^2)	Contact surface area/height (mm)	Probability of resting orientation
Orientation 1	5.92	38.75	6.55	0.36
Orientation 2	11.5	29.97	2.61	0.14
Orientation 3	8.58	31.54	3.67	0.2
Orientation 4	8.27	44.58	5.39	0.3
			$\Sigma A/y = 18.22$	

6 Experimental Determination of Natural Resting Orientation

The components were subjected to a drop test from different heights with random initial orientations to test the obtained theoretical results. Wooden and acrylic floors were used to determine the dependence of the resting orientation with the material of the base on which the component is dropped. The wooden floor permitted bouncing of the component which caused its toppling resulting in different orientations, and hence, it is a soft surface. However, in the case of the acrylic floor, the material absorbed the energy of the dropped component restricting its bouncing, and hence, it is a hard surface. Figures 4 and 5 show the setups used for the drop test involving wooden and acrylic floors, respectively.

Fig. 4 Setup for wooden floor

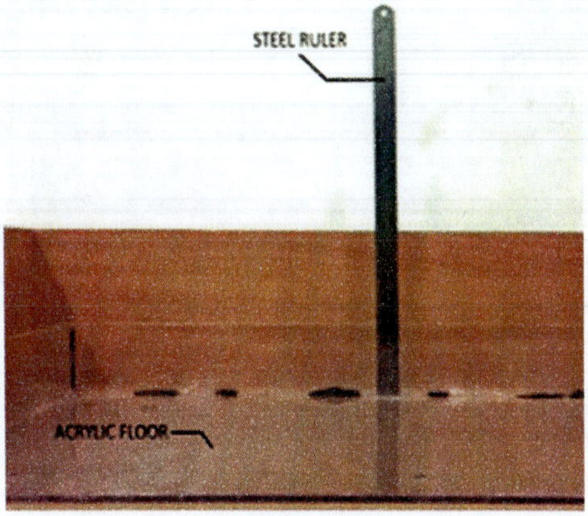

Fig. 5 Setup for acrylic floor

7 Procedures

The following steps were followed to conduct the drop test using component-1 on the wooden floor:

1. The steel ruler was fixed on the wooden floor on which the drop test was done.
2. With random initial orientations, the component was dropped from a height of 50 mm from the floor.
3. The final resting orientation of the component was recorded.
4. Steps 2 and 3 were repeated 100 times, and the final resting orientations of the component during each trial were observed and recorded.
5. Steps 2–4 were repeated at different heights of 100, 150, and 200 mm, and the final orientations at each stage were recorded as a table.
6. The number of occurrences of each orientation was obtained from the table, and the probabilities of occurrence of the orientations were calculated.

The same procedure was repeated for component-2 on the wooden floor to obtain the set of probabilities of occurrence of each resting orientation of the component. For carrying out the drop test on an acrylic floor, the steel ruler was removed from the wooden floor and was placed on the acrylic floor, and the test was carried out as before.

8 Results from Drop Test

Figures 6 and 7 show the graphs obtained from the drop test result of components-1 and 2, respectively, when dropped from different heights.

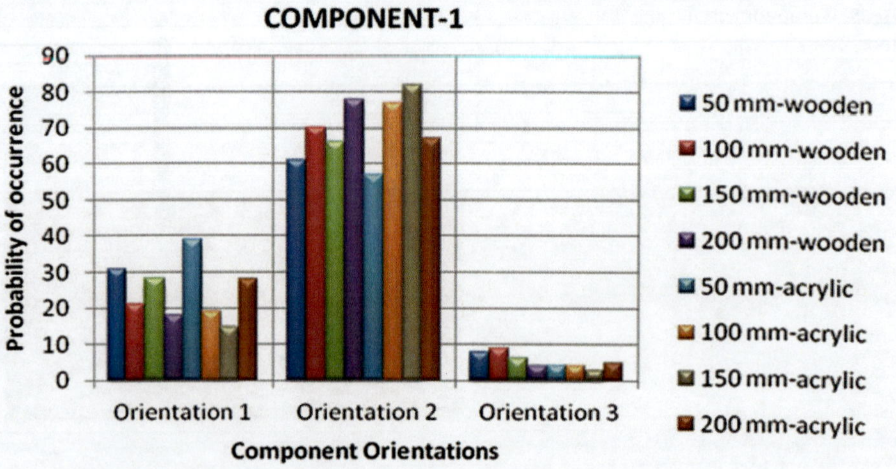

Fig. 6 Drop test result of component-1

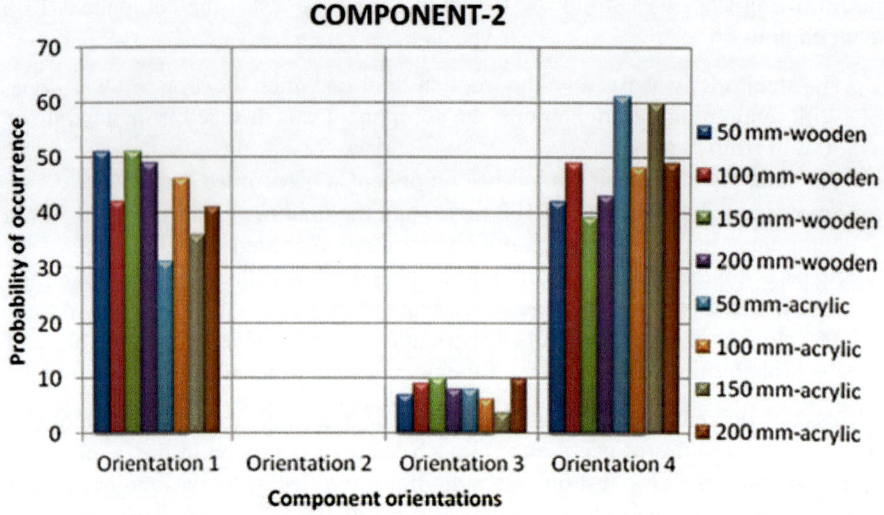

Fig. 7 Drop test result of component-2

9 Inference

From the drop test results, it was observed that in component-1, orientation 2 had the highest probability of occurrence in comparison with the other two orientations when tested on both wooden and acrylic floors from different heights. Similarly, in component-2, orientations 1 and 4 had higher probabilities of occurrence than the other orientations. In component-2, it was also observed that orientation 1 had a slightly higher probability of occurrence when dropped on the wooden floor, while orientation 4 was more probable when tested on the acrylic floor.

10 Comparison Between Theoretical and Experimental Results

The results obtained from the theoretical stability method and drop test were compared and analyzed to determine the accuracy of the theoretical method.

10.1 Experimental Study Using Component-1

The comparison of drop test and theoretical results obtained for component-1 when tested on wooden and acrylic floors is presented in Figs. 8 and 9, respectively. It was observed from the study that the theoretical results correlated with the observed experimental determinations. Though some deviations were observed, the most probable natural resting orientation of the component-1 was observed to be orientation 2 in both the theoretical stability method and experimental drop test conducted on wooden and acrylic floors.

In line with the extensive study, it was observed that the deviations between actual and theoretical values were minimal in orientation 3 than in orientations 1 and 2 though both theoretical and actual determinations concluded that the most probable natural resting orientation of component-1 was orientation 2.

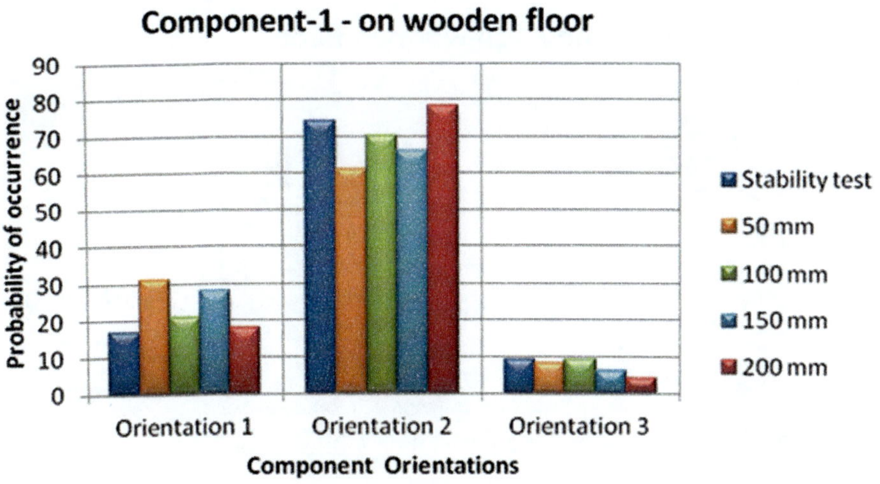

Fig. 8 Comparison of theoretical and experimental results of component-1 on the wooden floor

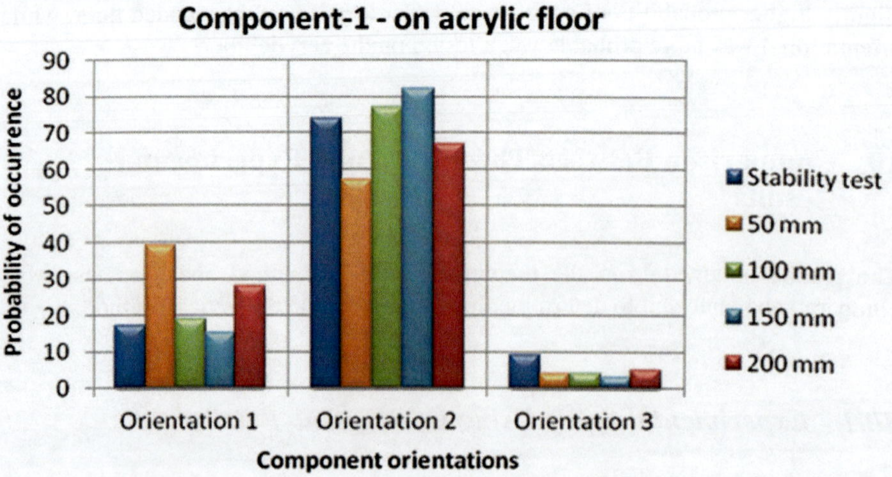

Fig. 9 Comparison of theoretical and experimental results of component-1 on acrylic floor

10.2 *Experimental Study Using Component-2*

Figures 10 and 11 show the comparison of the drop test and theoretical results obtained for component-2 when tested on wooden and acrylic floors, respectively. From Figs. 10 and 11, it was observed that orientations 1 and 4 were more probable than the other orientations. Also, it was observed that orientation 2 which did not have any occurrence during the drop test was found to have a probability of occurrence of about 14% through theoretical determination.

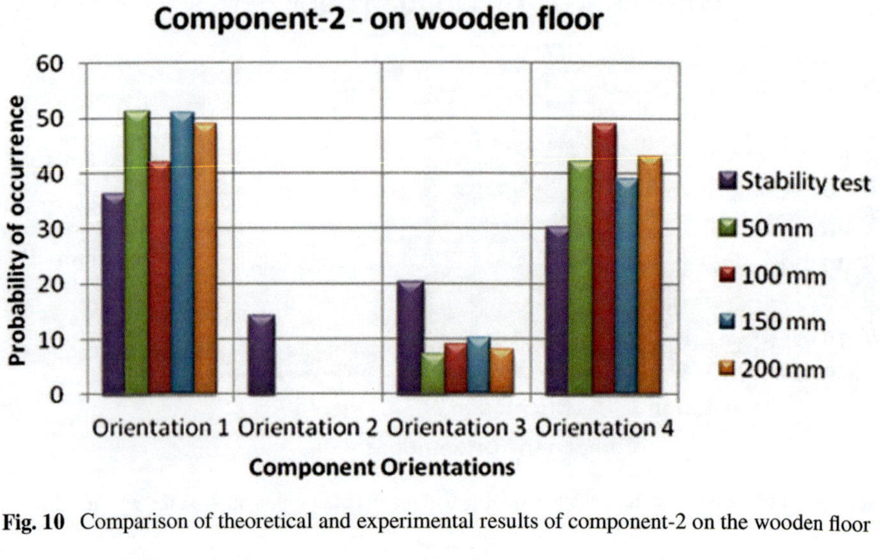

Fig. 10 Comparison of theoretical and experimental results of component-2 on the wooden floor

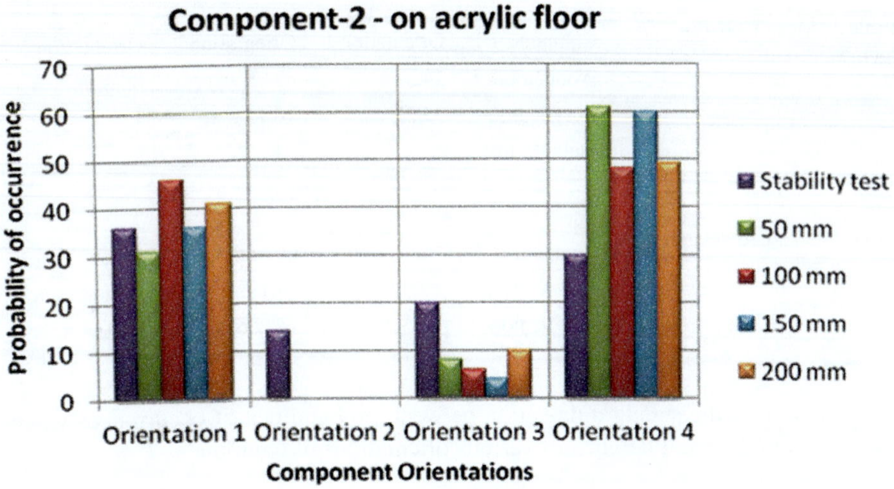

Fig. 11 Comparison of theoretical and experimental results of component-2 on acrylic floor

In addition, orientation 3 was found to have more probability of occurrence when determined through the theoretical stability method than what was observed during the drop test. In component-2, the theoretical results varied for all orientations of the component. Though the stability method predicted orientation 1 to be the most probable resting orientation of the component followed by orientation 4, the drop test results showed that both the orientations had almost equal probabilities of occurrence when dropped from different heights on wooden and acrylic floors.

11 Comparison Using Testing of Hypothesis

Null hypothesis

i. The probabilities of the orientation of components in the wooden floor and the stability test are the same.
ii. The probabilities of the orientation of components in the acrylic floor and the stability test are the same.

Alternate hypothesis

i. The probabilities of the orientation of components in the wooden floor and the stability test are not the same.
ii. The probabilities of the orientation of components in the acrylic floor and the stability test are not the same.

Table 3 gives the mean of probabilities of occurrences when dropped from different heights, of various orientations of component-1.

Table 3 Mean table for component-1

Component-1	Orientation 1	Orientation 2	Orientation 3
Wooden floor	24.5	68.75	6.75
Acrylic floor	25.25	70.75	4
Stability test	17	74	9

Table 4 Standard deviation table for component-1

Component-1	Orientation 1	Orientation 2	Orientation 3
Wooden floor	6.027713773	7.182153809	2.217355783
Acrylic floor	10.65754819	11.08677891	0.816496581

Table 4 gives the standard deviation between probabilities of occurrences when dropped from different heights, of various orientations of component-1.

11.1 Comparison of Probabilities on Wooden Floor and Stability Test for Component-1

Testing of hypothesis for the orientation 1 of component-1

Null hypothesis: $\mu = 17$
Alternate hypothesis: $\mu \neq 17$
$n = 4$
Test statistic,

$$\begin{aligned} Z &= (x - \mu)/\left(\sigma/(n)^{(1/2)}\right) \\ &= (24.5 - 17)/\left(6.027713773/(4)^{(1/2)}\right) \\ Z &= 2.4885 \end{aligned} \quad (2)$$

Considering the level of significance, $\alpha = 0.05$,

$$Z_{\alpha/2} = Z_{0.025} = -1.96, 1.96$$

From Fig. 12, it is found that the value of Z lies in the area of rejection, the null hypothesis is rejected. Hence, for orientation 1 of component-1, the probabilities in the wooden floor and the stability test are not the same. Similar comparisons are made between the probabilities of occurrences of each orientation and the stability test for both components-1 and 2. In all the comparisons, the mean of all the probabilities of occurrences while dropping the component from different heights is compared with the result of the stability test.

Henceforth, the null hypothesis states that the mean of the probabilities at different heights is closer to the result obtained in the stability value. A similar procedure is

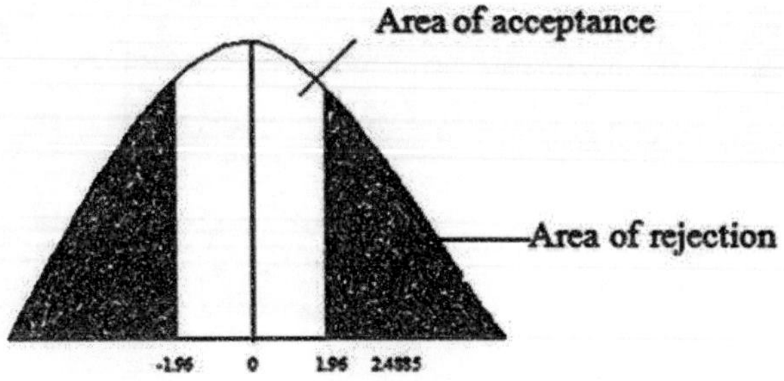

Fig. 12 Normal curve showing the rejection for orientation 1 of component-1 on the wooden floor

repeated for all the orientations of both components-1 and 2, and the results are discussed.

Testing of hypothesis for the orientation 2 of component-1

Null hypothesis: $\mu = 74$
Alternate hypothesis: $\mu \neq 74$
$n = 4$
Test statistic, $Z = -1.462$

$$Z_{\alpha/2} = Z_{0.025} = -1.96, 1.96$$

From Fig. 13, it is found that the value of Z lies in the area of acceptance, the null hypothesis is accepted. Hence, for orientation 2 of component-1, the probabilities in the wooden floor and stability test are the same.

Testing of hypothesis for the orientation 3 of component-1

Null hypothesis: $\mu = 9$
Alternate hypothesis: $\mu \neq 9$
$n = 4$
Test statistic, $Z = -2.0294$

$$Z_{\alpha/2} = Z_{0.025} = -1.96, 1.96$$

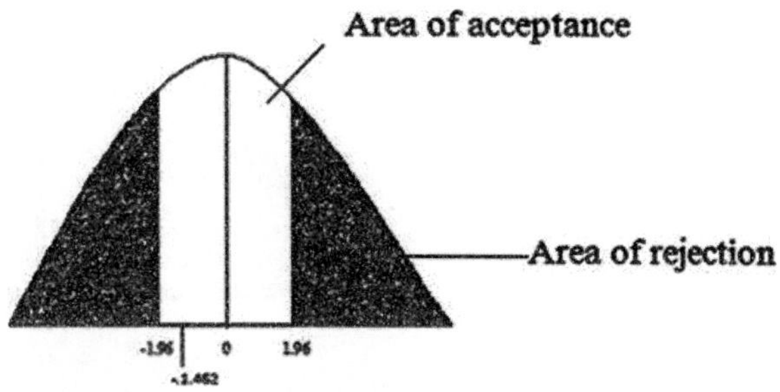

Fig. 13 Normal curve showing the rejection for orientation 2 of component-1 on the wooden floor

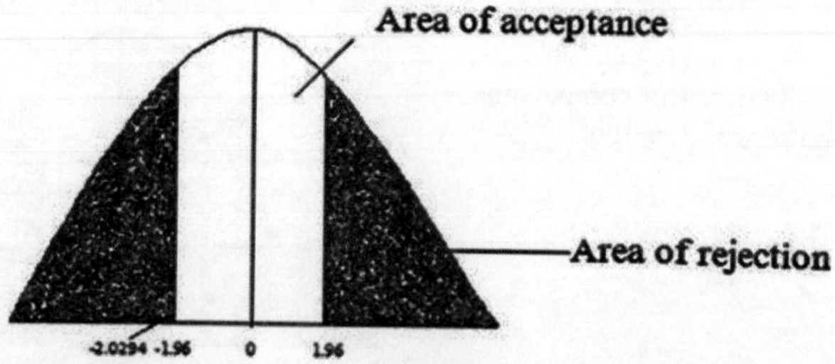

Fig. 14 Normal curve showing the rejection for orientation 3 of component-1 on the wooden floor

From Fig. 14, it is found that the value of Z lies in the area of rejection, the null hypothesis is rejected. Hence, for orientation 3 of component-1, the probabilities of occurrences in the wooden floor and stability test are not the same.

11.2 Comparison of Probabilities on Acrylic Floor and Stability Test for Component-1

Testing of hypothesis for the orientation 1 of component-1

Null hypothesis: $\mu = 17$
Alternate hypothesis: $\mu \neq 17$
$n = 4$
Test statistic, $Z = 1.5482$

$$Z_{\alpha/2} = Z_{0.025} = -1.96, 1.96$$

From Fig. 15, it is found that the value of Z lies in the area of acceptance, the null hypothesis is accepted. Hence, for orientation 1 of component-1, the probabilities of occurrences in the acrylic floor and stability test are the same.

Testing of hypothesis for the orientation 2 of component-1

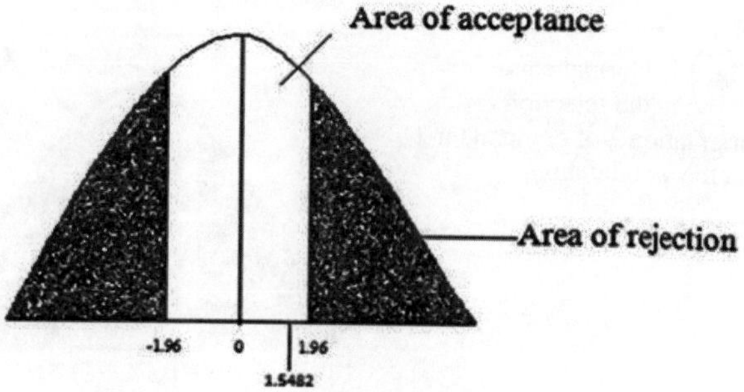

Fig. 15 Normal curve showing the rejection for orientation 1 of component-1 on acrylic floor

Null hypothesis: $\mu = 74$
Alternate hypothesis: $\mu \neq 74$
$n = 4$
Test statistic, $Z = -0.5863$

$$Z_{\alpha/2} = Z_{0.025} = -1.96, 1.96$$

From Fig. 16, it is found that the value of Z lies in the acceptance region, the null hypothesis is accepted. Hence, for orientation 2 of component-1, the probabilities of occurrences in the acrylic floor and stability test are the same.

Testing of hypothesis for the orientation 3 of component-1

Null hypothesis: $\mu = 9$
Alternate hypothesis: $\mu \neq 9$
$n = 4$
Test statistic, $Z = -12.247$

$$Z_{\alpha/2} = Z_{0.025} = -1.96, 1.96$$

From Fig. 17, it is found that the value of Z lies in the rejection region, the null hypothesis is rejected. Hence, for orientation 3 of component-1, the probabilities of occurrences in the acrylic floor and stability test are not the same.

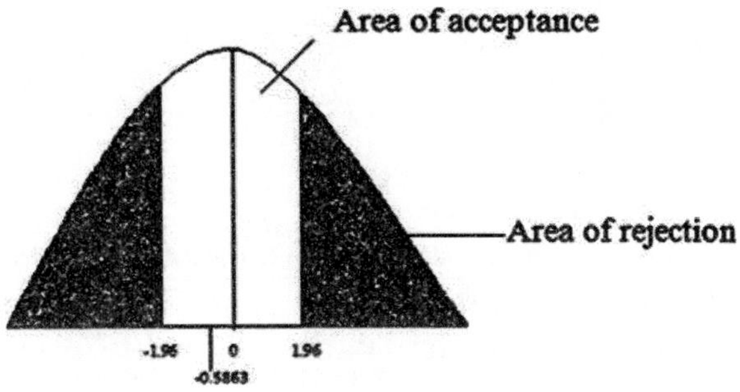

Fig. 16 Normal curve showing the rejection for orientation 2 of component-1 on acrylic floor

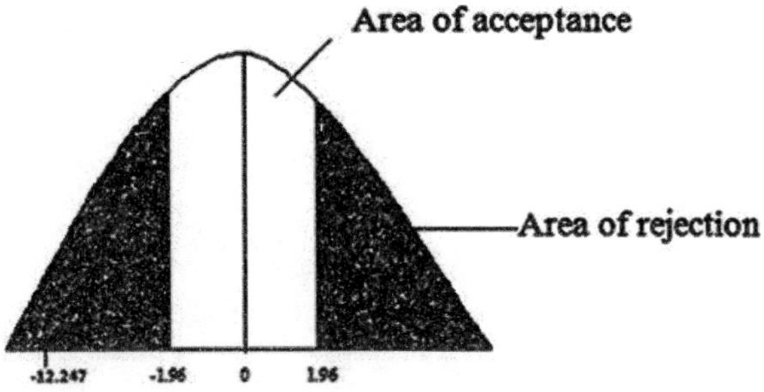

Fig. 17 Normal curve showing the rejection for orientation 3 of component-1 on acrylic floor

11.3 Comparison of Probabilities on Wooden Floor and Stability Test for Component-2

Table 5 shows the mean values of probabilities of occurrences while dropping from different heights for every orientation. Table 6 shows the values of standard deviation between the probabilities of occurrences from different heights for every orientation.

Testing of hypothesis for the orientation 2 of component-2

Null hypothesis: $\mu = 36$
Alternate hypothesis: $\mu \neq 36$
$n = 4$
Test statistic, $Z = 5.735$

$$Z_{\alpha/2} = Z_{0.025} = -1.96, 1.96$$

From Fig. 18, it is found that the value of Z lies in the rejection region, the null hypothesis is not accepted. Hence, for orientation 1 of component-2, the probabilities of occurrences in the wooden floor and stability test are not the same.

Table 5 Mean table of component-2

Part 2	Orientation 1	Orientation 2	Orientation 3	Orientation 4
Wooden floor	48.25	0	8.5	43.25
Acrylic floor	38.5	0	7	54.5
Stability test	36	14	20	30

Table 6 Standard deviation table

Part 2	Orientation 1	Orientation 2	Orientation 3	Orientation 4
Wooden floor	4.272001873	0	1.290994449	4.193248542
Acrylic floor	6.454972244	0	2.581988897	6.952217872

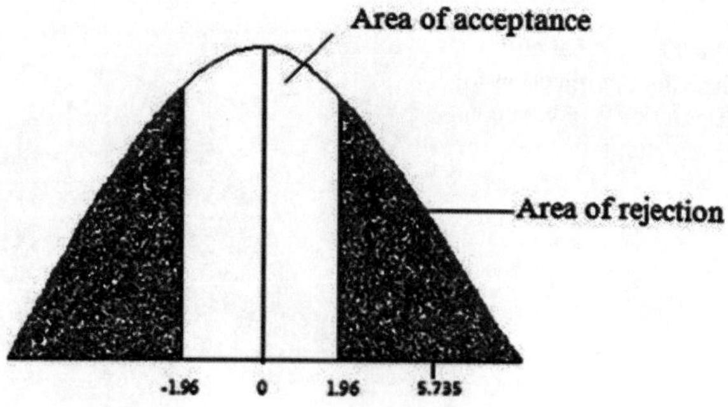

Fig. 18 Normal curve showing the rejection for orientation 1 of component-2 on the wooden floor

Testing of hypothesis for the orientation 2 of component-2

Null hypothesis: $\mu = 14$
Alternate hypothesis: $\mu \neq 14$
$n = 4$
$Z =$ Indeterminate

As the value of Z turns out to be indeterminate, it is not possible to proceed with the comparison of probabilities of occurrences in the wooden floor and acrylic floor for orientation 2 of component-2.

Testing of hypothesis for the orientation 3 of component-2

Null hypothesis: $\mu = 20$
Alternate hypothesis: $\mu \neq 20$
$n = 4$
Test statistic, $Z = -17.8157$

$$Z_{\alpha/2} = Z_{0.025} = -1.96, 1.96$$

From Fig. 19, it is found that the value of Z lies in the rejection region, the null hypothesis is rejected. Hence, for orientation 3 of component-2, the probabilities of occurrences in the wooden floor and stability test are not the same.

Testing of hypothesis for the orientation 4 of component-2

Null hypothesis: $\mu = 30$
Alternate hypothesis: $\mu \neq 30$
$n = 4$
Test statistic, $Z = 6.3197$

$$Z_{\alpha/2} = Z_{0.025} = -1.96, 1.96$$

From Fig. 20, it is found that the value of Z lies in the rejection region, the null hypothesis is rejected. Hence, for orientation 4 of component-2, the probabilities of occurrences in the wooden floor and stability test are not the same.

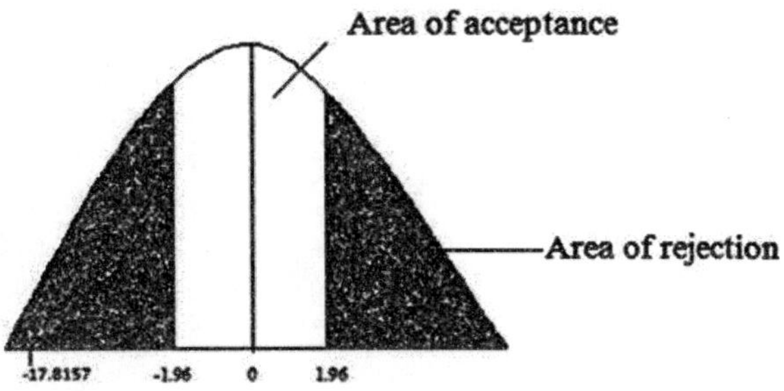

Fig. 19 Normal curve showing the rejection for orientation 3 of component-2 on the wooden floor

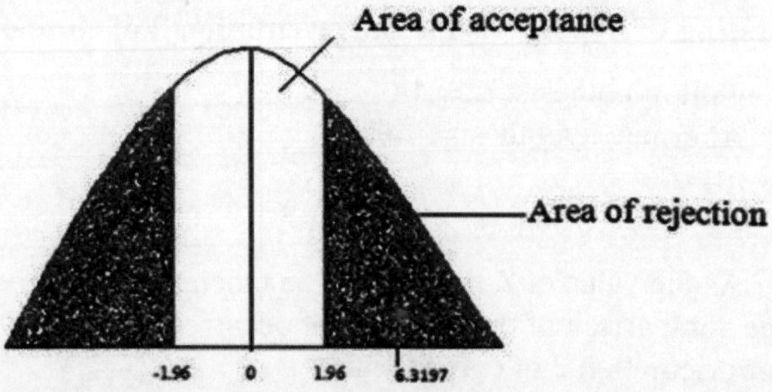

Fig. 20 Normal curve showing the rejection for orientation 4 of component-2 on the wooden floor

11.4 Comparison of Probabilities on Acrylic Floor and Stability Test for Component-2

Testing of hypothesis for the orientation 1 of component-2

Null hypothesis: $\mu = 36$
Alternate hypothesis: $\mu \neq 36$
$n = 4$
Test statistic, $Z = 0.7746$

$$Z_{\alpha/2} = Z_{0.025} = -1.96, 1.96$$

From Fig. 21, it is found that the value of Z lies in the acceptance region, the null hypothesis is accepted. Hence, for orientation 1 of component-2, the probabilities of occurrences in the acrylic floor and stability test are the same.

Testing of hypothesis for the orientation 2 of component-2

Null hypothesis: $\mu = 14$
Alternate hypothesis: $\mu \neq 14$
$n = 4$
Test statistic, $Z =$ Indeterminate

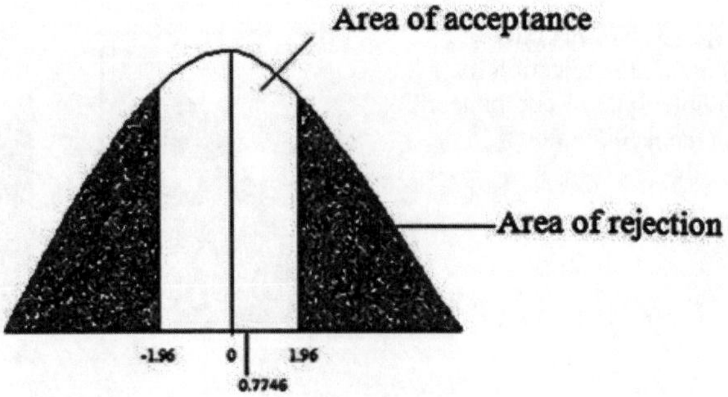

Fig. 21 Normal curve showing the rejection for orientation 1 of component-2 on acrylic floor

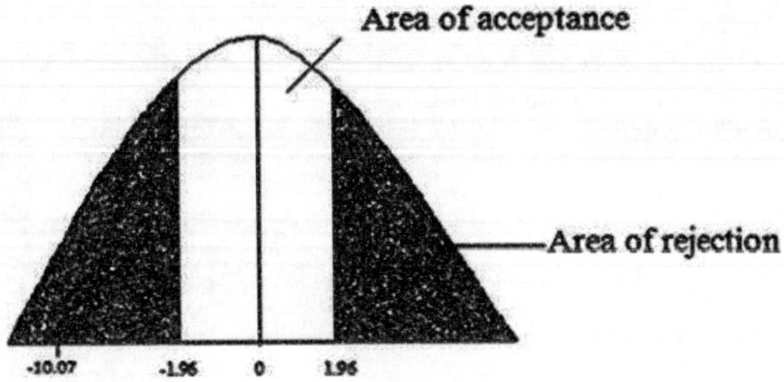

Fig. 22 Normal curve showing the rejection for orientation 3 of component-2 on acrylic floor

As the value of Z turns out to be indeterminate, it is not possible to proceed with the comparison of probabilities of occurrences in the acrylic floor and stability test for orientation 2 of component-2.

Testing of hypothesis for the orientation 3 of component-2

Null hypothesis: $\mu = 20$
Alternate hypothesis: $\mu \neq 20$
$n = 4$
Test statistic, $Z = -10.07$

$$Z_{\alpha/2} = Z_{0.025} = -1.96, 1.96$$

From Fig. 22, it is found that the value of Z lies in the rejection region, the null hypothesis is not accepted. Hence, for orientation 3 of component-2, the probabilities of occurrences in the acrylic floor and stability test are not the same.

Testing of hypothesis for the orientation 4 of component-2

Null hypothesis: $\mu = 30$
Alternate hypothesis: $\mu \neq 30$
$n = 4$
Test statistic, $Z = 7.0481$

$$Z_{\alpha/2} = Z_{0.025} = -1.96, 1.96$$

From Fig. 23, it is found that the value of Z lies in the rejection area, the null hypothesis is rejected. Hence, for orientation 4 of component-2, the probabilities of occurrences in the acrylic floor and stability test are not the same.

12 Conclusions

The most probable natural resting orientations of two different components from the LEGO robot kit were determined using the theoretical stability method whose results

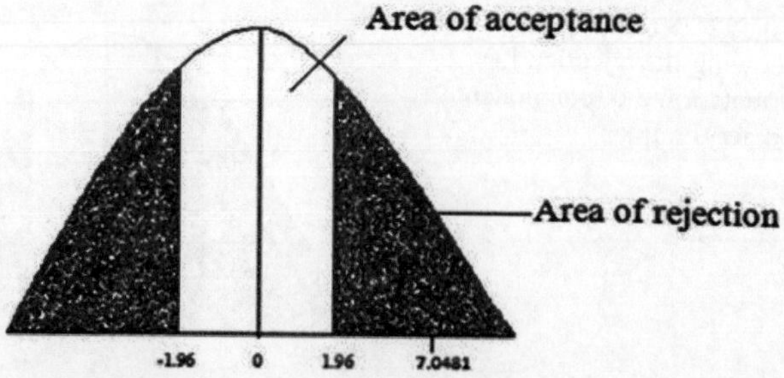

Fig. 23 Normal curve showing the rejection for orientation 4 of component-2 on acrylic floor

were verified using a drop test on wooden and acrylic floors from different heights. It is inferred from the hypothesis testing that the experimental drop test method on the wooden floor and stability method is same for the orientation 2 of component-1 and different for the other two orientations. The experimental method on acrylic floor and stability method is the same for orientations 1 and 2 of component-1 and different for orientation 3. The experimental method on the wooden floor and stability method is different for orientations 1, 3, and 4 of component-2. The experimental method on acrylic floor and stability method is the same for orientation 1 of component-2 and different for orientations 1 and 4. The test statistic is indeterminate for orientation 3 of component-2 on both wooden and acrylic floors. It was concluded that the results showed a good relation in the case of component-1, while the results from component-2 proved the theoretical conclusions to be insufficient to determine the most probable resting orientation of the component. In component-1, orientation 2 was the most probable natural resting orientation that can be made use of while designing the part feeder. In component-2, orientations 1 and 4 had equal probabilities of occurrence when subjected to the drop test, while theoretically, orientation 1 had a higher probability than orientation 4. The determination of the most probable natural resting orientations will help in the design of efficient part feeders.

References

1. Chua PSK, Tay ML (1998) Modelling the natural resting aspect of small regular shaped parts. Trans ASME 120:540–546
2. Udhayakumar S, Mohanram PV, Keerthi Anand P, Srinivasan R (2013) Determining the most probable natural resting orientation of sector-shaped parts. Assembly Autom 33(1):29–37
3. Boothroyd G, Redford AH, Poli CR, Murch LE (1972) Statistical distribution of natural resting aspects of parts for automatic handling. Manuf Eng Trans 1:93–105
4. Lee SSG, Ngoi BKA, Lim LEN, Lye SW (1997) Determining the probabilities of natural resting aspects of parts from their geometries. Assembly Autom 17:137–142
5. Ann NK, Woh LS, Jie C (1996) Analysing the natural resting aspect of a prism on a hard surface for automated assembly. Int J Adv Manuf Technol 11:406–412
6. Levy JC, Rosario LM, Predicting the natural resting aspects probabilities for a family of connectors: a comparison of theoretical and experimental results
7. Lo EK, Dick MD (1990) The solid modeling of part orienting tracks for automated assembly. Int J Prod Res 28(8):1513–1525

8. Nevins JL, Whitney DE (1980) Assembly research. Automatica 16:595–613
9. Ou-yang C, Maul GP (1993) A computer analysis of orientation devices for vibratory bowl feeders. Int J Prod Res 31(3):555–578
10. Yeong MY, de Vries WR (1994) A methodology for part feeder design. Ann CIRP 43:19–22

Lead Time Reduction and Quality Improvement in a Manufacturing Industry Using DMAIC Methodology—A Case Study

M. Gomathi Prabha, Theivanth Rajamohan, S. Manikandan, and Shashikiran Reddy Petluru

Abstract For the manufacturing sectors to survive and flourish in the competitive international market, constant improvement of small and medium firms is a must. The current scenario of any manufacturing industry focuses on the quality and on-time delivery of the products to satisfy their customers. This paper is the result of a study conducted in the manufacturing industry in south India, with a particular focus on the machining of several automobile components. The study aimed to identify and eliminate bottlenecks in the production line, as well as to increase production using techniques like define, measure, analyse, improve, and control (DMAIC), SIPOC, value stream mapping, IoT-enabled coolant concentration monitoring system, and IoT-enabled tool life monitoring. The overall equipment effectiveness (OEE) for the bottleneck operation was found to be 58.74%. By implementing Lean Six Sigma tools and techniques, the quality of the product has been increased and the lead time has been reduced and as a result, the OEE was improved by 4.88%. This in turn helps in improving the overall production in the industry.

Keywords Lean Six Sigma · Internet of Things · Tool life monitoring · Value stream mapping · OEE

1 Introduction

Motorola, Inc. developed LSS in 1986 to compete with Kaizen (lean manufacturing technique) business model developed in Japan. The LSS is a method used to reduce the waste and variation produced for a process/product. The use of LSS expanded into many sectors such as health care, supply chain, etc., in the early 2000s. In short, lean manufacturing focuses to reduce the process variation and Six Sigma aims to reduce the variation causing defects leading to a cycle of continuous improvement.

The LSS [1] tools such as the SIPOC chart, sigma calculation, and Pareto analysis help to understand the cause of defects. A standardized deployment diagram of the

M. Gomathi Prabha (✉) · T. Rajamohan · S. Manikandan · S. R. Petluru
Department of Mechanical Engineering, PSG College of Technology, Coimbatore 641004, India
e-mail: mgp.mech@psgtech.ac.in

U. S. Dixit et al. (eds.), *Advances in Forming, Machining and Automation*, Lecture Notes in Mechanical Engineering, https://doi.org/10.1007/978-981-19-3866-5_47

Table 1 Nomenclature

LSS	Lean Six Sigma
IoT	Internet of Things
TLM	Tool life monitoring
VSM	Value stream mapping
CSVSM	Current state value stream mapping
NVA	Non-value-added activities
OEE	Overall equipment effectiveness
A	Availability
P	Productivity
Q	Quality

analysed process was derived to understand the working of the process which will improve the actualization rate in a short time. Considering the three pillars of TBL [2], it was possible to identify a high positive influence of LSS on organizational sustainability. The impacts of LSS stand out as the most relevant to organizational sustainability: (i) reduce waste, (ii) reduce defect rate, (iii) increasing the quality of service, (iv) increasing product quality, and (v) reduce process variability. Among the five impacts most influential, the first two are related to the costs dimension and the other three are aligned to the quality dimension.

Further, failure mode and effect analysis [3] are used as a basic consideration in developing the continuous improvement programme. From the research, the NVA time and the process sigma level are determined and it is found that the implementation of LSS furthers the continuous improvement. The DPMO is determined as a prerequisite to sigma level calculation. The nomenclature is listed in Table 1.

Conducting Decision Making Trial and Evaluation Laboratory (DEMATEL) and developing a cause-and-effect model, it is evident that Industry 4.0 [4, 5] is important. The implementation of control and monitoring techniques of sustainable Industry 4.0 [6, 7] is beneficial. The methodology of the system can be formulated from sensor networks that yield big data, wherein it is analysed to either predict a solution/result, schedule a decision making or modelling scheme, or developed as a machine learning algorithm that produces a solution based on the stipulated information or from past experiences. As a whole, the sustainability of Industry 4.0 relies on reducing energy consumption. It is essential to understand and estimate the expected impacts from Industry 4.0 technologies [8, 9] on economic, environmental, and social aspects for sustainable development to enrich the productivity of the required sector. The Six Sigma methodology is a well-disciplined and structured approach used to enhance process performance and achieve high levels of quality [10, 11] and low levels of variability.

The lean methodology is proven to help organizations achieve on-time delivery [12] of the right quality and quantity to satisfy customers. Lean Six Sigma and lean manufacturing [13, 14] focus on achieving the quality levels of 3.4 parts per million defects, but the quality-related issues can be solved by the collection and analysis of data. The integration of Industry 4.0 and Lean Six Sigma focuses on data derived from the deployed tools will help apply Lean Six Sigma for obtaining faster and reliable decisions that contribute towards reducing lead time and producing better quality products. With the implementation of Industry 4.0, the steps taken will be as a result of customer feedback. The IoT employed can be optimized by the experience gained either as a supervised or unsupervised learning method. After the implementation of IoT, the industry needs to be followed by risk assessment to control the quality levels. Lean Six Sigma has an effective part when incorporated with new technologies into an operation in a meaningful way.

Lean Six Sigma and Industry 4.0 collectively known as Lean Six Sigma 4.0 [15, 16] is the data collection and analysis part of the technology that furthers the solution-driven model of LSS. It can be seen that integration serves as the next revolution and how it impacts the future scope of the industries. The benefits and advantages of the integration are an inclination in actualization rates as well as improved quality with decreased lead time. With productivity being the main objective for improvement, the parameter assessed to demonstrate the capability of a particular machine is OEE [17]. OEE is a measure constrained to an element/machine of a process and can be regarded concerning available time, performance rate [18], and quality quotient. Lean tools and techniques are often promoting a positive impact on the industry overall, which specifically shows the improvement of OEE [19].

From the literature survey, it is clear that the DMAIC methodology helps in improving the quality and also reduces the lead time of any product. In this paper, the product Rear Cover 515 has been selected based on high demand. The current process study for the selected component in the manufacturing industry was carried out, and the bottleneck operation has been identified. Using DMAIC methodology and suitable tools and techniques, the quality and lead time have been improved, and as a result, the OEE of the bottleneck machine has been increased.

2 Problem Background

The production rate is decreased due to high lead time and many non-value-added activities. Due to a lack of interaction between the manufacturer and the customer, customer satisfaction is not up to the mark. The quality of the products is decreased due to an increase in defective items, which are caused due to tool wear, human errors, improper functioning of the machine, and so on. The increase in defective items also leads to an increase in waste materials generated.

The company selected for study is a manufacturing company that receives its raw materials as casting from foundry. Two CNC machining centres and one VMC machining centre is required in the manufacture of Rear Cover 515 shown in Fig. 1. The VMC machining centre is selected for further analysis since its cycle time exceeds that of other machining centres. The difference in the coolant concentration influences the tool life of the machining centre. Tool changeover time caused by

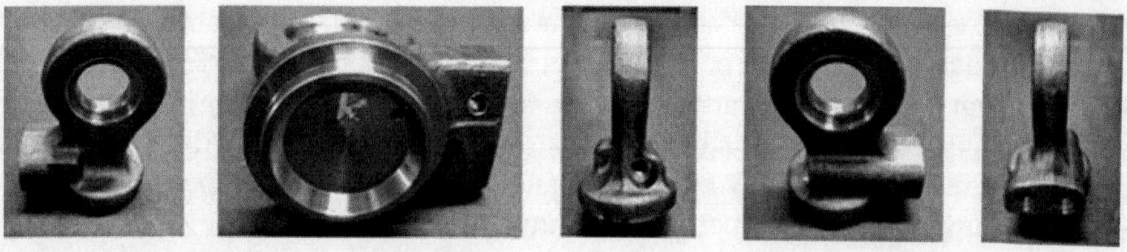

Fig. 1 Rear Cover 515 after machining

the coolant concentration reduces the uptime of the machine. The work proposes to address the enhancement of OEE of the machining centre.

The LSS implementation requires proper training of the management and staff, and the tool changeover time is addressed by introducing an IoT-enabled tool life monitoring (TLM) system which is used for continuous improvement (Kaizen). The coolant concentration is monitored by the use of live-refractometer apparatus linked to a mobile application.

3 Process Study

The manufacturing line consists of two CNC machines, one VMC machine, an inspection section, a quality control section, and a packaging section as shown in Fig. 2. In each process, the raw materials are machined according to the customer (automotive industries) requirements. If the components do not meet the inspection criteria, it is sent to rework. The satisfactory components are progressed to the further steps.

The current state value stream mapping (CSVSM) was determined using the initial data obtained, and bottleneck operation was identified. The OEE of the machine was analysed; corresponding to reduced uptime of 30 min and thereby enhancement of OEE is carried by the methodology shown in Fig. 3. IoT-enabled coolant concentration monitoring and tool life monitoring system were developed in parallel to be implemented in the machine.

4 Methodology

The methodology followed for this project is shown in Fig. 3. The initial data obtained are shown in Table 2.

The flow of the project sets off with the background study of LSS and IoT concepts along with the research papers to fix the suitable process parameters to attain the proposed objective.

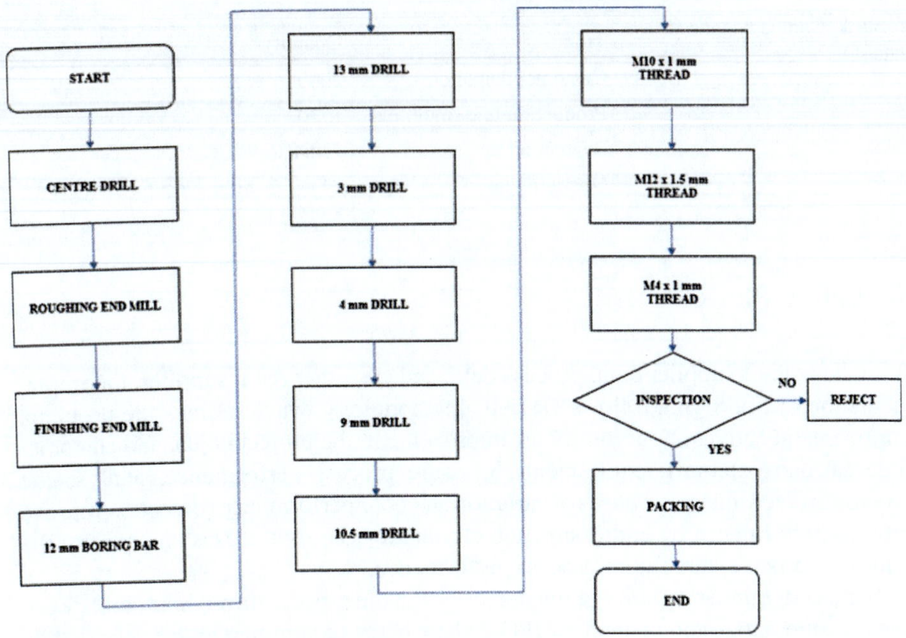

Fig. 2 Rear Cover 515 process map

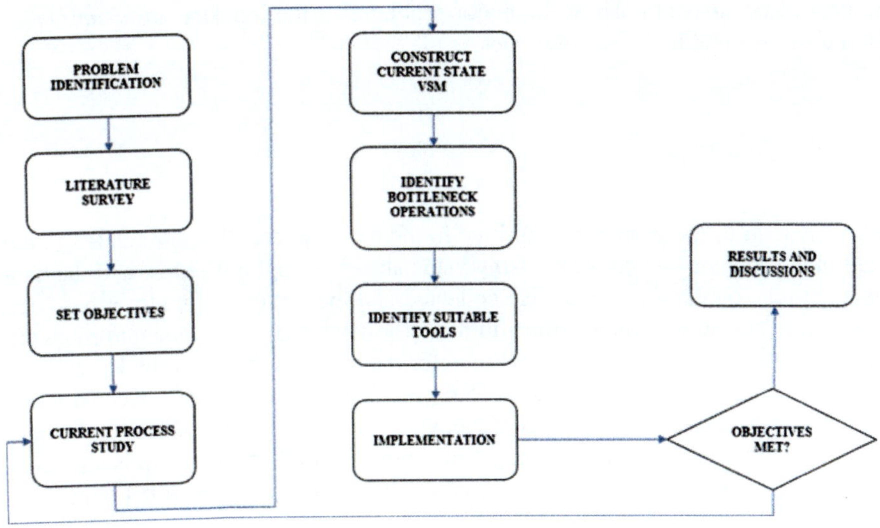

Fig. 3 Methodology adopted

Table 2 Company generic data

Title	Quantity
Customer demand	2000
Production rate (monthly)	1530
Rejection rate	17.64%
Operating hours	24 h—2 shifts—1 h break (except Sunday)

5 Define Phase

With the help of supplier data, process data, and customer data, supplier, the implementation of Six Sigma follows DMAIC methodology which defines the problem, improvement activity, opportunity for improvement, the project goals, and customer (internal and external) requirements, measure process performance, analyse the process to determine root causes of variation and poor performance (defects), improve process performance by addressing and eliminating the root causes and control the improved process, and future process performance.

For a better understanding of the operations in the process throughout, tools were used to summarize the contents. SIPOC chart helps to summarize the very details of the process to outline the functioning of the manufacturing line and is shown in Fig. 4. The defects produced have a striking difference between the most and least number of occurrences. Thus, the major problems in the industry are identified as high lead time and high rejection rates.

6 Measure Phase

With the help of the problems identified in the define phase, the data related to the lead time problem was collected using value stream mapping (VSM), and the data related to the quality issue was also collected and the current sigma level has been calculated. Takt time is the required duration for machining a component/product to meet the customer demand. The takt time for the production line of the Rear Cover 515 was calculated and found to be 15.6 min. The current VSM was constructed with the help of data like total cycle time, value-added time, non-value-added time, uptime, availability of work in process inventory, and takt time which is shown in Fig. 5. The total lead time for the selected product was found to be 2954 min.

$$\text{DPMO} = \frac{(\text{Number of Defects})}{(\text{Number of Units} * \text{Number of Defect Opportunities per unit})} \times 10^6 \tag{1}$$

For the quality issue, the types of defects, number of defects, total units produced, and the number of defects per opportunity were collected for one month and are

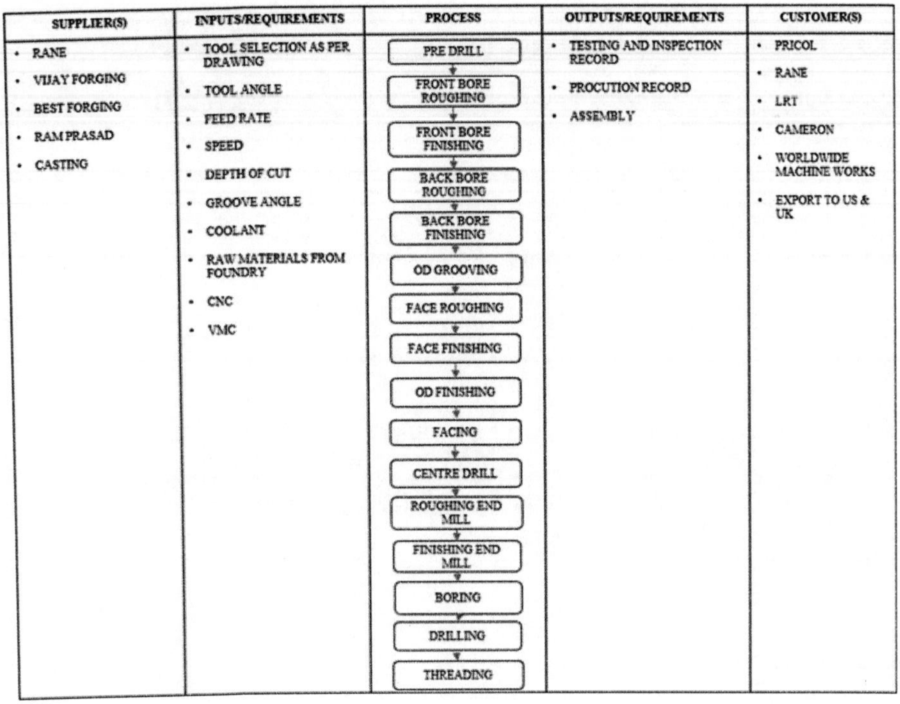

Fig. 4 SIPOC diagram

shown in Table 5. With the help of the data collected which is shown in Table 3, the defects per million opportunities (DPMO) was calculated using Eq. 1 and it was found to be 25,210. The current sigma level was found to be 3.456.

7 Analyse Phase

In this phase, the root cause for the problems mentioned in the define phase has been analysed. Using the current VSM constructed in the measure phase, a bar chart has been developed using the total cycle time data, which is shown in Fig. 6. The vertical milling centre (VMC) is the bottleneck operation since it exceeds the takt time.

Vertical machining centre has been selected for further improvements to reduce the overall lead time of the component Rear Cover 515. A time study was conducted in the VMC for 20 days which is presented in Table 4. The data collected was used to check the availability of the particular machine which is the bottleneck operation.

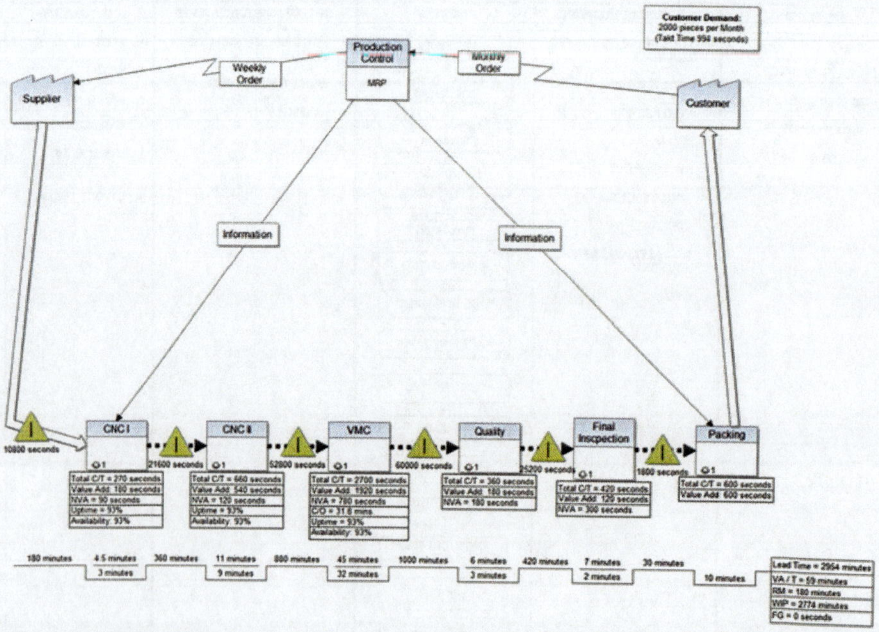

Fig. 5 Current state VSM

Table 3 Data for sigma calculation

Number of defects	270
Number of units produced per month	1530
Number of defect opportunities	7

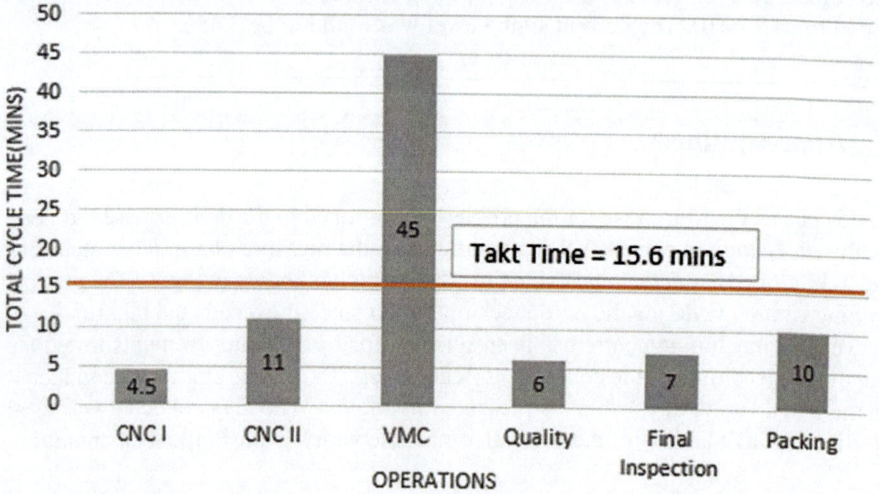

Fig. 6 Variation of the cycle time of all the workstations

Table 4 Time study data of VMC

Date	Planned production time = Shift length-break (h)	Stop time (h)	Run time = Planned production time − Stop time (h)
19-01-2021	11.5	1	10.5
20-01-2021	11.5	1	10.5
21-01-2021	11.5	1	10.5
22-01-2021	11.5	1	10.5
25-01-2021	11.5	2	9.5
26-01-2021	11.5	7	5
27-01-2021	11.5	2	9.5
28-01-2021	11.5	1.75	9.75
29-01-2021	11.5	2	9.5
30-01-2021	11.5	2	9.5
01-02-2021	11.5	1	10.5
02-02-2021	11.5	1.5	10
03-02-2021	11.5	1.5	10
04-02-2021	11.5	1	10.5
05-02-2021	11.5	1	10.5
06-02-2021	11.5	1	10.5
08-02-2021	11.5	1	10.5
09-02-2021	11.5	1	10.5
10-02-2021	11.5	1.5	10
11-02-2021	11.5	1	10.5

With the help of time study data, the overall equipment effectiveness (OEE) of VMC is calculated.

The formula to calculate OEE is as follows:

$$\text{OEE} = A \times P \times Q \tag{2}$$

where

A Availability
P Performance
Q Quality

$$\text{Quality} = \frac{\text{Good count}}{\text{Total count}} \tag{3}$$

From Eq. (3), quality is found to be 82.3%.
Run Time = 9.91 h/shift (average of 20 days)

Table 5 Type and number of defects of Rear Cover 515

Type of defects	Number of defects
Tool wear	93
Casting	61
Clamping	50
Wrong offset	29
Setting error	18
Inspection error	13
Others	6

$$\text{Takt Time} = \frac{\text{Run time}}{\text{Total demand}} \tag{4}$$

From Eq. (4), takt time is found to be 0.257 h

$$\text{Performance} = \frac{\text{Takt time} \times \text{Total count}}{\text{Run time}} \tag{5}$$

From Eq. (5), performance is found as 82.9%

$$\text{Availability} = \frac{\text{Run time}}{\text{Planned production time}} \tag{6}$$

From Eq. (6), the availability is found as 86.1%.

Using Eq. 2, OEE is calculated as 58.74%.

The current OEE of the vertical milling machining is found to be 58.74%. If the lead time and the quality are improved, the overall OEE of the VMC will be increased.

The type of defects and the respective quantity in VMC has been collected which is represented in Table 5 and it is analysed.

Pareto analysis is a technique used to select a limited number of tasks that produce an overall effect. It uses the Pareto principle, the idea that by doing 20% of the work, you can generate 80% of the outcome of doing the entire job. This technique is used to perform defect analysis as shown in Fig. 7.

From the Pareto chart, it is inferred that tool wear, casting, and clamping were identified as the "vital-few" defects. Implementation of tools and techniques for the reduction of these vital defects will increase the quality of the product. From the analysis, the tool wear defect needs to be addressed as it produces 93 defects in total with 78 defects solely in VMC.

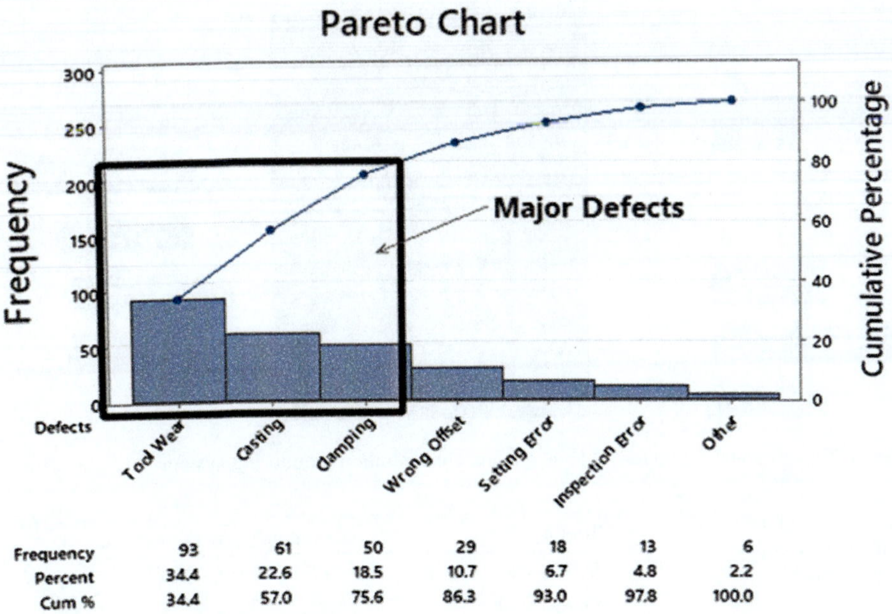

Fig. 7 Pareto chart for defect analysis

8 Improve Phase

At the end of analysing phase, tool wear was shortlisted as one of the major defects. It contributes the maximum consequence in the quality of the product selected.

The tool needs to be changed when it starts to wear but that is not properly monitored which leads to rejection of the component. To ensure this, IoT-enabled tool life monitoring system is used. It reduces the tool changeover time. The coolant needs to be maintained at a specific concentration to maintain tool performance and to prevent tool wear. For this, IoT-enabled coolant concentration monitoring system is used. It results in a better surface finish and ensures rust prevention.

The experimental set up for the tool and coolant concentration monitoring system is shown in Fig. 8. The dashboard installed specifies the tool to be changed. Tool(s) to be changed in Nth count will be displayed in the dashboard during $(N-1)$th count. The tools that need a change in the 250th step will be displayed during the 249th count (prior operation), thereby reducing the tool changeover time. The respected tool to be changed in the specific tool count by the tool life is shown in Table 6.

Coolant Cumi-Ferro 216 (5% by volume) is used. An increase in coolant concentration causes pronounced chipping of the tool's cutting edge. A decrease in coolant concentration increases the friction coefficient. Figures 9 and 10 show the circuit and block diagrams of the combined monitoring system.

Fig. 8 Experimental set up for tool and coolant concentration monitoring system

Table 6 Tool name and number of components manufactured

Tool name	Components manufactured
SCEM-Ø8MM	133
SCEM-039	200
INSERT-026	212
HSSD-236	240
TAP-005 TAP-164 TAP-186	250
INSERT-650	300
HSSD-003	320
SCEM-008	400
TB-0243	500
HSSD-121	600
HSSD-240	800
HSSD-180	996
SCD-379	1250
SCD-246 SCD-373 SCD-020	1790
HSSD-411	2490
Combination drill	2988

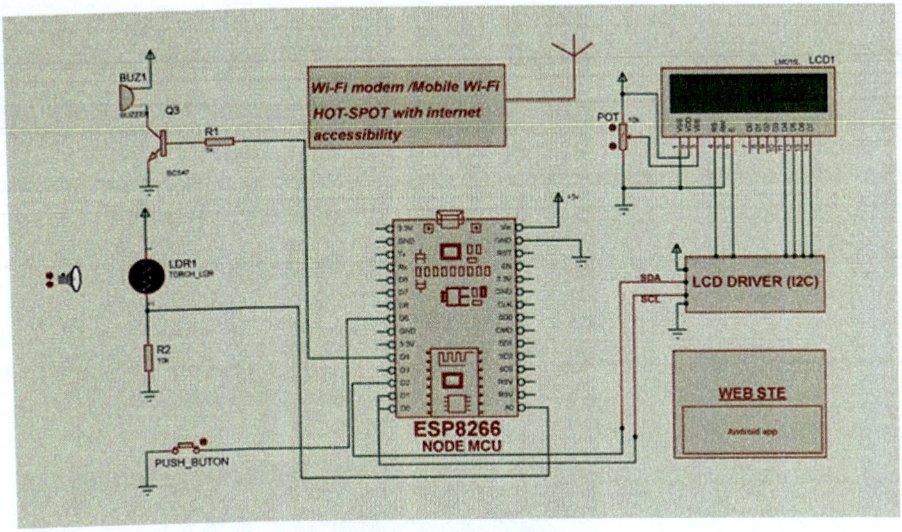

Fig. 9 Circuit diagram of the monitoring systems

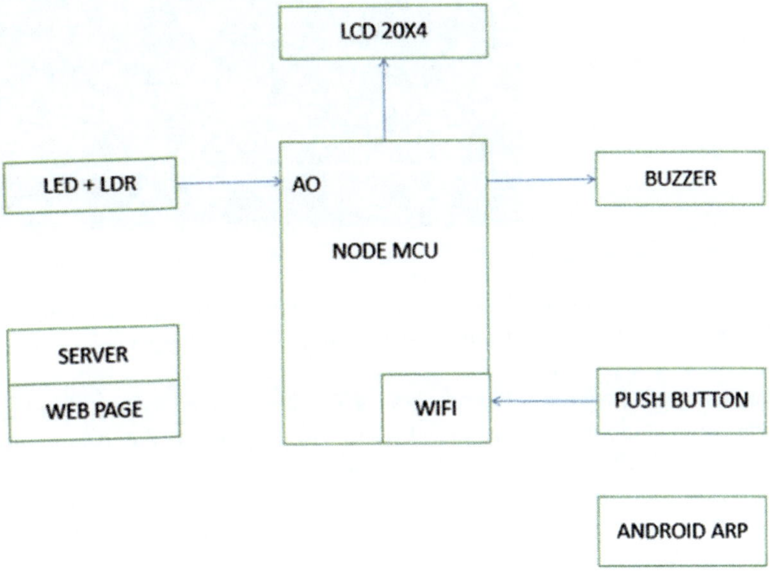

Fig. 10 Block diagram of the monitoring systems

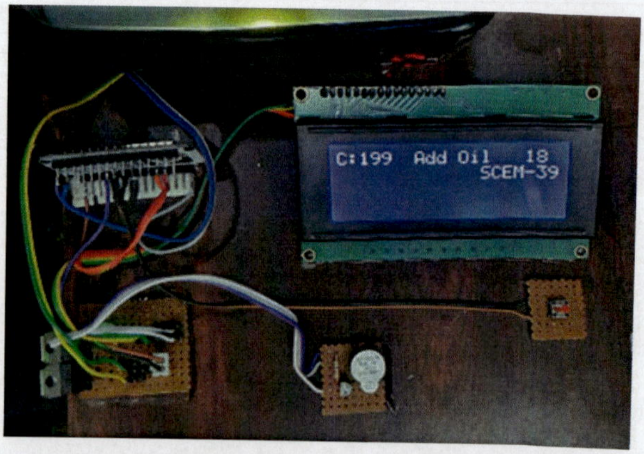

Fig. 11 Coolant concentration display

Fig. 12 Tool life display

The concentration and tool display occur in the same display as shown in Figs. 11 and 12. The concentration and tool change display are accompanied by a buzzer alarm.

The concentration and tool life monitoring system is linked to an android application via cloud computing as shown in Fig. 13 This step allows the live monitoring of the system, reducing the manual labour time.

The average tool changeover time reduces from 31.75 to 23.25 min with the help IoT-enabled tool life monitoring system. This in turn reduces the lead time. The future state VSM is developed and shown in Fig. 14.

9 Control Phase

It was determined to construct a control plan and checklist to keep a constant ability to check and monitor the process in terms of emerging difficulties. This is to ensure

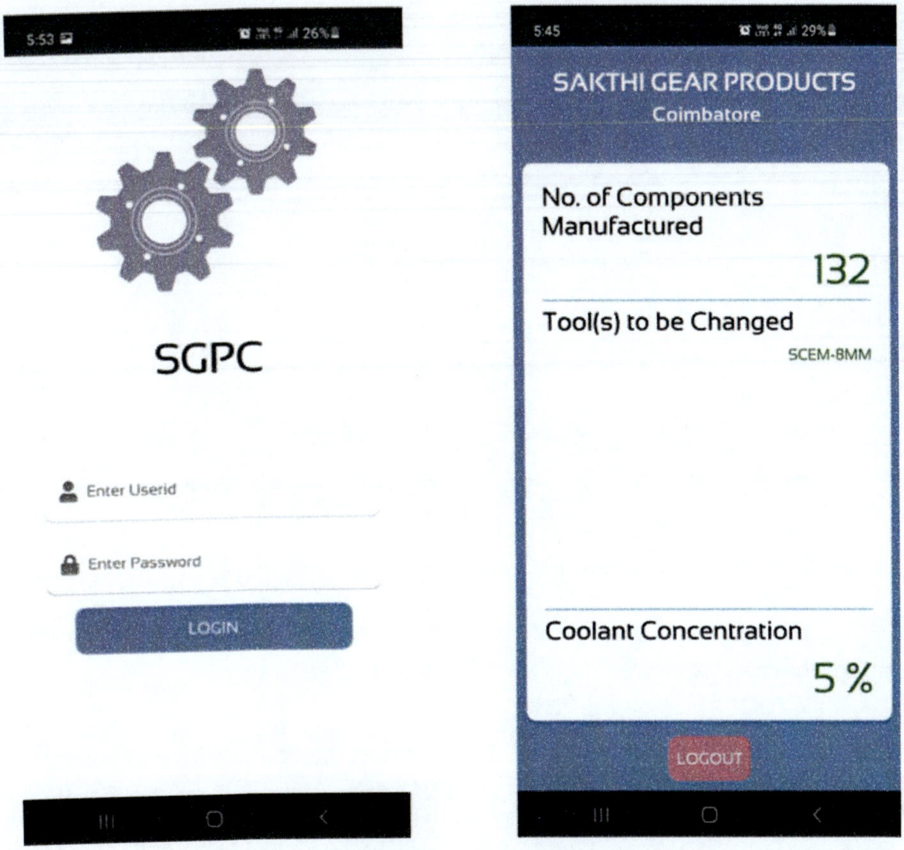

Fig. 13 IoT-enabled tool life monitoring system in an android application

that the foundation work process maintains a high degree of quality. The results were examined and measured for one month after these remedies were put into action.

10 Results and Discussion

With the help of IoT-enabled tool life monitoring system and IoT-enabled coolant concentration monitoring system, the tool wear defect has been eliminated. Therefore, the total rejection rate has been reduced which is shown in Fig. 15. The number of defects was found to be 192, the number of units produced and the number of defect opportunities is 1720 and 7, respectively. Using Eq. (1), DPMO is obtained as 15,946.

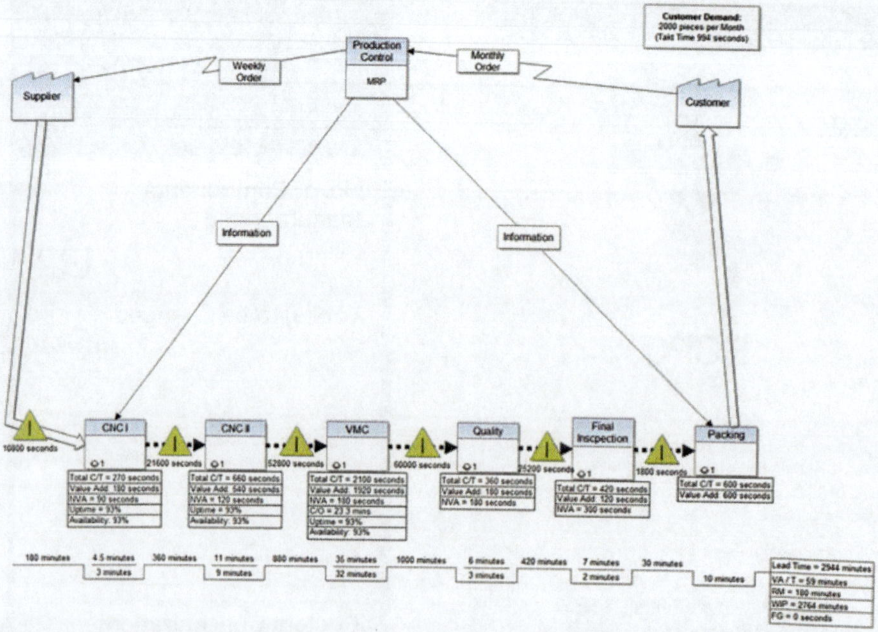

Fig. 14 Future state VSM

Fig. 15 Comparison of the number of defects before and after implementations

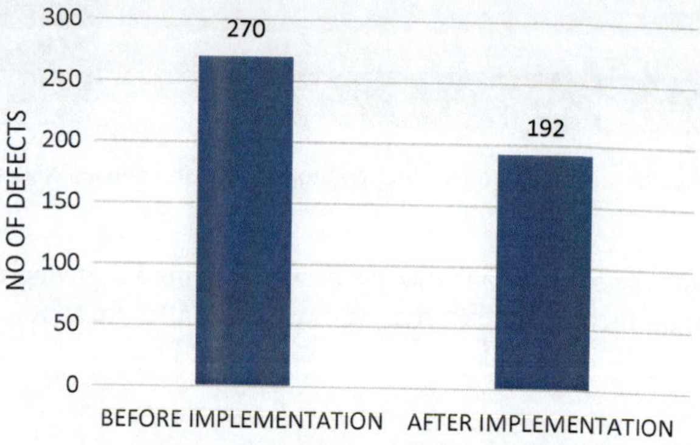

For the obtained DPMO value, the process sigma level is increased to 3.65 and the comparison of sigma level before and after implementations is shown in Fig. 16. As a result, the quality of the product has been increased.

Similarly, with the help of IoT-enabled tool life monitoring system, tool changeover time has been reduced from 31.75 to 23.25 min. This in turn lead time of the product has been reduced by 10 min.

The OEE of the vertical milling machine is increased following improvement in the subjective factors: availability, performance, and quality.

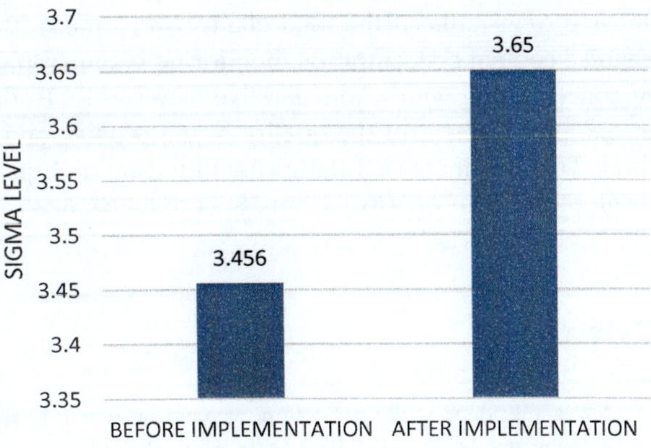

Fig. 16 Comparison of sigma level of the process before and after implementations

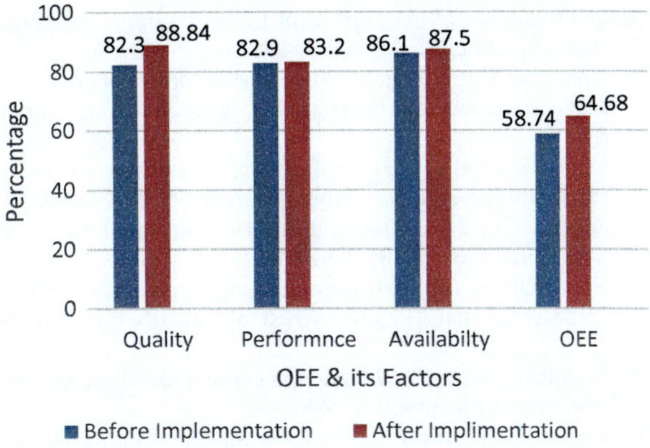

Fig. 17 Comparison of OEE and its factors before and after implementations

Quality is increased to 88.84%, with run time increased to 10.07 h/shift (avg of 20 days) and takt time 0.262 h, performance, and availability are increased to 83.2% and 87.5%, respectively.

Adhering to the increased factors, OEE after implementation is calculated as 64.68%. A comparison of OEE factors like quality, performance, and availability is shown in Fig. 17.

11 Conclusions

An integrated approach of Lean Six Sigma and the Internet of Things has been effectively adopted in the automotive manufacturing industry in this article. The case study addresses the issues such as high lead time and poor quality of the product which are faced by the manufacturing industry. Implementation of LSS and IoT improved productivity and enhanced the quality of the product. The IoT-enabled

TLM system reduced the total changeover time in the selected machine and IoT-enabled coolant concentration monitoring system replaced the manual measurement of concentration with automated live monitoring. Both the monitoring system helps in the improvement of the quality of the product and also in the reduction of lead time. This has increased the availability rate and quality rate of the product, as a result of the OEE of VMC from 58.74 to 64.68%.

References

1. Suryaprakash M et al (2021) Improvement of overall equipment effectiveness of machining centre using TPM. Mater Today Proc 46:9348–9353
2. Lia Z et al (2019) Data-driven smart manufacturing: tool wear monitoring with audio signals and machine learning. J Manuf Process 48:66–76
3. de Freitas JG (2017) Impacts of Lean Six Sigma over organizational sustainability: a survey study. J Cleaner Prod 156:262–275
4. Andersson C, Bellgran M (2015) On the complexity of using performance measures: enhancing sustained production improvement capability by combining OEE and productivity. J Manuf Syst 35:144–154
5. Kumar R et al (2020) Applications of industry 4.0 technologies in SMEs for ethical and sustainable operations: analysis of challenges. J Cleaner Prod 275:1–13
6. Scheffer C et al (2004) An industrial tool wear monitoring system for interrupted turning. Mech Syst Signal Process 18:1219–1242
7. Nara EOB et al (2020) Expected impact of industry 4.0 technologies on sustainable developments: a study in the context of Brazil's plastic industry. Sustain Prod Consumption 25:102–122
8. Scheffer C et al (2003) Development of a tool wear-monitoring system for hard turning. Int J Mach Tools Manuf 43:973–985
9. Kumar V et al (2020) Industry 4.0 implementation challenges in manufacturing industries: an interpretive structural modelling approach. Procedia Comput Sci 176:2384–2393
10. Sick B (2001) On-line and indirect tool wear monitoring in turning with artificial neural networks: a review of more than a decade of research. Mech Syst Signal Process 16(4):487–546
11. Indrawati S et al (2015) Manufacturing continuous improvement using Lean Six Sigma: an iron ores industry case application. Procedia Manuf 4:528–534
12. Singh K et al (2019) Selection of optimal cutting conditions and coolant flow rate (CFR) for enhancing surface finish in milling of aluminium alloy. Mater Today Proc 21:1520–1524
13. Di Bona G et al (2021) Implementation of industry 4.0 technology: new opportunities and challenges for maintenance strategy. Procedia Comput Sci 180:424–429
14. Bloj M-D et al (2020) Lean Six Sigma in energy service sector: a case study. Procedia Manuf 46:352–358
15. Abdul-Hamid A-Q et al (2020) Impeding challenges on industry 4.0 in circular economy: palm oil industry in Malaysia. Comput Oper Res 123:1–14
16. Arcidiacono G et al (2018) The revolution Lean Six Sigma 4.0. Int J Adv Sci Eng Inform Technol 8:141–149
17. Luthra S et al (2018) Evaluating challenges to industry 4.0 initiatives for supply chain sustainability in emerging economies. Process Saf Environ Prot 117:168–179

18. Moktadir MdA et al (2018) Assessing challenges for implementing industry 4.0: implications for process safety and environmental protection. Process Saf Environ Prot 117:730–741
19. Puvanasvaran AP et al (2013) Overall equipment efficiency improvement using time study in an aerospace industry. Procedia Eng 68:271–277

Low-Cost Automation Mechanism for Rotor Press Machine

Chaitanya Anil Koli and Nithin Tom Mathew

Abstract Pressing rotors on a high-pressure press are an important stage in the manufacturing of rotors for ceiling fans and other appliances. This pressing is done mainly to eliminate bending or distortion defects formed on its surface due to the aluminum die casting process taken place before. Due to this distortion or bending defects, rotors get rejected during their assembly. The available facility at the organization for this purpose was not very efficient and was causing a lot of production as well as financial loss to the organization. For this purpose, it was proposed to conceptualize, design, and develop a low-cost automation mechanism, which would eliminate this problem. The main task in hand was to make sure the entire diameter of the rotors gets pressed properly in one single stroke of the press machine. For this purpose, design changes in the layout of the machine were done. By using this project, the organization was able to improve its production efficiency, and human intervention in manufacturing activities was reduced to a minimum.

Keywords Manufacturing · Pneumatics · Automation

1 Introduction

For manufacturing of rotors of fans and other appliances, the last activity is pressing them under a high-pressure press. This activity is mainly carried out to eliminate bending or distortion defects produced in them due to the pressure die casting process. There are various kinds of parts produced at the organization according to customer specifications. Variants of products produced are shown in Table 1. During the pressure die casting process, aluminum is cast onto the outer surface of rotors. As the

C. A. Koli (✉)
Department of Mechanical Engineering, Birla Institute of Technology and Sciences, Pilani, Hyderabad Campus, Secunderabad 500078, India
e-mail: h20191060033@hyderabad.bits-pilani.ac.in

N. T. Mathew
Department of Mechanical Engineering, Birla Institute of Technology and Sciences, Pilani, Secunderabad 333031, India
e-mail: nithin.mathew@pilani.bits-pilani.ac.in

U. S. Dixit et al. (eds.), *Advances in Forming, Machining and Automation*, Lecture Notes in Mechanical Engineering, https://doi.org/10.1007/978-981-19-3866-5_48

Table 1 Specification of parts

Sl. No.	Inner diameter (mm)	Outer diameter (mm)	Weight (gm) before PDC	Height Min–Max (mm)
1	104.7	127.28	305–330	13.3–14.5
2	127.2	152.9	374–395	10.8–11.5

heating and cooling process takes place very rapidly during the rotor die casting process, surfaces of rotors usually have unevenness in them, and in some cases, bending or distortion can be easily observed. If such rotors are used, they can be easily rejected during the assembly process. To eliminate such rejection of parts, they are pressed onto a rotor pressing machine.

The use of basic, low-cost machinery is providing just the necessary functions which results in a relatively high manual operation effort. The preferred approach to reducing manual work in production is automation. Traditional automation of man–machine systems—especially in Western countries—tends to be comprehensive and thus often complex and expensive. Low-cost, lean automation intelligently being adapted to the individual task, as well as a decision method for choosing the tasks worth being automated, is required. The first step on the road toward a scientifically sound low-cost automation method for a manufacturing line is identifying and quantifying the different manual tasks which could potentially be automated [1]. Today, for the growth of any manufacturing unit, it must implement Industry 4.0. Highly connected cyber-physical systems (CPSs) are central elements for the digital factory of Industry 4.0 [2]. Hence, to grow and implement Industry 4.0 methodologies on shop floors, the organization wanted to start implementing automation in their manufacturing. Previous attempts for the implementation of low-cost automation mechanisms in assembly lines are represented in [3]. Automation is the creation and application of technology to monitor and control the production and delivery of products and services. In today's world, for any industry to survive in the competitive market, it must go for automation [4]. But, the use of automation and Industry 4.0 can be expensive, and small-scale industries cannot always go for it. Hence, the scope of low-cost automation increases in those scenarios. The case studies and industries' views on such scenarios are given in [5–9].

The labor cost has been one of the main reasons for the industry to move some of the production to so-called low-cost countries. Research has shown that this issue is more complex than calculating labor cost as the main driver. Organizational culture, research and development, and technical competence are also important drivers for a successful automation strategy [10]. The effectiveness and trends of implementation of such automation strategies have been discussed in [11–17]. The benefits of the implementation of automation in terms of tangible and intangible results for organizations are discussed in [18, 19].

The existing facility available at ASA industries, Noida, was operational for the past few years. The facility at the organization had a ram that would reciprocate in the vertical plane. This ram would press the rotor, while moving down and after pressing, it would move up to make room for the next part. The facility had an electric motor as the prime mover and the up and down motion of press were achieved through the pneumatic circuit. The organization wanted to ensure that the rejection of parts is brought down to as minimum as possible, and automation is implemented in daily manufacturing activities. This would ensure less human contact and better productivity for the organization. For this purpose, the task was to design an automated rotor feed mechanism that would ensure the entire diameter of the rotor gets pressed completely in a single stroke of the press. For this purpose, modifications were made to the existing layout of the machine. It was decided to stop the rotors under high-pressure press, and when the pressing is done, a pneumatic cylinder would actuate and push the rotor out of the pressing area through the exit chute provided, to make room for the next rotor.

2 Materials and Methods

The up and down motion of the ram of the press was achieved through pneumatic circuits. The pneumatic connections ensured the safe and easy operation of the ram. The electric motor was the prime mover which was used along with the pulley system to achieve primary rotational motion, and then, a pneumatic supply line was used which would ensure the up and down motion of the slide was achieved. The up and down moving slide of the ram was adjustable with its maximum height being 50 mm. This was to accommodate various heights of rotors being manufactured at the organization. Operators can adjust the required height of ram manually according to the quality requirements of customers to ensure zero rejections (Figs. 1 and 2).

2.1 Rotor Pressing Facility Available at the Organization

The existing facility at ASA Industries, Noida, for pressing the rotor was Kaushika made with a 50 mm maximum stroke of its slide and 7.7 hp electric motor as its prime mover. It had a layout such that it would make use of gravity for the movement of parts. At one end, the parts would be fed by an operator; they would slide onto the chute in a downward direction and would reach under press. A slight dip was provided after the press area to ensure easy removal of parts after pressing is completed. The layout was such that the parts would leave the machine in a direction parallel to the feed direction. The inlet and exit chutes were inclined to ensure the movement of parts (Table 2).

Fig. 1 Rotor press machine

2.2 *Proposed Design for Machine Layout*

The design proposed had a simpler guideway structure as shown in Fig. 3. The rotors would be fed by a human operator at the top end of the guideway as shown. Then, due to the smooth surface of the guideway and the effect of gravity, the rotor would slide down the inclined surface easily and would stop at end of the guideway. Here, the high-pressure press would press the rotor. The guideway design was fabricated such that it would make sure exact positioning of the rotor under rotor press, and no misalignment would take place.

At the end of the guideway, a stopper will be placed. The stopper would have a proximity sensor fitted into it. This would make sure that the electronic circuit gets the signal when parts have reached the rotor press area, and pressing has been completed. Once the pressing is completed, the delay timer fitted in the electronic circuit would send a signal into the pneumatic circuit, and by using air supplied to the pneumatic cylinder from a high-pressure compressor line, the pneumatic cylinder would actuate. This pneumatic cylinder would push out the rotor from the rotor press area and make room for the next rotor. This way cycle would continue, and smooth functioning of the mechanism can be obtained.

Fig. 2 Machine chute layout

Table 2 Press machine specifications

Model	Kaushika Make
Strike of slide	50 mm
Pressure	65 tons
Number of strokes	90 strokes per minute

The purpose of using automation in manufacturing is that it should aid the shop floor workers in making their tasks easier. With shop floor-oriented production support, human skills and automation create synergetic effects. The mastery of the manufacturing process is in the hands and brains of skilled workers; automation gives the necessary support to execute tasks and rationalize decisions. This represents a way of low-cost automation [20].

Design Calculations First, the effect of gravity was considered to make sure rotors do not pass at a very high speed into the pressing area. For this purpose, the inclination angle of the entry chute was decided such that it is just enough to cause the impending motion of the rotor.

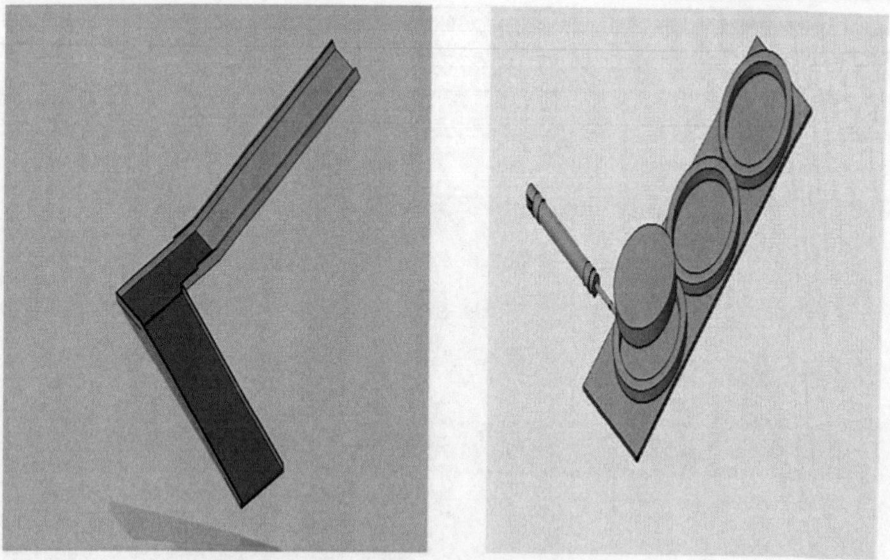

Fig. 3 Proposed design for machine chute and cylinder positioning

According to the proposed design, the part will be completely stopped under the rotor press area, and a pneumatic cylinder will be used to push the part out of the rotor pressing area. The choice of the pneumatic cylinder is dependent on many factors such as speed of motor available, up and down strokes of ram, the time required for forwarding travel and return of cylinder before ram reaches down for next stroke, the force required for removal of parts from the press area. The rotor press is driven by an electric motor. Hence, when the motor runs at 1440 rpm, the ram runs at approximately 90 strokes per minute.

$$\text{Maximum Weight of single rotor(after Pressure die casting)} = 640\ \text{gm}$$
$$= 0.64\ \text{kg}$$
$$\text{Inclination Angle} = \theta = 14^{\circ}.$$

At each instant, there will be three rotors on the inclined slide.
Therefore, the total force which causes downward motion due to inclination is

$$\begin{aligned} &= 0.64 \times 3 \times 9.81 \times \sin\theta \\ &= 18.84 \times \sin 14 \\ &= 4.54\ \text{N}. \end{aligned} \tag{1}$$

$$\text{Normal Reaction force} = N$$
$$= m \times g \times \cos\theta$$

where g is the acceleration due to gravity

$$= 18.84 \times \cos 14$$
$$= 18.28 \text{ N}.$$

Here, the coefficient of friction between rotor and surface is assumed as 0.6 (coefficient of friction between the aluminum rotor surface and steel chute surface).

Hence, total frictional force is

$$\begin{aligned} \text{Fr} &= \mu \times m \times g \times \cos\theta \\ &= 0.6 \times 18.28 \\ &= 0.6 \times 18.28 \\ &= 11 \text{ N}. \end{aligned} \tag{2}$$

Hence, from Eqs. (1) and (2), the minimum force which needs to be applied is

$$\begin{aligned} &= \text{Frictional force} - \text{Downward force due to inclination.} \\ &= 11 - 4.54 \\ &= 6.46 \text{ N}. \end{aligned} \tag{3}$$

The minimum force required for motion is 6.46 N. Also, as the ram is moving up and down at 90 strokes per minute, the cylinder would have approximately 0.667 s to move back and forth under the press area.

2.3 *Specifications of Pneumatic Cylinder*

Based on Eq. (3), the pneumatic cylinder chosen was FESTO made with its specifications as DSNU-16-125-P-A. The piston diameter of the pneumatic cylinder was 16 mm, and the stroke length was 125 mm. 5/2 double solenoid direction control valve was used for controlling the forward and reverse motion of the pneumatic cylinder. The inbuilt proximity sensors were provided to adjust the stroke length of pneumatic cylinders (Table 3).

Pneumatic Cylinder

Figure 4 shows the actual pneumatic cylinder. The pneumatic cylinder was supplied with air supply through tubing pipes from compressors. The proximity sensors for controlling the stroke of the cylinder are also mounted on its surface into their retainers. These sensors were connected to the electronic circuit. When an electric supply was given, these sensors would get a signal about the position of the pneumatic cylinder. When the cylinder has completed its full forward stroke, the forward stroke proximity sensor mounted on the cylinder would get a signal, and it

Table 3 Pneumatic cylinder specifications

Product	Specifications	Quantity
Pneumatic cylinder	DSNU-16-125-P-A	1
Push-in L-fitting	QSML-M5-4	2
Plastic tubing	PUN-H-4X0,75-BL	1
Foot mounting	HBN-12/16X2	1
5/2 Double solenoid control valve	JMFH-5-1/8 solenoid valve	1
Flange mountings		1
One-way flow control valve	GRLA-M5-QS-4-RS-D	2
Proximity sensors	SMT-8M-A-PS-24V-E-0,3-M8D	2
Retainers for sensors	SMBR-8-16	2
Connecting cable	NEBU-M8G3-K-5-LE3	2
Piston rod attachments	SG-M6	1

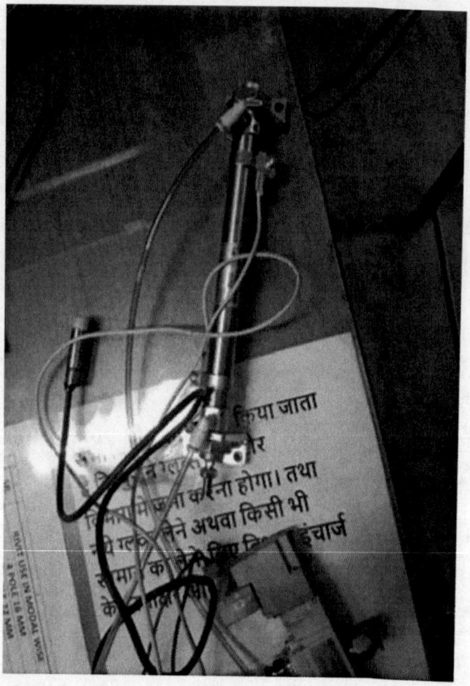

Fig. 4 Pneumatic cylinder

would send a signal to the direction control valve. Then, 5/2 double solenoid direction control valve in a pneumatic circuit would change the direction of airflow. Now, air would flow in opposite direction (Fig. 5).

Hence, the cylinder piston would retract and reach its initial position. Now, the proximity sensor placed at end of the cylinder would get a signal when the cylinder

Fig. 5 Block diagram of the actuation system

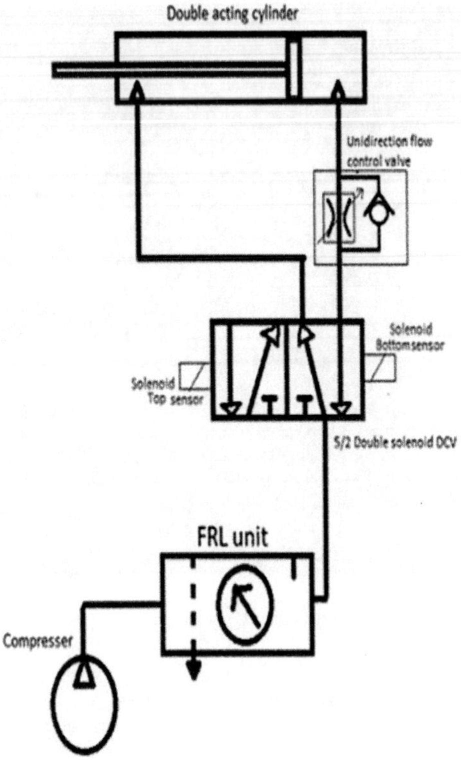

is fully retracted. Now, again, the sensor would send a signal to the direction control valve, and the flow direction would change, and the piston would move in the forward direction giving continuous cylinder operation.

Electronic Circuit

The electronic circuit was used for cylinder actuation. The electronic circuit operation starts with getting a signal from the proximity sensor about the position of the rotor. The proximity sensor mounted in the stopper of chutes would get the signal when the rotor has reached the rotor pressing area. Accordingly, the electronic circuit would get a signal acknowledging the position of rotors. Then, the ram of the press, which is moving up and down, would press the rotors. This pressing ensures the removal of the defects from the parts. Four 12 V of DC relays were used in the electronic circuit. These relays would send signals and ensure the execution of the cycle (Fig. 6).

After the pressing of rotors is completed and the ram is moving up, the pneumatic cylinder would move forward and push the rotors out of the rotor pressing area. This makes room for the next rotor to be pressed. The delay timer is used for this cylinder operation. The proximity sensor would send a signal into the electronic circuit as soon as the rotor reaches the rotor pressing area. Now, the cylinder actuation needs to be taken place after the rotor has been pressed, and the ram is moving up. To provide

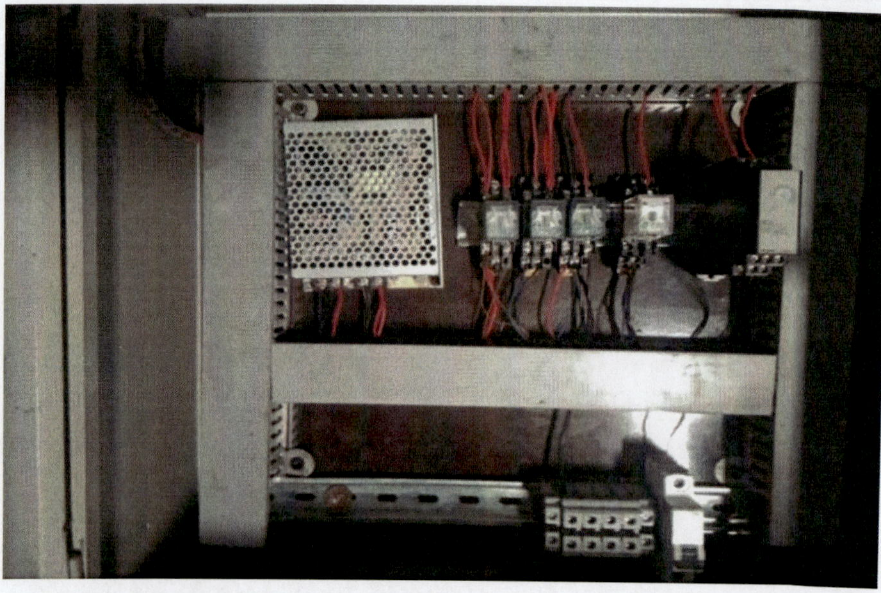

Fig. 6 Electronic circuit

this time-lapse Omron made, analog delay timer was used. A switched-mode power supply was used for a power source for the circuit, and a miniature circuit breaker was also provided.

3 Process Parameters

The manufacturers of machines had designed the layout of the machine such that the rotors will be moving in a straight path. The operator will feed the rotors onto the inclined platform; then, it will be pressed onto the rotor press, and it will pass down an inclined slope in a straight line. But, in this design of machine layout, the rotors were not getting properly pressed onto the rotor press machine. The bending or distortion defect originally present in rotors was not resolved. Hence, a modification in design was required. The main reason for the rotors not getting pressed onto rotor press was due to the inclination of bed rotors was moving down the incline at high speed, and incomplete pressing of the rotor was taking place. The inclination of the bed was not supposed to change because that change would avoid proper movement of rotors onto the bed.

To solve this problem, a low-cost automation mechanism was proposed. The mechanism was designed such that the complete downward motion of the rotor was restricted, and a sensor-actuated pneumatic cylinder was used to remove rotors from the rotor pressing area. For this purpose, a permanent stopper mechanism was placed onto a bed of the machine such that it will completely stop the motion of the rotor and bring it to a standstill. The stopper was positioned onto the bed such that the complete diameter of the rotor is getting pressed. To make sure, this is achieved several trials, and runs were carried out.

To actuate a pneumatic cylinder, a proximity sensor was used. The pneumatic cylinder was actuated by getting a signal about the position of the rotors. Air supply to actuate cylinder was taken from main supply lines available at the organization. Once the operator has fed the rotors into the rotor press machine, it will move down the inclined plane of the rotor press bed. When it has reached under rotor press, the stopper will restrict its downward motion, and it will come to a standstill position. After this, the ram of the rotor press will completely press the rotor such that the defects like bending or distortion present in the rotor due to pressure die casting will be removed. The press machine exerts a pressure of 60–65 tons onto the rotor due to which all defects are removed.

When the rotor has reached under the press, a proximity sensor placed on the stopper will detect the position of the rotor. As soon as the rotor reaches the under press, proximity sensor will get a signal about the positioning of the rotor, and it will pass the signal into the electronic circuit. The electronic circuit was designed such that it would actuate the cylinder 0.3 s after it has reached that position. This time-lapse was provided using a delay timer in an electronic circuit. The delay time used in the electronic circuit was an analog one which would provide time steps of 0.1 s up to 1 s. This 0.3 s time-lapse was provided for the rotor to get pressed under a high-pressure press. After the rotor has reached under the press, the proximity sensor will send a signal into the circuit indicating positioning of rotor accordingly circuit will generate a signal, and it will send this signal to the delay timer. This delay timer will send a signal into the pneumatic circuit 0.3 s after it has received signal from a proximity sensor, and by using air from the main supply line from compressor, pneumatic circuit will actuate the cylinder, and rotor will be removed out of the pressing area, making room for next rotor. The entire cycle would take a maximum of 0.6–0.7 s to press one rotor.

3.1 *Stopper for Sensor Positioning*

It can be seen in Fig. 7 due to the failure of previous trials for the mounting mechanism, we had designed a more rigid structure for mounting the sensor. The material used for this mounting was stainless steel, and welding and lathe machines were used in its fabrication. The middle part was drilled using a drill machine. This part was drilled to make room for the sensor. The size of the hole was made the same as that of the sensor. Hence, when the stopper was mounted on the bed of the rotor press, this will ease the positioning of the sensor. This front part will be where the rotor will come and stop due to striking, and the sensor mounted into the hole as shown in the figure will send a signal for positioning of the rotor.

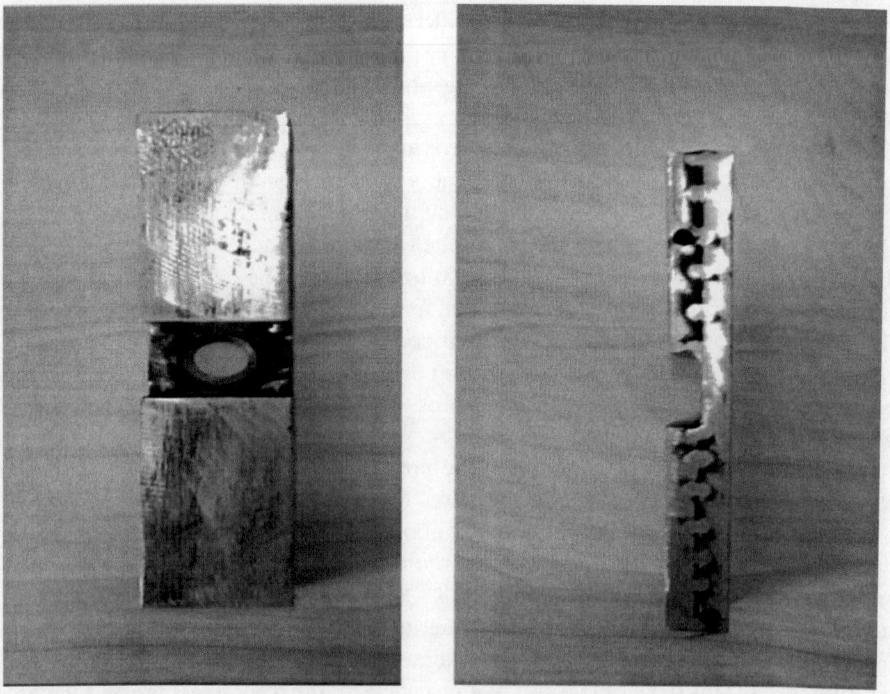

Fig. 7 Stopper for sensor positioning

3.2 Actual Positioning of Stopper Mechanism on Machine Bed

In Fig. 8, it can be seen how an actual mechanism was installed for positioning the sensor on the bed. The stopper mechanism will be permanently bolted to the bed by using Allen bolts of M6 * 80 size as shown. This would fix the mechanism firmly onto the bed and make sure it does not fail due to any kind of impact on it due to rotors or anything else. Then, the sensor which will be mounted at the center hole of the stopper mechanism will get the signal when the rotor is close enough. This would ensure that the rotor is at an exact position and does not slide anywhere else. The position of the stopper is such that it makes sure proper pressing of the rotor is taking place, and defects in it are removed.

Entire System Installation at the Machine
Figure 9 shows the actual installation of all the above-discussed components onto the machine. After successful installation at the machine, the trials were taken, and it was observed that this mechanism was working and acceptable.

Fig. 8 Positioning of stopper on machine bed

4 Results and Discussion

From the successful trial of the mechanism, it was observed that the mechanism can be implemented at rotor press machine, and commercial production of parts can be obtained from it. At the organization, there were many variants of rotors produced. The main difference was in their height and outside diameter. Accordingly, an adjustment in the slide of ram was made for each trial. The cylinder stroke was designed such that it would push out the minimum diameter rotor easily so that proper functioning for all kinds of rotor sizes can be obtained. From trials, it was observed that the time taken to press the rotor before the installation of machines and after the installation of the mechanism had only marginal change. Before installation of the machine, each part from feeding to getting pressed was taking 3–4 s. After installation of the mechanism as the time required for the downward motion of part was reduced, each part pressing was taking 2–3 s compared to 3–4 s from the previous usage. Also, if the part is trapped under the press, the operator manually had to push the part out of the press area before installation of the mechanism. Now, as cylinder and electronic circuits were used for this purpose, the human intervention was reduced. This improved the process efficiency. From the quality point of view, it was observed that comparatively better-quality parts were obtained after the successful implementation of the above mechanism, and the rejection rate was reduced. The rate of production was also unaffected with approximately 30–40 parts getting pressed per minute. The organization was benefitted as the rejection of the part was reduced which improved

Fig. 9 Actual installation on a machine

their production efficiency, and better-quality product free from distortion and any other defect was obtained. For the parts produced, a completely flat surface free from unevenness and satisfying customer requirements was obtained.

5 Conclusion

By the installation of a low-cost automation mechanism at the rotor press machine, it was observed that bending and distortion defects can be completely removed from the rotor surface, and better-quality products can be obtained without having a significant effect on the rate of production. By the use of automation in daily production activities, reduced human intervention was achieved. This resulted in better production efficiency for the organization. Hence, a low-cost automation mechanism implemented correctly and at a proper location can result in better production quality, better production rate, better efficiency, and better profit for an organization.

Acknowledgements The authors would like to acknowledge *ASA Industries, Noida* for their support and guidance during the work.

References

1. Seifermann S, Böllhoff J, Metternich J, Bellaghnach A (2014) Evaluation of work measurement concepts for a cellular manufacturing reference line to enable low-cost automation for lean machining, variety management in manufacturing. In: Proceedings of the 47th CIRP conference on manufacturing systems
2. Müller-Polyzou R, Meier N, Georgiadis A (2020) Case study analysis of laser-assisted low-cost automation assembly. IFAC Papers Online 53(2):10403–10410
3. Kihlman H, Ossbahr G, Engström M, Anderson J (2004) Low-cost automation for aircraft assembly. SAE technical paper 2004-01-2830
4. Gajmal SS, Sudhir GB (2014) Low-cost automation (LCA): a case study. Int J Adv Eng Glob Technol 2(12)
5. Frohm J, Granell V, Winroth M, Stahre J (2006) The industry's view on automation in manufacturing. In: Conference paper, May 2006
6. Erbe H-H (2003) Manufacturing with low-cost automation. IFAC Automatic systems for building the infrastructure in developing countries, Istanbul, Republic of Turkey
7. Gamberini R, Gebennini E, Rimini B, Spadaccini E, Zilocchi D (2009) Low-cost automation and poka yoke devices: tools for optimising production processes. Int J Prod Qual Manag 4(5–6)
8. Bjornsson A, Lindback J-E, Eklund D, Jonsson M (2016) Low-cost automation for prepreg handling—two cases from the aerospace industry. SAE Int J Mater Manuf 9(1)
9. Tamizharasi G, Kathiresan S, Balaji D, Jegathiesan S (2015) Low cost automation in a power press. Indian J Sci Technol 8(32)
10. Fast-Berglund Å, Salunkhe O, Åkerman M (2020) Low-cost automation—changing the traditional view on automation strategies using collaborative applications. IFAC Papers Online 53(2)
11. Bortolini M, Faccio M, Gamberi M, Pilati F (2020) "Station-sequence" parts feeding in mixed models assembly: impact of variations and industry 4.0 possible solutions. IFAC Papers Online 53(2)
12. Erbe H (2003) Introduction to low cost/cost effective automation. Robotica 21(3):219–221
13. Albertos P (1998) Trends in low-cost automation. IFAC Proc 31(25)
14. Karunakaran SS, Pushpa V, Akula S (2010) Low-cost integration of additive and subtractive processes for hybrid layered manufacturing. Robot Comput Integr Manuf 26(5)
15. Dev AKr (2019) Low-cost automation for small to medium shipyards. Paper, SNAME maritime convention, Tacoma, Washington, USA, Oct 2019
16. Minchala LI, Peralta J, Mata-Quevedo P, Rojas J (2020) An approach to industrial automation based on low-cost embedded platforms and open software. J Appl Sci
17. Daundkar MM, Savant BN (2016) Low-cost automation solution for paper bag production using commodity hardware. In: 2016 IEEE International conference on recent trends in electronics, information & communication technology (RTEICT)
18. Sathish T, Isaac Premkumar IJ, Saravanan R, Basker S, Parthiban A, Vijayan V (2020) Multiply of process speed, quality, and safety through low-cost automation—a case study. AIP Conf Proc 2283:020067
19. Kopacek P (1992) Low-cost factory automation. IFAC Proc 25(25)
20. Erbe HH (2002) Low-cost intelligent automation in manufacturing. In: 15th Triennial World congress, Barcelona, Spain, 2002 IFAC

Numerical Investigation on Phase Change Material (PCM) for Thermal Management Buildings Using Design-Builder Software

S. Syath Abuthakeer, D. Arunkumar, M. Ramu, and K. Sripriyan

Abstract The building sector consumes 35% energy for the cooling/heating of the buildings. To diminish the energy consumption of the buildings, it is necessary to advance the research on passive cooling/heating methods. The development of new and advanced materials for building components will provide a lesser environmental impact. The phase change materials (PCM) integrated with the building wall can store the latent heat thermal energy system. This will be reducing and shift the thermal load on buildings during the transition of phase change solid to liquids. In this present work, the real-time residential building is modeled and simulated using Energy Plus software with and without the use of PCM residential buildings. The building energy scenario for PCM and non-PCM buildings is discussed in warm and humid (Chennai), moderate (Bangalore), and composite (Delhi) climatic conditions. The PCM buildings are the major reduction of annual electricity consumption, annual energy use intensity, and annual energy demand are compared to non-PCM buildings. The PCM building HVAC's load is reduced by around 40% than that of non-PCM buildings. The PCM building is more suitable for providing indoor thermal comfort for the occupants.

Keywords Design-builder · Phase change materials · Energy savings in building · Energy demand

S. Syath Abuthakeer (✉) · D. Arunkumar · K. Sripriyan
Department of Mechanical Engineering, PSG College of Technology, Coimbatore 641 004, India
e-mail: ssa.mech@psgtech.ac.in

K. Sripriyan
e-mail: ksp.mech@psgtech.ac.in

M. Ramu
Department of Mechanical Engineering, Amirta Vishwa Vidyapeetham University, Coimbatore 641 112, India
e-mail: m_ramu@cb.amrita.edu

U. S. Dixit et al. (eds.), *Advances in Forming, Machining and Automation*, Lecture Notes in Mechanical Engineering, https://doi.org/10.1007/978-981-19-3866-5_49

1 Introduction

In the last decade, the cooling and heating energy demand for buildings is substantially increased in the world. The buildings sector consumes more than 30% of electricity in India being second to industries. It has been projected that the total built space in the country increases fivefold from 2005 to 2030. The commercial building space is increased with 60% of the building is to be air-conditioned [1]. The fast improvement of the economic growth around the world and high living standards like green living environment and reducing the carbon levels is imposing an increasing demand for energy-efficient buildings. To minimize the usage of energy is one of the most important factors and is being needed for the public. The faster reduction of greenhouse gas emissions, as well as protection of the environment [2], is very important. The indoor temperature and electrical energy consumption of the buildings are affected by various factors such as orientation, building shape, wall color, thermal insulation, glazing material, thermal mass, window size, shading devices, green roof system [3]. One essential method that can be used to reduce the temperature of indoor conditions and energy consumption of the building is to incorporate the phase change materials (PCMs) in the buildings [4]. The thermal energy storage system is used to store the energy in PCM through a solidification process during the nighttime. During the daytime, the ambient temperature increases, the PCM can be releasing the stored energy [5]. The simulation for PCM [6] and non-PCM ventilating system for space cooling during the summer season in Beijing was studied. They indicated that, higher the indoor set point, it required more cooling energy for PCM and the structural elements. The PCM building can reduce the cooling from 16.9 to 50.8% compared to conventional ventilated buildings. Experimental and Energy Plus simulation of PCM residential buildings were [7] analyzed. For the PCM, room wall temperature reduces 1.5 ºC compared to the reference building. The cooling load reduction was 63% in bedroom 2 and 28% for bedroom 1. The effect of PCM in light [8] weight official buildings with an HVAC system is used in hot-arid climatic conditions using a purge ventilation system.

The PCM is applied in various building elements. It significantly reduces the cooling load. Energy Plus simulation for residential buildings [9] with free and passive cooling methods of PCM integrated thermal energy storage system of the building is analyzed. During the summer season, free cooling results show that the reduction of indoor temperature is 1.8 °C compared with passive cooling methods. The experimental analysis of PCM integrated with the building walls [10] can make the comfort temperature for occupants so that the temperature fluctuation is stabilized and decrease the cooling load for the air-conditioning system. The incorporation of PCM in building walls can be decreased [11] the energy consumption of buildings in different cities in Australia. The Energy Plus simulation software can be used for the investigation of the effect of PCM in buildings. For each city, the optimum melting temperature of PCM has been identified. The results show that the PCM in buildings under cold temperature, mild temperature, and warm temperature is reduced the energy consumption of buildings. In the hot-humid climatic conditions, the effect of

PCM in buildings has minor energy consumption. The PCM effect on the buildings depends on PCM layer thickness, local climate, surface area, thermostat range, and application of the buildings. The selection of melting temperature of PCM is outside the comfort range, and it does not deliver the energy reduction irrespective of a large reduction in temperature variations. PCM incorporated in the building wall [12] can reduce the cooling load for air-conditioners in Singapore through the Energy Plus simulation tool. The PCM layer was used 10 mm thickness, and the melting temperature of PCM is selected for 28 °C is added to the external building surface of vertical wall. PCM buildings can effectively reduce the cooling load throughout the year of 21–32%.

The selection of suitable PCM transition temperature is critical. For the tropical climatic conditions, the PCM can be incorporated into the external wall surface shows the performance is better compared applied to the inner surface of the wall because fewer energy penalties are predictable from the solidification of PCM at nighttime. Thermal performance of [13] enhancement of PCM integrated composite building was analyzed with experimental and numerical simulation of Energy Plus software. The reduction of temperature is 3.4 °C for the PCM cold storage room, and this can be better by nighttime ventilation. Innovative double-layer PCM in buildings for the country of Iran [14] was simulated using Energy Plus software. This method can deliver thermal comfort and reduce the energy demand. The first layer of PCM was chosen for the melting point of PCM and the second layer of PCM was chosen for the average room air temperature. To attain the better energy performance of double-layer PCM, the first layer of the melting point of the PCM during the winter season in Iran, 1 °C should be lower for the desired indoor set point should be selected. During the summer season, the temperature of 2–3 °C is higher than the summer set point. The heating energy of 17.5% reduced in warm and dry climates and 10.4% in mild climates. The cooling load energy is decreased by 12.3% in cold climatic conditions and 9.8% in mild climatic conditions.

PCM integrated into the residential flat was constructed in its bedroom and the living room was modeled and simulated using Energy Plus in a subtropical city, Hong Kong [15]. The energy performance and thermal analysis of PCM integrated buildings were simulated. The effect of PCM residential buildings is a decrease in indoor temperature. The living room can attain the annual energy savings in electricity consumption of the air-conditioning system. The cooling load for the PCM incorporated bedroom can be increased as associated with the reference case. The PCM integrated into the building wall surface interior temperature always stays above 28 °C.

In this paper, a real-time south-facing residential building is situated in warm and humid climatic conditions (Chennai city13°4 N and 80°14 E) is modeled and analyzed with and without the use of PCM in the building envelope. The same has been followed for the Bangalore (moderate climate 12°58 N and 77°34 E), and New Delhi (composite climate 28°38 N and 77°13 E) with Energy Plus software tool used to simulate different climatic conditions in India. A comparative residential building model with and without the use of PCM was simulated for the above three climatic conditions.

2 Methodology

2.1 Description of a Simulated Building

A typical single-story double bedroom residential building was chosen for the effect of PCM enhancement in the building. The building model has created the size of 365.6 m^2 (9.14 m length × 12.19 breath and 3.3 m height) and simulated with the actual building model and exact floor as shown in Figs. 1 and 2. This building is occupied with six people for the simulated period. The building is designed with and without the use of PCM construction materials under three climatic conditions (warm-humid, moderate, and composite). The floor area of the building is 365 m^2. The double glazing window of size is 1.2 × 1.2 m and the distance of 0.9 m distance between each window's bottom from the floor. The main door is located on the southern wall. The roof and walls are exposed to environmental conditions. The ratio for the window to wall area in southern and northern sides are 18% and 11%, respectively. The selected ratio is within the civil building thermal design code range [16]. The window glass thickness of 3 mm was chosen and had the solar transmittance of 0.48, whereas the visible light transmittance and the thermal conductivity of the glass were 0.8 and 0.9 W/m K, respectively.

The PCM layer is arranged in all the external building walls of the residential buildings. The schematic diagram of PCM and non-PCM walls is shown in Figs. 3 and 4. The thermophysical properties of the PCM, details of the building layer,

Fig. 1 Building model

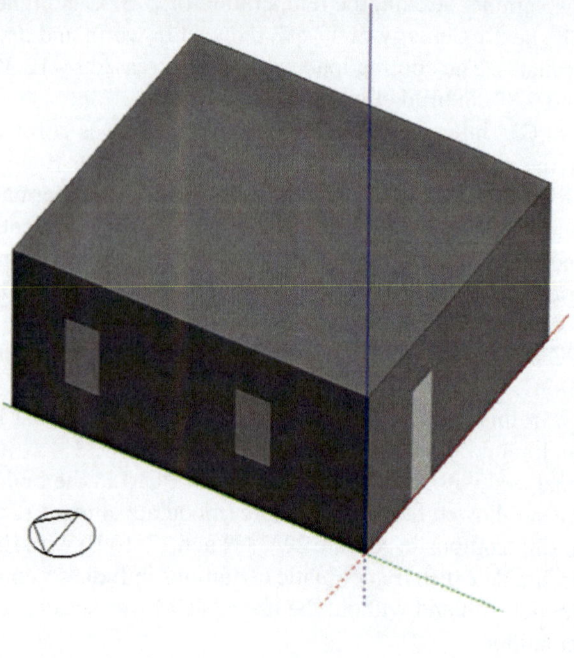

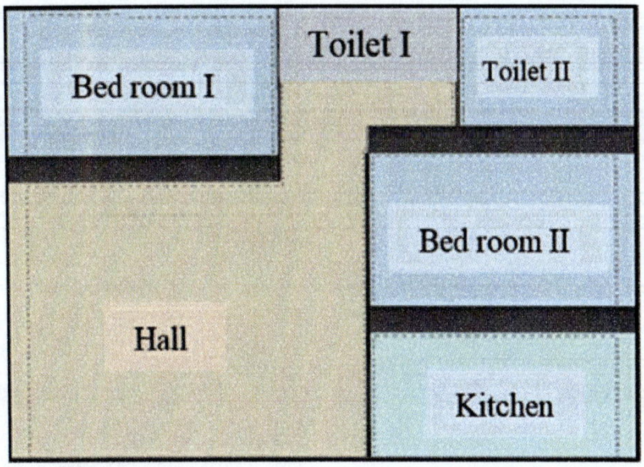

Fig. 2 Floor plan

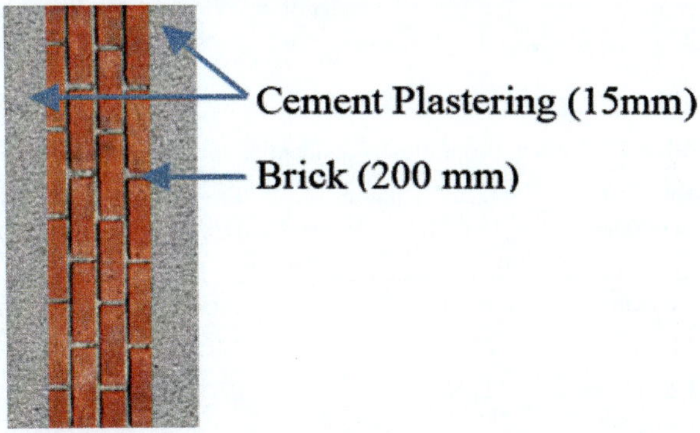

Fig. 3 Non-PCM building wall layer

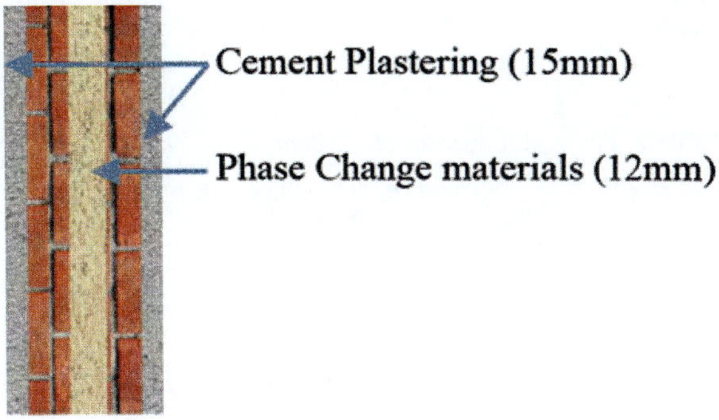

Fig. 4 PCM building wall layer

and building materials properties are given in Tables 1, 2, 3, and 4, respectively. The effects of PCM building for cooling/heating energy performances with thermal comfort chamber was studied. Two residential building models were used for the

Table 1 Properties of phase change materials (PCMs)

Parameter	Value
Melting point (° C)	29
Conductivity (W/m K)	0.172
Density (kg/m^3)	976
Specific heat (J/kg K)	1390
Latent heat (J/kg)	194

Table 2 Details of the residential apartment building envelope [17]

Envelope	Construction (inside to outside)
Floor	100 mm reinforced concrete
Wall	15 mm mortar, 200 mm brick, 15 mm mortar
Wall with PCM	15 mm mortar, 200 mm brick, 25 mm PCM,15 mm mortar
Roof	15 mm mortar,100 mm reinforced concrete, 15 mm mortar

Table 3 Physical properties of residential apartment building envelope

Building materials	Density (kg/m^3)	Specific heat (J/kg K)	Thermal conductivity (W/m K)	Thermal resistance (m^2 K/W)
Cement	1600	1050	0.84	0.15
Mortar	1800	1.05	0.93	0.01
Reinforced concrete	2500	0.92	1.74	0.02
Thermal insulating materials	1000	0.92	0.085	0.07

Table 4 Physical properties of the door

Materials	Thickness of the door (m)	Thermal conductivity *K* (W/m K)	Density (kg/m^3)	Specific heat (J/kg K)
Plywood door	0.0381	0.12	510	1380

simulation. One residential building without PCM in the wall is considered for reference and is compared to PCM integrated into the building wall. The PCM was incorporated in the brick wall with 25 mm thickness. Both the sides are done the cement plastering with the thickness of 15 mm as shown in Figs. 3 and 4.

2.2 Numerical Model

The numerical simulation of the residential building was carried out using Energy Plus v8.1 software for dynamic building energy simulation [18]. In this software, PCM can be added to the building wall layer by using conduction finite difference (CondFD) solution algorithm which will discretize the building wall envelope into different nodes and solve the heat transfer equation numerically using a finite difference method (FDM), which will be will be selected for Crank–Nicolson or fully implicit equation. In this present study, a fully implicit discretization technique can be used to simulate the PCM in buildings. To simulate the PCM layer, the specific heat change process is considered for the CondFD method. This method is integrated with the enthalpy-temperature function which gives the inputs of enthalpy at different temperatures as shown in Eq. 1 [19]. Each iteration, the enthalpies of each node can get updated and each time step specific heat is developed. The accuracy of the simulation is considered to use the time step of for one minute [20].

$$C\rho\Delta x\frac{T_i^{j+1}-T_i^j}{\Delta t}=\left[k_W\frac{(T_{i+1}^{j+1}-T_i^{j+1})}{\Delta x}+k_E\frac{(T_{i-1}^{j+1}-T_i^{j+1})}{\Delta x}\right] \tag{1}$$

where

q = density (kg/m^3),
Cp = specific heat capacity (kJ/kg K),
T = nodal temperature (k),
$j+1$ = new time step for calculation,
j = previous time step for calculation,
i = modeled node,
$i+1$ = adjacent node to interior,
$i-1$ = adjacent node to exterior,
k = thermal conductivity (kW/m K).

2.3 Validation

The PCM building and CondFD models of Energy Plus software are validated at different test conditions [20] which consist of the comparative test, systematic verification, and experiential validation [21]. The Energy Plus PCM algorithm was validated compared to experimental investigation data by other researchers [22]. Additionally, the experimental data and the PCM model of Energy Plus software were validated [21, 22]. Moreover, the Energy Plus simulation of the PCM layer model was validated compared to experimental data by other researchers [21].

2.4 Air-Conditioning System

A package air-conditioner was selected for providing cooling/heating to buildings. The supply fan and motor efficiency were selected as 0.7 and 0.9, respectively. The energy efficiency ratio (EER) of the cooling coil has 2.52, and the coefficient of performance (COP) of the heating coil is 2.75 as recommended by ASHRAE 90.1 standard [21, 22]. The HVAC system operates throughout the year. The indoor temperature set point is between 27 °C for cooling applications.

2.5 Results and Discussions

Table 5 shows the residential modeled building's annual energy use intensity (EUI). The EUI means the building's annual energy usage per unit area concerning the floor area of the buildings. The energy usage of the PCM is substantially reduced compared to the non-PCM buildings. The energy savings per unit area of the building in warm and humid (Chennai), moderate (Bangalore), and Composite (Delhi) weather conditions are 29%, 8%, and 32%, respectively, in PCM buildings. It is due to the lower thermal conductivity of PCM. This PCM is not allowing the solar heat to gain into the buildings during the daytime. The simulated building energy use intensity is summarized in Table 5 for the PCM and non-PCM buildings. The energy used in the three climatic conditions of PCM residential buildings is minimum compared to the non-PCM buildings. The net site energy reduced for warm and humid, moderate, and composite climates are 34.9%, 18.2%, and 32.5%, respectively. Similarly, annual energy savings for the warm-humid (Chennai), moderate (Bangalore), and composite climates (Delhi) are 3219.57, 820.25 and 2565.9 kWh/h, respectively.

Figures 5 and 6 show the annual electricity usage and month-wise percentage distribution of electricity utilization for several loads of PCM and non-PCM buildings. The annual electricity usage of PCM buildings is always lower than the non-PCM buildings for the cities Chennai, Bangalore, and Delhi as shown in the Fig. 5a, b, and c. The electricity consumption of PCM buildings in Bangalore city is minimum

Table 5 Energy use intensity of the modeled building

Climate conditions	Annual net site energy (kWh/year)		Annual specific energy consumption (kWh/m^2)	
	PCM buildings	Non-PCM buildings	PCM buildings	Non-PCM buildings
Chennai (warm-humid)	9204.6	5985	82	53.4
Bangalore (moderate)	4500.3	3680	40.18	32.85
Delhi (composite)	7895	5329.1	70.49	47.5

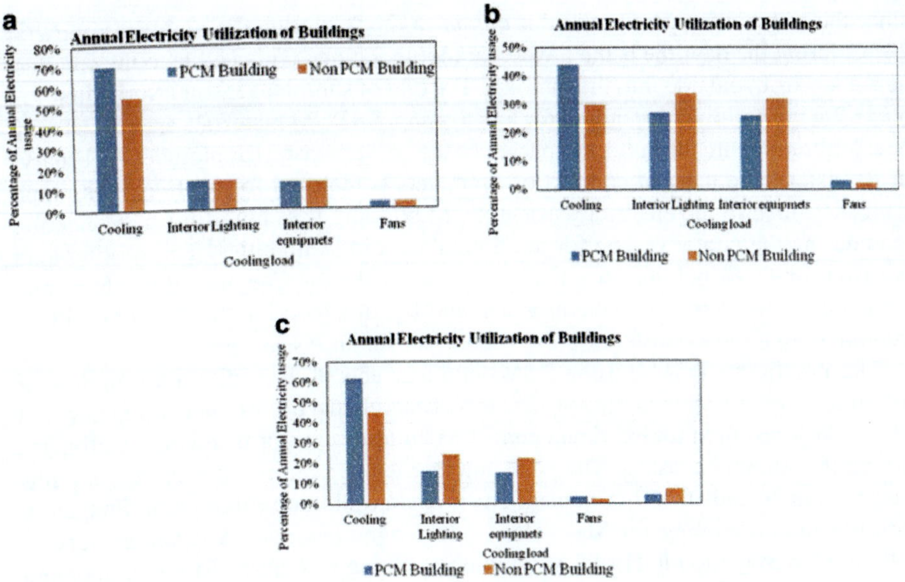

Fig. 5 **a** Residential building's annual electricity utilization for warm-humid climate (Chennai), **b** residential building's annual electricity utilization for moderate (Bangalore) climate, and **c** residential building's annual electricity utilization for composite climate (Delhi)

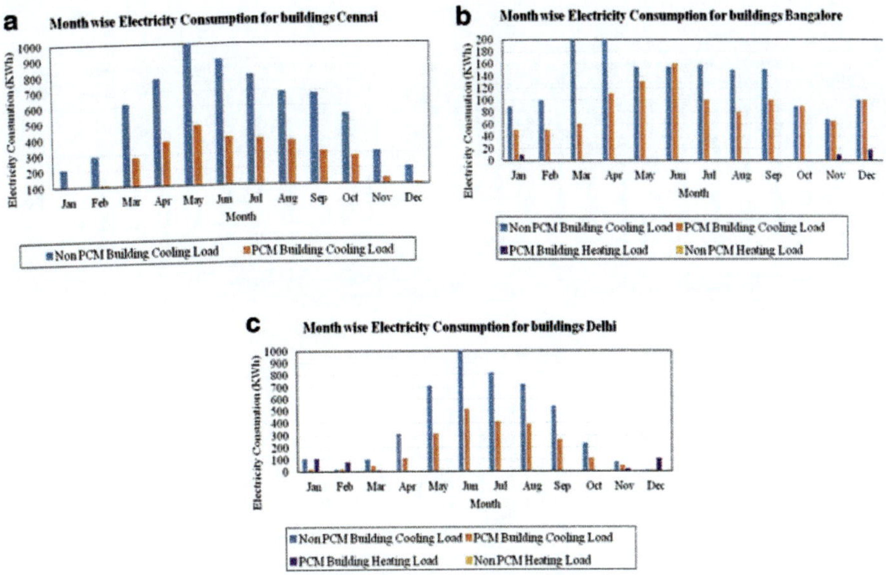

Fig. 6 **a** Month-wise building's annual electricity consumption of buildings for warm-humid climate (Chennai), **b** month-wise building's annual electricity consumption of buildings for moderate climate (Bangalore), **c** month-wise building's annual electricity consumption of buildings for composite climate (Delhi)

since the ambient temperature in this city is 15–20 °C during the nighttime and 26–33 °C during the daytime is the lowest for all other cities. This is due to the smallest heat entry from outside into the building. For end of summer season (April–June) in the PCM, the building is required for less heating. So PCM materials are allow lesser heat infiltration into the building space. For the winter season (September–February) in Bangalore, the thermal comfort of occupants is required for heating is essential which is added to the electricity usage. PCM building is required for a 3% heating load during the winter season for the Bangalore and Delhi cities. The cooling load is saved for PCM building for 13%, 14%, and 17% for Chennai, Bangalore, and Delhi cities, respectively. The equipment and lighting load for PCM and non-PCM buildings are almost constant for the winter and summer seasons.

The month-wise PCM and non-PCM buildings' annual electricity consumption for different cities as shown in Fig. 6a, b, c. It is observed that a constant cooling demand of 2.8 kW is required for PCM and non-PCM buildings in warm and humid climates during the summer seasons. The peak ambient temperature in the summer months ranges from 36 and 40 °C in Chennai. The moderate climatic condition in Bangalore city is required cooling for May. This is due to the peak ambient temperature of 34–36 °C in May month. For PCM buildings, heating is required for December and January due to lesser ambient temperature. Also, 3 kW demand for PCM and 4 kW demand for non-PCM are required for hot summer months as shown in Fig. 7a, b, c. Heating demand is also required for the winter season (November–February) due to lesser ambient temperature. The average annual electricity demand was saved for

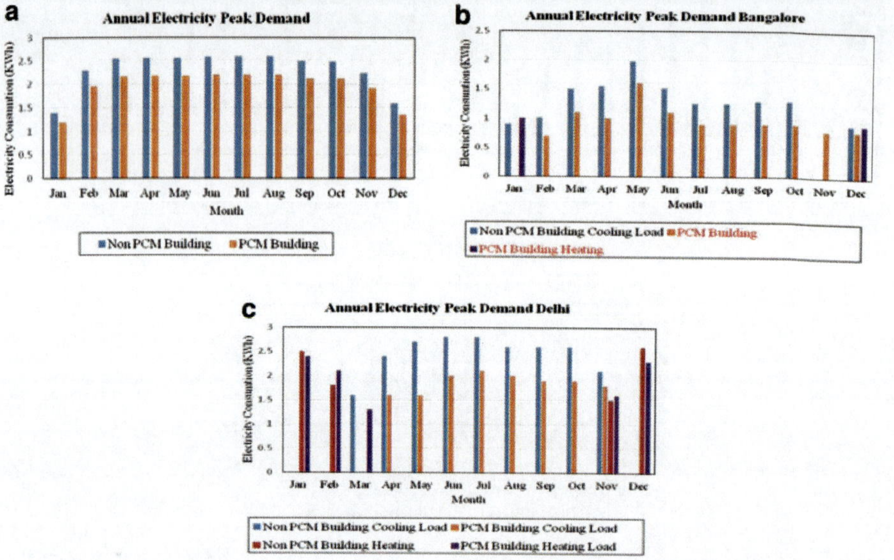

Fig. 7 Residential building's month-wise electricity peak demand. **a** residential building's month-wise electricity peak demand (Chennai), **b** residential building's month-wise electricity peak demand (Bangalore), and **c** residential building's month-wise electricity peak demand (Delhi)

warm-humid, moderate, and composite climate conditions of 884.23 W, 574 W, and 301.2 W, respectively, for summer months and the winter season 389 W, 321 W, and 289 W, respectively. The PCM room is significantly reduced the overall electrical consumption and annual electrical demand was observed as shown in Fig. 6a, b, c.

Figure 7a, b, c shows the residential building's month-wise electricity peak demand per annum for the PCM and non-PCM buildings. It is observed that a stable cooling demand of 2.5 kW is required for non-PCM buildings during the summer months from March to September. This is due to the high rise temperature in the summer season in the range of 35 and 39 °C in Chennai. This demand was reduced to 2 kW in the PCM building due to the lower thermal conductivity. The interior equipment, lightning, and fan power have not changed in PCM and non-PCM buildings. The maximum summer temperature during May in moderate climatic conditions of Bangalore city is observed. The heating demand is required for December and January.

Figure 8a, b, c shows the month-wise HVAC cooling load profile for PCM and non-PCM buildings for a whole year. The HVAC load for the PCM building has decreased considerably compared to the non-PCM buildings approximately half of the building of HVAC load for all climatic conditions. For warm and humid climatic conditions, peak cooling and peak heating are required for moderate and composite climates. The PCM buildings are transferred a lesser amount of heat into the buildings and also reduce the HVAC load.

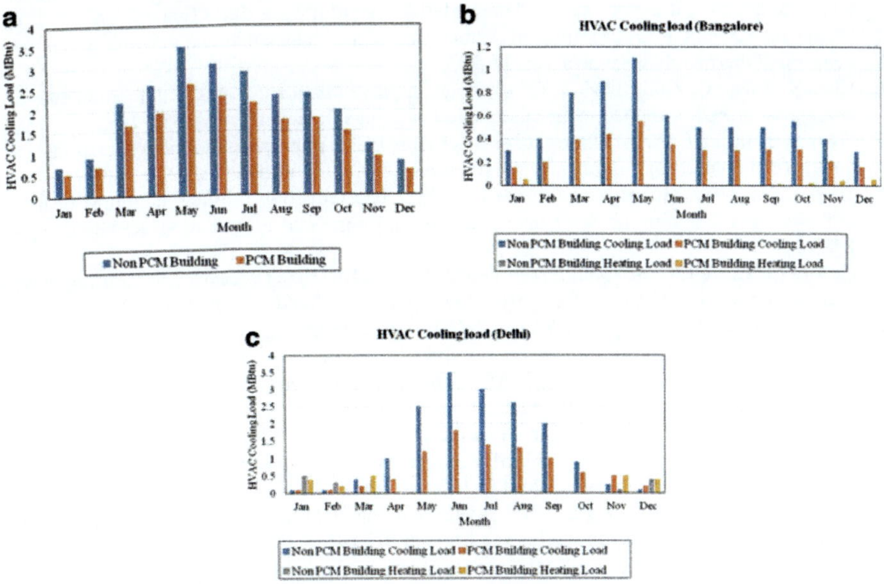

Fig. 8 **a** Residential building's month-wise HVAC load profile (Chennai), **b** residential building's month-wise HVAC load profile (Bangalore), and **c** residential building's month-wise HVAC load profile (Delhi)

3 Conclusion

The residential building integrated with PCM and without PCM in buildings is evaluated the annual energy consumption using Energy Plus software tool. A residential building is modeled and simulated using the Energy Plus software. The PCM and non-PCM buildings for warm-humid, moderate, and composite climate conditions are simulated. The annual energy utilization and electricity usage of residential buildings are compared with PCM buildings. The space heating loads are additionally added for the moderate and composite climate. A significant percentage of energy consumption was reduced in PCM buildings. From the simulation analysis, it might be decided that the PCM can be used for future residential building applications for providing thermal comfort for occupants.

References

1. Axaopoulos I, Axaopoulos P, Gelegenis J, Fylladitakis ED (2019) Optimum external wall insulation thickness considering the annual CO2 emissions. J Build Phys 42(4):527–544
2. McKinsey (2009) Environmental and energy sustainability: an approach for India 90
3. Sripriyan K, Ramu M, Thyla PR (2017) Effect of flat electrode and fuzzy logic model for the prediction of weldment shape profile in GMAW. J Mech Sci Technol 71(5):2477–2486
4. Vautherot M, Maréchal F, Farid MM (2015) Analysis of energy requirements versus comfort levels for the integration of phase change materials in buildings. J Build Eng 1:53–62
5. Pasupathy A, Velraj R (2008) Effect of double-layer phase change material in building roof for year-round thermal management 40:193–203
6. Chen X, Zhang Q, Zhai ZJ, Ma X (2019) Potential of ventilation systems with thermal energy storage using PCMs applied to air-conditioned buildings. Renew Energy 138:39–53
7. Nghana B, Tariku F (2016) Phase change material's (PCM) impacts on the energy performance and thermal comfort of buildings in a mild climate. Build Environ 99:221–238
8. Solgi E, Fayaz R, Mohammad B (2016) Cooling load reduction in office buildings of hot-arid climate, combining phase change materials and night purge ventilation. Renew Energy 85:725–731
9. Cabeza LF, Castell A, Martorell I, Medrano M, Pe G (2010) Experimental study of using PCM in brick constructive solutions for passive cooling. Energy Build. 42(4):534–540
10. Alam M, Jamil H, Sanjayan J, Wilson J (2014) Energy-saving potential of phase change materials in major Australian cities. Energy Build 78:192–201
11. Lei J, Yang J, Yang E (2016) Energy performance of building envelopes integrated with phase change materials for cooling load reduction in tropical Singapore. Appl Energy 162:207–217
12. Sripriyan K, Ramu M, Thyla PR, Anantharuban K, karthigha M (2022) Characteristic of weld bead using flat wire electrode in GMAW inline during the process: An experimental and numerical analysis. Int J Press Vessels Pip 196:104623
13. Gao Y, Gao W, Meng X, Long E (2017) Influence of the PCM layer location on the multilayer wall thermal performance. Open J Energy Effic 06(01):1–13
14. Pedersen CO (2007) Advanced zone simulation in energy plus: incorporation of variable properties and phase change material (PCM) capability. Build Simul 1341–1345
15. Tabares-Velasco PC, Christensen C, Bianchi M (2012) Verification and validation of energy plus phase change material model for opaque wall assemblies. Build Environ 54:186–196
16. Sripriyan K, Ramu M, Thyla PR, Anantharuban K (2021) Weld bead characterization of flat wire electrode in GMAW process part II: a numerical study. 35(5):1–8

17. Sheeja R, Kumar S, Chandrasekar P, Jospher AJ, Krishnan S (2020) Numerical analysis of energy savings due to the use of PCM integrated in lightweight building walls. IOP Conf Ser: Mater Sci Eng 923:012070
18. Sripriyan K, Ramu M, Palani PK (2016) Study and analysis of the macrostructure characteristics in FCAW with the use of a flat wire electrode and by optimizing the process parameter using the Taguchi method and regression analysis. High Temp Mater Process 20(3):197–224
19. Medrano M, Cabeza LF, Castello C, Leppers R, Zubillaga O (2007) Use of microencapsulated PCM in concrete walls for energy savings. 39:113–119
20. Sailor DJ (2014) Evaluation of phase change materials for improving thermal comfort in a super-insulated residential building. Energy Build 79:32–40
21. Sripriyan K, Ramu M (2017) An experimental investigation of flat wire electrodes and their weld bead quality in the FCAW process. High Temp Mater Process 21(1):65–79
22. Kuznik F, Virgone J (2009) Experimental assessment of a phase change material for wall building use. Appl Energy 86(10):2038–2046

Performance Measurement for Integrated Lean Six Sigma and Industry 4.0—A Case Study

Somishang A. Shimray and S. Vinodh

Abstract In this fast-paced environment, to be able to compete in the market, new technology and innovation play a vital role. To keep the companies competitive, many of them are adopting I4.0 technologies and innovative methodologies like lean six sigma (LSS). Integration of I4.0 and LSS and adopting it will bring innovation, increased response time, and will bring about continuous improvement to the company. In order to have better judgment of the situation in an organization, performance measurement is important. Here, the study is focused on performance measurement for integrated LSS and I4.0. To analyze the performance of the project, a model is developed consisting of two levels. The first level consists of 9 enablers and the second level of 36 criteria, and each enabler is associated with criteria. The fuzzy logic approach is used to analyze the performance of integrated LSS and I4.0 and found that the performance level is around 60% which shows that the performance level is strong. FPII is also calculated to find the ranking score of criteria. Weaker areas are identified.

Keywords Lean manufacturing · Six sigma · Lean six sigma · Performance measurement · Fuzzy logic

1 Introduction

In this competitive era, outdated technology and processes have become obsolete and cannot compete in a rapidly changing world. To compete in this kind of environment and keep the company as the frontrunner in the market, upgrading technology and bringing innovation to the company play a vital role. Technology advancement pushes companies to innovate processes, create new value chains, and help bring satisfaction to the company [1, 2]. Due to the rapid growth of information technology (IT), the amalgamation of physical and virtual entities is done through cyber-physical systems

S. A. Shimray · S. Vinodh (✉)
Department of Production Engineering, National Institute of Technology Tiruchirappalli, Tiruchirappalli, India
e-mail: vinodh_sekar82@yahoo.com

U. S. Dixit et al. (eds.), *Advances in Forming, Machining and Automation*, Lecture Notes in Mechanical Engineering, https://doi.org/10.1007/978-981-19-3866-5_50

(CPS), which initiated Industry 4.0. Utilization of CPS, I4.0 restructured manufacturing sector in industrial value creation [3]. The aim of I4.0 is to revolutionize industrial manufacturing by making use of new technology and digitalization. Thus, I4.0 permits customized products and personalized products as it is an adjustable production system. In recent times, the market has been pushed by global competition, and there is a constant demand for quick response type of production to meet constantly upgrading and changing demands. To meet this type of demand, one needs to upgrade the technologies [4–6]. The main focal point of lean is waste reduction and reducing lead time, while six sigma is a data-driven methodology that focuses on bestowing guidance to the organization and includes tools and techniques for maintaining continuous improvement and keeping the customer satisfied. Lean six sigma (LSS) integrates lean and six sigma concepts in a concerted way to facilitate process excellence. Lean and six sigma combined dictates value by inscribing efficiency and effectiveness. To assess the situation of the organization and to see how it is performing, performance measurement is needed. This allows the company to see how they are performing in the competition and see how they are performing with their company's objective. Better performance leads to profit for an organization and gives an advantage over others. All companies strive for better performance and profit.

The study has been done in an automotive component manufacturing industry where LSS methodology has been implemented, and there exists a need to integrate I4.0 technology to be more competitive in this ever-growing market. To analyze the project that integrates LSS and I4.0, a model is developed consisting of two levels, the first level being an enabler and the second level being criteria. An intensive literature review on performance measurement of lean, six sigma, LSS, and I4.0 is conducted, and nine enablers and 36 criteria are selected; performance level is assessed based on integrated LSS and I4.0 using the fuzzy logic approach. Based on the analysis, performance level has been found as stronger and weaker areas are analyzed for improvement.

2 Literature Review

Shah and Shrivastava [7] developed and validated performance measures in the Indian context for LSS. They made an instrument for helping the execution of processes in small and medium-scale enterprises (SMEs). They stated that manufacturing organizations could use this validated instrument of LSS factors to execute LSS in SMEs. It can also be used for evaluating the emphasis of management efforts. They stated that there are some limitations in their work as the questionnaire is subjective in nature, and ratings given by the interviewee are based on their own understanding and up to how much it is applicable to their own company. The authors also stated that there might be limitations as the phrasing of each item will have an influence on the results during construct verification based on factor analysis. More work needs to be done to overcome these kinds of limitations. Sangwa and Sangwan [8] proposed a framework for performance measurement of lean organizations. They discussed how to implement and practice the developed framework. They stated that the proposed framework would be helpful for managers, and implementation of lean in their organization will be favored. They stated that more work needs to be done to rationalize the concept and knowledge of lean performance measurement in other enterprises. Yin and Qin [9] investigated a smart design performance measurement approach to assist performance measurement execution in an integrated design process. They suggested a smart design performance measurement way to assist smart product design project management and also created an I4.0 technology framework. Raval et al. [10] developed LSS balanced scorecard (BSC) performance measurement system. The authors looked into important measures which are currently in use by Indian manufacturing companies. From this study, the authors showed that for LSS to succeed, performance measurement is required. Also, it is helpful in formulating performance metrics guidelines. The authors revealed that this study is done to help manufacturing companies find out their weaknesses and will serve as a benchmark. It will also help them in concentrating on the most important and suitable criteria and objectives. Ruben et al. [11] analyzed the performance of LSS systems combined with sustainability using fuzzy-based approaches. They designed a performance assessment model. From the analysis, they found out that LSS performance with sustainability concerns possessed strong performance. The authors stated that further work could be done by adding more criteria to make evaluation more exhaustive and to carry out the study in different manufacturing organizations. Kamble et al. [12] used an integrated method of investigative and experimental research design. The authors pointed out and verified performance measures for the evaluation of smart manufacturing system investments in auto-component manufacturing SMMEs based in India. They found out that there is more profit for an I4.0-enabled SMS than for a traditional manufacturing system. The authors further suggested investigating organizations having different lean levels.

Based on the literature review, it has been found that there are studies on LSS performance measurement and Industry 4.0 measurement; but concrete studies on integrated LSS and I4.0 performance measurement are not available to the best knowledge of authors. In line with industrial developments, performance measures for integrated LSS and I4.0 are to be developed and validated, which forms the motivation for the present study.

3 Methodology

The methodology of this work is shown in Fig. 1. Literature review is carried out on performance measurement of lean, six sigma, LSS, and I4.0. From the literature survey and expert opinion, a conceptual model for performance measurement is derived, as shown in Table 1. The model consists of two levels, the first level includes nine performance measurement enablers, and the second level includes 36 performance measurement criteria. Each enabler is connected with the second-level criteria.

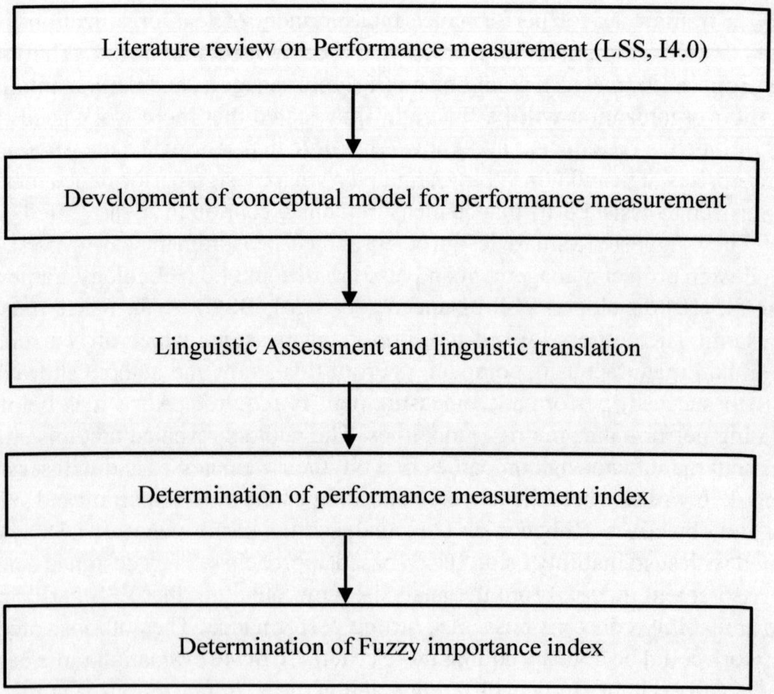

Fig. 1 Methodology

4 Case Study

The case study is done in an automobile component manufacturing company where they have integrated I4.0 technology with LSS methodology. To assess how the organization is performing, performance measurement is needed. The detailed case study done is presented below.

The linguistic scale is fixed for evaluating the performance ratings and importance weights of performance measurement after discussion with two experts from the company who have vast experience and knowledge. Table 2 shows the linguistic scale of performance ratings and importance weights [14].

To measure the performance ratings and importance weights, linguistic terms are used. The expert provided ratings which are in linguistic terms and are shown in Table 3. The linguistic term is transformed into fuzzy numbers using the relationship between linguistic terms and fuzzy numbers. Table 4 shows the approximated fuzzy numbers.

Table 1 Conceptual model for performance measurement

S. No.	Enablers	Criteria	Literature support
1	Defect	Prevalence of defect monitoring system	Kamble et al. [12], Ruben et al. [11]
		Ability to recognize CTQ measures	Ruben et al. [11]
		Incorporation of poka yoke mechanism	Sangwa and Sangwan [8], Ruben et al. [11]
		Defects of critical products/components	Kamble et al. [2], Bhasin [13]
2	Waste	Existence of waste analysis practice	Ruben et al. [11]
		Inclusion of environmental waste into the waste category	Ruben et al. [11]
		Costing system with a focus on value addition	Shah and Shrivastava [7], Ruben et al. [11]
3	Employee	Strong employee spirit and cooperation	Ruben et al. [11], Kamble et al. [12]
		Multi-skilled personnel	Ruben et al. [11], Shah and Shrivastava [7]
		Training effectiveness	Ruben et al. [11]
		Employees motivation and cultivating LSS culture	Raval et al. [10]
		Employee involvement	Yin and Qin [9], Raval et al. [10]
		No. of employee suggestions received	Shah and Shrivastava [7]
		Safety of employees	Kamble et al. [12], Bhasin [13]
		Build high morale within the team	Kamble et al. [12], Shah and Shrivastava [7]
4	Safety	Execution of corrective action	Ruben et al. [11]
		Compliance with safety standards	Ruben et al. [11]
		Health and safety per employee	Bhasin [13], Kamble et al. [12]
5	Business	Rate of incorporating customer response adoption	Ruben et al. [11]
		Prevalence of continuous improvement culture	Ruben et al. [11]
		Contribution to improving GDP	Kamble et al. [12], Ruben et al. [11]
		New technology development	Bhasin [13]

(continued)

Table 1 (continued)

S. No.	Enablers	Criteria	Literature support
6	Environmental impacts	Energy monitoring audit	Ruben et al. [11]
		Carbon footprint tracking	Ruben et al. [11]
		Adoption of life cycle assessment	Sangwa and Sangwan [8], Ruben et al. [11]
		Institutionalizing R3 and R6 cycles	Ruben et al. [11]
7	Customer perspective	Customer involvement	Raval et al. [10]
		Customer alignment	Raval et al. [10]
		Responsiveness	Bhasin [13]
		On-time delivery	Sangwa and Sangwan [8], Bhasin [13]
8	Process perspective	Innovation process	Yin and Qin [9], Raval et al. [10]
		Optimal resource utilization	Kamble et al. [12], Sangwa and Sangwan [8]
9	Supplier perspective	Supplier participation	Raval et al. [10]
		Supplier development	Raval et al. [10]
		Supplier quality management	Shah and Shrivastava [7], Raval et al. [10]

Table 2 Linguistic variables and fuzzy numbers [14]

Performance rating (*R*)		Importance weighting (*W*)	
Linguistic variables	Fuzzy number	Linguistic variables	Fuzzy number
Worst (W)	(0, 0.5, 1.5)	Very low (VL)	(0, 0.05, 0.15)
Very poor (VP)	(1, 2, 3)	Low (L)	(0.1, 0.2, 0.3)
Poor (P)	(2, 3.5, 5)	Fairly low (FL)	(0.2, 0.35, 0.5)
Fair (F)	(3, 5,7)	Medium (M)	(0.3, 0.5, 0.7)
Good (G)	(5, 6.5, 8)	Fairly high (FH)	(0.5, 0.65, 0.8)
Very good (VG)	(7, 8, 9)	High (H)	(0.7, 0.8, 0.9)
Excellent (E)	(8.5, 9.5, 10)	Very high (VH)	(0.85, 0.95,1.0)

Table 3 Linguistic performance ratings and importance weights (excerpt)

PC_i	PC_{ij}	W_i	W_{ij}	R_{ij}
PC_1	PC_{11}	H	V.H	G
	PC_{12}		H	F
	PC_{13}		H	G
	PC_{14}		FH	G
	PC_{15}		H	F
PC_2	PC_{21}	VH	H	G
	PC_{22}		FH	F
	PC_{23}		H	G
PC_3	PC_{31}	H	H	VG
	PC_{32}		VH	G
	PC_{33}		H	F
	PC_{34}		H	F
	PC_{35}		H	G
	PC_{36}		H	G
	PC_{37}		H	F
	PC_{38}		FH	F
PC_4	PC_{41}	FH	H	G
	PC_{42}		FH	G
	PC_{43}		FH	F

4.1 *Determination of Fuzzy Performance Measurement Index (FPMI)*

Fuzzy indices of the first level, i.e., criteria, are calculated, and they are extended to second-level enablers. Using Eq. 1, the fuzzy index of each enabler is calculated.

$$PC_{ij} = \sum_{k=l}^{n}\left(W_{ij}{*}PC_{ij}\right) \Big/ \sum_{k=l}^{n} W_{ij} \tag{1}$$

where PC_{ij} is the performance rating of second-level criteria, and W_{ij} represents the fuzzy importance weight of performance measurement criteria. Using the same equation, the fuzzy index of performance measurement enablers has been obtained. Table 5 represents fuzzy index of each enabler for performance measurement.

FPMI needs to be coordinated with proper performance measurement levels. Euclidean form is the most commonly used distance method. The natural level expression set LSSI4.0L = {extreme performance (E.P.), strong performance (S.P.), fair performance (F.P.), low performance (L.P.), poor performance (P.P.)}. By using Eq. 2, the Euclidean distance '*D*' from FPMI to each member set in PML is calculated.

Table 4 Linguistic terms approximated by fuzzy numbers (excerpt)

PC_i	PC_{ij}	W_i	W_{ij}	R_{ij}
PC_1	PC_{11}	(0.7,0.8,0.9)	(0.85,0.95,1.0)	(5,6.5,8)
	PC_{12}		(0.7,0.8,0.9)	(3,5,7)
	PC_{13}		(0.7,0.8,0.9)	(5,6.5,8)
	PC_{14}		(0.5,0.65,0.8)	(5,6.5,8)
	PC_{15}		(0.7,0.8,0.9)	(3,5,7)
PC_2	PC_{21}	(0.85,0.95,1.0)	(0.7,0.8,0.9)	(5,6.5,8)
	PC_{22}		(0.5,0.65,0.8)	(3,5,7)
	PC_{23}		(0.7,0.8,0.9)	(5,6.5,8)
PC_3	PC_{31}	(0.7,0.8,0.9)	(0.7,0.8,0.9)	(7,8,9)
	PC_{32}		(0.85,0.95,1.0)	(5,6.5,8)
	PC_{33}		(0.7,0.8,0.9)	(3,5,7)
	PC_{34}		(0.7,0.8,0.9)	(3,5,7)
	PC_{35}		(0.7,0.8,0.9)	(5,6.5,8)
	PC_{36}		(0.7,0.8,0.9)	(5,6.5,8)
	PC_{37}		(0.7,0.8,0.9)	(3,5,7)
	PC_{38}		(0.5,0.65,0.8)	(3,5,7)
PC_4	PC_{41}	(0.5,0.65,0.8)	(0.7,0.8,0.9)	(5,6.5,8)
	PC_{42}		(0.5,0.65,0.8)	(5,6.5,8)
	PC_{43}		(0.5,0.65,0.8)	(3,5,7)

Table 5 Fuzzy index of each grade for performance measurement

PC_i	R_i
PC1	(5.25, 5.90, 7.60)
PC2	(4.47, 6.06, 7.69)
PC3	(4.18, 5.97, 7.63)
PC4	(4.41, 6.03, 7.68)
PC5	(5.53, 6.89, 8.25)
PC6	(4.20, 5.83, 7.53)
PC7	(5.67, 7.00, 8.33)
PC8	(4.00, 5.50, 7.50)
PC9	(3.76, 5.54, 7.34)

$$D(\text{FLSSI4.0}I,\ \text{LSSI4.0LL}_i) = \left\{ \sum_{\text{xep}} (f_{\text{FPMI}}(x) - f_{\text{PML}}(x))^2 \right\}^{1/2} \quad (2)$$

where

$D(\text{FLSSI4.0}I, \text{LSSI4.0}L_i)$	Euclidean distance between FPMI and PMLi.
$\text{FLSSI4.0}I$	Fuzzy performance index.
$\text{LSSI4.0}L_i$	Related fuzzy number for natural language.
$f_{\text{FPMI}}(x)$	TFN of FPMI.
$f_{\text{PML}}(x)$	TFN of PML.

Euclidean distance calculated value is shown below:
D(FLSSI 4.0I, E) = 4.98.
D(FLSSI 4.0I, S) = 1.65.
D(FLSSI 4.0I, F) = 2.14.
D(FLSSI 4.0I, L) = 5.52.
D(FLSSI 4.0I, P) = 8.96.

5 Results

By matching the linguistic variable with minimum Euclidean distance (D), the performance measurement level of integrated LSS and I4.0 is found to be around 60%, and thus, the performance is strong. In order to identify weaker areas, a fuzzy performance importance index (FPII) is calculated. Value of FPII less than 1.63 shows weaker performance. FPII is calculated using the equation

$$\text{FPII}_{ij} = W'_{ij} \times R_{ij} \tag{3}$$

$FPII_{ij}$	is FPII for ijth attribute.
W'_{ij}	Complement of ijth attributes importance weight.

Where $W'_{ij} = \left[(1, 1, 1) - W_{ij}, W_{ij}\right]$ is the fuzzy importance weight of performance of integrated LSS and I4.0.

Fuzzy numbers do not always give a set order, unlike real numbers do. Thus, FPII has to be ranked [15].

$$\text{Ranking score} = \frac{a + 4b + c}{6} \tag{4}$$

The centroid method is considered for ranking the fuzzy number. The equation for calculation is represented in Eq. 4. Where a, b, and c denote lower, middle, and upper numbers of fuzzy triangular numbers. The fuzzy performance importance index of performance measurement of integrated LSS and I4.0 is shown in Table 6.

Table 6 FPII of criteria

PCij	R_{ij}	$W'_{ij} = [(1, 1, 1) - W_{ij}]$	FPII	Ranking score
PC_{11}	(5,6.5,8)	(0.0,0.05,0.15)	0.0, 0.32, 1.2	0.41
PC_{12}	(3,5,7)	(0.1,0.2,0.3)	0.3, 1, 2.1	1.06
PC_{13}	(5,6.5,8)	(0.1,0.2,0.3)	0.5, 1.3, 1.8	1.25
PC_{14}	(5,6.5,8)	(0.2,0.35,0.5)	1, 2.27, 4	2.34
PC_{15}	(3,5,7)	(0.1,0.2,0.3)	0.3, 1, 2.1	1.06
PC_{21}	(5,6.5,8)	(0.1,0.2,0.3)	0.5, 1.3, 1.8	1.25
PC_{22}	(3,5,7)	(0.2,0.35,0.5)	0.6, 1.75, 3.5	1.85
PC_{23}	(5,6.5,8)	(0.1,0.2,0.3)	0.5, 1.3, 1.8	1.25
PC_{31}	(7,8,9)	(0.1,0.2,0.3)	0.7, 1.6, 2.7	1.63
PC_{32}	(5,6.5,8)	(0.0,0.05,0.15)	0.0, 0.32, 1.2	0.41
PC_{33}	(3,5,7)	(0.1,0.2,0.3)	0.3, 1, 2.1	1.06
PC_{34}	(3,5,7)	(0.1,0.2,0.3)	0.3, 1, 2.1	1.06
PC_{35}	(5,6.5,8)	(0.1,0.2,0.3)	0.5, 1.3, 1.8	1.25
PC_{36}	(5,6.5,8)	(0.1,0.2,0.3)	0.5, 1.3, 1.8	1.25
PC_{37}	(3,5,7)	(0.1,0.2,0.3)	0.3, 1, 2.1	1.06
PC_{38}	(3,5,7)	(0.2,0.35,0.5)	0.6, 1.75, 3.5	1.85
PC_{41}	(5,6.5,8)	(0.1,0.2,0.3)	0.5, 1.3, 1.8	1.25
PC_{42}	(5,6.5,8)	(0.2,0.35,0.5)	1, 2.27, 4	2.34
PC_{43}	(3,5,7)	(0.2,0.35,0.5)	0.6, 1.75, 3.5	1.85

6 Conclusion

- Fuzzy logic approach is used for assessing the level of performance of the production line, which integrates LSS and I4.0.
- The performance level from the calculation is around 60%, and thus, the performance level is strong. FPII is calculated to identify the weaker performance criteria.
- A value less than 1.63 shows weaker performing criteria, and measures must be taken to improve the weaker criteria.
- The findings from this work can be used by researchers and industrial managers as a base for integrating LSS with I4.0.
- The case study carried out is only for automotive component manufacturing organizations. In future, studies could be done in different industries.

References

1. Klovienė L, Uosyte I (2019) Development of performance measurement system in the context of Industry 4.0: a case study. Eng Econ 30:472–482
2. Krishnan S, Gupta S, Kaliyan M, Kumar V, Garza-Reyes JA (2021) Assessing the key enablers for Industry 4.0 adoption using MICMAC analysis: a case study. Int J Prod Perform Manage
3. Huang Z, Kim J, Sadri A, Dowey S, Dargusch M (2019) Industry 4.0: development of a multi-agent system for dynamic value stream mapping in SMEs. J Manufact Syst
4. Aggarwal A, Gupta S, Ojha MK (2019) Evaluation of key challenges to industry 4.0 in Indian context: a DEMATEL approach. In: Advances in industrial and production engineering, pp 387–396
5. Sivakumar K, Mathivathanan D, Nishal M, Vimal KEK, Subhaa R, Pandian RS, Varela MLR (2020) Ensuring sustainability in industry 4.0: implementation framework. In: Sustainable manufacturing for industry 4.0, pp 215–252
6. Sodhi H (2020) When industry 4.0 meets lean six sigma: a review. Ind Eng J
7. Shah PP, Shrivastava RL (2013) Identification of performance measures of lean six sigma in small and medium-sized enterprises: a pilot study. Int J Six Sigma Compet Adv 8(1)
8. Sangwa NR, Sangwan KS (2017) Development of an integrated performance measurement framework for lean organizations. J Manufact Technol Manage
9. Yin Y, Qin SF (2019) A smart performance measurement approach for collaborative design in Industry 4.0. Adv Mech Eng
10. Raval SJ, Kant R, Shankar R (2019) Benchmarking the lean six sigma performance measures: a balanced scorecard approach. Benchmarking Int J 26(6):1921–1947
11. Ruben B, Vinodh S, Asokan P (2019) Application of multi-grade fuzzy and ANFIS approaches for performance analysis of lean six sigma system with sustainable considerations. Int J Serv Oper Manage
12. Kamble SS, Gunasekaran A Ghadge A, Raut R (2020) A performance measurement system for industry 4.0 enabled smart manufacturing system in SMMEs-a review and empirical investigation. Int J Prod Econ
13. Bhasin S (2008) Lean and performance measurement. J Manuf Technol Manag 19(5):670–684
14. Singh Patel B, Samuel C, Sharma SK (2017) Evaluation of agility in supply chains: a case study of an Indian manufacturing organization. J Manuf Technol Manag 28(2):212–231
15. Vinodh S, Vimal K (2011) Thirty criteria based leanness assessment using fuzzy logic approach. Int J Adv Manufact Technol

Predicting the Conveying Velocity of C-Shaped Parts on a Trap-Based Linear Vibratory Feeder

S. Udhayakumar, M. Saranya, and S. Reethika

Abstract Presenting the parts to the workstation in the required position is an important issue in the automated manufacturing system. For a long time, part feeders have been used in the industry to present parts in the desired position. The design of the industrial parts feeder is a complex process that involves trial and error methods, which may consume time in months. In this paper, the development of a trap-based parts feeds system that can handle C-shaped parts is discussed. The conveying velocity of the parts on the feeder is predicted using an artificial neural network (ANN) model. The model was developed through experimental results. The prediction model results were compared with the experimental results. The work shows that the ANN model can predict the conveying velocity of parts on the trap-based vibratory feeder without much deviation from the experimental results with a maximum average deviation of 3.7%. Such models are needed in industries to tune the input parameters according to the required conveying speed and maintain an uninterrupted flow of parts. In addition, the prediction of velocity is very important for adjusting the speed of the robot for material handling.

Keywords Trap · Vibratory feeder · Conveying velocity · Artificial neural network

1 Introduction

About 40% of labor and 50% of manufacturing costs are utilized for assembly and production [1]. Parts feeder can have parts in any position but provide output in a

S. Udhayakumar (✉)
Department of Mechanical Engineering, PSG College of Technology, Coimbatore, India
e-mail: suk.mech@psgtech.ac.in

M. Saranya
Department of I & C Engineering, PSG College of Technology, Coimbatore, India
e-mail: msa.ice@psgtech.ac.in

S. Reethika
Department of AMCS, PSG College of Technology, Coimbatore, India
e-mail: 20x031@psgtech.ac.in

U. S. Dixit et al. (eds.), *Advances in Forming, Machining and Automation*, Lecture Notes in Mechanical Engineering, https://doi.org/10.1007/978-981-19-3866-5_51

uniform position. The vibratory feeder is suitable for feeding parts in many industries such as machinery, electrical, pharmaceutical, bearings, optics, and fasteners. Berretty et al. discussed a series of mechanical devices, such as wipers and slots for filtering polygonal parts on the track [2]. They called them 'traps'. They had a series of mechanical barriers (gates), and they either reject parts that are mispositioned or repositioned the parts as per requirements. The former is called a passive trap, and the latter is called an active trap. Since active traps have an efficiency of 100%, they are preferred over passive traps [3]. These traps are installed at the exit of the vibratory feeder. Research work on trap-based vibratory feeders is limited [4–6]. Reznik et al. [7] discussed part movement on a part feeder with a longitudinally vibratory plate. Ramalingam and Samuel [8] analyzed the behavior of small parts on a linear vibratory feeder. The mathematical model developed to predict the conveying speed considered the impact of amplitude and frequency of vibration [8, 9], ignoring other parameters such as track angle and trap angle. The use of ANN to develop predictive models is extensively discussed in the literature [10–14]. From the literature survey, it could be found that work on models to predict the conveying velocity on a trap-based vibratory feeder for asymmetric parts is limited. The main objective of this paper is to determine the conveying velocity of parts on a trap using the ANN model. The full factorial experiment conducted on the trap-based vibratory parts feeder is discussed considering four factors at four levels. Using ANN, a model is developed to predict the conveying velocity of C-parts on the feeder taking into account the frequency of vibration (f), the amplitude of vibration (A), track angle (θ), and trap angle (α). The results of the predictive model were validated using experimental results.

2 Trap-Based Part Feeder

In a typical brake liner (C-shaped part) manufacturing industry, pellets and chemicals are preformed into sheets. A cutting machine cuts the sheet is cut into small pieces of brake liners which as C-shaped as in Fig. 1. The brake liners are then sent for subsequent processes such as grinding (internal and grinding), chamfering, and

Fig. 1 C-shaped part

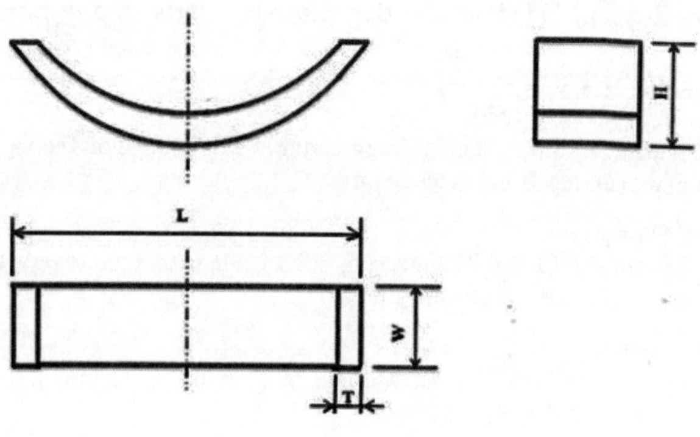

inspection. At each stage of operation, the parts are to be positioned properly and stacked for further processing. Without a proper parts feeding system, manpower will be wasted on positioning and stacking of parts. It is necessary to predict the conveying speed of parts at each stage of the process to avoid inventory and maintain an uninterrupted flow. To overcome this problem, Udhayakumar and Mohanram [15] developed a trap-based feeding and stacking system. The brake liner (Fig. 1) was considered to be conveyed in a trap-based parts feeder. The dimensions of the brake liner are shown in Fig. 1. The eight possible resting positions of the brake liners are shown in Fig. 2, numbered 1–8 are classified as 'a', 'b', and 'c', as shown in Fig. 2. The efficiency of the trap could be increased if it is designed in such a way that the part coming out of the feeder is always in the most probable naturally resting position. Udhayakumar et al. [16] concluded that, based on theoretical methods and experimental drop tests, positions 6 and 8 are considered favorable positions for part feeder design. Several configurations of traps have been designed and studied, the best of which this shown in Fig. 3. The gates used in traps are wiper, guide blocks,

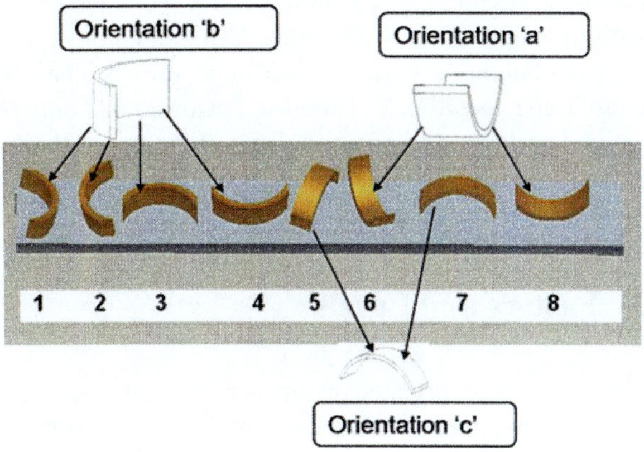

Fig. 2 Clubbing the positions of brake liner (C-shaped part)

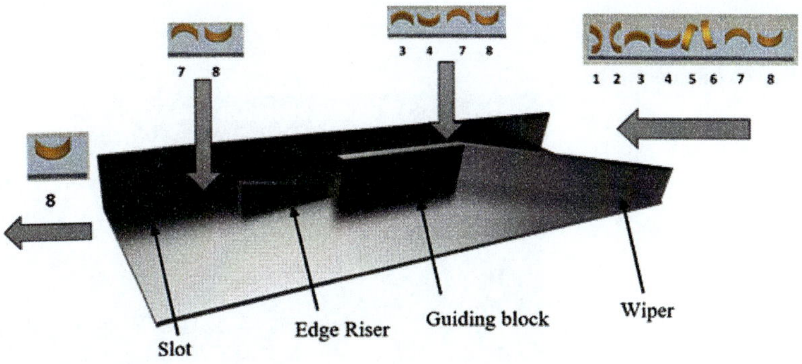

Fig. 3 Working of trap

edge risers, and slots. The movement of the parts on the trap is realized by the vibration provided by the vibratory feeder. The working of the trap (Fig. 3) is discussed as follows.

The entrance of the wiper blade is wider, but the exit is narrow. Whenever the parts with positions 1, 2, 5, and 6 (Fig.3) enter the wiper blade, they will be guided along the outline of the wiper blade and forced into a narrow area. To pass through the narrow area, positions 1, 2, 5, and 6 are converted to positions 3, 4, 7, and 8. The guiding block ensures that all parts whose positions are 3, 4, 7, and 8 reach the next gate without any change. The next gate is the edge riser. Parts with directions 7 and 8 are not disturbed by the edge riser because the gap between the wall and the edge riser is sufficient to allow them to pass. However, the parts in positions 3 and 4 cannot pass because their width is larger than the gap, so they must be moved above the edge riser. When the part with position 3 moves past the standing edge, its center of gravity falls outside the bottom, so it falls and changes to direction 8. Similarly, when the part in direction 4 moves past the edge riser, it is converted to direction 7. Output directions 7 and 8 move to the next step, which is the slot. The part with direction 7 falls into the slot when passing through the slot, and the part with direction 8 is not disturbed. When the conveyor belt is placed under the trap, the parts fall on the conveyor belt in position 8. The only undisturbed direction in the entire trap is position 8. Therefore, position 8 is only obtained at the exit of the trap, which is a favorable position. The efficiency of the trap is 100% because all input parts are switched to a favorable direction.

3 Factors and Levels

The trap was fabricated using acrylic material and installed on an electromagnetic vibratory feeder as shown in Fig. 4. LF II/059 type vibratory feeder and the microprocessor-based controller provided by M/s Elscint Automation, Pune were used for experimental research.

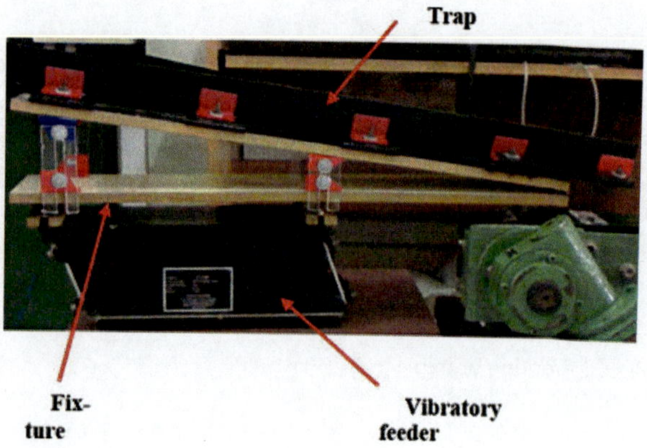

Fig. 4 Trap on vibratory feeder

The major parameters that influence the conveying velocity in a trap-based vibratory feeder are

1. Frequency of vibration, f
2. Amplitude of vibration, A
3. Track angle, θ
4. Trap angle, α

3.1 Frequency and Amplitude of Vibration

In the study of part motion on vibratory part feeders, the frequency and amplitude of vibration cannot be neglected.

3.2 Track Angle

To facilitate easy movement and effective positioning of parts, they should move along the track wall (Fig. 5). To achieve this, the track of the trap is slightly inclined relative to the track wall. Through initial experimental studies, a track angle (θ) of 10°–13° is recommended for effective repositioning.

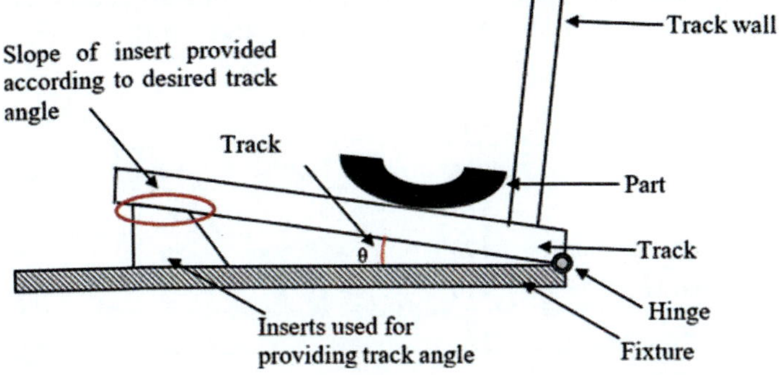

Fig. 5 Track angle

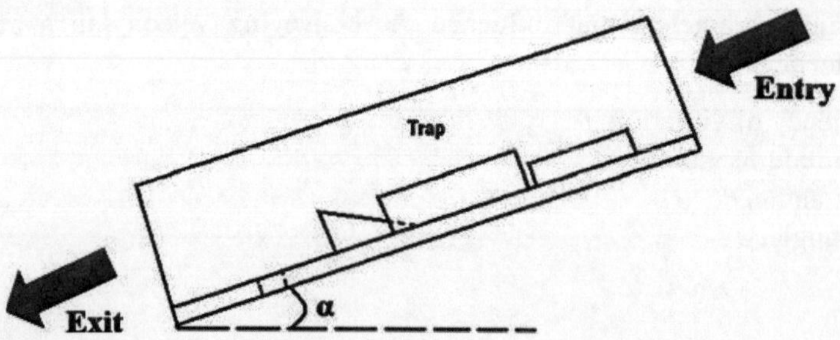

Fig. 6 Trap angle

3.3 Trap Angle

To move parts at a higher velocity, the trap has to be inclined as a whole as shown in Fig. 6. Trap angle (α) between 7° and 10° was found to be effective through experimental studies.

3.4 Levels are Chosen for Experimentation

The effect of friction between the trap and part was not taken into account since the material of the trap and the part was not changed during the experiment. Table 1 provides the details of factors and levels chosen for experimentation.

Four factors at four levels that led to 256 experiments were conducted. Since the trap handled parts in four different positions 3, 4, 7, and 8 (which were the positions at the exit of the wiper blade), the experiments were performed in each of these positions to find out the velocity. The video of part motion on the trap was recorded, and the time taken by the part from entry to exit of the trap was determined using a stopwatch. Under the influence of the above parameters, a model that predicts the conveying velocity of parts on the vibratory feeder is very important for determining throughput rates and maintaining an uninterrupted flow of parts at each stage of the process. The ANN model is developed based on the experimental results.

Table 1 Levels chosen

Parameter	Level-1	Level-2	Level-3	Level-4
Frequency (f), Hz	68	83	87	90
Amplitude (A), % of voltage input	70	75	80	85
Track angle (θ), degrees	10	11	12	13
Trap angle (α), degrees	7	8	9	10

4 Development of ANN Model

Considering the factors such as modeling ease and prediction accuracy, ANN techniques are proven to be more useful than statistical regression techniques. A neural network does not fall into a local minimum and can be trained faster to converge. Table 2 lists the parameters used to develop the ANN model to predict the conveying velocity of parts in directions 3, 4, 7, and 8. The program was developed using MATLAB software. The experimental results were used to train the network.

The learning rate of 0.01 and momentum of 0.95 were provided. The input combination of vibration frequency (f), vibration amplitude (A), track angle (θ), and trap angle (α) was provided into MATLAB software as input data. Taking the corresponding output values (conveying velocity in positions 3, 4, 7, and 8) as the target, the parameters listed in Table 3 were used to develop the network model. The ANN architecture is shown in Fig. 7. The number of neurons in the hidden layer is determined by the trial and error method [17] based on the R-value. The correlation between output and goal is represented by the R-value. The training, validation, and test data samples used to determine the number of neurons in the hidden layer are shown in Table 3. The hidden neurons were initially kept at 10 and then increased in increments of 10. The R-value obtained by changing the hidden neurons is shown in Fig. 8. It can be inferred from Fig. 8 that the hidden neuron size of 50 has the maximum R-value. Using this, the ANN network was trained again. Thus, an ANN model that can predict the conveying velocity based on input parameters has been developed.

Table 2 Parameters for developing ANN model

Structure and algorithm	Feed forward and back propagation
Type of training	Trainlm
Network structure	15, h and 1
Transfer function	TANSIG
Number of iterations	1000 (max epoch)
Performance function	MSE = 0.00001

Table 3 Samples for test and validation of data

S. No.	Description	Percentage of samples	Number of samples
1	For training	70	180
2	For validation	15	38
3	For testing	15	38

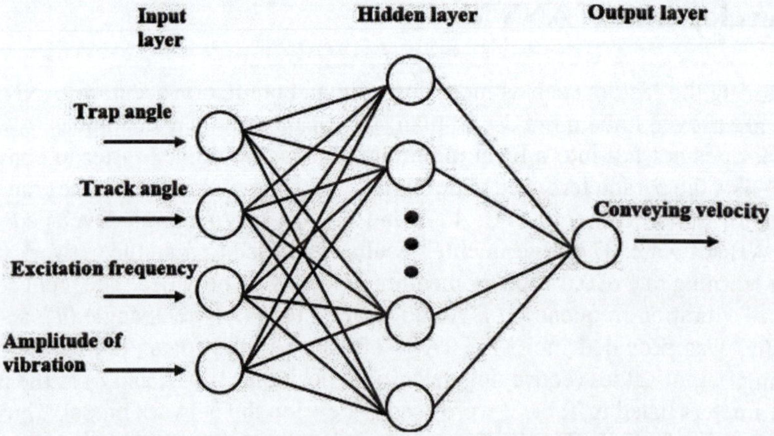

Fig. 7 ANN architecture

Fig. 8 Overall '*R*' versus no. of neurons in the hidden layer

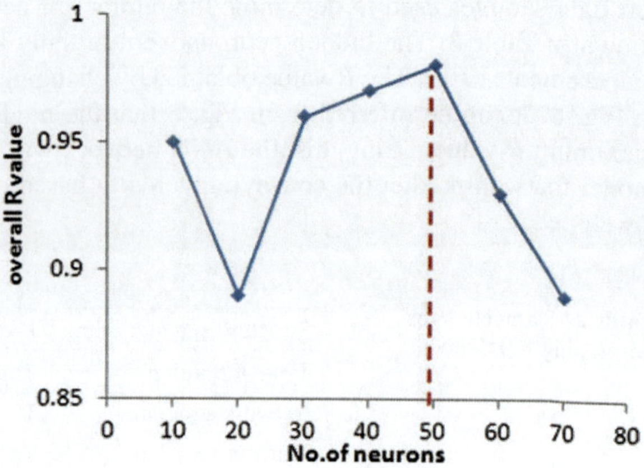

5 Validation of ANN Model

To validate the developed ANN model, twelve sets of inputs were randomly selected as listed in Table 4 and were compared.

Figures 9, 10, 11, and 12 show the comparison of conveying velocity for positions 3, 4, 7, and 8 determined experimentally and predicted by the ANN model.

It can be inferred from the results that the ANN model was able to predict the conveying velocity much closer to the experimental results. The average deviations for positions 3, 4, 7, and 8 predicted by the ANN model with experimental results are 2.5%, 2.8%, 3.72%, and 3.34%, respectively. ANN could predict more accurate results because they could make rules without any implicit formulae. They are non-parametric and do not make any assumptions about the distribution of data. They can 'let the data speak for itself'. In addition, they are robust to noise in training the

Table 4 Random set of input values considered for validation

Experiment no	Frequency, f (Hz)	Amplitude, A (% of voltage input)	Track angle, θ (degree)	Trap angle, α (degree)
E1	68	0.7	10	7
E2	90	0.8	10	9
E3	85	0.75	10	8
E4	66	0.7	11	8
E5	83	0.7	11	9
E6	90	0.7	12	7
E7	85	0.85	12	9
E8	87	0.8	12	9
E9	90	0.7	13	7
E10	87	0.85	13	8
E11	83	0.7	13	9
E12	70	0.85	13	7

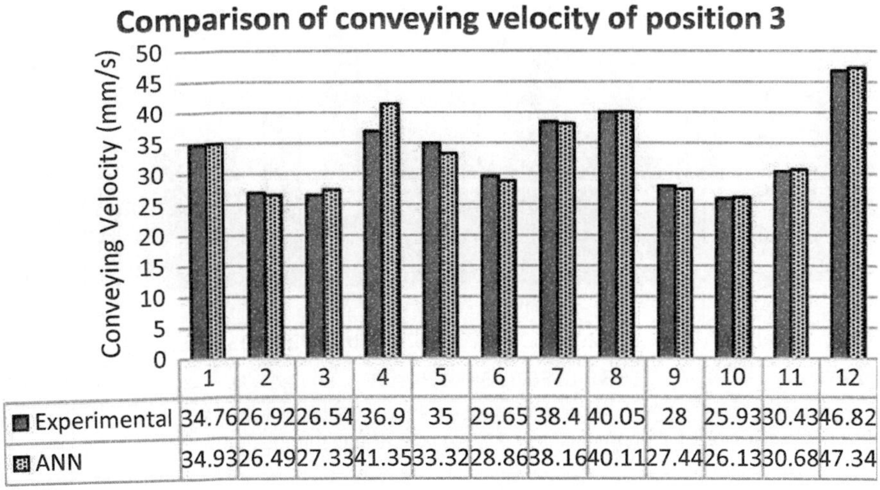

Fig. 9 Comparison of experimental and ANN model results for position 3

data and execute faster than many other models. Therefore, the developed artificial neural network model is suitable for predicting the conveying velocity of C-shaped parts on a trap-based vibratory feeder. This ANN model will be very useful in the industry to maintain uninterrupted flow by matching the conveying speed of parts with subsequent processes. However, for different sized/shaped parts, a different model has to be developed which is a major challenge for industries.

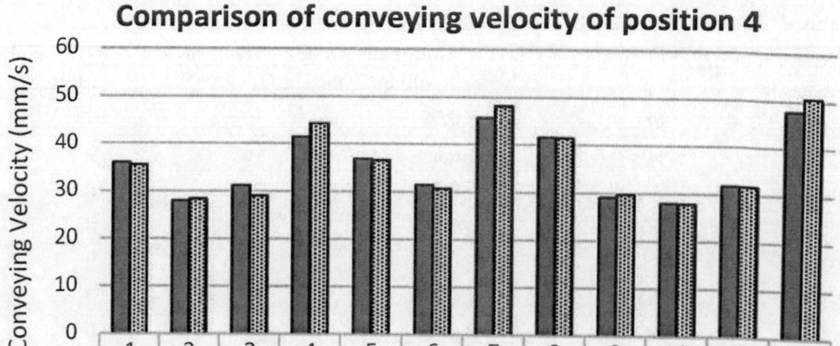

	1	2	3	4	5	6	7	8	9	10	11	12
Experiment	35.97	28	31.23	41.5	36.84	31.45	45.64	41.64	29.17	28	31.82	47.77
ANN	35.54	28.38	29.07	44.36	36.61	30.7	48.13	41.53	29.78	27.92	31.75	50.6

Fig. 10 Comparison of experimental and ANN model results for position 4

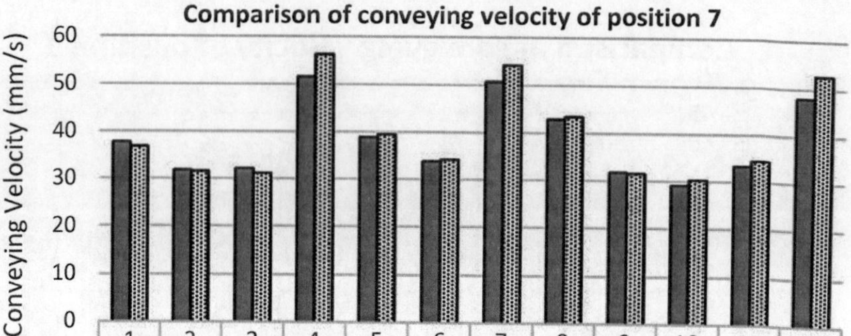

	1	2	3	4	5	6	7	8	9	10	11	12
Experiment	37.74	31.82	32.14	51.6	38.89	33.8	50.73	42.89	31.82	29.17	33.33	48
ANN	36.76	31.56	31.18	56.32	39.51	34.17	54.22	43.47	31.61	30.35	34.48	52.81

Fig. 11 Comparison of experimental and ANN model results for position 7

6 Conclusions

To enable parts to be presented in the most favorable position (position 8), an active trap was designed and developed. The trap included wiper blades, narrowing tracks, edge risers, and slots. The trap is fixed on the vibratory feeder, and a conveyor belt is installed below it. Full factorial experimentation was carried out to determine the conveying velocity of the C-shaped parts (positions 3, 4, 7, and 8) under the influence of frequency, vibration amplitude, track angle, and trap angle. The experimental results were used to develop an ANN model that predicts the conveying velocity of parts on the feeder. A predictive model considering the effect of excitation frequency

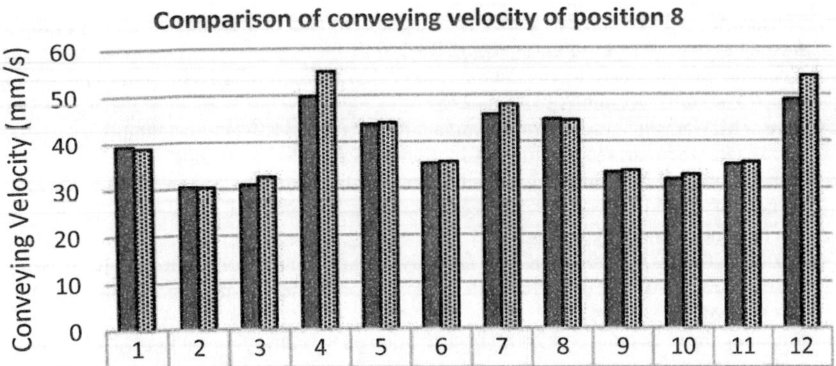

	1	2	3	4	5	6	7	8	9	10	11	12
Experiment	39.24	30.43	30.86	49.8	43.75	35.26	45.72	44.72	33.33	31.82	35	48.9
ANN	38.77	30.31	32.53	55.07	44.04	35.62	47.82	44.48	33.71	32.79	35.42	53.98

Fig. 12 Comparison of experimental and ANN model results for position 8

(f), amplitude (A), track angle (θ), and trap angle (α) was developed to determine the velocity of part positions 3, 4, 7, and 8. The results predicted by the model were compared with the experimental results. The average deviations between the ANN results and the experimental results of the part positions 3, 4, 7, and 8 were 2.5%, 2.8%, 3.7%, and 3.3%, respectively. It can be inferred from the results that the ANN model could predict the conveying velocity that is much closer to the experimental values. The accuracy in results could be attained because ANN is robust to noise and can formulate rules without any implicit formula. These types of models will be very useful in the industry to adjust the conveying speed of parts and maintain continuous flow. The major challenge would be developing a new model for different sized/shaped parts.

7 Scope for Future Work

The effect of part geometry (aspect ratio) on the conveying speed can be included to develop the predictive model. A much-generalized model to predict the conveying speed of parts with different geometric shapes may be more useful.

References

1. Boothroyd G (2005) Assembly automation and product design. CRC Press Taylor and Francis
2. Berretty RP, Kenneth Y, Goldberg K, Overmars MH, van der Stappen AF (2001) Trap design for vibratory bowl feeders. Int J Robot Res 20(11):891–908

3. Tay ML, Chua PSK, Sim SK, Gao Y (2005) Development of a flexible and programmable parts feeding system. Int J Prod Econ 98(2):227–237
4. Goemans OC, Goldberg K, van der Stappen FA (2006) Blades for feeding 3D parts on vibratory tracks. Assembly Autom 26(3):221–226
5. Suresh M, Narasimharaj V, Arul Navalan GK et al (2018) Effect of orientations of an irregular part in vibratory part feeders. Int J Adv Manuf Technol 94:2689–2702
6. Udhayakumar S, Mohan A, Gowthamachandran J, Prakash R, Shanmugam P (2021) Development of visionless flexible part feeder for handling shock absorbers. In: Lecture notes in mechanical engineering, pp 141–154
7. Reznik D, Canny J, Goldberg K (1997) Analysis of part motion on logitudinally vibrating plate. In: Proceedings of the IEEE/RSJ international conference on intelligent robots and systems (IROS), vol 1, pp 421–427
8. Ramalingam M, Samuel GL (2009) Investigation on the conveying velocity of a linear vibratory feeder while handling bulk-sized small parts. Int J Adv Manuf Technol 44:372–382
9. Lim GH (1997) On the conveying velocity of a vibratory feeder. Comput Struct 62(1):197–203
10. De Filippis LAC, Serio LM, Facchini F, Mummolo G (2017) ANN modelling to optimize manufacturing process 20th December 2017. https://doi.org/10.5772/intechopen.71237
11. Abiodun OI, Jantan A, Omolara AE, Dada KV, Mohamed NA, Arshad H (2018) State-of-the-art in artificial neural network applications: a survey. Heliyon 4(11):e00938. https://doi.org/10.1016/j.heliyon.2018.e00938
12. Ferrero Bermejo J, Gómez Fernandez JF, Olivencia Polo F, Crespo Marquez A (2019) A review of the use of artificial neural network models for energy and reliability prediction. A study of the solar PV, hydraulic and wind energy sources. Appl Sci 9:1844. https://doi.org/10.3390/app9091844
13. Asmara A (2020) A possibility of artificial neural networks to be applied in the predictive test: a systematic literature review and study case. J Phys Conf Ser 1456:012052
14. Africa ADM, Abaluna DAP, Abello AJA, Lalusin JMB (2020) Implementation of neural network control in a nonlinear plant using MATLAB. In: 2020 IEEE 12th international conference on humanoid, nanotechnology, information technology, communication and control, environment, and management (HNICEM), pp 1–6. https://doi.org/10.1109/HNICEM51456.2020.9400007
15. Udhayakumar S, Mohanram PV (2014) Design and development of an active trap for handling brakeliners. Ann Fac Eng Hunedoara–Int J Eng 9:151–156
16. Udhayakumar S, Mohanram PV, Keerthi Anand P, Srinivasan R (2014) Trap based part feeding system for stacking sector shaped parts. J Braz Soc Mech Sci Eng 36(2):421–431
17. Udhayakumar S, Mohanram PV, Keerthi Anand P, Srinivasan R (2013) Determining the most probable natural resting orientation of sector shaped parts. Assembly Autom 33(1):29–37

Review on Applications of Pneumatic Air Muscle

S. Udhayakumar, R. K. Bharath, N. Kowshik Santhakumar, and B. A. Mohamed Samsudeen Soofi

Abstract Pneumatic air muscles (PAMs) are lightweight, contractile, or extensional devices that were first developed in the 1950s under the name of McKibben Artificial Muscles for artificial limbs actuation. Commercially, available PAMs are generally in tubular structure, whereas other types of muscles have been developed in recent years. These muscles include wedge-shaped, flat nodular type, circular type, pleated type, etc. PAMs are available in different contraction ratios that can be used in various applications according to our requirements. They have greater power to weight ratio and as human muscles, work in an agonistic-antagonistic way, where compressed air will provide the necessary traction force. PAMs are also called short-stroke machines. Since human muscles work in agonistic-antagonistic pairs, pneumatic muscle-type actuators will play an important role in the development of an assistive rehabilitation robotics system. They can generate a great deal of force across a short distance. PAMs can contract by up to 30% with a minimal load. Due to these advantages, they are widely used for several applications. This paper provides a review of the applications of PAM.

Keywords Pneumatic air muscle · Rehabilitation · Medical robot

S. Udhayakumar · R. K. Bharath (✉) · N. Kowshik Santhakumar ·
B. A. Mohamed Samsudeen Soofi
Department of Mechanical Engineering, PSG College of Technology, Coimbatore 641 004, India
e-mail: 17m608@psgtech.ac.in

S. Udhayakumar
e-mail: suk.mech@psgtech.ac.in

N. Kowshik Santhakumar
e-mail: 17m622@psgtech.ac.in

B. A. Mohamed Samsudeen Soofi
e-mail: 17m628@psgtech.ac.in

U. S. Dixit et al. (eds.), *Advances in Forming, Machining and Automation*, Lecture Notes in Mechanical Engineering, https://doi.org/10.1007/978-981-19-3866-5_52

1 Introduction

Robotic systems are becoming increasingly prevalent today, and personal and mobile robots which assist humans are being developed. In recent years, pneumatic air muscle (PAM) plays a vital role in medical and delicate material handling applications. There are various types of muscles such as wedge-shaped pneumatic air muscles acting as sarcomeres that mimic the actual muscles in the human body and have the capability of bi-directional bending by regulating appropriate pressure levels. Xie et al. [1] designed a flat modular pneumatic muscle that can be lifted to an angle of more than 90° when pressurized. PAMs can provide better gripping. Al-Ibadi et al. [2] developed a circular muscle actuator that can lift a maximum load of 10.9 kg for the 6 cm height when pressurized. Manual rehabilitation exercise requires trained therapists. There are over 13.7 million new strokes each year of which 50% are affected by motor disabilities and 70% of people can be recovered by proper rehabilitation therapies. There is a need for rehabilitation robots that can be automated to restore the movement ability of hemiplegic patients. Since rigid motors lack compliance when getting in touch with patients, there is a need for rehabilitation robots by employing pneumatic muscle actuators. These muscles are soft and provide better compliance with human muscle patterns. PAM rehabilitation robots can be used to improve motor ability by providing rehabilitation exercises to various parts of the human body such as the wrist, upper limb, lower limb, and ankle. Due to the demand for a more compliant and interactive human–robotics system, there is a vast need in such fields in recent years. Therefore, this work discusses various types of PAMs, current rehabilitation orthoses techniques, and delicate material handling applications. Evaluation of control schemes, strategies, and theories has also been discussed.

2 Types of PAMs

Oguntosin and Akindele [3] designed a wedge-like pneumatic air muscle made of silicone rubber that would contract when loaded as shown in Fig. 1. To validate the design, isotonic and isometric exercises were carried out. The respective contraction parameters of both exercises were measured by visual processing. It also has the capability of bi-directional bending by regulating various pressure levels at each wedge unit of the proposed soft robotic muscle. The volume of the muscle can be increased by varying the wedge angle. The experimental validation showed that at zero loads, the velocity was maximum. The contractile strain of 67% was obtained. An increase in the number of wedge blocks would increase the contraction force thereby implemented in weight lifting or material handling applications where we need constant force with and without shortening the muscle.

To address the issue of friction and deformation of the muscle, Daerden and Lefeber [4] developed the braided pneumatic artificial muscles and pleated pneumatic

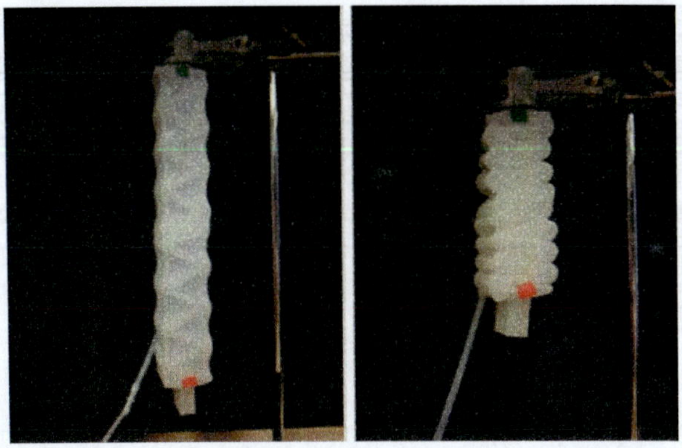

Fig. 1 Normal and contracted muscle [3]

artificial muscles (PPAM) which have an outer layer of aromatic polyamides that reduces friction. Veale et al. [5] developed a soft actuator Peano-fluidic muscle. As the muscle gets pressurized, it causes a reduction in length of about 15–30% which is flat before pressurizing. Belforte et al. [6] observed that textile soft actuators are used for active suits design with a small amount of force and contraction but braided pneumatic muscles are capable of producing more force than textile actuators. Textile muscles (Fig. 2) are operated at limited pressure to avoid the damage of repeatability

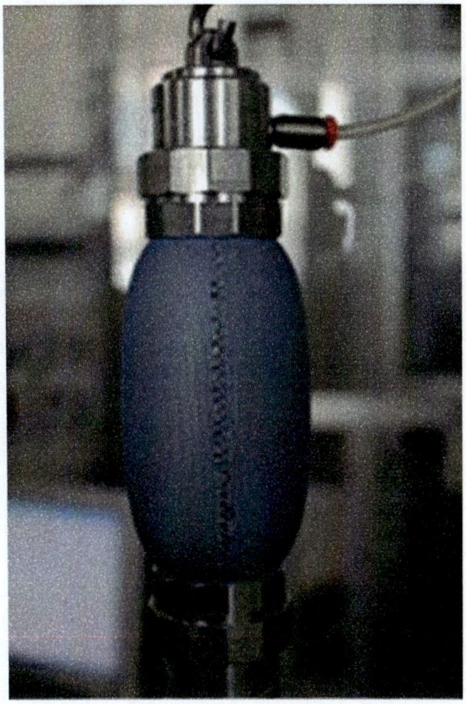

Fig. 2 Textile PAM under actuation [6]

properties of fabrics and are incapable of moving the human limb. To reduce the need for physiotherapists, different kinds of rehabilitation therapies can be availed with the help of robotized and automated equipment incorporating PAM as an actuator.

Xie et al. [1] presented a novel flat modular pneumatic artificial muscle. It has cylinder modules that do not expand along the radial direction but expand axially so that the length of the belt gets reduced on the pressurized side. Due to the decrease in length of the belt, it provides a bending motion such that there is output force that is generated. When pressurized, the arm can be lifted to an angle of more than 90° in less than 0.4 s.

3 Applications of PAM in Material Handling

Lau and Chai [7] designed a low-cost pneumatic air muscle actuated anthropomorphic robotic hand. It has a total of 16° of freedom. The actuator of the tendon-driven fingers was powered by 14 McKibben-style pneumatic air muscles. The power transmission between the tendon-coupled phalanges and the actuators was provided by the tendons. The return mechanism was achieved by installing passive springs at the back of each finger. To provide greater flexibility and dexterity for robotic hands in object manipulation. Instead of tactile feedback sensors, lineal potentiometers were used to keep track of the position and they were attached with pneumatic actuators. A simple fuzzy logic control algorithm was implemented for the control of the pneumatic actuators. The precision and power of the grasp were demonstrated by the experiments as shown in Fig. 3.

Scarfe and Lindsay [8] designed and developed an air muscle actuated low-cost humanoid hand. A total of 10° of freedom movements are possible in which the muscle is controlled by a computer GPI. It can hold the object and can also be used to place them in a different location. It can grip a water bottle comfortably but

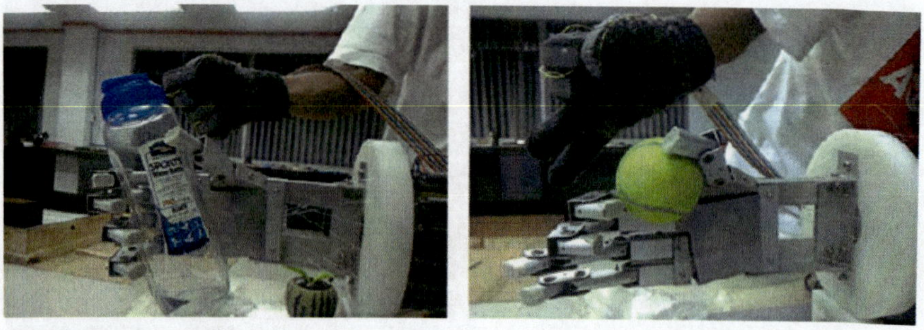

(a) (b)

Fig. 3 Grasping power of pneumatic air muscle actuated anthropomorphic robotic hand **a** holding cylindrical bottle to check power grasp and **b** holding tennis ball shows precision grasp [7] struggles to hold soft drinks which are comparatively small. Excluding the PC and air compressor, the total system costs around $1500.

Al-Ibadi et al. [2] proposed and developed a circular pneumatic muscle actuator (CPMA) which can be used to grip an object softly. The arrangement is such that when the muscle is pressurized, the actual diameter of the muscle gets increased while the inner diameter of the circular muscle gets decreased producing an inward radial force that can be used for lifting.

4 Rehabilitation Applications of PAM

Kawamura et al. [9] found that due to aging, there is a major need for walking assistance for elderly people as the muscle could not provide sufficient strength. Loss of muscle strength was one of the important factors for reduction in step length for elderly people. The step length was increased by the forward swing of the leg, and an advanced design of orthosis was also implemented to increase the forward swing phase of humans. Pneumatic air muscle has been employed to provide better flexibility and low rigidity. It was observed that the forward swing of the leg was assisted by sartorius muscle and the EMG experiments showed that the muscle strength decreased and the strength of semitendinosus muscle also decreases that helps in the restraining of legs against over assistance. Experiments result proven that further development needs to be done to overcome the aforementioned difficulties.

Al-Fahaam et al. [10] used pneumatic soft actuators to design a wearable splint that aids in the rehabilitation of wrist joints. In this design approach, extensor types of bending and also contraction types of muscles have been implemented. The biomechanics of the wrist joint has been studied and the bending angles for extension range from 0° to 70°, flexion movement ranges from 0° to 90°, radial deviation ranges from 0° to 20°, and ulnar deviation ranges from 0° to 50°. It was observed that the contraction and extension of 30% and 50% were obtained, respectively to produce a maximum generated output force of approximately 55 N.

Wang et al. [11] developed a tendon drive system that uses pneumatic air muscle to be implemented in the design of the soft and wearable robotic glove. The proposed structure of the cable drive system consists of a soft glove, cable guides, cables, and PAMs, and the human's skeleton has been used to transfer torque and force. The total weight of the glove part came around 120 g which was sufficiently light for stroke patients. When the actuators are pressurized, it does not give extra weight to the forearm which would be comfortable for a person's or any material handling applications for grasping a wide range of shapes and sizes of the object.

Choi et al. [12] developed the design of 2-DOF powered ankle–foot orthosis (PAFO) with talocrural and subtalar joints that are actuated by two pneumatic air muscles. The rotation of the inversion and eversion of the ankle was possible by the subtalar joint, which enables balanced walking for elderly people, and it is also considered to be one of the most important parts of balance training for people who face difficulty in walk and also for stroke rehabilitation. The sliding control method was used to experiment on the closed-loop frequency response. Rodrigues rotation formula was used to measure the length of the PAMs to avoid the complexity faced in the sliding control mode.

Petre et al. [13] developed pneumatic muscle actuated equipment for the rehabilitation of wrist muscle joints. It was constructed with a pair of bar mechanisms which exhibits Fin Ray-type structure like fish fins. Fins are the main thing that is helping fish to move further and ensuring stability and propulsion likewise, the same similarities were observed with the palm and fingers. The rehabilitation equipment was constructed with two Fin Ray-type mechanisms incorporating rack and pinion for linear movement with the help of pneumatic muscle. The experimental results have shown that the flexion and extension type motion of joints were achieved the angular limits of a healthy hand.

Polygerinos et al. [14] used pneumatic actuators to design a hand rehabilitation glove. The actuators were made up of elastomeric materials with integrated channels. These integrated channels function as pneumatic networks that are designed and analyzed for the comfort of human finger motion by producing bending motion. Experimental results on force generation showed that using a soft pneumatic actuator, one could ability to curl more than 320° and at the same time, it was capable of producing enough force for passive assistance of human fingers while closing. It was observed that further improvement in the manufacturing of actuators could provide high force and could able to withstand a huge amount of pressure.

Gao et al. [15] presented a new technology for elbow injury patients. A soft pneumatic elbow pad (SPEP). It was observed that SPEP attained a value of 9.22 N with a pressure of 70 kPa. The SPEP was able to reduce the forces of the biceps and the muscle in the upper arm by 31.33 N and olecranon by 22.12 N. The assistance provided is verified using electromyography. Wang et al. [16] designed and developed a soft rehabilitation glove with a double degree of freedom (DOF) soft pneumatic bending actuators (DPBA). Each finger and wrist can be controlled individually or in a combined manner by using this elbow pad. Results show that this developed glove could provide sufficient assistance as well as the potential for extension of hand for simple exercises and gripping of objects.

Zhong et al. [17] observed that the restoring movements of hemiplegic patients were done by rehabilitation robots, but motors were used as an actuator in these robots which lacks compliance with human body flexible movements when get touched. So, using pneumatic muscle actuator (PMA), they designed and developed a rehabilitation robot for the ankle, which shows similar compliance with human biological muscle, and the developed prototype as shown in Fig. 4 was analyzed for motion characteristics. Torque–angle relationship of the human ankle and structural dimensions was optimized with the help of robot dynamics. The driving torque was provided through the slider-crank mechanism for an antagonistically arranged pair of PMAs.

Waycaster et al. [18] developed an above-knee orthosis using pneumatic air muscle. Instead of the central support structure, which was a traditional one, a proposal of open frame structure has been implemented to accommodate the excessive larger profile expansion in the radial direction of the pneumatic air muscle during

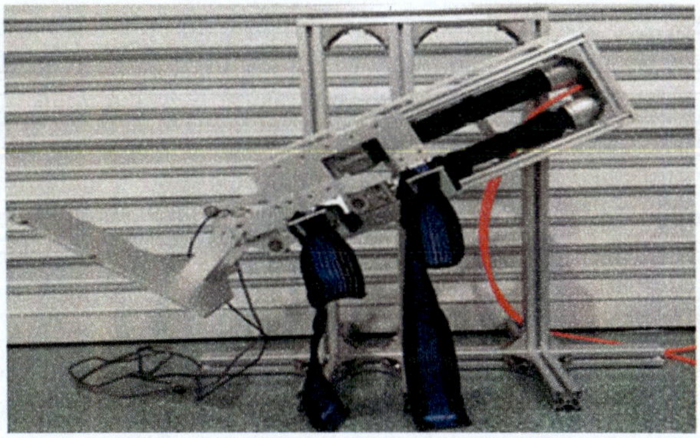

Fig. 4 Proposed ankle rehabilitation system [17]

its length shortening period. This design is capable of providing prosthesis to an individual of weight up to 75 kg and will be able to walk in a slow and normal cadence manner. A healthy test subject can walk on a treadmill wearing this prosthesis.

Wang et al. [19] presented a hand rehabilitation glove that doesn't have any joint actuated used pneumatic muscle actuation (PMA) which is used as a direct rotary actuator. The full flexion to horizontal position movements was achieved by optimizing 80 mm pneumatic air muscles. Since stroke-affected patients have dominant flexor hypertonia, the presented PMA could be served as a solution. The experimental result showed that using an 80 mm long PAM and pressurizing it up to 3.5 bar, we could able to achieve full rotation of finger joint.

Irshaidat et al. [20] proposed a novel exoskeleton soft robotic arm for rehabilitation of the human elbow for post-stroke patients, which is wearable, lightweight, and portable, and the pneumatic muscle is capable of bending. McKibben's muscle is used due to its force producing capacity. Two bending extensor pneumatic air muscles and one bending contractor PMAs were used to produce the necessary bending for the elbow rehabilitation. The prototype weighs only 0.35 kg, and it is capable of bending the forearm to weigh 1.5 kg which is the average mass for the forearm for both arms. From this paper, it is evident that more effective rehabilitation therapies can be done in-home at a lower cost.

Suzuki et al. [21] proposed a variable viscoelastic assistive suit that has a variable stiffness which has a magnetorheological fluid brake whose viscosity can be varied by applying a magnetic field. Two artificial muscles are connected in a manner of antagonistic pairs to a pulley of different diameters which produces the necessary output torque. It is controlled by a feedforward controller. Experiments were conducted to prove the assistive effect produced by the muscle.

Oliver-Salazar et al. [22] developed pneumatic air muscle-driven mechatronic fingers that show similar movements as compared to human fingers. A Manosil silicone rubber made pneumatic muscle was used to actuate the mechatronic fingers. A dsPIC-based, autonomous electronic board, and a graphic user interface was developed for conditioning the signal and processing it. A 1.5 kg, 4-DOF mechatronic finger was designed and constructed to validate the proposed model. It is found from the experiment that even when the medial phalanx was subjected to a load of 450 g mass, the finger can move to the required phalanx position smoothly.

Mendoza et al. [23] they created a low-profile vacuum-powered artificial muscle (LP-VPAM) with a 3D printed skeleton and flexible skin that has dimensions, contraction, and force generation suitable for actuating an infant leg in this study. They developed the LP-VPAM for rehabilitation exercises in a six-month-old infant's model leg. The LP-VPAM is appropriate for this application because it has a maximum force of 26 N at low vacuum pressure magnitudes (40 kPa) and can withstand weights of 0.3 kg, which is sufficient to move the leg. They used six-cell and 10-mm-height actuator used on the model leg was able to reach and exceed the target angle range of 54 in the side lying position, with a range of 61 in the prone position, demonstrating clinical relevance despite falling short of that target in the prone position due to the extra weight of the leg. In addition to its low-profile and low actuator pressure magnitudes, the LP-VPAM described in this paper generates forces and contraction percentages comparable to existing VPAMs. As a result, it has the potential to be useful in infant rehabilitation exercises.

Capace et al. [24] described the current stages of the mechatronic design of a biorobotic joint with controllable compliance, with the goal of realizing a wearable and soft exoskeleton for "as-needed" rehabilitation. Pneumatic artificial muscles (PAMs), which are soft actuators with variable stiffness, operate the biorobotic joint. As a result of the process of designing and validating the mechanism of joint compliance regulation, some experimental tests involving pressure and force measurements of the used PAMs were performed to characterize the performance of the mechanism of regulation influenced by the nonlinear and hysteretic response of the PAMs. The regulation mechanism allows for the implementation of a novel and effective control strategy that ensures safe and biomimetic performance for the optimal regulation of the mechanical compliance of biorobotic joints across their entire pressure and motion ranges. The proposed control law overcomes some of the limitations of existing strategies for controlling the pressure of the PAMs involved in the regulation of the mechanical compliance of rehabilitation robots actuated by pneumatic muscles.

Wang and Xu [25] designed and implemented a soft parallel robot for automated wrist rehabilitation. The presented wrist rehabilitation robot combines the benefits of both soft robot and parallel robot structures. Unlike traditional rigid-body-based rehabilitation robots, this soft parallel robot has a compact structure that is highly secure, adaptable, and flexible, making it a low-cost solution for personalized treatment. The proposed soft wrist rehabilitation robot is propelled by six evenly distributed linear actuators powered by pneumatic artificial muscles and one central linear electric motor. The introduced parallel-kinematic mechanism design enables the soft robot's output stiffness to be increased for practical use. To provide feedback signals for evaluating the rehabilitation training process, an electromyography sensor is used. A kinematic model of the designed robot is created, and a prototype is built for testing. The findings show that the developed soft rehabilitation robot can help the wrist perform all of the required training motions, such as abduction–adduction, flexion–extension, and supination–pronation.

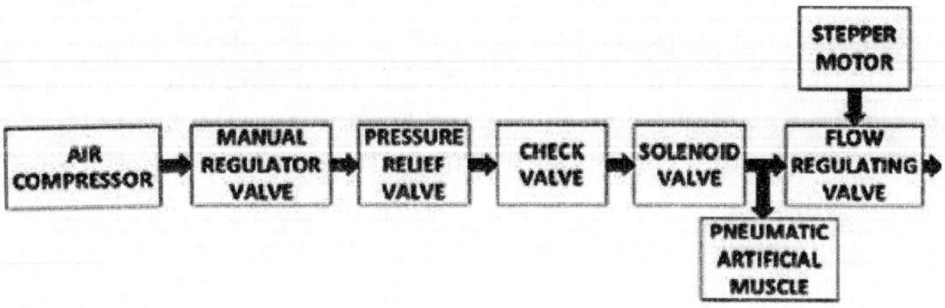

Fig. 5 Connection diagram of the PAM [27]

4.1 Control Strategies Followed in PAM

Al-Fahaam et al. [26] developed a mathematical model of open loop-controlled power augmented rehabilitation exoskeleton. A new approach was adopted in which the output force depends upon the pressure applied. Its assistance is verified by electromyography. An incremental adaptive Proportional-Integral-Derivative (PID) and a Proportional-Integral (PI) cascade were used to overcome the nonlinear behavior of PAM. PI and PID parameters were found in each stage by implementing the relay experiment method. The outer loop was responsible for the movement of the actuator and the inner loop controls the amount of flow. The proposed connection is as shown in Fig. 5.

Kalita and Dwivedy [28] studied the nonlinear dynamic behavior of pneumatically actuated artificial muscle considering the system to be parametrically excited. The governing equation of motion has been derived using Newton's law of motion to study the various responses in the system at principal parametric resonance conditions.

Ahn and Tu [29] analyzed the improvement of intelligent and inexpensive pneumatic artificial muscle (PAM) control system by incorporating proportional valves instead of costly servo valves, without changing external inertial loads. The proposed system uses a vector quantization neural network (LVQNN) to effectively contribute to the supervising switching controller. This proposed smooth switching algorithm proved that it was able to overcome the transient responses of the control system if the inertial load change is about 3000% also and able to reduce the steady-state error within 0.1°. Ahn et al. [30] also proposed the Magneto-Rheological brake for improving the performance of pneumatic muscle actuators (PMA). PMAs have certain disadvantages like air compressibility and less damping ability so, it was not easy to realize the transient response of PMAs with an additional factor of external inertial load change. A conventional PID controller and a variable damper Magneto-Rheological Brake (MRB) controlled by phase plane switching were equipped to conduct experiments and results were observed that the steady-state error was educed within + 0.05° and effectively control high gain with external inertial load variation up to 1000%. Chavoshianet and Taghizadeh [31] proposed and developed an adaptive neuro-fuzzy inference system (ANFIS), and the number of fuzzy rules was reduced without compromising the accuracy by implementing the subtractive clustering method. A conventional mathematical model was also constructed to compare the performance with the proposed model and observed that, forecasting error of the model was found to be 20% more than ANFIS. The modeling error for sinusoidal and trapezoidal inputs was about 5 and 3.5%, and there was a 0.6% increase in cluster error. By implementing this system for rehabilitations, we could able to reduce the dimension by two times.

Zhong et al. [32] used pneumatic air muscle (PMA) to analyze the ankle rehabilitation robot with a single neuron-tuned PID controller. The parameters of the PID controller own a constant value for P, I, and D because it was difficult to keep track of the nonlinearity and hysteresis of pneumatic air muscle. The single neuron PID controller was employed in this ankle rehabilitation robot. It shows excellent capacity in regulating structural parameters. The effectiveness of the controller was validated with the help of passive rehabilitation experiments.

Yang et al. [33] used pneumatic air muscle to develop a 3-DOF bionic joint. 3 connecting rods parallel mechanism was used in designing the bionic elbow joint. A parameter self-adjust model-free adaptive control (PSA-MFAC) strategy was proposed for the control of the system. It was based on the model-free adaptive control (MFAC) theory. Improved accuracy was found based on the experiment results. The tracking accuracy of the proposed PSA-MFAC controller was only 3.8% compared to conventional MFAC and PID controllers whose accuracies were 9.5 and 15%, respectively. Wu et al. [34] proposed a new parameter control of the PID controller to improve the precision of movements in rehabilitation. It is better than the conventional PID controller. This new control method uses a fuzzy PID which is implemented in a wearable rehabilitation robot hand. This new control method is proven to improve trajectory tracking.

Wang et al. [35] developed the design of a safe, flexible, and reliable finger rehabilitation robot (FRR) which was used for training with no human effort required, and a damper was used to provide the necessary damping for active training. To feel less when the subject is using the device, independent training of each joint is necessary by transferring the driving force to individual finger joints. The output provided by the damper is stable which makes it usable for active training.

Wang et al. [36] found out a new adaptive control for monitoring the upper limb rehabilitation training. To ensure adaptivity and stability during motion, an adaptive fuzzy backstepping control was proposed. Numerical simulations and Lyapunov stability were proved by the adaptive backstepping control. The aforementioned control was applied and expected dynamic and static performances were achieved with the help of a closed-loop system made up of the backstepping method, combining with the Lyapunov type adaptive law. The stability was maintained properly, due to external noises and variable inertia matrix. Real experiments and numerical simulations were performed and the results ensured the flexible motion of the arm via proposed control.

5 Summary

In this review article, evaluation of various rehabilitation systems and robotics applications for material handling and rehabilitation purposes using the pneumatic muscle-type actuators has been reviewed. The stiffness, stability, nonlinearity, and hysteresis were considered in the control algorithms and strategies. Although few advanced applications have been observed, the field of soft robotics is still rapidly evolving. One of the major issues faced in the design of an appropriate control system for pneumatic muscles to deal with the nonlinear behavior. Though the experimental analysis is costly and time-consuming, randomized controlled trials were necessary to identify suitable control algorithms for various applications.

References

1. Xie D, Zuo S, Liu J (2020) A novel flat modular pneumatic artificial muscle. Smart Mater Struct 29(6)
2. Al-Ibadi A, Nefti-Meziani S, Davis S (2018) A circular pneumatic muscle actuator (CPMA) inspired by human skeletal muscles
3. Oguntosin V, Akindele A (2019) Design and characterization of artificial muscles from wedge-like pneumatic soft modules. Sens Actuat A 297
4. Daerden F, Lefeber D (2001) The concept and design of pleated pneumatic artificial muscles Int J Fluid Power 2(3)
5. Veale AJ, Xie SQ, Anderson IA (2016) Modeling the peano fluidic muscle and the effects of its material properties on its static and dynamic behavior. Smart Mater Struct 25(6)
6. Belforte G, Eula G, Ivanov A, Sirolli S (2014) Soft pneumatic actuators for rehabilitation. Actuators 3(2)
7. Lau CY, Chai A (2012) The development of a low cost pneumatic air muscle actuated anthropomorphic robotic hand. Proc Eng 41
8. Scarfe P, Lindsay E (2006) Air muscle actuated low cost humanoid hand. Int J Adv Robot Syst 3(2)
9. Kawamura T, Takanaka K, Nakamura T, Osumi H (2013) Development of an orthosis for walking assistance using pneumatic artificial muscle. Int Conf Rehabil Robot
10. Al-Fahaam H, Davis S, Nefti-Meziani S (2017) Wrist rehabilitation exoskeleton robot based on pneumatic soft actuators
11. Wang L et al (2019) Soft robotics for hand rehabilitation. Intell Biomechatron Neurorehabilitat
12. Choi HS, Lee CH, Baek YS (2020) Design and validation of a two-degree-of-freedom powered ankle-foot orthosis with two pneumatic artificial muscles. Mechatronics 72
13. Petre I, Deaconescu A, Sârbu F, Deaconescu T (2018) Pneumatic muscle actuated wrist rehabilitation equipment based on the fin ray principle. J Mech Eng 64(6)
14. Polygerinos P et al. (2013) Towards a soft pneumatic glove for hand rehabilitation
15. Gao X, Sun Y, Hao L, Yang H, Chen Y, Xiang C (2018) A new soft pneumatic elbow pad for joint assistance with application to smart campus. IEEE Access 6
16. Wang J, Liu Z, Fei Y (2019) Design and testing of a soft rehabilitation glove integrating finger and wrist function. J Mech Robot 11(1)
17. Zhong J, He D, Zhao C, Zhu Y, Zhang Q (2020) An rehabilitation robot driven by pneumatic artificial muscles. J Mech Med Biol 20(09):2040008
18. Waycaster G, Wu SK, Shen X (2011) Design and control of a pneumatic artificial muscle actuated above-knee prosthesis. J Med Devi Trans ASME 5(3)

19. Wang B, Aw KC, Biglari-Abhari M, McDaid A (2014) Modelling of pneumatic air muscles for direct rotary actuation of hand rehabilitation glove. In: Lecture notes in computer science (including subseries lecture notes in artificial intelligence and lecture notes in bioinformatics), vol 8755
20. Irshaidat M, Soufian M, Al-Ibadi A, Nefti-Meziani S (2019) A novel elbow pneumatic muscle actuator for exoskeleton arm in post-stroke rehabilitation
21. Suzuki R, Okui M, Iikawa S, Yamada Y, Nakamura T (2019) Evaluation experiment of squat motion with variable viscoelastic assistive suit 'Airsist I'. IFAC-PapersOnLine 52(22)
22. Oliver-Salazar MA, Szwedowicz-Wasik D, Blanco-Ortega A, Aguilar-Acevedo F, Ruiz-González R (2017) Characterization of pneumatic muscles and their use for the position control of a mechatronic finger. Mechatronics 42
23. Mendoza MJ et al (2021) A vacuum-powered artificial muscle designed for infant rehabilitation. Micromachines 12(8)
24. Capace A, Randazzini L, Cosentino C, Romano M, Merola A, Amato F (2020) Design, realization and experimental characterisation of a controllable-compliance joint of a robotic exoskeleton for assist-as-needed rehabilitation
25. Wang Y, Xu Q (2021) Design and testing of a soft parallel robot based on pneumatic artificial muscles for wrist rehabilitation. Sci Rep 11(1)
26. Al-Fahaam H, Davis S, Nefti-Meziani (2018) The design and mathematical modelling of novel extensor bending pneumatic artificial muscles (EBPAMs) for soft exoskeletons. Robot Auton Syst 99
27. Giancarlo V, B C (2019) Control strategy of a pneumatic artificial muscle for an exoskeleton application. IFAC-PapersOnLine 52(Jan):281–286
28. Kalita B, Dwivedy SK (2019) Dynamic analysis of pneumatic artificial muscle (PAM) actuator for rehabilitation with principal parametric resonance condition. Nonlinear Dyn 97(4):2271–2289
29. Ahn KK, Tu DCT (2004.) Improvement of the control performance of pneumatic artificial muscle manipulators using an intelligent switching control method. KSME Int J 18(8)
30. Ahn KK, Thanh TUDC, Ahn YK (2005) Performance improvement of pneumatic artificial muscle manipulators using magneto-rheological brake. J Mech Sci Technol 9(3)
31. Chavoshian M, Taghizadeh M (2020) Recurrent neuro-fuzzy model of pneumatic artificial muscle position. J Mech Sci Technol 34(1)
32. Zhong J, Zhu Y, Zhao C, Han Z, Zhang X (2020) Position tracking of a pneumatic-muscle-driven rehabilitation robot by a single neuron tuned PID controller. Complexity 2020
33. Yang H, Xiang C, Hao L, Zhao L, Xue B (2017) Research on PSA-MFAC for a novel bionic elbow joint system actuated by pneumatic artificial muscles. J Mech Sci Technol 31(7)
34. Wu J, Huang J, Wang Y, Xing K, Xu Q (2009) Fuzzy PID control of a wearable rehabilitation robotic hand driven by pneumatic muscles
35. Wang D, Wang Y, Zi B, Cao Z, Ding H (2020) Development of an active and passive finger rehabilitation robot using pneumatic muscle and magnetorheological damper. Mech Mach Theory 147
36. Wang T, Chen X, Qin W (2019) A novel adaptive control for reaching movements of an anthropomorphic arm driven by pneumatic artificial muscles. Appl Soft Comput J 83